Hydraulics, Mechanics of Fluids,
Engineering Education

Selected
Writings of
Hunter Rouse

Hydraulics, Mechanics of Fluids, Engineering Education

Selected Writings of
Hunter Rouse

Edited by
JOHN F. KENNEDY and ENZO O. MACAGNO
INSTITUTE OF HYDRAULIC RESEARCH
UNIVERSITY OF IOWA

DOVER PUBLICATIONS, INC.

NEW YORK

Published in Canada by General Publishing Company, Ltd., 30 Lesmill Road, Don Mills, Toronto, Ontario.
Published in the United Kingdom by Constable and Company, Ltd., 10 Orange Street, London WC 2.

This Dover edition, first published in 1971, reproduces 57 papers by Hunter Rouse, unabridged and with minor corrections by the author. The Preface, Biographical Sketch, Complete List of Publications and Indexes were prepared especially for this edition.

International Standard Book Number: 0-486-60964-2
Library of Congress Catalog Card Number: 78-154483

Manufactured in the United States of America
Dover Publications, Inc.
180 Varick Street
New York, N.Y. 10014

PREFACE

The twentieth century has witnessed the evolution of hydraulic engineering from largely an art, in which the hydraulician's experience and intuition were his primary tools, to a precise, well-formulated discipline. Hunter Rouse has been in the forefront of the advance since the early 1930's. Rouse soon recognized that hydraulic engineering could only profit from an approach similar to that being applied by Ludwig Prandtl, Theodore von Kármán, Sir Geoffrey Taylor, and Stephen Timoshenko to other fields of engineering, in which the then unique blending of theoretical analysis, experimental verification, and practical application was producing advances at an unprecedented rate. Rouse undertook to apply this same approach in hydraulics, and to bring to bear upon hydraulic engineering the advances in fluid mechanics which were accompanying the rapid growth of aeronautics. Indeed, much of his remarkably productive research career has been spent at the interface between hydraulic engineering and fluid mechanics.

Throughout his professional life Hunter Rouse has been above all a teacher, and one who has excelled in all of the many avenues a teacher can use. He taught not only in the classroom, but also through his books, papers, lectures, films, participation in professional meetings, and then as Dean of the College of Engineering at The University of Iowa. Hence, when several of his colleagues began a few years ago to consider how his many contributions to engineering and education might suitably be recognized on the occasion of the 65th anniversary of his birthday, it was soon concluded this could be done most appropriately by publishing in a single volume a selection of his writings reflecting his many-faceted career as a teacher.

Rouse's interests have always transcended the specialized areas in which his research was concentrated. He is a university man more than an engineer, and he is more universal than most university men. Thus in the first paper included here, preceding even his first doctoral dissertation, one finds his observations on popular views about the United States in the Germany of the early 1930's. The paper presenting his views of the trends in hydraulics and its leading unsolved problems at the middle of the century was judged valuable not only to demonstrate his perception and foresight at that time, but also as an indication of what still remains to be accomplished. Between one of his papers on the Bernoulli theorem and one on pressure fluctuations, he found time to dwell on the amusing latitude existing in the pronunciation of the name of one of the great hydraulicians of the last century. (It is interesting to speculate if a similar article might be someday published, perhaps by a Japanese hydraulician, on the pronunciation of Rouse!) His address to the Latin American engineers in a meeting in Caracas on scientific and technological approaches has been included to illustrate his international influence as a counselor in matters of engineering, education, and research. The volume closes with Rouse's most recent articles, representative of the ever-widening breadth of his concern and interest. In the final article we find a man who contributed so much to Growth pondering now about what will be the principal ingredients of Progress in the future, and on what will be the role of the engineer in contributing to the newly defined Progress.

In spite of the inclusion of writings which tend to reveal Rouse as a man of many parts, his papers convey only glimpses of the human side of the man; perhaps it is inevitably difficult to judge this from a man's technical writings alone. In the Biographical Sketch we attempt to convey a broad view of his life, but still cannot resist the desire to say something more of his warmth by quoting from a recent letter from Vito A. Vanoni: "One time during the war he [Rouse] and John McNown arrived at our home on a Friday night when Edith [Vanoni] and I were tired out from a hard week. Hunter was in a festive mood, despite his own hard week and he soon had us all cheered up and ready for the big night that followed. A story or two, some reminiscing, all punctuated with laughter, and we were in the mood and away we went until the wee hours. The fun was not in the wine or the food or the place, but in enjoying each other."

As we worked on this project our enthusiasm for it continued to grow. It is our conviction that the many developments reported in the papers compiled in this volume will make it valuable to both present and future fluid mechanicists and hydraulicians. We hope also that engineering educators will find inspiration in Rouse's articles in the area of education. Moreover, we believe that Rouse's unique position as a leader in the recent progress of hydraulics will make the volume of interest for its historical value, for the selection of his papers appearing in this volume gives a concise overview of the development of hydraulics over the past four decades.

We are very pleased to have had the collaboration of Dover Publications Inc. in this undertaking; Mr. Hayward Cirker, President, was immediately agreeable to having his firm undertake publication, and Messrs. Thomas Reilly and Clarence Strowbridge were extremely cooperative and spared no effort in bringing the volume to realization in very fine style. Dr. Jung-Tai Lin of the Institute of Hydraulic Research provided invaluable assistance in retrieving from long dormant files many of the original drawings and photographs. The careful work of Mrs. Beth Buffum of the Institute, Miss Catherine S. Chambers of The University of Iowa Printing Service, and Mrs. Janice Arnio of the Iowa College of Engineering in typing and in recomposing some of the material used in the volume measurably enhanced its appearance.

ENZO O. MACAGNO
JOHN F. KENNEDY
IOWA CITY, IOWA
February, 1971

BIOGRAPHICAL SKETCH

Hunter Rouse traces his origin in America to about 1700, when the first of his line in the New World, a British conscript, is said to have jumped ship off the coast of New England. Others of the lineage lived in Cornwall, Connecticut; Grinnell, Iowa; and Toledo, Ohio; they included a judge, a minister, and several businessmen. Hunter was born in Toledo on 29 March 1906, the only child of Henry Esmond Rouse, a hardware dealer, and Jessie Rouse, née Hunter. He lost his father while still a child, and thereafter was reared by his mother (who lived in Iowa City for many years before her death, in 1970, at the age of 95) and Adolph de Clairmont, a French-born physician-scientist-inventor who was a close friend of the family and Hunter's godfather. De Clairmont, a linguist and a widely traveled man of broad interests, had a many faceted influence upon him.

In his youth Hunter was active in the Episcopal Church, in the Boy Scouts (where he advanced beyond the rank of double Eagle), and on the high school magazine and yearbook staffs. His boyhood dream was to become a civil engineer and build bridges in South America; toward this end he studied Spanish during his secondary schooling. In 1924 immediately following his graduation from high school, where he had won four gold medals for scholarship, he entered Toledo University. After one summer and a scholastic year he dropped out of the university and took a job as instrumentman on a county surveying crew to earn money to continue his schooling.

In the fall of 1926 he entered M.I.T. as a sophomore in civil engineering and was initiated into Sigma Chi Fraternity. To meet expenses he worked during the school terms through a Boston theatrical agency as an entertainer, utilizing his considerable talent as a puppeteer and ventriloquist which had its source in a boyhood hobby acquired from his godfather. During summer vacations he worked on surveying and highway construction crews. At M.I.T. he was an honor student; editor of the student newspaper (an activity "selected" at the command of fraternity brother James Killian, later President of M.I.T. and Chairman of the M.I.T. Corporation); winner of two Boit Prizes for composition; skit writer for the Tech Show; specialty actor for the Combined Musical Clubs; and a member of the senior honor society and two honorary fraternities: Tau Beta Pi and Chi Epsilon. Upon his graduation, in 1929, he was given the choice of becoming a graduate assistant to Karl Terzaghi, then relatively new at M.I.T., or going to Germany on an Institute Traveling Fellowship (which carried a stipend of $1000 per year for all expenses) to study under Theodor Rehbock at the Technische Hochschule Karlsruhe. De Clairmont, Hunter's godfather, had implanted in him not only the aim of becoming a civil engineer but also that of studying in Europe (preferably at the Sorbonne), so the choice was an easy one. This was an important juncture in his career; had he remained at M.I.T. under Terzaghi's guidance, his professional interests would surely have turned toward soil mechanics rather than hydraulics.

The departure from New York for Europe in the spring of 1929 was not uneventful; Hunter arrived at the pier in ample time for a midday sailing, but the ship—and his luggage—had already left the preceding midnight. The following week he was on board well before nightfall, and set out to seek his future—and his luggage—in Germany. Following his arrival in Karlsruhe he went, on Rehbock's suggestion, onward to Munich to learn German during the summer. As happy chance would have it, he moved into the same pension where Miss Dorothee (Doi) Hüsmert was to live while in a Munich photo school. Hunter had been told that the only way really to learn German was to date a German girl, and Doi had been given the corresponding information about learning English. (Three years later, after both had become reasonably bilingual and Hunter had earned a German degree—a condition set by her father—they would be married.) At the end of the summer he returned to Karlsruhe to live with a family of Dutch-German origin (as were the Rehbocks themselves). There he was told that he must obtain the German *Diplom* before proceeding to the doctorate, and accordingly he took a full load of classes during the winter term. By spring it was evident that the lectures were a needless repetition of previous work, and the ministry of education agreed that he might begin his dissertation project, energy distribution at the free overfall, with the understanding he would obtain the S.M. degree in civil engineering from M.I.T. before being admitted to examination for the doctorate. During this time he supplemented his income by conducting scour tests on various models, including that of the Zuyder Zee, under Rehbock's direction.

The summer of 1930 was spent at the outdoor hydraulics laboratory of the Kaiser Wilhelm Gesellschaft zur Förderung der Wissen-schaften on the Walchensee near the Bavarian Alps, assisting with various tests of instruments, writing about German laboratories for a proposed supplement to John R. Freeman's book, *Hydraulic Laboratory Practice,* and traveling in Germany, Switzerland, and Italy. In the fall he returned to Karlsruhe for continued work on the free overfall. Three months of the following spring were spent living with a young family in Paris to learn French. This period was followed by visits to hydraulics laboratories at Toulouse and Delft, and to the Zuyder Zee closure which he had helped investigate in model studies under Rehbock. Before returning to Karlsruhe he traveled through Austria, Hungary, and Rumania.

In July, following completion of most of the analysis for his dissertation, Hunter moved to Berlin to study for four weeks with Franz Eisner at the Preussische Versuchsanstalt für Wasserbau und Schiffbau. He returned to M.I.T. in September and enrolled as a candidate for the master's degree to fulfill the prerequisite that had been set for the doctorate. He was appointed Assistant in Hydraulics and worked in the River Hydraulics Laboratory that had been established by K. C. Reynolds, one of the first Freeman scholars. His co-workers during this period included John Drisko, who had held the Institute Traveling Fellowship for two years just before him, and Hugh MacDougall, who held it for a year just after. While at M.I.T. Hunter investigated flow over weirs of various relative heights, took courses and wrote a thesis ("Weir Flow and Spillway Design") for the required degree, taught classes in hydraulics, and did some additional work on his German dissertation under the guidance of Wilhelm Spannhake of T. H. Karlsruhe, who was then a visiting professor in mechanical engineering at M.I.T. He received his master's degree in June 1932.

Directly thereafter he returned to Germany to attend to several matters: a civil wedding ceremony in which he was married to Doi at Erfurt; a visit to the American Consulate at Berlin to arrange for his wife's visa; defense of his dissertation at Karlsruhe; a swim with Rehbock in the Rhine River (where he ruptured an eardrum on his first dive); another swim in the Rhine (during which he suffered a sunstroke) followed by a party with Spannhake; and finally—after being driven back to Erfurt by solicitous wedding guests—a formal church wedding and banquet. During their honeymoon Hunter recovered from his graduation swims, revisited the laboratory on the Walchensee, toured the hydraulic facilities at Hannover, and met Ludwig Prandtl and Albert Betz during a memorable two days at Göttingen.

In the fall of 1932 he returned to M.I.T. to teach and do further laboratory work on weir flow. He also served as Spannhake's assistant and prepared mimeographed notes for the latter's hydrodynamics class (which included the oceanographer Athelstan Spilhaus and the meteorologist Harry Wexler). Although Hunter assumed that he was established at M.I.T. for life, in 1933 Vannevar Bush, then Vice President of the Institute, did him the great favor (as Bush later claimed) of terminating his assistantship, for lack of funds. Hunter contacted Boris A. Bakhmeteff, whom he had met (along with O. G. Tietjens) at an ASME meeting where his paper commissioned by Freeman on the Walchensee laboratory had been presented, to inquire about a post at Columbia. A position was offered as instructor in civil engineering, and in the fall of 1933, following the birth of their first son, Richard Hunter, the Rouses moved to New York.

At Columbia Hunter taught courses on hydraulics, read papers for Bakhmeteff's course in fluid mechanics and, together with Arthur Matzke and Lincoln Reid, designed and built a small hydraulics laboratory in the basement bedrock of the engineering building. During this period he wrote several discussions of journal papers at Bakhmeteff's instigation, began actively to produce papers of his own, and started his first book, *Fluid Mechanics for Hydraulic Engineers*. His association while at Columbia with James Kip Finch, then Head of Civil Engineering (who was to

author *The Story of Engineering*, and whose office walls were lined with portraits of notable engineers and scientists), stimulated in Hunter an interest in the origins of his profession; this led eventually to publication of *History of Hydraulics* by Rouse and Ince in 1957.

In 1935 he received an offer from Robert T. Knapp to join the Sedimentation Laboratory of the Soil Conservation Service at California Institute of Technology as Associate Hydraulic Engineer. The offer was accepted and in January 1936 Hunter made the westward trek (Doi and Dick being momentarily in Germany), stopping for several days to visit the laboratories of the Iowa Institute of Hydraulic Research in Iowa City and the Bureau of Reclamation in Denver. At Caltech he worked under Knapp and Vito A. Vanoni on a variety of sedimentation problems; it was here that he developed his analysis of the distribution of suspended sediment in turbulent flows. (Arthur Ippen, already at Caltech from Germany by way of Iowa, had independently performed a parallel analysis of the same problem and obtained a comparable formulation; it appears that both were influenced by Professor Theodor von Kármán to undertake analyses based on the analogy between the turbulent diffusion of momentum and suspended material.) After Hunter had been at Caltech about a year, von Kármán and Knapp arranged for him to receive an appointment as Assistant Professor of Fluid Mechanics. He then taught several undergraduate and graduate courses in hydraulics and fluid mechanics and supervised one M.S. thesis (on the free overfall, of course). His colleagues at Caltech included—in addition to von Kármán, Knapp, Vanoni, and Ippen—Donald Barnes, Raymond Binder, Francis Clauser, Nephi Christensen, James Daily, Hans Einstein, Calvin Gongwer, William Sears, Merit White, and George Wislicenus. The manuscript for *Fluid Mechanics for Hydraulic Engineers* was completed, reviewed by Professor J. P. Den Hartog of Harvard University, and recommended for publication in the McGraw-Hill Engineering Societies Monograph Series;

it appeared in the summer of 1938.

The summer of 1938 was also highlighted by a trip to Germany to visit in-laws in Erfurt as well as Prandtl in Göttingen and other research authorities. On his way back to Pasadena Hunter participated in the memorable Fifth International Congress for Applied Mechanics at Harvard and M.I.T. and presented (with Bush presiding) a paper on his turbulence-jar experiments with sediment suspensions. At the ASCE national meeting that winter he was awarded the Norman Medal for his paper "Modern Conceptions of the Mechanics of Fluid Turbulence," and while there was offered jobs at Illinois Institute of Technology, the Corps of Engineers' Beach Erosion Board, and the University of Iowa. Largely because of Dean F. M. Dawson's expert undersalesmanship, he accepted the position at Iowa as Professor of Fluid Mechanics and consultant to the Iowa Institute of Hydraulic Research.

Following the birth of their second son, Allan Hüsmert (who spent his first weeks in the Pasadena home of Don and Thedia Barnes), the Rouses moved to Iowa in the fall of 1939. Shortly after renting a home that had belonged to the recently deceased Floyd Nagler, founder of the Iowa Institute of Hydraulic Research, Hunter ruptured his right Achilles' tendon while playing tennis and, as a result, spent the next three months on crutches. During his first year at Iowa he was offered the post of Head of the Department of Mechanics and Hydraulics, but elected to concentrate on laboratory research in fluids. In succeeding years he declined several other proffered academic positions that were primarily administrative, insisting that no research man in his right mind would even consider such a post—particularly that of dean!

An unprecedented Hydraulics Conference had been organized by F. T. Mavis (then Head of the Department of Mechanics and Hydraulics at Iowa) and held during the summer of 1939. Hunter had participated in it and had become convinced of the value of a continuing series of such meetings. The next year, while driving to Vicksburg to visit the Waterways Experiment Station, Hunter, Dawson (not only Dean but also Institute Director), E. W. Lane (Associate Director of the Institute), and J. M. Robertson (a graduate student, later a professor at Penn State and Illinois), discussed the matter at length, and the decision was made to hold a second conference in 1942. With the collaboration of J. W. Howe (who had succeeded Mavis as Head of M & H), C. J. Posey, and A. A. Kalinske, the series of triennial conferences was firmly established. It continued (with a one-year delay during the war) through the Seventh Conference, held in 1958, whereupon it was concluded that the function of these meetings was being fulfilled by the annual ASCE Hydraulics Division Specialty Conferences and that the Iowa series should hence be discontinued. Proceedings volumes of the Iowa Conferences are still in demand. The proceedings of the Fourth Conference (1949), planned and edited by Hunter and published in 1950 under the title *Engineering Hydraulics*, continues to be a standard reference in the field.

In 1942 Lane departed Iowa to pursue war-related work at T.V.A., and Hunter and Tony Kalinske were appointed Associate Directors of the Institute. Under their guidance and the exigencies of the war, the level of research activity rapidly increased and diversified, with most of the sponsorship coming from the War and Navy Departments. The research work ranged from problems of diffusion of fog and gases, to wind flow over mountainous terrain, underwater ballistics, ship resistance, grid turbulence, fire monitors, and development of electronic instrumentation. In 1943 the Rouses' daughter, Patricia Mary, was born. The following year Dawson relinquished the directorship of the Institute to Hunter, and John S. McNown eventually replaced Kalinske as Associate Director. Their leadership during the ensuing years established the Institute as one of the premier fluids research and educational organizations in the world. Hunter directed pioneering work on such diverse problems as cavitation, heat convection, roughness, free-turbulence shear flows, flow in open channels, sediment transport, density-stratified flows, boundary layers, turbulence, and turbulent diffusion; and members of his staff initiated under his stimulation research in a variety of other fields. His colleagues during the Iowa years included M. L. Albertson, L. M. Brush, F. R. Hama, P. G. Hubbard, L. Landweber, E. M. Laursen, E. O. Macagno, D. M. Metzler, E. Naudascher, E. M. O'Loughlin, W. Rand, and C. S. Yih. A particularly close relationship and collaboration developed between Hunter and Joe Howe, who for more than thirty years has been a sounding board for Hunter's ideas and a source of continuing counsel and encouragement. Another valuable association was that with Dale Harris, Institute shop foreman since the early forties, who transformed Hunter's ideas for apparatus and instrumentation into unexcelled laboratory equipment.

His second book, *Elementary Mechanics of Fluids,* was published in 1946, to be followed in 1953 by *Basic Mechanics of Fluids* (with Howe). Both texts had extensive favorable impact on the instruction of fluid mechanics in the United States and in many foreign countries. *History of Hydraulics,* co-authored with his doctoral student Simon Ince, was published first as a bilingual fold-in supplement to *La Houille Blanche* and then in book form by the Institute in 1957. *Advanced Mechanics of Fluids,* written by members of the Institute staff and edited by Hunter, appeared in 1959. Proceeds from the sale of the last two books were used to establish in the University of Iowa Library a collection of rare source material in hydraulics, now one of the finest of its kind anywhere. Hunter made the preliminary designs and supervised the construction of equipment for complete hydraulics laboratories at the Universities of Colombia, Venezuela, and the Philippines, and his 1961 monograph *Laboratory Instruction in the Mechanics of Fluids* led to the widespread adoption of instructional experiments that he had developed at Iowa. In 1960 he began work on a series of six thirty-minute educational motion picture films for fluid mechanics, an activity that took ten years to complete and involved the collaboration in turn of Lucien Brush and Emmett O'Loughlin. The films are characterized by the ingeniously conceived experiments and arguments and the excellence in detail that have long been the hallmark of his research and writing.

At the University of Iowa Hunter supervised 35 doctor's and 45 master's theses. He continued to teach his famous course Intermediate Mechanics of Fluids (using his *Elementary Mechanics of Fluids* as a text) until 1968; the course was sufficiently different from year to year and the lectures invariably so masterfully executed that many graduate students and some staff members would sit in on the course for one or more successive years after they had passed it for credit. He continued to travel extensively, lecturing in some 22 foreign countries and visiting 18 others. Many trips were made in the company of Art Ippen, Hunter's very close friend since their Caltech days together. In 1952-53 he spent 12 months as a Fulbright Research Scholar at the University of Grenoble, where he was Professeur d'Echange in the Faculté des Sciences. Thirteen months in 1958-59 as a National Science Foundation Senior Post-Doctoral Fellow were divided between the Max Planck Institut für Strömungsforschung in Göttingen, the University of Rome, Cambridge University, and the Ecole des Ponts et Chaussées and the Sorbonne (where he was examined for the French doctorate on the Fourth of July!). Much of this period was spent in European libraries continuing his study of the history of hydraulics. In 1960 he was sent by the State Department as educational consultant to the United Arab Republic; in 1969, by NSF to India; and in 1970, by UNESCO to Thailand. In 1961 he organized the first exchange visit between hydraulics laboratory directors of the United States and the Soviet Union, and in 1965 was co-organizer of the first NSF-sponsored U.S.-Japan Seminar, on instrumentation. He was also consistently active in professional societies and as consultant to various government agencies, engineering organizations, and industrial firms.

During his forty years of professional activity, much well-deserved recognition has been accorded him. He received the ASCE Norman Medal in 1939; in 1951 and 1961 he shared with a number of graduate students the ASCE Karl Emil Hilgard Prize for two joint papers; he received the ASEE George Westinghouse Award in 1948 and the Vincent Bendix Award in 1958, was elected a Fellow of the American Academy of Arts and Sciences in 1958, was recipient of the ASCE Theodor von Kármán award in 1963, and was elected a member of the National Academy of Engineering in 1966. Also in 1966 he was selected to give the first John R. Freeman Memorial Lecture to the Boston Society of Civil Engineers; Freeman's influence on his early professional life made this event particularly meaningful to him. The following year he became an Honorary Member of the ASME.

In 1966 Howard Bowen, then President of the University of Iowa, persuaded Hunter to accept the deanship of the College of Engineering, to relieve a difficult administrative situation that had developed (and, Hunter insists, to make room for a younger hand at the Institute helm). As Dean he demonstrated the same energetic style and exacting standards that had characterized his earlier career. He supported a complete revision of the undergraduate curriculum, upgraded teaching practices, and brought about a pronounced strengthening of the faculty and its scholarly productivity. Expression for his love of books was found in the superb new Iowa Engineering College Library which he conceived and brought to realization. He has continued during his deanship to produce research papers, although much of his recent writing has logically been concerned with engineering education. His latest paper included herein, it will be noted, even ventures into the still more esoteric field of ecologic demography.

The Rouses continue to live in the house—located next to City Park and straight north from the Hydraulics Laboratory, as many a former graduate student will recall—that they purchased in 1940. Their son Dick is now an associate professor in medieval history at UCLA; Allan is in charge of research for the Connecticut legislature at Hartford; and Trish is an army wife making her family's home in first one part of the world then another. Thus far there are no more engineers in the family, but with four grandchildren there is still great hope for the future.

TABLE OF CONTENTS

SELECTED WRITINGS

"Amerikanismus," *The Technology Review,* December 1930. [1]* 2

"Night Watch at Obernach," *The Technology Review,* December 1932. [4] 6

Verteilung der hydraulischen Energie bei einem lotrechten Absturz (doctoral dissertation, Technische Hochschule, Karlsruhe) Oldenbourg, Munich and Berlin 1933. [5] 10

"Model Research on Spillway Crests" (with L. Reid), *Civil Engineering,* Vol. 5, No. 1, 1935. [8] 47

"Discharge Characteristics of the Free Overfall," *Civil Engineering,* Vol. 6, No. 4, 1936. [11] 52

Discussion of Langbein's "Pressure Distribution and Acceleration at the Free Overfall," *Civil Engineering,* Vol. 7, No. 7, 1937. [12] 56

"Modern Conceptions of the Mechanics of Fluid Turbulence," *Transactions ASCE,* Vol. 102, 1937. (Excerpt from closing discussion.) [13] 57

"Experiments on the Mechanics of Sediment Suspension," *Proceedings Fifth International Congress of Applied Mechanics,* Cambridge, Massachusetts, 1938. [20] 62

"Criteria for Similarity in the Transportation of Sediment," *Proceedings Hydraulics Conference,* University of Iowa Studies in Engineering, Bulletin No. 20, 1940. [25] 67

"Suspension of Sediment in Upward Flow," *Investigations of the Iowa Institute of Hydraulic Research 1939-1940,* University of Iowa Studies in Engineering, Bulletin No. 26, 1941. [28] 84

"Evaluation of Boundary Roughness," *Proceedings Second Hydraulics Conference,* University of Iowa Studies in Engineering, Bulletin No. 27, 1943. [32] 93

Discussion of Moody's "Friction Factors for Pipe Flow," *Transactions ASME,* Vol. 66, No. 8, 1944. [36] 105

"A General Stability Index for Flow Near Plane Boundaries," *Journal of the Aeronautical Sciences,* Vol. 12, No. 4, 1945. [37] 107

Discussion of Bakhmeteff and Allan's "The Mechanism of Energy Loss in Fluid Friction" (with A. A. Kalinske), *Transactions ASCE,* Vol. 111, 1946. [38] 110

"Gravitational Diffusion from a Boundary Source in Two-Dimensional Flow" (Sixth International Congress for Applied Mechanics, Paris, 1946), *Journal of Applied Mechanics,* Vol. 14, No. 3, 1947. [40] 120

"Use of the Low-Velocity Air Tunnel in Hydraulic Research," *Proceedings Third Hydraulics Conference,* University of Iowa Studies in Engineering, Bulletin No. 31, 1947. [44] 124

*Bracketed numbers indicate chronological order of publication. *See* Complete List of Publications, p. 605.

"Diffusion of Submerged Jets" (with M. L. Albertson, Y. B. Dai, and R. A. Jensen), *Transactions ASCE*, Vol. 115, 1950. (Awarded Hilgard Prize, 1951.) [49] 139

"High-Velocity Flow in Open Channels —Design of Channel Expansions" (with B. V. Bhoota and E. Y. Hsu), *Transactions ASCE*, Vol. 116, 1951. [55] 165

"On the Growth and Decay of a Vortex Filament" (with H.- C. Hsu), *Proceedings First National Congress of Applied Mechanics*, 1951. [57] 182

"Experimental Investigation of Fire Monitors and Nozzles" (with J. W. Howe and D. E. Metzler), *Transactions ASCE*, Vol. 117, 1952. [58] 188

"Present-Day Trends in Hydraulics," *Applied Mechanics Reviews*, Vol. 5, No. 2, 1952. [61] 217

"Gravitational Convection from a Boundary Source" (with C. S. Yih and H. W. Humphreys), *Tellus*, Vol. 4, No. 3, 1952. [62] 219

"Free Convection over Parallel Sources of Heat" (with W. D. Baines and H. W. Humphreys), *Proceedings of the Physical Society*, B, Vol. 66, 1953. [63] 229

"Cavitation in the Mixing Zone of a Submerged Jet" (Eighth International Congress for Applied Mechanics, Istanbul, 1952), *La Houille Blanche*, Vol. 8, No. 1, 1953. [64] 236

"Measurement of Velocity and Pressure Fluctuations in the Turbulent Flow of Air and Water," *Riabouchinsky Anniversary Volume*, Ministère de l'Air, 1954. [66] 243

"Form Drag of Composite Surfaces" (with T. T. Siao), *Proceedings Second U.S. National Congress of Applied Mechanics* (1954), ASME, 1955. [68] 248

"Turbulent Diffusion across a Density Discontinuity" (with J. Dodu), *La Houille Blanche*, Vol. 10, No. 4, 1955. [70] 254

Discussion of Chow's "A Note on the Manning Formula," *Transactions AGU*, Vol. 37, No. 3, 1956. [71] 262

"Experiments on Two-Dimensional Flow over a Normal Wall" (with M. Arie), *Journal of Fluid Mechanics*, Vol. 1, Part 2, 1956. [73] 264

"Diffusion in the Lee of a Two-Dimensional Jet," *Proceedings 9th International Congress of Applied Mechanics* (Brussels, 1956), Vol. 1, 1957 [75] 277

"Characteristics of Flow over Terminal Weirs and Sills" (with P. K. Kandaswamy), *Journal of the Hydraulics Division*, ASCE, Vol. 83, No. HY 4, 1957. [77] 286

"Turbulence Characteristics of the Hydraulic Jump" (with T. T. Siao and S. Nagaratnam), *Transactions ASCE*, Vol. 124, 1959. (Awarded Hilgard Prize 1961). [78] 299

"Répartition de l'énergie dans des zones de décollement" (doctoral dissertation, University of Paris), *La Houille Blanche*, July 4, 1959. [84] 324

"On the Bernoulli Theorem for Turbulent Flow," *Tollmien Festschrift—Miszellaneen der Angewandten Mechanik*, Akademie-Verlag, Berlin, 1962. [86] 351

"Current Trends in American Hydraulics," *Physical Sciences, Some Recent Advances in France and the United States*, New York University Press, 1962. [87] 357

"Interfacial Mixing in Stratified Flow" (with E. O. Macagno), *Transactions ASCE*, Vol. 127, 1962. [91] 369

"Energy Transformation in Zones of Separation," General Lecture, *Proceedings IAHR 9th Convention*, Dubrovnik, 1961. [94] 396

"The Role of the Froude Number in Open-Channel Resistance" (with H. J. Koloseus and J. Davidian), *Journal of Hydraulic Research*, Vol. 1, No. 1, 1963. [96] 408

"On the Role of Eddies in Fluid Motion," *American Scientist*, Vol. 51, No. 3, 1966. *Science in Progress*, Yale University Press, 1966. [97] 411

"On the Art of Advancing the Science of Hydraulics," *Proceedings 1st Australasian Conference* (1962), Pergamon Press, 1963. [98] 457

Discussion of "Sediment Transportation Mechanics: Suspension of Sediment and Density Currents," *Journal of the Hydraulics Division*, ASCE, Vol. 90, No. HY 1, 1964. [100] 470

"Critical Analysis of Open-Channel Resistance," *Journal of the Hydraulics Division*, ASCE, Vol. 91, No. HY 4, July 1965. [105] 475

"Cavitation and Energy Dissipation in Conduit Expansions" (with V. Jezdinsky), *Proceedings of IAHR 11th Congress*, Leningrad, 1965. [106] 500

"The Bernoulli Theorem," *Proceedings JSME*, Vol. 68, No. 562, November 1965. [107] 506

"On a Matter of Latitude in Pronunciation," *La Houille Blanche*, Vol. 20, No. 6, 1965. [108] 512

"La Investigación y el Ingeniero," *Simposio el Ingeniero ante la Ciencia y la Tecnología Contemporáneas*, Colegio de Ingenieros de Venezuela, Caracas, 21-28 March 1965. [109] 513

"Fluctuation of Pressure in Conduit Expansions" (with V. Jezdinsky), *Journal of the Hydraulics Division*, ASCE, Vol. 92, No. HY 3, 1966. [110] 517

"Jet Diffusion and Cavitation," *Journal of the Boston Society of Civil Engineers*, Vol. 53, No. 3, 1966 [111] 529

"German-American Observations on Educational Reform" (with E. Naudascher), *Journal of Engineering Education*, Vol. 57, No. 1, 1966. [112] 545

"On the Use of Models in Fluids Research" (with E. O. Macagno), *Proceedings First International Conference on Hemorheology* (Iceland), Pergamon Press, London, 1968. [118] 552

Preface to *Hydrodynamics by Daniel Bernoulli and Hydraulics by Johann Bernoulli*, translated from the Latin by C. Carmody and H. Kobus, Dover, New York, 1968. [119] 557

"Iowa's Quest for Curricular Balance," *Engineering Education*, Vol. 59, No. 3, 1968. [120] 566

"Jet-Induced Circulation and Diffusion" (with C. Iamandi), *Journal of the Hydraulics Division*, ASCE, Vol. 95, No. HY 2, 1969. [121] 570

"Work-Energy Equation for the Streamline," *Journal of the Hydraulics Division*, ASCE, Vol. 96, No. HY 5, 1970. [125] 583

"International Heritage," *Engineering Education*, Vol. 61, No. 3, 1970. [126] 595

"Pierre Danel's Influence on American Hydraulics," *La Houille Blanche*, Vol. 25, No. 6, 1970. [128] 598

"Growth $\neq$ Progress?" *Mechanical Engineering*, Vol. 93, No. 4, 1971. [130] 600

COMPLETE LIST OF PUBLICATIONS 605

CO-AUTHOR INDEX 611

AUTHOR INDEX 613

SUBJECT INDEX 615

Hydraulics, Mechanics of Fluids, Engineering Education

Selected
Writings of
Hunter Rouse

AMERIKANISMUS

In Germany an Engineer Learns the Truth about Americans

By Hunter Rouse

LIESL, the German *Zimmermädchen* who cares for my room, just came in to turn down the bed-covers, lay out my pajamas, and wish me a good night's rest. The sudden question, "Liesl, what do you think of America?" rather took her unawares, accustomed as she is to my ungrammatical inquiry as to the next day's weather, but she must have caught my meaning, for her reply was exactly what I sought. "America is a land of money and money-chasing people; I'd like to work there — but live in Germany." That is the *Amerikanismus* concept in its simplest, commonest form.

We "money-chasers" are surely the most highly blessed race of all the world insofar as gratis criticism is concerned. From the native magazine writer, to the self-exiled American in permanent residence on the continent who sends his opinions home for publication — not to mention the European who is prolific in his statements after a four-day tour of the States — everyone who has a single thought on the matter gives it freely, till the burden of analyzed error would seem almost more than any shoulder could bear. But we are either inherently thick-skinned or so toughened by experience that we seldom bother even to smile tolerantly. To be sure, we have solved very easily and typically this problem of unending censure, but with it we have lost in large measure the ability to listen when some view of actual worth is presented — to distinguish valuable comment from sheer headline flourish.

A foreign country offers the most promising field for search; but just as one must go alone and live the life of the country to learn that country's life, so must one do to learn that country's genuine thoughts about our land of wealth and promise across the seas. A year of study as a hydraulic engineer in a different world has surely given me a new outlook upon American civilization — that is inevitable and entirely a self-contained process — but through the eyes of the German people has come a keener comprehension of the life we are wont to live.

To understand the German outlook, one must consider for a moment or so under just what conditions the German mind has to function. Twenty years ago the empire had reached almost the pinnacle of world prestige — *Kultur*, colonies, manufacture, education, science, engineering, the arts — in all these Germany held avowed leadership, with only one or two other powers who were her equal. She developed a national pride which became a byword to her people, literally a propaganda. Then came the Great War; who was to blame is quite beside the point; surely not the *Volk*, for when one listens to the serious comment of those who lived through the crisis, one even begins to question the Allied doctrines the slightest trifle!

Through Versailles, political revolution, inflation of currency, a decade and more of reparations payment with tremendous taxation and millions of inhabitants without employment, the country has been reduced to almost a shadow of the old empire. True, the people are the same, and their pride can never be conquered, be it the result of habit or genuine belief. But with empty purses, compulsory taxes that make accumulation utterly impossible, is it little wonder that the *Volk*, whether of Liesl's class or the comparatively well-to-do, look upon the neighbor across the ocean with a feeling ranging between envy and bitterness?

Simultaneous with this purely financial change came further alterations, each with its own peculiar influence upon the German mind and vision. The "Made in Germany" regime is at such a stand that American products are now widely used in Germany, and American enterprises are controlling many a German firm. Previous military stiffness is being rapidly superseded by a prevalent turn to gymnastic and competitive sport. The morals of the post-war generation have made an almost radical departure from the old-age decorum, with an accompanying freedom and naturalness that would abash even an American. And the experiment with smoothness and efficiency in republican rule has made a heated politician out of the most taciturn citizen of former years. All these points must one bear in mind if one will weight the German views to their true intrinsic worth.

Liesl gave us the *Amerikanismus* concept in its basic state. Let us advance a step to the group gathered perforce for the evening meal in a Münchener Pension, the most assorted collection of average human beings one could wish. The elderly *Pensionsmutter*, a hearty Royalist with a portrait of the ex-Kaiser above her bed, is a fair specimen of her order. She likes Americans on the whole, in spite of a resigned tolerance for their obvious poor breeding, for they pay well and are not so class-conscious as others, and she has fallen in love with each male guest in turn for countless years. Next in importance comes the equally elderly *Northgerman* spinster, who has spent a good two decades teaching languages in England, and returned to Germany with the typical English aloofness toward all things savoring of the American. (This, it must be noted, seems to be passed on to every student of the "real" English language on the continent.) But she is surprisingly well informed on all that exists on the other side — so well that she was once bluntly enlightened upon the sundry points in which an American considered her the most in error, and behold, in this former proudest of the proud of the old continental *Kultur* we now have a staunch supporter of the tremendous vitality and youth that exists in the new world.

From there we may pass on to the four or five young girl students, for the most part excellent sketches of the new generation (to our Lindsey is attributed this metamorphosis), with a lone example of the inexperienced pre-war type. The others are of every sort: an army officer or so at times, artists and students of several nationalities, and the general flow of two-day guests. Dinner-table conversation is generally lively enough with the *Northgerman* spinster at the helm and a Norwegian pessimist to spur her on; after an amused allusion to this and that American weakness, it takes a fairly international turn, especially when the unlikely happens and the Americans present can speak the language. But once let the room be free of all visitors from the land of wealth and promise, and not a fault of ours but what is brought to light, beaten, pummeled, stripped naked of excuses before the grim multitude — and then filed carefully away in memories, to be reviewed at the next round-table.

RECENTLY I attended a travel lecture by a professor who has world-wide repute in his field, after one of his trips through the States. He is a member of several American professional societies, follower of the best American engineering journals, and never hesitates to make friends with Americans who chance his way. However, this lecture was for the Germans, and consisted principally of pointed jokes at the expense of American fallacies and the proof that without German brains, American wealth would be useless. It is what the *Volk* seem to want, and it is only natural that he, as a popular figure, would give it to them.

Newspapers and magazines follow much the same trend: cartoons of the brain-pocketbook alliance, and ever-recurring citation of the case of this and that German scientist or engineer who designed this and that famous American project. An article in the latest number of a prominent weekly devoted two pages of illustrations to the topic "What a German would find highly inconvenient in America" — an example or so will suffice: the hotel night-table containing *only* a radio set and an unhandy form of telephone; the one-arm quick-lunch chairs "made purposely as uncomfortable as possible so that the guest will be glad to leave as quickly as he can and make room for the next!"

So much for the feeling that shows on the surface. It seems bitter, unfriendly, almost insulting at times, but if we consider once again our earlier discussion as to underlying causes, the whole resolves itself into what is, really, envy, jealousy that such an adolescent nation can without effort take her place in the world while she, Germany, is chained almost to a standstill. What she says and publishes is not for us here across the seas; it is for herself, an attempt at self-consolation. Yet we are still apt to take offense at this effrontery, we are prone to forget that the German may speak heatedly but very seldom emptily, and that there is a kernel of truth, whatever his underlying purpose may be, in every fallacy he finds in us. It is equally childish and useless for us to call bad names in return or to hold our fingers in our ears; these fallacies are after all our fault, and at that not beyond repair.

Our migration every summer to do Europe in eight or ten weeks has played an undue part in branding America as a country of unbred, common, money-made people. Our tourists may be typical of our nation — surely Europe has every cause to draw this conclusion — but heaven help our land if they are! Picture a crowd of loud-mouthed, carmine-lipped women thumbing over every bit of lace or jewelry in a shop — "No, we aren't buying, just looking!" Picture the throngs at the "American" bars, tanking up till the next trip across. Listen to the sniffs of disdain at an old cathedral that has withstood the test of centuries — "Who ever built the way *we* are doing?" Yes, indeed, Europe caters to us; we have made Oberammergau the commercialized production that it is; we keep scores and scores of towns in business by our annual visits. They cater to us — that is business; but respect is another matter — what barker at a county fair ever respected the throngs that filed from one exhibit to the next?

Behind this conscious and justifiable belittling and ridicule and wholesale criticism, there is a far more respectful feeling, *but for the land, and not the people!* Liesl admits she would go across immediately were the opportunity offered. Scores of students and young engineers have asked me what chances they would have were they to seek their fortune in our country. Not a family but what has relatives or friends already settled in the States, from whom come glowing tributes to the opportunities of this land of promise. But beneath it all is a trace of doubt — Liesl phrased it rather aptly when she ventured, "I'd like to work there — but live here."

No doubt Liesl was only repeating the ideas of another, and would find America a perfect paradise of amusement; we cater particularly to minds like hers. Therein lies what the German considers the distinguishing feature between our two worlds: we have developed Civilization, they, *Kultur.* We live a perfected, systematized, bodily and sensually luxurious existence at a tremendous pace, seeking with ever greater speed a goal whose nature we do not even bother to wonder about; that is Civilization. They feel they have long since found their goal — the finer cultivation of the mind — and they can enjoy their higher life without haste. The German visitor in America misses his accustomed leisure in which to appreciate life. The American in Germany curses the lack of conveniences.

So, be it strictly true or not, we have given the Germans good cause to believe that the United States is a land of wealth and wealth worshippers. How can they believe otherwise? Tourists fairly reek with money, or try to give that impression. Press reports are profuse in their tabulation of huge American financial transactions. An expenditure of hundreds of thousands is sheer waste and display in the light of the German's forced thriftiness. Luxuries to a German are things he has learned to do without, yet to an American they have become the very mainstays of existence. With reports of the recent market panic and unemployment menace in America, Germany almost gasped; the U. S. A. in financial trouble? The rumors continued to come, but so did the tourists and the reports of huge expenditure and business deals. Hence our money crisis is largely a myth to the average German mind and America is still the land of millionaires.

To the German it would seem that we have the horse and cart most assuredly in reverse order: had our wealth been the obvious result of leadership in education, science, and the arts; or even had the growth of all been more simultaneous, it would have been respected. But financial prestige came first, and what nation could not buy the rest once given sufficient resources? Perhaps the respect for more than our wealth will follow in the course of the years, but only when the German race begins to feel that America is producing from its *own* talent, and not buying from the rest of the world.

FROM the surface matter, the self-alleviation of her wounded pride by ridicule, we have gone a step deeper into Germany's feeling toward us; but there is yet greater depth more seldom encountered. In the number for March, 1930, of the "*Deutsche Rundschau*" I chanced upon an unusual article, "*Amerikanismus als Schlagwort und als Tatsache*" by Theodor Lüddecke. He has departed entirely from the usual attitude, and in unbiased writing, so candid with his own countrymen that America is discussed in almost flattering terms, he analyzes "Americanism" in its actual and hypothetical influence upon the German race. He too writes, not for us, but for his own people; yet, though his general theme as applied to Germany does not concern us at this time, many of his remarks are so pertinent that I quote a number as I translate their meaning.

He begins, "The term *Amerikanismus* is today almost at the center of every explanation of political, economic, or cultural questions. The commanding political position of America, which she has won directly through her economic powers" brings forces to bear which "influence the destiny of us all to an ever greater degree. . . . We have no longer the definite choice of acceptance or refusal — the American forms of life are, through economic channels, simply forced upon us."

But he notes rightly that what Germany terms *amerikanisch* is often far more "Americanized" than the original — modern youth, for instance — and develops the interesting thesis that "Americanization" is simply a process of national evolution, that the United States happened to reach this stage before the rest, and that "Americanization" would just as surely have come about had the U. S. A. never existed!

The American, as such, is in his life-force still younger than the European, but in his methods of living and working already far older. Why? Because we are not, like European states, the direct product of centuries of development; we took instead a strength and self-dependence from Europe to a land of endless resource, a land filled with the challenge to be conquered. "But no thought of self. No consideration for matters without direct practical value. There is not time for that. This part of the earth can only be conquered through toil. And thus did work become the leading principle of American religion and philosophy. This the German philosophers term Pragmatism and do their best to refute. But Pragmatism cannot be 'refuted'. . . ."

"America knocks at our door. What is this America anyhow with which we have to do? It is basically an industrial mass of extremely lively, adaptable youths, who — optimistic, humorous, athletic — seize every task within possible reach, are without limit in endurance, and feel themselves equal to every situation. Perhaps they are, in the words of a German professor, uneducated in the higher sense (*höchst ungebildet*), but they are, however, thoroughly trained (*gut durchgebildet*) — physically as well. They are at least well enough educated to buy up our German factories."

Our abnormal addiction to sport has not been without its marked result. He remarks that the Americans, from laborers to business directors, taken as a whole, are more in form than the inhabitants of almost any other land. An American 'stands out' when he strolls through Berlin. He is radiant with an energy that everyone notices. "*Wie kommt denn das?*"

"The American features are freer from care. In no land in the world is there so much laughter as in the U. S. A.

Granted — the American has it easier; the natural opportunities of his land offer an almost boundless field of activity. He permits no damming-up of his intellect. European visitors frequently affirm that America is problemless. Little wonder. When the whole land is 'in form' . . . when the people and the conditions are healthy, then even the unsolved problems are trifles. The people live intensively; they live not only with the head, but with the whole body. They have so great a surplus of life-force that it often takes the form of childish gayety. One is always surprised anew at the natural freshness that the American spirit shows.''

But he does not deny that there exists in America a certain monotony, and that directly on that account the higher classes seek closer connection with Europe, where a great warmth seems to be given forth. To him it is clear that the robust instinct of our early pioneer, his natural freshness, is already on the wane.

His views on comparative politics should be encouraging to us to say the least. One returning from America and viewing the German situation, he observes, is wont to quote Romain Rolland's ''Too many clever minds are busying themselves with the State.'' As a German voter once remarked to me, ''Every German wants to have his say in politics; you Americans are wise — you leave it to the politicians.'' Lüddecke continues, ''Men who otherwise understand each other very well, become irate once the discourse touches on politics. Talk begets neither potatoes nor automobiles. . . . America has developed herself through work.'' Nothing can satisfy all parties as well as a ''rational production of new products, through which all pull on the same rope. That is basically the secret of the American economic morale. . . .

''In other words, the economic attitude of mind has so taken hold of political thought, that politics have received an entirely different character. America maintains a strong army and a strong navy — however, when one talks in terms of peace, that is no empty way of speaking. This attitude is entirely an outgrowth of the American principle of predominant economic thought. This has nothing to do with Pacifism. Pacifism is — say what one will — surely no sign of strength. America is conquering the world, perhaps to a far greater extent today than any nation has ever done. She conquers, however, with different weapons — that is, without weapons, by means of economic superiority. Militarily she lies on the defensive; from the economic point of view she is absolutely on the offensive.''

If we are content to be typically American, and neglect to fill in the countless little ''exception'' sub-headings which Herr Lüddecke has left for us in his somewhat rosy outline or correct his obvious errors, we may now smile a bit triumphantly. ''Herr Lüddecke has found us the greatest nation of all times, and he's not so far wrong at that; we're showing the old world a thing or two in greater business methods.'' Ha — ''money-chasing people''! ''What,'' asks Lüddecke of his own nation, ''is the use of all the cheap patent ice-chests and automobiles, when the spine of the *Volk* is crushed? The economic moment can be for us only the means — never the end.''

But what is it to America? If only a means, then to what purpose? Is there a single soul in our nation who knows why we strive, why we build, why we live at our tremendous pace? Was Liesl far from the truth when she branded us as merely ''money-chasers'' — have we a greater goal in mind than what the dollar signifies? No doubt the reader will balk a bit at this point; ''It's not the dollar we're after; it's Progress!'' Good enough; but whither?

Lüddecke tells of the technically-inclined Englishman who read Goethe's ''Faust'' to the end, and then exclaimed triumphantly — ''Yes, but the German Faust died as an *engineer!*''

At the moment that is the most a great many of us would think to wish for. ''Decisiveness, magnitude of vision, the joy of doing, and cold calculation in economic matters — America teaches these,'' says Lüddecke. Perhaps we in turn can take a lesson or so from Germany. ''We have all that we need, and more than we can ever use. We have large bank balances, villas, autos, yachts — the one thing we lack is — Time! Our minds are fixed on production. And now the whole machine is dragging us along. When our nation was young, perhaps we would have willed it otherwise — now that it could be otherwise, we have lost the will to change.''

Lüddecke may be right, that it is only the will we have lost; and we are so imbued with our doctrine of Progress, and all the mass production, luxury, speed, and accompanying pursuit of the dollar that the term embodies, that we long for nothing more. Or it may be that we have lost, not only the will, but moreover the ability, to discover in life the intangible fineness that we have obscured with our maze of practical contraptions. The *Northgerman* spinster once told me that the *Kultur* of her country was built upon the foundation of the Latin and Greek languages; a late American engineer of world repute decried the impractical mind of the German professor who devoted a lifetime to translating the Latin Classics into ancient Greek. Both views are a trifle ludicrous, but they symbolize beautifully the conceptions of two narrow worlds. America is already the arrogant possessor of one-half the apple of greatness, yet blind to the fact that the other half is missing.

Northern end of the Experiment Station. In the distance a range of the Tyrolean Alps can be seen, and in the foreground is the bed of the Obernach River. Practically the entire tract owned by the Research Institute for Water Power and Hydraulic Engineering is visible

Night Watch

"The World's Largest Open-Air

By Hunter

NOT at all surprising is the fact that it was Oskar von Miller of Munich who first conceived the idea for an outdoor hydraulic experiment station in the foothills of the Bavarian Alps. He had already developed the Deutsches Museum into an object of international wonder, and united the power plants of Bavaria into a system known as the most perfect electrically of any on the globe. And he has now sponsored this latest endeavor so successfully that it has been recognized as the greatest open-air laboratory for hydraulic engineering in the world.

Despite Germany's previous success with small-scale experiments, there are still engineers who look with doubt upon the results of tests on models reduced a hundred-fold from natural proportions and often not even geometrically similar. A theory of hydraulic similitude is logical enough in its way, but without actual proof of the dependability of its conversion factors, not all will accept it unconditionally. Furthermore, there are experiments whose very nature prohibits scale reduction, yet which are vitally essential to industry and engineering.

Because of this need for large-scale tests, the Research Institute for Water Power and Hydraulic Engineering was formed in 1926 as a department of the Kaiser Wilhelm Society for the Promotion of Science. It was made a practically independent unit, financed by the national, Bavarian, and Munich governments, together with a number of private concerns interested in furthering the science.

After a careful search for a site fulfilling the qualifications of land, water supply, and proximity to Munich, the Board decided upon the valley south of the Walchensee, a mountain lake of Bavaria, where all three requirements were met to a high degree. At this point the diversion canal of the Isar River joins the Obernach, a stream emptying into the Walchensee, and provides the principal supply for the power plant at the lower end of the lake. The actual station grounds include an elongated area of some 25 acres, bounded on either side by the Obernach and the Mittenwald-Munich state highway. Thus the station has at its disposal ample space in a locality abounding in actual hydraulic developments, together with a steady water from which it may draw without charge as much as 425 second feet.

When I first visited this experimental paradise in the autumn of 1929, the plant was already in such a healthy state of development that I returned for three months the following summer. By then the initial installations had been completed and rated, with the opening series of experiments well under way. These involved a rigid comparative test of the four standard European and American methods of discharge measurement (weir, current-meter, salt-dilution, and salt-velocity), and the equipment was planned for these runs, yet made easily adaptable to future demands.

At the southern end of the station is a permanent distrib-

The nearby Walchensee Power Plant where important tests have been made

at Obernach

Hydraulic Experiment Station"

ROUSE

uting basin holding 88,250 cubic feet of water, of sufficient size to serve as many as four or five separate outlets to individual experiments; water is drawn from the Isar canal, and the surface level kept constant by a 30-foot spill weir wasting into the Obernach. From this basin leads a main canal about 1,900 feet in length and divided into four sections of varying pattern; the last of these sections of the experimental canal is rectangular and of concrete for the more important experiments.

At the end of the concrete flume (about 8.2 x 6.6 feet in cross section for a peak flow of 141 second feet) is located the large concrete measuring basin, for the actual volumetric check upon discharge through the flume. The care taken in calibrating this basin is typical of the thoroughness of the station methods: the entire structure was measured by tape and transit and the volume-depth relation calculated geometri-

cally; as a further check, the same function was tabulated by weighing successively volumes of water and introducing them into the basin, reading the gages in the basin pit after each addition; later the capacity was even measured photographically, by a method similar to aerial mapping. Thus the basin has been calibrated to its maximum capacity of 52,000 cubic feet with an exactitude that allows a total error in discharge measurement of $\pm$ 0.1%.

With this measuring basin as the key of the station's research accuracy, other experiments may be based either upon direct use of the basin or upon measurement by instruments which have already been so rated. Thus all experiments — whether in the permanent concrete channel or at any point in the area devoted to temporary model set-ups — may have a conclusive volumetric check.

A staff of eight or ten engineers is on duty throughout the summer months when active experimentation is in progress, a major run requiring the assistance of all, since the size of the layout necessitates constant attention at numerous points over a large area. A telephone system to all points keeps the staff members in close touch with each other during the run. The station has its own metal-working shop, where much of the apparatus is constructed, together with space for drafting and chemical analysis; in the headquarters cabin are living accommodations for the several men who are in permanent residence.

Delicate measuring instruments and clever error-trapping schemes all too numerous to describe are typical features of this unique outdoor laboratory; special tests for seepage through the concrete flume and basin walls; injection tanks providing constant discharge over great length of time; electric water gages; chronometers giving permanent ink records of experimental data and sending the same one-second impulses electrically to all parts of the station; and added to this a corps of men who have been with the Research Institute since the days when they were first putting the plans to paper.

Measuring the discharge of a mountain stream by the salt-dilution method

LONG will the memory of my first night watch with one of these assistants remain fresh in my mind. A ridge of the Tyrolean Alps was visible toward the south, but the night was black with the shadows of nearby pines. From the west came the dull roar of the Obernach rapids, while the sound of a steady downpour just beside the door of the little control house marked the lower end of the gray stretch of experimental canal fading into the darkness.

Monotonously ticklish business this, reading a dozen

depth gages at hourly intervals throughout the night; only a donning of headphones to note the buzz of contact when the gage point touched the surface, but — a meter of chilly mountain water below one's feet, hobnails catching on the tiniest irregularities of the concrete brink, and drowsiness playing tricks with one's eyelids. Yet when a run still lacks twenty-odd hours till completion, night is of little consequence. Rain — that surprisingly frequent and unforetellable factor of Upper-Bavarian summers — was the only matter of sufficient import to vary the routine of the station. With the advent of a shower, a run was immediately discontinued; but night meant only a change of shift, and pine shadows on Alpine foothills.

With the depth gages tabulated for another hour, I joined my comrade in his search for the elusive groundwater standpipes. Of very little concern, to be sure, when the water to be measured is flowing in a concrete flume — and of particularly little concern when one is sleepy and cold. But who knows — perhaps a fraction of one percent closer to the ultimately accurate if seepage losses are not ignored, and the station prides itself on excluding error; even a negative answer is better than none at all.

At midnight we pulled the levers that sent the discharge tumbling for several minutes into the large basin just behind the control house, where a chronometer recorded the exact instant of arrival and departure of the twin Tainter gates at their sills. The basin full and

the record dated and filed, we wheeled through the darkness a half mile to the cabin at the head gates of the canal to sup on beer and sausage and hardy gray bread, awaiting the quieting down of the rhythmic sway of the water in the measuring basin.

Beer is beer, and waves in such a basin do not exhaust themselves in five minutes — but perhaps at one o'clock I began my vigil over the basin gages, deep in a covered well at the side. Vigil? — and what a one! A glass standpipe before me, with a tiny light and mirror to facilitate noting the instant the knife-edge caught the water surface; a long sheet of graph paper on which I plotted the readings to $\frac{1}{20}$ of a millimeter against one-minute intervals of time of night; and an envelope of Austrian cigarettes that made me curse the tax that had sent common American varieties to $2 a pack.

This meant some hundred or more readings of water depth for only one filling of the

Left: Reading water elevation electrically in concrete section of experimental canal. The vertical pipes are the electrodes for the Allen salt-velocity method. Below: Bavarian laborers constructing model of Elbe River for Engels' large-scale experiments

measuring basin; again an apparent exaggeration of necessary care. But such pains are more comprehensible when one follows the waves in the graph — a breeze, and the water in the basin begins to sway again with accompanying variations in the readings; and only by a long watch at the gages may one accumulate enough data for a satisfactory average. Completely justifiable is this care when one realizes that the entire success of the run depends upon the volumetric records of this basin.

Still other night shifts with a book of logs and Rehbock's long weir formula before me, roughly checking the accuracy of the run before conditions changed; days by burettes and evaporating dishes analyzing samples of discharge into which . . . .

salt solution had been injected; hours with long paper chronographic ribbons tabulating current meter turns or enlarging curves of the Allen salt-velocity method — yet ever with the laborious, careful, error-chasing effort which characterized that first night shift.

German genius for thoroughness, for progress, for improvement in scientific methods. Yet Germans are human too — at least the Bavarians! For who could live in those mountains, sport around in leather shorts with a sky-high feather in the cap, and not be human? Could a cold-blooded hydraulician strip off his shorts of an evening and plunge into the measuring basin to practice the American crawl? Hardly — yet they did, till the drowned mice drove them out!

Little wonder that I returned again a year ago. Finally for the fourth year in succession the Walchensee proved too great an attraction for me to resist. True, the rain was more persistent than ever, but even rain must stop sometime, and up the valley a long model for a Chinese river, the Yellow River, neared completion. For Geheimrat Engels, father of small-scale laboratories, had come to Obernach a second summer to substantiate his earlier indoor tests on a large-scale reproduction in this greatest of outdoor laboratories.

VERTEILUNG DER HYDRAULISCHEN ENERGIE BEI EINEM LOTRECHTEN ABSTURZ

Theoretische und experimentelle Untersuchungen
der Wirkung gekrümmter Strombahnen, ausgeführt
im Flußbaulaboratorium der Technischen Hochschule
zu Karlsruhe

Von

Dr.-Ing. HUNTER ROUSE

Master of Science
Massachusetts Institute of Technology

Mit 20 Abbildungen und 3 Plänen

MÜNCHEN UND BERLIN 1933
VERLAG VON R. OLDENBOURG

10

A. Theoretische Behandlung der Druckverteilung beim Abfluß mit gekrümmten Strombahnen.

I. Überblick über die bisherige Art der Betrachtung des Abflusses mit gekrümmten Strombahnen.

Wenn man die Reibung vernachlässigt, so ist jede Strömung, die aus der Ruhe heraus unter der Einwirkung der Schwerkraft entsteht, als eine Potentialströmung zu behandeln. Wenn sie außerdem eine ebene (zweidimensionale) Strömung ist, so können alle für diesen Idealfall ausgearbeiteten graphischen oder rechnerischen Methoden auf sie angewandt werden. Von den rechnerischen käme insbesondere die Theorie der komplexen Funktionen in Frage (siehe z. B. die von Misessche Behandlung des Überfalles)[1].

Die Anwendung dieser Methoden erfordert in jedem Falle die Kenntnis der Randbedingungen. Unter diesen sind im vorliegenden Falle nicht nur kinematische (geometrische), sondern auch rein dynamische enthalten. Außer der geometrisch durch die Gerinnesohle festliegenden Begrenzung existieren nämlich mindestens eine, im Spezialfalle des freien Absturzes zwei freie Grenzflächen, deren Form man erst durch die Lösung mitbestimmen muß und von denen man zunächst weiter nichts weiß, als daß auf ihnen der Druck konstant ist (dynamische Grenzbedingung). Das Vorhandensein von freien Grenzflächen macht die Aufgabe besonders kompliziert.

Eine solche Lösung erfordert offenbar viel Mühe und Zeit. Ihre schließlich erreichbare theoretische Genauigkeit hängt bei Benutzung einer graphischen Methode von den Grenzen der zeichnerischen Genauigkeit und davon ab, wie oft die Berichtigung wiederholt wird. Die Ergebnisse aber bedürfen trotz aller theoretischen Strenge der Methode der Nachprüfung durch den Versuch, da ja die gemachten Voraussetzungen in Wirklichkeit nicht zutreffen. Dabei ergibt sich häufig eine verblüffende Übereinstimmung zwischen Theorie und Versuch.

Die sorgfältige experimentelle Erforschung der im offenen Gerinne durch gekrümmte Stromfäden hervorgerufenen Druck- und Geschwindigkeitsänderungen ist bisher noch recht unvollkommen, so daß noch fast keine praktischen Ergebnisse über die Übereinstimmung der theoretisch ermittelten Größen mit dem wirklichen Bewegungsvorgang bekannt sind.

Bazin[2] war einer der ersten, der es unternahm, die Energieverteilung im frei fallenden Wasserstrahl zu messen und die Ergebnisse seiner Messungen dazu zu benutzen, die Anwendbarkeit des Bernoullischen Theorems nachzuprüfen. In Koch-Carstanjens „Bewegung des Wassers"[3] ist eine theoretische Grundlage zur Berechnung der im allgemeinen Fall auftretenden Kräfte gegeben, aber nur mit wenigen Anwendungen auf praktische Beispiele. Bei mehreren Beispielen sind nur die Sohlendrücke angegeben, die in Kochs Laboratorium gemessen wurden. Lediglich für das scharfkantige Wehr wurde die Druckverteilung zwischen der Wehrscheide und der Wasseroberfläche bestimmt.

---

[1] von Mises, „Berechnung von Ausfluß- und Überfallzahlen". Zeitschrift des VDI, Mai 1917, Berlin.
[2] Bazin, „Ecoulement en Déversoir". Annales des Ponts et Chaussées, 1890, Bd. XIX.
[3] Koch-Carstanjen, „Von der Bewegung des Wassers und den dabei auftretenden Kräften". Berlin, Julius Springer, 1926.

Böß[4]) entwickelte im Jahre 1929 eine Gleichung für die Beziehung zwischen Wassertiefe und Unterdruck beim Überströmen des Wassers über Abstürze verschiedener Form, wobei er zur Vereinfachung der theoretischen Behandlung oberhalb der Absturzkante eine lineare Druckverteilung zwischen Oberfläche und Sohle annahm, die für die untersuchten Fälle auch mit genügender Genauigkeit zutraf, wie die gute Übereinstimmung zwischen dem Ergebnis der Versuche und den ermittelten Gleichungen bewies.

Auch R. Ehrenberger[5]), der Leiter der Versuchsanstalt für Wasserbau in Wien, der sich auf die von Michitaro Hasumi, Professor an der kaiserlichen Universität in Fukuoka, Japan, ausgeführten Versuche stützte, fand oberhalb des eigentlichen Absturzes eine geradlinige Druckverteilung.

Theoretische Überlegungen führten den Verfasser zu der Vermutung, daß die durch die Krümmung beeinflußte Druckänderung im allgemeinen nicht nach einer geraden Linie verläuft, sondern insbesonders bei starker Krümmung der Stromfäden beträchtlich davon abweicht. Diese Überlegungen gründen sich auf Kochs sogenanntes Stützkraftgesetz (das in Wirklichkeit der übliche Impulssatz ist) in seiner ursprünglichen allgemeinen Form und führen, wie im folgenden gezeigt wird, zu einer brauchbaren Formel.

II. Anwendung des Kochschen Stützkraftgesetzes.

1. Weitere Entwicklung des allgemeinen Gesetzes.

Kurz gesagt ist die Stützkraft der Momentan-Druck, der auf die Oberwasserseite einer plötzlich quer zum Wasserstrom gestellten Scheibe im Augenblick des Einsetzens wirken würde, d. h. der gesamte Druck von Gewicht, „Zusatzspannung" und Impuls des Wassers. Mathematisch ausgedrückt:

$$(1) \quad S = D_n - Z_n + K_n = \gamma\, b \int_0^n t\, dn - b \int_0^n z\, dn + b \int_0^n m\, v\, dn = P_n + K_n = b \int_0^n p\, dn + b \int_0^n m\, v\, dn.$$

(Die Integrale sind vektoriell zu nehmen.) Dieser Ausdruck enthält die Werte, die für den ganzen Querschnitt von der Breite b gelten. Zur Vereinfachung und zur leichteren Übersicht soll jedoch für die weiteren Betrachtungen die Einheit der Breite des Gerinnes zugrunde gelegt werden, d. h. b wird für die Folge gleich 1 gesetzt.

Die nachstehende Formel für die Stützkraft hat sich besonders in den Fällen als brauchbar erwiesen, in denen man zwei von der Krümmung unbeeinflußte Querschnitte vergleichen kann, z. B. beim Wechselsprung über einer ebenen Sohle. Hier sind die Orthogonaltrajektorien vertikal und daher ist $n = t$. Die Stützkraft S erhält dabei die Größe:

$$(2) \qquad S = \frac{\gamma\, t^2}{2} + 2\, \alpha_u\, \gamma\, t\, \frac{v_m^2}{2\, g} = \frac{\gamma\, t^2}{2} + 2\, \gamma\, t\, k\ ^6)\ ^7).$$

Auch im allgemeinen Fall des Abflusses mit gekrümmten Stromfäden wird n gewöhnlich gleich t gesetzt, obgleich man im Anschluß an die Kochsche Aussage n in Wirklichkeit als die Länge

[4]) P. Böß, „Berechnung der Abflußmengen und der Wasserspiegellage bei Abstürzen und Schwellen unter besonderer Berücksichtigung der dabei auftretenden Zusatzspannungen". Wasserkraft und Wasserwirtschaft, 1929, Heft 2—3.

[5]) R. Ehrenberger, „Versuche über die Verteilung der Drücke an Wehrrücken infolge des abstürzenden Wassers". Die Wasserwirtschaft, 1929, Heft 5.

[6]) α_u ist der Geschwindigkeitshöhen-Ausgleichswert (s. Th. Rehbock, „Die Bestimmung der Energielinie bei fließenden Gewässern mit Hilfe des Geschwindigkeitshöhen-Ausgleichswertes", Der Bauingenieur, 1922, Heft 15). Sein Kleinstwert beträgt nach bisherigen Messungen etwa 1,006 für sehr glatte Versuchsrinnen, während er für große Ströme im Mittel bis auf 1,20 anwächst. Der Wert darf aber nur verwendet werden, wo die Druckhöhe überall gleich der Tiefe unter der Oberfläche ist, so daß die Energielinienhöhe die Größe $H = t + \alpha_u\, v_m^2/2\, g$ besitzt. Dieser Wert ist daher für die weiteren sich mit den „Zusatzspannungen" befassenden Untersuchungen nicht ohne weiteres zutreffend, wie später ausführlich gezeigt wird.

[7]) Th. Rehbock, „Die Verhütung schädlicher Kolke bei Sturzbetten". Der Bauingenieur, 1928. — Th. Musterle, „Die Stützkraft und ihre Anwendung zur Berechnung von Staukurven". Die Wasserwirtschaft, 1929, Heft 6—7.

Die Stützkraft wird sich also mit dem Gesamtdruck P ändern. Daher beträgt der durch die lotrechte Tiefe t ausgedrückte Verlust an Stützkraft, der dem Reibungswiderstand zuzuschreiben ist:

$$(9) \qquad \Delta S_x = V_x = \Delta P = \gamma\, t\, \Delta H.$$

Die Kurven an den beiden Enden des abgeschnittenen Teiles des Wasserstromes in Abb. 3 können jetzt durch die Vertikale ersetzt werden. Durch Einsetzung von Gleichung (7) in Gleichung (8) erhält man:

$$(10\,a) \qquad P_1 + \frac{\gamma\, Q^2}{g\, t_1} - P_2 - \frac{\gamma\, Q^2}{g\, t_2} + \Sigma\,(p_b \cos \beta) - V_x = 0$$

$$(10\,b) \qquad \frac{\gamma\, Q^2\, \mathrm{tg}\, \alpha_1}{g\, t_1} - \frac{\gamma\, Q^2\, \mathrm{tg}\, \alpha}{g\, t_2} + G - \Sigma\,(p_b \sin \beta) - V_y = 0.$$

Gleichungen (10 a) und (10 b) können ganz allgemein auf den Abfluß zwischen parallelen Wänden Anwendung finden und bei genauer Beachtung des Wertes b bei veränderlicher Breite auf alle Fälle ausgedehnt werden. P_b enthält nicht nur den Sohlendruck, sondern auch den Druck auf quer zum Strom stehende Wandflächen. Es ist hier angenommen, daß die Komponenten der Druckwirkungen auf die Seitenwände, soweit sie quer zur Stromrichtung stehen, sich gegenseitig aufheben, so daß das Problem zweidimensional bleibt. Sowohl Gleichung (10 a) wie (10 b) können zur Lösung des Problems benutzt werden. Da aber der Wert G nur in (10 b) erscheint und P nur in (10 a), wird der Gebrauch von (10 a) gewöhnlich bevorzugt.

Der bei gekrümmten Stromfäden entstehende Unter- oder Überdruck stellt sich für alle praktischen Fälle ein, nachdem das Wasser den kritischen Querschnitt passiert hat. Dies ist streng genommen nicht richtig, denn infolge der Reibungsverluste beginnt die Senkungskurve wahrscheinlich schon etwas oberhalb des kritischen Querschnittes. Für eine reibungslose Flüssigkeit würde die Grenztiefe unendlich weit stromaufwärts von der Querschnittsänderung liegen. Formel (10 a) kann daher vereinfacht werden durch die Annahme, daß der erste Schnitt an der kritischen Stelle liegt, wo alle Werte bekannt sind, wenn zuvor der Durchfluß auf die Einheit der Breite des Gerinnes bestimmt wurde. Da also in diesem Schnitt:

$$t_{Gr} = \frac{2}{3}\, H_{Gr} = 2\, k_{Gr}$$

erhält man für die Stützkraft in diesem Schnitt:

$$S_{Gr} = \frac{\gamma\, t_{Gr}^{2}}{2} + 2\, \gamma\, t_{Gr}\, k_{Gr} = \frac{2}{3}\, \gamma\, H_{Gr}^{2}$$

und H_{Gr} kann ermittelt werden aus:

$$H_{Gr} = \frac{3}{2} \sqrt[3]{\frac{Q^2}{g}}.$$

2. Anwendung auf den einfachen Absturz.

a) Ermittlung des Gesamtdruckes in einem Querschnitte.

Für die experimentelle Überprüfung der Formel (10) wurde eine einfache Form des Abflusses in gekrümmter Bahn ausgewählt, indem ein Wasserstrom durch eine praktisch unbegrenzt lange Rinne mit waagerechter Sohle geleitet wurde, an die sich ein lotrechter Abfall scharfkantig anschließt. Der zwischen lotrechten Begrenzungen frei abfallende Strahl wurde voll belüftet. Es wurde der größte Abfluß ausgewählt, der bei der vorhandenen Gerinneausbildung mit Rücksicht auf die Versuchsgenauigkeit noch praktisch möglich war. Für die so ausgewählte Abflußmenge von 125 l/s/lfd. m wurde das Längenprofil der Wasseroberfläche von einer 1,5 m oberhalb der Absturzecke gelegenen Stelle aus bis in den Abfallstrahl hinein gemessen und auf Millimeterpapier aufgetragen. Hieraus wurden die Werte der Wassertiefe, die durchschnittliche Neigung der Stromlinien in den Querschnitten und der Abfall der Energielinie entnommen. Da der zuletzt genannte Wert ohne Kenntnis der gemessenen Druckverteilung nur oberhalb des kritischen Querschnittes berechnet werden konnte, wurde angenommen, daß das relative Energieliniengefälle $\Delta H : l$ von dieser Stelle aus abwärts bis zum Ende der waagerechten Sohle das gleiche bleibt und weiterhin im frei fallenden Strahl auf die Hälfte abnimmt (vgl. Tabelle I).

Da der Bodendruck lotrecht wirkt, nimmt Formel (10a) folgende sehr vereinfachte Form an:

$$(12) \qquad \frac{P}{\gamma} = \frac{2}{3} H_{Gr}{}^2 - \frac{Q^2}{t\,g} - t\,\Delta H.$$

Hier ist auf Berücksichtigung der ungleichförmigen Geschwindigkeitsverteilung verzichtet (Fußnote 6).

In Tabelle I sind schrittweise die durch Anwendung dieser Formel auf die verschiedenen Stellen vom kritischen Querschnitt bis in den abfallenden Strahl erhaltenen Werte eingetragen

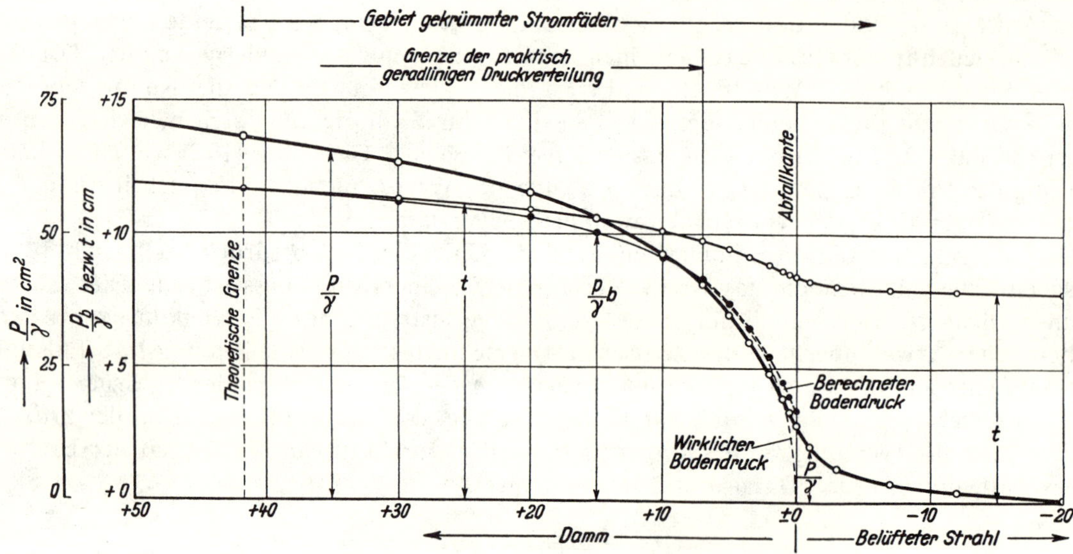

Abb. 4. Darstellung der Beziehung zwischen der lotrecht gemessenen Strahlstärke t und den berechneten Werten $\dfrac{P}{\gamma}$ und $\dfrac{p_b}{\gamma}$.

und die so gefundenen Inhalte der Druckflächen $\dfrac{P}{\gamma}$ in Abb. 4 dargestellt. Zu beachten ist, daß eine geringe Ungenauigkeit in der Tiefenangabe auf die Größe des Flächeninhaltes in der Gegend des Abfalles von großem Einfluß ist. Das gleiche gilt auch für den angenommenen Energieverlust[8]).

Wenn die Druckverteilung in der Lotrechten unter jeder Bedingung nach einer geraden Linie verläuft, dann könnte jeder Flächeninhalt durch die halbe Tiefe dividiert werden, um den resultierenden Druck auf den Boden zu erhalten. Diese Untersuchung ist ebenfalls in Tabelle I vom kritischen Querschnitt abwärts bis zur Absturzecke durchgeführt, wo der Bodendruck gleich Null werden muß. Der resultierende Bodendruck ist ebenfalls in Abb. 4 dargestellt. Man erkennt ohne weiteres, daß

---

[8]) Ein anderer Ausdruck für P kann folgendermaßen aus der theoretischen Energielinienhöhe abgeleitet werden:

$$H = \frac{t}{2} + \frac{P}{\gamma\,t} + \frac{Q^2}{2\,g\,t^2\cos{}^2\alpha}$$

$$\frac{P}{\gamma} = tH - \frac{t^2}{2} - \frac{Q^2}{2\,g\,t\cos{}^2\alpha}.$$

Hierin ist H die Höhe der Energielinie über dem untersten Punkte des betrachteten Schnittes. Daß diese Formel gleich Formel (12) ist, ist sehr schwer mathematisch zu beweisen. Jedoch durch Einsetzen der in Tabelle I gegebenen Werte liefert sie fast genau dieselben Werte bis zur Krone, von wo aus diese allmählich etwas kleiner werden als diejenigen nach Formel (12). Formel (12) hat sich für den Gebrauch als praktischer erwiesen.

einer zwischen der Sohle und der Oberfläche verlaufenden Kurve zu berechnen hat, die in jedem Punkte einen rechten Winkel mit der Strömungsrichtung bildet. Daher wird im allgemeinen Fall die oben betrachtete, plötzlich in den Strom einzusetzende Scheibe eine solche Form haben müssen, daß sie zu allen Stromlinien rechtwinklig steht. Es wird sich demnach im allgemeinen um eine gekrümmte Fläche handeln, die nur bei parallelem Strömungsverlauf in eine Ebene übergeht.

Da es praktisch einfacher ist, die Wassertiefe in der Lotrechten zu messen, ist die Länge der wirklichen Tiefenkurve n durch eine Näherungsmethode auf einen Äquivalentwert t zu reduzieren. Die Kurve soll zunächst als eine gerade unter dem Winkel α gegen die Lotrechte in ihrem tiefsten Punkte gerichtete Linie betrachtet werden. Die weitere Annäherung kann hiervon schrittweise abgeleitet werden.

Das Dreieck $a_n b c_n$ in Abb. 1 gibt pro Einheit der Gerinnebreite die in einem bestimmten Querschnitt des Wasserstromes über die ganze Tiefe n_0 wirksame Kraft an, soweit sie vom Gewicht des Wassers herrührt. Dabei ist der Querschnitt streng genommen rechtwinklig zur Strömung zu messen.

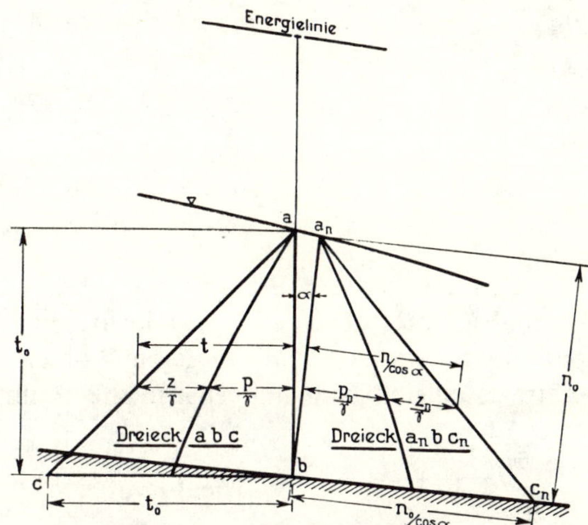

Abb. 1. Druckverhältnisse zwischen einem vertikal und einem normal zur Stromrichtung gelegten Querschnitte.

Er kann aber näherungsweise zunächst durch einen ebenen, unter einem gewissen Winkel α geneigten Schnitt ersetzt werden. Nach Koch wird diese Kraft jedoch verkleinert (oder vergrößert) durch den veränderlichen Unterdruck z_n (oder Überdruck $- z_n$). Daher lautet die Gleichung für den auftretenden Gesamtdruck im ganzen Querschnitt:

$$(3) \qquad P_n = D_n - Z_n = \gamma \frac{n\,t}{2} - \int_0^n z\,d n.$$

Entsprechend läßt sich in bezug auf den Vertikalschnitt folgende Gleichung aufstellen:

$$(4) \qquad P = D - Z = \gamma \frac{t^2}{2} - \int_0^n z\,d t.$$

Da nun der Druck an jeder Stelle nach allen Richtungen gleich ist, sind die Werte $\dfrac{p_n}{\gamma}$ bzw. $\dfrac{z_n}{\gamma}$ für jeden Punkt gleich $\dfrac{p}{\gamma}$ bzw. $\dfrac{z}{\gamma}$, und die lotrecht gemessene Tiefe unter dem Wasserspiegel ist an dieser Stelle

$$t = \frac{n}{\cos \alpha}.$$

Daher ist:

$$(5) \qquad P_n = \gamma \frac{t^2}{2} \cos \alpha - \int_0^{t \cos \alpha} z\,d t = P \cos \alpha.$$

Die in der Abweichung der beiden Tiefenlinien beruhende Ungenauigkeit bei dieser Näherung kann praktisch außer acht gelassen werden, wenn man sich die Linie n durch Aneinandersetzen kleiner, auf die Linie t gelegter Strecken zusammengesetzt denkt (s. Abb. 2). Der Winkel α ist dann ein Durchschnittswert. Diese Substitution ruft aber noch eine Druckkomponente hervor von der Größe $P \cdot \sin \alpha$ mit einer Wirkung quer zur Stromrichtung (s. Abb. 2).

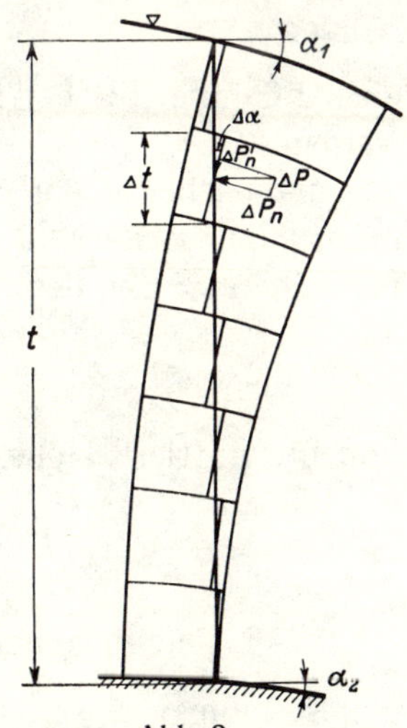

Abb. 2.
Darstellung der annähernden Genauigkeit der Gleichungen:
$t \cdot \sin \alpha = \Sigma\,(\varDelta t \cdot \sin \varDelta \alpha)$
$t \cdot \cos \alpha = \Sigma\,(\varDelta t \cdot \cos \varDelta \alpha),$
wo $\alpha = \dfrac{\alpha_1 + \alpha_2}{2}$
$\varDelta P = \sqrt{\varDelta P_n'^2 + \varDelta P_n^2}.$

Wird daher die gesamte Stützkraft durch die Größen im Vertikalschnitte ausgedrückt, so entsteht die Formel (6), worin S_s die Komponente in der Richtung der Strömung und S_n die Komponente normal zur Strömung sind:

$$(6) \qquad S_s = P \cos \alpha + 2 \gamma t k \cos \alpha \quad \text{und} \quad S_n = P \sin \alpha.$$

Mit

$$k = \frac{v_m^2}{2 g} = \frac{Q^2}{2 g n^2} = \frac{Q^2}{2 g t^2 \cos^2 \alpha}$$

geht Gleichung (6) über in:

$$(7) \qquad S_s = P \cos \alpha + \frac{\gamma Q^2}{t g \cos \alpha} \quad \text{und} \quad S_n = P \sin \alpha.$$

Das Kochsche Stützkraftgesetz besagt: „In einem durch zwei Normalschnitte begrenzten Stromabschnitt stehen die Stützkräfte im Gleichgewicht mit Eigengewicht, Wanddrücken und Reibungswiderstand." Mit anderen Worten: die Summe der horizontalen Komponenten all dieser Kräfte ist gleich Null und ebenso die Summe der vertikalen Komponenten.

Unter der Annahme eines Gerinnes mit gleichbleibender Breite ist also:

$$\Sigma X = 0 = \Sigma Y.$$

Daraus folgt:

$$(8) \qquad S_{1_x} - S_{2_x} + P_{b_x} - V_x = 0 = S_{1_y} - S_{2_y} + \\ + G - P_{b_y} - V_y,$$

worin V die dem Energieverlust zwischen den Querschnitten entsprechende Kraft ist und die Indizes 1 und 2 sich auf oberen und unteren Schnitt durch den Wasserstrom beziehen. Eine übersichtliche Darstellung der Kräfte zeigt die schematische Skizze in Abb. 3.

Die Werte P_b und G sind Funktionen, die noch immer zu verwickelt sind, um irgendwie kurz formuliert werden zu können. V kann dagegen durch

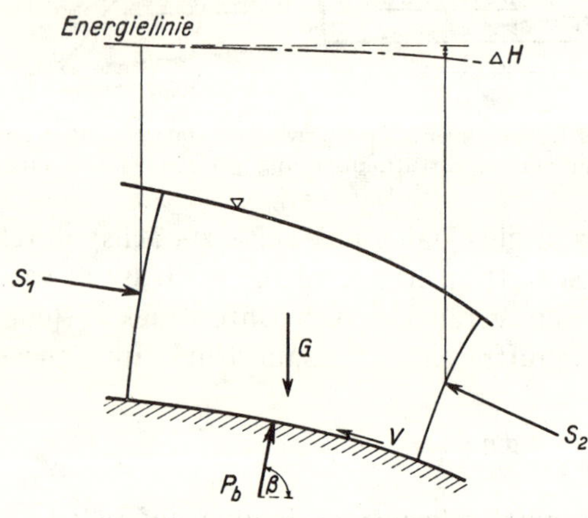

Abb. 3. Kraftwirkung auf einen isolierten Teil des Stromes.

die Abnahme ΔH der Höhe der Energielinie zwischen den beiden Vertikalschnitten ausgedrückt werden.

Auch bei einem Abflußvorgang, bei dem die Krümmung der Stromfäden eine von der Tiefe unter der Oberfläche abweichende Druckhöhe verursacht, kann die Höhe der Energielinie für einen beliebigen Punkt des Querschnittes angegeben werden zu:

$$H = h + \frac{p}{\gamma} + \frac{v_m^2}{2 g}$$

und die mittlere Höhe über der Sohle für den ganzen Vertikalschnitt zu:

$$H_m = \frac{1}{t} \int_0^t \left(y + \frac{p}{\gamma} + \frac{v^2}{2 g} \right) d y = \frac{t}{2} + \frac{P}{\gamma t} + \frac{v_m^2}{2 g} \quad \text{(s. S. 12)}.$$

Bei gleichbleibender Durchflußmenge und horizontaler Sohle für eine beliebige Tiefe t kann der Reibungsverlust mit dem Durchschnittswert der Druckhöhe $\frac{P}{\gamma t}$ ausgedrückt werden, folglich:

$$\Delta H = \frac{\Delta P}{\gamma t} \quad \text{und} \quad \Delta P = \gamma t \Delta H.$$

Entsprechend gilt für die Stützkraft bei konstantem Durchfluß und der gegebenen Tiefe t:

$$S_x = P + \frac{\gamma Q^2}{g t}.$$

die Kurve die offensichtlich falsche Bodendruckgröße von 3,42 cm an der Abfallkante erreicht, statt auf Null herunterzugehen. Mit anderen Worten, die lineare Druckverteilung darf in der Nähe der Abfallkante nicht mehr angenommen werden.

Unter der Annahme, daß die geradlinige Verteilung bis kurz vor die Abfallkante Gültigkeit hat, kann man eine wahrscheinliche Kurve des Bodendruckes ungefähr nach Gefühl eintragen. Eine merkbare Abweichung von dem berechneten Wert dürfte etwa bei Querschnitt $+7$ beginnen. Dadurch wird die Konstruktion aller Dreiecke stromaufwärts von dieser Stelle möglich, wobei also die Druckverteilung im Gebiete der letzten 7 cm der horizontalen Sohle noch unbestimmt ist.

Formel (10b) gestattet eine andere, wenn auch weniger geeignete Methode zur Berechnung des Sohlendruckes. Für ein Grundwehr mit waagerechter Sohle vereinfacht sich diese Formel zu:

$$\frac{G - P_b}{\gamma} = \frac{Q^2 \operatorname{tg} \alpha}{g\,t},$$

worin $\dfrac{G - P_b}{\gamma}$ den Flächeninhalt zwischen der Wasseroberfläche und der Sohlendruckkurve in Abb. 4 bezeichnet. Dieser Ausdruck wurde auf den betrachteten Fall angewendet und ergab einen Gesamtdruck über der Sohle, der kaum mehr als 1% größer war als der in der bereits beschriebenen Weise annähernd bestimmte Flächeninhalt. Mittels des obigen Ausdruckes kann die Druckverteilung durch schrittweise Berechnung des Flächeninhaltes jeweils für ein kleines Längenintervall vom kritischen Querschnitt bis zur Abfallkante hin allmählich als stetige Bodendruckhöhenkurve ermittelt werden. Der Rechnungsgang ist jedoch sehr mühsam und in diesem Falle gänzlich zwecklos, da eine viel einfachere Methode bereits vorhanden ist.

Tabelle I

Ausführung der Berechnungen nach den Formeln

$$\frac{P}{\gamma} = \frac{2}{3} H_{Gr}^2 - \frac{Q^2}{g\,t} - t\,\varDelta H \quad \text{und} \quad \frac{p_b}{\gamma} = \frac{2\,P}{\gamma\,t}$$

$$Q = 1250 \text{ cm}^3/\text{s/cm} \quad H_{Gr} = 17{,}52 \text{ cm}$$

$$\varDelta H = l \times 0{,}326\%, \quad l \times 0{,}163\%.$$

1	2	3	4	5	6	7	8
Querschnitt	t	$\varDelta H$	$\dfrac{2}{3} H_{Gr}^2$	$Q^2/g\,t$	$t\,\varDelta H$	$\dfrac{P}{\gamma}$	$\dfrac{p_b}{\gamma}$
cm	cm	cm	cm²	cm²	cm²	cm²	cm
$+41{,}67$	11,68	—	204,6	136,4	—	68,2	11,68
$+30{,}0$	11,31	0,038	204,6	140,8	0,43	63,4	11,20
$+20{,}0$	10,91	0,071	204,6	146,0	0,78	57,8	10,61
$+15{,}0$	10,57	0,090	204,6	150,6	0,95	53,0	10,03
$+10{,}0$	10,11	0,103	204,6	157,3	1,04	46,3	9,16
$+7{,}0$	9,78	0,113	204,6	162,9	1,11	40,6	8,30
$+5{,}0$	9,45	0,120	204,6	168,5	1,13	35,0	*7,41
$+3{,}5$	9,18	0,125	204,6	173,5	1,15	29,9	*6,52
$+2{,}0$	8,88	0,130	204,6	179,4	1,15	24,0	*5,41
$+1{,}0$	8,65	0,133	204,6	184,1	1,15	19,3	*4,46
$+0{,}5$	8,53	0,134	204,6	186,7	1,14	16,8	*3,94
$\pm 0{,}0$	8,42	0,136	204,6	189,1	1,14	14,4	*3,42
$-1{,}0$	8,25	0,138	204,6	193,1	1,14	10,4	—
$-3{,}0$	8,08	0,141	204,6	197,1	1,14	6,4	—
$-7{,}0$	7,97	0,147	204,6	199,8	1,17	3,6	—
$-12{,}0$	7,90	0,156	204,6	201,6	1,23	1,8	—
$-20{,}0$	7,85	0,169	204,6	203,0	1,33	0,3	—

* Weiterer Gebrauch der Gleichung $\dfrac{p_b}{\gamma} = \dfrac{2\,P}{\gamma\,t}$ ergibt falsche Werte, da die Fläche $\dfrac{P}{\gamma}$ nicht länger angenähert ein Dreieck ist.

17

b) Ermittlung der Druckhöhe für jeden beliebigen Punkt.

Wenn der Abfluß überhaupt mit der üblichen Methode der Darstellung der Energiekurve, der Geschwindigkeitshöhen und der Druckverteilung in einem gegebenen Schnitt behandelt werden kann, dürfte der folgende Weg zur Bestimmung der Druckverteilungskurve geeignet sein.

Q und t sind experimentell gemessen. Aus diesen Werten können mit obigen Formeln H, v_m, $\frac{P}{\gamma}$ und $\frac{p_b}{\gamma}$ — letzteres unter der erwähnten, gefühlsmäßigen Korrektur in der Nähe der Absturzecke — bestimmt werden. In Abb. 5 sind diese Werte beispielsweise für den Schnitt $\pm 0{,}0$ dargestellt. H wird nun im ganzen Querschnitt konstant angenommen; die Anteile $\frac{p}{\gamma}$ und k werden als horizontale Abszissen in jedem Punkt des Vertikalschnittes aufgetragen.

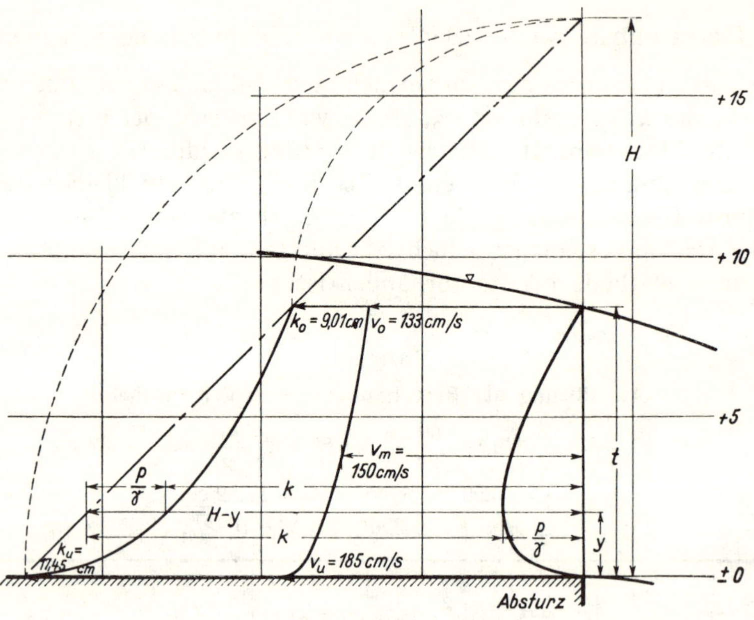

Abb. 5. Methode zur Bestimmung der Druckverteilung mittels der theoretischen Höhe der Energielinie.

Unbekannt ist noch der Verlauf entweder der Druck- oder der Geschwindigkeitshöhenkurve. Ist eine der beiden bekannt, so kann die andere daraus abgeleitet werden. Zunächst können die Endwerte der Geschwindigkeitshöhenkurve an der Sohle und der Oberfläche aus der Energielinienhöhe über diesen beiden Stellen bestimmt werden, da die Druckhöhen an beiden Stellen gleich Null sind. Mit Hilfe dieser Werte können die Sohlen- und Oberflächengeschwindigkeiten gefunden werden, die zusammen mit dem schon bekannten Wert v_m zwei feste und einen seiner Höhenlage nach zunächst noch unbestimmten Wert der Geschwindigkeitskurve ergeben.

Es gibt offenbar eine beliebige Anzahl von Kurven, die diesen Bedingungen genügen, während nur eine davon die Gesuchte ist. Es liegt nahe, diese durch einen Ansatz von der Form $y = c\,(v_u - v)^n$ anzunähern. Wie aus der späteren Untersuchung hervorgeht, liefert diese Gleichung hinreichend zuverlässige Ergebnisse. Mit dem Ansatz:

$$y = c \cdot x^n, \text{ worin } x = v_u - v \text{ und } dx = -dv$$

erhält man für den Flächeninhalt unter der Geschwindigkeitskurve:

$$t \cdot v_m = c \int_0^{v_u - v_o} x^n \, dx + t v_o = -c \int_{v_o}^{v_u} (v_u - v)^n \, dv + t v_o.$$

Da nun

$$y = 0, \quad x = 0 \qquad \text{folgt} \quad t = c\,(v_u - v_o)^n$$

$$y = t, \quad x = v_u - v_o \qquad \text{und} \quad c = \frac{t}{(v_u - v_o)^n}.$$

Durch Einsetzen:

$$t\,(v_m - v_o) = \frac{t}{(n+1)\,(v_u - v_o)^n}\,(v_u - v_o)^{n+1}$$

$$v_m - v_o = \frac{v_u - v_o}{n+1}$$

$$n = \frac{v_u - v_o}{v_m - v_o} - 1 = \frac{v_u - v_m}{v_m - v_o}.$$

Hiernach geht die ursprüngliche Gleichung über in:

$$y = \frac{t}{(v_u - v_o)^{\frac{v_u - v_m}{v_m - v_o}}} \cdot (v_u - v)^{\frac{v_u - v_m}{v_m - v_o}}$$

oder

$$v_u - v = (v_u - v_o) \left(\frac{y}{t}\right)^{\frac{v_m - v_o}{v_u - v_m}}$$

Mit $k = \dfrac{v^2}{2\,g}$ ist dann:

$$k = \frac{1}{2\,g}\left[v_u - (v_u - v_o) \left(\frac{y}{t}\right)^{\frac{v_m - v_o}{v_u - v_m}} \right]^2.$$

Nach der Theorie der Energielinie ist:

$$\frac{p}{\gamma} = H - k - y.$$

Also lautet der vollständige Ausdruck für die Druckverteilung in einem gegebenen Querschnitt:

$$(13) \qquad \frac{p}{\gamma} = H - y - \frac{1}{2\,g}\left[v_u - (v_u - v_o) \left(\frac{y}{t}\right)^{\frac{v_m - v_o}{v_u - v_u}} \right]^2.$$

Wenn auch diese Gleichung für die praktische Anwendung am einfachsten ist, kann ihr dennoch eine in bezug auf die Veränderlichen übersichtlichere Form gegeben werden, wobei H, v_o und v_m durch Q und t ausgedrückt werden können und v_u sich nach der Gleichung

$$v_u = \left[\left(H - \frac{p_b}{\gamma}\right) \cdot 2\,g\right]^{\frac{1}{2}}$$

ergibt.

Gleichung (13) enthält daher bei konstanter Durchflußmenge nur drei Variable: $\dfrac{p}{\gamma}$, y und t. Mangels eines brauchbaren theoretischen oder praktischen Ausdruckes für die Tiefe in Abhängigkeit von der Durchflußmenge, vom Abstand von der Abfallkante und vom Reibungsbeiwert an

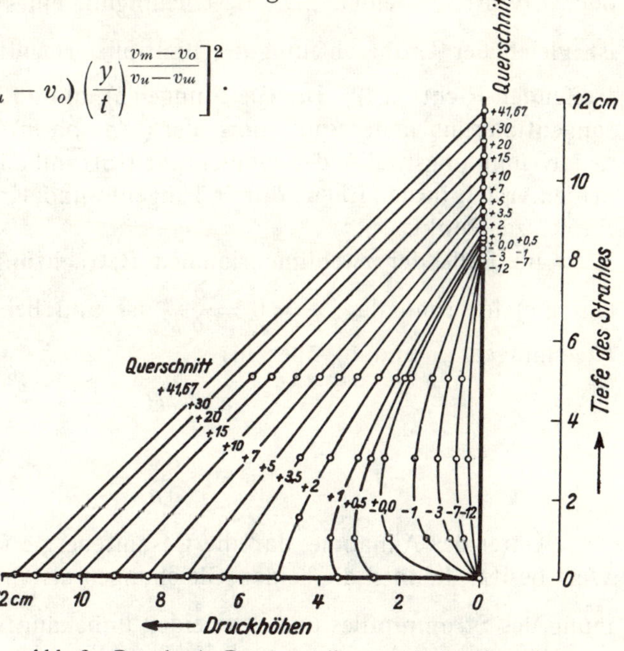

Abb. 6. Berechnete Druckverteilung nach der Formel

$$\frac{p}{\gamma} = H - y - \frac{1}{2\,g}\left[v_u - (v_u - v_o) \left(\frac{y}{t}\right)^{\frac{v_m - v_o}{v_u - v_m}} \right]^2.$$

Sohle und Wandungen muß t aus den Ergebnissen von Versuchsmessungen entnommen werden.

Aus den nach Formel (13) berechneten und den gemessenen Werten, die in Tabelle I eingetragen sind, wurde die Druckverteilung in den Lotrechten von dem kritischen Querschnitt an bis in den Abfallstrahl bestimmt. Diese Werte sind in Abb. 6 zusammengestellt, und zwar so, daß die Vertikalschnitte und deren tiefsten Punkte aufeinander liegen. Man erkennt daraus, daß der Übergang von einer Kurve zur nächsten einen stetigen Verlauf besitzt und daß vom kritischen Querschnitt bis zum Schnitt 7,0 eine praktisch lineare Druckverteilung vorhanden ist. Die so ermittelten Flächeninhalte stimmen recht gut mit den nach Formel (12) errechneten überein.

Infolge des im Exponenten stehenden Wertes $\dfrac{v_m - v_o}{v_u - v_m}$ verursacht schon eine geringe Ungenauigkeit in der Angabe der Tiefe eine beträchtliche Abweichung im Exponenten. Daher sind diese Werte für alle Schnitte sorgfältig aufgetragen und der gesuchte Exponent aus der durch die verschiedenen Punkte gezogenen ausgeglichenen Kurve entnommen.

III. Die Potentialströmungstheorie und ihre Anwendung zur Bestimmung der Druckverteilung mittels der Netzkonstruktion.

In der Hydromechanik werden die Beziehungen zwischen Druck und Geschwindigkeit in der Bewegung einer idealen Flüssigkeit mittels der Eulerschen Bewegungsgleichungen in folgender Weise dargestellt:

$$\text{a)} \quad \frac{\partial v}{\partial T} + \frac{\partial}{\partial s}\left(\frac{v^2}{2}\right) = - g \, \frac{\partial}{\partial s}\left(\frac{p}{\gamma} + h\right)$$

$$\text{b)} \quad \frac{\partial v_n}{\partial T} + \frac{v^2}{\varrho} = - g \, \frac{\partial}{\partial m}\left(\frac{p}{\gamma} + h\right)$$

$$\text{c)} \quad \frac{\partial v_m}{\partial T} = - g \, \frac{\partial}{\partial n}\left(\frac{p}{\gamma} + h\right)$$

oder in Worte gekleidet: „Die Beschleunigung eines Flüssigkeitsteilchens nach irgendeiner Richtung ist gleich der Erdbeschleunigung multipliziert mit dem Gefälle der Summe $\left(\dfrac{p}{\gamma} + h\right)$ in der betreffenden Richtung[9]“. Die Gleichungen a), b) und c) enthalten insbesondere die Beschleunigung in tangentialer, normaler und binormaler (s, n und m) Richtung zur natürlichen Stromlinie, während ϱ den Krümmungsradius der momentanen Strombahn bezeichnet. Von den Koordinaten liegen die beiden ersten in der Ebene durch Tangente und Krümmungsradius. Diese wechseln ihre Lage von Punkt zu Punkt.

Im Falle der zweidimensionalen Betrachtung eines stationären Abflußproblems fällt Gleichung c) fort und da $\dfrac{\partial v}{\partial T} = 0 = \dfrac{\partial v_n}{\partial T}$ ist und bei wirbelfreier Strömung H konstant ist, gehen Gleichungen a) und b) über in:

$$\text{a')} \quad H = \frac{p}{\gamma} + \frac{v^2}{2\,g} + h = \text{konstant}$$

$$\text{b')} \quad \frac{\partial v}{\partial n} = \frac{v}{\varrho} \quad \text{oder} \quad v = c \cdot e^{\int \frac{d n}{\varrho}}.$$

Unter der Annahme, daß die Gesamtenergie für jeden Punkt der Strömung einen konstanten Wert besitzt, können die beiden Gleichungen als Grundlage für eine graphische Lösung zur Bestimmung des Stromprofiles und der beiden unbekannten Größen v und $\dfrac{p}{\gamma}$ benutzt werden, wenn man

---

[9]) W. Spannhake, „Das Wichtigste aus der Hydromechanik“. 1. Kapitel von „Kreiselräder als Pumpen und Turbinen“, S. 28. Julius Springer, Berlin, 1931.

voraussetzt, daß das Absturzprofil und die Größe des Abflusses bereits bekannt sind. Zunächst werden nach Augenmaß geschätzte Stromlinien aufgezeichnet, die das ebenfalls geschätzte Längsprofil möglichst genau in eine bestimmte Zahl von Flächen gleicher Durchflußmenge teilen. An einer beliebigen, für die Untersuchung ausgewählten Stelle sind die Werte von n (Bogenlängen auf einer vom äußeren Rand des Stromes bis zum inneren laufenden Orthogonaltrajektorie der Stromlinien) bestimmt und mit diesen als Abszissen wird die Hilfskurve

$$q = \int e^{\int \frac{dn}{\varrho}} \cdot dn$$

aufgezeichnet. Für diese Konstruktion müssen die Werte von ϱ aus den in den Wasserstrom einskizzierten Stromlinien entnommen werden.

In diesem Diagramm wird der Endwert von q in ebenso viele gleiche Abschnitte zerlegt wie das ursprüngliche Längsprofil des Stromes. Parallele zur Abszissenachse, in diesen Abschnitten gezogen, liefern durch ihren Schnitt mit der q-Linie die Abschnitte auf der n-Abszisse, in denen die Stromlinien die Orthogonaltrajektorie schneiden[10]). Wenn die letzteren Punkte nicht mit den ursprünglich nach Augenmaß angenommenen übereinstimmen, muß der Arbeitsvorgang wiederholt werden. Mit dem schließlich so bestimmten Wert v erhält man aus der Gleichung a') den Wert $\frac{p}{\gamma}$ und nun muß kontrolliert werden, ob die beiden freien Stromlinien tatsächlich konstanten Druck ergeben. Ist dies nicht der Fall, so muß weiter probiert werden.

Wenn diese graphische Lösung für das ganze Längsprofil des Stromes durchgeführt ist, indem die n-Kurven so gelegt sind, daß die eingeschlossenen Flächen angenäherte Quadrate darstellen, wird das Resultat der Netzkonstruktion ähnlich sein, die auf Plan I gezeigt ist. Hierbei sind die theoretischen Geschwindigkeiten umgekehrt proportional den Längen der unterteilten Linien. Offenbar kann dieselbe Konstruktion vereinfacht werden, wenn die Druckverteilung aus Messungen bekannt ist, um die Stromlinien und Geschwindigkeiten an jeder Stelle übersichtlich darzustellen.

Vorstehende Methode zur Bestimmung und Aufzeichnung der Abflußverhältnisse hat verschiedene beachtenswerte Vor- und Nachteile. Sie gestattet eine theoretisch strenge Behandlung der Abflußbedingungen in jedem Punkte und ermöglicht daher, die Bewegung eines einzelnen Wasserteilchens längs seines ganzen Weges zu verfolgen. Obgleich die vorerwähnte Anwendung des Kochschen Stützkraftgesetzes auf hydrodynamische Vorgänge zulässig ist, ist dieser Satz immerhin von einem Vergleich zwischen zwei verschiedenen, einen Teil des ganzen Wasserstromes einschließenden Querschnitten abhängig.

Anderseits jedoch wird bei der Theorie der Potentialströmung das Wasser als ideale Flüssigkeit behandelt und der Reibungseinfluß auf die Geschwindigkeitsverteilung, der wiederum einen bemerkenswerten Einfluß auf den Druck haben könnte, nicht berücksichtigt. Weiterhin ist diese Theorie als graphische Methode zur Bestimmung der Druckverteilung äußerst mühsam und die praktisch möglichen Grenzen zeichnerischer Genauigkeit machen die Bestimmung des Druckes bestenfalls zu einem ungenauen Verfahren.

IV. Hydrodynamische Diskussion der verschiedenen Kraftwirkungen auf ein Wasserteilchen beim Abfluß mit gekrümmten Strombahnen.

1. Die Beziehung zwischen Schwerkraft, Druckgefälle und Beschleunigung.

Die oft zum Unterschied von der „Zusatzdruckhöhe" $\frac{z}{\gamma}$ und der „Gesamtdruckhöhe" $\frac{p}{\gamma}$ (siehe Abb. 1) gewählte Bezeichnung „statische Druckhöhe" ist in Wirklichkeit nur die Höhe des Wasserspiegels über einem gegebenen Teilchen und kommt nur bei geradlinig gleichförmiger Strömung oder bei ruhendem Wasser zur vollen Auswirkung. Was Koch „Zusatzspannung" nannte,

---

¹⁰) Spannhake, S. 34.

R o u s e , Energie.

ist einfach die Differenz zwischen der Höhe des Wasserspiegels über einem gegebenen Punkt und der wirklichen Druckhöhe an diesem Punkt. Mit anderen Worten ist die „Zusatzspannung" derjenige Teil des Gewichtes der gesamten Wassersäule über dem Teilchen, der eine positive nach unten gerichtete Beschleunigung anstatt eines Druckes zu erzeugen bestrebt ist, bzw. im Falle einer nach oben gerichteten Beschleunigung derjenige Teil der kinetischen Energie der Wassersäule, der in Druck umgewandelt diese Beschleunigung erzeugt.

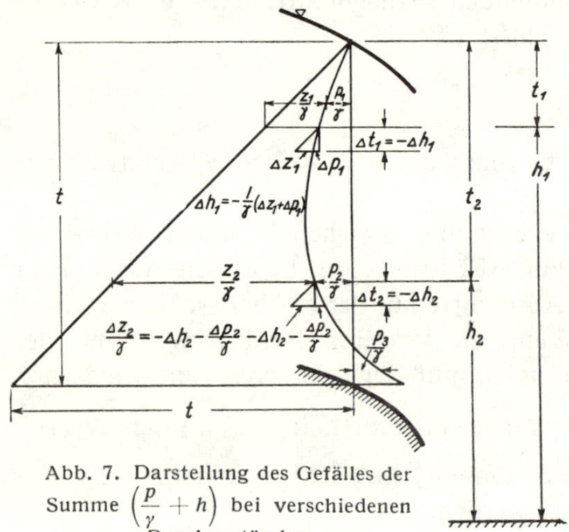

Abb. 7. Darstellung des Gefälles der Summe $\left(\dfrac{p}{\gamma} + h\right)$ bei verschiedenen Druckzuständen.

Die folgende Erörterung soll diese Beziehung näher beleuchten. Die Hydrodynamik der reibungslosen Flüssigkeit sagt: „Nach irgendeiner Richtung ist die Komponente der aus Druck- und Schwerewirkung zusammengesetzten (auf die Volumeneinheit bezogenen) Kraftwirkung gleich dem Gefälle der Summe $(p + \gamma h)$ nach dieser Richtung"[11].

Wird die vertikale Komponente dieser Kraft mit f_t bezeichnet, so ist

$$f_t = -\frac{d}{dt}(p + \gamma h),$$

worin h die Höhenlage des betrachteten Punktes über einem Nullniveau, z. B. der Gerinnesohle ist (siehe Abb. 7). Daher ist $dt = -dh$, so daß wir erhalten

$$f_t = -\frac{dp}{dt} + \gamma.$$

Dies bedeutet eine vertikal nach unten gerichtete, auf die Volumeneinheit bezogene Kraft. Wird nun der ganze Ausdruck einmal integriert, so gilt für eine beliebige Tiefe t_1 (siehe Abb. 7):

$$\int_0^{t_1} f_t\, dt = \gamma\, t_1 - p_1 = z_1$$

und man erkennt, daß in diesem Falle der Wert z_1, den Koch einfach mit „Zusatzspannung" bezeichnet, in Wirklichkeit die Differenz ist zwischen dem Gewicht des Wassers über einem gegebenen horizontalen Flächenelement und dem Druck an dieser Stelle, oder die gesamte vertikale Kraft in dieser Einheitswassersäule, welche die Beschleunigung in der Säule in vertikaler Richtung bewirkt.

Dort wo tatsächlich ein Druckgefälle existiert, d. h. wo $\dfrac{dp}{dt}$ negativ ist, wie im unteren Teil der Abb. 7, nimmt die Kraft sogar einen größeren Wert an als das Gewicht des Wasserteilchens. Das heißt, nicht nur das Gewicht eines Teilchens erzeugt seine Beschleunigung, sondern auch die durch den abnehmenden Druck verursachte Kraft.

Falls der Druck an einer Stelle einen negativen Wert annimmt $\left(\text{siehe } \dfrac{p_3}{\gamma} \text{ in Abb. 7}\right)$, so muß beachtet werden, daß dies eine Senkung des Druckes unter den atmosphärischen bedeutet, die den absoluten Betrag der Differenz zwischen dem augenblicklichen Atmosphärendruck und dem von der Temperatur abhängigen Dampfdruck des Wassers nicht überschreiten kann. Tritt irgendwo der Dampfdruck wirklich auf, so verlieren natürlich obige Betrachtungen ihre Gültigkeit. Im übrigen ist es in Fällen, wo Druck unter dem atmosphärischen auftritt, vielleicht bequemer, mit dem absoluten Vakuum als Nullniveau zu rechnen, um alle Drücke positiv bezeichnen zu können.

Die horizontale Komponente der Beschleunigungskraft lautet:

$$f_x = -\frac{dp}{dx},$$

---

[11] Spannhake, S. 9.

worin also der Wert γ verschwindet. Mit anderen Worten: die Beschleunigung eines Teilchens infolge der Gravitationskraft wirkt nur in vertikaler Richtung, während die weitere Beschleunigungskraft infolge des Druckabfalles im allgemeinen anders gerichtet ist, d. h. senkrecht zu einer Kurve gleichen Druckes, die durch diesen Punkt geht.

In Abb. 8 sind die Kurven gleichen Druckes aufgetragen, die aus den auf Plan III dargestellten Messungsergebnissen entnommen sind. Hieraus ersieht man die relativen Richtungen der Gravitationskraft und der durch das Druckgefälle verursachten Kraft, wobei deren Vektorsumme in jedem Falle Größe und Richtung der auf ein Teilchen wirkenden Gesamtbeschleunigungskraft ergibt.

Diese wichtige Tatsache kann in folgender Weise zusammengefaßt werden:

Die auf irgendeinem Wasserteilchen bei gekrümmtem Abfluß wirkende Beschleunigungskraft kann als die Vektorsumme zweier Komponenten betrachtet werden: einer Vertikalkomponenten, die gleich der Gravitationskraft auf das Teilchen ist und einer Komponenten, die die gleiche Größe und Richtung des maximalen Druckgefälles an diesem Punkte besitzt.

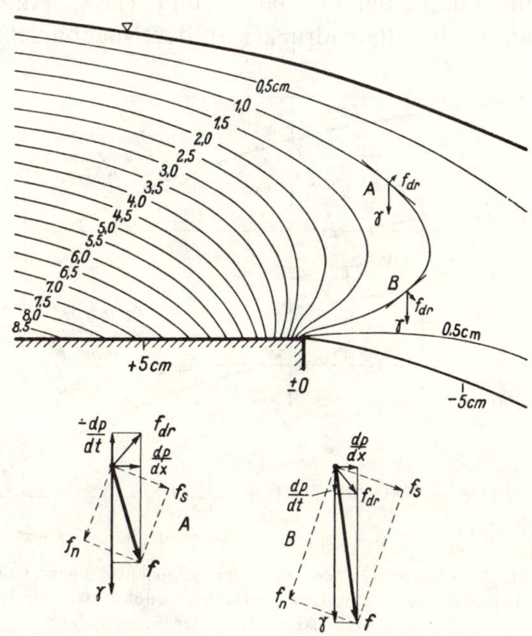

Abb. 8. Linien gleichen Druckes im Bereich des Absturzes mit Vektordiagrammen der Beschleunigungskräfte.

2. Zerlegung des Gesamtdruckes.

Die verschiedenen auf ein Wasserteilchen wirkenden Kräfte können in den folgenden beiden Komponenten zusammengefaßt werden: eine, die in der Richtung der Strömung beschleunigend wirkt und eine zweite, senkrecht zur Richtung der Strömung, die bestrebt ist, die Richtung der Bewegung des Teilchens zu ändern. Das heißt, diese beiden Komponenten erzeugen beziehungsweise eine tangentiael und eine zentripetale Beschleunigung.

In der vorigen Überlegung wurden die verschiedenen Kräfte von einem anderen Standpunkt aus behandelt, indem gezeigt wurde, daß an einem gegebenen Punkt die ganze zur Beschleunigung verwendete Kraft an einem Teilchen (auf die Volumeneinheit bezogen) der Vektorsumme von Schwerewirkung und Druckgefälle in diesem Punkt gleich ist. Natürlich ist die Resultierende dieser beiden Komponenten identisch mit der der eben erwähnten tangentialen und normalen Komponenten, man darf aber die einzelnen Komponenten der einen Darstellung nicht unabhängig voneinander mit denen der anderen vergleichen. Mit anderen Worten: Zentripetalbeschleunigung und Beschleunigung durch das Druckgefälle stehen in einem von der Schwerkraft beeinflußten Zusammenhange.

In einem Schnitt, wo die Stromlinien noch gerade und horizontal sind, wird das ganze Gewicht der Teilchen in einer vertikalen Einheitssäule zur Erzeugung eines Druckes verwendet, der unmittelbar von der Tiefe unter dem Wasserspiegel abhängig ist. Dies wirkt sich im tiefsten Punkt in einem Druck gleich dem ganzen Gewicht der Säule aus, dem durch einen vom Boden in der entgegengesetzten Richtung ausgeübten Druck das Gleichgewicht gehalten wird. Weiter stromabwärts nimmt dieser Druck allmählich mit der stetig zunehmenden Krümmung der Stromlinien ab, bis er am Absturz auf Null gesunken ist. Das heißt, die einzige außer der Schwerewirkung auf die Strömung wirkende äußere Kraft nimmt mit der Annäherung an die Absturzecke ab, wo sie ganz verschwindet. Trotzdem herrscht über der Absturzecke noch ein beträchtlicher Druck im Innern des Strahles.

Ein Blick auf die Netzkonstruktion in Plan I oder auf die Linien der gemessenen Geschwindigkeiten in Plan II und III (s. Anhang) zeigt, daß entsprechend der allmählichen Abnahme des Bodendruckes in der Nachbarschaft der Absturzecke die Bodengeschwindigkeiten entschieden größer sind als die im Wasserspiegel, während die letzteren allerdings schon eine nach unten gerichtete Komponente haben. Daher würde die freie Abflußparabel eines einzelnen Teilchens aus den tieferen Schichten des Strahles sich horizontal viel weiter erstrecken als eine für ein Teilchen aus den oberen Schichten (s. Abb. 9). Aber es liegt auf der Hand, daß die Teilchen nicht ihren natürlichen Abflußparabeln folgen können, weil sie sich dabei kreuzen müßten. Deshalb muß innerhalb der Strömung eine Kraft vorhanden sein, welche die Richtung und wahrscheinlich auch die Geschwindigkeiten der konvergierenden Wasserteilchen zu ändern bestrebt ist.

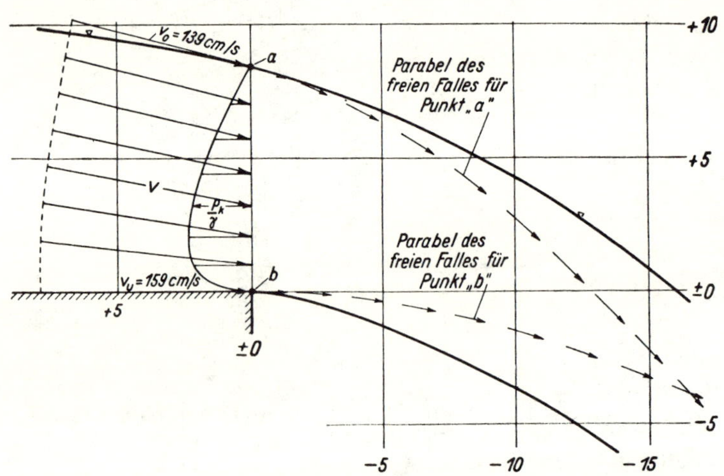

Abb. 9. Darstellung der von der ungleichmäßigen Konvergenz und Geschwindigkeitsverteilung herrührenden Druckverteilung an der Stelle größter Krümmung der Stromfäden.

In einem Schnitt wo der Boden noch einen Druck auf die Strömung ausübt, darf man den Druck innerhalb des Stromes als durch zwei Ursachen bestimmt ansehen: durch die Bodenreaktion, die mit großer Wahrscheinlichkeit eine lineare Funktion der Tiefe ist und die ihren Höchstwert in dem Bodendruck selbst hat (s. Abb. 10), und durch die gegenseitige Beeinflussung der konvergierenden Stromlinien, welche die Richtung eines jeden einzelnen Teilchens der allgemeinen Richtung des Stromes im ganzen anzupassen sucht.

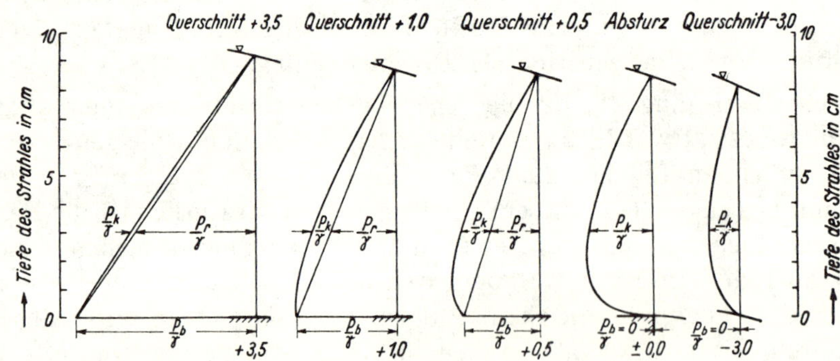

Abb. 10. Zerlegung des Gesamtdruckes in die von äußeren, statischen bzw. von inneren, kinetischen Kräften herrührenden Drücke.

Daher darf man die Vertikalkomponente der auf die Volumeneinheit bezogenen Kraft, um diese beiden Arten des inneren Druckes einzuführen, in folgender Weise umschreiben:

$$ f = \gamma - \frac{dp}{dt} = \gamma - \frac{dp_r}{dt} - \frac{dp_k}{dt} = \gamma - \frac{p_b}{t} - \frac{dp_k}{dt}. $$

Die vorausgegangene Betrachtung hat gezeigt, daß das Gewicht einer vertikalen Einheitssäule sich durch die Strömung hindurch teils in der Erzeugung von Bodendruck, teils in der Hervorrufung von vertikaler Beschleunigung auswirkt. Im frei fallenden Strahl erzeugt das gesamte Gewicht Beschleunigung; daher ist der innere Druck in diesem Teil des Strahles nicht dazu da, um die totale Beschleunigung zu vermindern — d. h. der Druck hängt nicht mit der Schwerkraft, sondern eben mit der relativen Richtung und Bewegungsgröße der Teilchen zusammen. Daher darf der Wert $\dfrac{dp_k}{dt}$ als unabhängig von γ angesehen werden.

Ein Studium des Planes III zeigt, daß trotz der allmählichen Abnahme des Bodendruckes vom kritischen Querschnitt ab bis zu einem Abstand von 7 cm von der Absturzecke die Druckverteilung praktisch noch eine lineare Funktion der Tiefe ist; bis zu diesem Schnitt ist die Krümmung der Stromlinien verhältnismäßig gering und ihr Konvergenzwinkel praktisch vom Spiegel bis zum Boden konstant. Später indessen nehmen nicht nur die Krümmung und die Bodengeschwindigkeiten zu, sondern es wird auch der Konvergenzwinkel in der Nähe des Bodens zunehmend größer, sobald sich der Bodendruck dem Wert Null nähert. Gleichzeitig wird auch die Druckverteilung mehr und mehr eine ausgeprägte Kurve, die am meisten gekrümmt ist in dem Schnitt über der Absturzecke, wo sowohl die Geschwindigkeit als auch der Konvergenzwinkel die größte Veränderlichkeit zwischen Spiegel und Boden zeigen (s. Abb. 9). An dieser Stelle ist der Wert p_b verschwunden, so daß der ganze verbleibende Druck p_k auf Rechnung der gegenseitigen Beeinflussung der konvergierenden Stromlinien kommt.

Diese Betrachtung kann wie folgt zusammengefaßt werden:

Im Falle gekrümmter Stromlinien darf die Kurve der Druckverteilung über einem Vertikalquerschnitt als die Summe zweier Kurven betrachtet werden; die erste ist wahrscheinlich eine lineare Funktion der Tiefe mit der größten Ordinate gleich dem Bodendruck in diesem Schnitt; die zweite ist eine nicht lineare Funktion der ungleichförmigen Geschwindigkeitsverteilung und ungleichförmigen Konvergenz der Stromlinien im Schnitt.

Der Verfasser hat vergeblich versucht, die zuletzt genannte Funktion in einfache mathematische Ausdrücke zu fassen, die die Bewegungsgröße und die relativen Winkel der Stromröhren enthalten. Es ist möglich, daß eine solche Formulierung schließlich notwendigerweise zu den früheren Ausdrücken für normale und tangentiale Beschleunigung zurückführen würde. Hierdurch würde natürlich die ursprüngliche Absicht dieser Untersuchung vereitelt werden, nämlich eine vereinfachte Behandlung einer Strömung mit gekrümmten Stromfäden unter Benutzung horizontaler und vertikaler Koordinaten zu finden.

B. Experimentelle Nachprüfung der Theorien.

I. Vorversuche.

Da eine genaue Kenntnis des Wasserspiegelverlaufes für die Anbringung der Druckwasserrohre an den geeignetsten Stellen der Seitenwand unbedingt erforderlich war, wurde zunächst eine Anzahl von Vorversuchen mit verschiedenen Abflußmengen durchgeführt. In die für die Durchführung der Versuche benutzte, beiderseits durch Glaswände begrenzte Rinne wurde auf die volle Breite von 50 cm eine glatte, ungefähr 3½ m lange Zementsohle eingebaut. Diese Sohle lag 35 cm über dem Boden der Rinne und erhielt am unteren Ende eine scharfe Ecke mit einer lotrechten Abfallwand. In diese Wand wurde eine Vorkehrung zur Belüftung des Strahles einbetoniert.

Abb. 11. Aufnahme der Versuchsrinne vom oberen Ende gesehen. Im Vordergrund sieht man das 50 cm breite Meßwehr unter einer Abflußmenge von 62,5 l/s. Weiter stromabwärts ist der Absturz deutlich erkennbar, wo der Abfluß in der auf 25 cm eingeengten Rinne einer Wassermenge von 125 l/s/lfd. m entspricht. Rechts vom Absturz steht das Gerät zur Messung der Wand- und Sohlendrücke.

Mittels eines Spitzenmaßstabes mit Noniusteilung, die ein Ablesen auf Zehntelmillimeter gestattete, wurde das Wasserspiegelgefälle für 12 verschiedene Durchflußmengen von 2 bis 100 l in der Sekunde je lfd. m Rinnenbreite bestimmt. In allen Fällen wurde sowohl die Oberfläche wie die untere Begrenzungsfläche des Strahles festgelegt und stromaufwärts das Wasserspiegelgefälle etwa fünfmal so weit gemessen, als der Strecke bis zum kritischen Querschnitt entsprach. Die vorgenommenen Messungen wurden in Rinnenmitte ausgeführt.

Die so aufgenommenen Längsprofile wurden gemeinsam unter Kennzeichnung der Stelle der theoretischen Grenztiefe zeichnerisch dargestellt. Aus dieser Auftragung wurde gefolgert, daß das Froudesche Ähnlichkeitsgesetz, wenn man die kleinen Wassertiefen von der Betrachtung ausschließt, mit genügender Sicherheit angewendet werden kann, d. h. daß aus den gemessenen Profilen andere Durchflußprofile abgeleitet werden können, vorausgesetzt, daß das Umrechnungsverhältnis nicht so groß ist, daß die Messungsgenauigkeit einen erheblichen Einfluß erlangen kann. Diese Ergebnisse wurden auf Grund der gemessenen Profile zuverlässig bestätigt. Beachtenswert ist, daß die Verbindung der bei jedem Längsprofil bezeichneten Punkte des kritischen Querschnittes annähernd eine gerade Linie durch die Abfallkante ist. Nur in der nächsten Nähe der Abfallkante selbst, d. h. bei kleinen Wassertiefen, lagen diese Punkte deutlich über der Geraden, so daß die wirkliche Tiefenkurve für die kritischen Querschnitte eine flache Parabel darstellt. Diese Abweichung von der Geraden dürfte darauf zurückzuführen sein, daß bei abnehmender Tiefe der Einfluß der Wand- und Sohlenreibung auf den Abfluß entsprechend größer wird. Auch dürften die Viskosität und die Oberflächenspannung bei kleiner werdenden Strahlstärken von wachsendem Einfluß sein.

Auf Grund dieser Folgerung aus den Meßergebnissen wurde das Durchflußprofil für eine Wassermenge von 125 l/s/lfd. m nach den Regeln der geometrischen Ähnlichkeit bestimmt. Diese Durchflußmenge wurde für die Hauptversuche gewählt, da sie genügend groß war, um eine sorgfältige Messung der Druckverteilung zu gestatten und eine gute Ausnutzung der Rinne bei noch glattem Abfluß ermöglichte. Auf diese Ergebnisse gestützt wurde das endgültige Modell hergestellt.

II. Hauptversuche.

1. Ausbildung der Versuchsrinne.

Die endgültige Ausbildung der Versuchseinrichtung zeigen die Abb. 11, 12 und 13. Die eigentliche Versuchsrinne ist am Meßwehr und auf den größten Teil ihrer Länge mit beiderseitigen Glaswänden versehen und trägt oben zwei Stahlschienen zur Führung der beweglichen Spitzenmaßstäbe und des Pitotrohrgerätes.

Am oberen Ende der Rinne befindet sich ein scharfkantiges Meßwehr von 50 cm Länge und 35 cm Höhe zur Messung von Wassermengen bis zu 60 l/s. Das Wasser wird dem Wehr aus einem Einlaufkasten zugeführt, in den das Zuleitungsrohr eingesetzt ist (s. Abb. 12). Für die Durchfluß-

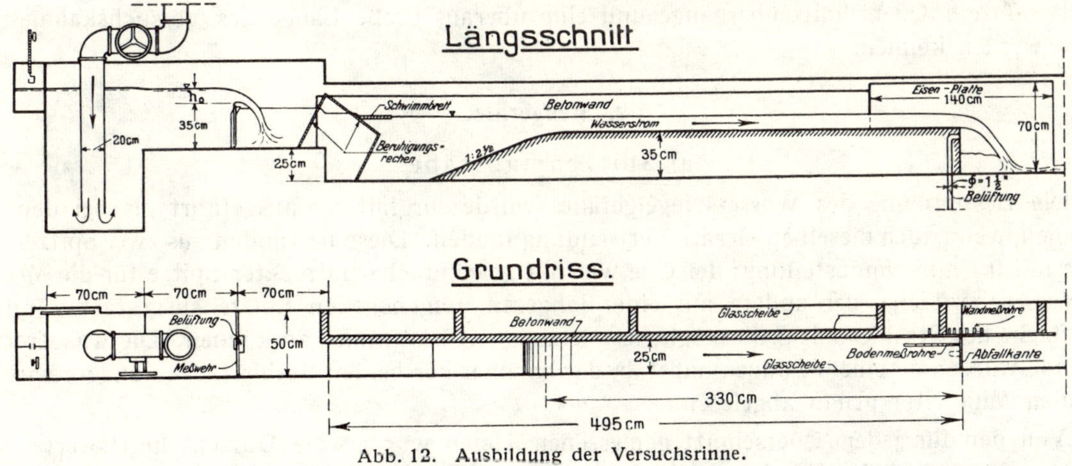

Abb. 12. Ausbildung der Versuchsrinne.

menge von 125 l/s für 1 lfd. m bzw. für die bei einer Rinnenbreite von 25 cm erforderliche Wassermenge von 31,25 l/s ergibt die Rehbocksche Wehrformel[12]) für das gegebene Wehr eine erforderliche Überfallhöhe von 10,32 cm.

____

[12]) Th. Rehbock, „Wassermessung mit scharfkantigen Überfallwehren". Zeitschrift des VDI, 1929, Nr. 24.

In der Rinne wurde der gleiche Zementboden beibehalten wie bei den Vorversuchen. Die am unteren Ende vorgesehene scharfe Abfallkante bestand aus einem polierten, sorgfältig einbetonierten Messingwinkel, dessen Höhenlage an verschiedenen Punkten eine äußerste Abweichung bis zu 0,03 cm zeigte. Zur Lüftung des Strahles war ein $1\frac{1}{2}$ Zoll starkes Rohr vorgesehen, das in die lotrechte Abfallwand eingebaut war. Die Sohle des Meßkanales lag wieder 35 cm über der normalen Rinnensohle.

Als rechte Seitenwand des Kanales wurden die Glasscheibe und die gestrichene Holzwand der eigentlichen Rinne beibehalten. Da der Kanal auf 25 cm Breite eingeengt war, um einerseits eine ausreichende Tiefe bei dem vorhandenen Durchfluß zu erhalten und anderseits den Einbau von Wasserdruckmeßrohren in eine der Seitenwände zu ermöglichen, ohne die Glaswand der Rinne entfernen zu müssen, wurde die linke Kanalwand aus Zementmörtel mit geglätteter Oberfläche hergestellt mit einer anschließenden dünnen Eisenplatte von 140×70 cm Größe, in deren Mitte die Abfallkante lag (s. Abb. 12). An der Eisenplatte wurden vor ihrem Einbau in die Rinne dünne Kupferrohre zur Druckmessung angebracht.

Alle Fugen in dieser Wand wurden sorgfältig geglättet und die Eisenplatte wasserseitig mit einem mehrfachen weißen Emaillelackanstrich versehen. An der Abfallkante war die Rinnenbreite auf genau 25 cm bemessen. Die größte Abweichung in der Breite lag 0,5 m stromauf und betrug weniger als — 0,5%.

Da auf der ganzen waagerechten Sohlenlänge von 3 m nahezu schießender, schon von geringster Störung stark beunruhigter Abfluß auftrat, wurde besondere Aufmerksamkeit auf die Wasserberuhigung verwendet. Am Ende des Sturzbeckens des Wehres wurde der Kanal plötzlich auf die Hälfte eingeschränkt, da die linke Seite vollständig abgeschlossen war. An dieser Stelle wurden zwei grobmaschige Metallrechen eingesetzt und durch ein hölzernes Floß für eine gute Beruhigung gesorgt. Der Boden fällt hier um ungefähr 25 cm ab, um dann wieder fast 1,0 m weiter stromab mit einer unter $1 : 2\frac{1}{2}$ geneigten Böschung zum Versuchsgerinne anzusteigen, wobei die Übergangsstelle zwischen Dammoberfläche und Böschung eine Abrundung erhalten hat.

Die Anordnung des Unterwasserbeckens und des Beruhigungsrechens sowie der allmähliche Böschungsübergang beseitigen alle merklichen Schwankungen im Abfluß, ohne aber das Auftreten der infolge der Querschnittseinengung sich einstellenden stehenden Reaktionswellen zu verhindern. Diese hätten nur durch allerdings praktisch schwer durchführbare, erheblich sanftere Querschnittsübergänge und eine überaus große Länge des Versuchskanales vermieden werden können.

2. Meßgeräte.

a) Spitzenmaßstäbe.

Die Bestimmung des Wasserspiegelgefälles wurde sorgfältiger ausgeführt als bei den Vorversuchen, wenn auch dieselben Geräte Verwendung fanden. Diese bestanden aus zwei Spitzenmaßstäben mit 0,1 mm Noniusteilung; der eine war mit gewöhnlicher lotrechter Spitze für die Messung der Wasseroberfläche, der andere mit einer längeren umgebogenen Spitze ausgestattet, um die Unterfläche des Strahles mit größtmöglicher Genauigkeit bestimmen zu können. Die waagerechten Abstände wurden an einem Längenmaßstab, der an einer der beiden Stahlschienen angebracht war, auf einen Millimeter genau abgelesen.

Von den für jeden Querschnitt gemessenen Tiefen wurden die Durschschnittswerte von 7 oder 8 Punkten längs des Schnittes gebildet, da durch die kleinen Diagonalwellen und die geringen zeitweiligen Schwankungen Abweichungen von ein oder zwei Millimetern verursacht wurden. Die stromaufwärts von der Abfallkante gelegenen Schnitte lagen in Abständen von 2,5 bis 10 cm und erstreckten sich auf einen Bereich von 155 cm. Soweit im Strahl mit genügender Genauigkeit gemessen werden konnte, betrugen hier die waagerechten Abstände der Meßpunkte 2,5 cm bis zu einer Gesamtlänge von 25 cm.

b) Wand- und Sohlendruckmeßrohre.

Die Druckmeßrohre in der linken Wand bestanden aus kurzen Kupferrohren von 5 cm Länge und 3 mm innerem Durchmesser. Sie waren in die Bohrlöcher der Eisenplatte eingelötet und mit der Plattenoberfläche genau bündig abgefeilt. Die Meßrohre wurden jeweils in lotrechten Reihen an 3 stromaufwärts, 4 stromabwärts und an einem unmittelbar an der Abfallkante gelegenen Querschnitt angebracht (s. Plan II). Im ganzen waren 43 Öffnungen vorhanden und zwar an den Stellen, die für die Betrachtung der Abflußvorgänge besonders zweckmäßig erschienen. Die spätere Erfahrung zeigte jedoch, daß es besser gewesen wäre, noch 25 weitere Öffnungen unmittelbar oberhalb und unterhalb der Abfallkante anzuordnen.

Als Ergänzung zu den seitlichen Meßrohren wurden in die Sohle im Abstande von 7,5 cm von der linken Wand 10 Öffnungen eingelassen. Diese bestanden aus Messingrohren mit dem gleichen inneren Durchmesser von 3 mm und wurden in einen Messingstreifen eingelassen, der sorgfältig in die Kanalsohle einbetoniert wurde. Die Meßpunkte erstreckten sich vom kritischen Querschnitt bis zu einem Abstand von 0,5 cm vor der Abfallkante.

Für die Ablesung wurde ein für diesen Zweck besonders gut bewährtes, im Karlsruher Flußbaulaboratorium entworfenes Gerät benutzt (Abb. 18), das aus einem vor einem Spiegel angebrachten Satz von 12 Glasröhren besteht, die durch Justierschrauben und Libelle genau lotrecht gehalten werden. Quer über alle Standrohre ist ein gemeinsamer beiderseits geführter Einstellfaden gespannt, an dessen Führung die Libelle angebracht ist. Seine Höhenlage kann an einer Noniusteilung auf 0,1 mm genau abgelesen werden. Der ganze Apparat war ursprünglich dazu konstruiert, über

Abb. 13. Blick von oben auf den Absturz. Die Absturzkante und Bodenpiezometerrohre kann man deutlich sehen, wie auch den Spitzenmaßstab, den Meßgeräträger und die Stahlschienen. Im Hintergrund bemerkt man die Kante des Meßwehres.

der Rinne aufgestellt zu werden, wobei die Wassersäulen durch ein gemeinsames Vakuum hochgesaugt werden sollten. Zur Vereinfachung und zur Erzielung einer besseren Übersichtlichkeit wurden oben offene Meßrohre in gleicher Höhe mit dem Absturz als zweckmäßiger angesehen.

Die Rohre wurden jedesmal vor dem Gebrauch mit Säure gereinigt, da ein unsauberes Rohr einen Meßfehler von mehr als einem Millimeter verursachen kann. Aus dem gesamten Meßrohrsystem wurde am oberen Ende der Standrohre stets so lange Wasser durch kräftiges Saugen herausgezogen, bis die Schlauchverbindungen frei von Luftblasen waren. Diese notwendige Voraussetzung für die Erzielung genauer Ergebnisse wurde nachgeprüft, indem die ganze Rinne bis zu einer über der höchsten Meßöffnung gelegenen Wasserspiegelhöhe aufgefüllt wurde, wobei jedes Standrohr die gleiche Ablesung ergeben mußte. Hiernach wurde die Höhe der Abfallkante als Nullage auf das Gerät übertragen. Dann wurde durch Öffnen des Rinnenauslasses mit dem Wasserdurchfluß allmählich begonnen, wobei darauf geachtet wurde, daß keine der Meßöffnungen auch nur vorübergehend dem Eintritt von Luft ausgesetzt war.

c) Pitotrohre.

Um dem oben erwähnten Mangel an Wandmeßstellen abzuhelfen, fanden Pitotrohre Verwendung, die gleichzeitig eine Prüfung und Ergänzung der Wanddruckmessungen ermöglichten. Die Anwendung eines normalen Pitotrohres, bei dem die Öffnungen für die statische Druckhöhe und die für die Geschwindigkeitshöhe plus statischer Druckhöhe an einem Rohr, aber in verschiedenen Querschnitten, angebracht waren, kam nicht in Frage, da die Abweichungen der Meßwerte zwischen den beiden Querschnitten selbst bei einem Abstand von nur einem Zentimeter zu große Fehler verursacht hätten. Für die Messungen mit jedem Teil des Rohres hätte dann eine Verschiebung des Gesamtgerätes vorgenommen werden müssen, was wiederum leicht zu Irrtümern und unnötigen Zeitverlusten Veranlassung gegeben hätte.

Es wurde daher zunächst ein besonderes Gerät hergestellt (s. Abb. 14), das zwei Parallelröhren in 1,85 cm Abstand erhielt, eine für die Ermittlung der Gesamthöhe (Druck- plus Geschwindigkeitshöhe), die andere für die reine Druckhöhenmessung. Im Laufe der Versuche konnte die Erfahrung gemacht werden, daß bei der recht kurzen Ausbildung dieser Rohre die lotrechte Tragstange des Gerätes infolge des sich unmittelbar davor einstellenden Staues das Meßergebnis sogar bei schießendem Abfluß beeinträchtigte.

Auf Grund dieser Feststellungen wurde eine andere Vorrichtung entworfen, die aus zwei langen, schmalen Röhren bestand, die abwechselnd an der gleichen Stange angebracht wurden. Für die Messung der Druckhöhen wurde ein Kupferrohr von 3 mm Durchmesser gewählt, das am vorderen, zugespitzten Ende mit einer scharfen Schneide versehen und auf die ganze Länge stark abgeplattet war. An jeder Seite war in 2,8 cm Abstand von der Spitze und 10,75 cm Entfernung vom unteren Befestigungspunkt eine Einlaßöffnung vorgesehen (s. Abb. 14). Das zur Messung der Gesamtdruckhöhe verwendete Messingrohr von 1,5 mm innerem Durchmesser besaß vom Befestigungspunkte bis zur Spitze eine Länge von 15 cm.

Dieses Hilfsgerät wurde an das übliche Gestänge angeschlossen, das eine lotrechte Millimeterteilung mit Ablesenonius besaß. Das Rohr wurde stets so lange im vertikalen Sinne gedreht, bis es parallel zur Stromlinie lag. Die zusätzliche Höhe infolge der Schrägstellung konnte an einem

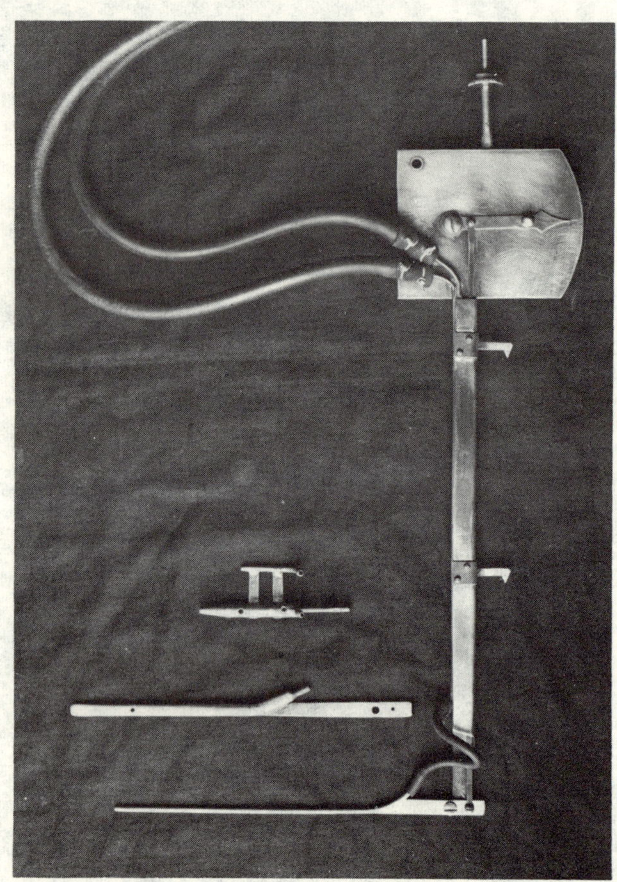

Abb. 14. Aufnahme der drei auswechselbaren Pitotrohre, mit denen alle Druck- und Staumessungen gemacht wurden. Das lange Staurohr ist bereits an der Stange festgeschraubt, deren oberes Ende an den Noniusmaßstab paßt. Der Zeiger vor der oberen Messingplatte dient zur Änderung der Richtung des Pitotrohres auch während der Messung.

Zifferblatt unmittelbar abgelesen werden. Es wurde jedoch festgestellt, daß dabei beträchtliche Fehler entstanden, so daß alle Bestimmungen der Höhe der Rohröffnung schließlich nur an einem kurzen bis zur Sohle des Kanales eingetauchten Millimetermaßstab ausgeführt wurden. Da die einzelnen Messungen für jeden Querschnitt in der gleichen Vertikalen vorgenommen wurden, mußte die in jedem einzelnen Falle entstehende waagerechte Abweichung durch Stromaufwärtsbewegung des ganzen Gestänges um einen der Neigung des Rohres entsprechenden Abstand ausgeglichen werden.

III. Ergebnisse der Versuche.

1. Wasserspiegellinien.

In Tabelle II sind die gemessenen Wasserspiegellagen wiedergegeben, aus denen das in Plan II dargestellte Längsprofil entnommen wurde. Schwierigkeiten bereitete nur die Messung der recht unregelmäßigen Strahlflächen an den Stellen, wo schon eine beträchtliche Richtungsänderung eingetreten war. Alle Punkte lagen jedoch genau genug, um sie durch eine stetige Kurve verbinden zu können. Die Lage der theoretischen Grenztiefe zwischen Strömen und Schießen wurde als Schnittpunkt der Profillinie mit der rechnerisch bestimmten Ordinatenhöhe ermittelt. Beachtenswert ist, daß dieser Punkt nur 1,3 cm weiter stromabwärts lag, als nach den Regeln der geometrischen Ähnlichkeit bei den 12 kleineren Durchflußmengen der Vorversuche ermittelt wurde, während die endgültige Wassertiefe an der Abfallkante nur rd. 0,1 cm oder 1,2% kleiner war. Nur in dem Gebiete des etwas unruhig fallenden Strahles konnte keine so gute Übereinstimmung festgestellt werden, da der Strahl bei den Hauptversuchen etwas steiler abfiel. Es ist dies daraus zu erklären, daß die Messungen bei den Vorversuchen nur in der Kanalmitte vorgenommen wurden und infolgedessen die Wirkung der Reibung nicht die gleiche war.

Tabelle II

Messungen des Stromprofiles

Stromaufwärts		Stromabwärts	
Abstand von der Kante in cm	Tiefe in cm	Abstand von der Kante in cm	Höhe der Oberfläche in cm
$+$ 0,0	8,45	$-$ 2,5	$+$ 7,67
$+$ 2,5	8,97	$-$ 5,0	$+$ 6,75
$+$ 5,0	9,45	$-$ 7,5	$+$ 5,65
$+$ 7,5	9,84	$-$ 10,0	$+$ 4,27
$+$ 10,0	10,11	$-$ 12,5	$+$ 2,73
$+$ 12,5	10,32	$-$ 15,0	$+$ 0,85
$+$ 15,0	10,57	$-$ 17,5	$-$ 1,21
$+$ 20,0	10,92	$-$ 20,0	$-$ 3,41
$+$ 25,0	11,11	$-$ 22.5	$-$ 6,01
$+$ 30,0	11,34	$-$ 25,0	$-$ 8,86
$+$ 35,0	11,47		
$+$ 40,0	11,63		
$+$ 45,0	11,79		
$+$ 50,0	11,94	Abstand von der Kante in cm	Höhe der Unterfläche in cm
$+$ 55,0	12,07		
$+$ 65,0	12,43		
$+$ 75,0	12,74	$-$ 2,5	$-$ 0,48
$+$ 85,0	13,02	$-$ 5,0	$-$ 1,31
$+$ 95,0	13,11	$-$ 7.5	$-$ 2,36
$+$ 105,0	13,02	$-$ 10,0	$-$ 3,57
$+$ 115,0	12,86	$-$ 12,5	$-$ 5,14
$+$ 125,0	12,82	$-$ 15,0	$-$ 7,05
$+$ 135,0	12,92	$-$ 17,5	$-$ 9,00
$+$ 145,0	13,24	$-$ 20,0	$-$ 11,29
$+$ 155,0	13,56	$-$ 22,5	$-$ 13,84

Als Unterlage für die Betrachtungen in Teil A dieser Abhandlung diente eine sorgfältige Auftragung auf Millimeterpapier in natürlicher Größe. Aus dieser Zeichnung wurden die Tiefen an den näher untersuchten Schnitten entnommen und ebenso die Winkel α der Stromrichtung, für die näherungsweise in jedem Querschnitt ein Mittelwert eingeführt wurde. Für diesen wurde oberhalb des Absturzes die halbe Neigung zwischen der Wasseroberfläche und der Waagerechten, im Gebiet des Strahles der Mittelwert zwischen den Neigungen der oberen und unteren Strahlfläche gewählt.

Um das Energieliniengefälle zu bestimmen, wurde, wie Abb. 15 zeigt, das ganze Meßprofil nochmals aufgetragen, und zwar diesmal in zehnfach überhöhtem Maßstabe. Hierbei wurden

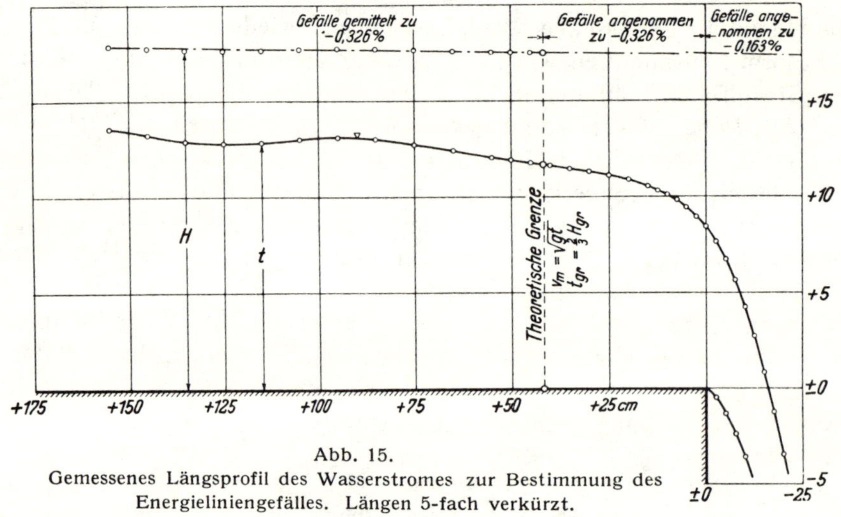

Abb. 15.
Gemessenes Längsprofil des Wasserstromes zur Bestimmung des
Energieliniengefälles. Längen 5-fach verkürzt.

zunächst die durch die lotrechte Querschnittseinengung sich einstellenden Reaktionswellen deutlich sichtbar. Für jede gemessene Tiefe oberhalb des kritischen Querschnittes wurde aus der Beziehung:

$$H = t + \alpha_u \frac{Q^2}{2\,g\,t^2}$$

die Energielinienhöhe H berechnet, unter der Annahme, daß α_u wegen der glatten Linienführung des Kanales gleich 1,0 ist. Diese geringe Ungenauigkeit hat keinen merklichen Einfluß auf das Energieliniengefälle. Die Wellen in der durch die so bestimmten Energielinienpunkte gezogenen Kurve wurden erheblich kleiner als die des Wasserspiegels, so daß als wahrscheinliches mittleres Energieliniengefälle eine gerade Linie durch die errechnete Energielinienhöhe im kritischen Querschnitt gezogen werden konnte. Hieraus wurde ein Energieliniengefälle von 0,326% ermittelt. Es wurde dann angenommen, daß dieses Gefälle auch für die anschließenden 41,67 cm bis zur Abfallkante fortgesetzt werden konnte und daß es weiterhin bis auf eine Strecke in den Strahl hinein in halber Größe, d. h. mit 0,163% in die Betrachtung eingeführt werden konnte. Es ist unwahrscheinlich, daß durch diese Annahme ein größerer Fehler begangen wurde als der Messungsgenauigkeit entspricht, da das Gesamtgefälle auf die Strecke vom kritischen Querschnitt bis zur Abfallkante (41,67 cm) bei 0,326% Gefälle nur 0,136 cm betrug.

2. Wand- und Sohlendruckmessungen.

Die bei den Wand- und Sohlendruckmessungen erhaltenen Druckhöhen sind in Plan II dargestellt. Die Kurve der Sohlendrücke ist wahrscheinlich sehr genau, da alle gemessenen Werte auf einer stetig verlaufenden Linie liegen (s. Abb. 18). Die Druckdiagramme der Wandmeßstellen ergeben stromabwärts bis zu Querschnitt —12,0 ziemlich stetige Kurven, doch waren die Dreiecksgrundlinien im allgemeinen etwas kleiner als es den Sohlendrücken entsprach. Die Abweichungen in den Meßwerten an den Stellen — 12,0 bis — 20,0 sind entweder der Unvollkommenheit der Öffnungen selbst, verstärkt durch die große Geschwindigkeit, oder den Ablenkungen des Wasserstromes nach der Strahlmitte hin zuzuschreiben. An den Seiten zeigt der Strahl nämlich die Neigung sich abzurunden, so daß hier die Stärke des Strahles beträchtlich kleiner war als in der Mitte.

Bemerkenswert ist, daß alle berechneten Sohlendrücke, die auf Tabelle I zusammengestellt sind, fast vollkommen mit den an der Sohle gemessenen Drücken übereinstimmen. Die nach Formel (12) berechneten Druckflächeninhalte stimmen verhältnismäßig gut mit den an den Wandmeßpunkten ermittelten überein, nur sind alle etwas größer als die wirklich gemessenen. Daß die Sohlendrücke und die Rechnungswerte ähnlich und etwas größer sind als die gemessenen Wanddrücke läßt vermuten, daß die Drücke im unteren Teil des Wasserstromes direkt an der Wand kleiner sind als der Mittelwert für den ganzen Querschnitt. Da die Meßwerte für die beiden mittleren Punkte im Querschnitt — 12,0 annähernd mit denen der Rechnung übereinstimmen, wurde mit ihnen das Diagramm für diesen Querschnitt entworfen.

3. Verbesserte Pitotrohrmessungen.

Es wurde bereits erwähnt, daß mit dem kurzen doppelten Pitotrohr fehlerhafte Ergebnisse erzielt wurden. Beachtenswert ist, daß das Druckmeßrohr sogar im kritischen Querschnitt an der Sohle zu hohe und an der Oberfläche zu geringe Werte ergab. Im Strahl kam diese Tendenz so stark zur Geltung, daß sich der größte Druck am unteren Ende des Schnittes zeigte, wo er den Wert Null hätte erreichen müssen. Es wurden daher Versuche mit Öffnungen oben und unten oder an beiden Seiten des Rohres unternommen, aber beide ergaben beträchtliche Abweichungen. Die so erhaltenen Ergebnisse wurden für Querschnitt ± 0,0 zum Vergleich mit dem längeren Pitotrohr in Abb. 16 gemeinsam dargestellt.

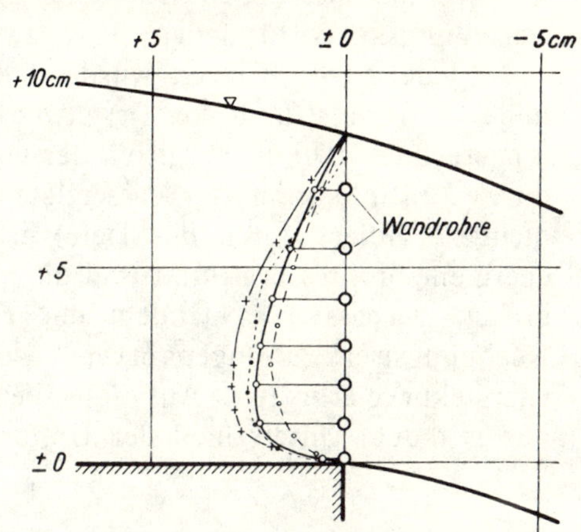

Abb. 16. Vergleich der Wanddruckmessungen mit den Pitotmessungen im Querschnitt ± 0,0. Wanddruckmessungen ⊙——⊙——⊙ Pitotmessung: 1. Kurzes Rohr mit a) seitlichen Löchern +——————+ b) Löchern oben und unten .————. 2. Langes Rohr ○——○——○

Die letztere, derart lang ausgebildete Röhre, daß das Vertikalgestänge keinen Einfluß mehr auf den Abflußvorgang an der Meßstelle ausüben konnte, ergab hingegen zu geringe Werte im Gebiete großer Geschwindigkeiten; ungefähr stromaufwärts von Querschnitt + 5,0 waren die Messungen ziemlich zuverlässig, nur unterhalb dieser Stelle wurden sie stets kleiner als die sowohl bei den Sohlen- wie auch bei den Wandmeßstellen abgelesenen Drücke. Wahrscheinlich ist dies der anwachsenden Geschwindigkeit zuzuschreiben, da die Abweichungen in dem unteren Teil des Wasserstromes, wo die Geschwindigkeiten schnell zunahmen, stets größer waren. Diese Ungenauigkeit ist wohl auf die durch die Rohrspitze verursachte Stromspaltung zurückzuführen, da hierdurch ein kleiner Unterdruck an der Öffnung erzeugt wurde.

Die Größe der Ungenauigkeit der zuletzt beschriebenen Druckmessungsmethode war jedoch unwesentlich im Verhältnis zu der beim kürzeren Rohr, und daher gab diese Methode schließlich ein Mittel, um ein relatives Maß für die Druckverteilung im Querschnitt zu erhalten, wenn man diese entsprechend den Wand- und Sohlendruckmessungen von in der Nähe gelegenen Schnitten verbesserte. Zum Beispiel ersieht man aus einem Vergleich der mittels Wanddruckrohres bzw. mit dem Pitotrohr bestimmten Druckflächen für Querschnitt + 5,0 und ± 0,0 die zulässigen größten Korrektionsfaktoren, die für die zwischenliegenden Querschnitte Verwendung finden dürfen. In ähnlicher Weise können die Verbesserungen zwischen ± 0,0 und — 3,0, zwischen — 3,0 und — 7,0 usw. näherungsweise ausgeführt werden. Im letzten Querschnitt (— 20,0) zeigte das Pitotrohr einen größten Druck von 0,11 cm in einer Gegend, wo auf Grund der Berechnungen der Druck fast vernachlässigt werden kann.

Mit dem Gesamtenergierohr (Staurohr) des kurzen doppelten Pitotrohres wurden sehr gute Kurven ermittelt, die aber vermutlich ebenfalls wie bei dem Druckhöhenmeßrohr durch das lotrechte Gestänge beeinflußt wurden. Die Meßergebnisse mit dem längeren Gesamtenergiemeßrohr lagen in der Regel bei Betrachtung der Verbindungskurve der Meßpunkte nicht ganz so gut, sie wurden dennoch als die wahrscheinlicheren vorgezogen, da anzunehmen war, daß das längere Rohr am unabhängigsten von äußeren Störungen arbeitet. Es muß jedoch betont werden, daß die Abweichungen beim kurzen Gesamtenergiemeßrohr im Verhältnis zum längeren an der Oberfläche zu große und an der Sohle zu kleine Werte lieferte, also gerade umgekehrt wie bei den Druckhöhenmessungen. Dies ist wahrscheinlich auf den Einfluß des längeren im Abstande von 1,85 cm liegenden Druckhöhenmeßrohres zurückzuführen (s. Abb. 14).

Die Messungen mit dem langen Druckhöhenmeßrohr wurden nur in der Mitte des Kanales ausgeführt, da in dem Teil des Kanales, der mit diesem Gerät erreicht werden konnte (bis zu 1 cm von der rechten und 2 cm von der linken Wand), nur kleine Druckhöhenschwankungen bemerkt wurden. Mit dem Gesamtenergiemeßrohr wurden an vier Stellen (+ 100; + 41,67; + 20,0; + 1,0) 21 über den ganzen Querschnitt verteilte Punkte bestimmt, so daß je sieben Punkte in drei waagerechten Reihen von Wand zu Wand lagen. Die Längsprofile wurden aber immer in Kanalmitte gemessen. Mittels einer kleinen, zwischen je zwei dieser Querschnitte notwendigen Interpolation wurden alle Messungen mit dem Gesamtenergiemeßrohr verbessert, um die mittlere vertikale Verteilungskurve für jeden der dazwischen liegenden Querschnitte zu erhalten.

Jede dieser Kurven wurde dann in ihre Komponenten — Druck- und Geschwindigkeitshöhe — zerlegt und die Geschwindigkeitshöhenkurve weiterhin in die Geschwindigkeitskurve umgerechnet. Durch Vergleich der von der Geschwindigkeitskurve umschriebenen Fläche mit dem durch Multiplikation der Wassertiefe mit der mittleren Geschwindigkeit (berechnet aus Durchflußmenge dividiert durch die Tiefe) erhaltenen Wert wurde der Umrechnungsfaktor des Gesamtenergiemeßrohres bestimmt und die gemessene Geschwindigkeitskurve proportional der Geschwindigkeit verbessert. Nachdem der Flächeninhalt der verbesserten Geschwindigkeitskurve dem Rechnungswert gut angepaßt war, wurde wieder die Geschwindigkeitshöhen- und dann die Gesamtenergiekurve ermittelt. Auf diese Weise wurde die Form der ursprünglichen Kurve wiedererhalten, nur mit dem Unterschied der Umformung auf die wirkliche Größe.

IV. Auswertung der Versuche.

1. Diagramm der Energieverteilung.

Plan III stellt ein übersichtliches Diagramm aller Ergebnisse dar. Grundsätzlich ist dieses Diagramm das gleiche wie dasjenige der Abb. 5, nur entsprechen die Werte besser den tatsächlichen Verhältnissen. Alle Querschnitte sind in dem Diagramm so übereinander gelegt, daß ihre beiden Achsen — die Rinnensohle und die Querschnittslinie — entsprechend zusammenfallen. Daher liegen alle stromaufwärts von der Abfallkante gelegenen Querschnitte vollkommen über der Sohle, während die weiter stromabwärts gelegenen Schnitte jeweils in der Höhe des Strahles in diesen Schnitten aufgetragen sind.

In den Diagrammen sind enthalten: erstens die Sohlen- und Wanddruckmessungen zusammen mit den verbesserten Pitotrohrablesungen; zweitens die theoretischen Hilfswerte der Gesamtdruckflächen $\frac{P}{\gamma}$; und drittens die geringe Verbesserung einiger Kurven, die dazu vorgenommen wurde, um einen stetigen Übergang im ganzen Meßbereich zu erzielen, ohne dabei auf die gegenseitigen Verhältnisse der Flächen untereinander und auf ihre Übereinstimmung mit dem Rechnungswert zu verzichten. Daher enthält Plan III nicht nur eine Zusammenstellung der Versuchsergebnisse oder der theoretisch ermittelten Werte, sondern eher eine verbesserte Kombination beider, um so eine gute Übersicht über den wirklichen Verlauf des Abflusses bei diesem Sonderfall der Wasserbewegung zu erhalten.

Bei näherer Betrachtung des Planes III lassen sich bezüglich der Energieumformungen eine Reihe wertvoller Schlüsse ziehen. Das Diagramm enthält drei veränderliche Größen: Druckhöhe, Geschwindigkeitshöhe und geodätische Höhe und außerdem die Summe aller drei: die Gesamtenergiehöhe. Jede Änderung einer dieser Größen entspricht entweder einer entgegengesetzten Änderung von einer oder beiden der anderen (nach dem Bernoullischen Gesetz) oder einer ähnlichen Änderung der Gesamtenergiehöhe (z. B. die Verminderung der Geschwindigkeit infolge des Reibungsverlustes, die nur eine Verminderung der Gesamtenergie hervorruft).

Stromabwärtsgehend tritt die erste Energieumsetzung infolge der Sohlenreibung ein, die die Geschwindigkeiten in den unteren Teilen des Wasserstromes vermindert. In der idealen Flüssigkeit würde einer Abnahme der Geschwindigkeit ein Anwachsen des Druckes entsprechen, weil die Gesamtenergie konstant bleibt. In Wirklichkeit aber bedingt die Reibung einen Verlust an Gesamtenergie und der Druck bleibt oberhalb der kritischen Grenze statisch verteilt, wodurch eine plötzliche Abnahme im unteren Teile der Gesamtenergiekurve entsteht. Die Messungen bei Querschnitt + 100 ergaben eine Gesamtenergiekurve, bei der dieser untere Abfall nicht annähernd so ausgeprägt war wie bei dem kritischen Querschnitt, und Plan III zeigt einen stetigen Abfall in der Gesamtenergie an der Sohle bis zur Abfallkante hin. Daß diese Verminderung nur unmittelbar an der Sohle wahrnehmbar ist, zeigt, daß die Reibungsverluste sich nicht gleich über den ganzen Querschnitt erstrecken, was vorausgesetzt wäre, wenn für jeden Punkt die gleiche Gesamtenergie angenommen wird.

Ein Vergleich der theoretisch erhaltenen, in Abb. 6 dargestellten Druckverteilungskurven mit denen in Plan III, die sich auf die gemessenen Werte beziehen, führt scheinbar zur Annahme, daß der Reibungseinfluß auf die Geschwindigkeitsverteilung sich nicht auch auf die Druckverteilung erstreckt. Die berechneten Kurven sind zwar in Wirklichkeit nach der empirischen Formel (13) bestimmt, jedoch stützt sich diese auf die wirklichen Abflußbedingungen bei einem vollkommen reibungslosen Kanal. Wenngleich die Reibung eine von diesem Grenzfall beträchtlich abweichende Geschwindigkeitsverteilung hervorrufen könnte, ist die unter Annahme fehlender Reibung erhaltene Druckverteilung praktisch die gleiche wie die beim Vorhandensein einer starken Reibungswirkung[13]). Zu großes Gewicht darf jedoch auf diesen Punkt nicht gelegt werden, da sich Formel (13) auf eine angenommene Kurve stützt, die selbst für den idealen Fall der reibungslosen Durchströmung eines Kanales vielleicht nicht ganz zutrifft.

Die weitere Betrachtung von Plan III dürfte ergeben, daß die wechselseitige Änderung zwischen Druck- und Geschwindigkeitshöhe bis zur Abfallkante hin keinen bemerkenswerten Einfluß auf die Verteilung der Gesamtenergie über den ganzen Querschnitt besitzt. Außerhalb der eben betrachteten, von dem Reibungsverlust herrührenden Umwandlung, hat die Gesamtenergiekurve praktisch die gleiche Form sowohl bei Querschnitt — 20,0 wie auch beim kritischen Querschnitt, obgleich jedoch die Geschwindigkeitshöhenkurve eine größere Änderung in Gestalt und Größe durch die Umformung potentieller und Druckenergie in kinetische Energie erfährt[14]).

Daher ist eine wichtige Annahme für den Abflußvorgang bei gekrümmten Stromlinien bestätigt: Mit Ausnahme der Wirkung der nicht zurückgewinnbaren Energieverluste bleibt die Gesamtenergie eines in Bewegung befindlichen Wasserteilchens konstant, und die Verschiebung der Energie von einem Teil des Querschnittes zum anderen kann höchstens durch Reibungswirkung erreicht werden, die wiederum weitere Verluste bedingt. Deshalb ist die andere häufige Annahme,

---

[13]) Es ist bemerkenswert, daß die berechneten Kurven nicht immer mit den gemessenen Flächen oder den nach Formel (12) gefundenen eng übereinstimmen. Dies ist teils auf die äußerste Empfindlichkeit des Gliedes $\dfrac{v_m - v_o}{v_u - v_m}$ in Formel (13) zurückzuführen, die einen kleinen Fehler in der Tiefe sehr übertreibt, teils vielleicht auch auf die Nichtberücksichtigung des Geschwindigkeitshöhen-Ausgleichwertes.

[14]) Die Gesamtenergiehöhenkurve ist in den alleruntersten Punkten der im Strahl gemessenen Querschnitte nicht ganz genau, da es unmöglich war, sehr nahe an der Unterfläche zu messen.

daß die Gesamtenergie gleichförmig über den Querschnitt verteilt sei, falsch, da sie eine sofortige Verteilung des Energieverlustes über den ganzen Querschnitt voraussetzt.

Vorstehende, sich auf sorgfältige experimentelle Messung stützende Erörterungen haben gezeigt, welche Beziehungen im allgemeinen zwischen Druck- und Geschwindigkeitsverteilung bei dem in vertikaler Richtung gekrümmten Abfluß bestehen: d. h., sobald die Wasserteilchen eine von der Geraden abweichende Bewegung besitzen, erzeugt die dadurch bedingte ungleichmäßige Beschleunigung eines jeden Teilchens im Strom eine nicht lineare Querschnittsdruckverteilung. Obgleich diese nichtlineare Druckverteilung streng genommen selbst für den Fall schwachgekrümmter Stromlinien gilt, ist sie nur dann von Bedeutung, wenn die relative Krümmung der von den Stromteilchen durchlaufenen Wege verhältnismäßig groß ist. Daher ist, im vorliegenden Beispiel des Abflusses über einem Absturz, obgleich die Druckverteilung bereits unmittelbar nach dem kritischen Querschnitt nach einer Kurve erfolgt, diese Abweichung von einer Geraden so unbedeutend, daß die Druckverteilung bis zu ungefähr 7 cm Abstand von der Abfallkante als praktisch linear betrachtet werden kann.

Überdies ist es aus der Theorie der Potentialströmung offensichtlich, daß die Geschwindigkeitsverteilung und die Druckverteilung in der Weise voneinander abhängig sind, daß eine Änderung in der Geschwindigkeit infolge äußerer Einflüsse sich auch auf den Druck auswirkt. Es kann, mit anderen Worten, nicht angenommen werden, daß die Druckverteilung unabhängig von der Änderung der Geschwindigkeitsverteilung infolge des Reibungsverlustes ist. Obgleich im vorliegenden Falle kein offensichtlicher Einfluß des Reibungsverlustes auf den Druck erkennbar ist, ist es dennoch wahrscheinlich, daß eine weitgehende Veränderung in der Geschwindigkeitsverteilung, die durch kleine Schwellen oder andere in die Sohle stromaufwärts von der Abfallkante eingebaute Hindernisse hervorgerufen wird, einen bemerkenswerten Einfluß auf die Druckverteilung im Gebiet der stärksten Krümmung haben würde.

2. Die wirkliche Höhe der Energielinie.

Nach dem Bernoullischen Gesetz ist der Gesamtgehalt an Energie in jedem Stromfaden konstant. Entspringen die Stromfäden in einem Gebiet durchweg konstanter Energie, so ist in idealer Flüssigkeit die Gesamtenergie auch in einem gegebenen Querschnitte durch den Wasserstrom konstant. Unter dieser Annahme kann die Größe der Energiehöhe in jedem Wasserteilchen durch eine einzelne Linie über dem Wasserstrom angegeben werden. In Wirklichkeit trifft dies nicht zu. Die Reibungsverluste äußern sich zunächst in einer Senkung der mittleren Energielinie von Querschnitt zu Querschnitt. Außerdem aber ist die Energie über jeden einzelnen Querschnitt nicht gleichförmig verteilt. Diese ungleiche Verteilung zeigt sich auch bei geradliniger Strömung durch eine ungleichmäßige Verteilung der Geschwindigkeiten.

Die Erkenntnis dieser Tatsache veranlaßte die Einführung eines Ausgleichswertes α_u[15]), der den durch die Annahme gleicher Geschwindigkeiten im ganzen Querschnitt begangenen Fehler ausgleichen soll. Durch Multiplikation des Wertes $\frac{v_m^2}{2g}$ mit dem Ausgleichswert α_u ergibt sich die tatsächliche mittlere Geschwindigkeitshöhe für den ganzen Querschnitt, d. h. die mittlere kinetische Energie, auf die Einheit des Durchflusses. Geschwindigkeitshöhe und Wassertiefe zusammen ergeben dann bei geraden Stromfäden die mittlere Höhe der Energielinie. Der Geschwindigkeitshöhen-Ausgleichswert α_u ist dem wirklichen Abfluß entsprechend unter stärkerer Berücksichtigung der größeren Geschwindigkeiten nach Th. Rehbock folgendermaßen abgeleitet[16]):

$$k = \frac{v^2}{2g}$$

---

15) Koch-Carstanjen, „Bewegung des Wassers".
16) Th. Rehbock, „Die Bestimmung der Lage der Energielinie bei fließenden Gewässern mit Hilfe des Geschwindigkeitshöhen-Ausgleichswertes".

$$k_m = \frac{1}{t} \int_0^t \frac{v}{v_m} \cdot \frac{v^2}{2\,g} \cdot d\,t = \alpha_u \frac{v_m^{\,2}}{2\,g}$$

$$\alpha_u = \frac{1}{Q\,v_m^{\,2}} \int_0^t v^3 \cdot d\,t \quad \text{wo} \quad Q = \text{m}^3/\text{s/m}$$

$$\text{oder} \; = \frac{1}{Q\,v_m^{\,2}} \int_0^F v^3 \cdot d\,F \quad \text{wo} \quad Q = \text{m}^3/\text{s}.$$

Ist die Änderung der Geschwindigkeit über den ganzen Querschnitt beträchtlich, dann muß der zweite Ausdruck benutzt werden, worin F den Flächeninhalt des Querschnittes bedeutet. Hat der Kanal eine große Breite und gleichbleibende Tiefe längs des Querschnittes, so daß die Seitenwände keinen merklichen Einfluß ausüben, so ist der erste Ausdruck anzuwenden.

Böß machte in der bereits erwähnten Schrift darauf aufmerksam, daß diese Methode der Berechnung der Höhe der Energielinie nicht anwendbar sei, wenn die Druckhöhe im Wasserstrom von der statischen Höhe verschieden ist[17]). Betrachtet man nämlich einen Punkt innerhalb des Wasserstromes, an dem die Druckhöhe beispielsweise nur halb so groß sei wie unter normalen Verhältnissen, so würde die zur Höhenlage des Punktes hinzuaddierte Druckhöhe nur in der halben Entfernung des Punktes von der Oberfläche liegen. Daher würde sich bei Hinzuaddieren der Geschwindigkeitshöhe zur Wassertiefe eine um die Größe des Unterdruckes zu hohe Lage der Energielinie ergeben. Dieser Fehler wird noch offensichtlicher, wenn man die Unterfläche des Strahles betrachtet. Hier wird die Geschwindigkeitshöhe nicht auf die obere Begrenzungsfläche des Strahles aufgesetzt, wie bei geradliniger Strömung, sondern auf die Höhe der Strahlunterfläche, da die Druckhöhe an dieser Stelle Null ist.

Daher muß im Falle gekrümmten Abflusses eine andere Methode angewandt werden, um die mittlere Energielinienhöhe zu erhalten, bei der die Druckverteilung mit berücksichtigt wird. Die einem Wasserteilchen zukommende Energiehöhe unter Voraussetzung eines sehr breiten Kanales mit ebener Sohle lautet:

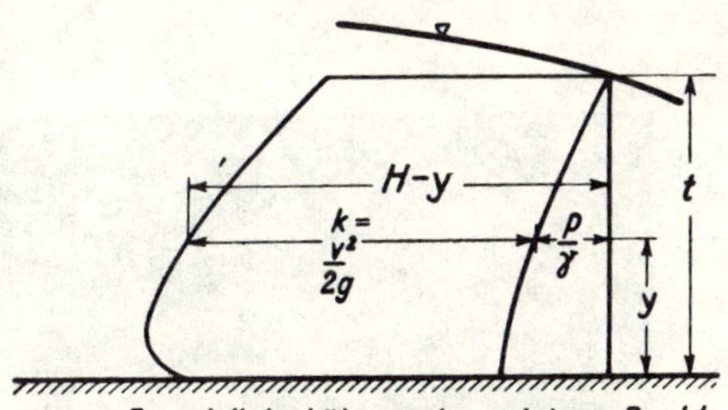

$$H = y + \frac{p}{\gamma} + \frac{v^2}{2\,g}.$$

Um einen Mittelwert für den Querschnitt zu erhalten, müssen in der zuvor beschriebenen Weise (s. oben) die den größeren Geschwindigkeiten angehörenden Geschwindigkeitshöhen stärker berücksichtigt werden. Daher nimmt die Gleichung für den gewissermaßen abgewogenen Mittelwert der Energielinienhöhe folgende Form an:

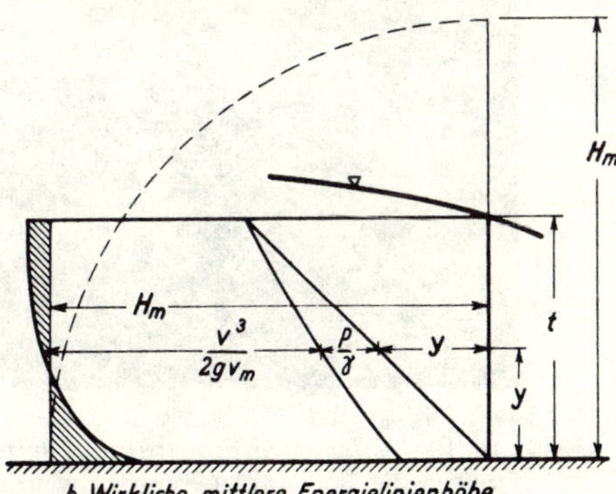

a. Energielinienhöhe an irgendeinem Punkt

b. Wirkliche mittlere Energielinienhöhe

Abb. 17. Bestimmung der wirklichen mittleren Höhe der Energielinie für Querschnitte im Bereich gekrümmter Stromfäden.

$$H_m = \frac{1}{t} \int_0^t \left(y + \frac{p}{\gamma} + \frac{v^3}{2\,g\,v_m} \right) d\,y$$

$$= \frac{1}{t} \left(\frac{t^2}{2} + \frac{P}{\gamma} + \frac{1}{2\,g\,v_m} \int_0^t v^3\,d\,t \right).$$

In Abb. 17 ist eine graphische Ermittlung dargestellt, die zur Auflösung dieses Ausdruckes führt. Abb. 17a zeigt die Methode, wie man die Gesamtenergie eines Punktes des Querschnittes

---

in üblicher Weise bestimmt, während Abb. 17b für den gleichen Querschnitt die Geschwindigkeits-höhenkurve unter stärkerer Berücksichtigung der größeren Geschwindigkeiten darstellt, so daß die mittlere Abszisse der Gesamtenergiefläche die wirkliche „abgewogene" Höhe der Energielinie für den ganzen Querschnitt angibt. Diesen Wert erhält man am besten durch Ausplanimetrieren und Dividieren der Fläche durch die Wassertiefe oder durch lotrechtes Abgleichen des linken Teiles der Abb. 17b, bis die kleinen schraffierten Flächen gleich sind.

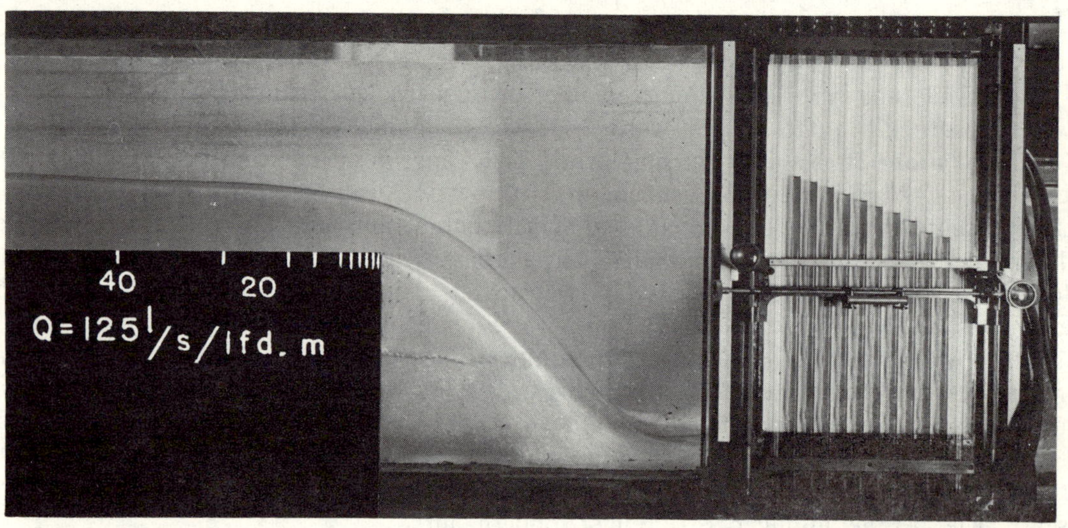

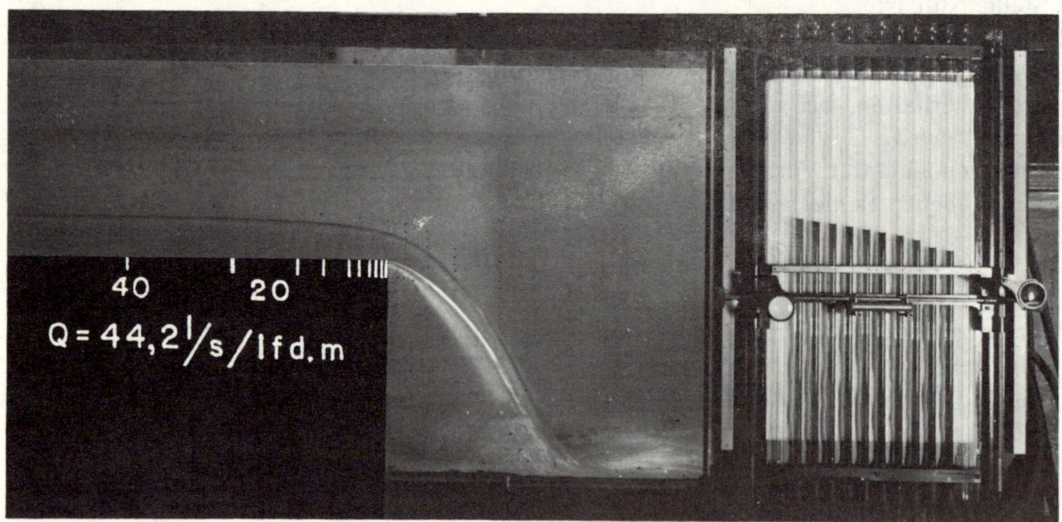

Abb. 18, 19. Aufnahmen des Absturzes unter einem Abfluß von 125 l/s/lfd. m bzw. 44,2 l/s/lfd. m. Die Piezometerrohre zeigen die Bodendrücke an den weiß markierten Stellen. Die Wandlöcher sind durch den Strahl erkennbar.

Das Glied $\dfrac{1}{2\,g\,t\,v_m}\displaystyle\int_0^t v^3\,dt$ in Formel (14) ist zwar identisch mit dem früher benutzten Wert $\alpha_u\dfrac{v_m{}^2}{2\,g}$, jedoch darf die so berechnete Geschwindigkeitshöhe in diesem Falle nicht unmittelbar auf den Wasserspiegel aufgesetzt werden, um die wirkliche Energielinienhöhe zu erhalten.

Diese Methode wurde auf die untersuchten Fälle, für den kritischen wie auch wieder für Querschnitt $\pm\,0{,}0$ angewandt, wobei die Ergebnisse 17,60 cm bzw. 17,43 cm erhalten wurden. Daß diese Werte mit den für die früheren Berechnungen benutzten Werten (17,52 und 17,38) so gut übereinstimmen, zeigt, daß die ursprüngliche Annahme ziemlich zutreffend war, und daß die angenommene Größe des Energieverlustes genügend genau war. Die geringe verbleibende Abweichung dürfte wenig Einfluß auf die Druckberechnungen haben und bei Gebrauch der Formeln (10) oder (12) im allgemeinen vernachlässigt werden können.

3. Ergänzungsversuche mit anderen Durchflußmengen.

Auf Grund des Ähnlichkeitsgesetzes kann zunächst geschlossen werden, daß alle bei dem Durchfluß von 125 l/s/lfd. m gemessenen Werte dazu verwendet werden können, die Verhältnisse bei jeder anderen Wassermenge zu bestimmen. Dies um so mehr, als die erfolgreiche Anwendung zur Bestimmung des wahrscheinlichen Längsprofiles bei der gegebenen Wassermenge auf Grund der bei den zwölf kleineren Durchflüssen der Vorversuche gemessenen Profile dies bereits bewiesen hat. Daher erstrecken sich alle in den vorhergehenden Erörterungen besprochenen Versuche nur auf diese eine Durchflußmenge.

Wenn auch die theoretische Berechnung mit den Messungen in diesem Falle hinreichend in Übereinstimmung war, so ist die Betrachtung dennoch unvollständig, wenn sie nicht auch noch auf andere Beispiele ausgedehnt wird. Um in dieser Hinsicht noch weitere Beweise zu haben, wurden noch bei drei anderen Abflußmengen (81,19; 44,19 und 8,74 l/s/lfd. m) Untersuchungen angestellt.

Die ersten beiden sind so gewählt, daß die eindimensionalen Größen $\left(t, \dfrac{p}{\gamma} \text{ usw.}\right)$ genau $\tfrac{3}{4}$- bzw. $\tfrac{1}{2}$ mal so groß waren als ursprünglich. Nach dem Ähnlichkeitsgesetz ist dann:

$$\left(\frac{t_a}{t_b}\right)^{\frac{3}{2}} = \frac{Q_a}{Q_b},$$

worin Q in l/s/lfd. m einzusetzen ist. Die Längsprofile, die Sohlendrücke und die Vertikaldruckverteilungen an der Abfallkante wurden wieder sorgfältig gemessen. Alle diese Werte können aus Abb. 20 ersehen werden.

Da die Druckmeßrohre für die ursprüngliche Durchflußmenge bestimmt waren, traten bei kleineren Durchflußmengen einige Schwierigkeiten auf, denn durch die Maßstabsverkleinerung wurde einmal die Zahl der verfügbaren Wandöffnungen vermindert und zweitens die in ihrer Konstruktion begründete Ungenauigkeit etwas vergrößert. Da die Versuche in der gleichen Rinne, also bei gleicher Sohlen- und Wandrauhigkeit, ausgeführt wurden, war der relative Wert der Rauhigkeit bei jeder Wassermenge ein anderer.

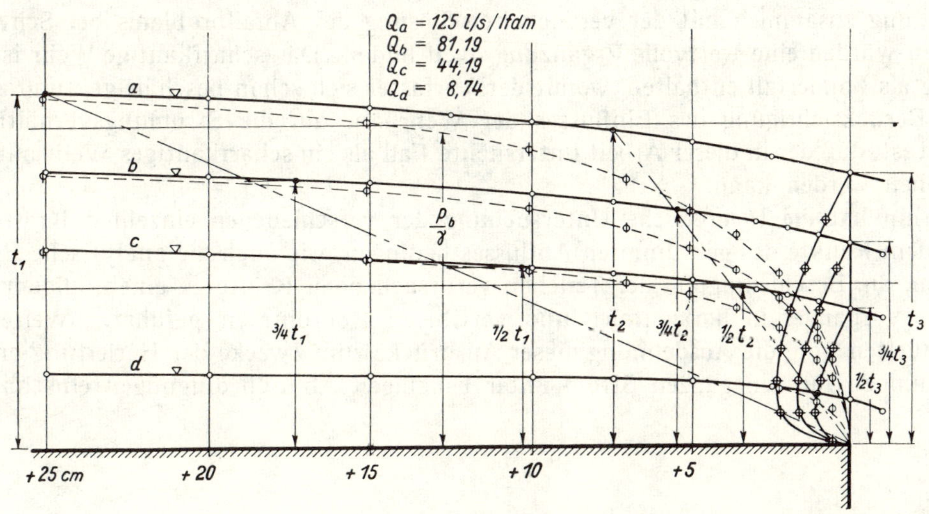

Abb. 20. Darstellung der Profillinien und Wand- und Bodendruckmessungen bei verschiedenen Abflußmengen.

Die Auftragung in Abb. 20 zeigt jedoch, daß trotzdem in jedem Falle eine hinreichende Übereinstimmung erzielt wurde. Die gemessene Tiefe an der Abfallkante zeigte eine Abweichung von weniger als 0,02 cm gegenüber der erwarteten. Die Formeln der Druckverteilungskurven an der Abfallkante sind einander sehr ähnlich, sie unterscheiden sich nur ein wenig in den Größenverhältnissen.

Die Längsprofile und die Kurven des Sohlendruckes sind, nach dem ersten Anschein zu urteilen, nicht in allen Fällen die gleichen, da sie bei geometrischer Ähnlichkeit alle durch die Abfallkante gelegten geraden Linien in proportionale Teile zerlegen müßten. Die wahren Teilpunkte liegen, wie Abb. 20 zeigt, in den beiden Oberflächen für die kleineren Durchflußmengen — entsprechend $\frac{3}{4}$ bzw. $\frac{1}{2}$ der ursprünglichen Tiefe — etwas rechts von den geraden Linien. Es ist dies ein Umstand, der bereits bei den Vorversuchen als auffallend bei den kritischen Grenzen erwähnt wurde. Dies ist bei den gleichbleibenden Kanalabmessungen und verschiedenen Abflußmengen dem veränderlichen Reibungsfaktor zuzuschreiben, zum Teil wohl auch dem verhältnismäßig größer werdenden Einfluß von Viskosität und Oberflächenspannung bei kleiner werdendem Durchfluß. Für die Tiefen t_1 und t_2 in Abb. 20 und die entsprechenden Tiefen bei kleineren Durchflußmengen stehen die zugehörigen Sohlendrücke in fast genau dem gleichen Verhältnis wie die entsprechenden Tiefen. Die Geschwindigkeitsverteilungskurven ähneln sich an der Abfallkante in ihrer Form und zeigen bei kleiner werdenden Durchflußmengen nur eine geringe Abnahme im proportionalen Größenverhältnis.

Daher ist die Annahme gerechtfertigt, daß das Ähnlichkeitsgesetz mit Ausnahme der Reibungswirkung am benetzten Umfang auf die Bestimmung der Druckverteilung bei anderen Durchflüssen nach den in dieser Abhandlung gegebenen Formeln angewendet werden darf.

4. Gebiet für weitere Forschung.

In den vorhergehenden Erörterungen ist die Energieverteilung insbesondere bezüglich der Druckenergie sowohl theoretisch wie experimentell für einen grundlegenden Fall gekrümmten Abflusses eingehend untersucht worden. Durch Gleichsetzung von Stützkraft und äußeren Kräften wurde eine praktische Methode zur Bestimmung des Druckes in einem beliebigen Querschnitt gefunden und am vorliegenden Beispiel nachgeprüft.

Diese Methode ist noch auf weitere Fälle anwendbar; so ist der Verfasser bereits zur Betrachtung anderer Beispiele übergegangen, bei denen stromabwärts der Abfallkante verschiedene Dammbegrenzungen vorgesehen sind, die zwischen einem sehr flachen Abfall und einem solchen schwanken, bei dem der Strahl sich an eine lotrechte Wand anschmiegt. Auch ein kreisförmiger Abfallrücken ist bereits untersucht worden. Ähnliche Untersuchungen über den Einfluß der oberwasserseitigen Wandausbildung zusammen mit der verwickelten Lösung des Abfallproblems bei Schwellen und Grundwehren würden eine wertvolle Ergänzung dazu bilden. Das scharfkantige Wehr ist in dieser Betrachtung als Sonderfall enthalten, womit der Verfasser sich schon beschäftigt, und zwar unter besonderer Berücksichtigung des Einflusses der Wehrhöhe auf die Strömungsverhältnisse. Bemerkenswert ist, daß der in dieser Arbeit untersuchte Fall als ein scharfkantiges Wehr mit der Höhe Null angesehen werden kann.

Weiterhin hat die theoretische Untersuchung der verschiedenen einzelnen Kräfte, die den Druck in jedem Punkte des gekrümmten Abflusses bedingen, wie auch die analytische Behandlung der einzelnen, die Beschleunigung der Teilchen verursachenden Kräfte zu einem allgemeinen Ausdruck dieses Vorganges in horizontalen und vertikalen Koordinaten geführt. Weitere Untersuchungen dürften wohl die Ausdehnung dieser Ausdrücke zum Zwecke der Herleitung brauchbarer Abflußgleichungen für den ganzen Strom unter beliebigen Abflußbedingungen einschließen.

C. Zusammenfassung.

Die des näheren in obiger Abhandlung erörterte Betrachtung des Wasserabflusses, bei dem eine senkrechte Krümmung des Wasserstromes auftritt, ergänzt und nachgeprüft durch Versuche an einem waagerecht überfluteten Damme mit belüftetem Abfallstrahl, können folgendermaßen kurz zusammengefaßt werden:

1. Sobald die Teilchen eines Wasserstromes von einer geradlinigen Bahn abweichen, ist der Druck nicht mehr statisch sondern im allgemeinen nach einer Kurve verteilt. Die Krümmung dieser Kurve ist allerdings nur bei stark gekrümmten Stromfäden merkbar und die Verteilungskurve kann sonst als geradlinige Funktion der Tiefe unter der Oberfläche angenommen werden.

2. Diese Änderung der normalen Druckverteilungskurve läßt sich folgendermaßen erklären:

   a) Die auf irgendein Wasserteilchen bei gekrümmtem Abfluß wirkende Beschleunigungskraft kann als die Vektorsumme zweier Komponenten betrachtet werden: einer Vertikalkomponenten, die gleich der Gravitationskraft auf das Teilchen ist und einer Komponenten, die die Größe und Richtung des maximalen Druckgefälles an diesem Punkte besitzt.

   b) Die Druckverteilungskurve ist die Summe zweier sekundärer Kurven, von denen die eine wahrscheinlich eine geradlinige Funktion der Tiefe ist und einen Maximalwert gleich dem Bodendruck in diesem Querschnitt besitzt, die andere eine nicht lineare Funktion der ungleichmäßigen Konvergenz und Geschwindigkeitsverteilung über den Querschnitt ist.

3. Der Gesamtdruck in einem beliebigen Querschnitt durch irgendeinen Wasserstrom kann nach den allgemeinen vom Impulssatz abgeleiteten Formeln berechnet werden:

$$\left(P_1 + \frac{\gamma Q^2}{g\,t_1}\right) - \left(P_2 + \frac{\gamma Q^2}{g\,t_2}\right) + \Sigma\,(p_b \cos\beta) - V_x = 0$$

$$\frac{\gamma Q^2}{g\,t_1}\,\mathrm{tg}\,\alpha_1 - \frac{\gamma Q^2}{g\,t_2}\,\mathrm{tg}\,\alpha_2 + G - \Sigma\,(p_b \sin\beta) - V_y = 0.$$

4. Für den Fall eines einfachen Wasserabsturzes mit belüftetem Strahl können diese Formeln vereinfacht werden zu:

$$P = \frac{2}{3}\,\gamma\,H_{Gr}{}^2 - \frac{\gamma Q^2}{g\,t} - \gamma\,t\,\Delta H$$

$$G - P_b = \frac{\gamma Q^2}{g\,t}\,\mathrm{tg}\,\alpha.$$

In dem Gebiet, wo die Druckänderung praktisch eine lineare Funktion der Tiefe ist, kann der Sohlendruck weiterhin auf Grund nachstehender Formel annähernd berechnet werden:

$$p_b = \frac{2\,P}{t}.$$

5. Mittels der empirischen Formel:

$$\frac{p}{\gamma} = H - y - \frac{1}{2g}\left[v_u - (v_u - v_o)\left(\frac{y}{t}\right)^{\frac{v_m - v_o}{v_u - v_m}}\right]^2$$

kann die Druckhöhe beim Abfluß über einen waagerechten Damm mit belüftetem Strahl für einen beliebigen Punkt mit guter Annäherung berechnet werden.

6. Die Höhe der wirklichen mittleren Energielinie in einem gegebenen Querschnitt ist im Falle gekrümmten Abflusses nicht gleich dem Ausdruck

$$H_m = t + \alpha_u \frac{v_m^2}{2\,g},$$

sondern muß ermittelt werden aus der Beziehung

$$H_m = \frac{1}{t} \int_0^t \left(y + \frac{p}{\gamma} + \frac{v^3}{2\,g\,v_m} \right) d\,y$$

$$= \frac{t}{2} + \frac{P}{\gamma\,t} + \alpha_u \frac{v_m^2}{2\,g}.$$

7. Unter hinreichender Berücksichtigung der Reibungsbeziehungen kann das Ähnlichkeitsgesetz auf die Verhältnisse des gekrümmten Abflusses mit Sicherheit Anwendung finden; ausgenommen sind solche geringen Wassermengen, bei denen die besonderen Eigenschaften des Wassers (Viskosität und Oberflächenspannung) merkbaren Einfluß auf den normalen Abflußvorgang haben.

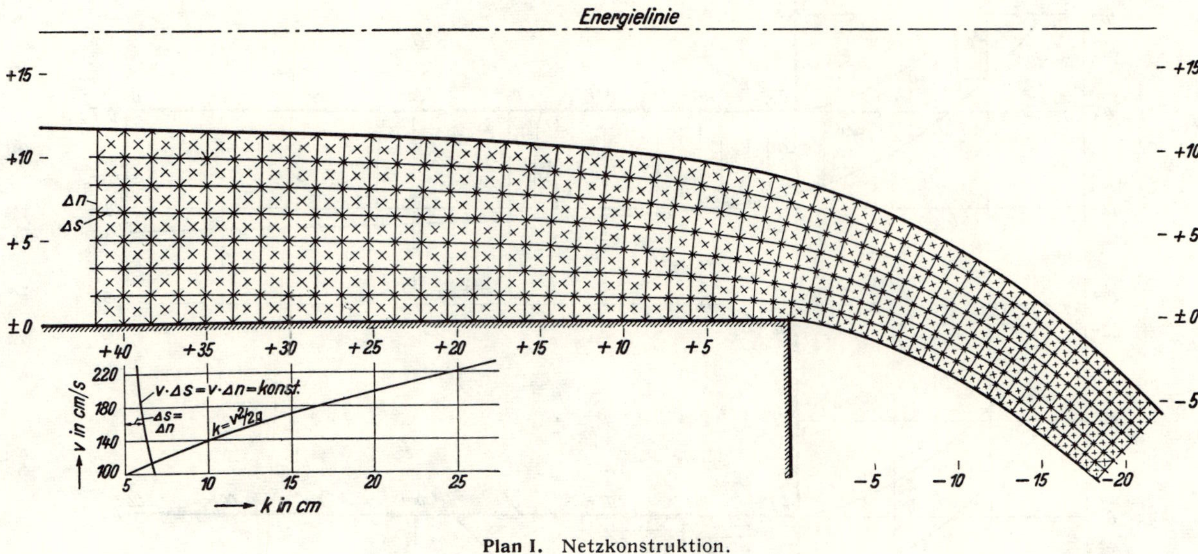

Plan I. Netzkonstruktion.

43

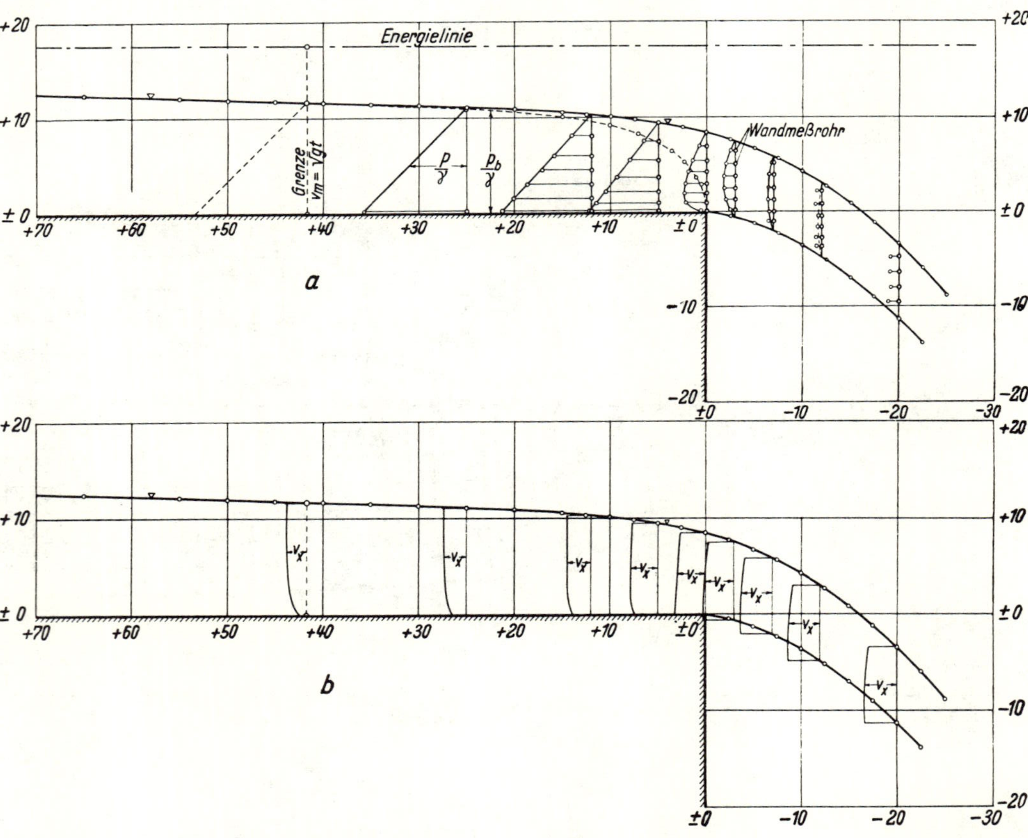

Plan II. a) Stromprofil im Bereich des Absturzes mit Darstellung der gemessenen Wand- und Bodendrücke.
b) Darstellung der horizontalen Komponenten der gemessenen Geschwindigkeiten Maßstab 1 cm = 1 m/s.

44

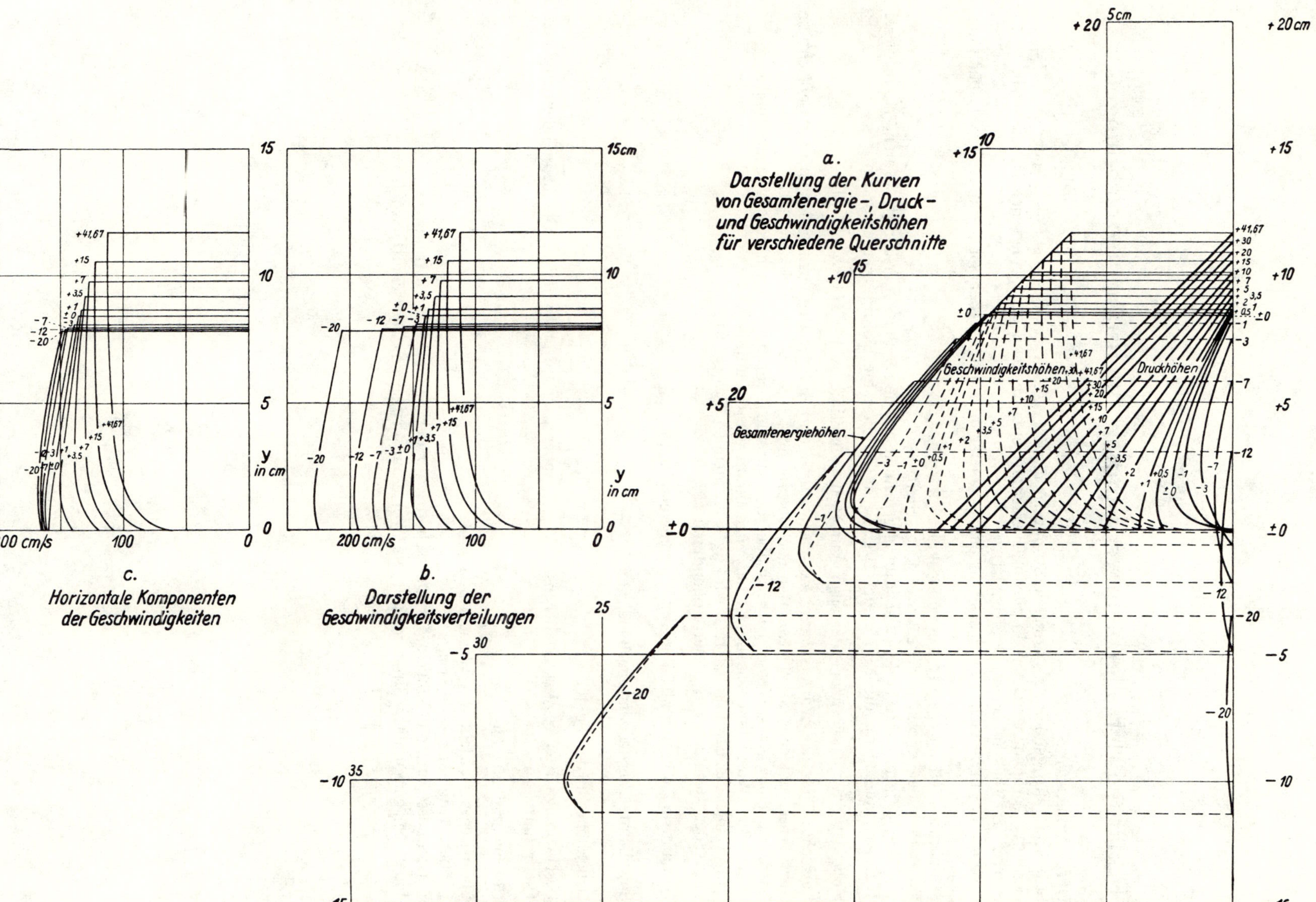

Plan III. Graphische Darstellung der Energieverteilung eines Stromes von 125 l/s/lfdm. ober- und unterhalb eines vollkommen belüfteten Absturzes.

Erklärung der allgemein verwandten Buchstabenbezeichnungen.

Q — Durchflußmenge.

S — Stützkraft an irgendeinem normalen Stromquerschnitte, gleich der Summe $P + K$ oder $D - Z + K$.

P — Gesamter vorhandener Druck in einem Querschnitt entsprechend den gemessenen Druckhöhen.

D — Statischer Druck in einem Querschnitt.

Z — Gesamter „Unter- oder Überdruck" in einem Querschnitt infolge der sogenannten Zusatzspannung, gleich $D - P$.

K — Stützkraftanteil infolge der Bewegungskraft.

P_b — Gesamter Bodendruck zwischen zwei Querschnitten.

    Die obigen Werte beziehen sich im allgemeinen auf den vollen Querschnitt, während sie in dieser Abhandlung für die Einheit der Breite der Gerinnesohle benutzt werden sollen.

H — Höhe der Energielinie über einer angenommenen Grundlinie, wobei $H = h + \dfrac{p}{\gamma} + \dfrac{v^2}{2\,g}$.

t — Lotrechte Tiefe des Wasserstromes zwischen oberer und unterer Begrenzungsfläche. Bei Betrachtung eines Wasserteilchens die Tiefe unter der Oberfläche.

y — Lotrechte Höhe eines Wasserteilchens über der unteren Begrenzungsfläche des Wasserstromes.

h — Höhe eines Wasserteilchens über einer angenommenen Grundlinie.

n — Länge einer die Stromlinien rechtwinklig schneidenden Trajektorie.

b — Breite des Gerinnes. In dieser Abhandlung $b = 1$.

$\dfrac{p}{\gamma}$ — Gesamtdruckhöhe (unmittelbar am Piezometerrohr abgelesen).

$\dfrac{z}{\gamma}$ — Sogenannte Zusatzspannungshöhe, gleich $t - \dfrac{p}{\gamma}$.

$\dfrac{p_b}{\gamma}$ — Druckhöhe an der Sohle.

v — Geschwindigkeit eines Wasserteilchens. v_o, v_u, v_m obere, untere und mittlere Geschwindigkeit in einem Querschnitt.

k — Geschwindigkeitshöhe, gleich $\dfrac{v^2}{2\,g}$.

γ — Spezifisches Gewicht des Wassers in kg/m³.

g — Erdbeschleunigung $= 9{,}81$ m/s².

m — Wassermasse, die die Einheit des Querschnittes in der Sekunde durchströmt.

Gr — Index bei geradliniger Strömung für die Werte im kritischen Querschnitt, wo $v_m = \sqrt{g\,t}$.

Model Research on Spillway Crests

A Study of Pressure Distribution and of Discharge as a Function of Crest Design

By Hunter Rouse
Instructor in Civil Engineering, Columbia
University, New York, N.Y.

and Lincoln Reid
Junior Engineer, U. S. Bureau of
Reclamation, Denver, Colo.

FOR determining the profile of an overflow spillway, recourse has been had to the rule-of-thumb, but faulty, method of making it coincide with the free path of a thin jet of water discharging into the atmosphere with an initial horizontal velocity corresponding to two-thirds the expected head on the spillway crest. Actually, no single filament of the nappe can be found which corresponds with the assumed jet discharging freely under atmospheric pressure. A more reasonable and exact method is to make the profile agree with the under surface of the ventilated nappe of a sharp-crested weir at a corresponding head. Investigations made by Messrs. Rouse and Reid and by others throw new light on the behavior of such a spillway under the expected head as well as under appreciable overload. Although additional data at large scale are needed to obtain in dimensionless coordinates the shape of the ventilated nappe over a complete range of ratios of operating head to depth of approach flow, certain conclusions can now be drawn. The actual head may exceed the designed head on a properly shaped crest by a considerable amount without the likelihood of the sheet springing loose, despite the presence of negative pressures under the sheet. Of primary importance in causing separation is an abrupt or rapid change in the curve of the spillway profile. This dangerous condition can be avoided by giving careful attention to the curves, especially in the immediate vicinity of the crest.

IN the design of spillways, there are three essential requirements to be met: the maximum expected discharge must be passed with the least possible increase in elevation of the reservoir surface; the structure must be not only stable under all conditions but economical of construction as well; and the hydraulic characteristic of flow over the crest should be known more definitely than is now possible on the basis of unverified assumptions in design. It is questionable whether these three requirements can be met fully without a more accurate knowledge of discharge conditions than has been available in the past.

Research on the distribution of pressure and velocity in such curvilinear flow as occurs over various types of weir sections led Mr. Rouse to make an analytical investigation of the design of spillway crests. This investigation was conducted by him at the Massachusetts Institute of Technology, and presented in the form of a thesis for an advanced degree, entitled "The Distribution of Hydraulic Energy in Weir Flow with Relation to Spillway Design." Through this method of attack, a number of fundamental points became evident that apparently are often neglected or misunderstood in current practice. Conclusions reached in this investigation have been tested by Mr. Reid in the River Hydraulics Laboratory of the Massachusetts Institute of Technology by experiments with models. The results of his tests were presented as a thesis for an advanced degree under the title, "An Experimental Study of a Spillway Conforming to the Measured Lower Surface Nappe of Flow Over a Rectangular, Suppressed, Sharp-Crested Weir." Such definite substantiation was obtained by Mr. Reid that both analytical and experimental results are presented briefly here with the hope that they may be of value as well to those engaged in the design of hydraulic structures as to those interested in the interpretation of research with hydraulic models.

American engineers as a rule base the curve of the spillway crest on the profile of the fully ventilated nappe of water flowing over the sharp-crested weir. For example, in his book, *Engineering for Masonry Dams*, W. P. Creager, M. Am. Soc. C.E., gives a number of dimensionless curves based on measurements by Bazin. On the other hand, various purely empirical formulas are often assumed to give a fairly accurate idea of the trajectory of the undermost water particles from the moment they leave the top of the spillway. Such methods are founded on the belief that the presence of a solid structure below the nappe, and in immediate contact with it, will not appreciably affect the course of the freely falling particles of water; that is, neither will the nappe tend to spring loose from the spillway face, nor will it exert pressure on it.

At once certain questions arise. Will the discharge really be uninfluenced by a solid structure filling the space formerly occupied by air, and is the pressure between spillway face and nappe actually atmospheric? Under what conditions could the falling sheet spring free? Do such methods of determining the profile of the spillway crest yield the maximum permissible coefficient of discharge without sacrificing either safety or structural economy? And how much leeway in design is possible without seriously changing the conditions of discharge?

Flow over a sharp-crested weir is governed by the geometrical proportions of channel and weir, and by three independent forces—gravitation, viscosity, and surface tension. If viscosity and surface tension did not

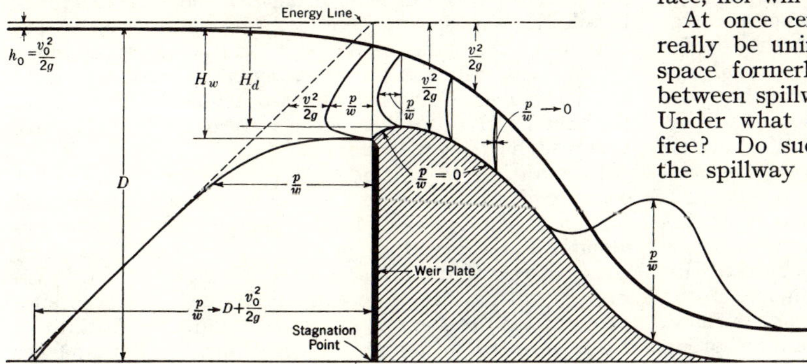

Fig. 1. Schematic View of Spillway Conforming to Ventilated Nappe of a Sharp-Crested Weir

Flow Profile and Expected Pressure Distribution

exist, geometrically similar flow conditions at different scales would also be dynamically similar; that is, the coefficient of discharge for any ratio of head to depth of approach, H/D, would have a constant value regardless of the actual magnitude of the head, and the nappe profiles would be exactly similar. Furthermore, since a frictionless fluid can exert no tangential force, the flow could not be changed by fitting the under surface of the nappe with a solid spillway section of exactly the same form. If this were done the pressure would be exactly atmospheric at all points of contact downstream from the former position of the weir crest, the same conditions of pressure and velocity would continue to exist at all points within the nappe, and the rate of discharge would remain unchanged.

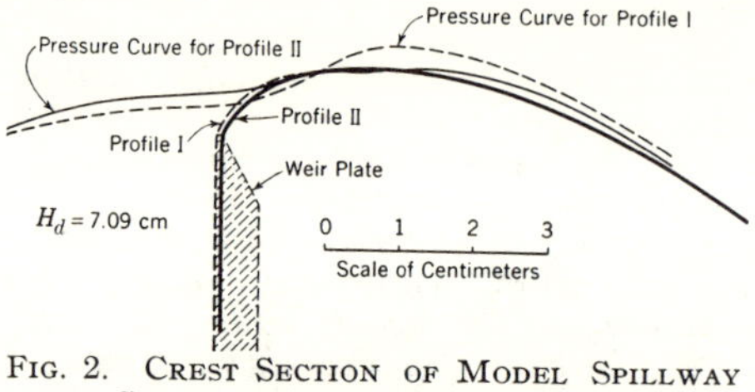

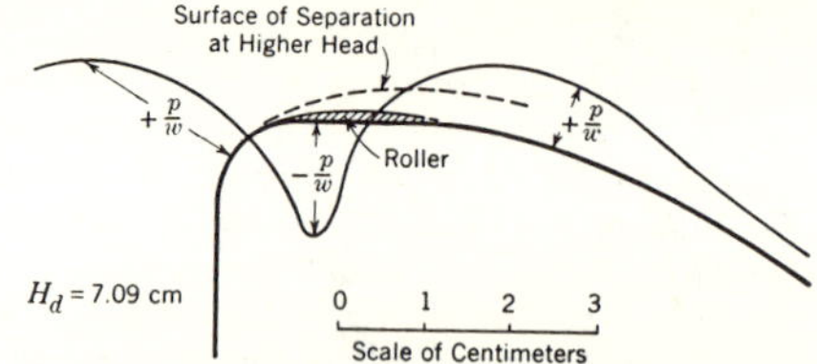

FIG. 2. CREST SECTION OF MODEL SPILLWAY WITH CURVES OF MEASURED PRESSURE DISTRIBUTION FOR PROFILES I AND II
Pressure Heads Plotted Normal to Spillway Face

FIG. 3. MODIFIED CREST SECTION OF MODEL SPILLWAY, PROFILE III
With Pressure Distribution at Designed Head and Surface of Separation at Higher Head

Although gravity is the primary force that acts in the flow of water over a weir, viscosity and surface tension play significant rôles in altering the magnitude of the discharge coefficient. The extent to which viscous forces affect the discharge is not as yet well known; however, it has been definitely established that the effect of surface tension varies approximately with the reciprocal of the head, thus being greatest for low discharges. Hence accurate results cannot be expected if the nappe profiles of small weirs are used to prophesy conditions for large spillways.

It is apparent that studies of small-scale models in the hydraulic laboratory cannot alone be depended upon to give accurate quantitative values for determining the behavior of much larger prototypes. Only a series of studies of the same structure at various scales will give the essential key to the interrelation of the three basic physical forces as the scale ratio changes. On this phase of the subject, the late Fr. Eisner's work, "Ueberfallversuche inverschiedener Modellgrösse," *Mitteilungen der Preussischen Versuchsanstalt für Wasserbau und Schiffbau* (Berlin, 1933), is of interest; also that of M. L. Escande, "Étude théorique et expérimentale sur la similitude des fluides incompressibles pesants," a thesis for the doctorate at the University of Toulouse (France, 1929). However, studies of small-scale models can provide very significant qualitative information. Furthermore, if a model spillway crest patterned after flow over a weir of the same dimensions behaves as expected, it is only logical to conclude that a large spillway will follow just as surely the nappe of a correspondingly large weir. Determination of the profile of very large nappes is not possible in the usual hydraulic laboratory. Therefore this would be a most desirable project to be undertaken by such an institution as the National Hydraulic Laboratory at Washington and would supply greatly needed data.

A hydrodynamical approach to the problem of curving flow, such as that over weir crests, shows that conditions at any point in the flow are dependent upon those directly upstream. Thus it has been demonstrated by Bazin and other investigators that any change in the crest of a weir will result in a change in the nappe profile. Logical as this reasoning may seem, designers are prone to determine carefully the downstream curve of the spillway face, and then proceed to round off, bevel, or prolong the short upstream curve to suit their fancy—despite the fact that the downstream curve most emphatically depends on this curve which they modify.

In order to study a spillway properly designed according to the profile of the nappe over a sharp-crested weir of the same proportions, and to illustrate the effect of a change in curvature in the immediate vicinity of the crest, experiments were made to determine the flow profile of discharge over a weir 40 cm (about 1.3 ft) in height under a head of 8 cm, and to investigate the pressure and velocity distribution throughout the flow. The spillway model is shown schematically in Fig. 1. Based on the measured curve of the lower surface of the nappe, a spillway section was accurately modeled in concrete and provided with 11 piezometer inlets at intervals over the vertical and curved surfaces in the region of the crest. In this way it was possible to measure discharge, profile, and pressure distribution during flow over this spillway under the same head, and compare these values directly with the previous results. Moreover, the crest of the spillway could be varied either by cutting away or by building up with plasticine to any desired form.

In the case of the weir, surface tension causes the nappe to spring loose a short distance below the crest rather than at the sharp upstream edge, due principally to the small width of the crest, which is never made absolutely sharp. However, the relative capillary effect diminishes as the head increases. Hence, some change was to be expected in the case of the spillway model, since the inserted body completely prevented the action of surface tension. Not only was the discharge over the spillway model actually lower than that over the weir by about 3 per cent, but the piezometers indicated a positive pressure at the top of the spillway curve, as shown in Profile I, Fig. 2. This influence was removed by rubbing down the vertical face of the model by an amount equal to the width of the weir crest, that is, 0.5 mm. While the differences in curvature were slight, the pressure variations were considerable. In this form, Profile II, the spillway gave a discharge only 0.7 per cent lower than that of the weir, and the pressure along the curves surface in the crest region did not vary from the atmospheric by more than a negligible fraction of the head. No further efforts were made to reduce these slight discrepancies.

In addition to this ideal form of crest, two other common designs were tested, one with a short horizontal section at the top (Profile III in Fig. 3), such as is frequently encountered in practice, preceded by a short curve of constant radius; and the other (Profile IV in Fig. 5) with a lip projecting upstream. Both profiles followed the ideal curve except as noted. Each of these crests resulted in a pressure distribution different from that for which the downstream curve was designed. As may be seen from Fig. 3, the departure of the first of these two from the ideal curve is by far the more marked, since the change in form lies more nearly in the region

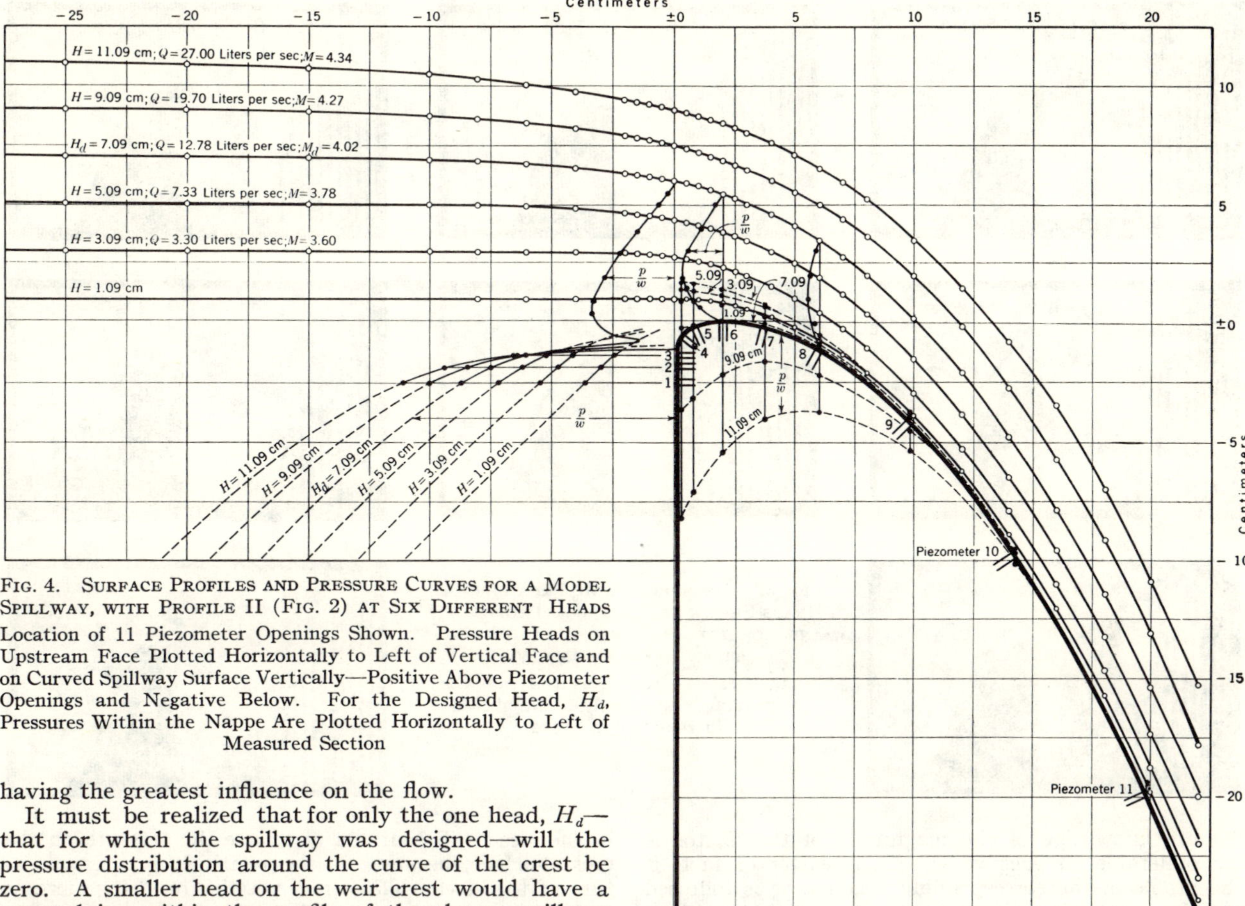

FIG. 4. SURFACE PROFILES AND PRESSURE CURVES FOR A MODEL SPILLWAY, WITH PROFILE II (FIG. 2) AT SIX DIFFERENT HEADS Location of 11 Piezometer Openings Shown. Pressure Heads on Upstream Face Plotted Horizontally to Left of Vertical Face and on Curved Spillway Surface Vertically—Positive Above Piezometer Openings and Negative Below. For the Designed Head, H_d, Pressures Within the Nappe Are Plotted Horizontally to Left of Measured Section

having the greatest influence on the flow.

It must be realized that for only the one head, H_d— that for which the spillway was designed—will the pressure distribution around the curve of the crest be zero. A smaller head on the weir crest would have a nappe lying within the profile of the chosen spillway section, whereas a larger discharge over the weir would give a curve lying outside the profile. The lower discharge, on passing over the spillway, would of necessity be forced to adapt itself to the established curve, and a certain positive pressure would exist between the sheet and the crest, forcing the nappe to follow a higher trajectory than that of free fall. Similarly, the greater discharge, the natural trajectory of which lies outside the spillway profile, would follow the established profile only through a reduction in pressure along the crest. That is to say, the sheet would be forced back against the spillway surface by the difference between the atmospheric pressure above the sheet and the less-than-atmospheric pressure beneath it. In all cases, however, the pressure would be positive a short distance past the crest.

In Fig. 4 profile and pressure curves are given for six different discharges over a crest shaped to Profile II, showing the variation in pressure conditions as the head varies from a small value to one considerably larger than H_d. A photographic record of profiles and pressure curves for the same range of H is presented in Fig. 5. The model spillway weir was set between two glass plates exactly one foot apart. All measurements were made with vernier gages reading to 0.1 mm (about 0.0003 ft). For photographic purposes the rear glass panel was covered with white tracing paper and the model illuminated from behind. Thus the water surface is visible as a black line caused by the meniscus formed along the glass wall. A few drops of potassium permanganate solution in the manometers made the water columns visible.

The pressures on the upstream face, plotted horizontally to the left in Fig. 4, deserve attention, since textbooks often refer erroneously to "impact" in this region. Hydrodynamically speaking, at the base of the upstream face—that is, at the intersection of the floor with the spillway—there is a point of stagnation, indicated in Fig. 1, where the velocity is reduced to zero because of the abrupt angle. If in this region the total head is the same as that of the average flow, this reduction in velocity head must be compensated by an equal increase in pressure head. Hence the normal static head, as given by the total depth of flow at a point where the surface of the approaching stream is still horizontal, is augmented by an amount equal to the velocity head of approach, and not by twice this velocity head as in the case of "impact." The fact that frictional losses tend to reduce the velocity along the floor only serves to reduce this stagnation pressure, as is also shown in Fig. 1. As the flow accelerates along the upstream face in the direction of the crest, the pressure drops accordingly, approaching zero as the crest is reached. The stagnation pressure becomes of appreciable magnitude with greater velocities of approach—that is, for low spillways under high heads. This was investigated over a great range of the ratio H/D.

Of primary interest to the designing engineer is the coefficient of discharge, M, in the equation $Q = MBH^{3/2}$. For a spillway this factor is generally assumed—3.8 being a common guess—and the design then proceeds by rule of thumb on the basis of this assumption. Obviously,

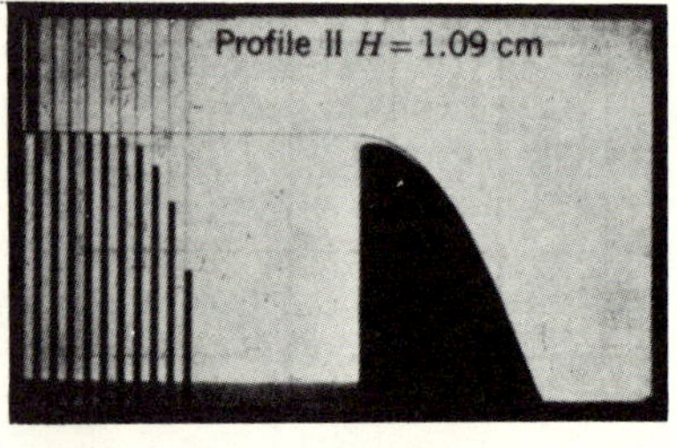

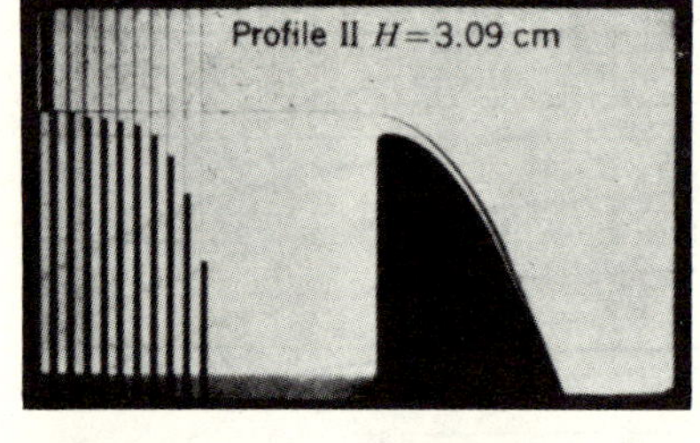

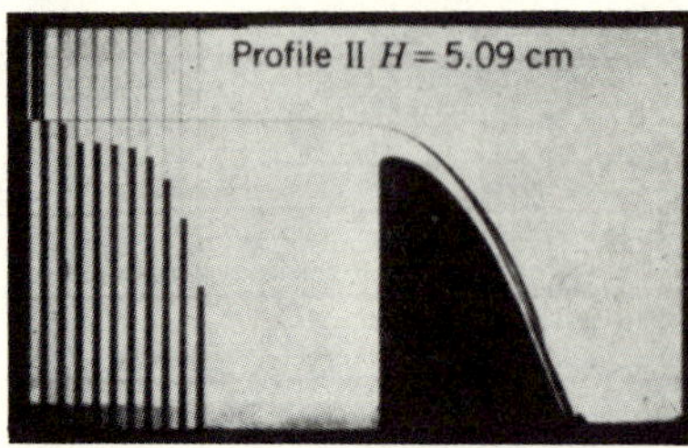

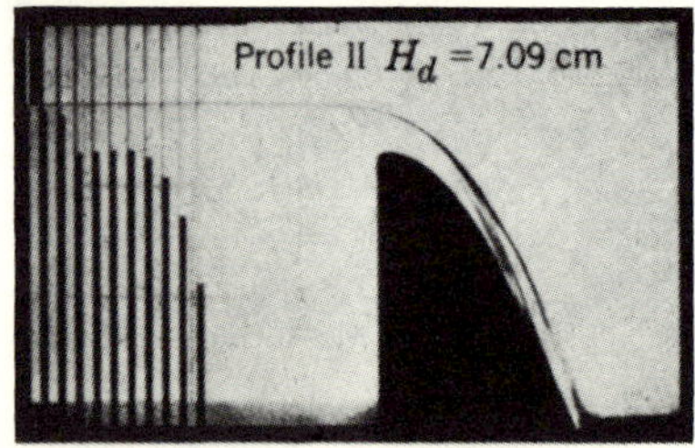

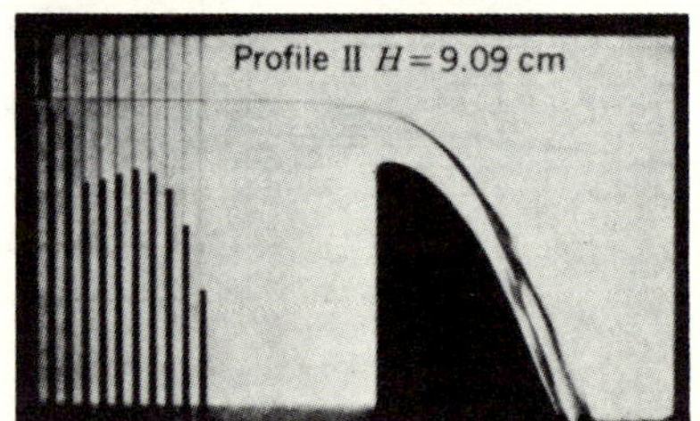

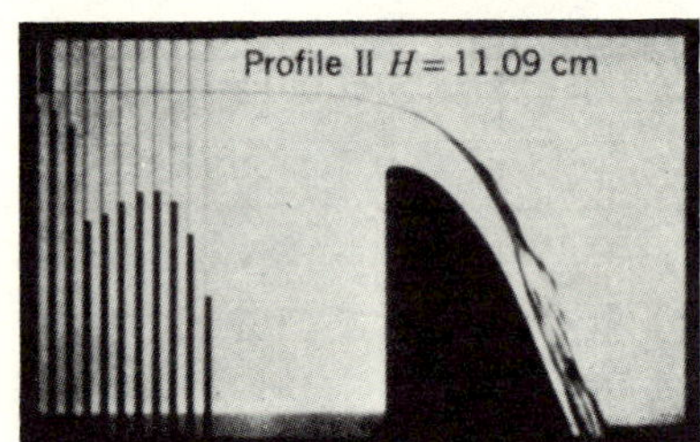

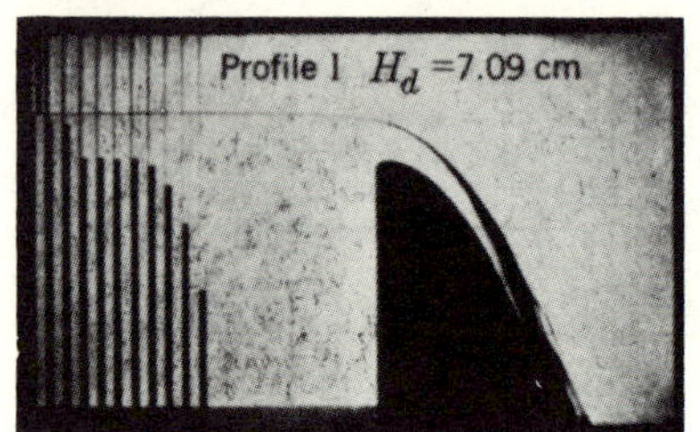

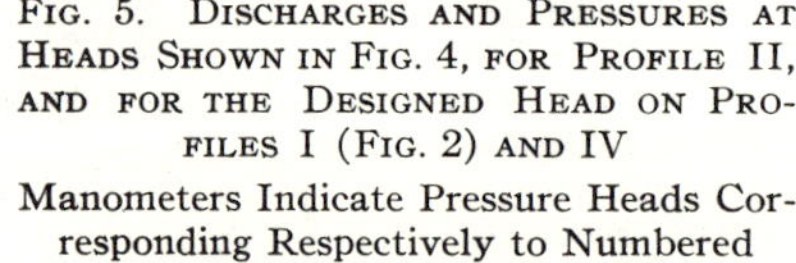

FIG. 5. DISCHARGES AND PRESSURES AT HEADS SHOWN IN FIG. 4, FOR PROFILE II, AND FOR THE DESIGNED HEAD ON PROFILES I (FIG. 2) AND IV

Manometers Indicate Pressure Heads Corresponding Respectively to Numbered Piezometers in Fig. 4

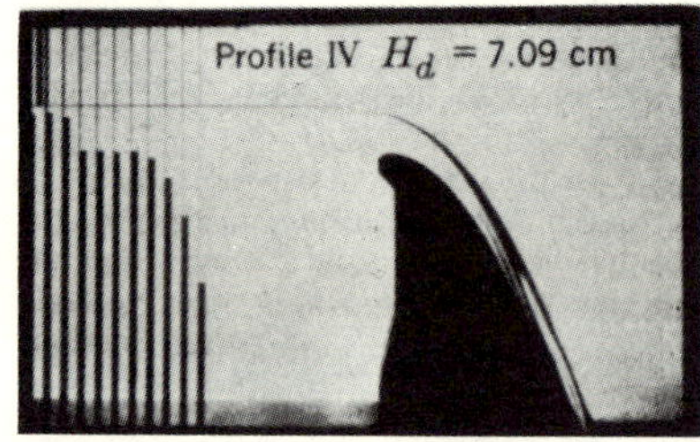

definite knowledge of the magnitude of this factor is far preferable to guesswork. As a matter of fact, if the profile of the corresponding weir nappe is followed exactly in designing the crest, not only can the discharge under the designed head be foretold with negligible error, but the complete rating curve for all heads is definitely established.

Measurements on the weir nappe for an 8-cm head gave a value of 3.39, in English units, for the coefficient M_w and the maximum elevation of the lower surface above the crest amounted to 0.91 cm. It will be noted that this is somewhat lower than Bazin's average value of 0.13 H, because of the increased capillary effect at low heads. Using the crest elevation as the zero level of the spillway itself, the head on the spillway would then be 7.09 cm, which gives the spillway coefficient, M_d, the value of 4.05. That the actual measurement of discharge over this spillway (Profile II) under the 7.09-cm head resulted in a value of 4.02 for M_d is ample proof of the accuracy which may be obtained by careful design. The measured coefficients for Profiles I, II, and IV are, respectively, 3.93, 3.89, and 3.97. A lip in the upstream side does not appreciably influence the discharge, since the effect of the slight reversal of flow is offset by the increased curvature of the filaments.

In general it may be said that negative pressure in the region of the crest is accompanied by an increase in the discharge coefficient, whereas a positive pressure signifies that the coefficient is reduced. This generalization holds not only for changes in profile under the given head, but for a variation in head as well. The plot of the discharge coefficient against the head for Profile II, given in Fig. 6, shows this relationship clearly. Of great importance is the fact that as the head increases beyond its design value, the rate of increase of the coefficient approaches a minimum. Further increase

in the head will continue to reduce the pressure on the crest surface, because of the relatively greater curvature of the stream filaments; yet the resulting increase in velocity is compensated by a gradual reduction in the rate at which the depth at the crest increases, since the sheet is deflected from its natural course with increasing abruptness.

The curve of H against M, as plotted in Fig. 6, applies only to the given value of H/D. This ratio will vary from nearly zero for very high spillways to unity for the free overfall; it will be found that the coefficient M for any designed head will increase with H/D, approaching its maximum value of $\sqrt{g}$, or 5.67, for extremely low spillways under great head. In this higher range, the coefficient will approach this constant value for all heads, since the discharge over the free overfall is at the critical point and is independent of the crest form.

From this discussion it would appear that the most economical head under which to operate a spillway would not be the designed head H_d, but somewhat higher on the curve, where M has attained a greater value. Yet designing engineers have not taken advantage of this fact through fear that the nappe would separate from the spillway face the moment a negative pressure occurred at any point. The curve of the discharge coefficient M, taken from Creager, becomes constant at 3.94 and ends abruptly when H becomes H_d (Fig. 6). Many texts advocate extending the downstream part of the spillway some distance into the normal profile of the free sheet as a sort of safety factor—but at the same time they unwittingly modify the upstream part of the curve in such a way that the pressure at the crest will often be less than atmospheric. It should be repeated, therefore, that it is not the downstream face but the entire upper region of the crest which influences the pressure distribution; and this region influences just as certainly the dangerous phenomenon of separation.

With each of the profiles investigated, every attempt was made to cause the nappe to spring loose from the spillway under the maximum available discharge head of $1.6\,H_d$. Air was blown into the flow through the piezometer tubes; piers were built in at both ends of the crest to allow free passage of the air along the sides; and the sheet was even disrupted by hand. Only in the case of Profile III was it possible to cause a surface of separation to form. In this case, under a head of about 9 cm, separation could be produced with ease, as shown in Fig. 3, the entire sheet then remaining free of the spillway face for an indefinite period. It will be noted from this illustration that even under the designed head the stream had left the face over a short distance, beginning with the abrupt change in curvature, the space below being filled with a small roller. The obvious conclusion is that sudden changes in curvature are to be avoided, for although a roller may at first fill the region of separation, negative pressure at this point will facilitate the entrance of air at either end of the crest to replace the roller.

As a further example of the formation of rollers at points of abrupt change in curvature, the case of the Puechabon spillway in France is cited, since this is a design often encountered in Europe. Results of measurements made by M. L. Escande at Toulouse, previously referred to, are reproduced in Fig. 7, taken from model studies of the 13-meter spillway at scales of 1:19.5 and 1:100. The illustration shows definitely the two surfaces of separation at the points of too-rapid change in the profile curve. The pressure curve is worthy of careful study in the light of the foregoing discussion.

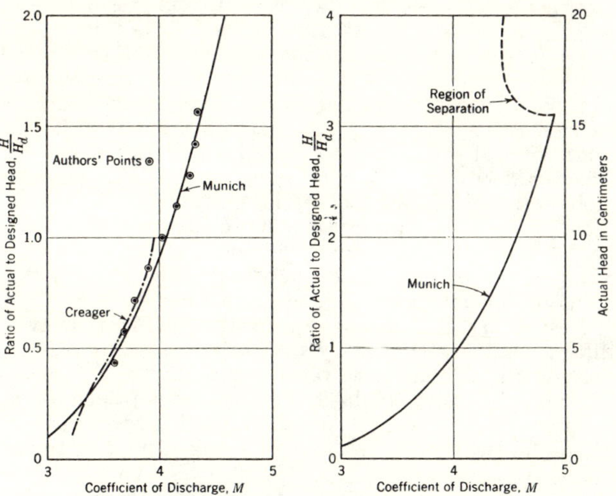

FIG. 6. VARIATION OF SPILLWAY DISCHARGE COEFFICIENT WITH RATIO OF ACTUAL TO DESIGNED HEAD FOR PROFILE II

Also Curves from Creager's "Masonry Dams" and Dillman's "Untersuchungen an Ueberfällen" (Munich). At the Right the Munich Curve Is Extended to and Beyond the Head at Which the Sheet of Water Separates from the Spillway Surface

At no point along the continuous curve of Profile II is it possible for such a roller to exist. For the same reason, if air is introduced at any point, it will immediately be carried downstream, because the steady change in curvature is sufficient to ensure against separation even under relatively high heads. Obviously, however, even for the ideal crest there is a limiting head under which the curvature becomes excessive, and for which a clinging sheet is impossible. Unfortu-

nately, laboratory facilities did not permit increasing the head on the model spillway to such an extreme magnitude, which would probably be several times the normal head, H_d.

Recently the laboratory of Dr. Thoma in Munich undertook an extensive investigation somewhat similar to the studies here described, determining the variation of pressure and discharge coefficient for spillways of various scales patterned after the general weir nappe. This work is reported by O. Dillmann in "Untersuchungen an Ueberfällen," Mitteilungen des Hydraulischen Instituts der Technischen Hochschule zu München, Heft 7 (Oldenbourg, Munich, 1933), and is referred to in Fig. 6. His crest section was made of hard-

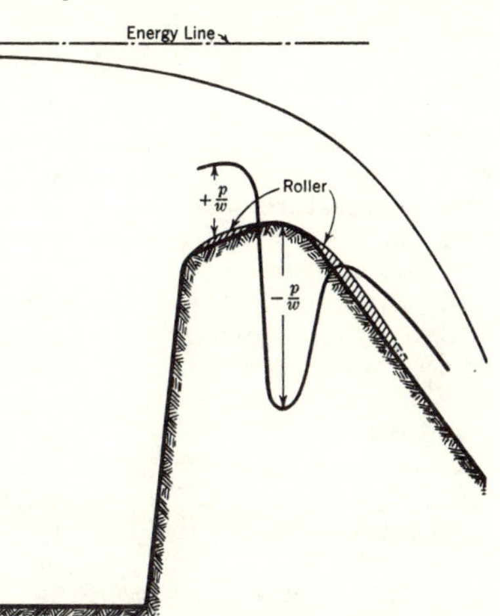

FIG. 7. PROFILE OF PUECHABON SPILLWAY, FRANCE

Pressure Distribution and Regions of Separation Under Normal Discharge, as Determined by Escande from Model Experiments at the University of Toulouse

wood, and the curve was approximated by a combination of ellipse and parabola which did not exactly correspond to the measured curve of the weir nappe.

He found that the head could be increased on his model spillway from the designed head of 5 cm to a magnitude of 15.5 cm without separation of the sheet at any point. As was to be expected, the curve of the coefficient M varied smoothly to this maximum value, as shown in Fig. 6. With a head of 16 cm, however, the entire discharge profile had changed, the sheet separating from the crest near the beginning of its curvature and the space underneath being filled by a roller of the type already discussed. The formation of this roller was accompanied by an increase in the radius of curvature of the nappe and hence by a decrease in pressure along the spillway face and a corresponding rise in the surface level. It will be seen that the coefficient of discharge drops abruptly at this point and then rises only slightly with increasing head.

Although such experiments on small models as are here discussed cannot be expected to give accurate quantitative results for the operation of large-scale spillways, qualitatively their indications are of direct importance in actual design. On none of these model tests was separation encountered with a head less than about three times that for which the crest was designed. Although such negative pressure may be possible in models, in the enlargement to full size for the case of a large spillway it may approach the vapor pressure of water—nearly absolute zero—under which conditions cavitation will result. In the Munich tests approximately 13 times the designed head, or over 4 times the maximum head, was discharged over the spillway without separation. While the resulting negative pressures can by no means be neglected, it is evident that the existing fear of separation on correctly designed spillways, even under an appreciable overload, is greatly exaggerated.

Discharge Characteristics of the Free Overfall

Use of Crest Section as a Control Provides Easy Means of Measuring Discharge

By Hunter Rouse, Assoc. M. Am. Soc. C.E.

Soil Conservation Service, Pasadena, Calif.

CONSIDERABLE attention has been given in engineering literature to the measurement of flowing water by means of critical-depth meters of one form and another. In this category fall such hydraulic devices as the broad-crested weir, the venturi flume, and similar contractions of the flow section, which presumably cause the critical depth to occur at some point along the flow profile. Numerous publications have cited efforts to control the longitudinal position of this critical point, in order that a single measurement of depth at one established section might permit a direct computation of the discharge.

As yet no satisfactory degree of success has been attained in these efforts. All measuring devices of this type require careful rating before accurate results may be expected, because a variable coefficient must be applied to the expression for critical flow; this is largely to compensate for the fact that the apparent critical section wanders upstream or down with changing discharge, so that the depth at the established point of measurement is seldom equal to the critical.

Three fallacies in reasoning lie at the root of the poor success of these investigations:

1. The shape of the flow profile is directly dependent upon the relation between a variable discharge and a constant geometrical form of the hydraulic device. Since the surface profile must then vary in shape as the discharge changes, geometrical similarity between any two discharges is impossible, and hence complete dynamic similarity is quite out of the question. Obviously no single section can be chosen at which the computed critical depth may be expected to occur for two different rates of flow, for no two flow profiles are either geometrically or dynamically similar. It is evident that the only type of meter which might have a constant discharge coefficient—that is, the only universal critical depth meter—must be of such form that the flow profile will always bear the same geometrical relationship to the flow boundaries, regardless of change in head. This condition is possible, for instance, in the case of a ventilated, sharp-crested, suppressed weir of infinite (that is, extremely great) height, sloping at any desired

REFINEMENTS are constantly being introduced in the methods of measuring flowing water. In recent years, many attempts have been made to avoid the necessity of calibrating each individual measuring device, by utilizing the principle of critical depth for parallel flow. Unfortunately, in the critical-depth meters thus far devised, it has been impossible to prevent the apparent control section from shifting indeterminately upstream or downstream as the discharge varies, and so the need for experimentally determined coefficients has not yet been eliminated. In the present article Dr. Rouse points out the possibility of using the free overfall as a flow meter which needs no calibration. Although the flow at the overfall is not parallel, the crest section is that of true minimum energy and hence is the actual control section. Furthermore, the crest depth is a constant percentage of the computed critical depth for parallel flow. Dr. Rouse's analysis has been verified by experiment, and should provide a dependable and simple means of determining discharge at points of overfall.

angle within reasonable limits; the free overfall at the end of a long channel is one specific case of such a weir, the weir face then being approximately horizontal.

2. The familiar expressions for critical velocity, depth, and discharge are based upon conditions of linear motion with static pressure distribution; once curvilinear flow occurs, these expressions cannot be expected to apply, for the sum of pressure head and elevation is no longer equal to the depth. Obviously, any abrupt change of section—the very basis of weirs and venturi flumes—is accompanied by appreciable curvature of the flow profile, so that a depth equal to the computed critical depth is seldom found at a point of parallel flow.

3. The true section of minimum energy is that at which the sloping energy line reaches its minimum elevation above the lower boundary of flow, regardless of the degree of curvature of the flow in this region. In the case of an overfall at the end of a mild slope, the true critical section is located at the overfall crest; in the case of a sharp-crested weir, on the other hand, it lies at that point of the free nappe at which the lower surface has attained its maximum elevation. Unfortunately, however, neither is the discharge always a constant function of this true critical depth, since it may vary with change in the boundary conditions, nor is it as yet possible to determine analytically the form of this function for any given case of curvilinear flow.

Restricting this discussion to the problem of two-dimensional flow (the side walls of the channel remaining parallel throughout), the phenomenon of the free overfall may be used to illustrate the three principles in question. The overfall is of distinct importance in hydraulic engineering, aside from its close relation to the broad-crested weir, for it forms the starting point in computations of the surface curve in non-uniform channel flow in which the discharge spills into an open reservoir at the downstream end.

According to the principle of minimum energy in parallel flow, a given rate of discharge carried by a long canal of mild slope ending in an abrupt fall will flow under conditions of minimum specific energy or total head—

Symbols Used in This Article

Q = rate of discharge
S = channel slope
A = cross-sectional area of flow
C = Chezy coefficient
c_q = discharge coefficient
c_c = contraction coefficient
c_v = velocity coefficient
H = specific energy of flow or total head above channel floor
V = average velocity over a vertical section
V_a = average velocity of approach

g = gravitational acceleration
b = channel width
d = depth of flow at any section
d_c = critical depth for parallel flow (hydrostatic critical depth) $= \sqrt[3]{\dfrac{Q^2}{b^2 g}}$
d_0 = depth at overfall crest (true critical section)
h = head on weir
p = pressure intensity
γ = specific weight of fluid

that is, at the critical depth for parallel flow—at some section a short distance upstream from the fall. This critical depth is given by the following relationships (Fig. 1):

$$d_c = \frac{2}{3}\,H_c = 2\,\frac{V^2}{2g} = \sqrt[3]{\frac{Q^2}{b^2 g}}$$

If the slope of the channel is such that the normal depth for uniform flow is exactly equal to the critical depth, that is, if

$$S = \frac{Q^2}{A^2 C^2 R} = \frac{V_c^2\,(b + 2\,d_c)}{C^2 b d_c},$$

then at no point upstream from the crest will the depth be greater than d_c. Since the critical slope decreases with decreasing rate of energy loss—that is, with increasing values of C—a minimum slope of zero would represent a state of flow without any loss whatever, and the energy line would then be horizontal. Under these limiting conditions the flow would remain at the critical depth— that is, the surface would be horizontal —so long as the conditions of hydrostatic pressure distribution were fulfilled.

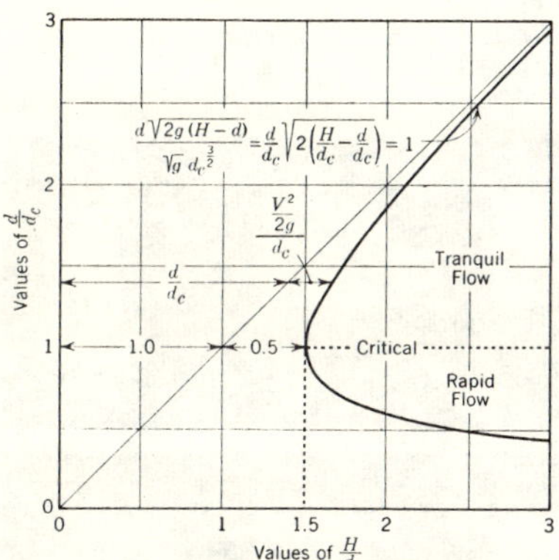

FIG. 1. DIMENSIONLESS SPECIFIC ENERGY DIAGRAM

PRESSURE DISTRIBUTION IN THE REGION OF THE CREST

At the lower end of the channel, however, the downward curvature of the discharging nappe of water produces a non-static distribution of pressure, as shown in Fig. 2(a) (drawn schematically after measurements by the author; see *Die Verteilung der hydraulischen Energie bei einem lotrechten Absturz*, H. Rouse, Oldenbourg, Munich, 1933). In the case of complete aeration, the pressure below as well as above the nappe must be atmospheric. Contrary to prevalent assumptions, however, it does not follow that the pressure within the nappe at the crest is also zero—in fact, there still exists considerable internal pressure in this region, caused by the convergence of the stream filaments between the upper and lower surfaces of the flow. As the filaments become more nearly parallel lower in the fall, this internal pressure gradually approaches zero as shown. Upstream from the crest there must be a gradual downward curve of the surface, since, in accordance with the energy principle, the decrease in pressure as the crest is approached signifies an increase in velocity, and the continuity principle in turn requires an accompanying decrease in depth. Mathematically speaking, this downward curve begins an infinite distance upstream from the

crest. For all practical purposes, nevertheless, it may be assumed that the depth is equal to the hydrostatic critical over the entire length of the channel, with the exception of a relatively short region near the crest.

Once resistance losses become appreciable, this basic picture of critical flow in a horizontal channel must be modified. As seen from the dimensionless energy diagram of Fig. 1, the critical depth for any given discharge under static pressure conditions occurs at the point of minimum specific energy or total head. As the channel remains horizontal, loss of energy causes the total head to decrease in the direction of flow, approaching this minimum value in the neighborhood of the crest. That is, the depth will increase in the upstream direction as shown by the upper part of the energy diagram. The maximum depth far upstream is limited only by the length of the channel and the rate of energy loss.

Since the surface curve is no longer horizontal, it is illogical to presume that the hydrostatic critical depth now occurs over even a small part of the channel length. Actually there is but one vertical section at which the computed critical depth may be found. At that section the surface slope resulting from wall and floor resistance joins the curve of transition to the fully ventilated nappe. Since for the case of negligible resistance the critical depth was an extremely great distance upstream, it might be reasoned that increasing resistance (hence increasing slope of the surface) would cause the computed critical section to move nearer and nearer to the crest; experimental evidence justifies this reasoning, although it is still impossible to predict the location with any accuracy. For relatively smooth channels it is a distance equal to approximately four times the critical depth upstream from the crest.

All curves of pressure distribution within the nappe in Fig. 2(a) apply with negligible change to the actual case of Fig. 2(b). As is shown by the curve of measured floor pressures, at the computed critical section the pressure is still statically distributed, so that the critical relationships already cited are still applicable. Using this section as the upper limit, it is possible through use of the momentum principle to determine the vertical dimension of the nappe at point 1, where internal pressure no longer exists. Neglecting the small boundary resistance over this short channel length, the only

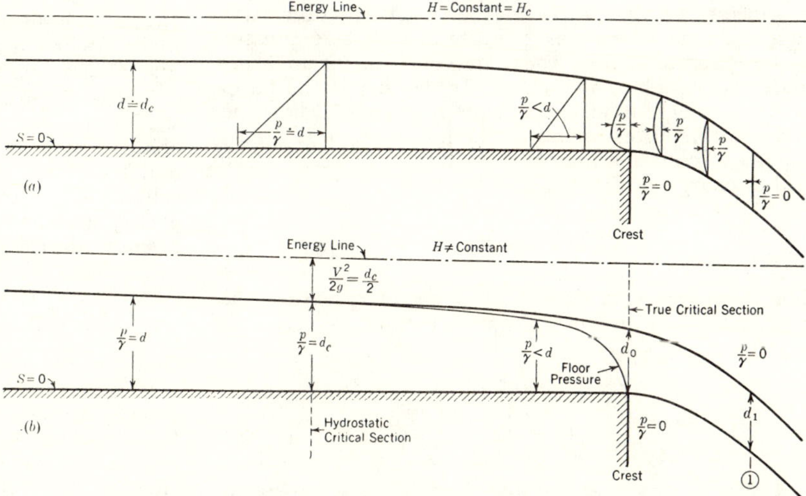

FIG. 2. SURFACE PROFILES AND PRESSURE DISTRIBUTION AT A FREE OVERFALL
(a) Assumed Case of Constant Specific Energy. (b) An Actual Profile; Pressure Distribution in the Nappe Is Similar to That in (a)

force causing the rate of change of momentum in the horizontal direction is the static pressure over the critical section. For all practical purposes this is equal to the mass discharge past either section multiplied by the change in the average horizontal component of velocity:

$$\frac{\gamma d_c^2 b}{2} = \frac{Q\gamma}{g}\,\Delta V = \frac{Q^2\gamma}{gbd_1} - \frac{Q^2\gamma}{gbd_c}$$

Now since $d_c^3 = \dfrac{Q^2}{gb^2}$, we can write this equation as

$\dfrac{d_c^2}{2} = \dfrac{d_c^3}{d_1} - \dfrac{d_c^3}{d_c}$, and, solving, $d_1 = \dfrac{2}{3}d_c$.

Experiments have shown that below the level of the crest, where the pressure within the nappe has become practically atmospheric, the vertical dimension of the sheet is only very slightly greater than two-thirds the critical depth (because of the neglected frictional loss), remaining practically constant through the downward course of the sheet.

DEPTH OF FLOW
AT THE CREST

Since the pressure distribution at the crest section as shown in Fig. 2(a) cannot be found without rather laborious hydro-mechanical methods, the principle of momentum is not sufficient to determine the depth of the flow at this point. (The hydromechanical methods are described in the thesis already referred to.) Recourse may be had, nevertheless, to a method which has a certain schematic significance—although it is still without concrete physical justification. The basic weir formula,

$$Q = c_c \frac{2}{3}\sqrt{2g}\,b\left[\left(h + \frac{V_a^2}{2g}\right)^{3/2} - \left(\frac{V_a^2}{2g}\right)^{3/2}\right]$$

in which c_c represents the so-called coefficient of contraction, assumes horizontal, parallel flow at zero pressure at a section similar to the vena contracta of an orifice under very high head; obviously, neither is the nappe of a weir a case of either horizontal or parallel flow, nor is the internal pressure atmospheric. Be that as it may, the above equation may be rewritten in terms of a general discharge coefficient, c_q, so that

$$Q = c_q \frac{2}{3}\sqrt{2g}\,bh^{3/2}$$

The coefficient c_q is then the product of c_c and another coefficient, c_v, which includes in dimensionless terms the velocity of approach:

$$c_v = \left(\frac{h + \dfrac{V_a^2}{2g}}{h}\right)^{3/2} - \left(\frac{\dfrac{V_a^2}{2g}}{h}\right)^{3/2}$$

Assuming now that the free overfall is a sharp-crested weir of zero height, and that the discharge head is the critical depth ($h = d_c$), the discharge coefficient c_q may be computed from the critical discharge relationship:

$$Q = \sqrt{g}\,bd_c^{3/2} = c_q \frac{2}{3}\sqrt{2g}\,bd_c^{3/2}$$

$$c_q = \frac{3}{2\sqrt{2}} = 1.061$$

c_v may also be computed from the relationship between hydrostatic critical depth and velocity head:

$$c_v = \left(\frac{1 + \frac{1}{2}}{1}\right)^{3/2} - \left(\frac{\frac{1}{2}}{1}\right)^{3/2} = 1.485$$

From these two values, there results for the coefficient of contraction c_c:

$$c_c = \frac{c_q}{c_v} = \frac{1.061}{1.485} = 0.715$$

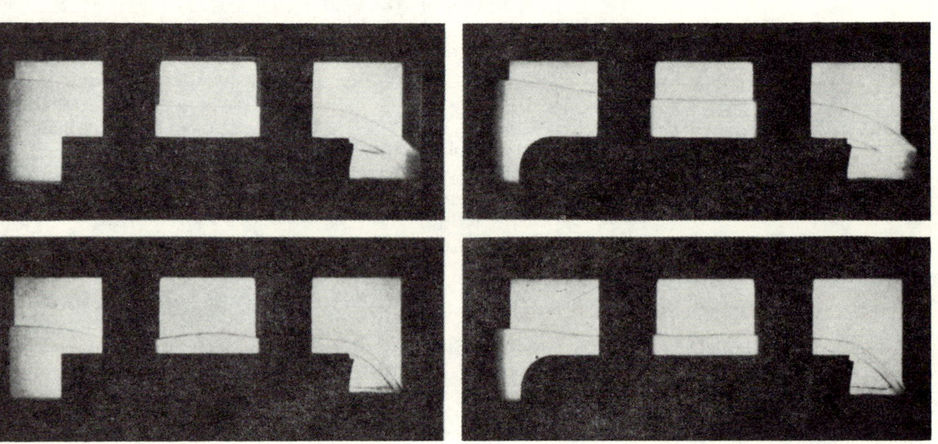

EFFECT ON SURFACE CURVE PRODUCED BY CHANGING THE PROFILE OF A BROAD-CRESTED WEIR
The Measured Heads in Views on Right Are the Same as Those in Corresponding Views on Left

Numerous measurements by the author have shown that this contraction coefficient is equal to the ratio between crest depth and computed critical depth, regardless of rate of discharge or channel width. (For first mention of this fact see "The Distribution of Hydraulic Energy in Weir Flow with Relation to Spillway Design," by H. Rouse, submitted in partial fulfillment of the requirements for the degree of master of science in civil engineering, Massachusetts Institute of Technology, 1932.)

Thus, despite the fact that the hydrostatic critical section for parallel flow has no definite location, but changes position with rate of discharge and channel slope and roughness, it is evident that the crest section itself, in a free overfall, is the true section of minimum energy, and as such is a suitable control point at which a single measurement of depth should permit a simple and direct determination of the rate of discharge:

$$Q = \sqrt{g}\,bd_c^{3/2} = 1.654\sqrt{g}\,bd_0^{3/2} = 9.4\,bd_0^{3/2}$$

The success of this determination, of course, would depend upon complete aeration of the nappe and a continuation of the channel sides a short distance beyond the crest. Needless to say, such depth measurements cannot be made by floor or wall piezometers, for decidedly non-static pressure conditions prevail in this region.

At this point reference must be made to a paper by P. Böss, which has been translated in this country and is often quoted in engineering literature ("Berechnung der Abflussmengen und der Wasserspiegellage bei Absturzen und Schwellen...," *Wasserkraft und Wasserwirtschaft*, Heft 2–3, 1929). In this article the discharge is computed on the assumption of linear pressure distribution at a vertical section of curvilinear flow; as this signifies zero pressure over the entire crest section of a ventilated overfall, the equation yields a value of

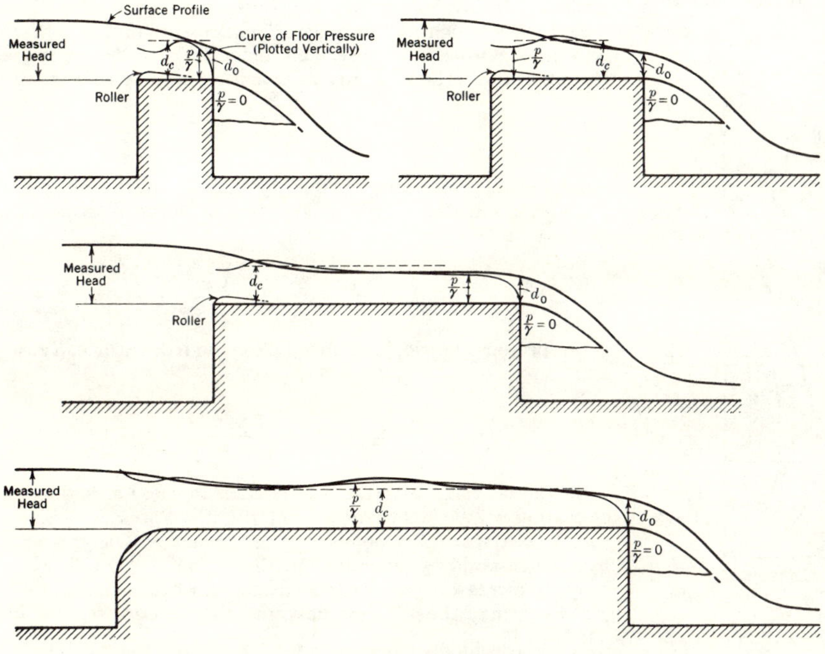

critical, and the surface thereafter is either highly undular (for very broad weirs) or of the general rising profile typical of such rapid motion, depending upon the ratio between the head, h, and the weir breadth. If the upstream end of the weir is rounded sufficiently, this roller and the accompanying rapid flow will no longer exist. The surface will remain undular, however, unless the rounding be exceedingly gradual, but the critical depth will not occur until the transition region of the crest is reached.

It must be noted that the position and number of the surface undulations depend entirely upon the geometrical relation between the head on the weir and the weir dimensions. Any undulation for a given weir profile will move downstream with increasing discharge, until it becomes absorbed in the lower transition curve. At the same time it is apparent that the position of these reaction waves, as determined by the magnitude of the head and the form and length of the weir, will influence the lower transition curve to such an extent that the hydrostatic critical section will be moved upstream or downstream from its corresponding position for the free overfall. A narrow weir, moreover, will have no section at which the critical depth for parallel flow may properly be located, because of curvature of the filaments over the entire profile.

Whether the crest depth itself remains unchanged in its relation to the computed critical depth, as the breadth of the weir varies, may be determined by experiment alone. Obviously a very narrow weir must at once be

FIG. 3. TYPICAL SURFACE PROFILES AND PRESSURE DISTRIBUTIONS FOR BROAD-CRESTED WEIRS OF VARIOUS PROPORTIONS

$\dfrac{d_0}{d_c} = 0.656$, which differs by more than 8 per cent from the value of 0.715 derived herein and substantiated experimentally.

EFFECTS OF THE BREADTH OF THE WEIR

Considering now the broad-crested weir, it is evident that the surface curve of the free overfall will be transformed by another local transition curve a relatively short distance upstream from the crest. That is, a short distance upstream from the weir the water will flow at great depth and low velocity, and at the upper end of the weir there will be a surface drop and an abrupt acceleration. If the weir is very broad, it is reasonable to expect the first transition curve to end some distance upstream from the transition of the crest region. A narrow weir, however, may result in a mutual interference of the two transition curves, in which case neither attains its full independent form. See Fig. 3 for typical profiles. The sharp-crested weir is the limiting case of such interference, the two curves here being blended into a single smooth profile.

If the upstream end of the weir is a sharp corner—the intersection of the vertical face with the floor—a region of discontinuity forms between the flow and the floor of the weir, this region being filled with a roller of water as shown in Fig. 3. Since the roller is not a part of the live flow, the discharge passes over it as though the roller were a small spillway built upon the weir. Because of the effective height of this roller above the weir floor, the flow reaches its toe at a velocity somewhat above the

eliminated from this consideration, for the sharp-crested weir as a limiting case surely does not display this constant relationship. It remained for a recent investigation of the broad-crested weir, conducted in the Fluid Mechanics Laboratory at Columbia University, to supply the necessary experimental test of this point ("The Hydraulics of the Broad-Crested Weir," by T. H. Prentice, a thesis submitted in partial fulfillment of the requirements for the degree of master of science, Department of Civil Engineering, Columbia University). The experiments were conducted on a series of weirs 1, 2, and 3 ft in breadth, with both square and rounded upstream ends, under heads varying from 0.2 to 0.6 ft. While the progression of undulations resulted in an average deviation from the mean of $\pm 1\frac{1}{2}$ per cent, the general average value of $\dfrac{d_0}{d_c} = 0.712$ for all weirs except the shortest (25 different profiles) is strikingly close to the value of 0.715 derived from the weir equation. On eight different heads for the free overfall, the author had previously found a mean value of 0.716, with an average arithmetical departure from 0.715 of 0.5 per cent.

While further large-scale experiments are now in order, including channels with mild slopes and various values of the roughness coefficient, the author believes that the use of the crest section as the control point for measurement of discharge will not only further the laboratory search for a satisfactory critical depth meter, but will provide a simple and dependable method of determining discharge in rectangular channels that end in a free overfall.

Pressure Distribution and Acceleration at the Free Overfall

To the Editor: Walter B. Langbein, Jun. Am. Soc. C.E., is to be commended for his desire to develop a general analysis of flow conditions at the free overfall for all velocities of approach, as evidenced by his article on "Discharge Over a Free Fall," in the May issue, for falls at the end of supercritical slopes are frequently encountered in practice. Not only has this general case lacked a

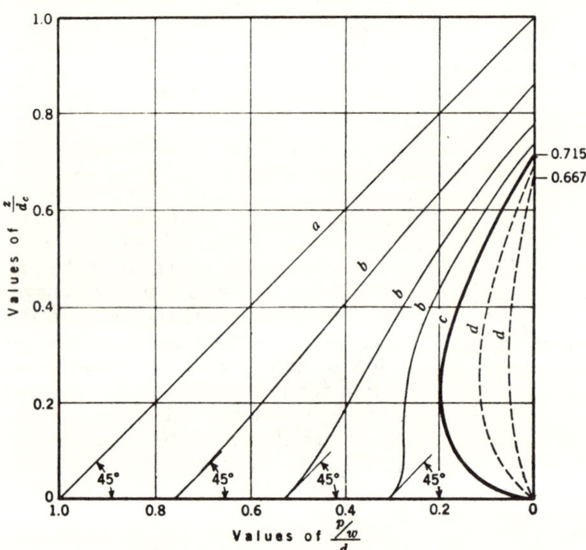

FIG. 1. DISTRIBUTION OF PRESSURE HEAD AT THE FREE OVERFALL
Showing Conditions (*a*) at the Critical Section Some Distance Upstream from the Crest; (*b*) at Intermediate Sections; (*c*) at the Crest; and (*d*) Within the Free Nappe.

satisfactory analysis, but adequate experimental data upon which to base such an analysis are nowhere to be found.

Since Mr. Langbein's equations were not formulated on the basis of experimental evidence, they may not properly be termed empirical. And since certain of the assumptions upon which his developments depend are contrary to fact, the significance of his derivation is open to considerable question. That his equations yield the ratio $\frac{d_o}{d_c} = 0.71$ as a limiting condition, in close agreement with my measurements, is hardly sufficient verification.

The fallacy seems to lie with the initial assumptions as to pressure distribution and acceleration. Mr. Langbein's Fig. 1 shows a pressure diagram of concave form—a condition that is seldom found in such curvilinear motion. Actually, the convergence of the stream lines at the free fall produces a higher pressure intensity within the flow, with a pressure diagram that is convex over practically the entire section.

In the accompanying Fig. 1 are summarized, in dimensionless form, my measurements of pressure distribution at successive vertical sections of an overfall with horizontal approach. Both pressure head, $\frac{p}{w}$, and elevation above the lower boundary, z, are plotted in their ratio to the critical depth, d_c. These curves were obtained for a channel of zero slope, but they apply very closely to mild slopes as well. Moreover, the individual curves should be of the same general form for any velocity of approach above the critical, varying only in relative proportions.

In cases of such rapid acceleration, pressure and weight are the only acting forces of appreciable magnitude. The component of force per unit volume in any direction, y, may be written

$$f_y = -\frac{d(p + wz)}{dy}$$

Since the force per unit volume, f_y, divided by the mass per unit volume, ρ, must equal the acceleration in the corresponding direction,

$$a_y = \frac{f_y}{\rho} = -g\frac{d\left(\frac{p}{w} + z\right)}{dy}.$$

If y represents vertical distance below the free surface, $dy = -dz$, and

$$a_y = g\left(1 - \frac{d(p/w)}{dy}\right)$$

So long as the pressure head increases in the y-direction, the acceleration will be less than g; similarly, if the pressure gradient becomes negative at any point, the acceleration will be greater than g by a corresponding amount. In the region of the fall, evidently, a_y must increase from the free surface downwards, because of the convex form of the pressure diagram. At the crest section, indeed, $d\frac{(p/w)}{dy}$ is decidedly negative in the lower portion of the flow, indicating a vertical acceleration considerably in excess of g. Nevertheless, at the floor itself the vertical acceleration must always have a magnitude of zero, since the lowermost stream line remains horizontal as far as the crest section. Hence, the slope of each pressure diagram must approach unity as a limit at the fixed boundary, as indicated in the illustration. The resulting concave portion of the diagram lies close to the floor, however, and becomes increasingly smaller in the direction of flow; at the crest this reversal of curvature entirely disappears.

It is reasonable to expect that the coefficient C in the expression

$$Q = Cbd_o^{3/2}$$

will increase with the velocity of approach, attaining the limit unity as the velocity becomes infinitely great. Despite unsound original assumptions, Eq. 2 of Mr. Langbein's paper thus indicates the proper trend. Should experimental data be found to substantiate this equation, hydraulics will add one more empirical paradox to the many which it now displays. It is only to be regretted that such occurrences tend to belie the desirability of sound physical analysis.

Hunter Rouse, Assoc. M. Am. Soc. C.E.

Pasadena, Calif.
May 24, 1937

Excerpt from

MODERN CONCEPTIONS OF THE MECHANICS OF FLUID TURBULENCE

BY HUNTER ROUSE

Similar conclusions will result from the following reasoning: In the Navier-Stokes equations the three components of the shearing stress, τ, are simply those due to viscous action, regardless of whether the flow is laminar or turbulent:

$$\tau_{xy} = \tau_{yx} = \mu \left(\frac{\partial v_x}{\partial y} + \frac{\partial v_y}{\partial x} \right) \quad \ldots \ldots \ldots \ldots \ldots (107)$$

$$\tau_{yz} = \tau_{zy} = \mu \left(\frac{\partial v_y}{\partial z} + \frac{\partial v_z}{\partial y} \right) \quad \ldots \ldots \ldots \ldots \ldots (108)$$

and,

$$\tau_{zx} = \tau_{xz} = \mu \left(\frac{\partial v_z}{\partial x} + \frac{\partial v_x}{\partial z} \right) \quad \ldots \ldots \ldots \ldots \ldots (109)$$

Once the instantantous velocity gradient is replaced by the mean velocity gradient, if the flow is turbulent there must exist additional terms to express the high viscous stresses in the turbulent eddies—that is, the effective shear of the momentum transport; introducing Boussinesq's turbulence coefficient, η, these become:

$$\bar{\tau}_{xy} = \bar{\tau}_{yx} = \mu \left(\frac{\partial \bar{v_x}}{\partial y} + \frac{\partial \bar{v_y}}{\partial x} \right) - \rho \, \overline{v'_x v'_o} = (\mu + \eta) \left(\frac{\partial \bar{v_x}}{\partial y} + \frac{\partial \bar{v_y}}{\partial x} \right) \quad ..(110)$$

$$\bar{\tau}_{yz} = \bar{\tau}_{zy} = \mu \left(\frac{\partial \bar{v_y}}{\partial z} + \frac{\partial \bar{v_z}}{\partial y} \right) - \rho \, \overline{v'_y v'_z} = (\mu + \eta) \left(\frac{\partial \bar{v_y}}{\partial z} + \frac{\partial \bar{v_z}}{\partial y} \right) \quad ..(111)$$

and,

$$\bar{\tau}_{zx} = \bar{\tau}_{xz} = \mu \left(\frac{\partial \bar{v_z}}{\partial x} + \frac{\partial \bar{v_x}}{\partial z} \right) - \rho \, \overline{v'_z v'_x} = (\mu + \eta) \left(\frac{\partial \bar{v_z}}{\partial x} + \frac{\partial \bar{v_x}}{\partial z} \right) \quad ..(112)$$

It must be realized that Equations (107) to (109) and (110) to (112) are merely different ways of stating the same basic effect of viscous shear.

Reducing Equation (110) to the case of flow in a pipe:

$$\bar{\tau} = \mu \, \frac{d\bar{v}}{dy} - \rho \, \overline{v'_x v'_y} = (\mu + \eta) \, \frac{d\bar{v}}{dy} \quad \dots\dots\dots\dots (113)$$

In the paper it was shown that with increasing Reynolds number the magnitude of η becomes increasingly greater than that of μ, and that above the approximate Reynolds number of 100 000, the magnitude of μ is comparatively insignificant. In the light of the foregoing discussion, the Reynolds number might also be considered to denote the relative magnitude of η and μ, the former being averaged over the flow section.

Since in highly turbulent motion $\mu \, \dfrac{d\bar{v}}{dy}$ is generally negligible when compared with $\eta \, \dfrac{d\bar{v}}{dy}$, Equation (113) may then be written in the form,

$$\bar{\tau} = - \rho \, \overline{v'_x v'_y} = \eta \, \frac{d\bar{v}}{dy} \quad \dots\dots\dots\dots\dots (114)$$

Introducing the expression, $|v'_x| = l \, \dfrac{d\bar{v}}{dy}$ (refer to the development of Equation (42)):

$$\frac{\eta}{\rho} = |v'_y| \, l \dots\dots\dots\dots\dots\dots (115)$$

in which the vertical bars denote mean absolute magnitude, regardless of sign.

Equation (115) is of very basic importance, for it not only lends added clarity to Boussinesq's coefficient of turbulence, but is a direct measure of the transporting power of the mixing process. The writer stressed only the transport of momentum by this means, but the fact must be appreciated that it is the same mechanism which results in a transportation of heat, salinity, or suspended matter from one region of turbulent flow to another. The rate of transport of momentum has been shown to depend upon only the factor, $\dfrac{\eta}{\rho}$, and the mean velocity gradient, thus producing an effective shearing stress:

$$\bar{\tau} = \rho \, | v'_y | \, l \, \frac{d\bar{v}}{dy} \quad\ldots\ldots\ldots\ldots\ldots\ldots\ldots\ldots(116)$$

The problem of suspended load in a stream may be attacked in similar fashion. If at any depth of flow there are n particles per unit volume of fluid, the concentration gradient being written $\dfrac{dn}{dy}$, then, due to the transverse fluctuations, fluid masses bearing n particles per unit volume will be carried the mean distance, l, across the flow to regions where the concentration differs by the amount, $l \, \dfrac{dn}{dy}$. Thus, the temporal rate of passage of sediment per unit area, N, will be the product of the rate per unit area of transverse flow, $| v'_y |$, and the difference between the sediment concentration in the traveling fluid mass and that of the region into which it comes, $- l \, \dfrac{dn}{dy}$ —in other words, the product of $\dfrac{\eta}{\rho}$ and the sediment gradient:

$$N = - \, | v'_y | \, l \, \frac{dn}{dy} \quad\ldots\ldots\ldots\ldots\ldots\ldots\ldots(117)$$

Identical reasoning applies to the transportation of salinity, and similar considerations show that the rate of heat transfer per unit area, q, will depend only upon $\dfrac{\eta}{\rho}$, the specific heat, c, and the mean temperature gradient across the flow, $\dfrac{d\theta}{dy}$:

$$q = - \, | v'_y | \, l \, c \, \frac{d\theta}{dy} \quad\ldots\ldots\ldots\ldots\ldots\ldots\ldots(118)$$

Such heat transfer is purely convective, and must be distinguished from the process of conduction in the laminar film.

The parallel development and structure of Equations (116) to (118) was shown to excellent advantage by von Kármán in a résumé of turbulence theories[49]. Since the problem of suspended load in a stream is of paramount interest in river hydraulics, amplification of this method of treatment should be of value at this time. If the sediment transportation of the stream as a whole is in a uniform state of equilibrium, the rate of upward transfer by the mixing process must equal the downward movement due to settling. Designating the settling velocity of a given particle size by w, then N, the temporal rate of transportation per unit area, must equal the number of particles per unit volume multiplied by their rate of fall:

$$N = w \, n = - \, | v'_y | \, l \, \frac{dn}{dy} \quad\ldots\ldots\ldots\ldots\ldots(119)$$

---

[49] "Some Aspects of the Turbulence Problem", *Proceedings,* Fourth International Congress for Applied Mechanics, Cambridge, England, 1934.

This expression may be integrated in the form,

$$\log_e \frac{n}{n_a} = - w \int_a^y \frac{dy}{|v'_y| \, l} \quad \dots\dots\dots\dots (120)$$

in which $\frac{n}{n_a}$ is the relative concentration at any point, referred to some arbitrary elevation, a, above the bottom. The evaluation of the last term in the expression depends only upon knowledge of the kinematic turbulence factor, $\frac{\eta}{\rho} = |v'_y| \, l$, as a function of the depth.

As shown by von Kármán, introduction of Equation (116) yields an integrable expression,

$$\frac{dy}{|v'_y| \, l} = \frac{dy}{\dfrac{\tau}{l \dfrac{d\bar{v}}{dy}}} = \frac{l \dfrac{d\bar{v}}{dy}}{\tau_0 \left(1 - \dfrac{y}{d}\right)} \, dy \dots\dots\dots\dots (121)$$

in terms of the mean velocity gradient. If the latter is of the universal logarithmic type, then from Equation (59),

$$\frac{d\bar{v}}{dy} = \frac{1}{\kappa} \sqrt{\frac{\tau_0}{\rho}} \frac{1}{y} \quad \dots\dots\dots\dots (122)$$

and Equation (120) becomes:

$$\frac{n}{n_a} = \left[\frac{1 - \dfrac{y}{d}}{\dfrac{y}{d}} \times \frac{\dfrac{a}{d}}{1 - \dfrac{a}{d}} \right]^z = \left[\frac{1 - \dfrac{y'}{d'}}{1 + \dfrac{y'}{a}} \right]^z \dots\dots (123)$$

in which $z = \dfrac{w}{\kappa \sqrt{\dfrac{\tau_0}{\rho}}}$. Two pertinent facts are to be noted in Equation (123):

The relative sediment distribution, for a given particle size (that is, for a given settling velocity), is a function of relative depth only. It cannot, however, be extended to the channel bottom, but only to some arbitrary depth, $d' = d - a$, and then only if the concentration at that depth is not so great as to change either the density of the fluid mixture or the assumed velocity distribution. If the bottom is smooth, the logarithmic velocity curve does not extend to the bottom, and there is no mixing in the final layer of fluid at the boundary. If the bottom is rough, knowledge of the total suspended load would permit complete solution of the sediment distribution (provided the logarithmic law could be extended to the free surface), but this expression in no way permits determination of the total transporting capacity of the stream. The latter probably depends upon conditions in the transition region between movable bed-load and completely suspended matter, a problem still awaiting a satisfactory physical analysis.

On the assumption that the logarithmic velocity curve actually extends to the free surface of a very wide stream, Equation (123) enables one to construct a plot of sediment distribution curves (Fig. 30) for various values of the suspended load parameter, $z = \dfrac{w}{\kappa \sqrt{\dfrac{\tau_0}{\rho}}}$, indicating, for a given state of flow, the relative distribution of the different particle sizes based upon their settling velocities. Such a diagram was first developed by Arthur T. Ippen, Jun. Am. Soc. C. E., at the suggestion of Professor von Kármán. An interesting comparison may be made with a group of measured curves published by L. G. Straub[50], Assoc. M. Am. Soc. C. E.*

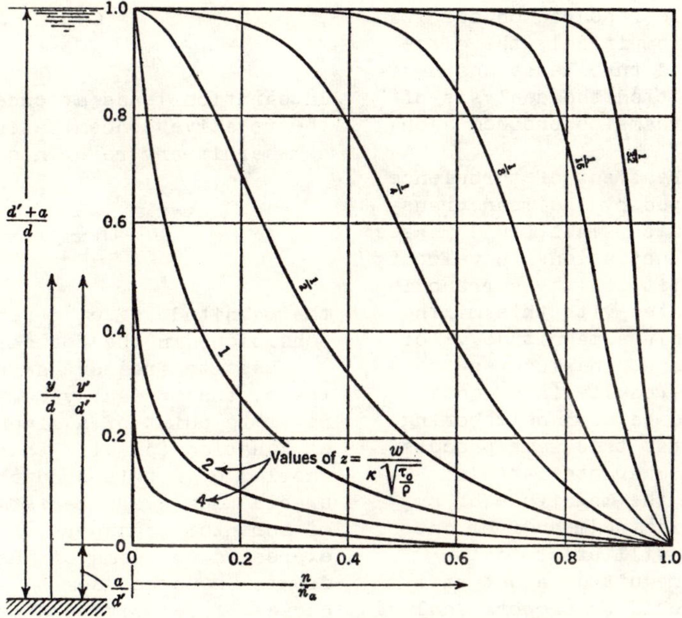

FIG. 30.—DIMENSIONLESS PLOT OF SUSPENDED LOAD
DISTRIBUTION.

---

[50]"Transportation of Sediment in Suspension", *Civil Engineering*, May 1936.

*For further discussion of this topic, the reader is referred to "Sediment Transportation Mechanics: Suspension of Sediment," p. 470 in this volume.

HUNTER ROUSE

EXPERIMENTS ON THE MECHANICS OF SEDIMENT SUSPENSION

FLUID TURBULENCE AS A MEDIUM OF LATERAL TRANSPORT: For many decades the problem of sediment transportation has claimed the attention of the hydraulic engineer, particularly in connection with river regulation, irrigation, and the protection of arable land. Laboratory and field investigations of this problem have dealt primarily with the movement of rock debris along the bed of a stream, and only in recent years has the study of suspended matter come into prominence. Yet while research on bed load has remained essentially empirical, the close relationship between fluid turbulence and sediment suspension has permitted the analysis of the suspended load of streams to proceed along more rational lines.

In the statistical treatment of turbulence, an important rôle is played by the mean transverse distance through which small fluid masses are carried by lateral fluctuations in velocity before losing their identity in the surrounding fluid; intimately associated with this mixing length l is the mean absolute magnitude v' of the transverse fluctuations. Inasmuch as the mixing due to turbulence results in a continuous interchange of fluid between neighboring regions, it is evident that this same process must involve every local characteristic of the fluid. Therefore, if the magnitude of any characteristic m varies in the mean from one region to another, there will exist in the direction of decreasing magnitude a net rate of transport M, which should be proportional to the gradient of m and to the intensity $v'l$ of the mixing process:

$$M = -\beta v'l \frac{dm}{dy} \qquad (1)$$

This equation is commonly used to express the rate of transport of momentum (ϱu) per unit fluid volume. But if the reasoning is generally valid, it should apply equally well to the transport of vorticity, salinity, heat, or matter carried in suspension, although the proportionality factor β would not necessarily be the same in each of the several cases[1]. Needless to say, the relationship is restricted to regions of fully developed turbulence in which the influence of fluid viscosity may be neglected.

BASIC EQUATION OF SUSPENSION: In the case of sediment suspension, it is customary to assume a state of equilibrium between the mean vertical rate of sediment transport due to turbulence and the normal rate of settling of the sediment due to its own weight. Thus, if w denotes the settling velocity, and c the sediment concentration per unit volume of the suspension

$$cw = -\beta v'l \frac{dc}{dy}. \qquad (2)$$

Integration leads at once to an expression for the relative concentration at any point above some arbitrary reference level a ,

$$\ln \frac{c}{c_a} = -\frac{w}{\beta} \int_a^y \frac{dy}{v'l}, \qquad (3)$$

the magnitude of c_a depending ultimately upon conditions in the boundary region.

Despite frequent assumptions to the contrary, the product $v'l$ will generally vary from point to point of a given turbulent flow; use of Equation (3) will thus depend upon prior knowledge of this product as a function of measurable flow characteristics. From the equation of momentum transport, for instance, $v'l$ may be expressed in terms of the mean velocity gradient - a method sometimes used in constructing curves of relative sediment concentration from known velocity measurements in rivers. On the other hand, if one assume von KÁRMÁN's universal law of velocity distribution to apply to open channels as well as pipes, it is possible to determine analytically the corresponding distribution function[2]. Unfortunately, existing data on the distribution of velocity and sediment are not sufficiently well correlated to prove beyond question the validity of the foregoing theory.

EXPERIMENTAL APPARATUS: Prior to undertaking an extensive study of the transportation of suspended load at the Pasadena laboratory of the Soil Conservation Service, definite quantitative information was sought on the mechan-

1. KÁRMÁN, TH. von, "Some Aspects of the Turbulence Problem", *Proceedings of the Fourth International Congress for Applied Mechanics*, 1935, p. 66.

2. ROUSE H., "Modern Conceptions of the Mechanics of Fluid Turbulence", *Proceedings of the American Society of Civil Engineers*, vol. 102, 1937, p. 536.

ics of suspension, with particular reference to the influence of sediment characteristics. The apparatus developed for this preliminary investigation (somewhat similar to that used by HURST[3]) consisted, in brief, of a cylindrical glass tank in which a lattice structure oscillated vertically in simple harmonic motion (see Figure 1), thus producing essentially the same

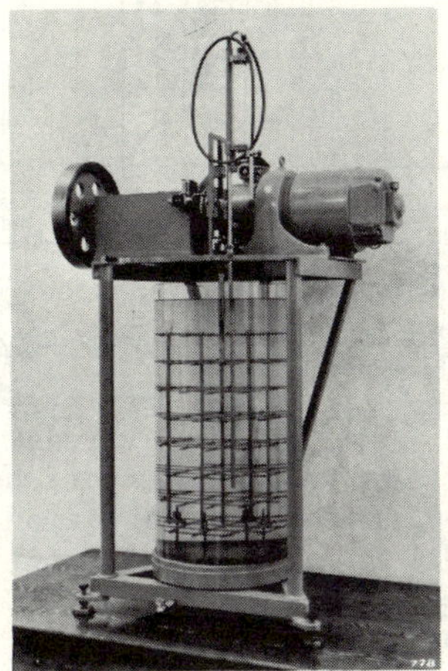

Fig.1 General view of experimental apparatus.

degree of mixing ($v'\ell$ = constant) over a considerable fluid volume, and avoiding the necessity of movement of the fluid as a whole past a stationary boundary as in the case of flow through a pipe or open channel. It should be evident that apparatus of this nature is quite as adaptable to the study of heat transfer as to the investigation of sediment suspension. Indeed, it offers extensive possibilities for research on the mechanism of the turbulence itself, through variation of the amplitude, frequency, scale, and form of the agitator, and the density and viscosity of the fluid, the distribution of suspended matter then providing a method of determining the turbulence characteristics. In the present investigation, however, the turbulence was regarded primarily as a means to an end; the form, scale, and amplitude of the agitator were therefore kept the same throughout the initial experiments, water at constant temperature served as the medium of suspension, and the intensity of mixing was regulated solely by varying the frequency of the agitator.

SEDIMENT CHARACTERISTICS: The sediment chosen for the experiments were nominally 1/4, 1/8, 1/16, and 1/32 mm in grain diameter. The first three sizes were separated by both air and sieve classification from Yuma dune sand, while the fourth was obtained from commercial white Nevada foundry sand by air and water sepa-

3. HURST, H.E., "The Suspension of Sand in Water", Proceedings of the Royal Society, vol. 124, 1929, p.196.

ration.Settling velocities were determined as arithmetic averages by timing the fall in water of a hundred or more representative particles of each sediment(see Table I and Fig.2).

Table I Sediment Characteristics

Nominal diameter d	Size range	Sieve mesh	Settling velocity	Reynolds number
1/4 mm	.351 -- .246 mm	42 -- 60	3.69 cm/s	11.0
1/8 mm	.175 -- .124 mm	80 -- 115	1.75 cm/s	2.6
1/16 mm	.088 -- .061 mm	170 -- 250	0.675 cm/s	0.5
1/32 mm	.053 -- .037 mm	----	0.131 cm/s	0.06

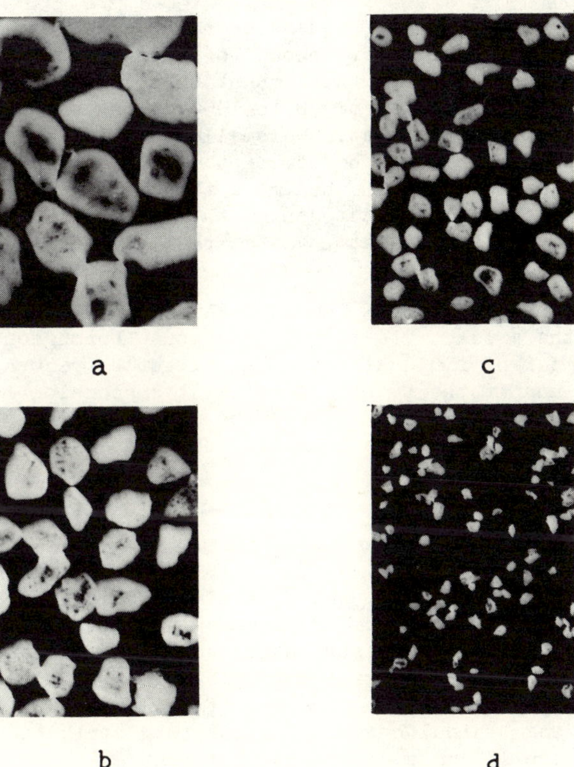

Fig.2 Photomicrographs of sediments used in experiments. Each picture includes a 1x1 1/2 mm field.

A special pipette with a capacity of 150 cubic centimeters (Figure 3) permitted accurate volumetric samples of the suspension to be withdrawn at any desired level. In each individual run a sampling traverse was made from top to bottom of the tank, followed immediately by a traverse from bottom to top at intermediate points. The dry weights of the resulting 16 samples were then determined on an analytical balance sensitive to one-tenth milligram. Such

Fig.3 Details of sampling mechanism and agitator drive.

position.

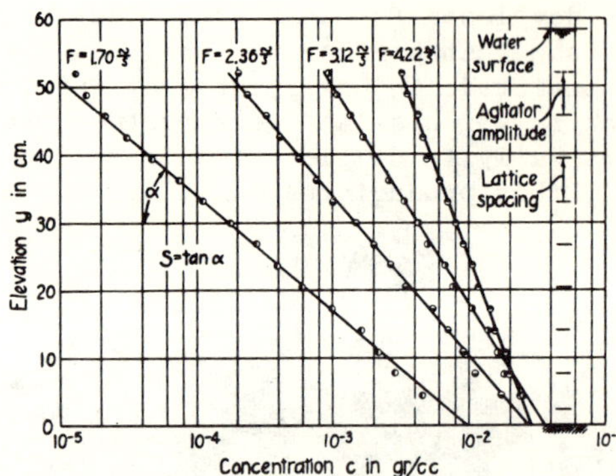

Fig.4 Typical semilogarithmic curves of sediment distribution.

runs were made at frequencies ranging between 0.8 and 5 cycles per second for each of the four sediments, in total quantities varying in three-fold steps from 50 to 1350 grams. With the agitator at rest, the smallest quantity barely covered the bottom of the tank, while the largest formed a bed more than a centimeter in thickness - just touching the lowermost lattice when the agitator was at the end of its stroke.

VERIFICATION OF THE EXPONENTIAL FUNCTION: If the product $v'l$ is truly constant throughout the jar of fluid at any given agitator frequency, and if the theory of sediment suspension is valid for such conditions, the distribution should follow the exponential relationship

$$\frac{c}{c_a} = e^{-\frac{w}{\varepsilon}(y - a)} \tag{4}$$

in which $\varepsilon = \beta v'l$, and c_a again represents the concentration at some reference level a . In other words, the distribution curve should be linear on a semilogarithmic plot. That this is actually the case may be seen from Figure 4, showing typical curves of absolute distribution for the 1/8-mm sediment at different agitator frequencies but with the same total quantity of material in the tank. Although conditions differ in the neighborhood of the upper and lower boundaries, it is evident that each series of points closely follows a straight line for more than 40 cm. With a smaller amount of sediment in the tank, all curves were shifted to the left - with a larger amount, to the right, even at the highest concentration (4% by weight) no systematic deviation from a linear plot was apparent. Invariably, however, the same frequency of agitation produced the same slope of the distribution curve, regardless of absolute

For a given sediment, in other words, all curves for the same frequency would be superposed if related to the concentration at an arbitrary reference level a . The experiments therefore indicate that the form of Equation (4) is mathematically correct.

INFLUENCE OF THE SETTLING VELOCITY: That the slope S of the semilogarithmic distribution plot should be directly proportional to the ratio ε/w may be seen by writing Equation (4) in the following form:

$$S = \frac{y - a}{\log \frac{c_a}{c}} = 2.3 \frac{\varepsilon}{w} . \tag{5}$$

That is, the greater the intensity of mixing, the more uniform will be the sediment concentration - i.e., the more nearly vertical the distribution curve; similarly, the greater the settling velocity, the more c will vary from one level to another. The term 2.3 is simply the conversion factor between natural and decimal logarithms. Since the frequency of agitation was the sole means by which the intensity of mixing was governed, the characteristic slope of each distribution curve was computed from this equation and plotted logarithmically against the measured frequency F (in cycles per second), as shown in Figure 5. It is evident that the points fall in four distinct groups, according to the settling velocities of the sands. Moreover, it will be found that the horizontal distance between any two groups is equal to the logarithm of the ratio of the respective settling velocities. It then follows that if each value of S is multiplied by the corresponding value of w , all points for any given agitator frequency should be superposed. According to Equation (5), division of these values by 2.3 should then yield the characteristic magni-

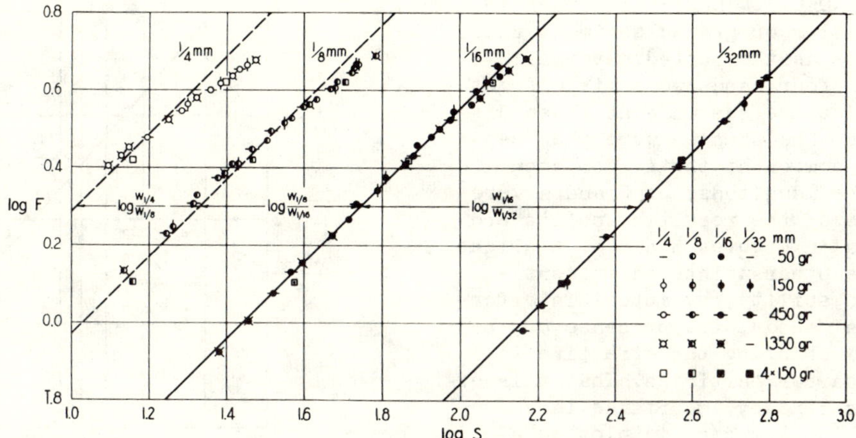

Fig.5 Logarithmic plot of the agitator frequency F against the slope S of the semilogarithmic distribution curves.

tude of ε for the given frequency, as shown in Figure 6. For all practical purposes these points could be represented by a single curve, although a systematic deviation from this average curve would be noted for each of the four sediments.

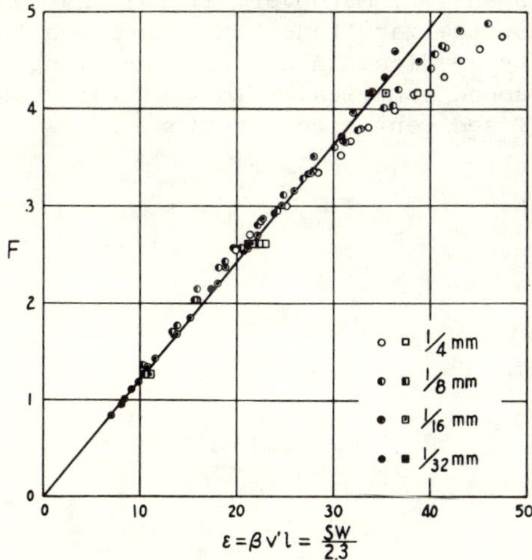

Fig.6 Relationship of the mixing coefficient ε to the agitator frequency F .

RELATIVE MIXING OF SEDIMENT AND FLUID: It will be recalled that ε is the product of a proportionality factor β and the mixing characteristics v' and l of the fluid. Owing to the plate-like form of the lattice bars, one would expect v' to vary directly with the agitator frequency (the stroke remaining the same), and l , a geometrical parameter, to remain constant so long as the geometry of the agitator was not changed. While β would then have a constant value for the fluid alone, it might be expected to vary in magnitude for matter in suspension (a) if the mixing process for the sediment is not practically identical with that of the fluid, or (b) if the presence of the sediment alters the fluid turbulence. Since the data for the 1/32 and 1/16-mm sands follow a straight line passing through the origin in Figure 6, it is evident that the product $\varepsilon = \beta v' l$ is directly proportional to F and that β is the same for both cases. The 1/8-mm sand deviates somewhat from this line, while the departure for the 1/4-mm sand is quite marked - indicating that the factor β is no longer constant. The individual trends of the four sediments are quite evident in Figure 5.

For the moderate concentrations investigated, it is unlikely that the presence of the sediment caused any essential change in the turbulence of the fluid, since in Figure 5 no systematic deviation with total amount of sediment is apparent. The variation in β must therefore be due to the fact that the mixing of the larger sediment particles is no longer governed wholly by the turbulence of the fluid, the inertia of the particles now having an appreciable effect. The degree of deviation from the linear plot might conceivably be a function of the ratios w/v' and d/l . On the other hand, it so happens that the 1/16 and 1/32-mm sands lie within the Stokes range of settling, while the larger sands do not. It is thus a moot question whether this line of demarcation corresponds to a critical ratio of sediment and turbulence characteristics, or to a critical Reynolds number for the settling alone. More definite information can be obtained only through further experimentation, using (a) a larger or smaller but geometrically similar agitator with the same fluid and sediment, and (b) the same agitator and fluid with a sediment of different density.

BEHAVIOR OF GRADED SEDIMENT: In order to determine the behavior of graded sediment under the same conditions as the sorted material, equal parts of all four sands were mixed together in the tank and traverses made at three different frequencies. The samples were then separated into their component parts, for each series of which the magnitudes of S and ε were determined as before; the resulting points are plotted as squares in Figures 5 and 6. A slight deviation from the other points is evident - caused, in all probability, by mutual interference of the several sands, the presence of the very fine material changing the effective fall velocity of the coarser particles. That this effect was also of secondary importance is shown by Figure 7, in which log c/c_a is plotted against $w(y - a)/\varepsilon$ for the constituent parts of each sample, using a mean value of ε for each value of F , and retaining the velocities of fall determined in clear water. The experimental values for all four sediments are seen to be in good agreement with the theoretical curve.

Definite verification of the suspension theory has been provided by the experiments herein described, with regard both to the exponential nature of the distribution function and to the rôle played by the settling velocity of the sediment. Some question still exists as to the strict similarity between the mixing characteristics of the sediment and those of

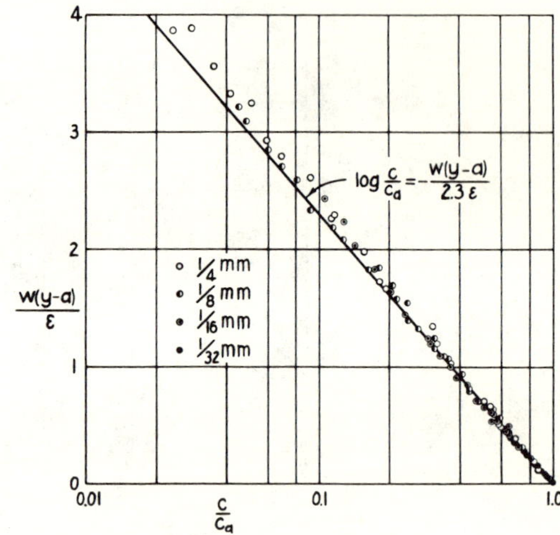

Fig.7 Dimensionless distribution function for the component parts of a sediment mixture at various agitator frequencies.

the fluid, for the factor ε appears to vary somewhat with the relative size of the particles in suspension. Moreover, it remains to be seen whether the magnitude of ε determined from the velocity gradient in a given turbulent flow will correspond numerically to that for the gradient of sediment concentration.

CRITERIA FOR SIMILARITY
IN THE TRANSPORTATION OF SEDIMENT

by

Hunter Rouse
Associate Hydraulic Engineer
Coöperative Laboratory, Soil Conservation Service
California Institute of Technology
Pasadena, California

SIMILITUDE AS A SUBSTITUTE FOR FUNCTIONAL ANALYSIS

Any phenomenon of fluid motion may be described mathematically by means of a functional relationship among a group of dimensionless parameters so constituted as to include the one dependent factor to be studied and all the independent quantities which govern its variation. Were the forms of these functional relationships known for all phases of hydraulics, there would be no need whatever of investigating the behavior of projected engineering works by means of laboratory models. However, in most cases these complex functions are at best only qualitatively understood, and one must generally resort to the principle of similitude as a practical substitute for rigorous or even approximate functional analysis. This principle merely embodies the fact that flow in a model will be dynamically similar to that in its prototype if all dimensionless parameters describing the flow in the model are numerically equal to the corresponding parameters for flow in the prototype. There is obviously no necessity of knowing how the several parameters are functionally interrelated, but for complete similarity it is essential that they include every variable which has an influence upon the fluid motion.

It is already a truism of hydraulic model testing that complete similarity is a goal which may often be approached, but seldom attained so long as one is forced to use the same fluid in the model as that which flows in the prototype. The discrepancies between perfect similarity and that actually obtained become the greater as one begins to deal with the movement of sediment by flowing water. Indeed, sediment transportation is perhaps the most complex problem of hydraulics, combining as it does all the difficulties of unsteady, non-uniform flow of water in open channels, with temporal change of the boundary configuration. Thus, any river displays not only a general downstream movement of sediment which varies with the seasonal rate of discharge, but phases of local scour and undercutting of banks on the one hand and phases of local deposition on the other. While fundamental research on sediment transportation is still restricted to the movement of bed load and suspended load in steady, uniform flow, model studies must seek to reproduce all major variations taking place in the prototype. It is often necessary, however, to sacrifice even approximate similarity in some respects in order to obtain more satisfactory results in specific phases of the motion; unfortunately, the validity of these results can be ascertained only by conducting model studies at various scales and extrapolating to prototype size, or by eventual comparison with prototype behavior. This unhappy situation will not be improved, it would appear, so long as it remains necessary to regard the grain size of the sediment as a geometrical characteristic subject to the same scale reduction as the other linear dimensions.

For the formulation of more general principles of sediment similarity than now exist, three lines of attack are at hand: (1) One may use as a guide the various empirical equations already available, as in the case of bed-load movement. (2) One may turn to theoretical analyses, still only partially tested experimentally, as in the case of sediment suspension by fluid turbulence. (3) For cases of motion as yet untouched either experimentally or analytically, one may obtain the essential results by a combination of dimensional and physical analysis, supplemented by a systematic experimental check. In the following pages each of these methods will be illustrated.

Sediment Characteristics

Parameters governing the transport of sediment by flowing water must necessarily include all effective geometric, flow, and fluid characteristics, and in addition the properties of the sediment itself. The characteristics of the fluid motion have been discussed elsewhere in considerable detail, but those of the sediment warrant amplification at this point. These include the size distribution, shape, and density of the individual grains, and the state of the bed as a whole.

Sediment density needs no further comment. Particle shape will likewise receive little mention herein, owing to the difficulty of measuring and expressing shape factors in a simple, representative fashion. It must suffice to note that in general the sphericity and roundness of natural sediment grains decrease with decreasing grain size, and that the crushed materials often used in the laboratory possess shape factors quite different from the natural product. The degree of compaction of a sediment bed is only partly dependent upon the sediment characteristics, for a given sediment may display different degrees of compaction for different loads or different manners of deposition. However, it must be assumed at present that all materials under discussion are deposited in the same manner, and that the porosity of the bed depends only upon the other sediment properties.

Grading characteristics of sediments are probably of far more importance in transportation studies than is usually believed, and it is imperative that significant parameters be adopted for the size distribution. Mechanical analyses of size[1] are now customarily plotted in the familiar histogram, with geometric (logarithmic) gradation of nominal diameter (Fig. 1a). As the class intervals become smaller,

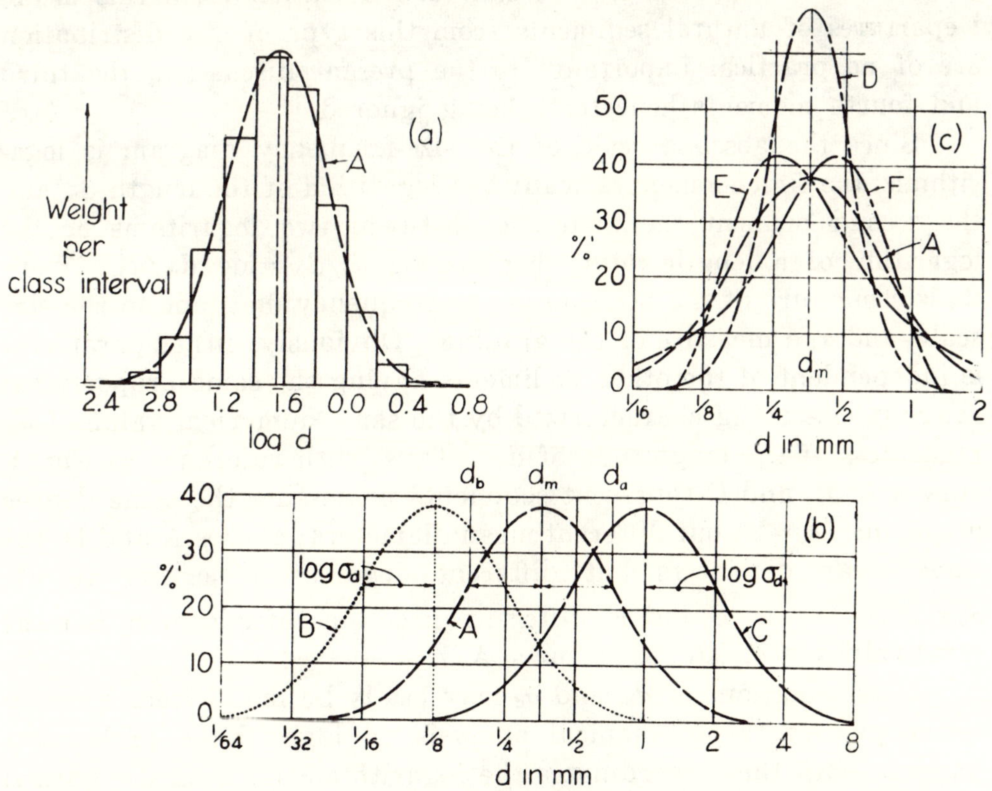

Fig. 1.—Definition Sketches for the Size-Frequency Characteristics of Sediment.

the histogram approaches the smooth curve of the size-frequency diagram shown as a broken line, the vertical scale of which (per cent per class interval, designated by %′ in Fig. 1b) is so adjusted as to make the enclosed area equal to 100%. The geometric mean diameter of the corresponding grains is obtained from the first moment of this area about an arbitrary vertical axis; this primary parameter is indicated by the abscissa of a vertical line passing through the center of gravity of the area under the smooth curve. The standard (root-mean-square) deviation from this mean—i.e., the radius of gyration of the area about the mean line—is obtained from the second moment. Similarly, the skewness (asymmetry) of the curve is obtained from the third moment, kurtosis (foreshortening of the peak) from the fourth moment.

Natural sediments deposited under constant conditions are characterized by size-frequency diagrams approaching the logarithmic form of the normal error curve. Since the error curve possesses neither skewness nor kurtosis, it is evident that it may be fully defined by its geometric mean and by the standard deviation from this mean. Departures of natural sediments from this type of size distribution are of no practical importance to the present discussion, the third and fourth moments henceforth being ignored.

Since the abscissa scale of the size-frequency diagram is logarithmic, the first moment is really the logarithm of the length d_m and the second moment the difference between two logarithms or the logarithm of a length ratio: $\log \sigma_d = \log d_a/d_m = \log d_m/d_b$. Thus, d_m is a measure of the position of the frequency diagram on the size scale, and σ_d a measure of the grading. Obviously, either parameter is independent of the other, sediments having the same proportional range in size being characterized by the same numerical value of σ_d, regardless of the magnitude of d_m. Thus, with reference to Fig. 1, curves A, B, and C represent sediments possessing the same degree of sorting ($\sigma_d = 2$) but different mean diameters; curves A and D, the same mean diameters but different degrees of sorting ($\sigma_d = 2$, $\sigma_d = \sqrt{2}$); curves E and F, different directions of skewness but the same values of d_m and σ_d as curve A.

Determinations of d_m and σ_d may easily be made from a cumulative plot of the mechanical analysis on logarithmic probability paper. With these coordinates the logarithmic form of the normal error curve becomes a straight line (Fig. 2), its position varying with d_m, its slope with σ_d.   The magnitude of σ_d is given by the ratio of d_m to the intercept at the 15.9 per cent line or the ratio of the 84.1 per cent intercept to d_m.   It so happens that many skewed curves (E and F of Figs. 1 and 2) may be approximated in form by two terms of a logarithmic Gram-Charlier series, the latter function invariably coinciding at the 15.9 per cent and the 84.1 per cent points with the normal error function having the same values of d_m and σ_d.

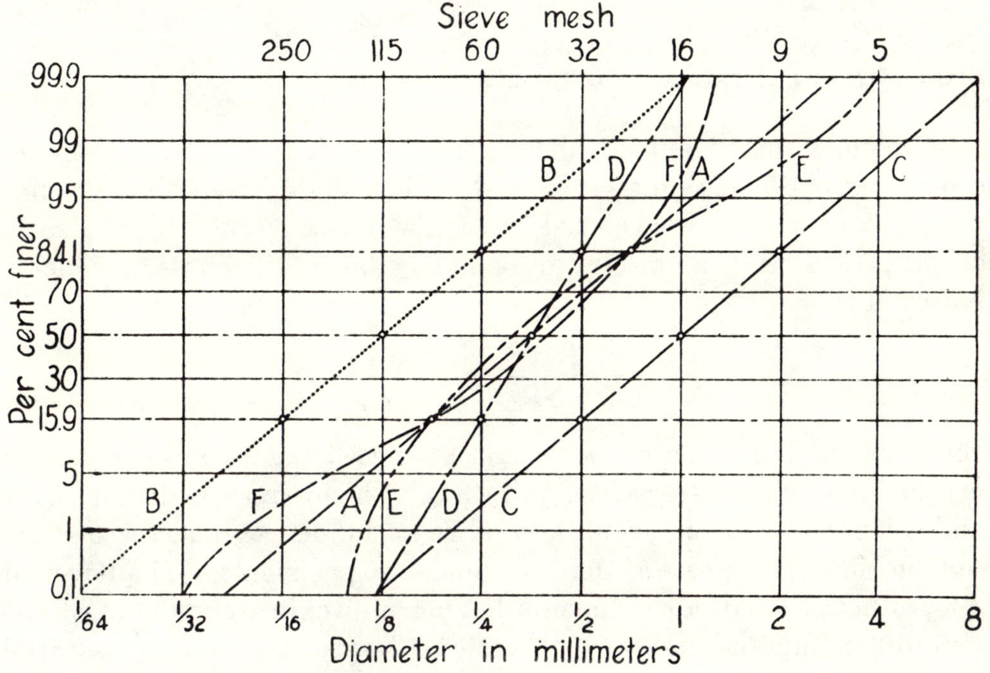

Fig. 2.—Cumulative Probability Plots Corresponding to the Frequency Diagrams of Fig. 1.

The mean and the standard deviation of skewed curves may therefore be obtained with sufficient accuracy for most practical purposes by passing a straight line through the 15.9 per cent and 84.1 per cent on the skewed plot, d_m and σ_d then being found from the straight line as before.   Not only are these parameters of great importance in the study of actual sediments found in nature, but they may well serve as two fundamental parameters to be varied systematically in the laboratory study of sediment transportation.

Such characteristics as permeability and angle of repose of the sediment grains cannot be regarded as independent characteristics, for they are governed wholly by the basic properties of sediment and fluid. Although the fall velocity of the individual grains is likewise a dependent characteristic varying with many factors of sediment and fluid, it is often treated as a basic variable in sediment research. Such use of the fall velocity in place of the other sediment characteristics is permissible when the latter have no individual effect upon the phenomenon, but this substitution should invariably be subjected to careful study before one accepts the simplified mathematical results which the method permits.

SIMILARITY IN BED-LOAD MOVEMENT

The most important advancement in the study of bed load in recent years is Shields' analytical and experimental proof that the initial movement from a level bed of sediment is governed by the ratio of mean sediment diameter to boundary-layer thickness.[2] Shields reasoned that

$$\frac{T_c}{(\gamma_s - \gamma)\, d_m} = \phi\left(\frac{d_m}{\delta}\right)$$

the form of the function ϕ being obtained from experiments with crushed amber, lignite, barite, and sand. Shields reasoned that shape and compaction factors should also have an influence upon the motion, but no such influence was detected in the experiments. The effect of σ_d was not studied, all sediments having approximately the same degree of sorting; nevertheless, this factor is probably a very essential characteristic in cases of different size distributions.

A purely dimensional analysis might have yielded the expression

$$\frac{T_c}{\gamma d_m} = \phi\left(\frac{\gamma_s}{\gamma}, \frac{d_m}{\delta}, \sigma_d\right)$$

(not to mention various others of even more complex form), indicating that for similarity of initial movement γ_s/γ, d_m/δ, and σ_d must be numerically identical in model and prototype. Shields' analytical and experimental studies showed, however, that certain of these factors could be combined, so that for sediments of similar grading characteristics the ratio d_m/δ appears to be the sole criterion for similarity of initial movement from a level bed. That d_m must be a basic factor in this case is evident from the fact that the grain size governs the absolute bed roughness so long as the bed remains level. On the other hand, Shields' experiments indicated that the magnitude of d_m/δ determines as well the form of the initial bed undulations, ripples appearing at small values of d_m/δ and long bars at large values—a point of fundamental importance in the attainment of bed similarity.

Since the rate of bed-load movement has not as yet been subjected to rational analysis, only purely empirical expressions are at hand upon which to base similarity criteria. Any such empirical relationship, it must be noted, is dependable for this purpose only within the range of experimental check, and cannot be generalized with safety. Using Shields' transport equation

$$\frac{G}{Q} = 10 \, S \frac{T - T_c}{(\gamma_s - \gamma) \, d_m}$$

for the sake of convenience, it would seem that the ratio of sediment discharge by weight (measured under water) to the water discharge by weight should depend only upon the parameter $S(T-T_c)/(\gamma_s-\gamma)d_m$. Again, however, the expression was developed from studies on uniform materials, and was restricted to cases in which viscous shear at the bed was negligible in comparison with turbulent shear. Therefore, in order for such a criterion to hold, it would seem essential that σ_d be approximately the same in model and prototype and that the Reynolds number be of the same order of magnitude. Needless to say, the Reynolds number in many model studies is far too low to eliminate the effect of viscosity in the boundary region.

While d_m is obviously a primary sediment characteristic in initial phases of bed movement, its rôle in advanced stages of transportation is less apparent—in particular since it is not the grain size, but that of the bed irregularities, which governs the bed roughness. It is a noteworthy fact that all experimental studies of sediment movement have been arbitrarily restricted to cases in which little or no material is carried into suspension—the latter being a phenomenon in which, for the most part, d_m is definitely not a pertinent factor.

Similarity in the Transportation of Suspended Load

Since the suspension of sediment in flowing water is governed by the turbulence of the flow, it is possible to express the relative form of the sediment-distribution function in terms of the fall-velocity characteristics of the sediment and the distribution of turbulence over the vertical section; the latter, in turn, is a function of the velocity distribution, while the fall-velocity characteristics may be defined in terms of w_m, the geometric mean, and σ_w, the geometric standard deviation of w from this mean—weight frequencies being used exactly as in the case of the sediment diameter. The distribution of sediment over the vertical section is then a function of the following form:[3]

$$\phi\left(C_a,\ \frac{a}{D},\ \frac{\sqrt{T/\rho}}{w_m},\ \sigma_w\right)$$

Herein, C_a is the weight concentration of the sediment at some arbitrary level a, while $\sqrt{T/\rho}$ is the so-called friction velocity, equal to $V\sqrt{f/8}$. The form of ϕ depends upon the distribution of turbulence, which varies with the velocity distribution, which (for fully developed boundary roughness) depends in turn upon f, or upon the relative roughness k/D, and upon the mean velocity of flow.

Since C_a is an independent variable in this function, the total amount of material transported in suspension is still indeterminate. If one regards the advanced state of sediment movement as the result of fluid turbulence over the entire flow section, it would seem reasonable to replace C_a by C_0, the bed concentration, and to make a proportional to k, the absolute roughness of the bed. $\sqrt{T/\rho}$ may be replaced by V alone, since f varies only with k/D. The total rate of transport would then have the form

$$\frac{G}{Q} = \phi\left(C_0,\ \frac{k}{D},\ \frac{V}{w_m},\ \sigma_w\right)$$

This expression has a reasonable, though hypothetical, basis and has yet to be verified experimentally. Were it to prove valid, the terms C_0, k/D, V/w_m, and σ_w would be the sole criteria for similarity of sediment discharge under moderate concentrations, provided only that the bed resistance did not depend upon the Reynolds number in either prototype or model.

SIMILARITY IN SCOUR PHENOMENA

Although the phenomenon of localized scour has been subjected to surprisingly little systematic study, it depends upon essentially the same geometric, flow, fluid, and sediment characteristics as any general problem of sediment transportation in unsteady, non-uniform flow. As in other cases involving the movement of sediment by water, it is not safe practice to combine all pertinent variables by simple dimensional analysis, for the elementary manner of applying the Π-theorem presumes that the problem involves the motion of a homogeneous fluid, which is far from the case; as a result, this method will invariably lead to a parameter of the form d_m/D, making the sediment diameter once and for all a basic geometrical factor subject to exact scale reduction in the model.

As a matter of fact, the most general phase of sediment transportation is a combination of two distinct types of motion—that of a fluid relative to geometrical boundaries, and that of granular solids relative to the moving fluid, the latter motion producing in addition a gradual change in the boundary geometry. The actual function, therefore, must be a composite one, the dimensional analysis of which has not yet been rationally approached. Under such circumstances, one is tempted to assume that the relative motion between sediment and fluid is proportional to the normal settling velocity of the individual sediment grains, this velocity of fall then being used in the customary application of the Π-theorem, and all related sediment characteristics being ignored.

Consider, for instance, a simple case of scour in which such fluid characteristics as weight and viscosity have no effect upon the flow itself, and upon the sediment movement only insofar as they vary the magnitude of w_m. It is desired to express the relative depth of scour s as a function of time, mean velocity of inflow, and the sediment characteristics w_m and σ_w. It follows that

$$\frac{s}{a} = \phi \left(\frac{w_m\, t}{a}, \ \frac{V}{w_m}, \ \sigma_w \right)$$

a being some length characteristic of the boundary geometry. The case is, to all appearances, conveniently simple; however, final acceptance of such a hybrid product of dimensional analysis must remain subject to experimental studies in which each of the individual factors is varied over a considerable range.

Experimental Procedure

In order to test the validity of the foregoing relationship, the boundary conditions shown in Fig. 3 were chosen arbitrarily. Water is fed under constant head from a portable supply system to a small,

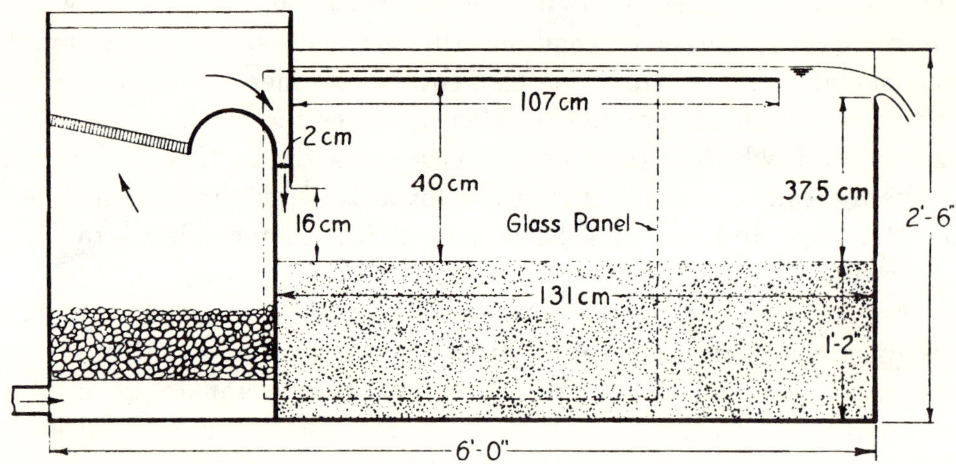

FIG. 3.—LONGITUDINAL SECTION THROUGH GLASS-WALLED TANK, SHOWING THE GEOMETRICAL BOUNDARY CONDITIONS AT THE BEGINNING OF EACH RUN.

glass-walled tank 6 inches in width, the inflow being given the form of a vertical jet impinging on an originally level bed of prepared sand. The course of the scour with time is mapped with wax pencil on one glass panel, the profiles finally being traced full scale on coordinate paper for permanent reference. Runs of from 3 to 24 hours duration (depending upon the rate of scour) have been made to date on three different sands of essentially the same σ_w, prepared by a combination of water, a i r, and sieve classification (sand characteristics are shown

FIG. 4. GENERAL VIEW OF EXPERIMENTAL EQUIPMENT.

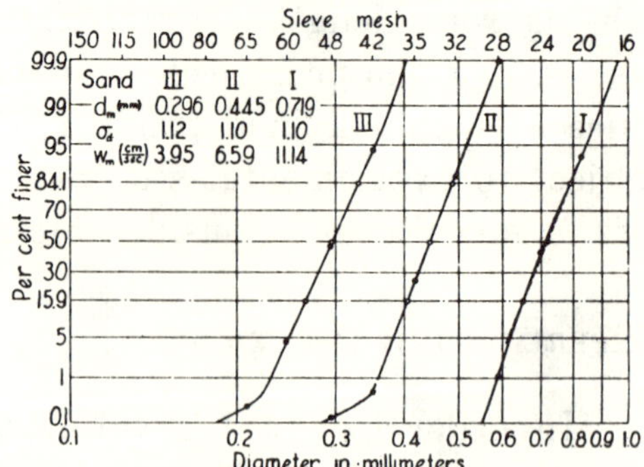

FIG. 5. GRADING CHARACTERISTICS OF SANDS I, II, AND III.

in Fig. 5). In addition, each sand has been studied under a ½-scale reduction of all geometrical proportions except flume width. Through successive changes in discharge, every series of runs has included as great a variation in the ratio V/w_m as operating conditions would permit.

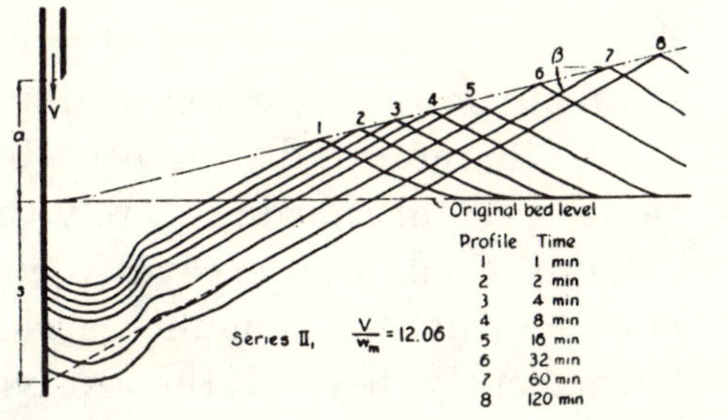

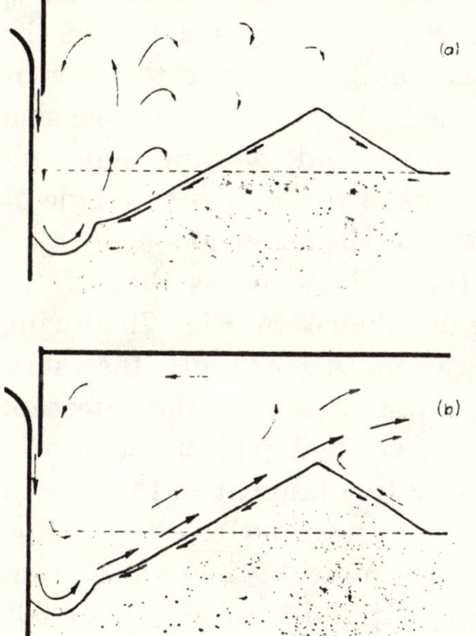

Fig. 6.—Progressive Development of the Scour Profile as a Function of Time.

Fig. 7. Flow Patterns for (a) Maximum and (b) Minimum Jet Deflection.

A family of profiles obtained with Sand II at full scale and constant jet velocity is shown schematically in Fig. 6. Evidently, the linear characteristics of the scour hole vary by approximately constant increments as the successive time intervals double in magnitude, suggesting at once an exponential relationship between scour and time. However, it will be noted that during this particular run the length increments abruptly increase in magnitude, indicating some break in the continuity of the function. Investigation of this discontinuity disclosed the fact that two distinct regimes of flow are possible, the jet either being deflected through nearly 180° (Fig. 7a) or following the boundary of the scour as far as the crest of the dune (Fig. 7b). The depth of scour at which a transition from the former regime to the latter will occur was found to vary with the velocity of the jet, the fall velocity of the material, and the boundary scale, the jet pendulating between its two limiting forms as the critical stage is approached.

For either regime, the actual removal of material from the scour hole is purely a phenomenon of suspension. Any material falling upon the upstream side of the dune gradually slides back toward the zone of excavation, only to be carried again into suspension as it meets the deflected jet. In the regime of Fig. 7b (minimum jet deflection) the actual transporting velocities of the fluid are far higher than in the case of Fig. 7a, despite the fact that the velocity of inflow is somewhat lower; as a result, although considerably less material is in suspension, more is carried over the crest before settling out. In selecting a characteristic depth of scour s for purposes of comparison, one must therefore distinguish between a factor showing merely the amount of material excavated and one showing the amount carried permanently beyond the excavation zone. The depth of the scour hole itself is of the former category, for at a given position of the dune crest it varies with V, w_m, and the boundary scale. On the other hand, whenever the flow is stopped, all material in suspension settles out and the dune slope adjusts itself to the natural angle of repose of the material—a slope essentially the same as that of the regime shown in Fig. 7b during operation. Accordingly, the variable s was chosen as the intercept on the vertical jet boundary of a sloping line tangent to the face of the dune near its crest (see Fig. 6); its angle of inclination was slightly greater than 29° for Sands I and II, and somewhat over 30° for the more angular Sand III.

FIG. 8. TYPICAL SCOUR REGIMES AT TWO SLIGHTLY DIFFERENT RATES OF DISCHARGE: ABOVE, MAXIMUM JET DEFLECTION; BELOW, MINIMUM JET DEFLECTION.

DISCUSSION OF RESULTS

All data obtained thus far with Sands I, II, and III at both full and half scale are reproduced in Fig. 9, s/a being plotted against log $w_m t/a$ for various values of V/w_m (points shown as squares refer to the regime of maximum jet deflection, circles to that of minimum deflection). At once apparent is the systematic variation of the slope

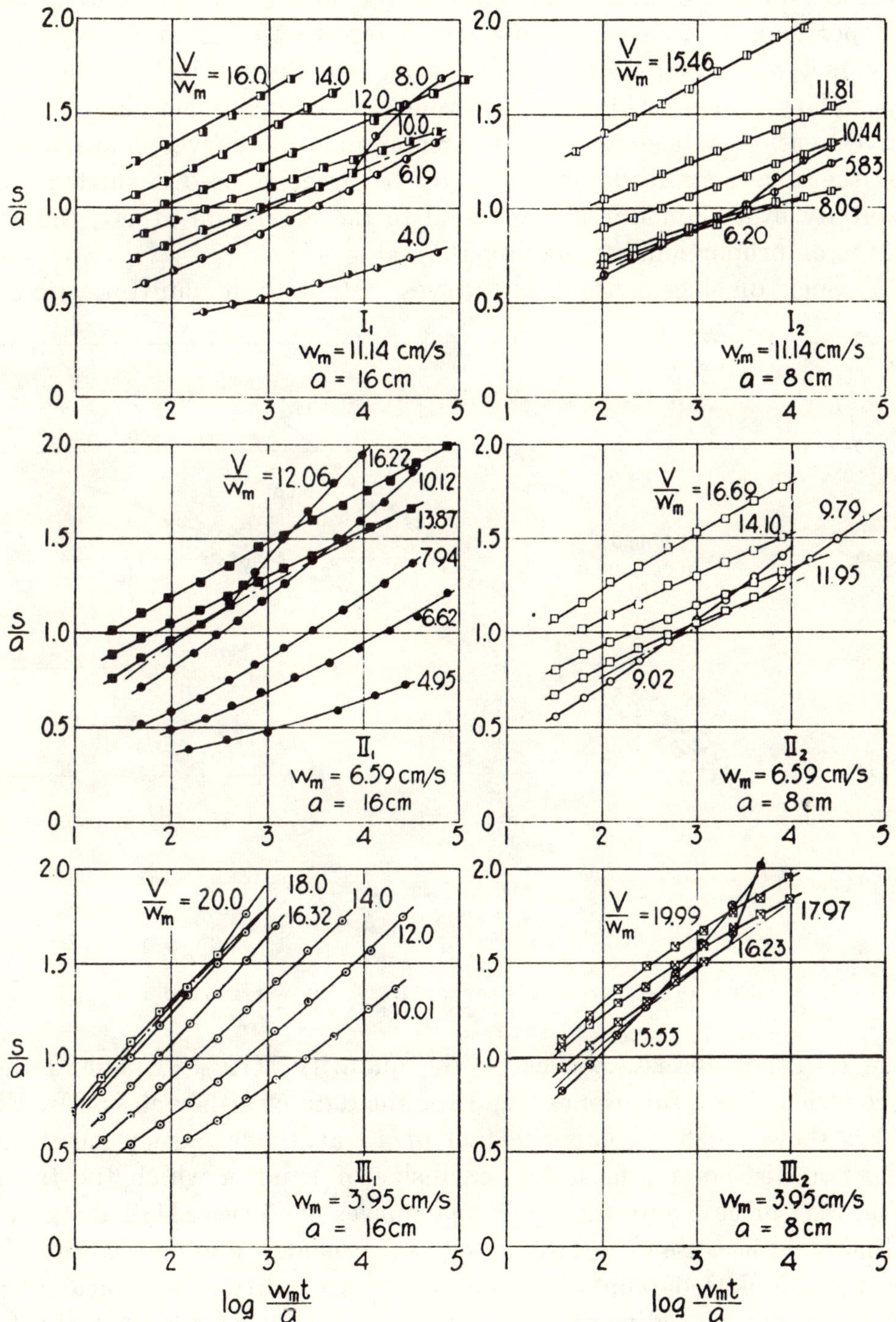

Fig. 9.—Experimental Measurements of Localized Scour.

of the curves of either regime with V/w_m; in fact, the slope is directly proportional to $V/w_m - 1$, indicating a zero rate of scour for $V/w_m = 1$ —a fact which is quite reasonable, since it is the jet itself, rather than secondary eddies, which accomplishes the scouring action. Further investigation will show that in each regime all curves characterized by the same magnitude of V/w_m will be superimposed if shifted horizontally by an amount proportional to the logarithm of w_m, the constant of proportionality having the value 3.

Such displacement of all curves, followed by multiplication of

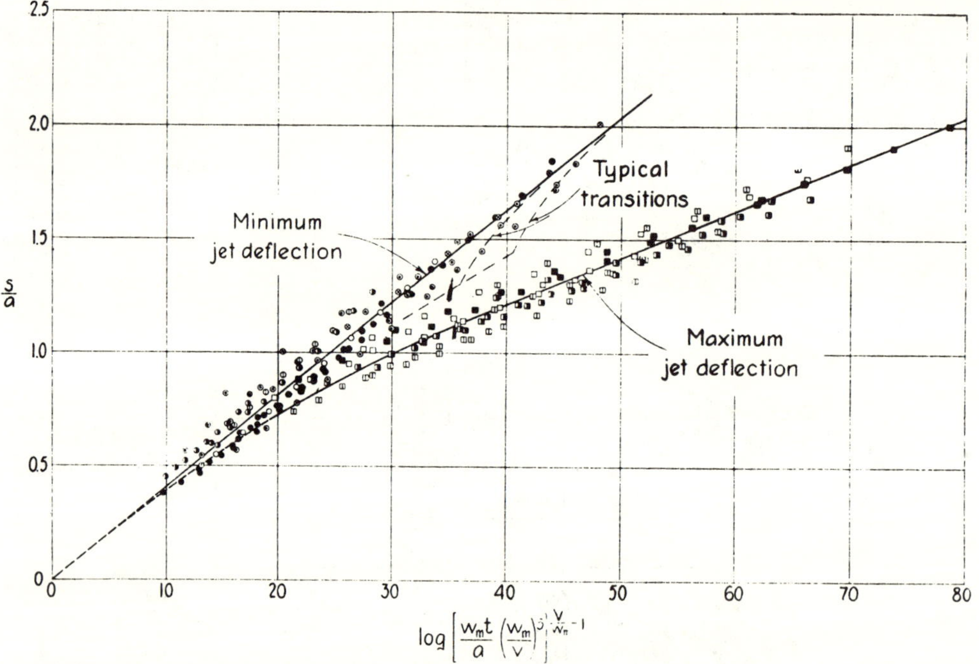

FIG. 10.—COMPOSITE PLOT OF DATA SHOWN IN FIG. 7.

the resulting abscissa values by the quantity $V/w_m - 1$, should then produce a single functional trend for the data of either class. In Fig. 10 is shown such a composite plot of all available measurements for the two distinct regimes (i.e., exclusive of runs in which the transition took place), only the upper two curves of Series III_2 having been eliminated because of unsatisfactory agreement. While a considerable scatter of plotted points is apparent, this condition is typical of all experiments on sediment transportation, and it is felt that the two curves passed through the points indicate the functional trends with good approximation. As a matter of fact, the actual forms of these functions are of no further concern in the selection of similarity criteria, for a glance at the ordinate and abscissa scales of Fig. 10 will show that the conclusions reached through the dimensional analysis of the problem are fully verified—with one exception: the abscissa term includes a dimensional factor v which was not foreseen in the preliminary analysis. This factor necessarily has the dimension of a velocity, in order that the ratio w_m/v may be properly dimensionless, and it must have the magnitude 5.25 for all three sands if the functional curves are to pass through the origin; while its significance is not fully clear, it might be regarded as the ratio of a to t_o, the latter being some initial time factor dependent upon the model scale—or, perhaps, v is proportional to $\sqrt{p/\rho}$, in which p is a characteristic bearing or shear capacity of the material which tends to oppose the scouring action.

As has already been mentioned, the value of s/a at which the transition from one regime to the other takes place varies with V, w_m, and a, the locus of the transition points for any one series of experiments being indicated in Fig. 9 by a broken line. While further study of the transition phenomenon would be of considerable academic interest, in the organization of similarity criteria it is necessary to define only the approximate transition zone so that model studies may lie entirely within the proper flow regime. This zone is shown in Fig. 11 for each of the two scales already investigated, the points being obtained from Fig. 9 and the width of the shaded areas from exploratory tests for each of the sands and boundary scales.

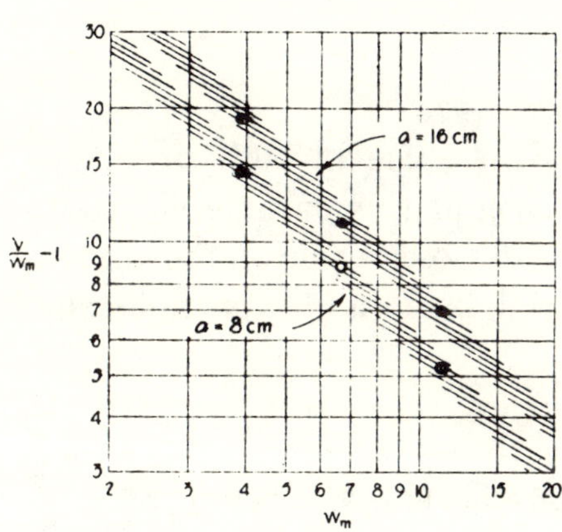

FIG. 11.   APPROXIMATE LIMITS OF THE TRANSITION ZONE.

As yet no detailed investigation has been made of the effect of sediment grading. From Fig. 5 it was seen that the standard deviation of particle size was essentially the same for all three sands, the assumption being made that with such narrow sorting σ_w could then be treated as a constant. For material so nearly uniform in fall velocity, the experiments have shown that no equilibrium of scour is to be expected at any depth, the removal of material continuing as an

81

exponential function of the time with only the fixed boundaries governing the ultimate limit of excavation. Preliminary experiments indicate, on the other hand, that a process of selective sorting will take place if σ_w is increased appreciably in magnitude, the finer material being carried over the dune, and the coarser being deposited repeatedly on the upstream slope and sliding back into the scour hole. In this way the magnitude of w_m for the material lining the hole will steadily rise, the resulting tendency to approach a state of equilibrium obviously increasing with the magnitude of σ_w for the original bed material.

Whatever the rôle of σ_w may be, the present experiments with constant σ_w indicate that for a given fall-velocity distribution of sediment the sole criterion for similarity of the scour pattern under the existing boundary conditions is embodied in the quantity

$$\left[\frac{w_m\, t}{a} \left(\frac{w_m}{v} \right)^3 \right]^{\frac{V}{w_m}-1}$$

provided only that the regime of flow is the same in both prototype and model. Although the mathematical form of this parametric group will probably vary with the boundary conditions, each of the dimensionless ratios should still play an essential part in the phenomenon.

Conclusions

In selecting criteria for similarity in the transportation of sediment, one must choose, in addition to the usual geometric, flow, and fluid parameters, those sediment characteristics having a definite influence upon the rate at which material is moved. So long as the size of the material relative to some length parameter of the flow is a governing factor, it is only reasonable to use the size-frequency properties as similarity criteria, the geometric mean diameter and the geometric standard deviation from this mean being the most significant size factors available. Thus, when the mean diameter is of the order of magnitude of the boundary-layer thickness, the ratio d_m/δ is a fundamental similarity parameter. Once d_m becomes sufficiently large to produce boundary turbulence, however, neither does viscosity directly influence sediment movement at the boundary nor does the size of moving grains determine the relative roughness of the bed. On the other hand, it must be noted that bed materials of pulverized pumice, amber, and lignite often used in the laboratory not only far exceed in size the nominal boundary-layer thickness, but often approach the depth of flow itself; under such circumstances the size again becomes a significant parameter—though for the model bed alone.

In by far the majority of actual canals and rivers, the bed material is very small in comparison with the depth of flow, while the depth, mean velocity, and bed roughness are such that the motion is wholly turbulent, viscous effects upon the resistance then being quite negligible. At sufficiently advanced stages of sediment movement the sediment diameter probably influences the rate of transport only insofar as it governs the velocity of fall. In such cases, the fall velocity would seem the only logical sediment characteristic for similarity criteria, the ratio V/w_m and the factor σ_w being of basic importance; needless to say, however, the model scale should not be so far reduced as to prevent the full development of turbulence.

Experiments now in progress on an arbitrary condition of localized scour indicate that the mean fall velocity and the standard deviation about this mean are again significant, the ratio V/w_m being a primary similarity parameter so long as the model scale is not so small as to yield a grain diameter approaching the order of magnitude of other flow dimensions. In this connection it is to be noted that the customary use of coarse, low-density materials in model studies of scour produces a condition of transport akin to the bed-load phase of uniform flow, whereas in nature scour is often predominantly a phenomenon of suspension.

Bibliography

(1) Krumbein, W. C., and Pettijohn, F. J. *Manual of Sedimentary Petrography*, Appleton-Century, New York, 1938.

(2) Shields, A. "Anwendung der Aehnlichkeitsmechanik und der Turbulenzforschung auf die Geschiebebewegung," *Mitteilungen der Preussischen Versuchsanstalt für Wasserbau und Schiffbau*, Berlin, Heft 26, 1936.

(3) Rouse, H. "An Analysis of Sediment Transportation in the Light of Fluid Turbulence" (Mimeographed), *U. S. Department of Agriculture, Soil Conservation Service, SCS-TP-25*, 1939.

SUSPENSION OF SEDIMENT IN UPWARD FLOW

by

Hunter Rouse

Professor of Fluid Mechanics
Consulting Engineer, Iowa Institute of Hydraulic Research

Engineers have long been aware of the fact that the percolation of a fluid through a granular material will displace the material in the direction of flow if restraining forces are not sufficiently great. The quicksand phenomenon is a well known instance of such conditions as found in nature, while the method of washing rapid-sand filters by reversing the normal direction of flow represents an engineering application of the same principle; in either case the drag exerted by upward flow through an initially compact bed tends to exceed the immersed weight of the bed material, thereby producing an increase in bed porosity and a corresponding reduction in drag. Although in such examples of upward flow the material is rarely expanded to a point at which the grains are no longer in mutual contact, it is not unreasonable to presume that ever higher degrees of expansion would be obtained with increasingly higher rates of flow — the particles ultimately becoming so dispersed as to be in a state of complete suspension within the rising fluid. It is with the analysis of this phase of the phenomenon that the present paper is concerned.

Once bed material is displaced by such flow, it is apparent that the coefficient of permeability k in the Darcy relationship $Q = k A \, dh/dz$ will no longer equal that of the initially compact, uniformly mixed material, but will increase considerably as the interstices within the material grow in size. Moreover, as the expansion of the bed reduces the frictional restraint upon the individual particles, they will not only begin to go into suspension, but will at the same time assume positions commensurate with the resistance which they offer to the flow. In other words, finer or lighter particles will be carried higher than coarse or heavy ones, until a state of complete suspension displays almost perfect stratification of material according to size and density. If the upward flow is then abruptly stopped, the material will settle to yield an even more perfectly stratified bed.

Since the Darcy coefficient k is at best an empirical factor, known values of which apply specifically to the case of pure percolation, a more rational approach to the case of complete suspension is to be preferred. Let it be assumed, therefore, that the immersed weight of the particles in suspension is exactly balanced by the drag exerted upon them by the upward flow. The external forces upon the elementary ''free body'' of the fluid-solid mixture shown in Fig. 1 must then also be in equilibrium, such forces involving the

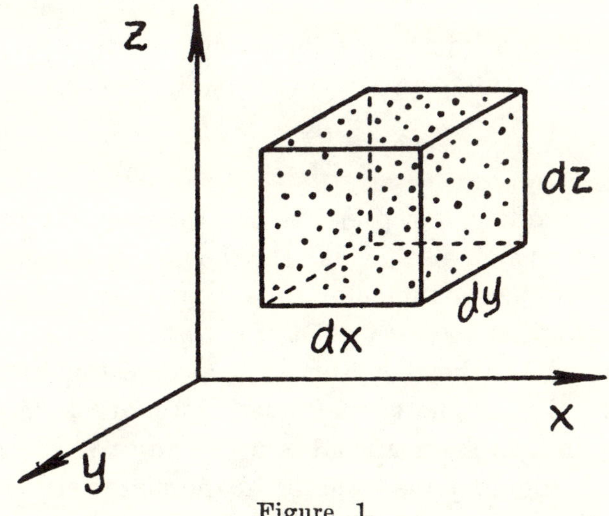

Figure 1.

weight of the mixture and the normal and tangential stresses upon the faces of the element. Designating by γ_s the specific weight of the solid matter, by γ the specific weight of the fluid itself, and by e the ratio of the volume of solids to volume of mixture, it follows that the weight of the element will have the magnitude $[e(\gamma_s - \gamma) + \gamma]\, dx\, dy\, dz$. If conditions are statistically similar at all points in any horizontal plane, the mean tangential stresses will be equal to zero over each of the faces. Normal stresses in the horizontal directions will then in themselves yield a state of equilibrium, while in the vertical direction the weight of the element will be balanced by the difference in pressure between the lower and upper faces. That is, expressing the change in pressure intensity with elevation as dp/dz, it will be seen that

$$- [e(\gamma_s - \gamma) + \gamma]\, dx\, dy\, dz - \frac{dp}{dz}\, dz\, (dx\, dy) = 0$$

whence

$$-\frac{dp}{dz} = \gamma + e\,(\gamma_s - \gamma)$$

Evidently, the rate of decrease of pressure intensity with elevation at any point is equal to the specific weight of the suspension at that point. Or, defining the concentration of the suspension as the difference between its specific gravity and that of the fluid alone (i.e., $c = e(\gamma_s/\gamma - 1)$),

$$-\frac{dp}{dz} = \gamma\,(1 + c)$$

In case the fluid is a liquid, the foregoing equation may be written in the more significant form

$$-\frac{dh}{dz} = c \tag{1}$$

in which the quantity h represents the manometric head $p/\gamma + z$ within the suspension — that is, the height to which the liquid would rise in an open manometer tube connected to a piezometer at the point in question. Eq. (1) therefore states that the rate of decrease of manometric head in the vertical direction is equal to the local concentration of the suspension. More practically expressed, the difference in manometric head between any two levels is directly proportional to the immersed weight of the material held in suspension between these two levels.

Although the foregoing relationship was obtained by assuming a state of static equilibrium to prevail between the immersed weight of the material and the drag exerted by the flow, this effectively presumes a kinematic balance between the settling velocity w of the material and the local velocity v of the upward flow. The settling velocity of the individual granules probably varies somewhat with the concentration of the suspension, but as a first approximation it may be considered proportional to the velocity of fall of a single representative particle in still liquid; thus, $w = K_1 v$. The local velocity of flow may likewise be expected to vary with concentration (or vice versa), but in a manner which is more readily evaluated. Representing by Q the rate of upward flow past a horizontal section of total area A, the local upward velocity should be directly proportional to the rate of flow and inversely proportional to the area of the reduced flow section. If the mean particle diameter is designated by d, and the mean spacing of the particles by L, the area of the total section may be expressed as $K L^2$ and that of the reduced section as $K (L^2 - K_2 d^2)$. Therefore, introducing the nominal velocity of flow $V = Q/A$ and combining the several proportionalities, it appears that

$$w = K_1 v = K_1 \frac{Q}{K(L^2 - K_2 d^2)} = K_1 \frac{KVL^2}{K(L^2 - K_2 d^2)} = \frac{K_1 V}{1 - K_2 \dfrac{d^2}{L^2}}$$

Similarly, the concentration c may be expressed in terms of the parameters d and L, in combination with the specific weights of fluid and solid, through the following proportionality:

$$c = e \left(\frac{\gamma_s}{\gamma} - 1 \right) = K_3 \frac{d^3}{L^3} \left(\frac{\gamma_s}{\gamma} - 1 \right)$$

Therefore,

$$\frac{d^2}{L^2} = \left(\frac{c}{K_3 \left(\dfrac{\gamma_s}{\gamma} - 1 \right)} \right)^{2/3}$$

Substitution of this expression in the foregoing relationship for w then yields the general equation

$$c = K_3 \left(\frac{\gamma_s}{\gamma} - 1 \right) \left(\frac{1 - K_1 V/w}{K_2} \right)^{3/2}$$

Finally, combining constants and introducing Eq. (1).

$$c = -\frac{dh}{dz} = A \left(1 - B \frac{V}{w} \right)^{3/2} \tag{2}$$

This equation is evidently in accord with the earlier statement that the finer material is carried toward the top of a zone of suspension, since c can decrease with elevation only if w also becomes smaller. Indeed, since the gradient dh/dz is a measure of c, one might now conclude that measurements of manometric head at various levels in such upward flow would provide a quantitative means of determining the fall-velocity characteristics of the suspended material. The import of such a conclusion cannot be over emphasized, for fall-velocity characteristics are comparable — and in some cases preferable — to size characteristics obtained by physical analysis. Certain features of the derivation, however, must not be overlooked. The coefficient A is governed by the shape and relative density of the material and perhaps even by viscous effects upon the pattern of flow around the individual particles; only if these factors are essentially the same for all particle sizes may A be expected to retain a constant magnitude. The factor B, on the other hand, embodies the proportionality between the rate of settling and the local velocity of the upward flow; since turbulence can produce a state of suspension without any upward flow whatever, the magnitude of B must be expected to vary with the degree of turbulence present in the flow.

To what extent such influences will affect the practicability of this relationship can only be determined through experimental tests, a series of which were conducted at the University of Iowa in 1939-40 by Warren DeLapp and described in "Sediment Behavior in Upward Flow," a thesis submitted in partial fulfillment of the requirements for the degree of Master of Science in the Department of Mechanics and Hydraulics. Duplicate studies were made in a glass cylinder 12 in. in diameter and 19 in. high and in a Lucite cylinder $1\frac{3}{4}$ in. in diameter and 54 in. high, each provided with means of measuring the manometric head over practically the entire height of cylinder. The granular material investigated consisted of quartz sand artificially graded to yield a size-frequency distribution closely following the normal error curve, 2% being finer than 0.07 mm. and 2% coarser than 0.8 mm.

A typical series of measurements made in the larger cylinder is shown in Fig. 2, the ordinate scale representing elevation above the fine-mesh screen which served to support the material at zero discharge, and the abscissa scale the difference between the manometric head within the water-sediment mixture and the head within the clear water above. Prior to the beginning of manometer readings, the entire bed was brought into suspension, and thereafter allowed to settle in stratified form. Measurements at a very low rate of flow then yielded the bottommost curve, the form of which is typical of pure percolation through a bed of decreasing grain size — that is, the Darcy permeability k grows smaller as the interstices are reduced in size, the slope $-dh/dz$ attaining its maximum value at the bed surface. At the next higher rate of flow, however, a reversal in curvature is noted in this region, indicating that the topmost material has been carried into suspension. Continued increase in the discharge brings more and more material into suspension, at the same time increasing the drop in head through that portion of the bed which has not yet begun to expand. Finally, however, the form of the curve indicates that the entire bed has been carried into sus-

% cumulative
0 20 40 60 80 100
z in feet
1.2
1.0
0.8
0.6
0.4
0.2
0
V = 0.0515 ft/sec
0.0197
0.0150
0.0403
0.0105
0.0284
0.154
0.114
0.074
0.00356
0.00141
V = 0.00111 ft/sec
Δh in feet
0 0.1 0.2 0.3 0.4

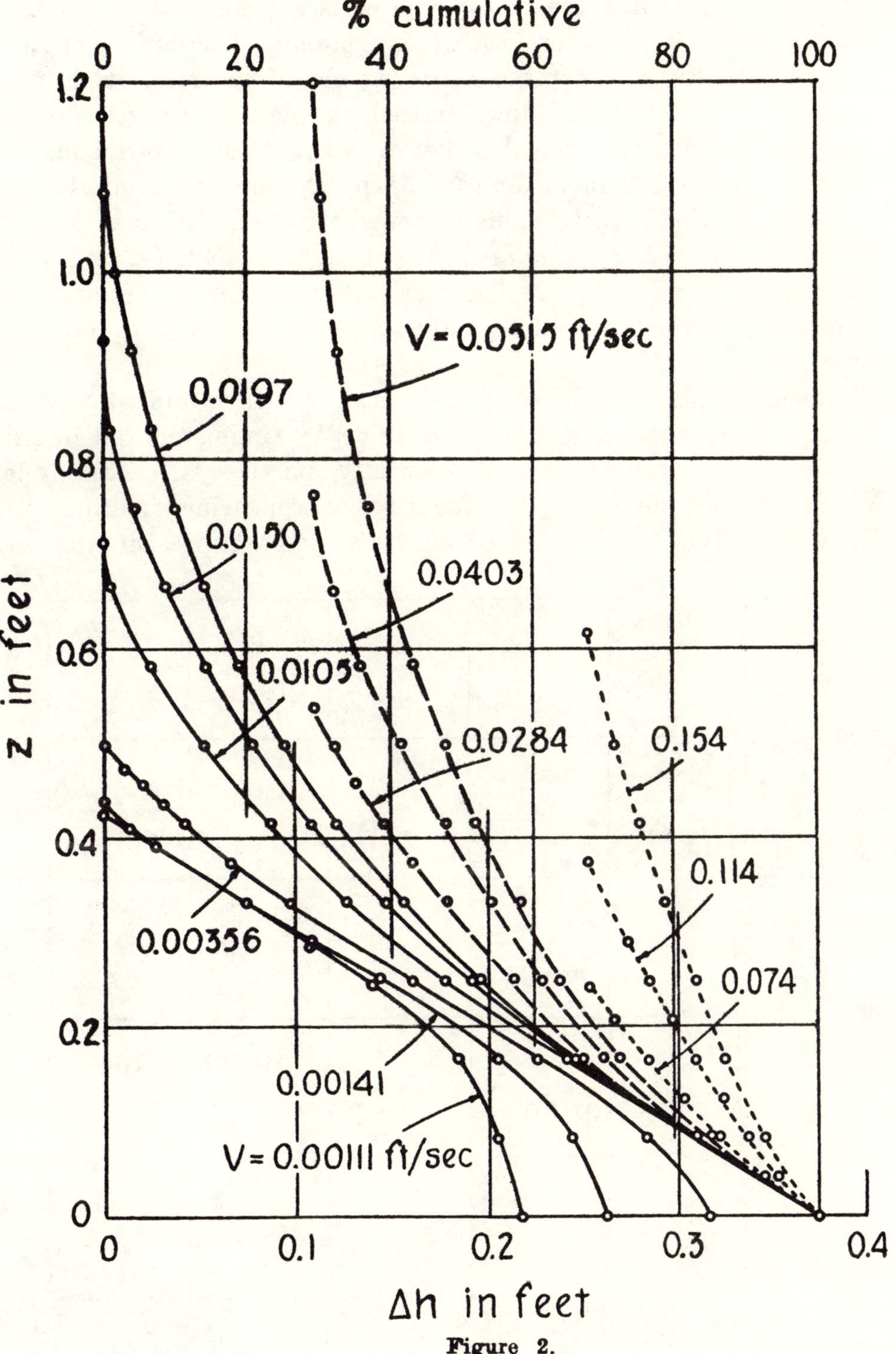

Figure 2.

pension, further increase in the rate of flow producing no further change in the head at the base of the column. This is in complete accord with Eq. (1), for the entire weight of the material is now supported by the upward flow. Indeed, supplementary tests proved that the addition or removal of known amounts of material invariably changed the manometer reading by the amount computed on the basis of Eq. (1). For this reason the broken curves in Fig. 2, obtained at such high rates of flow that the finer material had to be removed to avoid being carried over the top of the container, have arbitrarily been displaced to make the readings at zero elevation coincide.

From samples withdrawn at various elevations and at various rates of flow, it was possible to determine, by timing the fall in still water of a hundred or more representative particles of each sample, characteristic values of the quantity w appearing in Eq. (2). Measurement of the slope of $h{:}z$ curves at points represented by the

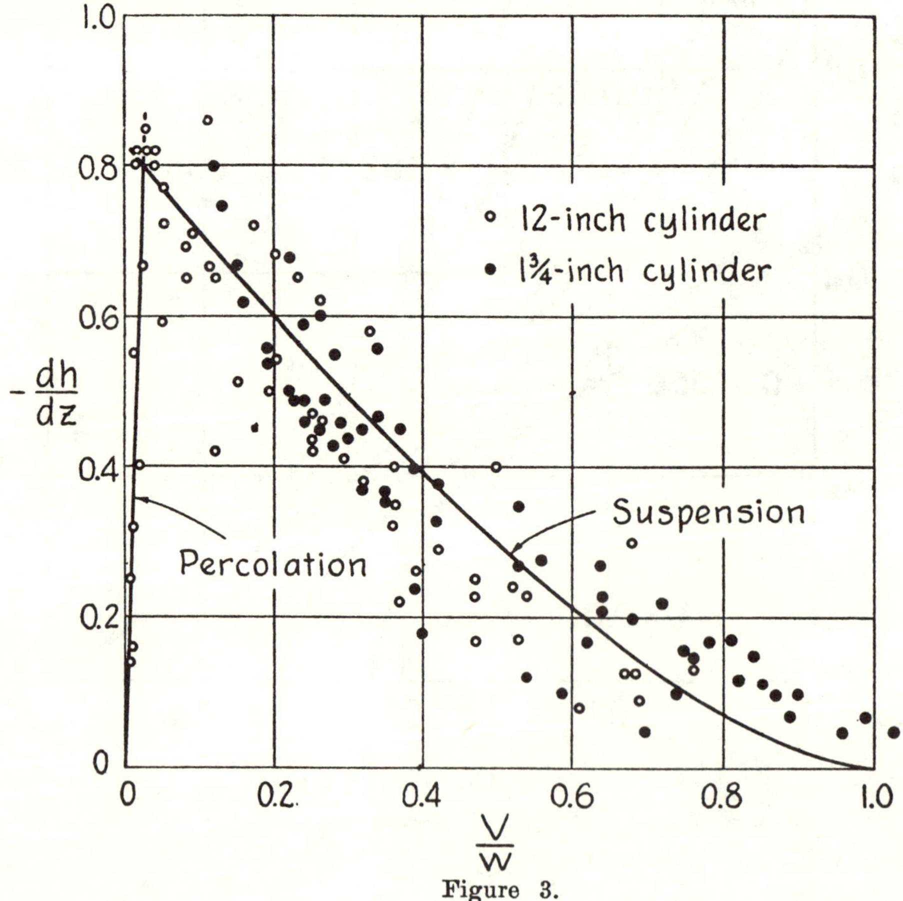

Figure 3.

samples then yielded the corresponding gradient — dh/dz, a plot
of which against the ratio V/w is shown in Fig. 3. The straight line
passing through the origin averages points obtained under conditions of pure percolation, its upper limit indicating that expansion
of the material begins at a value of — dh/dz of about 0.8, regardless of grain size. Once the material is in suspension, however, this
gradient rapidly decreases in magnitude, finally approaching the
limit zero as the nominal velocity of flow approaches the normal
settling velocity of each different size of grain. Although there is
appreciable scatter of points, they follow the general trend of the
plotted curve (refer to Eq. (2)), for which $A = 0.83$ and $B = 1.0$.
Nevertheless, it was found that the points lying above this curve
invariably corresponded to the coarser material, and points below
to the finer; in other words, the coefficient B, arbitrarily chosen as
unity, actually varies with the fall velocity of the various grain
sizes. Reference to the derivation of Eq. (2) will show that this
indicates incomplete proportionality between w and v, a variation
attributed to the effects of turbulence upon the suspension. Indeed,
a definite pattern of eddies was noticeable through the transparent
walls of the cylinders, apparently due neither to imperfect stilling
of the approaching flow nor to the wake behind each individual

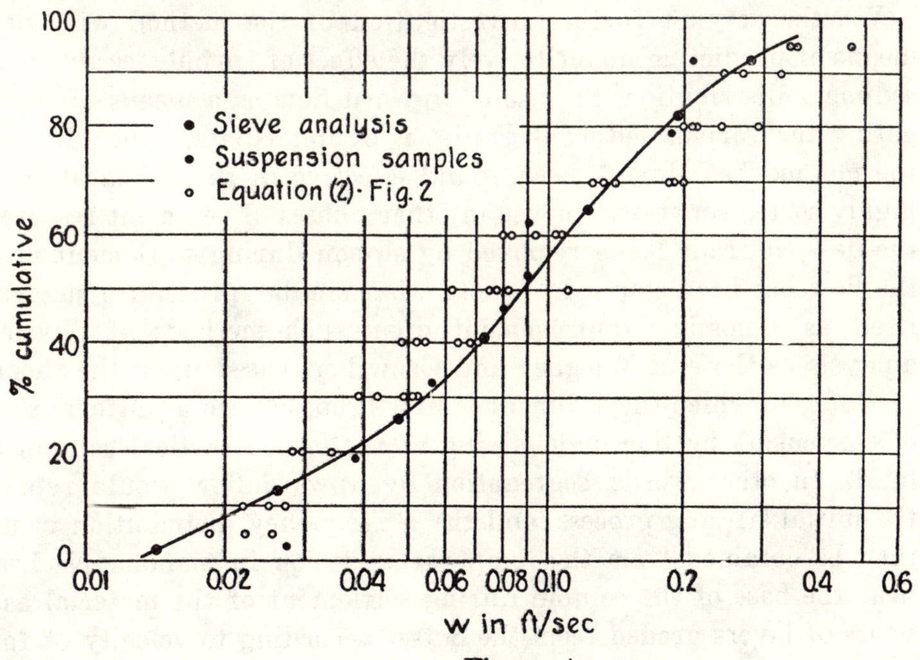

Figure 4.

grain, but rather to characteristic momentary fluctuations in the density of the water-sediment mixture.

Despite the probable lack of precision which such tendencies would produce if the method were used for sediment analysis, an effort was made to compare the results so obtained with the known characteristics of the original material. Using the size of sieve openings as a measure of the grain diameter and assuming the grains to be spherical in shape, points on the cumulative fall-velocity curve shown in Fig. 4 were computed. A second series of values was obtained by correlating the measured fall velocities of the several suspension samples with the corresponding points on the cumulative scale of Fig. 2. The third series resulted from measuring the slope $-dh/dz = c$ of the various curves of Fig. 2 for successive cumulative values, the corresponding magnitudes of w being obtained from the following form of Eq. (2).

$$w = \frac{V}{1 - 1.13\left(-\dfrac{dh}{dz}\right)^{2/3}}$$

Although considerable deviation of the points from one another is again evident, the consistent trend of the measurements as a whole indicates that Eq. (2) embodies the basic principles of sediment suspension in upward flow.

Whether or not further investigation of the method will yield means of predicting quantitatively the effect of turbulence upon the sediment distribution, the use of upward flow as a means of segregating the various sediment grades is of immediate value. Indeed, the method has already been found effective in the preparation of highly sorted separates for use in other phases of sediment research, the desired grade being removed by siphon during settlement after the flow has been stopped. A somewhat similar procedure suggests itself as a possible improvement upon such methods of physical analysis as those of Wiegner and Crowther, based upon the theory of Odén, in which the sediment is first brought into a uniform state of suspension by thorough mixing of the entire vertical column of fluid. In other words, segregation by upward flow would replace the initial stirring process, and the w-frequency distribution would then be obtained from the temporal variation in manometric head near the base of the column during settlement of the material as a series of layers graded from the outset according to velocity of fall.

EVALUATION OF BOUNDARY ROUGHNESS

by
Hunter Rouse
State University of Iowa
Iowa City, Iowa

Somewhat over a decade ago Professor von Kármán published an extremely significant analysis of the velocity distribution and resistance of turbulent flow past smooth and rough boundaries. This analysis has had a profound influence upon fluid mechanics throughout the world, but, despite considerable publicity in engineering journals, it has seen little application to hydraulic design. Current literature, for instance, continues to reflect adherence to exponential resistance formulas, and evidences at times a complete misunderstanding of the roles played by viscosity and boundary roughness.

There is probably a three-fold reason for this situation. First, the apparent complexity of the analysis does not of itself attract the interest of the practical mind. Second, the experiments substantiating the analysis involved artificial boundary roughness which was not geometrically similar to the roughness of commercial material. And, third, the analysis did not originally include that phase of resistance which is most frequently encountered in practice, in which the boundary is, in effect, neither completely smooth nor completely rough. In this paper the writer seeks to overcome each of these three difficulties, in the hope of providing information of immediate value to the engineer.

It matters little whether the resistance equation for uniform flow is written in the Darcy-Weisbach form

$$h_f = f \frac{L}{4R} \frac{V^2}{2g} \tag{1}$$

or in the Chezy form

$$V = C \sqrt{RS} \tag{2}$$

for the resistance coefficient f and the discharge coefficient C are evidently related through the expression

$$\frac{1}{\sqrt{f}} = \frac{C}{\sqrt{8g}} \qquad (3)$$

The factor f (and therefore C as well) has long since been shown to depend only upon the Reynolds number of the flow and the geometry of the conduit boundaries. In the case of uniform flow through a conduit of circular cross section, the relative roughness of the boundary is the only geometrical parameter involved. If such roughness is very slight—as is true of glass or drawn metal tubing—the Reynolds number $\mathbf{R} = VD/\nu$ is the sole resistance parameter, and above the critical limit $\mathbf{R} \approx 2000$ the coefficients f and C will vary according to the relationship

$$\frac{1}{\sqrt{f}} = -0.8 + 2 \log \mathbf{R}\sqrt{f} = \frac{C}{\sqrt{8g}} \qquad (4)$$

This (see Fig. 1) is the von Kármán resistance function for smooth pipes [1], the constants having been determined by Nikuradse

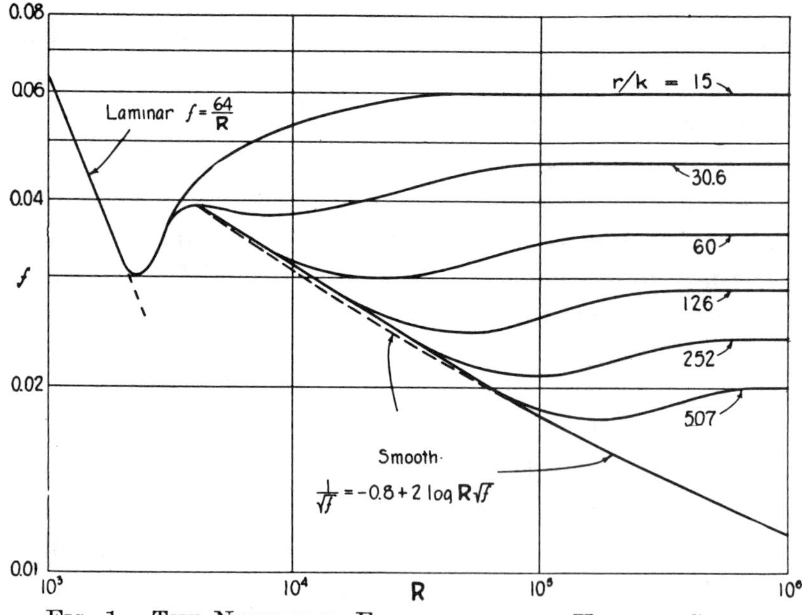

Fig. 1.—The Nikuradse Experiments on Uniform-Sand Roughness.

[2] and closely checked by other experimenters [3]. If the roughness is relatively great, on the other hand, viscous effects in the

boundary zone will be negligible at high Reynolds numbers, and the coefficients f and C will depend only upon the relative roughness—i.e., the ratio of a linear measure k of the boundary irregularities to the radius r of the pipe:

$$\frac{1}{\sqrt{f}} = 1.74 + 2 \log \frac{r}{k} = \frac{C}{\sqrt{8g}} \tag{5}$$

This is the von Kármán resistance function for rough pipes [1], the constants having been determined by Nikuradse [4] by means of pipes artificially roughened through coatings of uniform sand grains of diameter k. Since the Reynolds number is not involved in this equation, it evidently refers to the horizontal portions of the resistance curves shown in Fig. 1.

The validity of Eq. (5) for Nikuradse's artificial roughness was shown by plotting against r/k on semi-logarithmic paper the values of $1/\sqrt{f}$ for the horizontal portions of the curves of Fig. 1, the six points falling along the straight line of the equation as seen

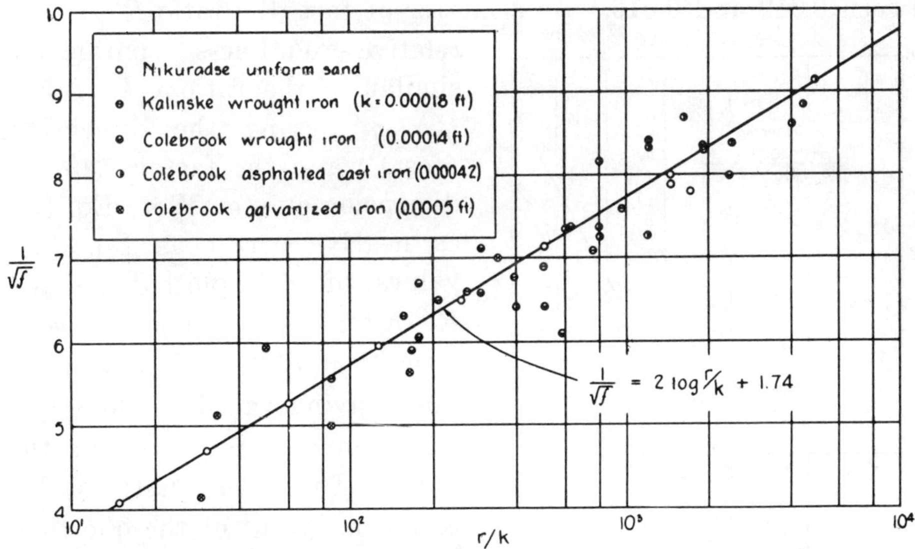

Fig. 2.—Correlation of Tests on Artificial and Natural Roughness.

in Fig. 2. The validity of the same equation for the case of natural roughness was first indicated in this manner by Kalinske [5] through an analysis of Kessler's tests [6] on new wrought-iron pipe of various sizes (see Fig. 2). As further indication of the correctness of this function for commercial roughness of comparable types, the writer has adapted to this form of diagram values obtained by Colebrook [7] in an analysis of measurements by a great number of experimenters on new wrought-iron, galvanized-iron, and asphalted cast-iron pipe. Despite the appreciable scatter of the results shown in Fig. 2 (probably due in large measure to variations in fabrication, experimentation, and evaluation of test results), the points unquestionably follow the trend of the original functional relationship.

If the familiar Manning formula is written in the Chezy form, it will be seen that

$$\frac{1}{\sqrt{f}} = \frac{1.49}{\sqrt{8g}}\frac{R^{1/6}}{n} = \frac{C}{\sqrt{8g}} \tag{6}$$

The factor n is evidently a measure of boundary roughness, since it is invariably tabulated solely as a function of boundary composition; in the case of wood-stave penstocks, for instance, n is normally assumed to have a value between 0.010 and 0.013, and for concrete between 0.012 and 0.016. As a matter of fact, the ratio $R^{1/6}/n$ is a relative-roughness parameter similar to the ratio r/k of Eq. (5), n having the dimension (feet)$^{1/6}$ and the factor 1.49 the dimension of $\sqrt{g}$. Were Eq. (6) as nearly correct as Eq. (5), values of r/k plotted against $R^{1/6}/n$ for the same magnitude of f or C would yield a straight line having a 6:1 slope on logarithmic paper. From the nature of the two equations, this is obviously out of the question, and the deviation of the Manning formula from the actual function may be judged from Fig. 3. Although empirical formulas of the Kutter-Manning-Bazin type may well continue to fill a useful purpose in

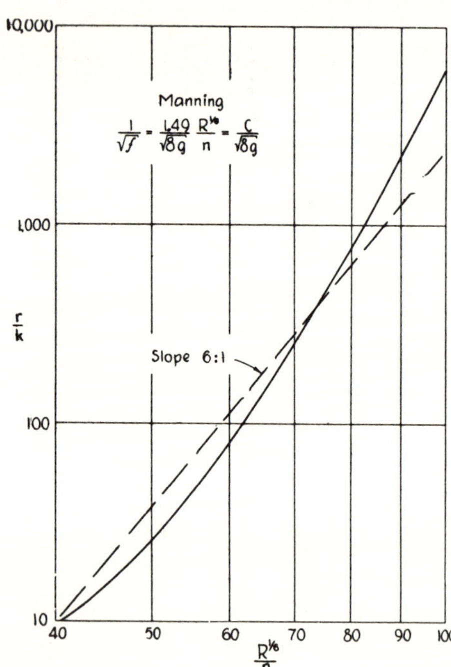

FIG. 3.—RELATIONSHIP OF THE VON KARMAN AND MANNING ROUGH-NESS EQUATIONS.

the field, two facts should now be apparent: Such formulas from their very nature embody the influence of boundary roughness alone, and hence do not apply to that zone of motion in which viscous effects are of appreciable magnitude. And since they are far from exact except over a limited relative-roughness range, they may be expected, when used in connection with research in allied fields, to introduce inherent errors not present in Eq. (5).

In reducing measurements of commercial pipe resistance to the form of Eq. (5), one must evaluate an effective absolute roughness k in terms of Nikuradse's sand-grain diameter, extrapolating when necessary to the zone in which the function becomes independent of $\mathbf{R}$. This has often been considered impracticable because of apparent differences in functional form in the intermediate zone. In the case of the uniform-sand roughness, to be sure, each of the family of curves shown in Fig. 1 is essentially similar to the others, beginning to deviate from the smooth-pipe curve as the nominal thickness δ of the laminar boundary film ($\delta = 65.6 \, r/\mathbf{R}\sqrt{f}$) approaches the order of magnitude of the boundary roughness k. Evidently, the boundary irregularities contribute in no way to the resistance so long as they are fully contained within the boundary zone of laminar motion. This is clearly seen by plotting the quantity $1/\sqrt{f} - 2 \log r/k$ against the ratio of $\mathbf{R}\sqrt{f}$ to r/k for all curves of Fig. 1, as shown in Fig. 4; the horizontal portions of these

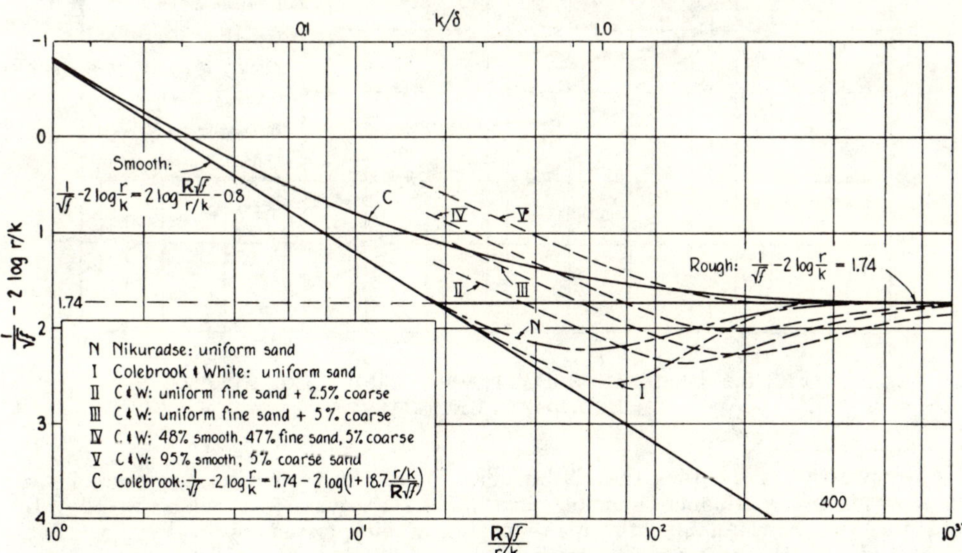

FIG. 4.—Transition Functions for Different Roughness Patterns.

curves are superposed at the ordinate 1.74, from which limit a single curve extends to the sloping line representing the following adaptation of Eq. (4), the smooth-pipe function :[1]

$$\frac{1}{\sqrt{f}} - 2 \log \frac{r}{k} = -0.8 + 2 \log \frac{R\sqrt{f}}{r/k} \qquad (7)$$

On the other hand, experiments by Colebrook and White [8] on artificial roughness of non-uniform character yield a different transition curve for each roughness pattern, as shown by curves I—V on the same plot. It is only reasonable, of course, to expect the larger boundary irregularities to disturb the laminar motion at the wall long before the smaller ones become effective, leading to the conclusion that each statistical combination of surface protuberances should produce its own characteristic transition function. Nevertheless, it was shown by Colebrook [7] that all three types of commercial pipe listed in Fig. 2 are characterized by essentially the same transition curve, for which he obtained the relationship

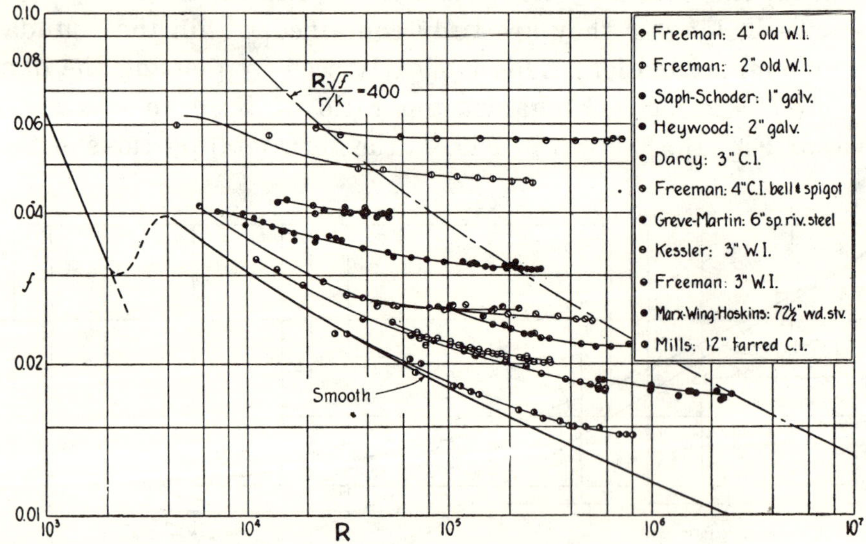

FIG. 5.—RESISTANCE DATA FOR VARIOUS COMMERCIAL BOUNDARY MATERIALS.

---

[1] Noteworthy is the fact that Nikuradse found the transition curves to have as asymptote the Blasius equation $f = 0.3164/R^{1/4}$ when plotted as in Fig. 1, and the von Kármán smooth-pipe equation when plotted as in Fig. 4. The magnitude of the discrepancy may be judged from the deviation of the broken line from the full line in Fig. 1.

$$\frac{1}{\sqrt{f}} = 1.74 - 2 \log \left(\frac{1}{r/k} + \frac{18.7}{\mathbf{R}\sqrt{f}} \right) \tag{8}$$

This expression will be found to approach Eqs. (4) and (5) as limits as r/k and $\mathbf{R}\sqrt{f}$, respectively, become very large; it is plotted in Fig. 4 in the alternative form

$$\frac{1}{\sqrt{f}} - 2 \log \frac{r}{k} = 1.74 - 2 \log \left(1 + 18.7 \, \frac{r/k}{\mathbf{R}\sqrt{f}} \right) \tag{9}$$

These equations are obviously too complex to be of practical use. On the other hand, if the function which they embody is even approximately valid for commercial surfaces in general, such extremely important information could be made readily available in diagrams or tables.

Contrary to Colebrook's conclusion, O'Brien, Folsom, and Jonassen [9]—far from finding the same transition curve for different boundary materials—asserted that they could not obtain a represenative curve for even the same material in different pipe sizes. In the effort to show that such variation is beyond practical significance, however probable its existence may be, the writer has plotted in Fig. 5 a series of measurements by various authorities on various types of commercial conduit [3, 6, 10, 11, 12, 13, 14, 15], the data having been selected on the one hand for range

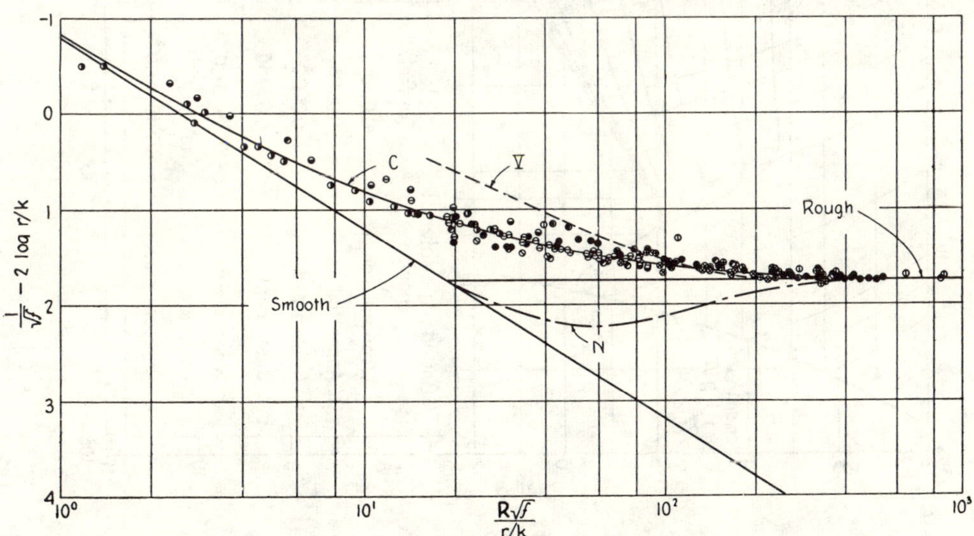

Fig. 6.—Transition Function for Data of Fig. 5.

of Reynolds number and on the other for range of f. On this familiar f:R type of diagram, the several curves show to some extent the variations in form so characteristic of the maze of

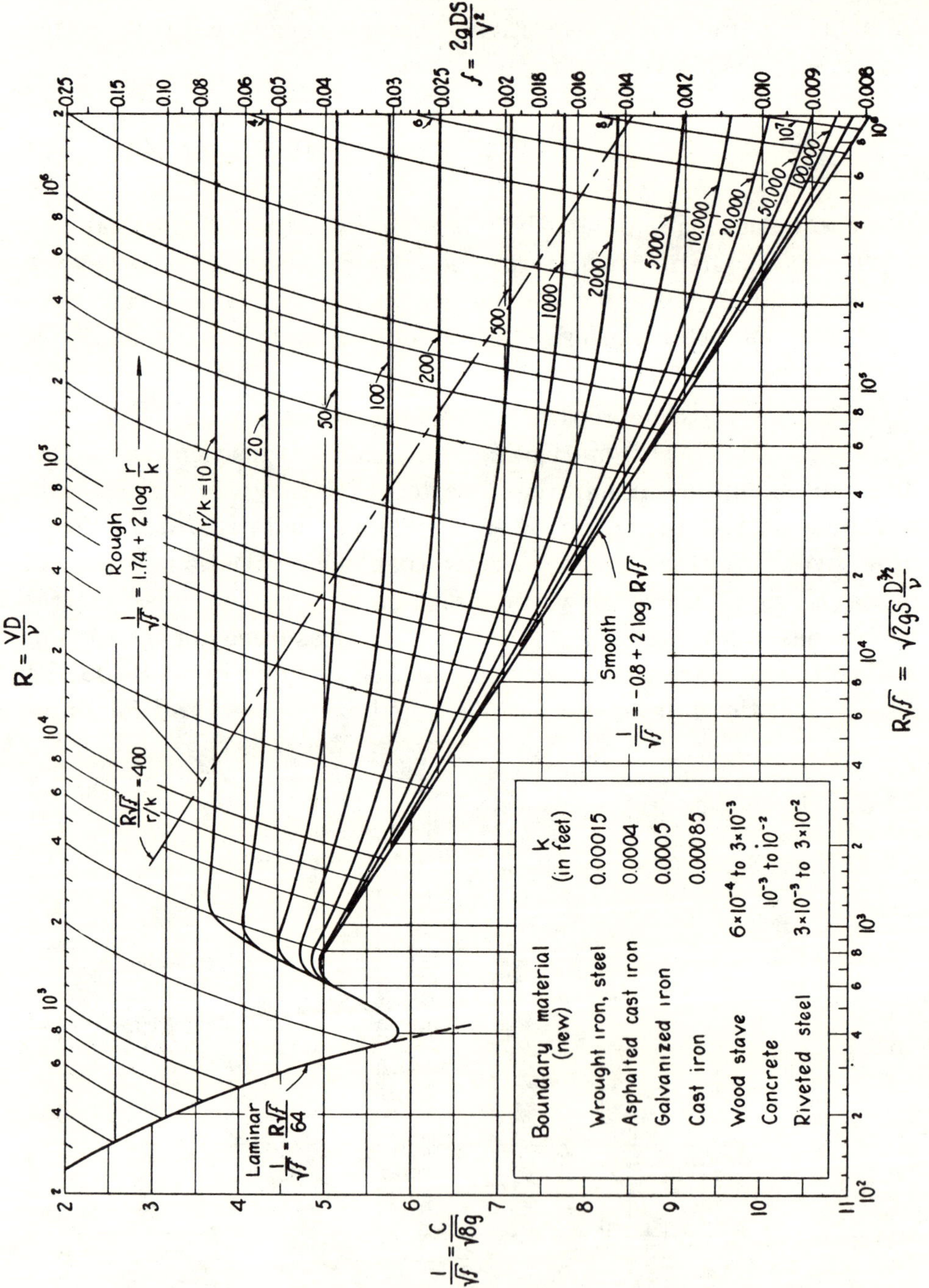

FIG. 7.—PROPOSED RESISTANCE DIAGRAM FOR COMMERCIAL PIPE.

experimental values often published in the past. However, it is at once evident that all curves tend to become more and more nearly horizontal as they approach the line $\mathbf{R}\sqrt{f}\ k/r = 400$, which corresponds to the value on the abscissa scale of Fig. 4 at which the rough-pipe law may be considered to begin. Indeed, when the several series of points are reduced to the form of Fig. 4 through extrapolation according to Eq. (8), their deviation from the plotted curve of Colebrook is evidently not much greater than the experimental scatter of the individual measurements in any one series, as may be seen from Fig. 6. Only in the case of the spiral-riveted pipe, which is of a basically different nature from the other boundary materials, is a somewhat different transition (such as Curve V) apparent, yet even here the departure from the general trend is not serious.

In the light of such results, the writer recommends the use of Eq. (8) as a close approximation to the actual resistance law. And in the light of the roles played therein by the parameters $1/\sqrt{f}$ and $\mathbf{R}\sqrt{f}$ of von Kármán's original analysis, the writer further recommends the use of these parameters as graph-paper co-ordinates in place of the customary f and $\mathbf{R}$. A diagram of this nature is shown in Fig. 7, in which the latter scales have been included for convenience, the basic vertical and horizontal scales nevertheless determining the form of the system of transition curves. These, it will be noted, are all geometrically similar—an obvious advantage over the f :$\mathbf{R}$ type of plot. Moreover, as shown by the alternative forms of the ordinate and abscissa parameters, values on the left-hand scale are directly proportional to (in American units about 1/16 of) the Chezy C, while in the bottom scale the velocity of flow does not appear. Such a diagram therefore permits the evaluation of either the velocity of flow or the hydraulic gradient in either the Darcy-Weisbach or the Chezy equation. Involving, if not the actual transition function, at least a close approximation thereto, this diagram may hence be used for design purposes with greater confidence than a purely empirical system of curves (such, for instance, as that formerly proposed by Pigott [16]). In addition, its efficacy in determining the magnitude of k from experimental data on new types of pipe will be apparent from the fact that it is no longer necessary either to extrapolate or to reduce the data to the forms of Figs. 2 and 6—but simply to plot the points on Fig. 7 and read directly the magnitude of r/k.

The engineer, in considering the general applicability of such a diagram, will probably ask at least one of the following questions: What value of k should be used for this or that specific surface? Does the diagram apply only to continuous sections of conduit, or is the effect of joints included? Is the transition function valid for pipes which have long been in service? And what application can the diagram have to conduits of other than circular cross section— in particular to open channels? Although full and precise information is by no means yet at hand, qualitative indications safely permit the following encouraging answers to be given at this time.

As may be noted from the table on Fig. 7, values of k (except for steel, approximately after Colebrook [7]) have been determined for five common types of pipe surface in new condition. Listed thereafter are three boundary materials which are likewise in common use, but which, unlike the others, vary considerably in absolute roughness with method of fabrication. More precise tabulation of the corresponding values of k is not consistent with existing experimental data [17]. However, lest the engineer, accustomed to the relatively small variation in the Kutter-Manning n for such surfaces, regard the tabulated ranges of k as exorbitant, the fact must be emphasized that a three-fold change in n (say from 0.01 to 0.03) corresponds roughly to a thousand-fold change in k. In other words, appreciable inaccuracies in the evaluation of k will not seriously affect either f or C.

So far as the influence of joints is concerned, it must be noted that connections of various types and frequencies are represented in the data of Figs. 2, 5, and 6, without marked variation in the form of the transition function. Both spacing and form of coupling may conceivably alter the effective magnitude of k to an appreciable degree, however, and systematic experimental study of this phase of the resistance problem is highly desirable.

Aging of the pipes, on the other hand, generally involves a continuous increase of the effective relative roughness k/r, corresponding to a continuous upward progression across the family of curves of Fig. 7 at a rate depending upon the conduit material and the fluid transmitted (it should be noted that such a trend will follow a vertical course if the hydraulic gradient remains constant, but will follow the Reynolds-number curves in the case of constant discharge). For evaluating the effect of age upon pipe capacity, Colebrook and White [18] have proposed an exponential expression which, in terms of boundary roughness, reduces to the following simple form [7]:

$$k = k_0 + \alpha t \tag{10}$$

The term k_0 represents the initial roughness of the new conduit, k the roughness to be expected after the time interval t, and α the rate of roughness growth. Determination of k_0 and α evidently involves at least two successive series of resistance measurements, their evaluation proceeding most conveniently according to the method herein recommended.

For reasons not at once obvious, the foregoing pages of this paper have dealt entirely with flow through uniform conduits of circular cross section. As shown by Keulegan [19], nevertheless, the von Kármán relationships for smooth and rough boundaries are fully as applicable to both closed and open conduits of other cross-sectional forms, provided only that the proper adjustment for form effect is made in the numerical coefficients. Such adjustment is, to be sure, never considered in using a formula of the Manning type, and is actually very small, unless the section departs considerably from the circular. The diagram of Fig. 7 should therefore yield as a first approximation the resistance characteristics of channels of moderate width-depth ratios, and should in any event indicate the general form of the resistance law for all conduits of uniform section.*

REFERENCES

[1] Kármán, Th. von, ''Mechanische Aehnlichkeit und Turbulenz,'' *Proceedings, Third International Congress for Applied Mechanics,* Stockholm, 1930; see also ''Turbulence and Skin Friction,'' *Journal of the Aeronautical Sciences,* Vol. 1, No. 1, 1934; for critical reviews of the von Kármán analysis see Miller, B., ''Fluid Flow in Clean Round Straight Pipe,'' *Transactions, American Institute of Chemical Engineers,* Vol. 33, 1937, and Mises, R. von, ''Some Remarks on the Laws of Turbulent Motion in Tubes,'' *Theodore von Kármán Anniversary Volume,* California Institute of Technology, 1941.

[2] Nikuradse, J., ''Gesetzmässigkeiten der turbulenten Strömung in glatten Rohren,'' *VDI Forschungsheft* 356, 1932.

---

*The reader is referred to the next paper in this volume for further discussion of this topic.

[3] Freeman, J. R., ''Experiments upon the Flow of Water in Pipes and Pipe Fittings,'' *American Society of Mechanical Engineers*, 1941.

[4] Nikuradse, J., ''Strömungsgesetze in rauhen Rohren,'' *VDI Forschungsheft* 361, 1933.

[5] Kalinske, A. A., ''A New Method of Presenting Data on Fluid Flow in Pipes,'' *Civil Engineering*, Vol. 9, No. 5, 1939.

[6] Kessler, L. H., ''Experimental Investigation of Friction Losses in Wrought Iron Pipe When Installed with Couplings,'' *University of Wisconsin Engineering Experiment Station Bulletin* 82, 1935.

[7] Colebrook, C. F., ''Turbulent Flow in Pipes, with Particular Reference to the Transition Region between the Smooth and Rough Pipe Laws,'' *Journal, Institution of Civil Engineers, February,* 1939.

[8] Colebrook, C. F., and White, C. M., ''Experiments with Fluid-Friction in Roughened Pipes,'' *Proceedings of the Royal Society*, Vol. 161, 1937.

[9] O'Brien, M. P., Folsom, R. G., and Jonassen, F., ''Fluid Resistance in Pipes,'' *Industrial and Engineering Chemistry*, Vol. 31, April, 1939.

[10] Darcy, H., *Recherches experimentales*, Paris, 1857 (see reference 14).

[11] Greve, F. W., and Martin, R. R., ''Flow of Water in Spiral Riveted Steel Pipe,'' *Purdue Engineering Experiment Station Bulletin* 8, 1921.

[12] Heywood, F., ''The Flow of Water in Pipes and Channels,'' *Minutes of Proceedings, Institution of Civil Engineers*, Vol. 219, 1925.

[13] Marx, C. D., Wing, C. B., and Hoskins, L. M., ''Experiments on the Flow of Water in the Six-foot Steel and Wood Pipe Line of the Pioneer Electric Power Company at Ogden, Utah,'' *Transactions, American Society of Civil Engineers*, Vol. 40, 1898.

[14] Mills, H. F., ''Flow of Water in Pipes,'' *Memoirs, American Academy of Arts and Sciences*, Vol. 15, No. 11, 1924.

[15] Saph, A. V., and Schoder, E. W., ''An Experimental Study of the Resistance to the Flow of Water in Pipes,'' *Transactions, American Society of Civil Engineers*, Vol. 51, 1903.

[16] Pigott, R. J. S., ''The Flow of Fluids in Closed Conduits,'' *Mechanical Engineering*, August, 1933.

[17] Hotes, F. L., ''Correlation of Experimental Data and Rational Equations on Boundary Roughness and Resistance,'' *Master's Thesis, Department of Mechanics and Hydraulics*, State University of Iowa, 1941.

[18] Colebrook, C. F., and White, C. M., ''The Reduction of Carrying Capacity of Pipes with Age,'' *Journal, Institution of Civil Engineers*, 1937.

[19] Keulegan, G. H., ''Laws of Turbulent Flow in Open Channels,'' *Journal of Research, National Bureau of Standards*, Vol. 21, 1938.

Friction Factors for Pipe Flow

By LEWIS F. MOODY, PRINCETON, N. J.

Discussion

HUNTER ROUSE.[15] Important results of laboratory research frequently do not reach the hands of practicing engineers until many years after the original papers have been published. A salient case in point is the discovery by Blasius in 1913 of the dependence of the Darcy-Weisbach resistance coefficient f upon the Reynolds number $\mathbf{R}$, which did not come into general engineering use until perhaps a decade ago. It often happens, however, that once engineers have accepted a new idea they are loath to modify it in any way. The paper under discussion is a very commendable endeavor to make recent experimental findings immediately useful to the engineer, but the writer feels that it still caters to a regrettable degree to the engineer's innate conservatism.

If the writer's belief is correct, this paper is intended to fulfill the same purpose as that which prompted the writer to present a somewhat similar paper (12) and resistance chart at the Second Hydraulics Conference in 1942. The author claims that in this chart, which is reproduced herewith in slightly modified form,[16] the writer adopted co-ordinates inconvenient for ordinary engineering use. Such criticism resulted from the writer's deliberate advancement beyond the now familiar Blasius f-versus-$\mathbf{R}$ notation in the belief that both greater convenience and greater significance would be attained thereby. Since these two papers of identical purpose thus differ in their basic method of approach, a criticism of the one point of view must necessarily involve a defense of the other.

Although Blasius' original dimensional analysis of the variables involved led to his adoption of the form VD/ν as the most significant grouping of terms upon which f should depend, it must be realized that the following three different combinations of the same variables are all equally valid for the basic case of a smooth pipe:[17]

$$f = \frac{2gh_f D}{LV^2} = \varphi_1 \left(\frac{VD}{\nu}\right) = \varphi_1(\mathbf{R})$$

$$f = \frac{2gh_f D}{LV^2} = \varphi_2 \left(\frac{2gh_f D^3}{L\nu^2}\right) = \varphi_2'(\mathbf{R}\sqrt{f})$$

$$f = \frac{2gh_f D}{LV^2} = \varphi_3 \left(\frac{2gh_f Q^3}{L\nu^5}\right) = \varphi_3'(\mathbf{R}f^{1/5})$$

The combination now most familiar to the engineer, of course, is the first, although it has long since been proved that it will yield a linear plot on logarithmic paper for only the laminar zone. The second, on the other hand, is the basis of the Kármán-Prandtl analysis of the turbulent zone, the general functional relationship simply being written in the specific form

$$1/\sqrt{f} = A + B \log (\mathbf{R}\sqrt{f})$$

Despite the author's indication to the contrary, f is not inextricably embodied in the second term of this relationship, as may be seen from the identity $\mathbf{R}\sqrt{f} = \sqrt{(2gh_f D^3/L\nu^2)}$. If

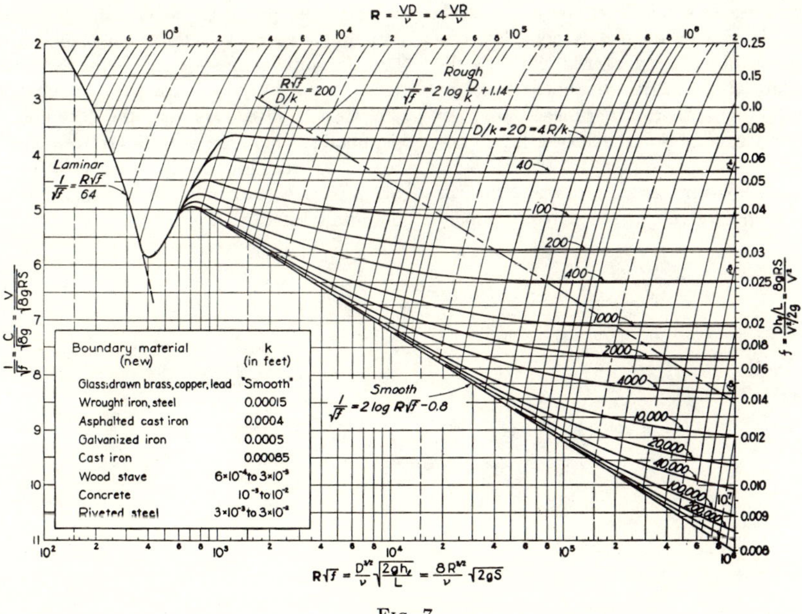

FIG. 7

[15] Director, Iowa Institute of Hydraulic Research, University of Iowa, Iowa City, Iowa.

[16] "Elementary Mechanics of Fluids," by Hunter Rouse; John Wiley & Sons, Inc., New York, N. Y. (in Press).

[17] "Solving Pipe Flow Problem With Dimensionless Numbers," by A. A. Kalinske, Engineering News-Record, vol. 123, 1939, p. 23.

the Kármán-Prandtl parameters are chosen as the basis of a semilogarithmic chart, as in the accompanying figure, not only will the smooth-pipe relationship plot as a straight line, but all transition curves from the smooth to the rough relationship will be geometrically similar. It would appear to the writer that this combines ease in interpolation (and hence convenience) with greater significance than the Blasius plot will permit. This, therefore, is one of the writer's two reasons for continuing to recommend the newer type of chart in preference to that of the author.

The writer's second reason will be evident after further inspection of the foregoing functional relationships. The first relationship will be directly useful in graph form only if the velocity or rate of flow is known; otherwise the desired coefficient may be evaluated from the graph only through the inconvenient process of trial and error. If the velocity or rate of flow is not known, on the other hand, a graph of the second functional relationship will make the desired coefficient immediately available. In order to provide a single chart which would satisfy both sets of requirements, the writer supplied ordinate scales of both f and $1\sqrt{f}$ (the latter being proportional to the Chezy C) and abscissa scales of both $\mathbf{R} = VD/\nu$ and $\mathbf{R}\sqrt{f} = \sqrt{2g\,h_f/L}\ D^{3/2}\nu$. Since the parameters $1/\sqrt{f}$ and log $(\mathbf{R}\sqrt{f})$ were selected by the writer for the primary ordinate and abscissa scales, the alternative abscissa parameter log $\mathbf{R}$ is necessarily represented by curved lines over a portion of the writer's chart. Had log f and log $\mathbf{R}$ been chosen as the primary parameters, log $(\mathbf{R}\sqrt{f})$ would still have required sloping lines in the grid; such choice therefore involves no particular advantage over the writer's but rather defeats the writer's purpose owing to the accompanying distortion of the entire system of transition curves. The author's graph, of course, contains no secondary grid system simply because it permits direct solution for only one of the several variables.

Brief mention might be made of the third possible combination of variables, which is evidently applicable to problems in which the diameter is the unknown quantity. So long as the pipe is smooth, such a plot will be of use, but the adoption of a similar function for the case of rough surfaces will still require a trial-and-error solution, unless the graph is made hopelessly complex, owing to the fact that for a given boundary material the pipe diameter must be known before the relative roughness may be evaluated. Solution by trial might therefore proceed just as well from either of the other two functional relationships contained in the writer's diagram.

The writer commends the author's presentation in graph form of the values of absolute roughness given in the writer's paper, but notes with interest that this plot is consistent with the writer's rather than the author's choice of basic parameters. Such a graph would therefore have its greatest value when prepared as a marginal extension of the writer's resistance chart, for then no relative-roughness computations would have to be made.

So far as the author's discussion of open-channel resistance is concerned, the writer takes exception to two points of fundamental importance: First, the author states that such relationships as the Manning formula should be used in open-channel computations in preference to values derived from pipe tests, implying that the familiar empirical open-channel formulas are inherently more valid. It is known, however, that the Manning formula (not to mention those of Bazin and Kutter) is not in accordance with the logarithmic law of relative roughness upon which the author's paper is based. So far as the writer can ascertain, the only reason pipe tests could not generally be used in evaluating open-channel resistance lies in the fact that few open-channel boundary surfaces are suitable to testing in pipe form. Aside from the moot question of the effect of cross-sectional shape (which the empirical open-channel formulas in no way answer), it would appear to the writer that a general resistance graph for uniform open channels should differ little from that for pipes, except in that the familiar parameters C and S might conveniently be included in the co-ordinate scales; this has been done in the present form of the writer's chart.

The writer's second objection to the author's closing section is in regard to his implication that the Froude number should replace the Reynolds number as the fundamental resistance parameter for open-channel flow. So far as boundary resistance is concerned, the writer can see no possibility of the Froude number playing a comparable role. It is true that viscous shear is of little significance in comparison with boundary roughness in most open-channel problems, but it is also true that the effect of surface waves upon the internal resistance to flow has not yet been ascertained. The open-channel problem is, in fact, quite analogous to that of ship resistance, in which the matter of surface drag is considered wholly independent of the Froude number. If, to be sure, the phenomena of slug flow, atmospheric drag, and air entrainment prove to govern the resistance in the comparatively infrequent case of supercritical flow in open channels, then the Froude number may well become an appropriate resistance criterion, as it already is for cases of channel transition. But to imply that it should replace the Reynolds number as a resistance parameter whenever a free surface exists seems rather untimely to the writer, in that it could lead to serious misinterpretation of those few principles of boundary resistance which have been definitely established.

A General Stability Index for Flow Near Plane Boundaries

HUNTER ROUSE*

University of Iowa

INTRODUCTION

SINCE THE TIME of Osborne Reynolds, the stability problem in fluid mechanics has consistently withstood every effort toward a general solution. The mathematical method of superposing a periodic disturbance upon a parallel flow has, to be sure, yielded qualitative results in specific cases, and experimental studies have supplied quantitative information in terms of characteristic Reynolds Numbers wherever required. However, a general criterion for stability is not yet at hand (as evidenced by the wide numerical range of critical Reynolds Numbers even for related types of flow), nor have existing analyses provided a physical picture of the controlling factors which is within the grasp of the average engineer. The stability index proposed in this paper does not have the justification of rigorous mathematical derivation, nor has it evolved from extensive laboratory study. Yet it does stem from logical and almost self-evident dimensional and physical concepts, and its general significance is indicated by experimental data already at hand for several basically different types of fluid motion.

FORMULATION OF THE STABILITY INDEX

A popular belief among engineers is that turbulence will develop at any point in laminar flow at which the velocity gradient becomes sufficiently high—this despite the fact that the gradient is invariably highest at the boundary itself and becomes still higher in the stable laminar sublayer between the boundary and the turbulent zone.

Quite inconsistently, however, disturbances in high-velocity zones far from the boundary are often blamed for the onset of turbulence, even though the velocity gradient is lowest in that region. Although the velocity gradient is assuredly a factor that tends toward instability fully as much as the viscosity of the fluid tends toward stability, the far greater effect of the boundary proximity is seldom given credit. As a matter of fact, unless the normal distance y from the boundary is considered as essential a variable as the velocity gradient dv/dy and the kinematic viscosity ν, the formulation of a general criterion of stability will be out of the question. If, however, these three variables are assumed to be the essential ones involved in the basic

Received January 18, 1945.

* Director, Iowa Institute of Hydraulic Research.

stability problem they may be related in the general functional form

$$\varphi(y,\ dv/dy,\ \nu) = 0$$

for which only one dimensionless combination is obtainable:

$$\varphi'[y^2(dv/dy)/\nu] = 0$$

Since dv/dy is invariably a maximum along a plane boundary when y is zero and since, with increasing values of y, the gradient decreases more and more rapidly, in any case of laminar motion the product $y^2(dv/dy)$—and hence its ratio to ν—should attain a maximum value at some point within the flow. From the structure of the dimensionless ratio of the three variables, it is evident that this optimum point represents the zone of minimum stability for the flow in question. Since, however, only one dimensionless parameter is involved in the functional relationship, it must necessarily be a constant for the conditions assumed. In other words, if the parameter is designated as the stability index χ, the limit of stability should correspond to a constant critical magnitude of this index:

$$\chi_{cr} = [y^2(dv/dy)/\nu]_{cr} = \text{constant}$$

Evidently, if the maximum value of χ for a given state of flow is less than this critical magnitude, then the flow should be inherently stable to all disturbances. If, on the contrary, the maximum value of χ exceeds this critical magnitude, then at the corresponding point within the flow the onset of turbulence should be expected once the necessary disturbance occurs.

FLOW ALONG A FLAT PLATE

Consider, for instance, flow in a laminar boundary layer along a flat plate, the well-known velocity distribution of which is plotted dimensionlessly in Fig. 1a.[1] In Fig. 1b is shown the corresponding plot of χ in its ratio to the square root of a Reynolds Number based upon the velocity v_0 at a point far removed from the plate and the distance x from the leading edge. The stability index evidently varies from zero at the boundary to a maximum value of $1.44\sqrt{v_0 x/\nu}$ at the boundary distance $y = 3.1\sqrt{x\nu/v_0}$, beyond which it again approaches zero. Then this point of maximum χ is the location at which the onset of turbulence is to be expected, provided that the absolute magnitude of the

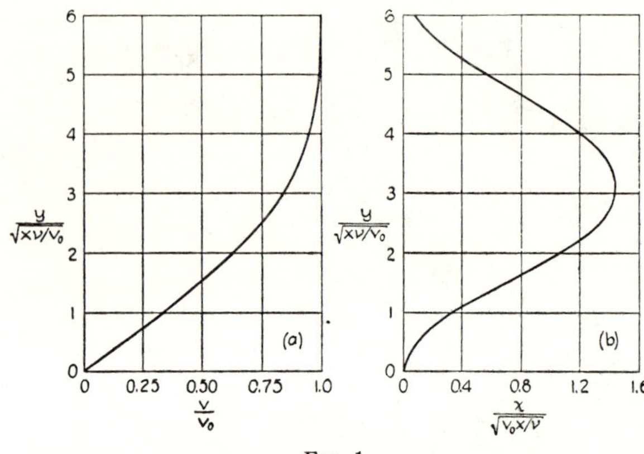

FIG. 1.

index exceeds its critical value and the necessary disturbance occurs. If one takes the Reynolds Number $u_0x/\nu = 1.5 \times 10^5$ as the lowest probable limit for boundary-layer turbulence,[2] the critical value of the index will be found to be approximately $\chi_{cr} = 560$.

FLOW THROUGH PIPES

A related, although considerably different, type of flow is that through a long pipe of uniform diameter, which is generally considered to be stable to all disturbances if a Reynolds Number based upon the average velocity V and the pipe radius r_0 is lower than 1,000. Since, according to the Poiseuille relationship,[3] the velocity gradient may be expressed in terms of the intensity of boundary shear τ_0, the dynamic viscosity μ, and the relative distance from the boundary,

$$dv/dy = (2\tau_0/\mu r_0)(r_0 - y)$$

and since, therefore,

$$\tau_0 = 2\mu V/r_0$$

it will be found that

$$\chi = \frac{y^2(dv/dy)}{\nu} = \frac{4Vr_0}{\nu}\left(\frac{r_0y^2 - y^3}{r_0^3}\right)$$

Placing the derivative of χ with respect to y equal to zero,

$$\frac{d\chi}{dy} = \frac{4Vr_0}{\nu}\frac{2r_0y - 3y^2}{r_0^3} = 0$$

from which it follows that $\chi = \chi_{max.}$ when $y = 2r_0/3$. This, then, should be the zone of initial instability in a long pipe, provided that the index exceeds its critical value and that a disturbance occurs. If, furthermore, the χ criterion is as general as assumed, its critical value for the pipe should be numerically equal to that for the boundary layer along a plate. From the foregoing development it will be found that

$$\chi_{max.} = (16/27)(Vr_0/\nu)$$

whence, for the value $Vr_0/\nu = 1,000$, $\chi_{cr} = 590$.

FLOW BETWEEN PARALLEL BOUNDARIES

Although little experimental information is at hand for the critical limit of laminar flow between stationary parallel boundaries or of laminar free-surface flow down a sloping boundary, such flow should obviously be subject to the same general stability criterion. Since the velocity distribution for laminar flow in either instance follows a parabolic curve similar to that of the pipe, the point of maximum χ should again be located two-thirds of the distance y_0 from the boundary to the midsection between the boundaries or to the equivalent free surface. If, as is customary, the critical limit of either type of flow is estimated from tests on pipes in terms of the so-called hydraulic radius (i.e., the ratio of the cross-sectional area of flow to the "wetted" boundary perimeter), the Reynolds Number based on the mean velocity V and the distance y_0 is found to have a magnitude of 500. However, experimental evidence for open-channel flow,[4] as well as the location of the intersection of the dimensionless resistance curves for the corresponding laminar and turbulent regimes, indicates that this value should be at least as high as 800 to be fully comparable. Under such circumstances

$$\chi_{max.} = (4/9)(Vy_0/\nu) \text{ and } \chi_{cr} = 380$$

As a final test, flow that is produced by the motion of one plane boundary past another should be considered. Experimental data are available only for the case of rotating cylinders, but it appears from extrapolation[5] that cylinders of infinite radius would yield, for stability, the condition

$$\Delta v \,\Delta r/\nu \approx 2,000$$

While the velocity gradient is constant between the inner and outer boundaries, variation in y would make χ a maximum midway between them, at which point

$$\chi_{max.} = \left(\frac{\Delta r}{2}\right)^2\left(\frac{\Delta v}{\Delta r \nu}\right) = \left(\frac{1}{4}\right)\left(\frac{\Delta v \,\Delta r}{\nu}\right)$$

Under such circumstances the value of $\chi_{max.}$ at which instability should exist is found to be $\chi_{cr} = 500$.

GENERAL SIGNIFICANCE OF THE STABILITY INDEX

Although four somewhat different values (560, 590, 380, and 500) have been obtained herein for an index that was asserted to be a constant, it should be noted that three of these values have been computed from commonly accepted critical limits without any attempt at readjustment and that the lowest is based upon a limit that itself is not yet fully established. In every case $\chi_{max.}$ is directly proportional to the Reynolds Number or is, in fact, a special form of the Reynolds Number, just as the Richardson Number is a special form of the Froude Number. Nevertheless, only by arbitrary manipulation of the length term in the Reynolds Number would it be possible to obtain comparable critical values in the several cases, while the χ criterion at once yields values that depart so little from the average $\chi_{cr} \approx 500$ that it would appear to have both physical and numerical significance.

Essentially different from the foregoing case of transformation from laminar to turbulent flow is the question of the limit of stability of the laminar sublayer—the zone of demarcation between laminar and turbulent motion occurring simultaneously in neighboring regions. The latter problem nevertheless involves the same three variables comprised in the χ parameter, which may readily be evaluated from existing measurements on smooth pipes.[6] According to these measurements, the nominal thickness δ of the laminar sublayer may be expressed as

$$\delta = 11.6\nu/\sqrt{\tau_0/\rho}$$

in which τ_0 is the intensity of boundary shear and ρ the fluid density. Since, in obtaining this value, it is necessarily assumed that

$$(dv/dy)_{y=\delta} = v_\delta/\delta = \tau_0/\mu$$

it follows that

$$\chi_{cr} = \frac{\delta^2 \tau_0}{\nu\mu} = (11.6)^2 = 134$$

Because of the essential difference between these two phases of the stability problem, it is hardly to be expected that the respective values of χ_{cr} should be numerically equal. That they are nonetheless of the same order of magnitude is therefore probably of greater interest than significance. As a matter of fact, the constancy of χ_{cr} indicated by the initial dimensionless relationship is wholly dependent upon the similarity of flow conditions tacitly assumed in the analysis. Whether or not the slight variation in χ_{cr} for the four cases previously discussed is also due to inherent differences in the flow conditions must remain subject to further investigation. On the other hand, if such additional variables as boundary curvature or density gradient were involved in the problem, the same stability index would then obviously assume the role of a dependent variable, its magnitude for critical conditions becoming a function of the boundary geometry or of either the Froude Number or the Richardson Number.

REFERENCES

[1] Blasius, H., *Grenzschichten in Flüssigkeiten mit kleiner Reibung*, Z. Math. und Physik, Vol. 56, 1908.

[2] Kármán, Th. von, *Skin Friction and Turbulence*, Journal of the Aeronautical Sciences, Vol. 1, No. 1, 1934.

[3] Lamb, H., *Hydrodynamics*, p. 585, Cambridge University Press, London, 1932.

[4] Robertson, J. M., and Rouse, H., *On the Four Regimes of Open-Channel Flow*, Civil Engineering, Vol. 11, March, 1941.

[5] Taylor, G. I., *Fluid Friction Between Rotating Cylinders*, Proc. Roy. Soc., Vol. A157, Nos. 892–893, 1936.

[6] Nikuradse, J., *Gesetzmässigkeiten der turbulenten Strömung in glatten Rohren*, Forschungsheft, Vol. 3, No. 356, September-October, 1932.

Excerpt from the discussion following

THE MECHANISM OF ENERGY LOSS IN FLUID FRICTION

BY BORIS A. BAKHMETEFF AND

WILLIAM ALLAN

HUNTER ROUSE,[41] M. AM. SOC. C. E., AND A. A. KALINSKE,[42] Assoc. M. AM. SOC. C. E.—Energy dissipation in turbulent flow is at best an extremely complex phenomenon, and a successful investigation of the mechanics of such a process requires more ability, intuition, and courage on the part of the investigator than perhaps any other phase of fluid motion. It is for this reason that the writers have only commendation for the authors of the present paper in undertaking so broad and fundamental a treatment of the problem. No fault can be found with the mathematical derivations, and portions of the interpretive matter are considered well worthy of reading and rereading by every one interested in fluid mechanics. Section 10, for instance, which contains a detailed description of the turbulence mechanism, is particularly lucid and significant.

Because of the very complexity of fluid turbulence, a paper of this nature can be of value to only a limited number of readers if the treatment is not in itself a simple one, devoid of all ambiguous or superfluous material. The authors presumably sought to clarify the problem for more than such a limited group, through use of analogies, new concepts, and numerous charts; but the

---

[41] Director, Iowa Inst. of Hydr. Research, Univ. of Iowa, Iowa City, Iowa.
[42] Associate Director, Iowa Inst. of Hydr. Research, Univ. of Iowa, Iowa City, Iowa.

writers are inclined to question whether the goal of clarity has actually been achieved.   Indeed, since the essential justification of the new concepts introduced by the authors was not properly noted, the application of these concepts to laminar and turbulent flow alike served to conceal rather than to emphasize the fundamental difference between the two types of motion.

Perhaps the writers' thesis may best be explained by reverting to the authors' use of analogy, to emphasize the fact that analogies are, to say the least, frequently misleading in their oversimplification.   Instead of the pulley system to illustrate the general flow process, one might more logically consider a uniform pile of lumber.   If an equal force is applied to the end of each plank to simulate a constant pressure gradient, movement apparently comparable to that of laminar flow will result in the longitudinal direction.   Thus, assuming a constant coefficient of friction (that is, "viscosity") between planks, the intensity of shear will increase linearly with depth of pile, just as—although for a different reason—the shear increases linearly with distance from the water surface or the center line of a pipe.   The work done by the force acting on the end of the top plank, therefore, will evidently not be equal to the generation of thermal energy through friction with the next plank, since both planks are moving and a portion of the force will be transmitted via shear to the lumber below.   However, to assume that the forces exerted upon the plank ends represent a form of potential energy (just as the pressure within a fluid is often improperly assumed to be potential energy of flow), which is then "transferred" by the shear process from one plank to another until finally dissipated in the form of heat, is hardly in accordance with accepted methods of mechanics.   Moreover, it obviously would be impossible to extend either the plank analogy or the pulley analogy to the case of turbulent motion, since there is no conceivable intermixing process that the planks could undergo to simulate the well-recognized mechanism of turbulent diffusion.   Neither analogy (nor, for that matter, any other) is thus capable of truly elucidating the problem under consideration.

In order to be rigorously acceptable in the mechanics sense, the authors' concept of energy borrowing and transmission must involve an actual mass exchange between neighboring zones of flow.   Such an exchange is the very essence of turbulent motion, to be sure, but it appears at first glance to be no more characteristic of laminar motion than of the plank analogy.   It must be recalled, nevertheless, that the large-scale or molar transfer of fluid matter produced by the mixing process of turbulence has its counterpart in the infinitesimal molecular mixing which takes place in turbulent and laminar flow alike.   As a matter of fact, the kinematic molecular viscosity ν and the kinematic eddy viscosity ϵ are both products of mixing lengths and velocities, differing effectively only in scale.   Just as the mixing process of turbulence is considered to produce a shear between neighboring zones of flow which is proportional to the lateral rate of momentum transfer, the viscous shear of laminar motion may be resolved into the lateral transfer of momentum through mixing on a molecular scale.   Indeed, one frequently compares the rates of molecular and molar diffusion of salinity, or of heat, so that the molecular transfer of kinetic energy is actually just as real a process in laminar flow as the molar transfer is in turbulent flow.

The authors' concepts of energy transmission in both laminar and turbulent flow are therefore fully justifiable, although for reasons not brought to the attention of the reader.   In the light of this justification, however, the essential difference between the two types of flow still must be properly clarified.   The writers seek to accomplish the necessary distinction by assembling certain of the curves presented by the authors, together with additional functions which they have discussed only casually, in the schematic chart of Fig. 31.

In the uppermost diagrams are included the primary characteristics of laminar and turbulent flow in pipes: (1) The pressure gradient, which is constant across both sections; (2) the intensity of shear, which increases linearly with the radius in both cases; and (3) the velocity, which varies parabolically in laminar flow, and approximately logarithmically in turbulent flow.   Neither these nor the remaining curves, it must be emphasized, are drawn to a particular scale, since it is the form of each function, rather than its absolute location, which is essential to the argument.

The diagrams next in order depict the mixing characteristics of the secondary motion—molecular in the laminar case and molar in the turbulent.   Curve $4(c)$ represents the mean molecular velocity u'_m, and $4(d)$ the mean velocity $\sqrt{\overline{(u')^2}_t}$ of the turbulence (to which $\sqrt{\overline{(v')^2}}$ is assumed proportional); curve $4(c)$ is necessarily constant across the section, whereas curve $4(d)$ decreases markedly toward the center line.   Curve $5(c)$ represents the mean free path l_m of the molecules, which is likewise constant across the section; but its molar counterpart curve $5(d)$, the mixing length or mean eddy size l, increases toward the center line.   The remaining curves $6(c)$ and $6(d)$ for the mixing coefficients ν and ϵ are simply the products of the velocity and length characteristics of the respective mixing processes.   Noteworthy is the fact that ν is constant across the section while ϵ reaches a maximum approximately midway between center line and wall.   In either case, nevertheless, the mixing functions are related to the shear and velocity curves of the upper diagrams through the parallel expressions $\tau = \nu\,\rho\,\dfrac{du}{dy}$ and $\bar{\tau} = (\nu + \epsilon)\,\rho\,\dfrac{d\bar{u}}{dy}\cdot$

The third pair of schematic diagrams portray the transfer and the spending of kinetic energy by the mean laminar and turbulent motion.   The local rate of spending, rather than the cumulative emphasized by the authors, is used herein for the same reason that local conditions of velocity and shear rather than their cumulative effects are considered more significant in the present argument.   Likewise, instead of the authors' cumulative transmission function W_t, the writers have chosen the ratio $W_t/2\,\pi\,r$, since it is the latter quantity which is directly proportional to the mixing coefficient and the velocity gradient of the foregoing diagrams.   In other words, the rates of transfer of kinetic energy of the mean motion toward the walls are, in terms of the respective mixing processes:

$$\left(\frac{W_t}{2\,\pi\,r}\right)_{\text{lam}} = \nu\,\frac{d\left(\dfrac{\rho\,u^2}{2}\right)}{dy} \quad\text{and}\quad \left(\frac{W_t}{2\,\pi\,r}\right)_{\text{turb}} = (\nu + \epsilon)\,\frac{d\left(\dfrac{\rho\,\bar{u}^2}{2}\right)}{dy}\cdot$$

Although it is evident from these diagrams that both the transfer and the spending of the energy of the mean motion in turbulent flow are greater in the wall region, owing to the different nature of the mixing process, the true sig-

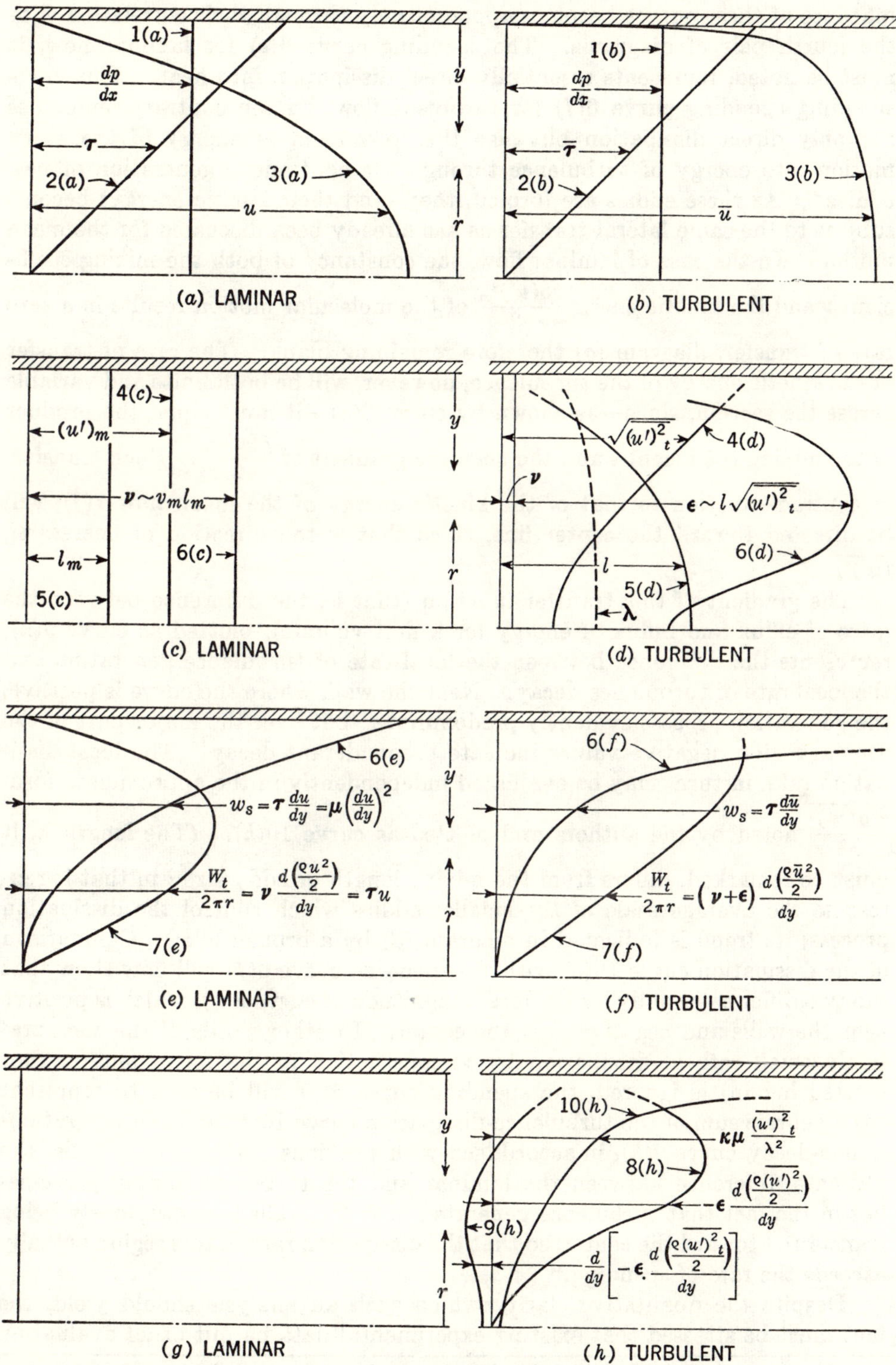

FIG. 31.—COMPARISON OF CHARACTERISTIC FUNCTIONS FOR LAMINAR AND TURBULENT FLOW IN PIPES

nificance of these curves becomes apparent only through their correlation with the fourth pair of diagrams. The spending curve 6(e) for laminar flow, it must be noted, represents essentially direct dissipation into heat. The corresponding spending curve 6(f) for turbulent flow, on the contrary, embodies not only direct dissipation but also the conversion of energy of the mean motion into energy of turbulence through the continuous generation of new eddies.[43] As these eddies are formed, they (and their kinetic energy) become subject to the same lateral transfer as has already been discussed for the mean motion. In the case of laminar flow, the constancy of both the mixing coefficient ν and the kinetic energy $\dfrac{\rho \overline{(u')^2}_m}{2}$ of the molecular motion results in a zero rate of transfer, diagram (g) therefore remaining blank. The rate of transfer of the kinetic energy of the turbulence, however, will be both finite and variable across the section, since—as shown by curve 8(h)—it must equal the product of the mixing coefficient ϵ and the negative gradient of $\dfrac{\rho \overline{(u')^2}_t}{2}$. Such transfer, in contradistinction to that of the kinetic energy of the mean flow 7(f), will be directed toward the center line, since that is the direction of decreasing $\overline{(u')^2}_t$.

The gradient of this transfer function (that is, the difference between the rates of efflux and influx of energy for a unit volume), plotted as curve 9(h), represents the difference between the local rate of turbulence generation and the local rate of turbulence decay. Near the wall, where the curve is positive, the generation process evidently predominates, but over the major part of the central region negative values indicate preponderant decay. The local dissipation rate, in turn, may be evaluated independently in the approximate form $\dfrac{\kappa\,\mu\,\overline{(u')^2}}{\lambda^2}$ noted by the authors and plotted as curve 10(h). (The length λ, it must be remarked, differs from the mixing length or eddy size l in that it represents the average scale of the smaller eddies which control the dissipation process; its trend is indicated in diagram (d) by a broken line.) Comparison of the dissipation curve 10(h) with the spending curve 6(f) will now show that the two differ essentially by the local magnitude of curve 9(h), which is positive near the walls and negative near the center. In other words, if the comparatively small rate of dissipation by viscous action in the mean motion (not plotted herein) is ignored, the spending curve 6(f) will be seen to represent the algebraic sum of the turbulence-dissipation curve 10(h) and the generation-minus-decay curve 9(h), in accordance with previous statements. Thus, the ultimate difference between the laminar and the turbulent mixing processes lies in the fact that turbulence generated near the walls is continuously being transferred toward the center, so that the dissipation rate in this region actually exceeds the rate of spending.[44]

Despite the qualitative clarity which such an analysis should yield, the fact must be stressed that existing experimental data permit exact evaluation

---

[43] "Statistical Theory of Turbulence," by Theodor von Kármán, *Journal of the Aeronautical Sciences,* Vol. 4, No. 4, 1937, p. 135.

[44] "Statistical Theory of Turbulence," by G. I. Taylor, *Proceedings,* Royal Soc. of London, Series A, Vol. CLI, 1935, pp. 462–464.

of only part of the functions shown schematically in the several diagrams. The laminar curves, of course, are accurately known. Completely lacking, however, are systematic measurements of $\overline{(u')^2}_t$, l, λ, and ϵ, for various boundary and flow conditions. Even the curve of velocity distribution in turbulent flow becomes indeterminate in the neighborhood of rough walls, as a result of which all curves for turbulent flow in Fig. 31 have advisedly been broken in the wall vicinity. Just what happens in the neighborhood of the wall, and at the wall itself, must therefore remain for the present a matter of complete conjecture. It is possible, however, to perform the following bulk evaluation of the relative parts played by the wall and the central regions, which may shed further light upon the dissipation problem.

In the case of smooth pipes, of course, the method of analysis discussed in the foregoing pages can be carried to the wall itself, for reasonably precise information as to the velocity distribution from boundary to center line is at hand. If one were to assume, arbitrarily, that the nominal thickness of the laminar sublayer $\delta = \dfrac{11.6 \, \nu}{\sqrt{\tau_o/\rho}}$ represents[45] the border between the central and the wall regions, it becomes possible at once to evaluate independently W_c, W_w, and W_o, the rates of energy spending per unit length of pipe in the central region, the wall region, and over the entire cross section, respectively. For example, since the velocity gradient in the sublayer is very nearly constant, as a close approximation (except at very low Reynolds numbers) the rate of dissipation in this wall zone will be $2 \pi r \delta \mu \, (du/dy)^2$, which reduces to

$$W_w = 11.6 \, \pi \, D \, \tau_o \, u_f \dots\dots\dots\dots\dots\dots (101a)$$

while the total rate of energy expenditure for the pipe becomes, in similar units,

$$W_o = \pi \, D \, \tau_o \, u_f / \sqrt{f/8} \dots\dots\dots\dots\dots\dots (101b)$$

The following ratio then results:

$$\frac{W_w}{W_o} = 11.6 \sqrt{\frac{f}{8}} \dots\dots\dots\dots\dots\dots (102)$$

Since the resistance coefficient f is a known function of the Reynolds number, a plot of W_w/W_o versus $\mathbf{R}$ may be made directly.

A diagram of a somewhat similar nature[46] is reproduced in Fig. 32. The over-all ordinate AD of this diagram represents the total rate of energy expenditure as given by Eq. 101b, and the ordinate AB the rate of dissipation within the laminar sublayer approximately in accordance with Eq. 101a. The spending rate in the central region, shown by the ordinate remainder BD, has been further subdivided into part BC, the rate of dissipation due to the mean motion, and part CD, the rate of dissipation within the superposed turbulence.

---

[45] "Modern Conceptions of the Mechanics of Fluid Turbulence," by Hunter Rouse, *Transactions, Am. Soc. C. E.*, Vol. 102 (1937), p. 498.

[46] "Sur la répartition entre le mouvement moyen et le mouvement d'agitation de l'énergie dissipée dans l'écoulement turbulent d'un fluide incompressible," by J. Kampé de Feriet and A. Martinot-Lagarde, *Comptes rendus des seances de l'Académie des Sciences*, December 13, 1937.

Immediately evident is the fact that the wall region plays a relatively large part in the total dissipation process, although the turbulence mechanism becomes more and more effective with increasing Reynolds number (that is, decreasing thickness of the laminar sublayer). Equally apparent is the very small rate of dissipation attributable to the mean motion in the central region, which likewise diminishes with increasing R.

Comparable representation of the relative dissipation rates for rough pipes would require a separate diagram for each successive degree of roughness. It

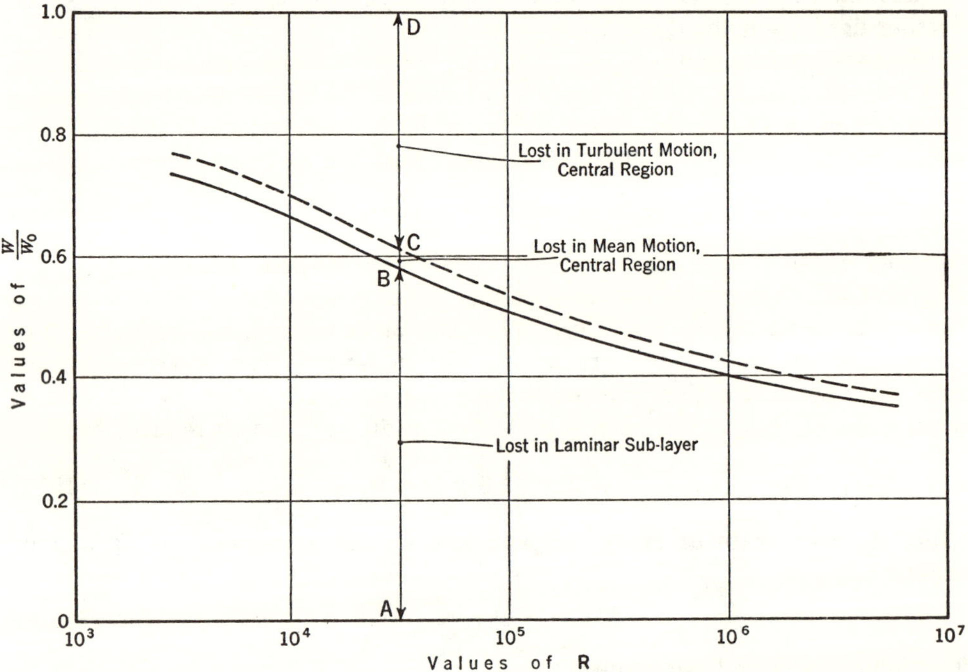

Fig. 32.—Relative Rates of Energy Loss in Different Zones of Flow for Smooth Pipes, after Kampé de Feriet and Martinot-Lagarde

would appear, moreover, that preparation of such diagrams must entail equally explicit information as to flow conditions near the boundary, which information is notably lacking in the case of boundary roughness. However, it will be recalled that the ratio W_w/W_o is equal to $1 - W_c/W_o$, which can be evaluated independently for all conditions. The authors, for instance, utilize this fact in interpreting the similar gradients of the velocity curves for different roughnesses in Fig. 16. That is, by assuming the same hydraulic slope for flow through pipes of the same diameter but different relative roughness, they show that the energy expenditure in the central region will be identical, whereas the total expenditure—and hence that in the wall region—will decrease with increasing roughness. The writers believe, however, that a far more general and significant interpretation may be made.

As must be concluded from Fig. 16, the dimensionless gradient $\dfrac{d(\bar{u}/u_f)}{d(r/r_o)}$ is the same for all velocity-distribution curves at any particular value of r/r_o in the central zone, regardless of the magnitude of the Reynolds number or the relative roughness. Therefore, the rate of energy expenditure in this central zone, up to that value of r/r_o at which the type of boundary begins to influence the given dimensionless gradient, may be expressed in completely general terms. In other words, letting $x = \bar{u}/u_f$ and $z = r/r_o$, it may be shown that

$$W_c = \pi\, D\, \tau_o\, u_f \int_0^{z'} z^2 \left(-\frac{dx}{dz}\right) dz \dots\dots\dots\dots(103)$$

in which z' is the limiting value of r/r_o for the central zone under discussion. The integral may readily be evaluated from the logarithmic velocity equation, for which $dx/dz = -\dfrac{1}{\kappa\,(1-z)}$, its magnitude evidently depending solely on the value of the limit z'. Upon division of Eq. 103 by Eq. 101b, and subtraction

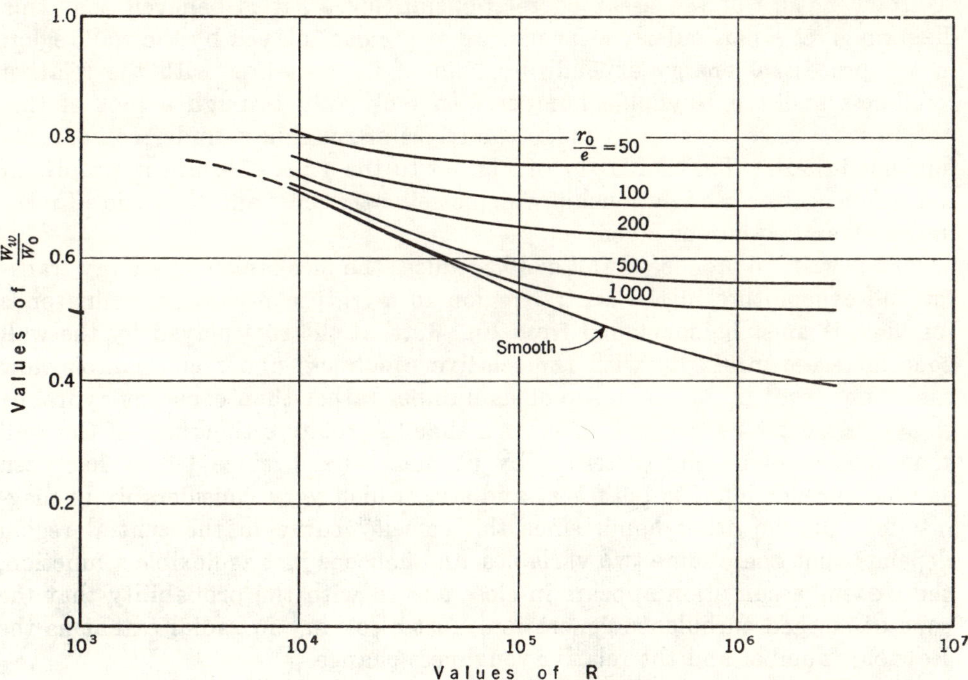

Fig. 33.—Relative Rates of Energy Loss in Wall Region for Smooth and Rough Pipes

from unity, the desired ratio of the rates of wall and total energy expenditure is seen to become simply a function of z' and f:

$$\frac{W_w}{W_o} = 1 - \frac{W_c}{W_o} = 1 - \sqrt{\frac{f}{8}} \int_0^{z'} z^2 \left(-\frac{dx}{dz}\right) dz \dots\dots\dots(104)$$

The question still must be answered as to what magnitude of z' to specify in determining values of W_w/W_o to plot against $\mathbf{R}$ and e/r_o. In the case of smooth pipes, of course, use of $z' = 1 - \dfrac{\delta}{r_o}$ will reduce the result to that of Eq. 102. In the case of rough pipes, it would then appear reasonable to seek a function of the roughness e which would involve a comparable portion of the pipe section; it would be desirable, moreover, to incorporate in this function the effect of viscosity in the practically important transition range from the smooth to the rough phase of the resistance phenomenon. Comparison of the intercepts on the smooth-pipe and the rough-pipe velocity-distribution curves indicates that a value of $3\,e$ is approximately the counterpart of the value δ used in Eq. 101a; and further consideration indicates that the relationship $1 - z' = \dfrac{\delta + 3\,e}{r_o}$ is the only simple function approaching the assumed limits of $\dfrac{\delta}{r_o}$ and $\dfrac{3\,e}{r_o}$.

Based upon this function, the writers have prepared from Eq. 104, together with a diagram of f versus $\mathbf{R}$ for smooth and rough commercial pipes,[47] the system of curves of W_w/W_o versus $\mathbf{R}$ shown in Fig. 33. Despite the admittedly arbitrary nature of the assumed relationship for z', it is believed that this diagram gives a particularly clear picture of the part played by the wall region in the process of energy expenditure, and of its variation with the relative roughness and the Reynolds number. In fact, only through a plot of this nature is it truly possible to judge the effect of roughness under comparable flow conditions. The similarity of Fig. 33 to the $f : \mathbf{R}$ diagram is simply an indication of the close relationship which exists between f and the ratio W_w/W_o in the several equations given.

If the writers' premise that Eq. 103 reduces the mechanism of energy transfer and expenditure in the central region to a truly common denominator is justified, it must be concluded from Fig. 33 that the rôle played by the wall zone increases markedly with the relative roughness under comparable flow conditions (that is, constant Reynolds number rather than constant hydraulic slope). It must be realized, of course, that the relative thickness of the wall zone considered by the writers is by no means fixed, since $1 - z'$ has been assumed proportional to both δ/r_o and e/r_o, which vary considerably in magnitude. On the other hand, since the velocity curve in the central region depends upon these same two variables, and hence is just as flexible a function, the writers' assumption appears in close accord with the probability that the zone of marked turbulence generation must also vary in radial extent as the Reynolds number and the relative roughness change.

Since, in the case of fully developed roughness action, there is no longer a laminar film at the wall in which direct dissipation of energy can occur, one is also led to conclude that the increase in the proportional rate of energy expenditure in the wall region must be due to the relative violence of the turbulence generated by the wall irregularities. It must not be assumed, however, that this large rate of expenditure represents an equally large rate of conversion into energy of turbulence which is subject to transfer toward the central region. On the contrary, as shown by the authors, a considerable

---

[47] "Evaluation of Boundary Roughness," by Hunter Rouse, *Proceedings*, 2d Hydr. Conference, June 1–4, 1942, State Univ. of Iowa, Iowa City, *Bulletin No. 27*, Studies in Eng., 1943.

portion of the actual energy dissipation probably occurs locally during the generation process.   This is due to the extremely high instantaneous stresses produced within the eddies as they are formed.   The energy required to maintain equilibrium of the turbulence structure, in other words, while eventually dissipated in its entirety, does not all pass through the intermediate stage of transferable energy of turbulence.

In this discussion the writers have consistently sought to add to, rather than detract from, the value of the authors' detailéd presentation.   As a new approach to a problem of vital importance is invariably subject to modification and improvement, the writers are hopeful that the interpretation presented herewith may likewise be altered and extended by the authors in their closure, and by those who later use the paper as a reference.   Indeed, quite aside from the fundamental material which the paper has provided, it is probable that it will have lasting worth as a stimulus to creative thought on a very stubborn problem.

Gravitational Diffusion From a Boundary Source in Two-Dimensional Flow

By HUNTER ROUSE,[1] IOWA CITY, IOWA

Convection currents of heated air downwind from a continuous line of gasoline burners at ground level provided an important means of dispersing fog over airfields during the recent war. A study of heat requirements as a function of burner location and cross-wind velocity, conducted by the Iowa Institute of Hydraulic Research in 1943, yielded results which are considered applicable to all similar problems of gravitational diffusion from a boundary source—such, for instance, as the mixing through induced convection of sediment-laden water introduced at the upper surface of a stream. The author presents herewith a general analysis of the problem, based upon solution of the differential equations of energy and diffusion for the assumption of (1) an error distribution of the specific-weight difference over any normal section and (2) a direct proportionality between the mixing length and the standard deviation of the error curve. Experimental results are shown to verify the analysis with good approximation.

EXISTING theories of turbulent diffusion are tacitly restricted to the condition of constant specific weight throughout the moving fluid. It is presumed, in other words, that the turbulence of the approaching flow provides the sole mixing action, and that such turbulence is neither augmented nor reduced as a result of the diffusion process. As a matter of fact, not only is the specific weight of fluid matter in zones of diffusion often variable over a considerable range, but such variation can in itself be the primary cause of the mixing phenomenon.

Consider, for purposes of illustration, the generation of heat along a line source at right angles to a uniformly turbulent flow. So long as the rise in temperature remains small, the heat will be diffused symmetrically about a longitudinal plane passing through the source; the transverse distribution will follow the probability function at all successive sections, the lateral spread increasing with distance from the source and the peak value decreasing accordingly. The geometry of the diffusion pattern, under such circumstances, will depend solely upon the relative intensity of the turbulence.

If now the rate at which heat is generated at the source is gradually increased, there will evidently be a steadily growing tendency for the heated fluid as a whole to rise. If the line source and the direction of flow are horizontal, the effective buoyant force will necessarily result in a continuous vertical displacement of the heated fluid as it moves downstream, with the result that the distribution of heat can no longer be symmetrical about the horizontal plane of the source. For reasons of two-dimensional continuity, however, a net vertical displacement of the heated fluid can take place only by a convective mixing process which is roughly akin to the original turbulence of the flow but quite independent thereof. Such convective mixing will obviously augment the diffusive capacity of the flow by an amount varying solely with the rate at which heat is generated by the source. Indeed, if the rate of generation is sufficiently great, the diffusion due to the initial turbulence of the flow may be quite small in comparison with that due to thermal convection.

Since the motivating force in such convective action is evidently gravitational, it should make little actual difference in the manner of analysis whether the fluid is heated or cooled as it passes the source. In fact, essentially the same convective phenomenon should result from the introduction at the line source of a lighter or a heavier fluid, or even a fluid suspension of finely dispersed solid, liquid, or gaseous matter. The dependent variable common to all such phases of the problem is simply the local increment (or decrement) in the specific weight of the diffusing fluid, regardless of whether the difference is due to mechanical, chemical, or thermal causes.

The specific-weight difference $\Delta\gamma$ may, in the elementary case illustrated by Fig. 1, be expected to depend upon the following

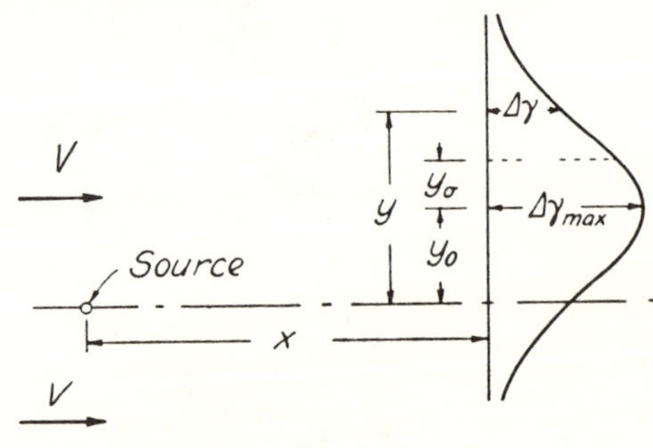

Fig. 1

variables: (1) the horizontal distance x from the line source in the direction of flow; (2) the vertical distance y from the longitudinal plane of the source; (3) the uniform velocity V of the ambient fluid; (4) the density ρ of the ambient fluid; and (5) the differential weight flux G of the diffusing fluid past any vertical section. The latter quantity, defined as

$$G = V \int \Delta\gamma \, dy \dots\dots\dots\dots\dots [1]$$

is seen to be proportional to the rate of heat output per unit length of source, or to the unit rate at which salinity or suspended matter is introduced. All six of the foregoing variables may readily be combined, by means of the Π-theorem of dimensional analysis, to yield (among others) the following arrangement of terms

$$\varphi_1 \left(\frac{y}{x}, \ \frac{\Delta\gamma x}{\rho V^2}, \ \frac{\rho V^3}{G} \right) = 0 \dots\dots\dots\dots [2]$$

The first two parametric groups evidently embody in dimensionless form the ordinate and abscissa scales of the vertical distribu-

[1] Director, Iowa Institute of Hydraulic Research, State University of Iowa. Mem. A.S.M.E.

Presented at the Sixth International Congress for Applied Mechanics in Paris, France, September, 1946.

Discussion of this paper should be addressed to the Secretary, A.S.M.E., 29 West 39th Street, New York, N. Y., and will be accepted until October 10, 1947, for publication at a later data. Discussion received after the closing date will be returned.

NOTE: Statements and opinions advanced in papers are to be understood as individual expressions of their authors and not those of the Society.

tion curve; the third is a number of the Froude type, characterizing the relative magnitude of the gravitational influence.

Such analysis obviously provides no clue as to the form of the indicated function. It may be assumed of course that the variation of $\Delta\gamma$ with y follows the probability equation, so that the distribution curve will be completely defined (see Fig. 1), by the centroidal distance y_0, the standard deviation y_σ, and the maximum specific-weight increment $\Delta\gamma_{max}$. Little will be gained thereby, however, unless y_0 and y_σ (measures of the weight displacement and the weight diffusion, respectively) can both be related to G. For lack of experimental evidence in regard to the general problem, it is necessary herein to eliminate the variable y_0 by considering the line source to lie in the plane of a horizontal boundary (such, for instance, as the ground or the free surface of flowing water). The limit $\Delta\gamma_{max}$ of the vertical distribution curve should then correspond at all sections to the elevation $y = 0$, as shown in Fig. 2. In other words, although there continues

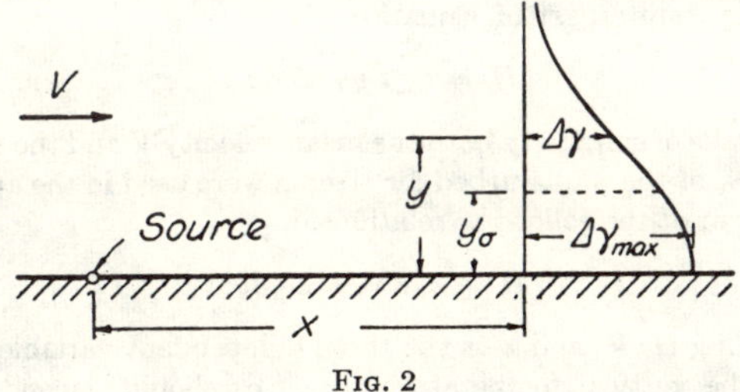

FIG. 2

to be a net vertical displacement of fluid weight from section to section, the presence of the boundary (or the absence of convection across the longitudinal plane of the source) now restricts such displacement to the change in the standard deviation (or the elevation of the centroid) of the distribution curve; y_σ thus becomes a measure not only of the weight diffusion but also of the weight displacement.

Under these conditions the variation in $\Delta\gamma$ with y at any section may be expressed by the following simple form of the probability function

$$\frac{\Delta\gamma}{\Delta\gamma_{max}} = e^{-\frac{y^2}{2y_\sigma{}^2}} \dots\dots\dots\dots\dots [3]$$

From the characteristics of this probability curve Equation [1] may at once be replaced by the identity

$$G = 1.25 V y_\sigma \Delta\gamma_{max} \dots\dots\dots\dots\dots [4]$$

Moreover, since the essential variables are now y_σ, $\Delta\gamma_{max}$, x, ρ, and V, the functional relationship of Equation [2] may likewise be replaced by

$$\varphi_2\left(\frac{y_\sigma}{x}, \ \frac{\rho V^2}{y_\sigma \Delta\gamma_{max}}\right) = 0 \dots\dots\dots\dots [5]$$

The first parameter of this relationship will be seen to indicate the relative scale of the convection pattern, while the second, upon introduction of G from Equation [4], will become identical in form to the Froude parameter of Equation [2]. If the Froude number is defined specifically as

$$\mathbf{F} = \sqrt{\frac{\rho V^3}{G}} \dots\dots\dots\dots\dots [6]$$

Equation [5] will take the significant form

$$\varphi_3\left(\frac{y_\sigma}{x}, \ \mathbf{F}\right) = 0 \dots\dots\dots\dots [7]$$

Assumption of the probability equation for the distribution curve, and restriction of the problem to the case of a boundary source, have thus reduced the terms of the original functional relationship to two simple parameters, but the function itself still remains to be determined.

A peculiarity of the diffusion phenomenon under discussion lies in the fact that it is a wholly self-contained process, that is, the variation in specific weight with elevation at any section represents a tendency toward instability in the form of unconstrained potential energy, which is gradually transformed into kinetic energy of convective mixing as the centroid of the weight-distribution curve rises. Only through the convective mixing thus produced, however, can the vertical displacement of weight essential to the work of energy transformation be accomplished. Evidently the convection mechanism is self-induced and progressively self-supporting, so that the energy of the mixing process may be expected to increase at a steady rate from section to section. In view of these conditions, it should be possible to write two related equations, one of work-energy and one of diffusion, to determine the form of the still unknown function.

The rate of change of potential energy, i.e., the rate at which work is done by the buoyant forces, within an elementary fluid column may be expressed as the integral of the product of the local specific-weight difference and the corresponding rate of vertical displacement. From Equation [1], and from the necessary constancy of G with x, it is apparent that the total vertical force upon a column of infinite height will also remain constant with x. For the assumed probability distribution, the quantity $V\partial y_\sigma/\partial x$ is proportional to the rate of vertical displacement of the centroid of the weight-distribution curve; hence the effective displacement rate at any elevation should differ therefrom in direct proportion to the ratio y/y_σ. Writing the rate of change of kinetic energy in terms of a quantity v' representing the vector magnitude of the convection velocity, the equation of work-energy then becomes

$$V\frac{\partial y_\sigma}{\partial x}\int_0^\infty \Delta\gamma \frac{y}{y_\sigma}\,dy = V\frac{\partial}{\partial x}\int_0^\infty \frac{\rho(v')^2}{2}\,dy \dots\dots [8]$$

The general equation of molar diffusion is of the form $N = -\epsilon\,dn/dy$, in which n represents the volume concentration of the characteristic under discussion, ϵ the diffusion coefficient, and N the net rate at which the characteristic is transported in the y-direction per unit normal area. In the problem under consideration, the net rate of vertical transport of fluid weight should be equal to the local specific-weight increment times the local rate of vertical displacement already discussed. The diffusion equation therefore becomes

$$\Delta\gamma \frac{y}{y_\sigma} V \frac{\partial y_\sigma}{\partial x} = -\epsilon\frac{\partial(\Delta\gamma)}{\partial y} \dots\dots\dots\dots [9]$$

which may, for the sake of similarity of form, be expressed in integral terms comparable to those of Equation [8]

$$V\frac{\partial y_\sigma}{\partial x}\int_0^\infty \Delta\gamma \frac{y}{y_\sigma}\,dy = -\int_0^\infty \epsilon\frac{\partial(\Delta\gamma)}{\partial y}\,dy \dots\dots [10]$$

Through comparison with Equation [1] it will be seen that the left sides of Equations [8] and [10] are proportional to products of the quantity G, which is necessarily constant with x, and the gradient $\partial y_\sigma/\partial x$. The right sides, however, cannot be evaluated without further knowledge as to the distribution of the convection characteristics. On the other hand, mean values of these characteristics over any vertical section may conveniently be defined as follows:

$$(v')_m{}^2 = \frac{1}{y_\sigma} \int_0^\infty (v')^2 \, dy \quad \dots \dots \dots [11]$$

$$\epsilon_m = -\frac{1}{\Delta\gamma_{max}} \int_0^\infty \epsilon \frac{\partial(\Delta\gamma)}{\partial y} \, dy \dots \dots [12]$$

Since the proportionality factors differ little from unity, Equations [3] and [10] may thus be simplified to

$$\frac{\partial y_\sigma}{\partial x} = \frac{\rho V}{2G} \frac{\partial}{\partial x} [y_\sigma (v')_m{}^2] \dots \dots \dots \dots [13]$$

$$\frac{\partial y_\sigma}{\partial x} = \frac{\epsilon_m \Delta\gamma_{max}}{G} \quad \dots \dots \dots \dots \dots [14]$$

If it is now considered that the mean diffusion coefficient is equal to the product of a mean length scale and the square root of the mean-square velocity of convection, i.e.

$$\epsilon_m = l_m \sqrt{(v')_m{}^2} \quad \dots \dots \dots \dots [15]$$

and if the reasonable assumption is made that the mean length scale varies from section to section in direct proportion to the standard deviation of the distribution curve, i.e.

$$l_m = C y_\sigma \dots \dots \dots \dots \dots [16]$$

the following significant relationships will at once be obtained

$$\sqrt{\frac{(v')_m{}^2}{V}} = \sqrt{2}\, \frac{1}{\mathbf{F}} \quad \dots \dots \dots \dots [17]$$

$$\frac{\epsilon_m}{Vx} = \frac{2C^2}{1.25} \frac{1}{\mathbf{F}^2} \quad \dots \dots \dots \dots [18]$$

$$\frac{y_\sigma}{x} = \frac{\sqrt{2}\,C}{1.25} \frac{1}{\mathbf{F}} \quad \dots \dots \dots \dots [19]$$

As is evident from these equations, the characteristics of the diffusion pattern when expressed in dimensionless form are dependent solely upon the Froude number of the flow. However, while the velocity characteristic is necessarily constant from section to section, the diffusion coefficient is a linear function of distance from the source. Herein lies the essential distinction between diffusion due to initial turbulence of the ambient fluid and that due to gravitational convection, that is, while the turbulence is subject to gradual decay as the result of viscous action, the convection process will continuously increase with distance downstream. Of particular significance in the present analysis is the third of these equations, for it not only states that the diffusion zone expands at a linear rate but represents in explicit form the functional relationship indicated by Equation [7]. Through use of Equation [19], in other words, it is now possible to generalize the distribution function of Equations [2] and [3] as follows

$$\frac{\Delta\gamma x}{\rho V^2} \mathbf{F} = \frac{1}{\sqrt{2}\,C} e^{-\left(\frac{0.63}{C}\frac{y}{x}\mathbf{F}\right)^2} \quad \dots \dots \dots [20]$$

With the exception of a single coefficient, the analytical solution of the basic problem may therewith be considered complete.

Experimental data of the type required to evaluate the coefficient C were obtained in 1943, by the Iowa Institute of Hydraulic Research in the course of heat-diffusion studies relating to the dispersal of fog over military airfields. Sponsored by the Office of Scientific Research and Development, these studies utilized a specially constructed air tunnel of rectangular cross section 4 ft high, 6 ft wide, and 20 ft long, with a velocity range from 1 to 25 fps. A burner manifold was set into the tunnel floor across the entire 6-ft width, through which a mixture of propane

and air could be discharged under constant pressure to produce a continuous blue flame of controlled intensity. Copper-constantan thermocouples in combination with a precision potentiometer permitted temperature measurements to be made at any point in the air stream. Vertical distribution curves could thus be obtained at various air speeds, rates of heat output, and distances downwind from the source.

Owing to the rapidity with which an approximate solution of the problem had to be obtained, experimental refinements were out of the question. For example, the thermocouples were not completely shielded against radiation from the open flame, with the result that an appreciable (though systematic) temperature correction was required at each section. Moreover, since the tunnel walls and floor were only partially insulated against heat transmission, the measurements at successive sections of the air stream did not satisfy completely the criterion of thermal continuity. Finally, although the rate of heat output per unit length of burner should, strictly speaking, be computed from the following counterpart of Equation [1]

$$H = c_p \int v\gamma \,\Delta T \, dy \dots \dots \dots \dots [21]$$

for the sake of simplicity both the mean velocity V and the specific weight γ_0 of the undisturbed air stream were used in the approximate form of the following relationship

$$H \approx c_p V \gamma_0 \int \Delta T \, dy \dots \dots \dots \dots [22]$$

Selecting H, V, and x as the three independent variables upon which the vertical temperature traverses should depend, three systematic series of measurements were made to determine the influence of each variable in turn upon the distribution curve. The results of these measurements, reproduced as Fig. 3, served as the basis of an empirical analysis, in which, for want of the solution presented herein, the data were reduced to a single composite plot by a rather arbitrary process of curve-fitting. Needless to say, all values of ΔT and H, plotted in Fig. 3, may be con-

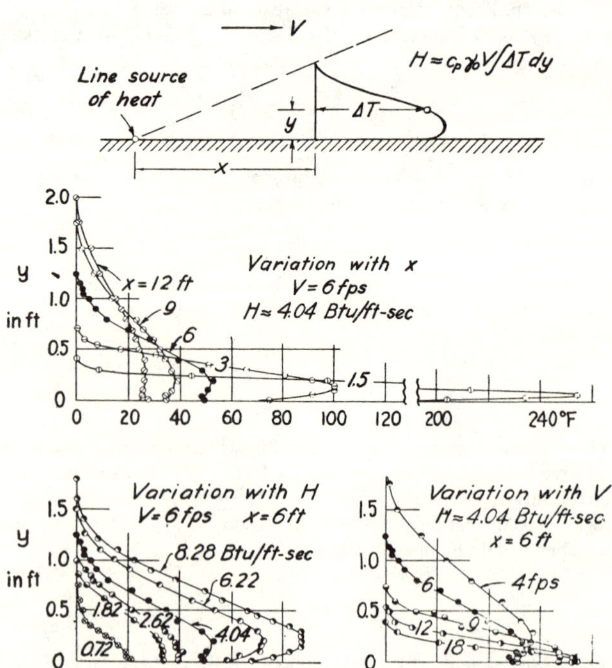

Fig. 3

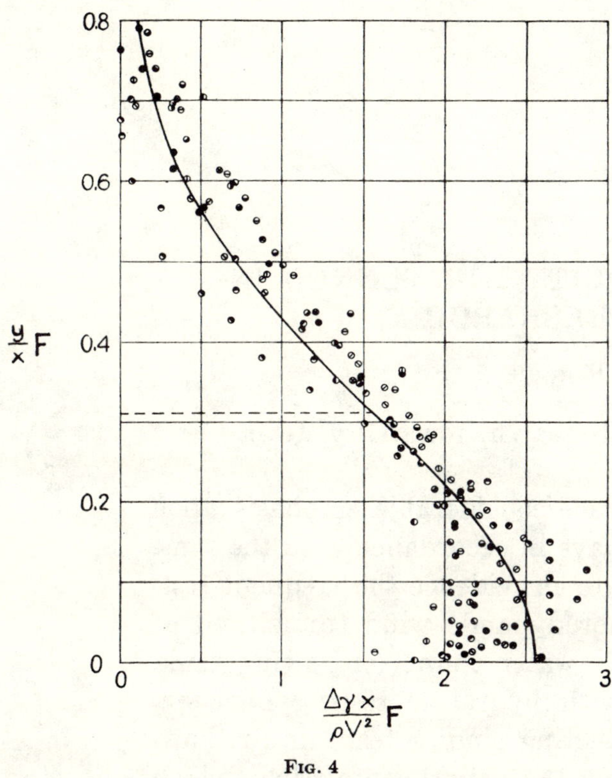

FIG. 4

verted to the more general terms $\Delta\gamma$ and G, discussed in the foregoing, through use of the approximate relationships

$$\Delta\gamma \approx \gamma_0 \frac{\Delta T}{T_0} \dots\dots\dots\dots\dots [23]$$

and

$$G \approx \frac{H}{c_p T_0} \dots\dots\dots\dots\dots [24]$$

the latter deriving from the definition equations of G and H. All data in Fig. 3 have thus been replotted in Fig. 4, according to the parameters of Equation [20], together with the probability curve of this equation which best fits the experimental points. Although, for reasons already enumerated, the scatter of the points is appreciable (note, for example, the consistent evidence of heat loss at the lower boundary), the close agreement between the mathematical curve and the general trend of the measured values is believed to provide ample verification of the foregoing analysis.

With the distribution function thus defined, it is now possible to evaluate the constant of Equation [16] and therewith to complete the expression for the mean characteristics of the diffusion pattern. Since, when $y/y_\sigma = 1.0$, the ordinate scale in Fig. 4 has the magnitude 0.31, it follows from Equation [19] that $C = 0.31 \times 1.25/\sqrt{2} = 0.27$. Therefore one may finally write Equation [20] in the specific form

$$\frac{\Delta\gamma x}{\rho V^2}\mathbf{F} = 2.6\, e^{-5.4\left(\frac{y}{x}\mathbf{F}\right)^2} \dots\dots\dots\dots [25]$$

The problem of gravitational convection from a boundary source in two-dimensional flow may herewith be regarded as solved to at least the same degree of approximation as the experimental data themselves display.

In the light of the foregoing discussion, further analysis of the convection phenomenon might logically proceed in several ways. On the one hand, it remains to be shown by statistical treatment of the motion of a fluid particle to what extent the distribution function actually deviates from the probability curve; however, this would very likely require assumptions as to the variation of $\sqrt{\overline{(v')^2}}$ and l with y as well as with x, which would obviously limit the rigor of such an approach. On the other hand, the approximate treatment described herein might next be applied to the problem of convection from a line source remote from a boundary, and then to the three-dimensional counterpart of each of these two-dimensional conditions. The results of such analyses, even though inexact, should provide information of use in fields as widely varied as meteorology, sanitary engineering, chemical engineering, and oceanography.

USE OF THE LOW-VELOCITY AIR TUNNEL IN HYDRAULIC RESEARCH

Hunter Rouse

Iowa Institute of Hydraulic Research, Iowa City, Iowa

Instructors in fluid mechanics place considerable emphasis upon the fact that both air and water behave in accordance with the same fundamental laws of motion, and cite as evidence the frequent testing of under-water bodies in the aerodynamic wind tunnel and of aircraft parts in the hydrodynamic water tunnel or towing tank. Not until investigators familiar with hydraulic testing have experienced the relative ease of conducting similar tests in air, however, does the actual broadening of the experimental field which this entails become directly apparent. Staff members of the Iowa Institute were initiated perforce into the technique of air-flow measurement early in the recent war, with results which were pleasantly surprising; since then they have been staunch proponents of the air tunnel as an essential tool of the hydraulics laboratory.

Five advantages which at once recommend the use of air instead of water for test purposes are as follows: (1) owing to the low density of air (approximately one eight-hundredth that of water), structural requirements are held to a minimum; (2) for the same reason, power demands are reduced many fold; (3) absolute air-tightness of a conduit is by no means as essential as absolute water-tightness, for wetting problems do not exist; (4) with the atmosphere serving at once as a supply reservoir and a catch basin, storage tanks and water costs are eliminated; finally (5) many phases of instrumentation become greatly simplified.

Apparently opposed to these advantages are factors stemming from the three major characteristics of liquids as distinguished from gases: the relatively high elastic modulus, the tendency to become discontinuous when the vapor pressure of the liquid is reached, and the ability to maintain a free surface. It would therefore seem that air could not replace water for experimental purposes under conditions involving the effects of compressibility, cavitation, or gravitational attraction. Such, however, is not entirely the case. On the one hand, air may undergo velocity changes of several hundred feet per second without the influence of its compressibility becoming apparent; on the other hand, there is a close analogy between compression waves in gases and gravity waves in liquids, as was emphasized by Dr. Ippen at the Second Hydraulics Conference. At that Conference, moreover, Mr. Mousson told of studying the pressure distribution in a model turbine as a step in reducing cavitation, air being used because of its convenience; during the recent war, similarly, the tendency for certain underwater projectile forms to produce cavitation was determined on models in aerodynamic wind tunnels. Finally, Dr. Knapp discussed at the same Conference the so-called density currents produced by silt-laden flow in fresh-water reservoirs, a gravitational effect closely akin to the flow of either dust-laden or merely colder air in the atmosphere; each of these gravitational phenomena is as definitely governed by the Froude criterion of similarity as any liquid flow with a free surface.

The writer is by no means recommending the construction of the familiar wind tunnel of aeronautics laboratories in every hydraulics laboratory, for the term ''low-velocity'' is used advisedly in the title of the paper. This term is, of course, wholly relative, but is intended to form a distinguishing category similar to ''supersonic'' and ''aerodynamic''. High-velocity wind tunnels producing speeds beyond that of sound in air will probably never have their liquid counterpart—although recent news items on the projection of molten metal cite velocities comparable to that of sound in the corresponding medium. Aerodynamic wind tunnels, on the other hand, are so devised as to attain prototype Reynolds numbers without reaching sonic speed, and even with appreciable compression of the air they generally involve velocities greater than several hundred feet a second. The low-velocity tunnel herein discussed is intended to cover the remaining range—with an air speed of perhaps one hundred feet per second as the practical maximum.

There at once arises a very essential question as to the usefulness of such relatively low air speeds when much higher values are normally required to produce dynamic similarity of the Reynolds type, for the velocity should then vary directly with the viscosity and inversely with the linear scale. (The kinematic viscosity of air, it must be noted, is some fifteen times as great as that of water, while the scale is invariably reduced for model study.) However, the fact remains that the majority of the problems confronting the hydraulic engineer are wholly lacking in the highly streamlined elements demanded by the aeronautical engineer, and hence many of the boundary conditions investigated in the hydraulics laboratory are completely independent of the viscous phenomena for which the Reynolds number is the essential criterion. In other words, geometrical effects rather than surface effects still predominate in hydraulics; since these are relatively free from velocity, scale, and viscous influence, for the study of such problems the low-velocity air tunnel is a most convenient tool.

Fig. 1—View of Simple Demonstration Tunnel for Two-dimensional Phenomena.

Just how simply equipment of this nature may be constructed is indicated by the portable, two-dimensional tunnel shown in Fig. 1, which was built in 1943 to fill a temporary need at the Institute laboratory and still finds considerable use. Air speeds up to 75 feet per second may be obtained in the 2x20x30-inch transparent test section by means of suction from a 5-horsepower centrifugal fan. A somewhat more elaborate experimental tunnel, of the three-dimensional open-throat type, is that sketched in Fig. 2, which was designed by the Institute for instructional purposes at the National University of Colombia and will be built essentially in duplicate for use in the student laboratory of the Institute. The larger of the two tunnels constructed by the Iowa Institute for NDRC research during the recent war was of the closed-throat type, having a test section 4 feet high, 6 feet wide, and 20 feet long (Fig. 3). Owing to the necessity of introducing either heat or gas into the flow, fresh air was drawn in through windows at the north end of the laboratory and exhausted by the 7½-horsepower fan through windows on the east and west sides; maximum air speeds of ap-

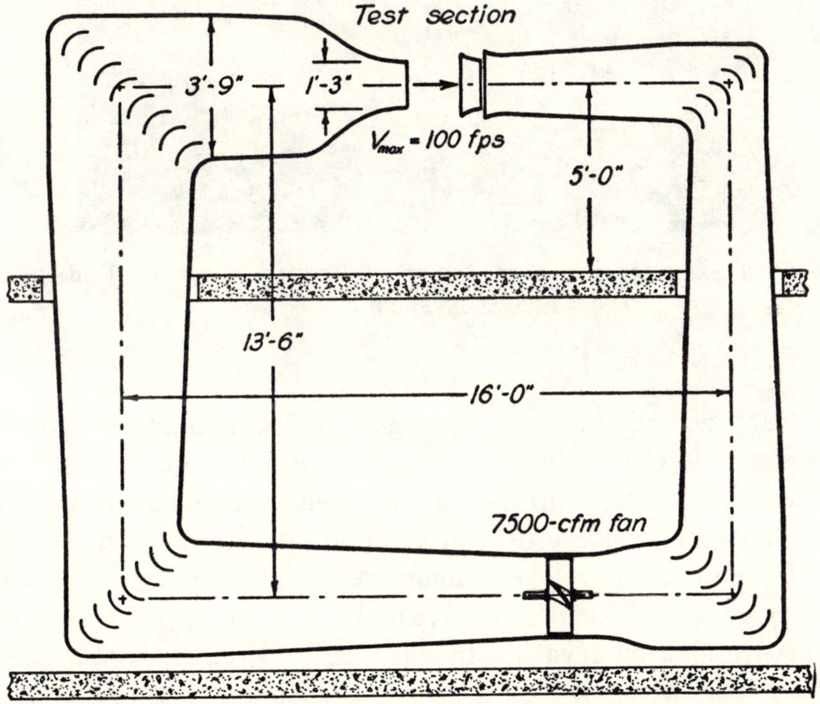

FIG. 2—SKETCH OF OPEN-THROAT AIR TUNNEL FOR STUDENT LABORATORY.

proximately 25 feet per second were attainable. With the forthcoming expansion of the Institute, it is planned to rebuild the latter tunnel as a two-dimensional closed-throat unit having a test section 2x3x6 feet in size with velocities up to 100 feet per second, and to install a recirculating, three-dimensional, closed-throat tunnel having a test section in the form of a 5-foot octagonal prism 20 feet long with a 100-foot-per-second maximum speed. Needless to say, additional fans and blowers of various capacities will be available for research requiring independent air supply.

FIG. 3—TEST SECTION OF LARGER INSTITUTE TUNNEL SHOWING URBAN MODEL USED IN GAS-DIFFUSION ANALYSIS.

The wartime uses of the larger Institute tunnel are pertinent to the present discussion in that they indicate at one and the same time the great range of application of such an experimental facility and the bearing of all test results upon fundamental problems of hydraulic engineering. The tunnel was constructed specifically for the study of gas diffusion in urban districts, a typical pattern of structures under observation being seen in Fig. 3. In the case of boundary angularity of this nature, it was found possible to express all measurements of gas concentration for a particular structural arrangement (i.e., regardless of wind speed, scale, and rate of gas flow) in terms of dimensionless concentration contours, through use of an index involving the local volume concentration c in its ratio to $c_0 = Q/Vh^2$, in which Q is the volume rate of gas release at the source, V is the mean wind speed above the structures, and h is the structural scale. Typical concentration contours are shown in

Fig. 4 for a schematic grouping of cubic forms. Needless to say, the diffusion of salinity or very fine sediment in a liquid under comparable boundary conditions could be evaluated in essentially the same manner, for the injection of gas into an air stream provides a most convenient means of simulating and measuring the diffusion process.

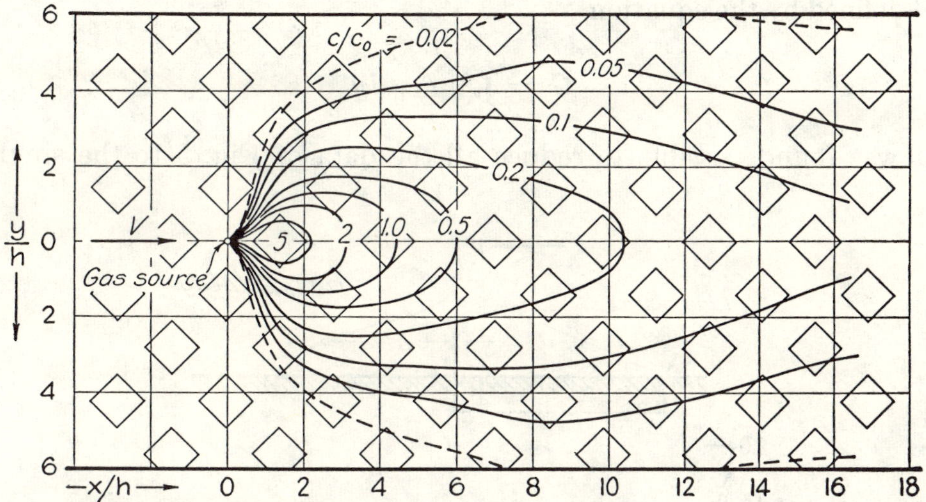

Fig. 4—Dimensionless Representation of Gas Distribution in Model Village Similar to that of Fig. 3.

The second phase of war research for which this facility was developed involved the problem of determining heat requirements for fog dispersal over airplane runways—the method known under the code term FIDO. A line source of heat simulating a continuous gasoline burner was produced by means of a jet manifold burning butane gas and extending across the tunnel floor. Since the rate of heat diffusion varied with the relative intensity of the convection currents produced by the local heating of the air, it was to be expected that the distribution of heat downwind from the source would depend upon the distance from the source, the wind speed, and the rate of heat release. Vertical temperature traverses were therefore made with a sensitive thermocouple circuit in such manner that the effect of each variable in turn could be evaluated, as indicated in Fig. 5. Since thermal effects were actually involved only to the extent that the temperature variation resulted in a proportional buoyancy of the heated air, the resulting convection was purely a gravitational phenomenon akin to many problems involving the relative motion between slightly lighter and heavier liquids and gases. The Froude criterion for similarity should therefore be expected to govern the conversion of experimental data to prototype conditions. In fact, by expressing the temperature differences in terms of the resulting difference $\triangle\gamma$ in specific weight, and by writing the rate of heat release in terms of a net weight flux G defined by the equation

$$G = \int v \triangle\gamma \, dy$$

it was found possible to reduce all the data of Fig. 5 to the single

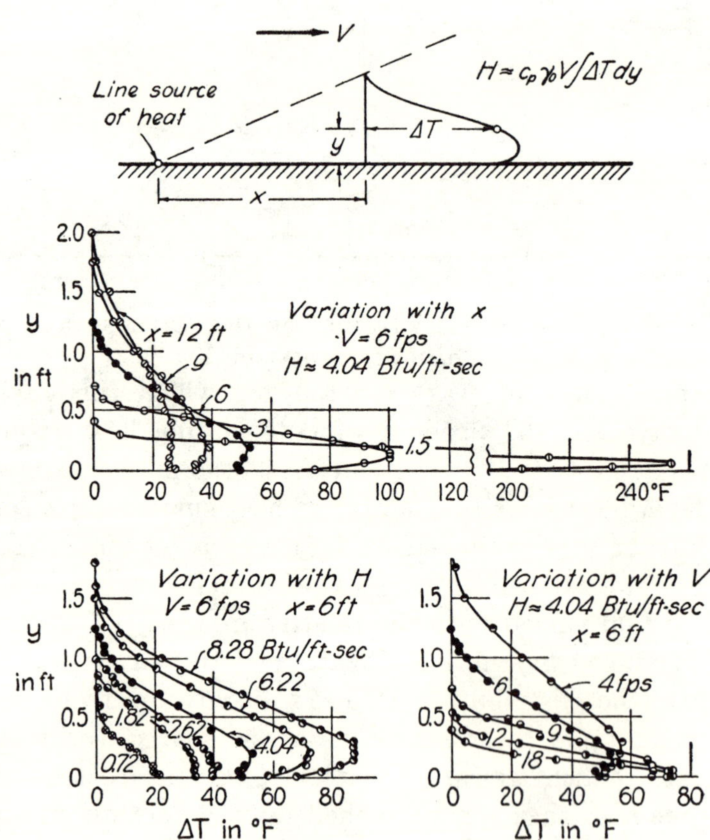

Fig. 5—Measurements of Temperature Distribution Downwind from a Line Source of Heat.

dimensionless plot of Fig. 6, in which the quantity $\varrho V^3/G$ is a form of the Froude number. The full line corresponds to an error distribution satisfying the elementary equations of energy and diffusion. It is to be noted that the rate of spread of the diffusion zone is inversely proportional to the Froude number. Evidently, information thus obtained in the air tunnel should be applicable as well to the flow of heated or cooled water, and convection produced in this manner by temperature differences should also characterize convection phenomena due to the presence of either finely dispersed air bubbles or suspended sediment.

The last of the wartime projects involving the low-velocity air tunnel dealt with the determination of wind patterns over models of mountainous terrain. For this purpose the sensitive direction-

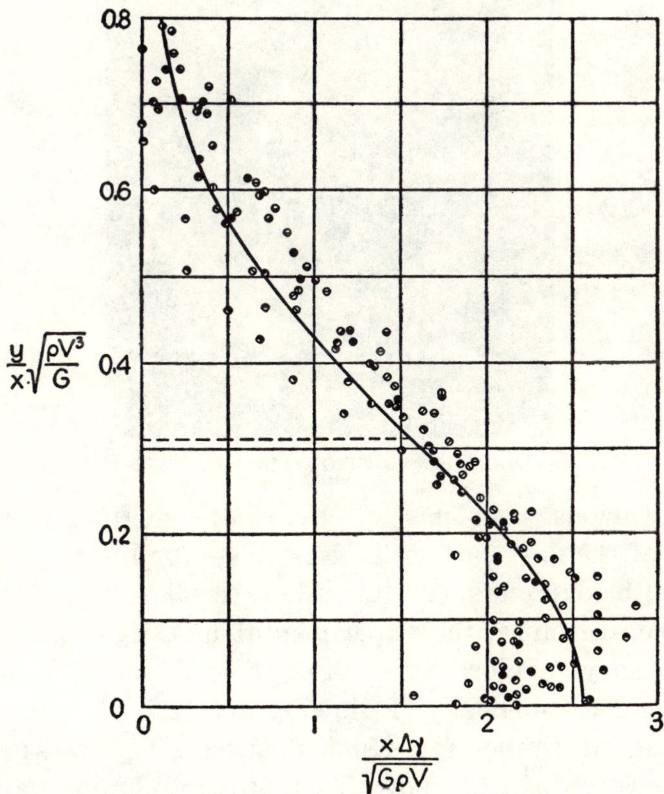

FIG. 6—DIMENSIONLESS PLOT OF GRAVITATIONAL DIFFUSION FOR THE DATA OF FIG. 5.

indicating Pitot tube shown in Fig. 7 was constructed, six fine tubes being so formed and mounted that, in combination with a differential gage of the Wahlen type reading to 0.001 inch of alcohol, the magnitude of the velocity vector and its horizontal and vertical inclinations could be measured at any point above the model. Aside from its immediate application to meteorology (and, under water, to oceanography), the same experimental technique is directly use-

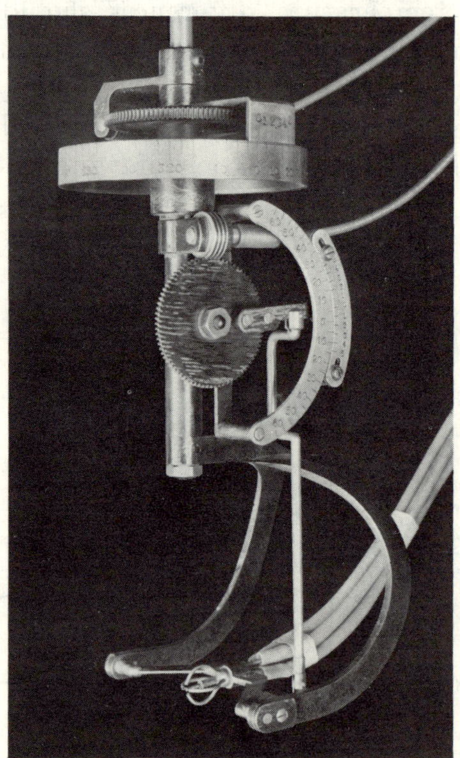

Fig. 7—Directional Pitot Tube for Air Flow.

ful in the various problems of hydraulic engineering in which knowledge of the flow pattern is desirable—for instance, the design of baffle and bridge piers, conduit inlets, scroll cases, or draft tubes. In every such instance, the experimental investigation—or at least the preliminary exploration—may proceed more rapidly, conveniently, and economically in flowing air.

Use of the air tunnel for demonstration purposes—particularly in the observation of eddy patterns requiring contamination of the fluid by a coloring agent—is of considerable importance in both instruction and research. During the war years the Institute prepared a number of motion-picture sequences on turbulence and diffusion for use in training personnel, typical frames of which are reproduced in Figs. 8, 9, and 10. Although these studies dealt specifically with such atmospheric problems as the diffusion of smoke and gas under stable and unstable conditions and in wooded and urban districts, and the diffusion of heated air over fog-bound landing fields, both the flowing medium and the demonstration technique are readily adaptable to the illustration of similar effects as they occur in any gas or liquid. The complete transparency of

FIG. 8—EDDY STRUCTURE IN THE LEE OF A SCHEMATIC BUILDING.

fresh air, the ready production of dense smoke plumes with either titanium tetrachloride or oil fog, and the extreme economy with which contaminated air may be wasted are all factors in favor of the air tunnel for such purposes.

Although the small tunnel of Fig. 1 was constructed wholly for visual and photographic studies of this nature, it was soon found to be very convenient in exploratory tests on pressure distribution around various two-dimensional boundaries characterized by either pronounced angularity or rapid convergence. Typical of such tests are Figs. 16 and 17 of Dr. Steinman's paper on page 159 of this volume and the two series of measurements reproduced herein as Figs. 11 and 12. The first of these was taken from preliminary studies for a graduate thesis on cavitation at gate slots in a high-velocity conduit or sluice. A schematic slot was constructed of sheet plastic in such manner that the proportions could readily be changed. The particular system of curves shown in Fig. 11 repre-

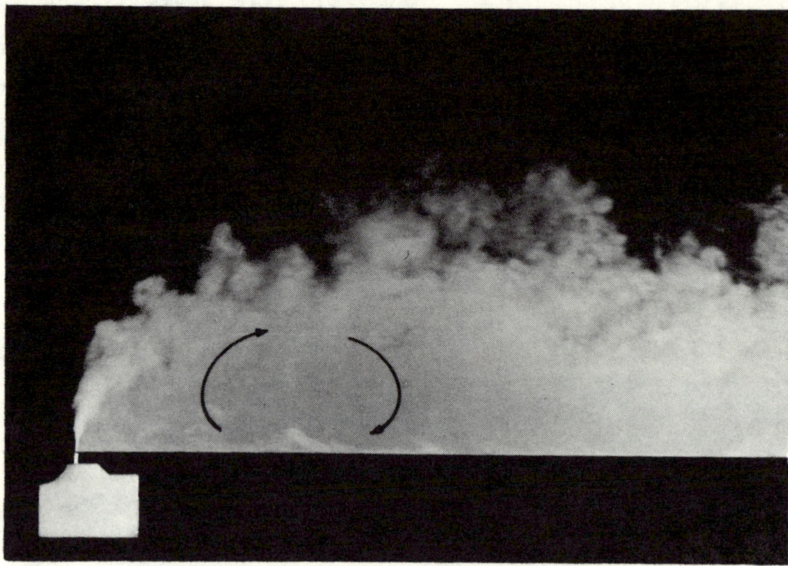

FIG. 9—FORMATION OF A STABLE EDDY BEHIND A WIND CURTAIN.

sents the variation in pressure distribution which is caused by relative displacement of the downstream wall in either direction from its normal plane. The great reduction in minimum pressure (the primary factor involved in the elimination of cavitation) which results from a slight depression of the wall is at once apparent.

FIG. 10—FLOW PATTERN IN THE WAKE OF A VERTICAL WALL.

Studies of actual cavitation effects obviously require the use of water, but exploratory tests in air not only reduce the necessary extent of the final investigation but do so at a few percent of the cost which would otherwise be involved.

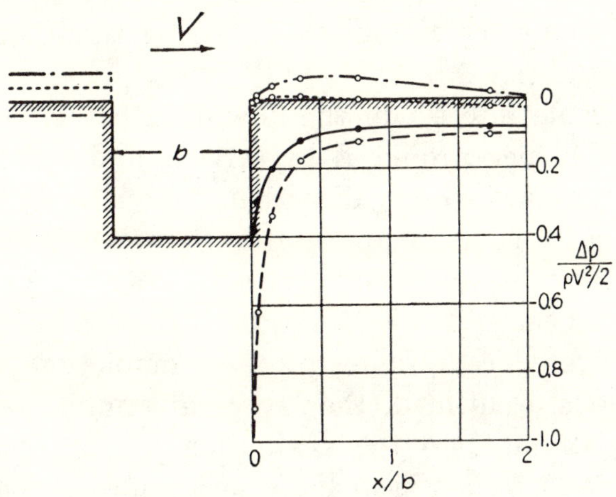

FIG. 11—DISTRIBUTION OF PRESSURE AT A GATE SLOT.

Fig. 12, in turn, illustrates the effect of radius of curvature upon the pressure distribution at a two-dimensional conduit inlet. According to the theory of potential flow, continued decrease in the ratio of radius to wall spacing should result in a continued increase in the pressure drop just upstream from the point of final tangency. This trend is well indicated by the sequence of the curves in the plot at the left and by the minimum-pressure function at the right. A zero radius would evidently result in a pressure intensity of negative infinity as the limit of this function—were it not for the minor effects of either compressibility in air or cavitation in water and for the major effect of separation which invariably occurs when the curvature of the boundary becomes too abrupt. The optimum degree of separation, of course, is found in the case of the sharp corner, and it is pertinent to note that the minimum pressure for this condition is not extreme. In other words, the greatest pressure drop does not coincide with the minimum rounding of the entrance, for the reduction in the separation tendency with increasing radius of curvature causes the minimum-pressure function to approach that of the potential theory more and more closely. So far as boundary pressure is concerned, a slight degree of rounding is therefore worse than none at all. To what extent such conclusions bear upon cavitation during the three-dimensional flow of water at high Reynolds numbers is discussed in another paper of the Conference Proceedings. The present tests were conducted in perhaps a week's time by M. L. Albertson, one of the Institute staff, simply to illustrate the simplicity of the air-tunnel method in such exploratory studies.

Of considerable interest in the flow of either air or water is the degree of turbulence produced by screens and baffles and its rate of decay in the downstream direction, for many types of research equipment (such as wind and water tunnels) involve the use of such stilling devices, and many types of research project (such as sediment and heat diffusion) require the production of known degrees of turbulence. Following the completion of the NDRC studies previously outlined, the larger air tunnel of the Institute was at once allocated to the evaluation of turbulence characteristics downstream from a systematic series of lattice and perforated

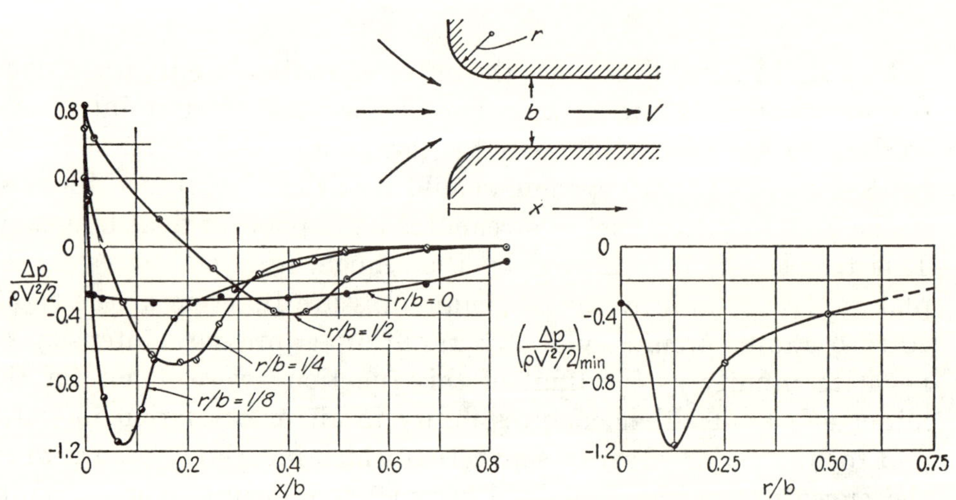

Fig. 12—Distribution of Pressure at a Conduit Inlet.

screens under the sponsorship of the David Taylor Model Basin of the Navy Bureau of Ships. The relative ease of measuring both the intensity and the scale of turbulence in air by means of either the gas-diffusion or the hot-wire process permitted data of the type plotted in Fig. 13 to be obtained. The results are, needless to say, equally applicable to the flow of air, water, or any other fluid of comparable viscosity.

It is eventually planned to utilize the same or a similar air tunnel in studying the effect of turbulence upon the rate of evaporation from both solid and liquid surfaces. Preliminary tests in this program were made by Mr. Albertson in the demonstration tunnel of Fig. 1 to ascertain practical aspects of evaporation measurement, the relative wall height of a schematic land pan being used as the geometric variable. As indicated in Fig. 14, a systematic trend of evaporation rate with air speed is obtained despite the relatively crude instruments and controls assembled for the initial tests and the obviously incomplete dimensional treatment of the variables involved.

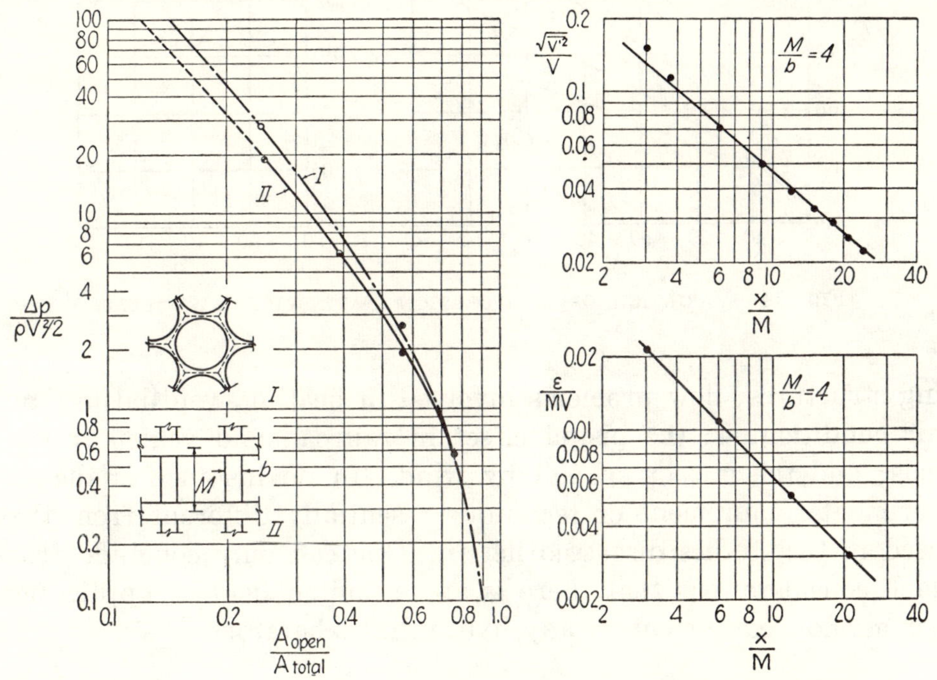

FIG. 13—ENERGY LOSS AND TURBULENCE PRODUCED BY SCREENS.

There are, of course, many instances such as this in which air flow plays a direct part in problems of hydraulic engineering: the entrainment of air by flowing water; the generation of waves by wind; the ventilation of weirs, siphons, and gates; and the general question of drag at an air-water interface. Practically every one of these has already been investigated in one or another hydraulics laboratory using some form of air duct or tunnel. In addition, however, there are various air-flow problems in allied fields of civil, mechanical, and marine engineering which must now be sent to already overtaxed aerodynamic wind tunnels for experimental analysis: pressure distribution on buildings in high winds; methods of smoke abatement; air-induced vibration of suspended or rotat-

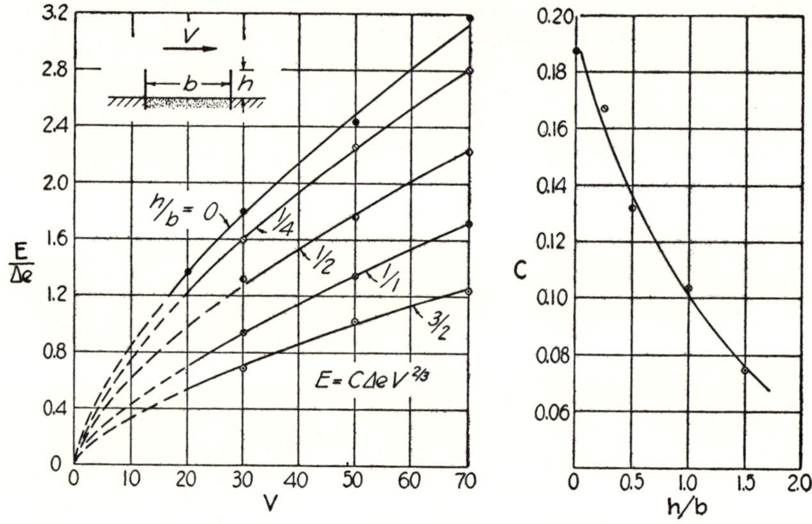

FIG. 14—VARIATION OF EVAPORATION RATE WITH AIR SPEED.

ing structures; flow problems involved in heating, ventilating, and air conditioning; the piston effect in railroad and vehicular tunnels; and the erosion of soil by wind. In no instance is the experimental equipment or technique essentially different from that used in the studies discussed herein. One can only conclude from such considerations that there is an unlimited field of application for air-flow equipment in any hydraulics laboratory.

Paper No. 2409

DIFFUSION OF SUBMERGED JETS

By M. L Albertson,[1] Jun. ASCE, Y. B. Dai,[2] R. A. Jensen,[3] and Hunter Rouse,[4] M. ASCE

With Discussion by Messrs. James S. Holdhusen, Duilio Citrini, Stanley Corrsin, W. Douglas Baines, Abraham Streiff, Harold R. Henry, and M. L. Albertson, Y. B. Dai, R. A. Jensen, and Hunter Rouse.

Synopsis

As the direct result of turbulence generated at the borders of a submerged jet, the fluid within the jet will undergo both lateral diffusion and deceleration, and at the same time fluid from the surrounding region will be brought into motion. The approximate characteristics of the corresponding mean flow pattern are derived analytically, with the exception of a single experimental constant, through assumptions that: (1) The pressure is hydrostatically distributed throughout the flow; (2) the diffusion process is dynamically similar under all conditions; and (3) the longitudinal component of velocity within the diffusion region varies according to the normal probability function at each cross section. Experimental data are presented which justify the analysis and provide the necessary coefficients for flow from both slots and orifices. All results are reduced to a form immediately useful for design purposes.

Statement of the Problem

As far as most engineers are concerned, high-velocity flow from a submerged outlet represents merely an irrecoverable loss of power, for a basic axiom of hydraulics states that the entire kinetic energy of such a jet will be dissipated through reaction with the surrounding fluid. In more explicit terms, the difference in velocity between a jet and the region into which it is discharged will give rise to a pronounced degree of instability, the kinetic energy of the oncoming flow steadily being converted into kinetic energy of turbulence, and the latter steadily decaying through viscous shear. However, such conversion is by no means restricted to the vicinity of the outlet, for any reduction in kinetic energy necessarily represents a decrease in the velocity of flow, and even the

Note.—Published in December, 1948, *Proceedings*. Positions and titles given are those in effect when the paper or discussion was received for publication.

[1] Associate Prof., Civ. Eng. Dept., Colorado Agri. and Mech. College, Fort Collins, Colo.
[2] Associate Prof., National Hunan Univ., Changsha, Hunan, China.
[3] Research Engr., Fuels Div., Battelle Memorial Inst., Columbus, Ohio.
[4] Director, Iowa Inst. of Hydr. Research, State Univ. of Iowa, Iowa City, Iowa.

most elementary considerations of continuity indicate that the area of the flow section must become great as the velocity becomes small. In view of the Newtonian principle of action and reaction, moreover, it will be realized that deceleration of the fluid in the jet can occur only through simultaneous acceleration of the surrounding fluid, so that the total rate of flow past successive sections of the jet will actually increase with distance from the outlet.

The phenomenon of "exit loss" so simply treated in hydraulics evidently involves far more than a local energy conversion. Indeed, a detailed knowledge of the velocity distribution within such an expanding stream is essential to many problems of diffusion. Flow of a liquid suspension from a submerged pipe into a settling basin is a case in point, although a closely related phase of mixing is found in the efflux of heated or cooled air from a ventilation duct in the wall of a room; in either case the engineer is interested in the shape and the "carry" of the jet, and in the amount of entrainment by (and hence dilution of)

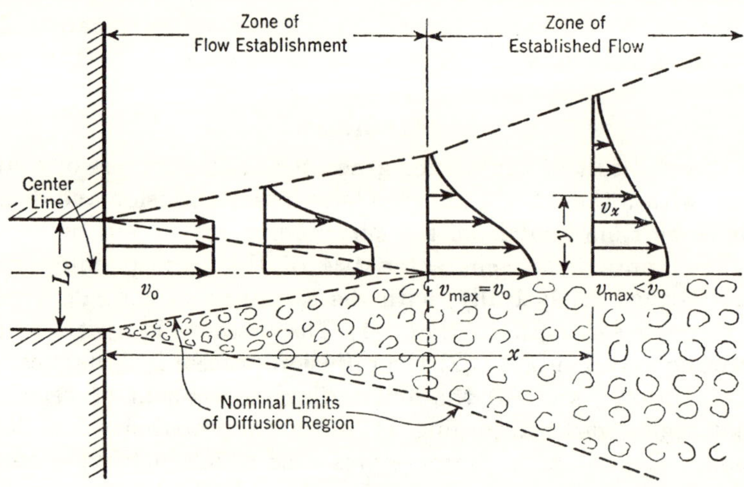

FIG. 1.—SCHEMATIC REPRESENTATION OF JET DIFFUSION

the jet from section to section. Directly related, although more closely restricted by boundary (that is, free-surface) conditions, is the diffusion of plunging nappes from spillways or of submerged jets from sluice gates, because information as to the accompanying rate of deceleration is often necessary in the prevention of scour beyond the apron. Finally, since the slip stream of a propeller differs little from any other type of jet, the velocity distribution in the wake of aircraft and watercraft (a matter of vital importance, for example, in submarine detection) should be subject to the same method of analysis.

With reference to Fig. 1, it will be seen that an initial zone of flow establishment must exist beyond the efflux section of either a two-dimensional or a three-dimensional submerged jet. Since the fluid discharged from the boundary opening may be assumed to be of relatively constant velocity, at the efflux section there will necessarily be a pronounced velocity discontinuity between the jet and the surrounding fluid. The eddies generated in this region of high shear will immediately result in a lateral mixing process which progresses both inward and outward with distance from the efflux section. Such lateral mixing produces a necessarily balanced action and reaction: On the one hand, the fluid within the jet is gradually decelerated; on the other hand, fluid from the surrounding region is gradually accelerated or entrained. As a result, the constant-velocity core of the jet will steadily decrease in lateral extent, whereas both the rate of flow and the over-all breadth of the jet will steadily increase in magnitude, with distance from the efflux section. The limit of this initial zone of flow establishment is reached when the mixing region has penetrated to the center line of the jet.

Once the entire central part of the jet has become turbulent, the flow may be considered as fully established, for the diffusion process continues thereafter without essential change in character. Further entrainment of the surrounding fluid by the expanding eddy region is now balanced inertially by a continuous reduction in the velocity of the entire central region. Such variation approaches (but never actually attains) a limit as the center-line velocity becomes of negligible magnitude a very great distance from the original efflux section, the lateral extent of the eddy region—and hence the quantity of fluid within the jet—then being wholly out of proportion to conditions at the initial section.

Just as the theoretical limit of the zone of established flow is evidently nonexistent, the border section between this zone and the zone of establishment (indicated by the vertical broken line in Fig. 1) can be considered sharply defined only to the extent to which one may neglect the zone of transition which necessarily exists between two distinct flow regimes. There is, in fact, no precise point at which the eddies from opposite sides of the jet can be said to meet, for the statistical nature of the mixing process makes it impossible to place more than an arbitrary limit on its lateral range. The boundaries of the diffusion region, like the border between the two zones, must therefore be accepted merely as convenient nominal designations.

Conditions within the zone of flow establishment and the zone of established flow were first investigated theoretically by W. Tollmien in 1926,[5] on the assumptions that: (a) The sole effective force was the tangential shear expressed in terms of the lateral momentum transport of the mixing process; (b) the mixing length varied with the first power of the longitudinal distance from the efflux section; and (c) the velocity of the turbulence was proportional to the product of the mixing length and the mean velocity gradient. With the exception of a numerical coefficient which had to be evaluated experimentally, Mr. Tollmien was able to derive a stream function which indicated with good approximation the actual pattern of the mean flow in each region.[6] Later studies by other investigators[7,8,9] introduced either the concept of vorticity transport or the assumption of constant eddy viscosity across any section of the diffusion zone,

[5] "Berechnung turbulenter Ausbreitungsvorgänge," by W. Tollmien, *Zeitschrift für angewandte Mathematik und Mechanik*, Vol. 6, 1926, p. 468.

[6] "Turbulente Ausbreitungsvorgänge im Freistrahl," by P. Ruden, *Die Naturwissenschaften*, Vol. 21, 1933, p. 375.

[7] "Investigations of the Turbulent Mixing Regions Formed by Jets," by A. M. Keuthe, *Journal of Applied Mechanics*, September, 1935, p. A-87.

[8] "Application of the Modified Vorticity Transport Theory to the Turbulent Spreading of a Jet of Air," by S. Tomotika, *Proceedings*, Royal Soc. of London, Vol. 165, Series A, March–April, 1938, p. 65.

[9] "Concerning the Velocity and Temperature Distribution in Plain and Axially Symmetrical Jets," by L. Howarth, *Proceedings*, Cambridge Philosophical Soc., Vol. 34, 1938, p. 185.

but did not add materially to the agreement with experimental measurements. As a matter of fact, the sensitivity of the mean velocity-distribution function to the type of turbulence structure assumed is not great; the comparison between measurement and theory has, as a result, never been sufficient to provide a conclusive check on the accuracy of the assumed mixing characteristics. Indeed, not until detailed measurements of the turbulence within two-dimensional and three-dimensional jets were made by S. Corrsin[10] and by H. W. Liepmann and J. Laufer[11] was the considerable discrepancy between assumption and fact fully realized.

Since an approximate analysis of the mean velocity distribution within either zone actually requires no assumption as to the distribution of turbulence, and since the characteristics of jet diffusion have not yet been published in a form readily usable by engineers, this paper seeks to fulfil this twofold purpose. The dimensional aspects of the phenomenon are first discussed, followed by an elementary physical analysis of the mean flow pattern. Experimental data covering a considerable range of each independent variable for both two-dimensional and three-dimensional flow are then presented to justify the analysis and to provide the single numerical coefficient still required for each case. All results are finally summarized in a convenient dimensionless form.

Analysis of the Mean Flow Pattern

General Considerations.—If the Reynolds number for fluid efflux from a submerged boundary outlet is not too low, the mean velocity v at any point (see Fig. 1) should depend only on the coordinates x, y, and z, on the efflux velocity v_0, and on a linear dimension L_0 characterizing the particular outlet form. These variables may be grouped in the dimensionless relationship:

$$\frac{v}{v_0} = f_1\left(\frac{x}{L_0}, \frac{y}{x}, \frac{z}{x}\right) \dots\dots\dots\dots\dots(1)$$

This relationship must be considered to involve the magnitude and the direction of the vector v, the components of which may further be related through the differential equation of continuity:

$$\frac{\partial v_x}{\partial x} + \frac{\partial v_y}{\partial y} + \frac{\partial v_z}{\partial z} = 0 \dots\dots\dots\dots\dots(2)$$

The rate of flow or volume flux Q past successive normal sections may be written as the integral of the differential flux $v_x\, dA$ over any normal section. Since, because of entrainment, Q will vary with the longitudinal distance x from the efflux section, its ratio to the efflux rate Q_0 may be expected to have the functional form:

$$\frac{Q}{Q_0} = \frac{\int_0^\infty v_x\, dA}{v_0 A_0} = f_2\left(\frac{x}{L_0}\right) \dots\dots\dots\dots\dots(3)$$

---

[10] "Investigation of Flow in an Axially Symmetrical Heated Jet of Air," by S. Corrsin, *Advance Confidential Report No. 3L23*, National Advisory Committee for Aeronautics, Washington, D. C., 1943 (*Wartime Report W-94*).

[11] "Investigations of Free Turbulent Mixing," by H. W. Liepmann and J. Laufer, *Technical Note No. 1257*, National Advisory Committee for Aeronautics, Washington, D. C., August, 1947.

in which A_0 is the cross-sectional area of the outlet. Similarly, since the momentum flux M may be written as the integral of the volume flux $v_x\,dA$ times the longitudinal component of momentum per unit volume $\rho\,v_x$, ρ being the fluid density, the ratio of M for any section to M_0 for the efflux section should be

$$\frac{M}{M_0} = \frac{\displaystyle\int_0^\infty (v_x)^2\,dA}{(v_0)^2\,A_0} = f_3\!\left(\frac{x}{L_0}\right) \dots\dots\dots\dots\dots (4)$$

Finally, since the energy flux E is expressible as the integral of the volume flux $v_x\,dA$ times the kinetic energy per unit volume $\rho\,v^2/2$, the ratio of E for any section to E_0 for the boundary outlet should take the form:

$$\frac{E}{E_0} = \frac{\displaystyle\int_0^\infty v^2\,v_x\,dA}{(v_0)^3\,A_0} = f_4\!\left(\frac{x}{L_0}\right) \dots\dots\dots\dots\dots (5)$$

Such general considerations will yield no further clue to the actual functional relationships without specific statements as to the dynamics of the flow. For example, on the basis of experimental evidence it may safely be presumed that the pressure distribution is essentially hydrostatic throughout the zone of motion—in other words, that the sole force producing the deceleration of the jet and the acceleration of the surrounding fluid is the tangential shear within the mixing region. Because this process is wholly internal, it follows at once that the momentum flux must be a constant for all normal sections of a given flow pattern:

$$\frac{M}{M_0} = \frac{\displaystyle\int_0^\infty (v_x)^2\,dA}{(v_0)^2\,A_0} = 1 \dots\dots\dots\dots\dots (6)$$

If, moreover, viscous action is presumed to have no influence on the mixing process, the diffusion characteristics—and hence the characteristics of the mean flow—should be dynamically similar under all conditions. Thus, in effect, the same velocity function must characterize every section within the diffusion region. As a matter of fact, experimental data follow the general trend of the Gaussian normal probability function:

$$\frac{v_x}{v_{\max}} = \exp\left(-\frac{y^2}{2\,\sigma^2}\right) \dots\dots (7)$$

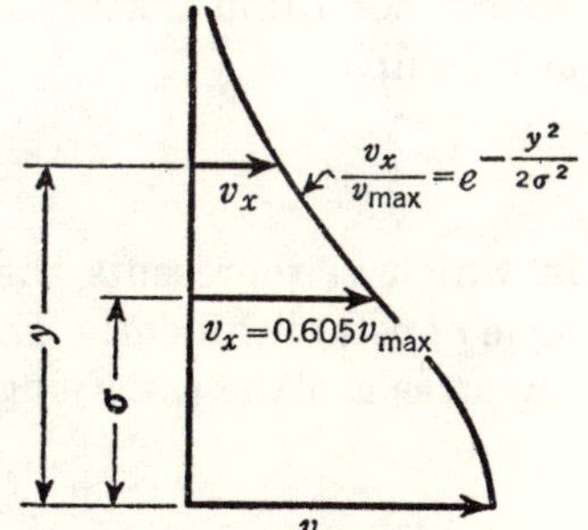

Fig. 2.—Characteristics of the Normal-Probability Curve

in which the right-hand member expresses the value e^n, the quantity in parentheses being the value of n. (A similar simplification applies subsequently in Eqs. 12, 19, 24, and 29.) The use of Eq. 7 in the analysis permits the characteristics of the entire flow pattern to be expressed in terms of the two parameters which define the proportions of the curve: the velocity $v_{\max}$ and the standard or root-mean-square deviation σ (see Fig. 2). Moreover, not only does Eq. 1 then reduce to

$$\frac{v_{\max}}{v_0} = f_5\left(\frac{x}{L_0}, \frac{\sigma}{x}\right) \dots\dots\dots\dots\dots\dots(8)$$

but the constancy of the momentum flux together with the similarity of the velocity profiles at successive sections will be found to require that

$$\frac{\sigma}{x} = C \dots\dots\dots\dots\dots\dots\dots\dots(9)$$

That is, the jet will spread at a linear rate defined by the constant C.

Through use of Eqs. 6, 7, and 9, all characteristics of the mean flow pattern for any specific boundary condition may now be determined analytically, with the exception of a single experimental coefficient. Such characteristics include the variation in the magnitude and direction of the velocity vector with coordinate space, and the variation of volume flux and energy flux with distance from the outlet section. In the following pages these characteristics are evaluated for the zone of flow establishment and for the zone of established flow under both two-dimensional and three-dimensional conditions.

Zone of Flow Establishment.—In the case of efflux from a long slot, it will be noted from Fig. 3 that $v_{\max}$ will correspond to v_0 as long as the eddies have not penetrated to the central plane of

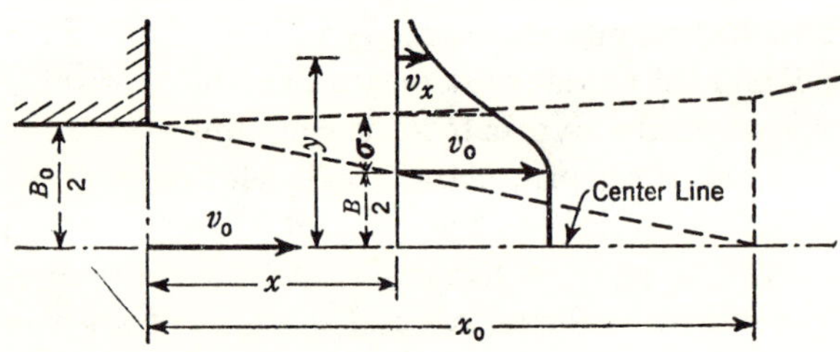

Fig. 3.—Definition Sketch for Zone of Flow Establishment

the jet. Therefore, the distribution of v_x across any normal section within this zone may be represented as a first approximation by the two symmetrical halves of the probability curve connected with a straight line through the constant-velocity core. Evaluation of the momentum-flux integral of Eq. 6, for the condition that $\sigma/x = C_1$, will be found to yield at once the specific relationship:

$$\frac{x_0}{B_0} = \frac{1}{\sqrt{\pi}\,C_1} \dots\dots\dots\dots\dots\dots\dots(10)$$

in which x_0 represents the distance from the efflux section to the end of the zone of establishment. Evidently, the nominal inner boundary of the diffusion region is a plane satisfying the expression:

$$\frac{B}{B_0} = 1 - \frac{x}{x_0} \dots\dots\dots\dots\dots\dots\dots(11)$$

The distribution function for v_x in the diffusion region then takes the explicit form:

$$\frac{v_x}{v_0} = \exp\left[-\frac{\left(y + \sqrt{\pi}\,C_1\frac{x}{2} - \frac{B_0}{2}\right)^2}{2\,(C_1 x)^2}\right] \dots\dots\dots\dots(12)$$

Similar evaluation of the volume-flux integral of Eq. 3, after introduction of v_x from Eq. 12, will result in the relationship:

$$\frac{Q}{Q_0} = 1 + \sqrt{\pi}\,(\sqrt{2} - 1)C_1\,\frac{x}{B_0}\dots\dots\dots\dots\dots(13)$$

Inasmuch as the rate of change in volume flux with longitudinal distance dQ/dx must equal twice the velocity of entrainment (that is, the velocity v_y at a considerable distance y from the jet), it follows that

$$\operatorname*{Limit}_{y\to\infty}\frac{v_y}{v_0} = -\tfrac{1}{2}\sqrt{\pi}\,(\sqrt{2} - 1)C_1\dots\dots\dots\dots(14)$$

The same result would be obtained as a limit through solution of the differential equation of continuity for v_y as a function of x and y; however, since the latter function must be shown graphically rather than algebraically, it is not introduced at this point. (All detailed evaluations of this nature have been deposited with the Engineering Societies Library for reference.)

Although the energy-flux integral properly involves the magnitude of the velocity vector in the term for kinetic energy per unit volume, the vector magnitude may be expected to differ little from that of its longitudinal component except in the outer regions, and the contribution of these regions to the total flux should be relatively small. The energy-flux ratio may therefore be approximated in terms of v_x, as follows:

$$\frac{E}{E_0} = 1 + \sqrt{\pi}\,(\sqrt{\tfrac{2}{3}} - 1)C_1\,\frac{x}{B_0}\dots\dots\dots\dots\dots(15)$$

Flow from a circular orifice should differ from the foregoing only in the substitution of the orifice diameter D_0 for the slot width B_0 and the substitution of the radial distance r for the lateral distance y in the basic equations, the integration being performed over the corresponding circular areas for the condition that $\sigma/x = C_2$. At the limit $x = x_0$, the momentum relationship at once results in

$$\frac{x_0}{D_0} = \frac{1}{2\,C_2}\dots\dots\dots\dots\dots\dots\dots\dots(16)$$

which is a counterpart of Eq. 10. However, the general solution of the momentum relationship yields an expression which—in contradistinction to Eq. 11 for the two-dimensional case—indicates that the diameter of the nominal inner border of the diffusion region for three-dimensional flow is not a linear function of x; that is,

$$\frac{D}{D_0} = \sqrt{1 + (\pi - 4)(C_2)^2\,\frac{x^2}{(D_0)^2}} - \sqrt{\pi}\,C_2\,\frac{x}{D_0}\dots\dots\dots\dots(17)$$

Fortunately, the difference between this curvilinear function and the linear approximation

$$\frac{D}{D_0} = 1 - \frac{x}{x_0}\dots\dots\dots\dots\dots\dots\dots(18)$$

is probably no greater than the discrepancy between the actual velocity distribution and the assumed probability curve. For all practical purposes it is sufficient, therefore, to write the velocity distribution within the region of diffusion in the simplified form:

$$\frac{v_x}{v_0} = \exp\left[-\frac{(r + C_2 x - D_0/2)^2}{2 (C_2 x)^2} \right] \dots\dots\dots\dots (19)$$

In terms of this simplification, the equation for the variation in the volume-flux ratio with distance from the orifice then becomes

$$\frac{Q}{Q_0} = 1 + 2 (\sqrt{2\pi} - 2)C_2 \frac{x}{D_0} + 4 (3 - \sqrt{2\pi})(C_2)^2 \frac{x^2}{(D_0)^2} \dots\dots (20)$$

Although, as in the two-dimensional case, the rate of entrainment may now be expressed as the derivative of the volume flux with longitudinal distance, for reasons of continuity the radial velocity v_r beyond the diffusion region will vary inversely with radial distance from the center line; in other words, the product $r\, v_r$ rather than v_r itself approaches a finite limit in such three-dimensional flow:

$$\operatorname*{Limit}_{r \to \infty} \frac{v_r}{v_0} \frac{r}{D_0} = -\tfrac{1}{4}(\sqrt{2\pi} - 2)C_2 - (3 - \sqrt{2\pi})(C_2)^2 \frac{x}{D_0} \dots\dots (21)$$

The variation of v_r within the diffusion zone may be evaluated from the differential equation of continuity; since this function must again be shown graphically rather than algebraically, it is not presented herein. Finally, as in the

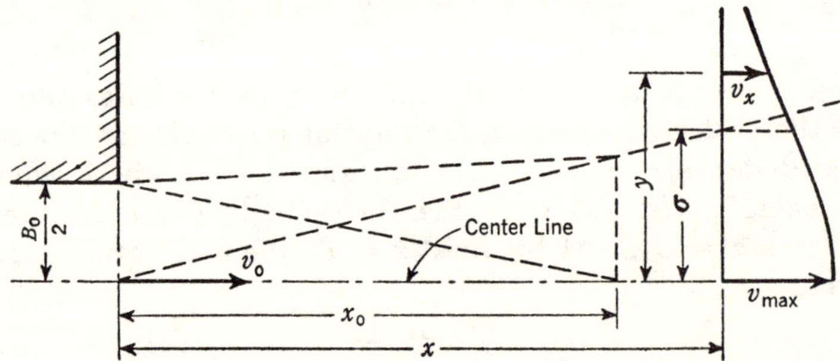

Fig. 4.—Definition Sketch for Zone of Established Flow

case of flow from a slot, it may be assumed that the difference between the magnitude of the velocity vector and that of its longitudinal component will have little effect on the expression for the energy flux past successive sections beyond an orifice, which thus becomes

$$\frac{E}{E_0} = 1 + 2 \left(\sqrt{\frac{2\pi}{3}} - 2 \right) C_2 \frac{x}{D_0} + 4 \left(\frac{5}{3} - \sqrt{\frac{2\pi}{3}} \right) (C_2)^2 \frac{x^2}{(D_0)^2} \dots (22)$$

Zone of Established Flow.—As will be noted from Fig. 4, at all sections beyond the zone of flow establishment the probability curve indicating the velocity distribution will be continuous across the line of symmetry, and the ratio σ/x will therefore become directly indicative of the rate of outward spread of the eddy region. Assuming, for the two-dimensional case, that $\sigma/x = C_1$, evaluation of the momentum flux will lead at once to the expression:

$$\frac{v_{max}}{v_0} = \sqrt{\frac{1}{\sqrt{\pi}\,C_1}\frac{B_0}{x}} \dots\dots\dots\dots\dots\dots (23)$$

in which the quantity $B_0/(\sqrt{\pi}\,C_1)$ represents the same parametric distance x_0 as that in Eq. 10. In other words, determination of the single constant C_1 from experimental measurement will suffice for both zones. The distribution of the longitudinal velocity component in the zone of established flow may then be written as

$$\frac{v_x}{v_0} = \sqrt{\frac{1}{\sqrt{\pi}\,C_1}\frac{B_0}{x}}\,\exp\left[-\frac{1}{2(C_1)^2}\frac{y^2}{x^2}\right] \dots\dots\dots\dots (24)$$

From the equation for volume flux it will be found, for the established zone, that

$$\frac{Q}{Q_0} = \sqrt{2\sqrt{\pi}\,C_1\,\frac{x}{B_0}} \dots\dots\dots\dots\dots\dots (25)$$

and differentiation with respect to x will yield the rate of entrainment and the corresponding lateral velocity at a considerable distance from the jet:

$$\operatorname*{Limit}_{y\to\infty}\frac{v_y}{v_0} = -\sqrt{\frac{\sqrt{\pi}\,C_1}{8}\frac{B_0}{x}} \dots\dots\dots\dots\dots (26)$$

Eq. 26 will also be the limit of the general function for v_y obtained by graphical solution of the equation of continuity and reproduced at a later point. Evaluation of the energy flux in terms of $v_x \approx v$ finally leads to the approximate expression:

$$\frac{E}{E_0} = \sqrt{\frac{2}{3\sqrt{\pi}\,C_1}\frac{B_0}{x}} \dots\dots\dots\dots\dots\dots (27)$$

Once more replacing y and B_0 by r and D_0, and integrating over the corresponding areas, the equivalent expressions for the three-dimensional case may be evaluated for the condition that $\sigma/x = C_2$. The momentum equation at once reduces to

$$\frac{v_{max}}{v_0} = \frac{1}{2\,C_2}\frac{D_0}{x} \dots\dots\dots\dots\dots\dots\dots (28)$$

in which the quantity $D_0/(2\,C_2)$ corresponds to the same parametric distance x_0 as that in Eq. 16. Evidently, determination of the single experimental constant C_2 will suffice to define all characteristics of both zones, just as in the two-dimensional case. The remaining expressions then become

$$\frac{v_x}{v_0} = \frac{1}{2\,C_2}\frac{D_0}{x}\,\exp\left[-\frac{1}{2\,(C_2)^2}\frac{r^2}{x^2}\right] \dots\dots\dots\dots (29)$$

$$\frac{Q}{Q_0} = 4\,C_2\,\frac{x}{D_0} \dots\dots\dots\dots\dots\dots\dots (30)$$

$$\operatorname*{Limit}_{r\to\infty}\frac{v_r}{v_0}\frac{r}{D_0} = -\frac{C_2}{2} \dots\dots\dots\dots\dots\dots (31)$$

and

$$\frac{E}{E_0} = \frac{1}{3\,C_2}\frac{D_0}{x}\dots\dots\dots\dots\dots\dots\dots(32)$$

Characteristics of the Mixing Process.—The foregoing analysis of the pattern of mean flow was made with complete disregard for the detailed characteristics of the diffusion pattern on which the phenomenon depends. If the assumed distribution of mean velocity and the resulting diffusion rate are in even approximate agreement with actual flow conditions, however, it should be possible to obtain at least a rough indication of the diffusion characteristics.

Since the pressure distribution was assumed to be hydrostatic, the only force to be considered in analyzing the interaction of the jet and the surrounding fluid must be the shear exerted along any longitudinal plane. The cumulative effect of such shear will necessarily disappear as the distance from the jet axis becomes great (that is, a zero net force corresponds to a zero longitudinal gradient of the total momentum flux), but within the zone of diffusion the intensity of shear τ may be expected to vary considerably. In fact, its local magnitude may be evaluated in terms of the change in momentum flux which occurs as fluid passes through an elementary part of the diffusion zone.

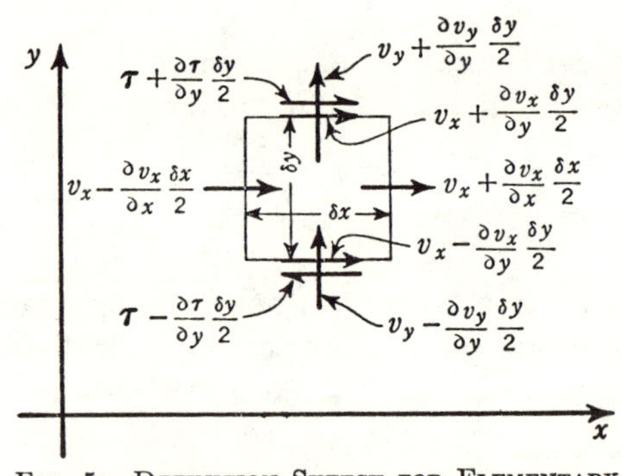

FIG. 5.—DEFINITION SKETCH FOR ELEMENTARY SHEAR AND MOMENTUM FLUX

With reference to Fig. 5, which represents the longitudinal cross section of a cubic element of space within the expanding jet, the flux of longitudinal momentum may be expressed as the product of the elementary volume flux at right angles to any face and of the corresponding longitudinal component of momentum per unit volume. The net longitudinal component of force on the faces of the element should then equal the net flux of longitudinal momentum through all faces of the element. For two-dimensional flow (that is, zero longitudinal shear on the $x-y$ faces), the momentum equation will be

$$\left(\tau + \frac{\partial \tau}{\partial y}\frac{\delta y}{2}\right)\delta x - \left(\tau - \frac{\partial \tau}{\partial y}\frac{\delta y}{2}\right)\delta x = \rho\left(v_x + \frac{\partial v_x}{\partial x}\frac{\delta x}{2}\right)^2\delta y - \rho\left(v_x - \frac{\partial v_x}{\partial x}\frac{\delta x}{2}\right)^2\delta y$$

$$+ \rho\left(v_x + \frac{\partial v_x}{\partial y}\frac{\delta y}{2}\right)\left(v_y + \frac{\partial v_y}{\partial y}\frac{\delta y}{2}\right)\delta x - \rho\left(v_x - \frac{\partial v_x}{\partial y}\frac{\delta y}{2}\right)\left(v_y - \frac{\partial v_y}{\partial y}\frac{\delta y}{2}\right)\delta x \dots (33)$$

which reduces to the following statement of conditions at a point:

$$\frac{\partial \tau}{\partial y} = \rho\left(v_x \frac{\partial v_x}{\partial x} + v_y \frac{\partial v_x}{\partial y}\right)\dots\dots\dots\dots\dots(34)$$

Evidently, the local intensity of shear will involve the integral of Eq. 34 with respect to y from the axis to the point in question:

$$\tau = \rho \int_0^y \left(v_x \frac{\partial v_x}{\partial x} + v_y \frac{\partial v_x}{\partial y} \right) dy \dots\dots\dots\dots\dots (35)$$

Because of the difficulty of expressing v_y by other than graphical methods, evaluation of this relationship must also proceed graphically, the results being shown in plotted form subsequently.   It is at once possible, on the other hand, to express certain characteristics of the turbulence in terms of this function.

From the Reynolds equations of acceleration, the intensity of longitudinal shear at any point in a turbulent fluid may be written as the sum of a mean viscous stress and an apparent turbulent stress:

$$\tau = \mu \frac{\partial v_x}{\partial y} - \rho \, \overline{v'_x v'_y} \dots\dots\dots\dots\dots\dots (36)$$

in which μ is the dynamic viscosity of the fluid.   Under conditions of such pronounced mixing as exists within the eddy region of a spreading jet, it appears reasonable to presume that the term for mean viscous shear will be relatively unimportant.   Indeed, this was prerequisite to the basic assumption of dynamic similarity.   The necessarily finite magnitude of the mean product of the turbulent velocity components $\overline{v'_x v'_y}$, on the other hand, indicates that a finite degree of correlation must exist between the components v'_x and v'_y. The turbulence must be definitely nonisotropic, therefore, and in all probability the root-mean-square values of these components will differ from one another at every point.   Because of the arbitrary nature of the assumed mean velocity distribution, it cannot be expected that the resulting mean flow pattern will yield any useful clue as to the relative magnitudes of these root-mean-square terms, and as a rough approximation it must necessarily be assumed that they are proportional.   Further assuming that the Prandtl concept of proportionality between these components and the product of a so-called mixing length l and the mean velocity gradient is not greatly in error, it follows that, qualitatively,

$$\tau = \rho \, \sqrt{\overline{(v'_y)^2}} \, l \frac{\partial v_x}{\partial y} = \rho \, l^2 \left(\frac{\partial v_x}{\partial y} \right) \left| \frac{\partial v_x}{\partial y} \right| \dots\dots\dots\dots (37)$$

in which the product $\sqrt{\overline{(v'_y)^2}} \, l$ represents the diffusion coefficient ϵ (that is, the so-called kinematic eddy viscosity).   Once Eq. 35 has been evaluated, it should be possible to express the approximate variation of l, $\sqrt{\overline{(v'_y)^2}}$, and ϵ as dimensionless functions of coordinate space, just as in the case of such other variables as v, Q, and E.   The results are shown graphically in the following section.

Experimental Results

Equipment and Test Procedure.—An experimental program resulting from a wartime fog-dispersal project of the Iowa Institute of Hydraulic Research (at the State University of Iowa, Iowa City, Iowa) involved the measurement of the mean velocity distribution in both two-dimensional and three-dimensional jets over a considerable range of each independent variable. Purely for convenience, all measurements were made in air, but there is no reason to presume that comparable results would not be obtained in any other fluid—whether liquid or gas—of moderately low viscosity. Essentially the same

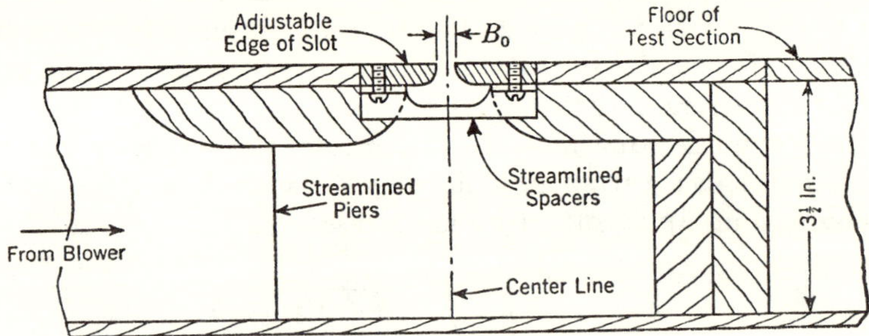

Fig. 6.—Details of Slot and Approach

experimental equipment was used in each phase of the project, the flow being produced by a 2-hp centrifugal fan with suitable ducts and stilling devices. The two-dimensional slot (Fig. 6) was formed from two 6-ft strips of 1¼-in. by ⅜-in. cold-rolled steel, carefully rounded along the inner edges and accurately adjustable in spacing from $\frac{1}{32}$ in. to ¼ in. This slot was set flush with the surface of a 6-ft by 12-ft level floor, and was bounded at the ends by 6-ft by 12-ft vertical walls. (Because attempts to measure the velocity distribution in the zone of establishment for even the ¼-in. slot proved futile, a special 1-in. slot

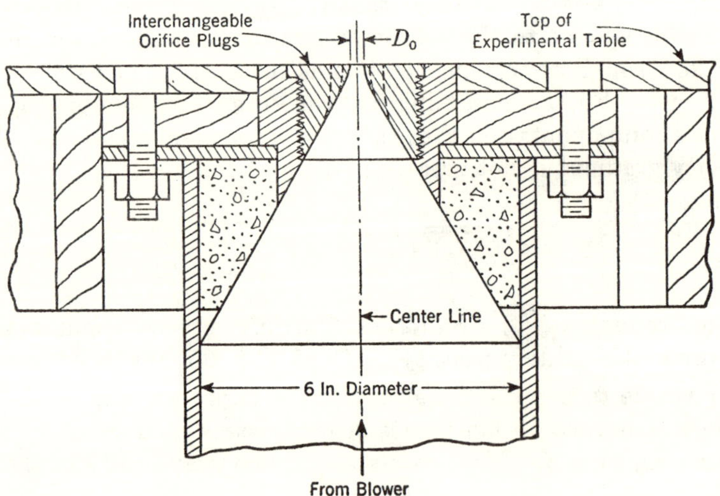

Fig. 7.—Details of Orifice and Approach

was made for this purpose.) The circular orifice (Fig. 7) was set into the top of a 4-ft by 4-ft table, and consisted of three interchangeable brass sections having outlet diameters of ¼ in., ½ in., and 1 in. with well-rounded transitions from the same 60° convergent approach section. In both cases, gage carriages permitted motion of measuring instruments in three coordinate directions, scales with verniers which read to 0.001 ft indicating the coordinate position.

Because of the extreme variation in velocity with location, it was necessary to utilize several different methods of velocity measurement. The efflux velocity of the jet and the velocity distribution in the outlet vicinity were determined with a No. 20 hypodermic needle used as a stagnation tube and connected directly to a differential water manometer read by point gage to 0.001 ft. In zones of moderate velocity the water manometer was replaced by a

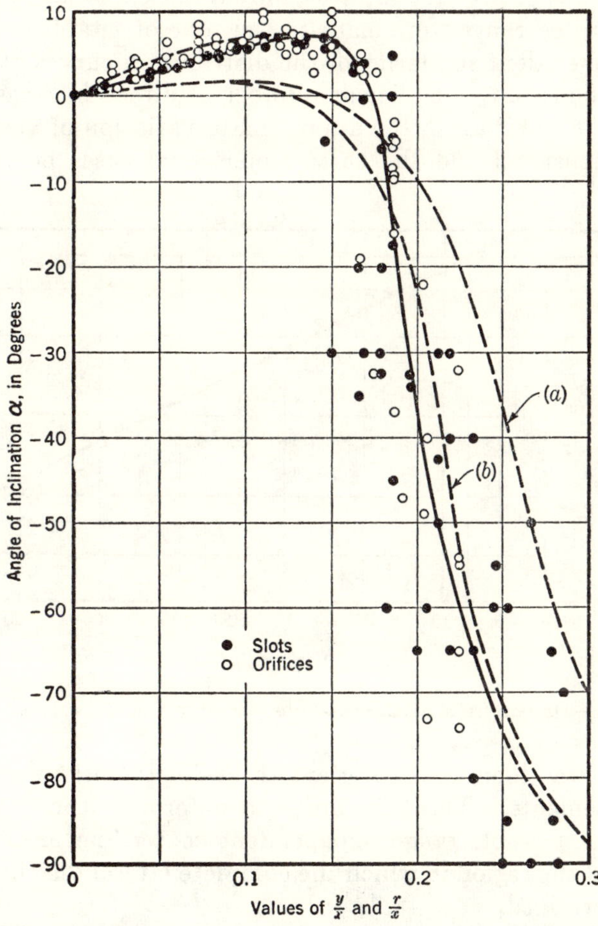

FIG. 8.—VARIATION IN DIRECTION OF FLOW WITHIN JETS FROM BOTH SLOTS AND ORIFICES

sensitive two-liquid gage, which was read by micrometer to 0.001 in. of alcohol. The lowest velocities were measured by a midget spoke-vane anemometer with jeweled bearings, which had been calibrated in an air tunnel. Except in the zones of flow establishment, the velocity instruments were oriented with respect to flow direction in accordance with prior indications of a directional vane consisting of a thin plate with a stagnation tube mounted on each side a short distance behind the leading edge. The vane was rotated about an axis through the leading edge until a zero differential reading was obtained on a manometer

connected to the two stagnation tubes.   In zones of very low velocity the vane
was replaced by a fine jet of smoke for visual observation.

Since the distribution of velocity at any cross section was assumed to be
governed solely by the efflux velocity, the outlet dimension, and the longitudinal distance to the section, these three independent variables were generally
so controlled in the tests that the maximum practicable range of each was
investigated in systematic steps as the other two were held constant.   There
were three exceptions to this general procedure: First, because the zone of
established flow was of primary interest, only a few exploratory measurements
of necessarily limited range were made in the zone of establishment.   Second,
as a result of the evident similarity of the distribution curves at all sections of
the established flow, only a sufficient number of direction measurements (see
Fig. 8) were made to establish the approximate variation of vector inclination
for the two-dimensional and the three-dimensional cases beyond reasonable

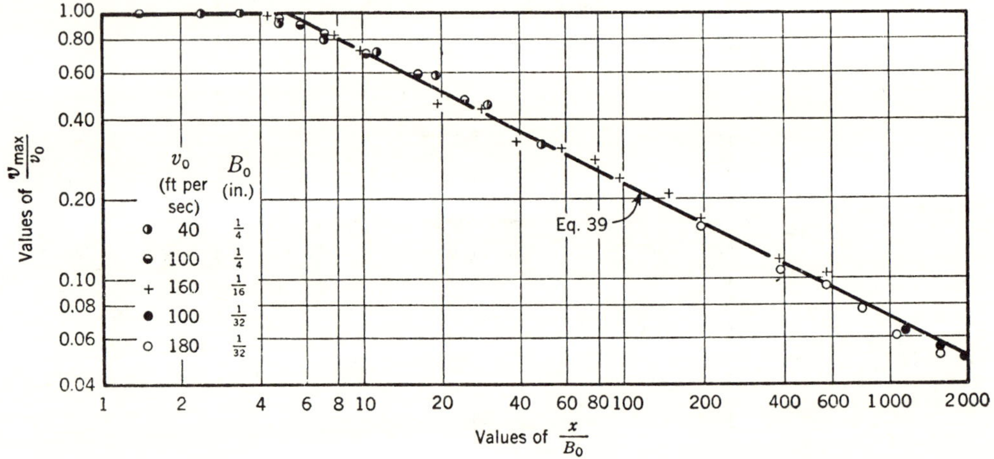

FIG. 9.—DISTRIBUTION OF CENTER-LINE VELOCITY FOR FLOW FROM SLOT

doubt, and this was then used to orient the velocity instruments for all subsequent measurements.   Third, in order to determine the variation in v_{max}
over the greatest possible range, independent center-line measurements were
made far beyond the region in which the complete lateral distribution could accurately be determined.

Results for the Two-Dimensional Jet.—Measurements of the center-line
velocity from the efflux section over a longitudinal distance equal to several
thousand times the slot width are shown as a dimensionless logarithmic plot
in Fig. 9.   Except for a limited zone of transition, the experimental points for
all slot widths and efflux velocities follow closely either a horizontal line with the
ordinate magnitude of unity for the zone of flow establishment, or a 1:2 sloping
line for the zone of established flow.   Since these two lines intersect at the
point $x_0/B_0 = 5.2$, it follows at once that the constant C_1 for the equations of
the zone of establishment of any two-dimensional jet will have the magnitude
$C_1 = B_0/(\sqrt{\pi}\, x_0) = 0.109.$

Introduction of this value into Eq. 12 yields the specific relationship for the velocity distribution in the diffusion region of the two-dimensional zone of establishment

$$\log_{10} \frac{v_x}{v_0} = -18.4 \left(0.096 + \frac{y - B_0/2}{x} \right)^2 \dots\dots\dots\dots (38)$$

which is plotted in Fig. 10, using the lateral location of the slot boundary as the zero reference on the ordinate scale. Superposed on the analytical curve are the measured values obtained in the exploratory study of the velocity distribution in this zone. The agreement between analysis and experiment is quite satisfactory.

For the zone of established flow, introduction of the same experimental coefficient C_1 = 0.109 into Eq. 23 leads to the specific expression for the sloping line of Fig. 9,

$$\frac{v_{\max}}{v_0} \sqrt{\frac{x}{B_0}} = 2.28 \dots (39)$$

Eq. 39 represents the limiting form of Eq. 24 after introduction of the constant C_1:

$$\log_{10} \frac{v_x}{v_0} \sqrt{\frac{x}{B_0}}$$

$$= 0.36 - 1.84 \frac{y^2}{x^2} \dots (40)$$

Eq. 40 is plotted in Fig. 11, together with all measured values for the longitudinal velocity component. Despite appreciable experimental scatter, the analytical curve is in reasonable agreement with the measurements over the entire range of the independent variables. The graphical solution of the continuity relationship for v_y, in terms of the constant C_1, is shown in the same general manner in Fig. 12, together with experimental values obtained by multiplying the measured velocities by the sines of the corresponding angles of inclination of Fig. 8. Although the data follow the general trend of the theoretical curve, the values computed from the measurements are at least 100% greater than those indicated by the plotted function—the discrepancy also

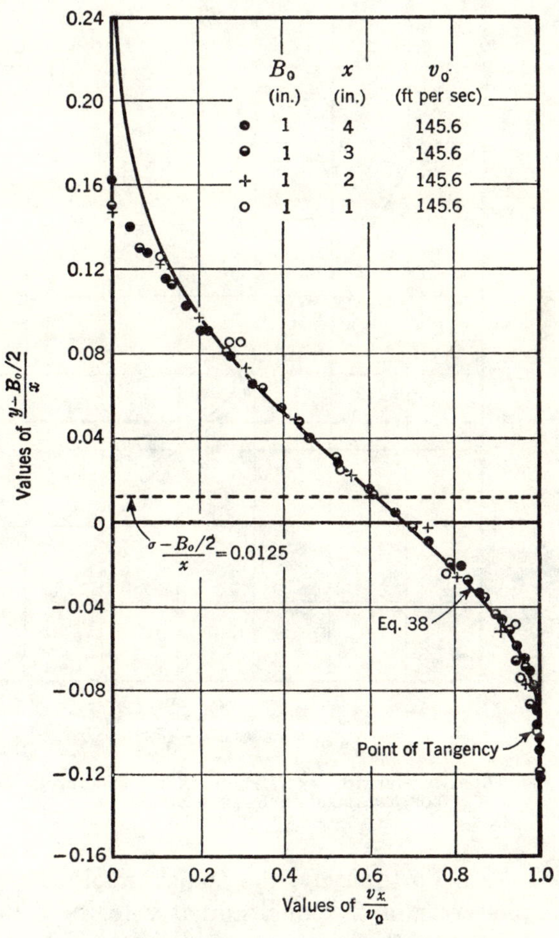

FIG. 10.—DISTRIBUTION OF LONGITUDINAL VELOCITY IN ZONE OF ESTABLISHMENT OF FLOW FROM SLOTS

being apparent from comparison of the analytical curve (curve (a)) of Fig. 8 with the measured values. No explanation can be given for this disagreement; neither do the computations disclose any likelihood of mathematical error, nor have repeated measurements by different observers led to essentially different

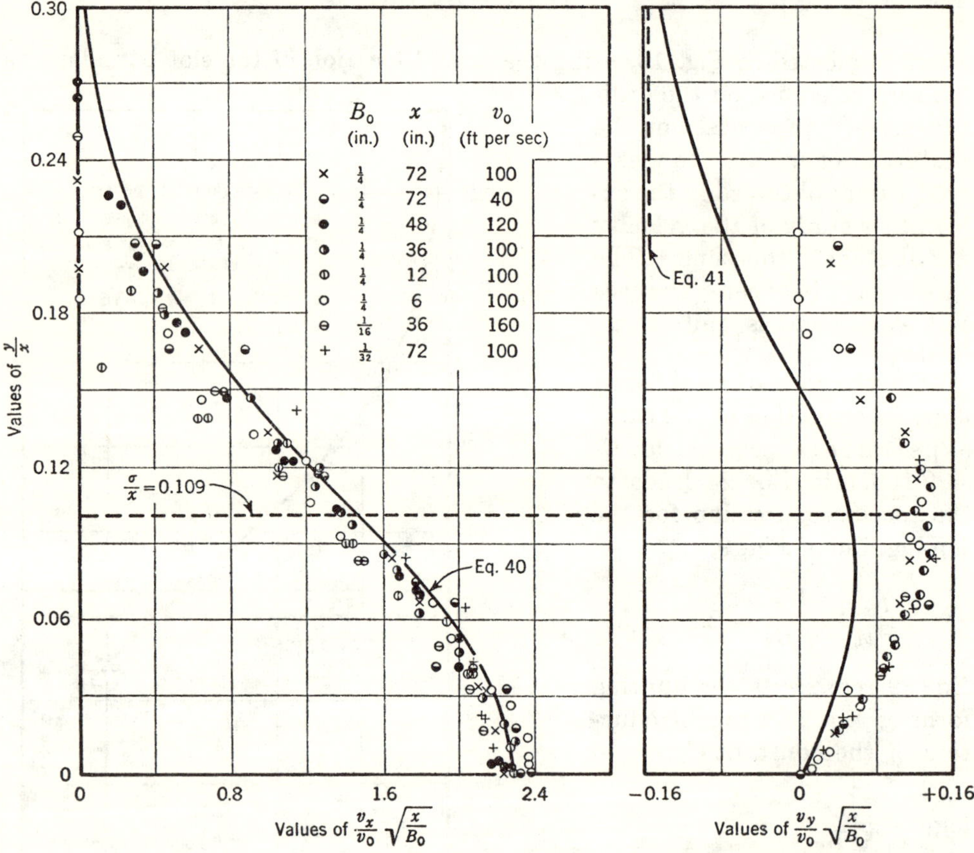

FIG. 11.—DISTRIBUTION OF LONGITUDINAL VELOCITY IN ZONE OF ESTABLISHED FLOW FROM SLOTS

FIG. 12.—DISTRIBUTION OF LATERAL VELOCITY IN ZONE OF ESTABLISHED FLOW FROM SLOTS

results. Fortunately, the angles involved were so small as to result in a negligible error in the longitudinal velocity. Because of the lack of dependable velocity measurements in the outer region, it cannot be stated whether or not the experimental data approach the velocity-of-entrainment limit given by the following specific form of Eq. 26:

$$\operatorname*{Limit}_{y \to \infty} \frac{v_y}{v_0} \sqrt{\frac{x}{B_0}} = -0.155 \dots\dots\dots\dots\dots(41)$$

Eqs. 13 and 15 for the volume-flux ratio and the energy-flux ratio in the zone of flow establishment become, in terms of the constant C_1,

$$\frac{Q}{Q_0} = 1 + 0.080 \frac{x}{B_0} \dots\dots\dots\dots\dots(42)$$

and

$$\frac{E}{E_0} = 1 - 0.035 \frac{x}{B_0} \dots\dots\dots\dots\dots\dots\dots (43)$$

Likewise, Eqs. 25 and 27 for the zone of established flow will reduce to

$$\frac{Q}{Q_0} = 0.62 \sqrt{\frac{x}{B_0}} \dots\dots\dots\dots\dots\dots\dots (44)$$

and

$$\frac{E}{E_0} = 1.86 \sqrt{\frac{B_0}{x}} \dots\dots\dots\dots\dots\dots\dots (45)$$

Eqs. 42, 43, 44, and 45 (as well as the basic expression $M/M_0 = 1$) are plotted in Fig. 13, together with values obtained by integration of the measured velocity-distribution curves for typical runs with x as the independent variable.

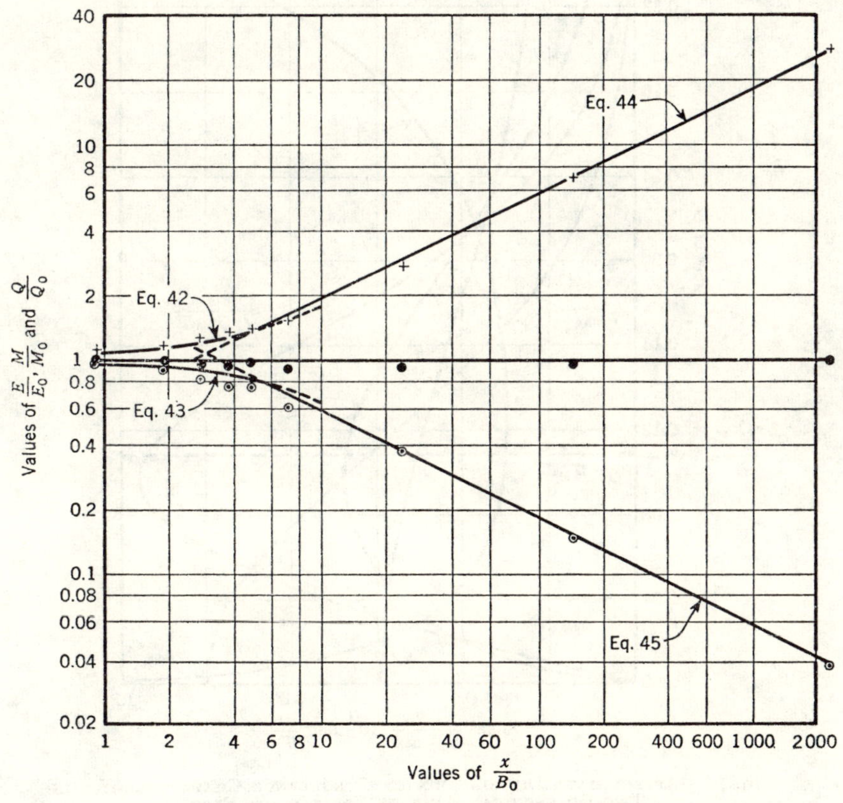

Fig. 13.—Distribution of Volume, Momentum, and Energy Flux Downstream from Slot

Although no measurements of turbulence were made in this experimental program, it is nonetheless pertinent to note the results of the approximate analysis of the eddy pattern. Fig. 14 includes the simultaneous graphical solutions of Eqs. 35 and 37 as dimensionless curves for τ, $\sqrt{(v'_y)^2}$, l, and ϵ at any

section of the zone of established flow, the assumed curve for v_x being included
for purposes of orientation. As will be noted therein, the zone of maximum
intensity of shear, and hence maximum intensity of turbulence, lies somewhat
inside the point of inflection or maximum slope of the velocity curve (that is,
where $y = \sigma$). The scale of the eddies is a minimum in this general zone,
whereas the mixing coefficient attains a maximum at the center line. The
indications are, of course, merely qualitative, but nevertheless quite significant.

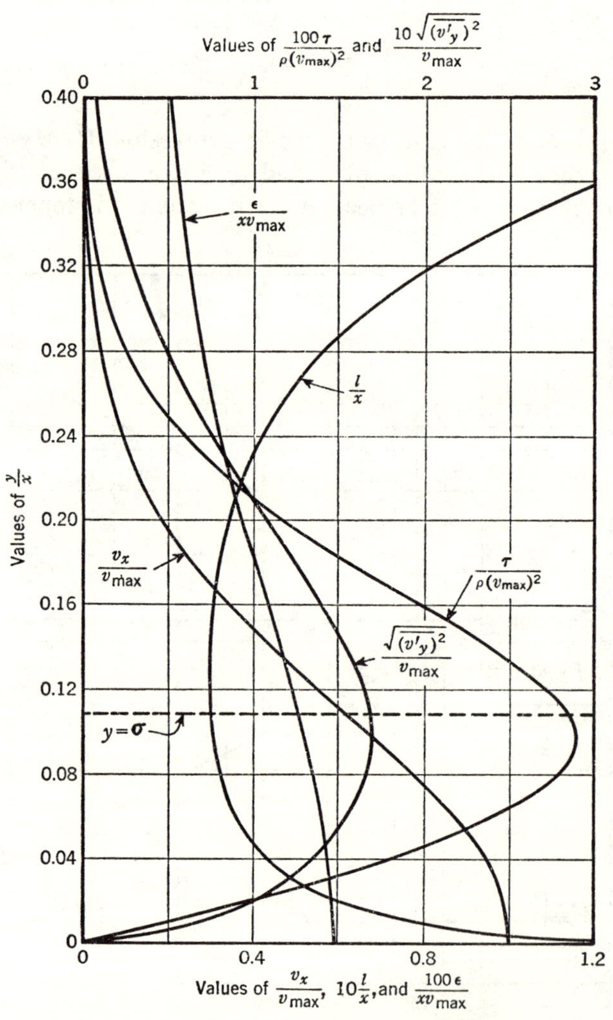

Fig. 14.—Approximate Distribution of Turbulence Characteristics in
Two-Dimensional Zone of Established Flow

Results for the Three-Dimensional Jet.—As in the two-dimensional case
just discussed, a logarithmic plot of all center-line velocity measurements
(Fig. 15) yields a horizontal and a sloping line for the two zones of flow, that
for the established zone now having a 1:1 slope. The intersection of these
lines at the point $x_0/D_0 = 6.2$ at once yields $C_2 = D_0/(2\,x_0) = 0.081$. In

terms of this constant, Eq. 19 for the distribution of the longitudinal velocity component in the zone of establishment assumes the specific form:

$$\log_{10} \frac{v_x}{v_0} = -33 \left(0.081 + \frac{r - D_0/2}{x} \right)^2 \dots\dots\dots\dots (46)$$

which is plotted, together with measured data, in Fig. 16. Unlike the case of the slot, the agreement between analysis and measurement is not wholly satisfactory. As shown by the broken line, considerable improvement would result through use of the smaller border value $x_0/D_0 = 4.5$ (that is, $C_2 = 0.111$).

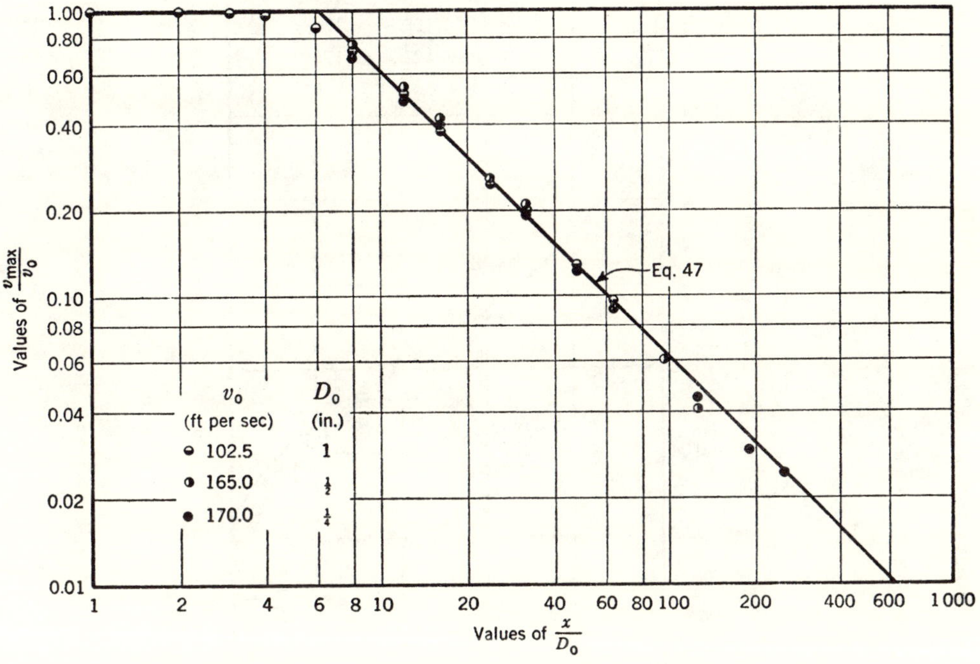

FIG. 15.—DISTRIBUTION OF CENTER-LINE VELOCITY FOR FLOW FROM ORIFICE

However, since close agreement in the limited zone of establishment is considered less significant than evaluation of both zones (as in the two-dimensional case) in terms of the same coefficient, the value $C_2 = 0.081$ is used throughout. Thus, Eq. 28 is seen to take the specific form:

$$\frac{v_{max}}{v_0} \frac{x}{D_0} = 6.2 \dots\dots\dots\dots\dots (47)$$

which corresponds to the sloping line of Fig. 15. This is simply the limiting form of the general relationship for v_x of Eq. 29, which accordingly becomes

$$\log_{10} \frac{v_x}{v_0} \frac{x}{D_0} = 0.79 - 33 \frac{r^2}{x^2} \dots\dots\dots\dots (48)$$

and is plotted in Fig. 17, together with measured data. Despite considerable experimental scatter due to the great range of the independent variables, the points will be seen to follow the probability curve with good approximation. Fig. 18 gives the graphical solution of the continuity equation for v_r, supplemented by experimental results obtained through multiplying the measured vector magnitudes by the sines of the corresponding angles of inclination. As in the two-dimensional case, the data follow the trend of the

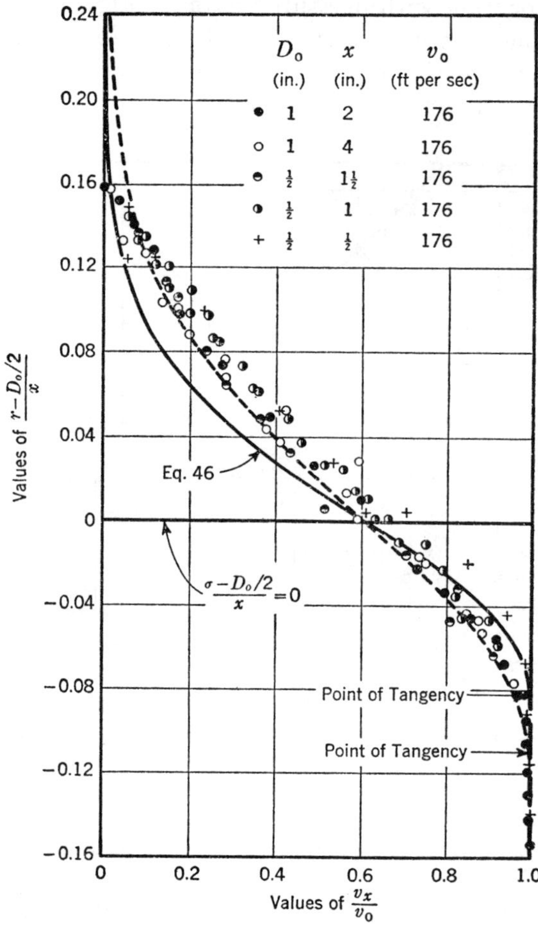

Fig. 16.—Distribution of Longitudinal Velocity in Zone of Establishment of Flow from Orifices

theoretical curve, but the values computed from the velocity measurements are at least twice as great as those of the plotted function (see also curve (*b*), Fig. 8). It is evidently impossible to discover whether or not they finally approach the hyperbolic curve corresponding to the following specific form of Eq. 31 for the radial velocity of entrainment:

$$\operatorname*{Limit}_{r \to \infty} \frac{v_r}{v_0} \frac{x}{D_0} = -\,0.040\,\frac{x}{r}\dots\dots\dots\dots\dots(49)$$

Introduction of the experimentally determined coefficients in Eqs. 20 and 22 for the volume-flux and energy-flux ratios in the zone of flow establishment gives

$$\frac{Q}{Q_0} = 1 + 0.083\,\frac{x}{D_0} + 0.0128\,\frac{x^2}{(D_0)^2}\dots\dots\dots\dots(50)$$

and

$$\frac{E}{E_0} = 1 - 0.090\,\frac{x}{D_0} + 0.0058\,\frac{x^2}{(D_0)^2}\dots\dots\dots\dots(51)$$

Similar evaluation of Eqs. 30 and 32 for the zone of established flow yields

$$\frac{Q}{Q_0} = 0.32\,\frac{x}{D_0}\dots\dots\dots\dots\dots\dots(52)$$

and

$$\frac{E}{E_0} = 4.1\,\frac{D_0}{x}\dots\dots\dots\dots\dots\dots(53)$$

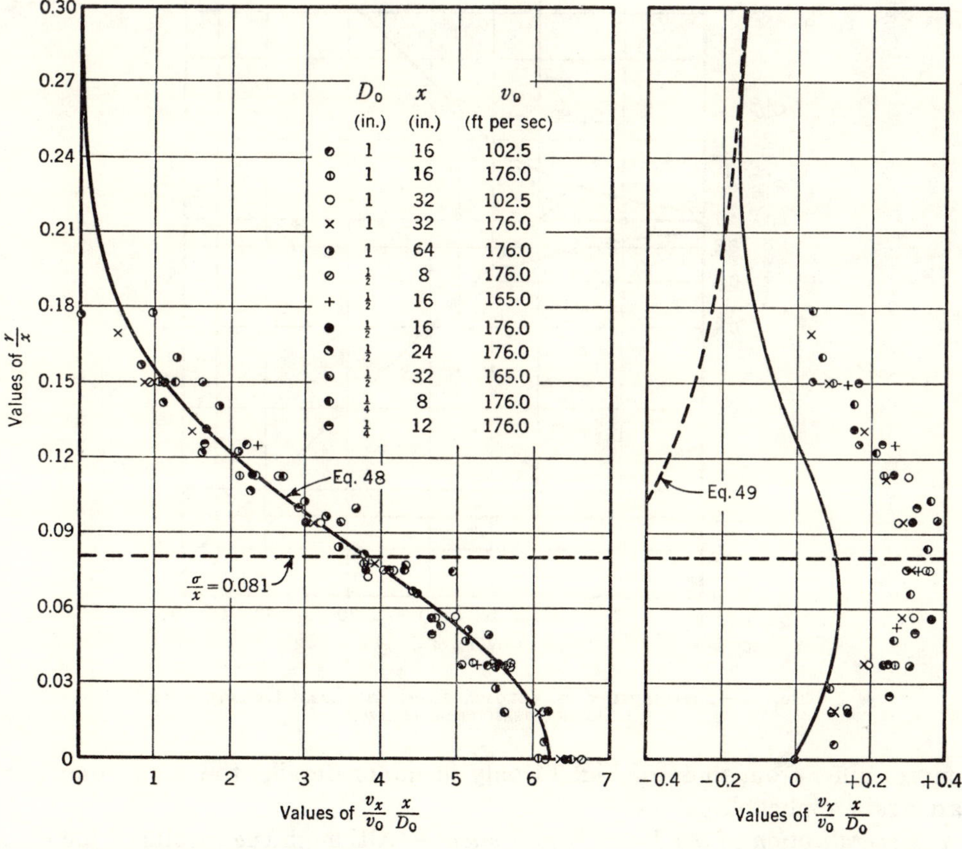

FIG. 17.—DISTRIBUTION OF LONGITUDINAL VELOCITY IN ZONE OF ESTABLISHED FLOW FROM ORIFICES

FIG. 18.—DISTRIBUTION OF RADIAL VELOCITY IN ZONE OF ESTABLISHED FLOW FROM ORIFICES

Eqs. 50, 51, 52, and 53 are plotted in Fig. 19, together with points representing the corresponding integrals of measured velocity-distribution curves for typical runs in which x was the only independent variable.

Except for the fact that the development leading to Eq. 35 must be revised according to the distinction between the cubic element of the two-dimensional case and the cylindrical element of the three-dimensional case, evaluation of the approximate characteristics of the turbulence proceeds in essentially the same manner. Solution of the corresponding differential equations results in the functions for τ, $\sqrt{\overline{(v'_r)^2}}$, l, and ϵ plotted in Fig. 20 in dimensionless form together with the function for v_x for purposes of orientation. Inasmuch as the

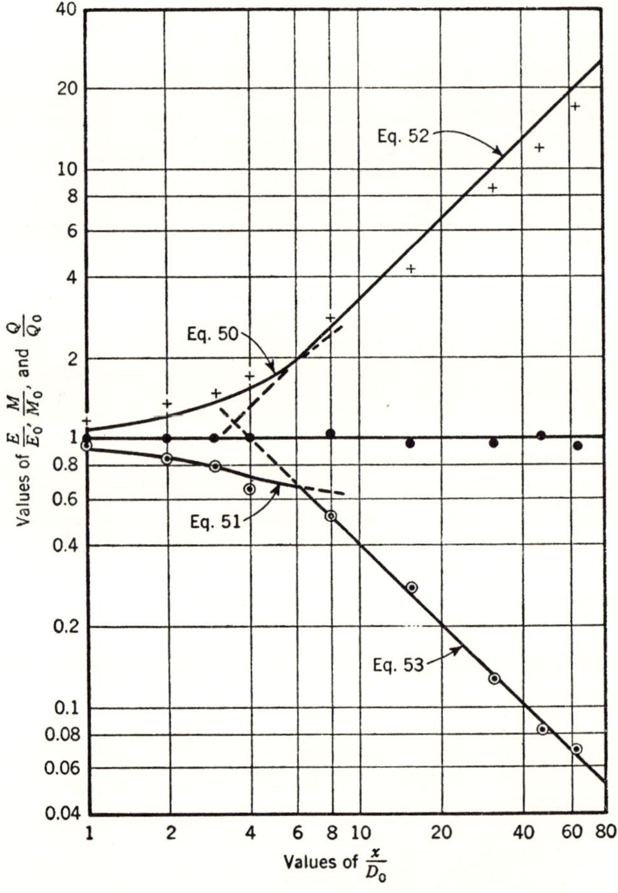

Fig. 19.—Distribution of Volume, Momentum, and Energy Flux Downstream from Orifice

curves differ from those of Fig. 14 only in minor details, the same comments are again applicable.

Generalization of the Mean Flow Pattern.—Although the results of the combined analytical and experimental investigation presented in the foregoing must be considered as approximate rather than as rigorous, the series of specific equations and plotted curves are nevertheless sufficiently close to actual fact to be used with confidence in general problems of design. The equations or diagrams themselves indicate no limit of applicability. In fact, no systematic deviation from the respective functions was noted with change in either length or velocity scale. Two probable limits of applicability should therefore be mentioned. At great distances from the efflux section—however accurately the measured curves may be extrapolated—the velocities of flow become so low that extraneous effects rather than the dynamics of the jet itself will begin to govern the flow pattern. Although no significant deviation of the center-line velocities from the predicted function was found in experiments at distances

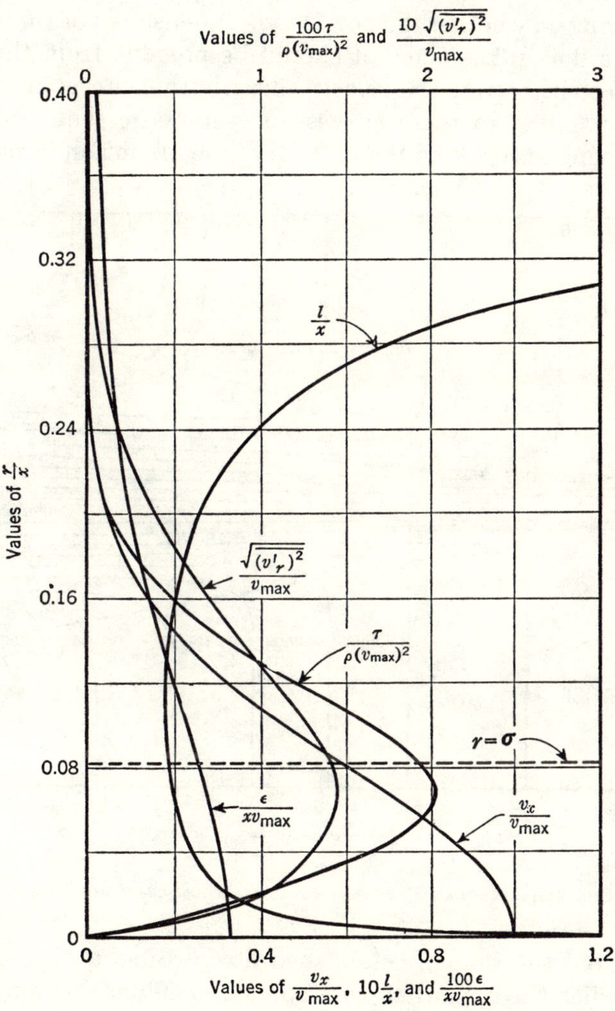

Fig. 20.—Approximate Distribution of Turbulence Characteristics in Three-Dimensional Zone of Established Flow

from the efflux section equal to several hundred times the orifice diameter and to several thousand times the slot width, it must be emphasized that extreme care was taken to eliminate all possible disturbance of the jet. At low Reynolds numbers for the efflux section, moreover, it is to be expected that viscous influences will prevent the full development of turbulence in the region of diffusion. The lowest Reynolds number, $\mathbf{R} = v_0 L_0/\nu$, attained in the tests described herein was about 1,500. There is, of course, no reason to expect any difference between the flow of a gas and the flow of a liquid at comparable Reynolds numbers. The results may be considered applicable, therefore, to the efflux of either type of fluid—provided only that gas velocities do not become so high that acoustic effects begin to influence the flow pattern and that liquid velocities do not become so high that appreciable cavitation occurs within the eddies.

If a streamline or stream surface of the mean motion is defined as a line or surface (depending on whether the flow is two-dimensional or three-dimensional) across which no flow takes place, it should be possible from the foregoing results to develop—either from the velocity-distribution curves or from the stream function corresponding to these curves—a systematic sequence of such lines or surfaces showing at a glance the pattern of mean motion in both the zone of

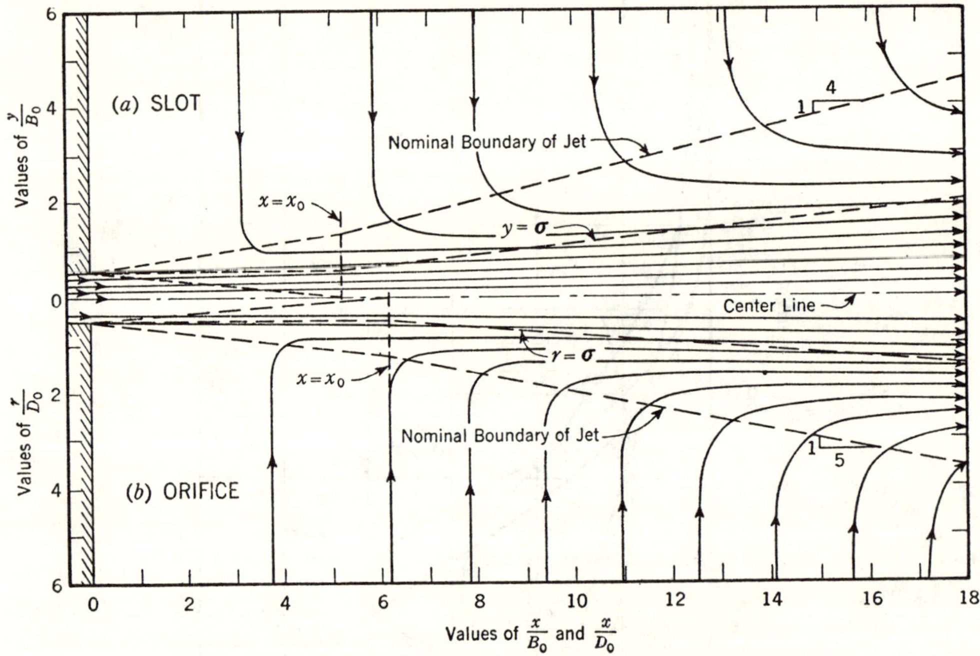

FIG. 21.—GENERAL PATTERNS OF FLOW FROM SLOT AND ORIFICE

establishment and the zone of established flow. Such a diagram is presented in Fig. 21, in which the upper half corresponds to efflux from a long slot and the lower half to that from a circular orifice. In the two-dimensional case, the spacing of the streamlines is inversely proportional to the local magnitude of the velocity, so that the speed as well as the direction of flow is readily apparent. In the three-dimensional case the spacing of the stream surfaces is inversely proportional to the velocity times the radial distance to the point in question, so that—although the direction of motion is immediately evident—comparison of velocity magnitudes by eye is not readily possible. Each of the diagrams contains the locus of the point of inflection (that is, the maximum slope) of the error curve for the longitudinal velocity distribution, which corresponds roughly to the region of maximum intensities of shear and turbulence. Also indicated is the nominal boundary of the jet—that is, the zone in which the longitudinal velocity is only a small percentage of that at the center line. The nominal boundary of the jet from a slot will thus have a slope of approximately 1:4, whereas that for an orifice is approximately 1:5. As previously indicated, this diagram should be characteristic of all jets under comparable flow conditions.

Not until after the completion of the present paper did the authors receive a copy of the very excellent and comprehensive treatise by H. Reichardt,[12] which discusses submerged jets as one of a series of related diffusion phenomena. Noteworthy in connection with the pattern of mean flow just presented is Mr. Reichardt's conclusion that the lateral streamlines of the two-dimensional jet—in contradistinction to the radial streamlines of the three-dimensional jet (and to the lateral streamlines as shown at the top of Fig. 21)—must be asymptotic to a family of parabolas having a common focus at the origin of the coordinate system, in order that the velocity of the lateral flow, like that of the radial flow, may properly be zero at infinity. This would accordingly modify the indicated outskirts of the two-dimensional pattern, but would affect the streamlines of the central flow to a wholly imperceptible degree.

CONCLUSIONS

Through assumptions of hydrostatic pressure distribution, dynamic similarity of the diffusion process, and normal-probability variation of the velocity, analytical expressions have been developed for the patterns of mean flow within submerged jets from both slots and orifices. Measurements conducted on the flow of air at different velocities from slots and orifices of different sizes have been shown to be in substantial agreement with the analytical relationships, and have provided the single experimental coefficient required to complete the analysis in the two-dimensional and in the three-dimensional case. Algebraic expressions and plotted curves have been presented in dimensionless form for the distribution of velocity, volume flux, and energy flux in the zone of flow establishment and the zone of established flow for each boundary condition. In addition, a qualitative treatment of the variation in mixing characteristics and a generalized diagram of the mean flow pattern have been given. All results are considered applicable for design purposes to the flow of any liquid or gas at moderate to high Reynolds numbers.

ACKNOWLEDGMENTS

This paper represents an original contribution of the Iowa Institute of Hydraulic Research of the State University of Iowa, all measurements having been conducted in the Institute laboratories and all analytical developments having been carried to completion by members of the Institute staff. The experiments on efflux from slots—with the exception of measurements in the zone of flow establishment contributed by D. E. Metzler—were performed by

---

[12] "Gesetzmässigkeiten der freien Turbulenz," by H. Reichardt, *Forschungsheft 414*, Verein Deutscher Ingenieure, 1942.

Messrs. Albertson and Jensen as a seminar project in the Department of Mechanics and Hydraulics of the University. The experiments on efflux from circular orifices were performed by Mr. Dai as a thesis investigation in partial fulfilment of the requirements for the degree of Master of Science in the Graduate College of the University. Because of its direct application to certain studies being conducted by the Institute for the David Taylor Model Basin (Carderock, Md.), Bureau of Ships, United States Navy, part of the analysis presented herein was supported by Navy research funds. J. S. McNown, Assoc. M. ASCE, acted frequently as consultant during the preparation of the paper, and W. D. Baines, Jun. ASCE, evaluated many of the more complex integrals. All phases of the project were under the direct supervision of Professor Rouse.

DESIGN OF CHANNEL EXPANSIONS

By Hunter Rouse,[26] M. ASCE, B. V. Bhoota,[27] Assoc. M. ASCE, and En-Yun Hsu[28]

Synopsis

Following an introductory discussion of supercritical flow in divergent channels, the matter of channel design is discussed under three sequent headings: (1) Surface configuration at abrupt expansions; (2) efficient curvature of expanding boundaries; and (3) elimination of disturbances at the end of transitions. The extent of the agreement between elementary wave theory and experimental measurement is shown, and the results are presented in the form of generalized diagrams convenient for rapid exploration and preliminary design.

Introduction

In the design of hydraulic structures it is often necessary to provide for the lateral expansion of flow emerging at high velocity from a closed conduit, sluice gate, spillway, or steep chute. If such a transition section is made to diverge too rapidly, the major part of the flow will fail to follow the boundaries; if the divergence is too gradual, on the other hand, waste of structural material will result; and, finally, if local disturbances are produced by incorrect boundary form, either at these points or farther downstream the walls may fail to confine the flow. Any particular problem of this nature, to be sure, may be subjected to cut-and-try investigation through model tests, with results which are necessarily restricted to the specific model form. An exact and general analytic solution, unfortunately, is still a matter for the future, and may be approached only through application of sound physical principles as discussed in the first Symposium paper. However, a step in this direction has been made in the development of reasonably general relationships among the several major variables involved, by means of which an approximate solution may be obtained for a great range of boundary conditions.

As in other problems of steady flow in an open channel of nonuniform cross section, the variation in velocity and depth through a channel expansion will depend on the geometry of the channel boundaries, the rate of flow, and the fluid properties. Under boundary geometry must be considered the form of the channel walls, the slope and form of the floor, and the surface roughness of floor and walls. In a strict sense, under fluid properties one should consider the density, specific weight, viscosity, and surface tension; except in small models, however, or under conditions in which boundary shear is of particular

---

[26] Director, Iowa Inst. of Hydr. Research, State Univ. of Iowa, Iowa City, Iowa.
[27] Engr., Foreign Div., The Dorr Co., Bombay, India.
[28] Research Associate, Iowa Inst. of Hydr. Research, State Univ. of Iowa, Iowa City, Iowa.

moment, both surface tension and viscosity are of very minor importance, and the two remaining properties then reduce to their ratio $\gamma/\rho = g$, the gravitational acceleration.

If these different independent variables are combined by the Π-theorem [29] of dimensional analysis into a series of dimensionless ratios, as many length ratios will be obtained as are necessary to describe the relative geometrical proportions of the boundary, together with a flow parameter of the Froude type. The latter is generally written in the form:

$$\mathbf{F} = \frac{V}{\sqrt{g\,h}} \dots\dots\dots\dots\dots\dots\dots\dots\dots (44)$$

in which V is the mean velocity and h is the mean depth of the approaching flow.

For given boundary conditions, the relative form of the free surface and the relative velocity distribution will depend solely upon the magnitude of the Froude number. As in all cases of open-channel flow, the critical magnitude $\mathbf{F} = 1$ marks the border between two wholly different types of surface configuration and velocity distribution. For Froude numbers less than unity, the depth then being greater than the critical, a gradual enlargement of the cross section will result in a gradual increase in mean surface elevation and a corresponding reduction in mean velocity. For Froude numbers greater than unity, the depth then being less than the critical, the same gradual enlargement of the cross section will result in a gradual reduction in mean surface elevation and a corresponding increase in mean velocity. However, only if the divergent boundaries are continuous planes (which is physically impossible if the transition is to begin and end with other than zero and infinite cross-sectional areas) will the depth of flow and the magnitude of the velocity be constant over any normal section. In other words, at the beginning and at the end of the transition the local curvature or angularity of the walls and floor will produce disturbances which make it impossible to handle such a problem satisfactorily on the elementary basis of mean velocity and mean depth. For Froude numbers less than unity (which are not the concern of the present paper), the boundary may be designed and the flow pattern may be evaluated in much the same manner as for the corresponding transition in a closed conduit. On the other hand, for Froude numbers greater than unity, the problem of design and evaluation becomes one of gravity-wave analysis, since each increment of the boundary deflection may be considered to generate an incremental surface wave which crosses the flow at an angle depending upon the Froude number and the boundary form; only through determination of the cumulative effect of all such waves may the depth and velocity at each and every point be predicted.

As has been described in the first Symposium paper, there is at hand a graphical method which permits the direct construction of streamlines, "isovels," and water-surface contours for any boundary form, provided that: (1) The channel walls are vertical and the floor is horizontal, (2) the energy

---

[29] "Fluid Mechanics for Hydraulic Engineers," by Hunter Rouse, McGraw-Hill Book Co., Inc., New York, N. Y., 1938, pp. 13–18.

loss due to boundary resistance is negligible, and (3) the pressure is hydro-statically distributed. These provisions may at first glance appear decidedly restrictive, but they become less so as they are considered individually. Vertical channel walls are common; indeed, sloping walls are generally to be avoided in nonuniform high-velocity flow because of their tendency to exaggerate surface disturbances. Channel floors are seldom horizontal, however, and the boundary resistance is never completely negligible; on the other hand, slight to moderate slopes of either the floor or the total-head line generally have an influence which is secondary to that of the wall expansion, and such effects are in fact compensative rather than additive. Moreover, only in the case of relatively abrupt curvature at the beginning or the end of the expansion is the existence of nonhydrostatic zones to be expected, and these may be effectively eliminated by proper easing of the transition curve. Only two factors, in actuality, tend to limit the graphical method in its use for the complete design of a well-proportioned expansion. First, its application depends upon prior knowledge or assumption of the boundary geometry, so that determination of the best form of transition involves the tedious process of trial and error. Second, if (as is usually the case) a hydraulic jump is to form at the end of the expansion, the method offers no clue as to the inherent stability (or, more likely, instability) of the phenomenon. As a matter of experience, the formation of a jump or standing wave by other than the boundary curvature (for instance, by backwater from a downstream control) may lead to an asymmetric pattern of flow within the transition which is still wholly unpredictable.

Since the purpose of this paper is the provision of general rather than specifically detailed information on the behavior of high-velocity flow in any channel expansion and on the preliminary design of particular expansion structures, primary attention (once the agreement between theory and experiment has been shown to be satisfactory) is focused upon the reduction of all experimental data to a few composite diagrams from which the basic details of design may be determined. Both the experiments and the generalization of the experimental results group themselves naturally into three subdivisions of the problem—first, the characteristics of a high-velocity jet expanding upon a level floor; second, the effects of boundary curvature in the zone of divergence; and, third, phenomena accompanying the return to uniform flow at the end of the transition.

All experiments described herein were conducted at the Iowa Institute of Hydraulic Research of the State University of Iowa under a project sponsored by The Engineering Foundation and the Committee of the Hydraulics Division, ASCE, on Hydraulic Research. The first part of the project, including the construction of equipment, was undertaken as a doctoral dissertation by Mr. Bhoota,[30] and the second part as a master's thesis by Mr. Hsu,[31] who then completed the investigation as a staff member of the Iowa Institute. Messrs. C. H.

---

[30] "Characteristics of Supercritical Flow at an Abrupt Open-Channel Enlargement," by B. V. Bhoota, thesis presented to the State University of Iowa, at Iowa City, Iowa, in December, 1942, in partial fulfilment of the requirements for the degree of Doctor of Philosophy.

[31] "Characteristics of Supercritical Flow at a Gradual Open-Channel Enlargement," by En-Yun Hsu, thesis presented to the State University of Iowa, at Iowa City, Iowa, in February, 1946, in partial fulfilment of the requirements for the degree of Master of Science.

Hsia and M. M. Hassan, Assoc. M. ASCE, assisted in various phases of the analysis. The entire project was under the direction of Mr. Rouse.

Essentially the same equipment (see Fig. 50) was used for all experiments. Water was supplied from constant-level tanks, through 4-in. lines and 8-in. lines containing calibrated elbow meters, to a pressure tank 2.5 ft in diameter and 5 ft long. One end of the pressure tank was provided with three interchangeable nozzles yielding rectangular jets 0.4 ft by 0.4 ft, 0.3 ft by 0.6 ft, and 0.25 ft by 1.0 ft in cross section—that is, having width-depth ratios of 1, 2, and 4. Rates of discharge were such that flow at any Froude number from 1 to 8 could be established. Flush with the bottom of the nozzle outlet sections was a level table, normally horizontal but adjustable to a maximum slope of approximately 10° and provided with a suitably hooded waste trough. This

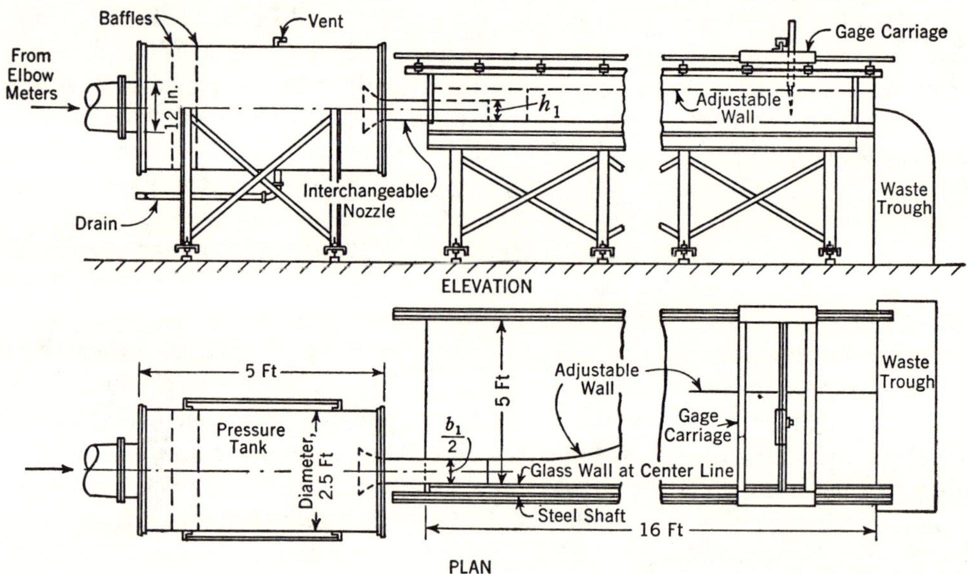

FIG. 50.—SCHEMATIC REPRESENTATION OF EXPERIMENTAL APPARATUS

table was originally 5 ft wide and 8 ft long, and was covered with oiled hardboard except for a plastic section with floor piezometers. It was later doubled in length and paved with finished concrete throughout. A gage carriage traveling on steel shafts above the table permitted three-directional movement of a point gage or pitot tube to any part of the test section. Through the earlier experiments one edge of the table was alined with the outer edge of each nozzle, a glass wall extending down the assumed line of symmetry of the transition for purposes of observation. That such elimination of one half of the flow pattern introduced negligible error was shown when the full transitions were later tested.

CHARACTERISTICS OF FLOW AT AN ABRUPT EXPANSION

The extreme case of a channel expansion is represented by the abrupt termination of the side walls, the channel floor continuing at the same slope.

If h_1 and V_1 represent the depth and mean velocity of the approaching flow, b_1 is the channel width, x and y are the longitudinal and lateral coordinates (measured from the outlet section and the center line, respectively) of a point of depth h, and if no other factors than the acceleration of gravity are assumed to influence the flow, these variables may be combined into the following dimensionless relationship:

$$\frac{h}{h_1} = f_1\left(\frac{x}{h_1}, \frac{y}{h_1}, \frac{b_1}{h_1}, \mathbf{F}_1\right) \dots\dots\dots\dots\dots\dots(45)$$

Evidently, the relative depth at any point of the flow should depend upon the relative coordinate location, the relative width of the channel outlet, and the Froude number of the approaching flow. The form of this functional relationship, of course, cannot be predicted through dimensional considerations, but must depend upon either physical analysis or experimental measurement.

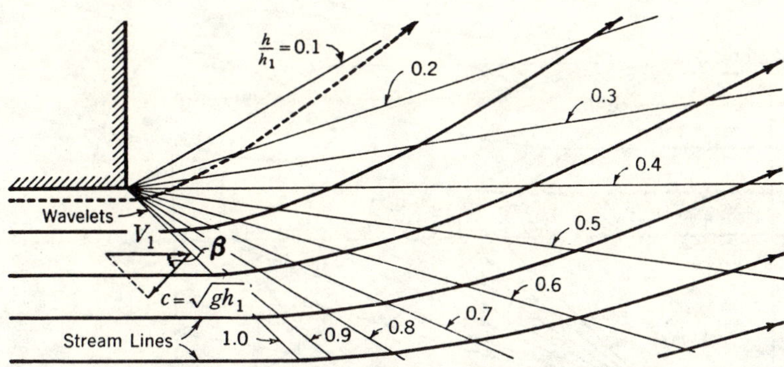

FIG. 51.—PATTERN OF FLOW NEAR THE ABRUPT END OF A CHANNEL WALL

The elementary wave theory indicates that the flow in the neighborhood of the end of either wall will begin to change in direction only as it passes the first negative wavelet (see Fig. 51), which lies at the angle $\beta = \sin^{-1}\sqrt{g\,h_1}/V_1 = \sin^{-1} 1/\mathbf{F}_1$ to the initial flow direction. From then on the streamlines may be considered to continue deviating through a series of infinitesimal steps, the angle of each succeeding wavelet depending upon the local magnitude of the continuously changing ratio of V to $\sqrt{g\,h}$—that is, the local Froude number. Each wavelet represents, in effect, a line of constant depth, so that proper selection among the infinite series of wavelets will yield the systematic series of surface contours shown in Fig. 51.

If now the zone is investigated in which the wavelets from the two opposite sides of the outlet begin to intersect, it will be seen (Fig. 52) that the resulting pattern of interference will yield a rather complex variation in depth and velocity of flow. The surface contours may again be determined by a rather laborious analysis of each element of the pattern in accordance with the elementary wave theory, but a far more rapid solution may be obtained by the graphical method of characteristics outlined in the first Symposium paper. Three such solutions, for different values of $\mathbf{F}_1$, are shown in Fig. 53.

The method of characteristics, in effect, reduces the functional relationship of Eq. 45 to the form:

$$\frac{h}{h_1} = f_2\left(\frac{x}{b_1}, \frac{y}{b_1}, \mathbf{F}_1\right) \dots\dots\dots\dots\dots\dots (46)$$

by combining the relative coordinate terms x/h_1 and y/h_1 with the initial width-depth ratio, b_1/h_1. This entails the inherent assumption of hydrostatic pressure distribution at all points—that is, the absence of appreciable vertical acceleration. As a matter of fact, at the abrupt end of either channel wall the pressure is far from hydrostatically distributed, as the water surface is practically vertical in such a zone. The extent to which this lack of fulfilment of

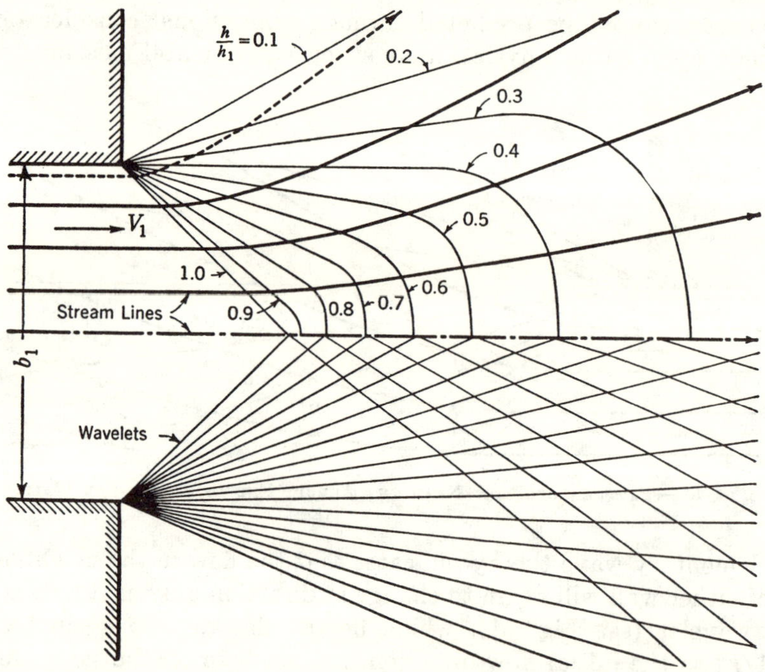

FIG. 52.—EFFECT OF WAVE INTERFERENCE FROM OPPOSITE SIDE OF CHANNEL

the assumption causes the actual surface configuration to differ from the theoretical evidently depends upon the magnitude of the ratio b_1/h_1. In wide, shallow channels the zone of disagreement is of relatively small extent; in narrow, deep channels, on the other hand, the pressure distribution will be markedly nonhydrostatic from wall to wall.

In illustrating the variation to be expected, experimentally measured surface contours for three different width-depth ratios are plotted in Fig. 54 for the same Froude numbers as those in Fig. 53. The deviations with b_1/h_1 are appreciable, but nevertheless secondary to the variation with the Froude number. In other words, using an average system of contours in preliminary design is quite in order. Even for the widest channel, however, it will be found that there is also a discrepancy between the measured contours and those

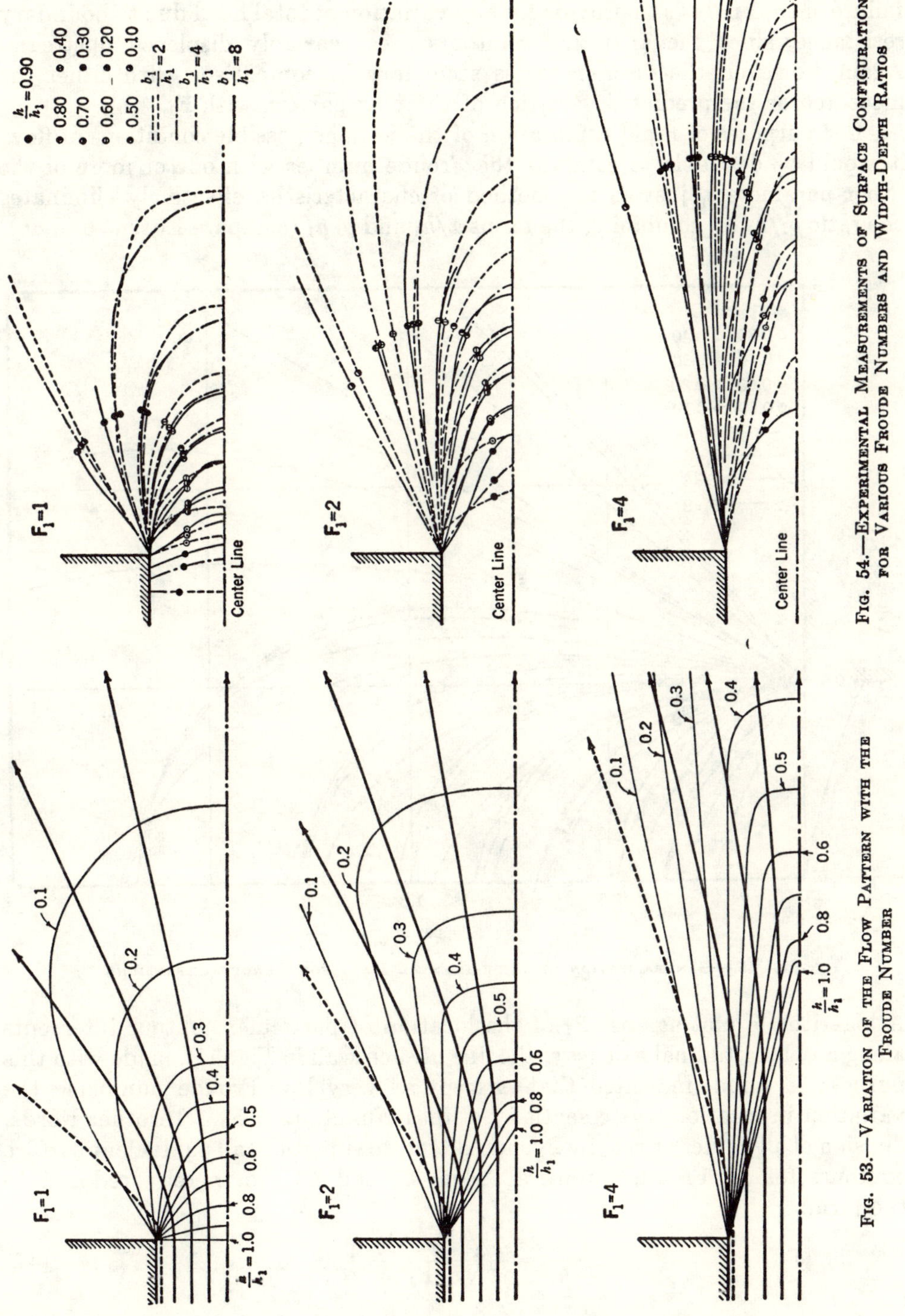

Fig. 54.—Experimental Measurements of Surface Configuration for Various Froude Numbers and Width-Depth Ratios

Fig. 53.—Variation of the Flow Pattern with the Froude Number

obtained by the graphical method of analysis. This may be attributed to the failure of the analysis to provide for any variation of total head due to boundary resistance, since the analytical contours are invariably displaced upstream. Again, however, the discrepancy is secondary in comparison with either the measured or the predicted variation of the flow pattern with F_1.

For purposes of rapid exploration of the various possible conditions of flow, it would be desirable to combine the Froude number with one or more of the other parameters, just as the method of characteristics effectively eliminates the ratio b_1/h_1 by combining the terms x/h_1 and y/h_1 (compare Eqs. 45 and 46).

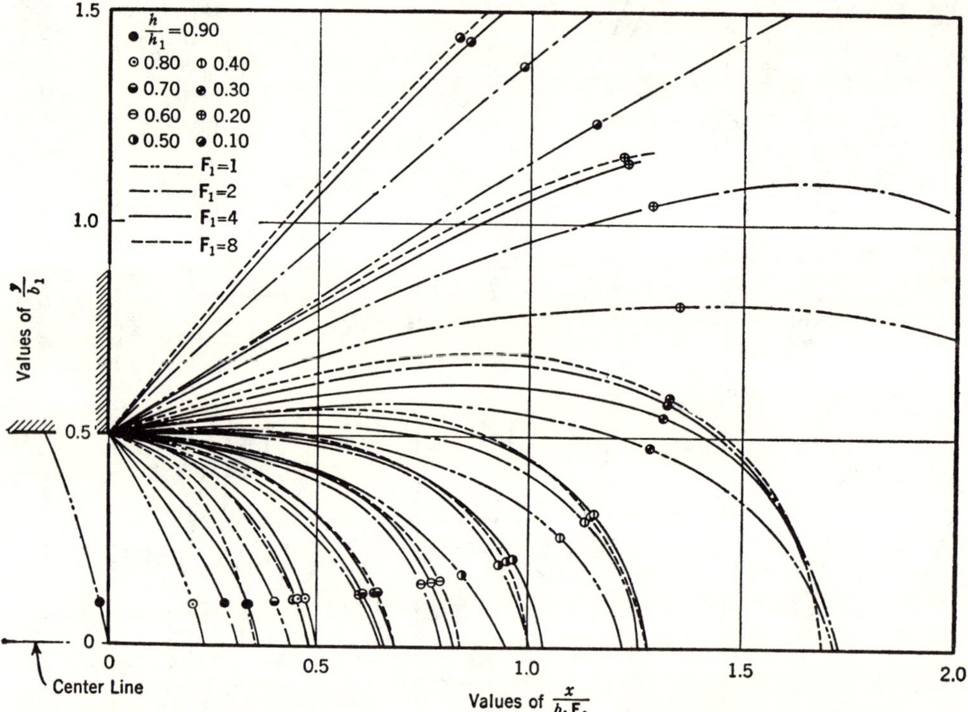

FIG. 55.—GENERALIZATION OF EXPERIMENTAL DATA FOR ABRUPT EXPANSIONS

A logarithmic plot against F_1 of the location of particular contour intercepts along any longitudinal axis (say, the line of each wall in Fig. 54), made with this purpose in mind, indicated that—except at very low Froude numbers—the variation in location was essentially a linear function of F_1. In other words, division of all values of x/b_1 by F_1 should tend to superpose all equivalent surface contours for all Froude numbers. Eqs. 45 and 46 are thereby reduced to the form:

$$\frac{h}{h_1} = f_3 \left(\frac{x}{b_1 \, F_1}, \frac{y}{b_1} \right) \dots\dots\dots\dots\dots\dots(47)$$

As may be shown analytically, this is certainly not a rigorous generalization. However, as will be seen from inspection of Fig. 55, when replotted in this manner the deviation of the mean contours taken from Fig. 54 is not great.

Although single average lines on this generalized diagram would evidently represent means of means, their departure from the contours of twenty-one runs at different values of F_1 and b_1/h_1 is considered sufficiently small to permit the use of this diagram for the preliminary analysis of abrupt expansions at practically any value of either parameter.

Efficient Curvature of Expanding Boundaries

There is a rule of thumb for the design of divergent boundaries in high-velocity flow which arbitrarily fixes the angle of divergence at $\theta = \tan^{-1}\frac{1}{6}$ to $\theta = \tan^{-1} = \frac{1}{9}$, regardless of the depth and the velocity of flow. As is apparent from the foregoing discussion of the expansion of flow without lateral constraint, the angle of divergence of any two neighboring streamlines is constant neither with the Froude number at a particular longitudinal distance nor with the longitudinal distance for a particular Froude number. At the abrupt beginning of such a uniformly divergent section the flow itself cannot abruptly change direction, and local separation as well as a concentration of negative wavelets will result; on the other hand, at a distance downstream which varies with the Froude number. the flow would naturally diverge more rapidly than the constant boundary angle will permit, thereby producing positive wavelets. Thus, as indicated by either of the contour maps of Fig. 56, obtained by the method of characteristics, such a divergent section is invariably inefficient at its beginning (and again before its end) unless the Froude number is so high that recovery from the initial zone is not accomplished before the transition ends.

It is obvious from such reasoning that an efficient boundary expansion should display a continuous change in curvature, and should have different proportions for every Froude number. The latter requirement suggests at once that, for purposes of preliminary design, the best form of boundary, as well as the resulting surface configuration, should be reducible to a generalized diagram such as that of Fig. 55. As a matter of fact, Fig. 55 was initially used in the arbitrary selection of a number of boundary curves for experimental and graphical investigation, the curves being formulated algebraically to approximate streamlines of the unconfined flow which enclosed about 90% of the total discharge.

The boundary equation eventually found to be most satisfactory was of the form:

$$\frac{y}{b_1} = \frac{1}{2}\left(\frac{x}{b_1\,F_1}\right)^{\frac{3}{2}} + \frac{1}{2}\dots\dots\dots\dots\dots\dots(48)$$

which is plotted in Fig. 57, together with surface contours for a mean value of b_1/h_1 and various values of F_1.

As will be noted from this composite plot, the beginning of the transition is sufficiently gradual to reduce effects of nonhydrostatic pressure distribution to a minimum, so that the factor b_1/h_1 is no longer an essential variable. The gradual increase in boundary angle, moreover, is sufficient to prevent the formation of positive waves—yet not so great as to cause an undue change in depth across any normal section. In fact, using circular arcs to approximate normalcy to the streamlines at successive sections, it will be found that the variation in depth from wall to wall does not exceed 30% of the center-line value.

It is, of course, possible to reduce such depth variation between wall and center line by decreasing the rate of flare—that is, by decreasing the coefficient

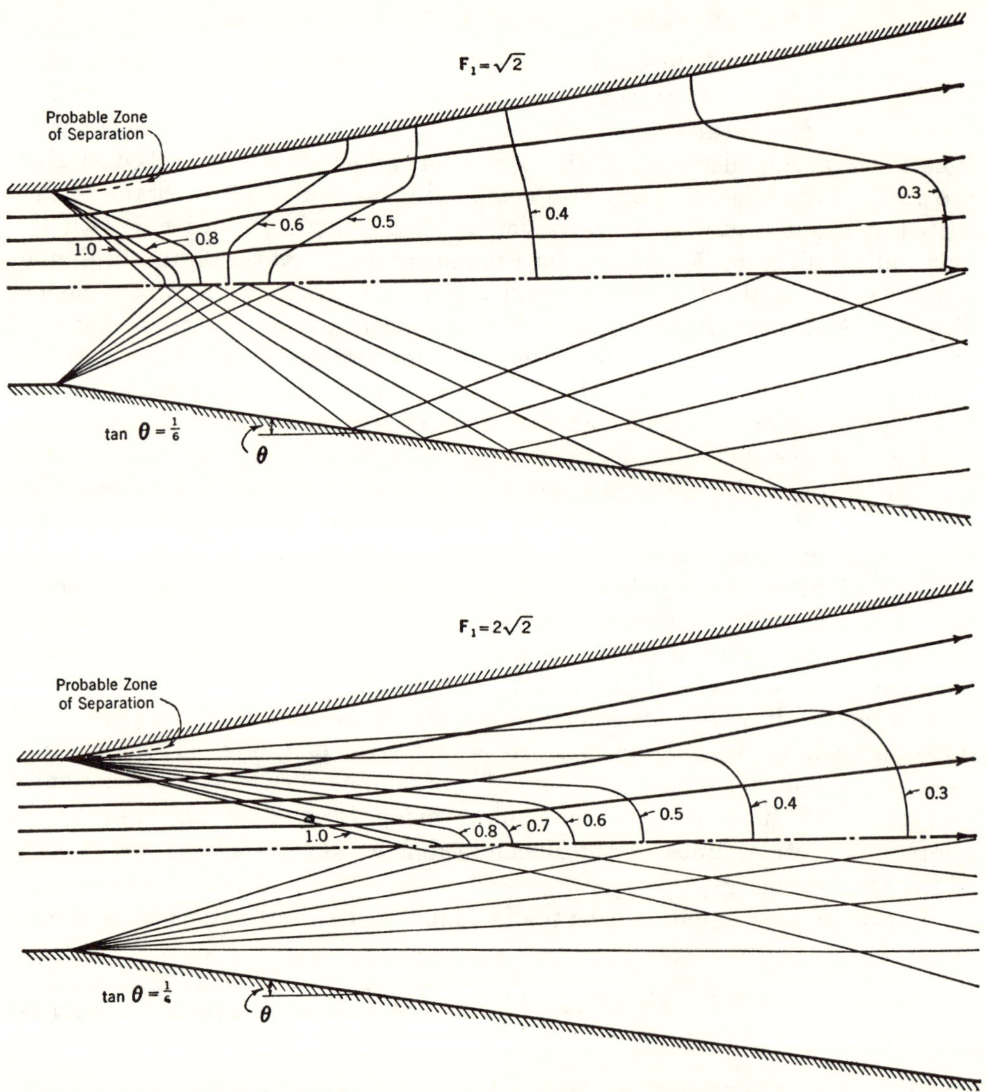

FIG. 56.—PATTERNS OF FLOW IN A UNIFORMLY DIVERGENT CHANNEL AT DIFFERENT FROUDE NUMBERS

of Eq. 48. This reduction, however, will result in a longer (and, hence, more costly) expansion for a given ratio of initial and final widths. Although the decision as to the greatest permissible depth variation is a matter either of judgment or of outlet requirements, it is believed that the curve shown in Fig. 57 will provide a satisfactory average basis for design.

The several curves reproduced in Fig. 57 typify the twenty or more which were determined experimentally for various Froude numbers and width-depth ratios and then checked by the graphical method of characteristics. Although a good general agreement was always obtained, the experimental results invariably yielded contours which were displaced downstream (from 0 to 35%, depending upon the ratio h/h_1). As this was attributed to the failure of the graphical method to take into account the gradual loss in total head due to boundary resistance, the same measurements were repeated on bed slopes varying from 4% to 10%. This, however, resulted in little displacement of the contours longitudinally, but in a considerable displacement laterally—that is, toward the center line, because the maximum bed slope was necessarily in the longitudinal direction rather than in the direction of each individual streamline.

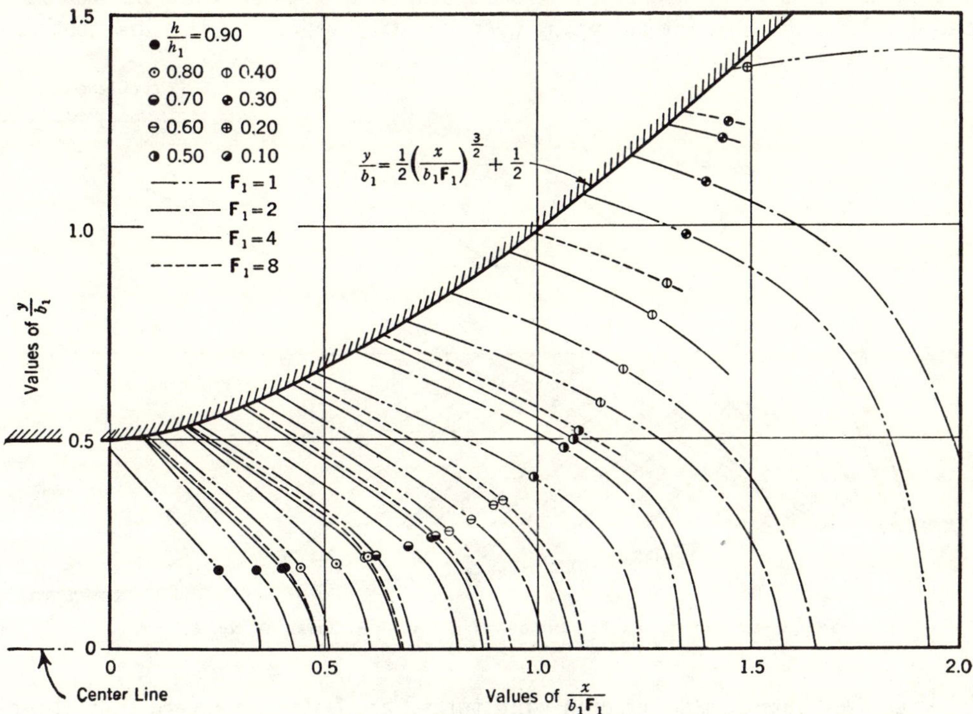

FIG. 57.—GENERALIZATION OF EXPERIMENTAL DATA FOR GRADUAL EXPANSIONS

From these considerations it would appear that the boundary form and surface contours for a level bed, shown in Fig. 57, may also be considered applicable to such moderate slopes as are normally encountered in open-channel design. Great slopes, on the other hand, would require a warped bed to prevent the major part of the flow from tending to follow the direction of maximum slope parallel to the center line. At present, the bottom surface can be warped satisfactorily only by trial and error at model scale. In the latter connection, however, it is to be noted that an effort was made to equalize both the surface elevation and the unit rate of flow across all normal sections by molding the bottom in conformity with the surface contours of Fig. 57. In other words, the initially horizontal bed was arbitrarily made higher at the center and lower at the walls in exact conformity to the indicated change in surface elevation across each normal section. Although full equalization of flow rate and surface elevation was not attained, conditions were improved perhaps 50%. Since application of this method of partial correction requires no further trial-and-error experimentation, its consideration is recommended where the added expense of bottom contouring is warranted.

Elimination of Disturbances at the End of an Expansion

Just as the analysis of supercritical flow of water is closely related to that of supersonic flow of gases, an open-channel transition for such flow should satisfy essentially the same general requirements as the test section of a supersonic wind tunnel—a variation in cross-sectional area such that the velocity and depth (or pressure intensity) are evenly distributed across the final section.

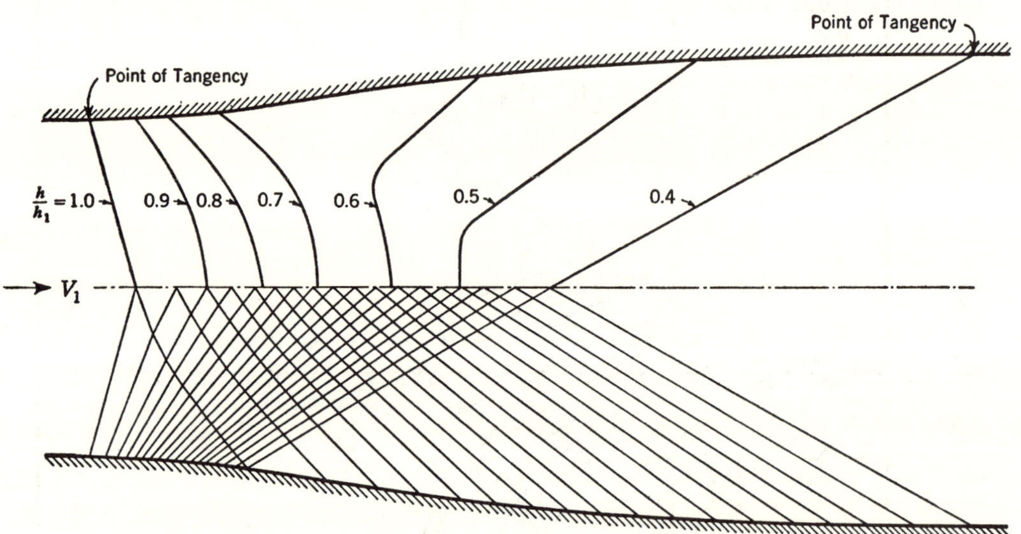

FIG. 58.—DESIGN OF AN EXPANSION WITH UNIFORM OUTFLOW FOR $F_1 = 2$, BY THE METHOD OF CHARACTERISTICS

The requirements of a supersonic wind tunnel are far more severe than those of an open-channel expansion, to be sure, even though experimental flow conditions are usually subject to arbitrary control; in fact, it is necessary to vary precisely the wall curvature of the tunnel test section from run to run in accordance with particular velocities of operation. Nevertheless, the same basic principles of boundary design could be applied quite generally to the case of open-channel expansions were it not for three practical limitations: First, under many circumstances the Froude number of the flow must be expected to vary over a considerable range; second, the length of transition required for either high Froude numbers or great expansion ratios will frequently be many times that permitted by structural economy; and, third, no method is provided thereby of stabilizing the hydraulic jump.

For the particular condition that an expansion represents merely a desirable increase in the width of a continuously paved channel, however, the same design procedure will yield at least the first approximation to an efficient design for a particular Froude number. As indicated in Fig. 58, the basis of the design technique is the control of wall curvature in such manner that the negative waves formed by successive elements of the outward curve just offset the positive waves formed by successive elements of the subsequent inward curve, so that the flow is restored to complete uniformity at the end of the transition. The procedure is, unfortunately, one of trial and error, and the resulting expansion ratio cannot be accurately foretold. A generalized series of boundary curves for successively greater expansion ratios is therefore presented in Fig. 59, as determined by interpolation from a series of solutions for various Froude numbers and expansion ratios by the method of character-

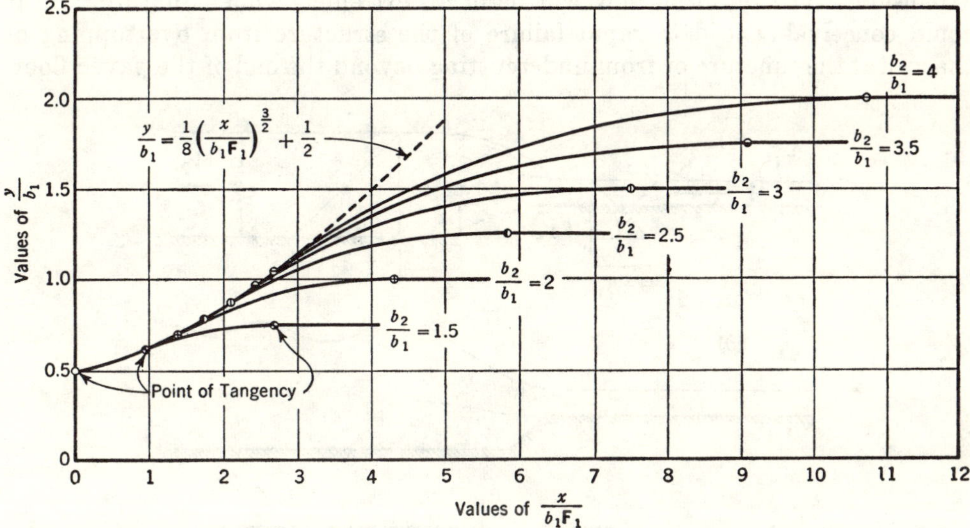

Fig. 59.—Generalization of Boundary Curves Determined by the Method of Characteristics

istics. These curves must be regarded merely as guides in preliminary design, for the following reasons: (1) Since the primary purpose was the generalization of results, each curve represents the average form of several somewhat different curves for different Froude numbers; (2) since the initial outward curve was chosen to yield without change the greatest practicable expansion range, it is probable that a somewhat shorter expansion curve could be devised for a particular condition; and (3) since the length of any transition is far in excess of that for which the drop in total head could be ignored, the assumption of zero loss in applying the method of characteristics leads to a predicted outlet depth which is considerably smaller than that which will actually prevail. For example, experiments on several expansions constructed in the laboratory on the basis of Fig. 59 resulted in outlet depths as much as from 20% to 40% in excess of that indicated by the simple wave theory, even though the flow was essentially uniform at the exit.

If, on the other hand, the channel expansion is intended to reduce the Froude number of the flow just prior to the formation of the hydraulic jump, it will be possible to hold the expansion to the shorter form of Fig. 57. If the divergent walls are followed by parallel walls with either an abrupt or a gradual transition, a positive wave will be formed at each wall junction and will extend diagonally across the flow at an angle varying with the local Froude number. Such waves would persist a considerable distance down the channel through repeated reflection if no jump were formed. If the toe of the jump lies at or near the end of the expansion, however, the diagonal waves will no longer form. On the other hand, should the jump as a whole advance even slightly into the expanding section, the smaller depth at each side will result in a progressively greater advancement of the jump along the walls, any slight asymmetry of the divergent flow finally giving rise to a deflection of the whole stream along one wall as the jump advances along the other almost to the upstream end of the expansion. The resulting flow will be of an extremely violent nature, and it could conceivably lead to rapid failure of the structure from overtopping of the wall at the juncture or from undercutting beyond the end of the paved floor.

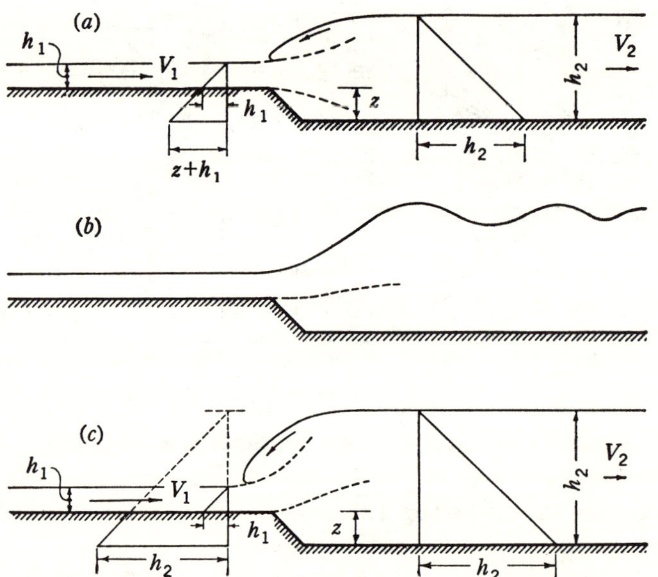

FIG. 60.—ALTERNATE FORMS OF THE HYDRAULIC JUMP AT AN ABRUPT DROP

In order to stabilize the jump at the end of such an expansion it appears necessary to provide a drop in floor level at the beginning of the parallel section. The relative magnitude of the change in elevation should depend primarily upon the Froude number of the flow as it leaves the expanding section. With reference to Fig. 60, it will be seen that the relationship between the Froude number, the relative change in depth, and the relative size of drop may be determined, like the equation of the jump itself, by the momentum and continuity relationships. There are, however, two different types of jump which may form, depending upon whether the downstream depth is below or above that which produces the standing wave indicated by Fig. 60(b). For the condition of Fig. 60(a), the pressure on the face of the drop will be determined by the upstream depth; for the condition of Fig. 60(c), the downstream depth will govern. The relationships for cases (a) and (c) are as follows:

$$\mathbf{F}^2{}_1 = \frac{1}{2}\frac{h_2/h_1}{1 - h_2/h_1}\left[\left(\frac{z}{h_1} + 1\right)^2 - \left(\frac{h_2}{h_1}\right)^2\right]\dots\dots\dots(49a)$$

and

$$\mathbf{F}^2{}_1 = \frac{1}{2}\frac{h_2/h_1}{1 - h_2/h_1}\left[1 - \left(\frac{h_2}{h_1} - \frac{z}{h_1}\right)^2\right]\dots\dots\dots\dots(49b)$$

Curves for Eqs. 49 will be seen in Fig. 61, the right-hand (or lower) series corresponding to Eq. 49a and the left-hand series to Eq. 49b. The critical zone for the formation of the standing wave—case (b)—cannot be foretold therefrom,

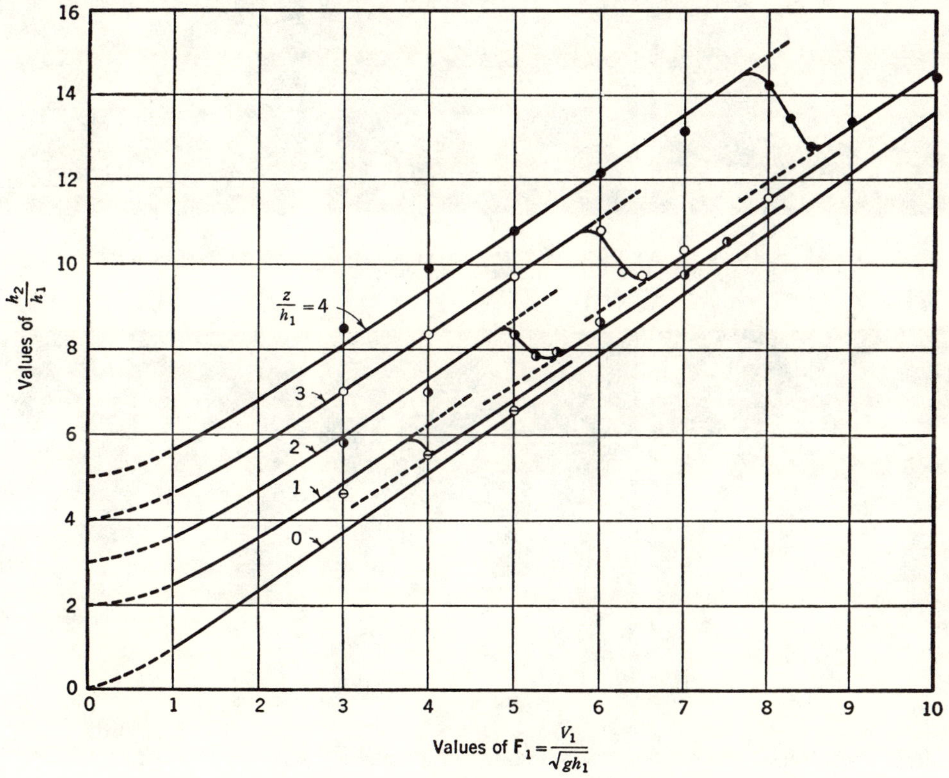

Fig. 61.—Analytical and Experimental Characteristics of the Hydraulic Jump at an Abrupt Drop

however, and recourse must be had to experimental measurement. Tests on both abrupt and sloping drops resulted in the points plotted in the figure, which not only verify the approximate analysis but indicate a systematic trend of the transition between the two regimes of flow.

From this diagram the magnitude of the drop for a given tailwater depth, or vice versa, may be determined once the average depth and the Froude number of the flow at the end of an expansion have been established. For protection of the structure, the design should be such that at the maximum expected Froude number the tailwater depth will be the minimum required to produce a jump. Figs. 62(a) and 62(b) are photographs of conditions for expansion ratios of 8 and 4, respectively, the details of the expansions following the recommendations presented herewith.

(a)　　　　　　　　(b)　　　　　　　　(c)

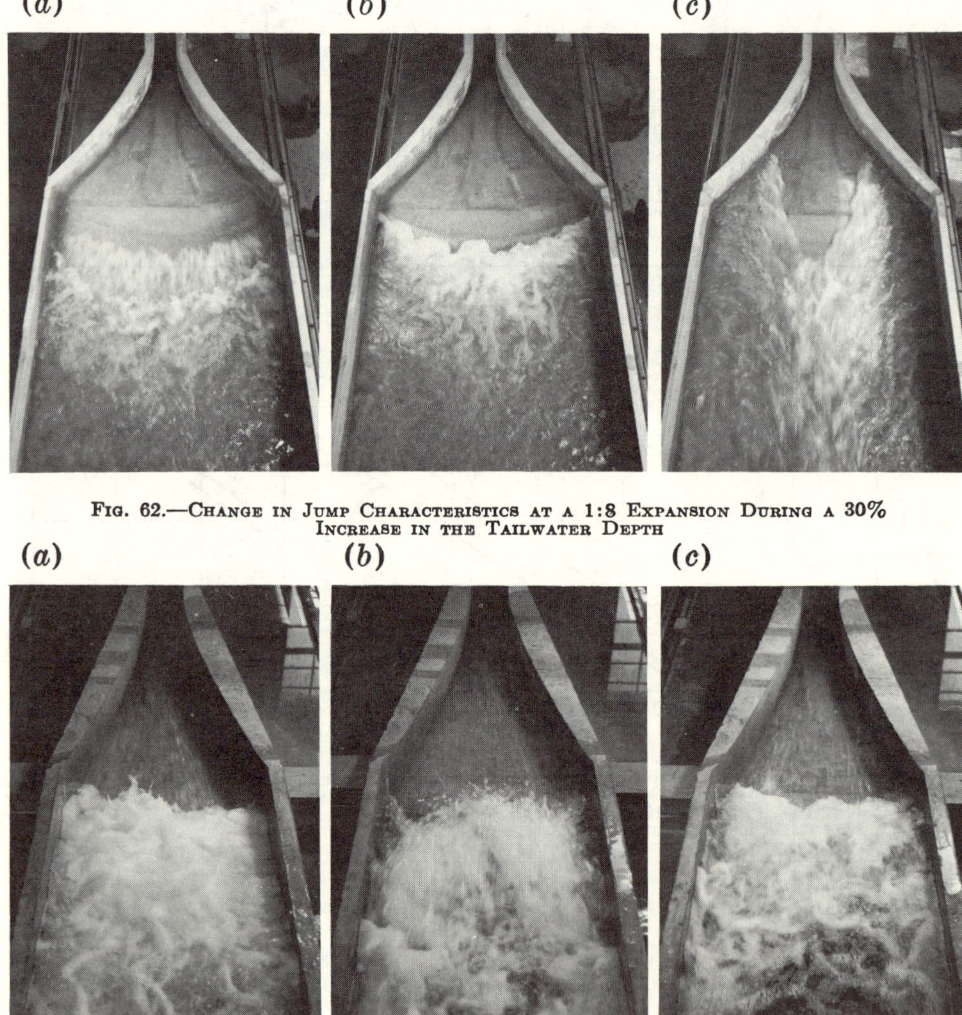

Fig. 62.—Change in Jump Characteristics at a 1:8 Expansion During a 30% Increase in the Tailwater Depth

(a)　　　　　　　　(b)　　　　　　　　(c)

Fig. 63.—Change in Jump Characteristics at a 1:4 Expansion During a 30% Decrease in the Rate of Flow

If the tailwater depth is increased beyond the value for minimum jump requirements, the undular regime will first appear (see Fig. 62(b)), followed by the second form of the jump and then by an uneven penetration of the expansion (Fig. 62(c)). An asymmetric pattern may eventually result, but the presence of the drop makes this far less likely than would otherwise be the case.   On the other hand, if the Froude number increases beyond the design value, or if the tailwater depth decreases, the jump will be carried downstream.   This phenomenon again is rendered less sensitive by the drop; the condition shown in Fig. 62(*a*), for example, will prevail during a 10% change in the downstream depth, whereas the three different stages shown in Fig. 62 represent a 30% change.   Finally, if the Froude number is decreased, essentially the same sequence will follow as for the increase in tailwater depth (see Figs. 63(*b*) and 63(*c*)).   In this event, however, the decrease in discharge will correspond to an equivalent decrease in harmfulness of the flow, and—although asymmetry may eventually develop—the structure planned for higher flows should then be safe.

Conclusions

Application of the elementary wave theory to the analysis of high-velocity flow in open-channel expansions may be expected to yield results in essential agreement with experiment as long as the assumptions involved in the theory are approximately satisfied.   For purposes of design, however, it is convenient to reduce all measured data for abrupt expansions to a single generalized plot of surface contours as a function of the initial Froude number and the relative coordinate location.   A similar procedure for gradual expansions permits selection of an efficient wall form for any initial Froude number and width-depth ratio.   To avoid dangerous asymmetry of the flow at the end of such an expansion, the hydraulic jump should be stabilized by a drop in the channel floor, the proper magnitude of which may be determined by the momentum equation.

by Hunter Rouse
Hsieh-ching Hsu

ON THE GROWTH AND DECAY OF A VORTEX FILAMENT

Iowa Institute of Hydraulic Research

ABSTRACT

Proceeding from the equations of motion of a viscous fluid, an investigation is made of the temporal variation of a single vortex of arbitrary initial characteristics: (1) The velocity field produced by the action of a vortex filament of infinitesimal diameter is expressed as a function of the strength of the filament, the viscosity of the fluid, and the time of generation. (2) The modification of this field after elimination of the generating filament is described in terms of the three parameters of the generation phase and the time interval thereafter. (3) The resulting equation is rewritten in a form involving three reference parameters which may be related to any three characteristics of an actual vortex at a particular instant; for example, the maximum circulation, the core diameter, and the center-line vorticity. An expression is thus obtained for the prediction of the subsequent change of the vortex characteristics with passage of time in a fluid of any viscosity.

STATEMENT OF THE PROBLEM

KNOWLEDGE of the variation in velocity and pressure throughout the field of a vortex filament is of basic importance in various phases of fluid mechanics. On the one hand, the formation of vortices in the wake of a cylindrical body is directly related to the variable force exerted upon the body, and the subsequent diffusion and dissipation of the vortex energy are essential factors in the study of wakes in general. On the other hand, although present-day analysis of fully developed turbulence no longer emphasizes the vortex nature of the turbulence structure, evaluation of instantaneous pressure fluctuations from a temporal velocity record may well be facilitated by consideration of the elementary vortex pattern. This possibility becomes the more worthy of attention with recognition of the fact that success in predicting incipient cavitation in zones of turbulence generation depends primarily upon the accuracy with which minimum local pressures can be foretold.

Since the time of Rankine (1)* the velocity and pressure fields in the vicinity of a vortex filament have been expressed in terms of his so-called "combined vortex." As indicated in Fig. 1, this idealization consists of a central rotational zone (or "forced" vortex) of constant vorticity, in which the velocity varies directly with radial

---

*Numbers in parentheses refer to the references at the end of the paper.

$v = \Gamma/2\pi r$, $\Gamma =$ const.

$v = \tfrac{1}{2}\zeta r$, $\zeta =$ const.

$p = 0$ at $r = \infty$

$-p = \frac{\rho}{2}\left[2\left(\frac{\Gamma}{2\pi r_c}\right)^2 - \left(\frac{\zeta r}{2}\right)^2\right]$

$-p = \frac{\rho}{2}\left(\frac{\Gamma}{2\pi r}\right)^2$

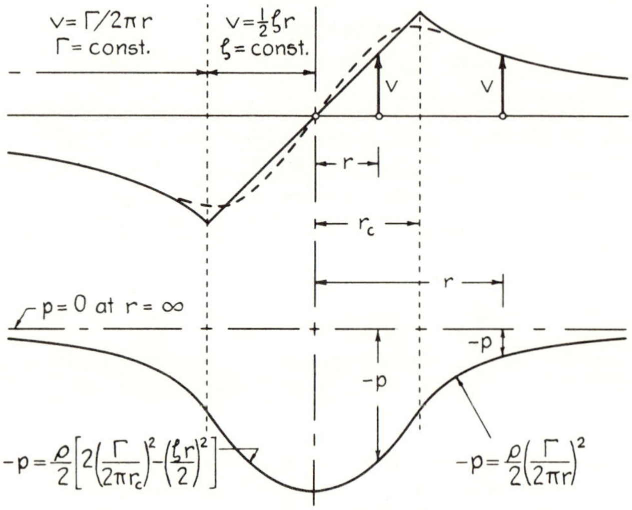

Fig. 1. VELOCITY AND PRESSURE FIELDS EXPRESSED IN TERMS OF RANKINE'S "COMBINED VORTEX."

distance from the axis, and a surrounding irrotational zone (or "free" vortex) of constant circulation, in which the velocity varies inversely with radial distance. As described in most textbooks on the subject (2), the resulting distribution of pressure — or elevation of a free liquid surface to which the axis is normal — will have the form shown in the figure, the maximum pressure intensity or surface elevation being that of the undisturbed fluid at an infinite radius and the minimum occurring at the centerline and depending upon the diameter and vorticity of the central filament.

Evident from examination of the structure of Rankine's combined vortex is the partial independence of the two zones of flow. In other words, while the circulation of the irrotational zone and the diameter and vorticity of the rotational zone are uniquely interrelated, the same irrotational field may accompany filaments having an infinite variety of diameters and vorticities. A question hence arises as to just what governs the size of an actual vortex — and, moreover, to what extent such a vortex differs from Rankine's idealization. Although it is indicated in a few references (3) that the existence of viscous shear necessarily modifies the velocity pattern somewhat in accordance with the heavy broken lines in Fig. 1, systematic measurements of the phenomenon appear to be wholly lacking. In fact, the only example of vortex motion which is conveniently subject to measurement is that which occurs over an open drain in a shallow tank; this, however, depends for its stability upon the existence of a radial as well as a tangential component of flow, which makes it inapplicable except in a general way to the problem at hand.

If, as suggested by the broken line in Fig. 1, one considers the development of a vortex to result from viscous shear, two existing analyses for rather idealized boundary conditions warrant at least preliminary examination. The first, by Lamb (4), assumes an irrotational field of constant circulation to exist initially, and the subsequent alteration of the velocity field is expressed as a function of time through use of the analogy between conduction of heat and diffusion of vorticity. The second, by Goldstein (5),·assumes a fluid body initially at rest but containing an isolated filament of constant diameter and vorticity, the effect of which upon the surrounding velocity field is likewise expressed as a function of time, although in the form of a differential equation of which the solution is not expressible in terms of elementary functions.

Consideration of the two analyses as means of describing vortex behavior at once reveals serious departures from the actual state of motion. In the first, a question immediately arises as to the original source of the irrotational field; this becomes the more problematic when it is realized that the initial kinetic energy of such a field is of infinite magnitude, and that the ultimate dissipation must hence also be infinitely great. The method has nevertheless been used by Hooker (6) as a basis for studying the diffusion of the Kármán vortex trail. In the second analysis, the postulation of a finite core of constant vorticity (in effect, a solid cylinder of unchanging rotational speed) is likewise highly artificial, except perhaps as merely the initial phase of development. In the latter event, no solution has been found in the literature which might characterize the further behavior of such a filament once its vorticity was permitted to change.

ιn view of these circumstances, an attempt has been made by the writers to formulate in accordance with the Navier-Stokes equations a reasonable expression for the growth and decay of an isolated vortex filament. Three specific requirements were given due consideration in this analysis: First, continuity of the function with respect to both time and space; second, agreement with actual vortex characteristics beyond some reference stage; and, third, expressibility in terms of significant parameters. As described in the following sections, these ends were attained through three successive steps. The velocity field produced in an initially still fluid by the action of a single line vortex was first expressed as a function of vortex strength, fluid viscosity, and passage of time. The modification of this field occurring after abrupt elimination of the generating vortex was then evaluated in terms of the original strength and length of generation, the viscosity, and the continued passage of time. Finally, the resulting function was rewritten in terms of parameters depending upon the centerline vorticity, core diameter, and circulation of an actual vortex at some instant of its existence, the viscosity, and the passage of time beyond this instant.

DEVELOPMENT OF THE PRIMARY FUNCTION

For motion in concentric circles about the origin of polar coordinates, the Navier-Stokes equations for the tangential and the radial directions reduce to the following linear forms:

$$\frac{\partial v}{\partial t} = \nu \left(\frac{\partial^2 v}{\partial r^2} + \frac{1}{r} \frac{\partial v}{\partial r} - \frac{v}{r^2} \right) \qquad [1]$$

$$\frac{v^2}{r} = \frac{1}{\rho} \frac{\partial p}{\partial r} \qquad [2]$$

In the case of an infinitely long generating cylinder of radius r_g which is suddenly brought into rotation with the peripheral velocity v_g, the instantaneous fluid velocity at any radial distance from the axis will be, according to Goldstein (5),

$$v = \frac{v_g r_g}{r} + \frac{2v_g}{\pi} \int_0^\infty \exp(-\nu t x^2)$$
$$\times \frac{J_1(xr) Y_1(xr_g) - Y_1(xr) J_1(xr_g)}{J_1^2(xr_g) + Y_1^2(xr_g)} \frac{dx}{x} \qquad [3]$$

in which J_1 and Y_1 are first-order Bessel functions of the first and second kind, respectively, and x is any variable.

Let it first be assumed that r_g approaches zero in such a manner that the product $v_g r_g$ maintains the constant value $\Gamma_g / 2\pi$, the limiting magnitude of the integral then being evaluated. In determining the limit of the integrand, the following relationships are useful:

$$J_1(0) = J_1(\infty) = Y_1(\infty) = 0 \qquad Y_1(0) = \infty \qquad [4]$$

$$Y_1(z) = \frac{1}{\pi} \left[2 \left(\gamma + \ln \frac{z}{2} \right) J_1(z) - \frac{2}{z} \right.$$
$$- \sum_{m=0}^{\infty} \frac{(-1)^m (z/2)^{1+2m}}{m! (m+1)!}$$
$$\left. \times \left(\frac{1}{1} + \frac{1}{2} + \ldots \frac{1}{m} + \frac{1}{1} + \frac{1}{2} + \ldots \frac{1}{1+m} \right) \right] \qquad [5]$$

in which γ is the Euler constant. The limit of $r_g Y_1(xr_g)$ as r_g approaches zero may be expressed as follows:

$$\lim_{r_g \to 0} r_g Y_1(xr_g) = \frac{1}{\pi} \lim_{r_g \to 0} \left[2 r_g \ln \left(\frac{xr_g}{2} \right) J_1(xr_g) - \frac{2r_g}{xr_g} \right] = -\frac{2}{\pi x} \qquad [6]$$

Hence, in view of Equation [4],

$$\lim_{r_g \to 0} \frac{v_g J_1(xr) Y_1(xr_g) - v_g Y_1(xr) J_1(xr_g)}{J_1^2(xr_g) + Y_1^2(xr_g)} = \lim_{r_g \to 0} \frac{r_g v_g J_1(xr)}{r_g Y_1(xr_g)}$$
$$= -\frac{\Gamma_g x J_1(xr)}{4}$$

Upon substitution of this limiting value into Equation [3], it is found that

$$v = \frac{\Gamma_g}{2\pi r} - \frac{\Gamma_g}{2\pi} \int_0^\infty \exp(-\nu t x^2)\, J_1(xr)\, dx$$

which through use of the formulas

$$J_1(z) = -\frac{d}{dz} J_0(z) \qquad [7]$$

and

$$\int_0^\infty J_0(ax)\, \exp(-p^2 x^2)\, x\, dx = \frac{1}{2p^2}\, \exp\frac{-a^2}{4p^2} \qquad [8]$$

may be integrated to yield

$$v = \frac{\Gamma_g}{2\pi r}\, \exp\frac{-r^2}{4\nu t} \qquad [9]$$

If the factor t be considered to approach the limit of infinity, Equation [9] will, paradoxically, become identical to that for the line vortex in an irrotational field of constant circulation. Evidently, this corresponds to an infinite input (as well as dissipation) of energy through everlasting rotation of the generating filament. The resulting velocity distribution, of course, is that indicated for the irrotational zone in Fig. 1, which approaches the limit of infinity as the radius approaches zero. If, on the other hand, the factor t is held to a finite value, the field surrounding the filament becomes restricted in its practical extent — even though the energy input is still infinite because of the infinite velocity at the axis.

Unlike the limiting case of the field of constant circulation, the flow characteristics of this field of decreasing circulation involve two parametric values: the circulation Γ_g of the generating filament and the time t_g during which it is assumed to operate. If, at the end of the generation period, the hypothetical filament suddenly ceases to exist, the parameters Γ_g and t_g will still characterize the subsequent dissipation. The analysis of the dissipation process over a particular time t_d is then reduced to finding a solution of Equation [1] which satisfies the condition

$$v = \frac{\Gamma_g}{2\pi r}\, \exp\frac{-r^2}{4\nu t_g} \qquad [10]$$

for $t_d = 0$, and the condition $v = 0$, $r = 0$ for $t_d > 0$.

One may show, through separation into two ordinary differential equations, that Equation [1] has a solution of the type

$$v = \exp(-\lambda^2 \nu t)\, J_1(\lambda r)$$

Since Equation [1] is linear, any linear combination of similar terms will also represent a solution. It is known, moreover, that any function subject to suitable restrictions may be expressed as a definite integral of the form

$$F(r) = \int_0^\infty \lambda\, d\lambda \int_0^\infty F(x)\, J_1(\lambda x)\, J_1(\lambda r)\, x\, dx$$

Thus, if $F(x)$ is replaced by $(\Gamma_g/2\pi x)\exp(-x^2/4\nu t_g)$, there results

$$\frac{\Gamma_g}{2\pi r}\, \exp\frac{-r^2}{4\nu t_g} = \frac{\Gamma_g}{2\pi} \int_0^\infty \lambda\, d\lambda \int_0^\infty \exp\frac{-x^2}{4\nu t_g}\, J_1(\lambda x)\, J_1(\lambda r)\, dx$$

The solution which satisfies the required conditions for $t_d = 0$ and $t_d > 0$ is

$$v = \frac{\Gamma_g}{2\pi} \int_0^\infty \lambda\, \exp(-\lambda^2 \nu t)\, J_1(\lambda r)\, d\lambda \int_0^\infty \exp\frac{-x^2}{4\nu t_g}\, J_1(\lambda x)\, dx$$

which, upon integration in accordance with Equations [7] and [8], assumes the desired form

$$v = \frac{\Gamma_g}{2\pi r}\left(\exp\frac{-r^2}{4\nu(t_g + t_d)} - \exp\frac{-r^2}{4\nu t_d}\right) \qquad [11]$$

a generalized plot of which is shown in Fig. 2.

This equation indicates the radial distribution of velocity throughout the zone of motion at any time t_d for

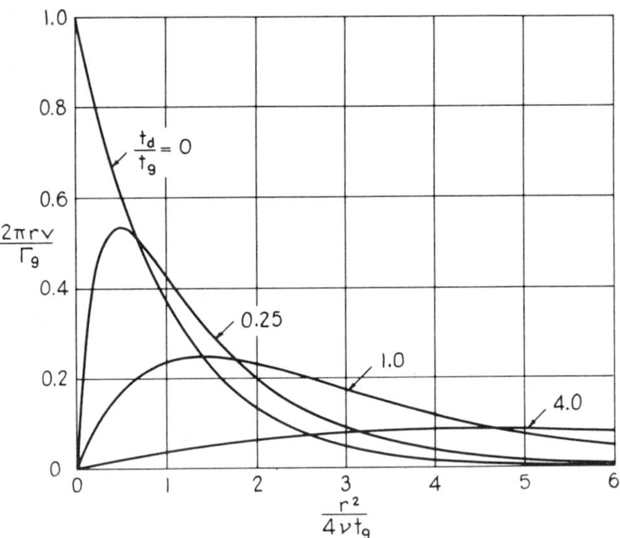

Fig. 2. GENERALIZED PLOT INDICATING THE RADIAL DISTRIBUTION OF VELOCITY THROUGHOUT THE ZONE OF MOTION AT ANY TIME t_d FOR GIVEN VALUES OF Γ_g, t_g, AND ν.

given values of the original parameters Γ_g, t_g, and ν.

From it may be derived the corresponding functional relationships for the distributions of circulation, vorticity, and pressure, as well as the nominal size of the vortex core and the total kinetic energy of the motion. As such, Equation [11] represents the primary dissipation function, which

184

serves as the basis of the subsequent interpretative reasoning.

METHOD OF APPLICATION TO SPECIFIC CONDITIONS

Although the generation of an actual vortex is surely triggered by viscous action, and although dissipation just as surely begins with the onset of the generation process, in its early stages the latter may still be susceptible to approximate analysis as a case of unsteady potential flow — at least for fluids of small viscosity. For example, the characteristics of a vortex shortly after its generation would appear to depend upon three independent factors: the scale of the generating mechanism (such as a paddle or a wing tip), the relative velocity of the motion, and either a time interval or a displacement. (Whether this would indicate three, or only two, degrees of freedom in the vortex characteristics must remain for the present a moot question.)

Under such circumstances, Equation [11] seems at first glance to represent little more than a hypothetical case of flow, for the parameters Γ_g, t_g, and t_d are too artificial to have counterparts in the actual generation process. However, other types of flow have been described successfully in terms of parameters which have no physical significance beyond their usefulness as reference parameters. In other words, although the actual generation process may well differ radically from the sequence of conditions embodied in the primary function, the flexibility of this function may equally well permit it to represent with satisfactory approximation the state of motion at some later time. All subsequent phases of the motion would then be predictable through an extension of this function.

The proposal is therefore made that — until a more exact solution becomes available — Equation [11] be considered to represent the velocity distribution of a vortex at some instant after its generation, and that the three parameters Γ_g, t_g, and t_d be expressed in terms of three conveniently measurable and significant characteristics. Those proposed are the centerline vorticity ζ_0, the maximum circulation Γ_0 and the nominal core radius r_0.

If, now, the primary function is rewritten in the form

$$v = \frac{A}{2\pi r}\left(\exp\frac{-r^2}{ak} - \exp\frac{-r^2}{a}\right) \qquad [12]$$

it will be seen that three arbitrary constants — A, a, and k — may be used to define the initial velocity distribution. These may be related to r_0, Γ_0, and ζ_0 as follows. Designating the radius of the core as that at which $\Gamma = \Gamma_{max}$ (as in the case of the Rankine combined vortex), since $\Gamma = 2\pi r v$,

$$\frac{\partial \Gamma}{\partial r} = 0 = \frac{2Ar}{a}\left[\exp\frac{-r^2}{a} - \frac{1}{k}\exp\frac{-r^2}{ka}\right]$$

and

$$r_0 = \left(\frac{ka}{k-1}\ln k\right)^{1/2} \qquad [13]$$

It follows at once that

$$\Gamma_0 = A(k-1)k^{\frac{k}{1-k}} \qquad [14]$$

Finally,

$$\zeta = \frac{1}{r}\frac{\partial(vr)}{\partial r} = \frac{A}{\pi ka}\left[k\exp\frac{-r^2}{a} - \exp\frac{-r^2}{ka}\right]$$

whence, for $r = 0$,

$$\zeta_0 = -\frac{A}{\pi ka}(k-1) \qquad [15]$$

To be noted is the fact that these three parameters combine to yield a dimensionless ratio expressible in terms of the single factor k:

$$\frac{\Gamma_0}{\zeta_0 r_0^2} = \frac{\pi(k-1)k^{\frac{k}{1-k}}}{\ln k} \qquad [16]$$

For particular values of the parameters r_0, Γ_0, and ζ_0 (and hence of the factors A, a, and k) the subsequent growth and decay of the vortex filament may be expressed in the form

$$v = \frac{A}{2\pi r}\left(\exp\frac{-r^2}{ak + 4\nu t} - \exp\frac{-r^2}{a + 4\nu t}\right) \qquad [17]$$

Three characteristics of the dissipation process are of primary interest: the size of the core at any time, the pressure intensity at the axis, and the total kinetic energy of the field. The first again involves the condition that $r = r_c$ when $\partial \Gamma/\partial r = 0$, with the result that

$$r_c^2 = a\frac{(k + 4\nu t/a)(1 + 4\nu t/a)}{k - 1}\ln\frac{k + 4\nu t/a}{1 + 4\nu t/a} \qquad [18]$$

The second and third characteristics, being integral quantities, must be evaluated with the aid of the following equations:

$$\int_0^\infty \frac{(e^{-px} - e^{-qx})^2}{x}\,dx = \ln\frac{p+q}{2p} + \ln\frac{p+q}{2q} \qquad [19]$$

$$\int_{x_1}^\infty \frac{(e^{-px} - e^{-qx})^2}{x^2}\,dx = \frac{(e^{-px_1} - e^{-qx_1})^2}{x_1}$$

$$+ 2q\int_{2px_1}^{2qx_1}\frac{e^{-y}}{y}\,dy$$

$$- 2(p+q)\int_{2px_1}^{(p+q)x_1}\frac{e^{-y}}{y}\,dy \qquad [20]$$

$$\int_0^\infty \frac{(e^{-px} - e^{-qx})^2}{x^2}\, dx = 2q \ln \frac{q}{p} - 2(p + q) \ln \frac{p + q}{2p} \qquad [21]$$

If the pressure at infinity is considered to have the reference magnitude of zero, the pressure at any radius is determined from Equation [2] as

$$p = -\rho \int_r^\infty \frac{v^2}{r}\, dr$$

$$= -\frac{\rho A^2}{8\pi^2} \int_r^\infty \frac{\left(\exp \dfrac{-r^2}{ka + 4\nu t} - \exp \dfrac{-r^2}{a + 4\nu t}\right)^2}{r^4}\, 2\pi r\, dr$$

Through use of Equation [20] this is reduced to the form

$$p = -\frac{\rho v^2}{2} - \frac{\rho A^2}{4\pi^2} \frac{1}{a + 4\nu t} \int_{\frac{2r^2}{ka + 4\nu t}}^{\frac{2r^2}{a + 4\nu t}} \frac{e^{-y}}{y}\, dy + \frac{\rho A^2}{4\pi^2} \left(\frac{1}{ka + 4\nu t}\right.$$

$$\left. + \frac{1}{a + 4\nu t}\right) \int_{\frac{2r^2}{ka + 4\nu t}}^{\frac{-r^2}{ka + 4\nu t} + \frac{r^2}{a + 4\nu t}} \frac{e^{-y}}{y}\, dy \qquad [22]$$

The minimum pressure, which occurs at the axis, is then

$$p_{min} = -\frac{\rho A^2}{4\pi^2 a}\left[\frac{1}{1 + 4\nu t/a} \ln \frac{k + 4\nu t/a}{1 + 4\nu t/a} - \frac{1}{1 + 4\nu t/a}\right.$$

$$\left. \times \left(1 + \frac{1 + 4\nu t/a}{k + 4\nu t/a}\right) \ln \frac{1 + \dfrac{k + 4\nu t/a}{1 + 4\nu t/a}}{2}\right] \qquad [23]$$

The kinetic energy per unit length of vortex is likewise expressible as

$$E = \pi\rho \int_0^\infty v^2 r\, dr = \frac{\rho A^2}{8\pi} \int_0^\infty \frac{1}{r^2}\left(\exp \frac{-r^2}{ka + 4\nu t}\right.$$

$$\left. - \exp \frac{-r^2}{a + 4\nu t}\right)^2 2r\, dr$$

which, through use of Equation [19], becomes

$$E = \frac{\rho A^2}{8\pi} \ln \frac{[1 + (k + 4\nu t/a)/(1 + 4\nu t/a)]^2}{4(k + 4\nu t/a)/(1 + 4\nu t/a)} \qquad [24]$$

So far as the underlying assumptions are valid, Equations [17], [18], [23], and [24] thus represent the pertinent characteristics of vortex dissipation.

There are, unfortunately, no experimental data yet at hand with which the validity of these equations can be checked. In fact, the only published measurements of the velocity field of a single vortex known to the writers are those made by Bourot (7) over a single cross section of a tip vortex downwind from a wing. Not only are these results obviously insufficient for verification purposes, but the relationship between dissipation with time and dissipation with axial displacement in the wake of the wing could in any event only be approximated. For lack of other evidence, however, the factors A, a, and k have been evaluated from the characteristics of Bourot's velocity traverse, with the results tabulated in Fig. 3. Merely for purposes of illustration, the subsequent temporal variation indicated by Equation [17] for the vortex field having

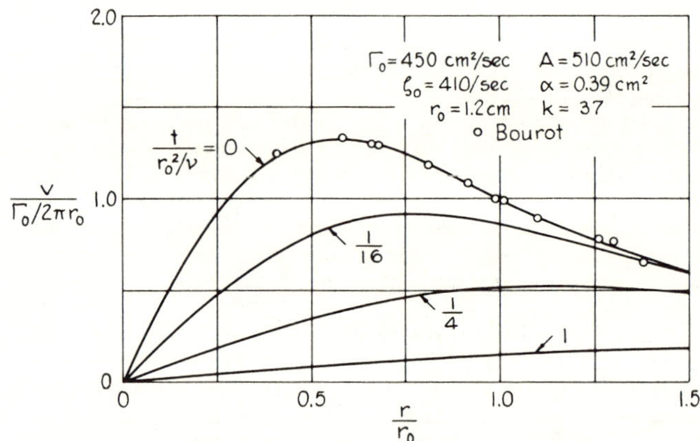

Fig. 3. GROWTH AND DECAY OF THE VORTEX FILAMENT PLOTTED FROM BOUROT'S VELOCITY TRAVERSE OF A TIP VORTEX.

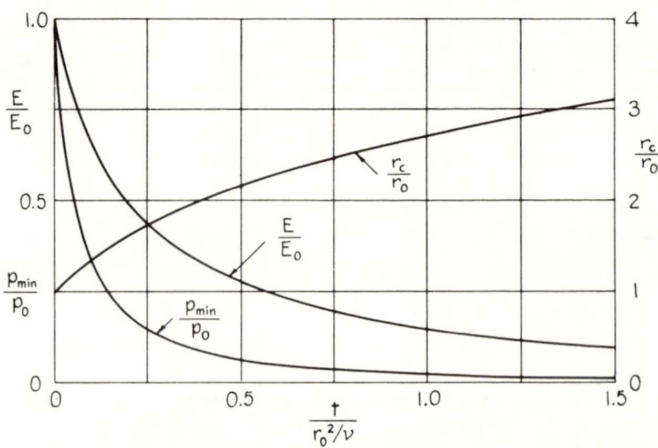

Fig. 4. VARIATION IN RELATIVE CORE DIAMETER, MINIMUM PRESSURE AND KINETIC ENERGY DERIVED FROM BOUROT'S MEASUREMENTS.

these reference characteristics is plotted in the same figure, while Fig. 4 shows the corresponding variation in relative core diameter, minimum pressure, and kinetic energy in accordance with Equations [18], [23], and [24].

ACKNOWLEDGMENTS

Based upon a suggestion by Dr. Francis H. Clauser of Johns Hopkins University in connection with the problem of eddy cavitation, the foregoing analysis was carried out in detail by the junior author under the senior author's general direction as part of the turbulence and cavitation program of the Iowa Institute of Hydraulic Research under Contract N8onr-500 with the Office of Naval Research.

REFERENCES

1. "Hydrodynamics," by H. Lamb, Cambridge, 1932, p. 29.

2. "Fluid Mechanics for Hydraulic Engineers," by H. Rouse, McGraw-Hill Book Company, New York, 1938, p. 76.

3. "Fundamentals of Hydro- and Aero-mechanics," by L. Prandtl and O. G. Tietjens, McGraw-Hill Book Company, New York, 1934, p. 208.

4. "Hydrodynamics," by H. Lamb, Cambridge, 1932, p. 592.

5. "Some Two-dimensional Diffusion Problems with Circular Symmetry," by S. Goldstein, *Proc. of the London Mathematical Society,* ser. 2, v. 34, 1932, pp. 51-88.

6. "On the Action of Viscosity in Increasing the Spacing Ratio of a Vortex Street," by S. G. Hooker, *Proc. Royal Soc. London,* A-154, 1936, pp. 67-89.

7. "Sur la Visualisation du Tourbillon Marginal d'une Aile, au Moyen de Lamelles d'Aluminium, en Soufflerie Aérodynamique," by J. M. Bourot, *C. R. Acad. Sci. Paris,* v. 228, May 1949, pp. 1567-1569.

Paper No. 2529

EXPERIMENTAL INVESTIGATION OF FIRE MONITORS AND NOZZLES

By Hunter Rouse,[1] M. ASCE, J. W. Howe,[2] M. ASCE,
and D. E. Metzler[3]

With Discussion by Messrs. G. Halbronn; Pierre Oguey, Marcel Mamin, and François Baatard; John H. Arnold; and Hunter Rouse, J. W. Howe, and D. E. Metzler

Synopsis

Since the inefficiency of fire streams is directly attributable to the initial turbulence of the free jet, attention was given at the Iowa Institute of Hydraulic Research to the design of a fixed monitor, a portable monitor, and a nozzle that will reduce the turbulence of the flow to a practicable minimum. The underlying principles upon which the designs were based are discussed in this paper, the test facilities are described, the efficiencies of existing and improved units are compared, and design details and performance curves are presented for the recommended forms.

Introduction

In 1888 John R. Freeman published a prize-winning paper on fire streams[4] that is still the most widely accepted treatise on this subject, despite great advances in nearly every other phase of hydraulics. Lack of further progress in this particular field has been due in part to the excellence of Mr. Freeman's experiments and in part to apparent—but fallacious—corollaries of perfectly sound experimental findings.

The most frequently quoted of Mr. Freeman's results are his coefficients of nozzle discharge, determined with a precision that would be acceptable even under present-day standards. For well-faired nozzles yielding a negligible contraction of the free jet, he found the coefficient C in the customary discharge equation to be within less than 2% of unity. Repeated efforts during the intervening years to improve flow conditions by changing the boundary curvature,

---

Note.—Published in October, 1951, as *Proceedings-Separate No. 92*. Positions and titles given are those in effect when the paper or discussion was received for publication.

[1] Director, Iowa Inst. of Hydr. Research, State Univ. of Iowa, Iowa City, Iowa.

[2] Head, Dept. of Mechanics and Hydr., State Univ. of Iowa, Iowa City, Iowa.

[3] Associate Prof., Dept. of Mechanics and Hydr.; Research Engr., Iowa Inst. of Hydr. Research, State Univ. of Iowa, Iowa City, Iowa.

[4] "Experiments Relating to Hydraulics of Fire Streams," by John R. Freeman, *Transactions*, ASCE, Vol. 21, 1889, p. 303.

lengthening the taper, or polishing the inner surface have resulted in no significant increase in the discharge coefficient.

Unfortunately, two related (and equally false) presumptions have been the outgrowth of this situation. First, the most efficient nozzle is generally considered to be that for which the discharge coefficient is nearest unity—despite the facts that a contracting jet is known to yield a somewhat higher efficiency, even though the coefficient C is then well below unity, and an expanding jet yields a considerably lower efficiency, even though the value of C may then be greater than unity. Second, since nozzles that are now considered the best have coefficients close to unity, their high efficiency as conduit outlets is taken as an indication of an equally high efficiency as fire-stream producers—despite the fact that both good and bad jets are known to result from nozzles having identical values of C.

A stream from any standard nozzle is, needless to say, a notoriously poor counterpart of the ideal jet that has its parabolic trajectory based upon the physics of free fall, particularly in a cross or adverse wind. Discrepancies between theory and practice are attributed almost entirely to the unavoidable drag of the air upon the water that has left the nozzle. Indeed, it was not until 1943, when it was discovered that lengthening the barrel of a fire-boat monitor increased the carry of the stream some 30%, that attention was finally focused upon the true source of fire-stream inefficiency: The turbulence of the flow.

Since the investment of the United States Coast Guard for fire-fighting equipment during World War II involved a very considerable sum, it was decided before the end of hostilities to invest as well in a study of possible improvements in monitor and nozzle design. The Iowa Institute of Hydraulic Research (hereinafter called "the Institute") was commissioned in 1944, as one phase of its contract with the Navy Bureau of Ships, to conduct an extensive series of investigations of both fixed and portable monitors and nozzles, culminating in complete performance characteristics of the recommended designs.

Underlying Principles of Flow

Basic Flow Equations.—For a clear understanding of fire-monitor performance, reference must be made to the following elementary principles of flow, the development of which is to be found in most textbooks.[5] The simple equation of continuity,

$$Q = A_1 V_1 = A_2 V_2 \dots\dots\dots\dots\dots\dots\dots(1)$$

states that the rate of flow Q, equal to the product of the cross-sectional area A of a conduit and the mean velocity V at that section, must be the same at all successive sections. The equally simple equation of energy (known as the Bernoulli theorem),

$$H_1 = H_2 + H_L \dots\dots\dots\dots\dots\dots\dots(2)$$

states that the total head H_1 at one section is equal to the total head H_2 at the next section plus the intervening losses H_L. The total head H at any

---

[5] "Elementary Mechanics of Fluids," by Hunter Rouse, John Wiley & Sons, Inc., New York, N. Y., 1946.

section consists of velocity head $V^2/2\,g$ and piezometric head h

$$H = \frac{V^2}{2\,g} + h \dots\dots\dots\dots\dots\dots\dots\dots (3)$$

in which g is the acceleration of gravity. More specifically, the total head H represents the height to which water will rise in an open manometer column connected to the tip opening of a Pitot tube, the piezometric head h the height in a column connected to the side opening of a Pitot tube (or to a piezometer in the pipe wall), and the velocity head $V^2/2\,g$ the difference between these two. The piezometric head is in turn equal to the sum of the pressure head p/γ and the elevation z:

$$h = \frac{p}{\gamma} + z \dots\dots\dots\dots\dots\dots\dots\dots (4)$$

in which p is the pressure intensity relative to that of the atmosphere (as read on a Bourdon gage at the level z of the point in question) and γ is the unit weight of the water. The datum above which z is measured is arbitrary but must be the same for all sections. Likewise the same consistent units of measure (such as feet, seconds, and pounds) must be used to express all quantities involved.

The loss in head H_L between any two neighboring sections may invariably be expressed in terms of a coefficient C_L and a velocity head as follows:

$$H_L = C_L \frac{V^2}{2\,g} \dots\dots\dots\dots\dots\dots\dots\dots (5)$$

The magnitude of the coefficient C_L at moderate to high velocities depends almost entirely upon the geometrical form of the intervening conduit that produces the loss, and hence may be considered a constant for a given conduit form, such as a fixed contraction ratio or fixed proportions of an elbow, regardless of the conduit dimensions. It follows at once that the loss H_L through a given type of fitting under a given discharge may be reduced by increasing the size of the fitting and thereby reducing $V^2/2\,g$ in accordance with Eq. 1 or by streamlining the fitting to reduce C_L in Eq. 5 as far as possible toward its ideal limit of zero.

In a free jet the pressure intensity is the same as that of the atmosphere ($p = 0$), so the Bernoulli equation becomes

$$\frac{(V_1)^2}{2\,g} + z_1 = \frac{(V_2)^2}{2\,g} + z_2 + H_L \dots\dots\dots\dots\dots (6)$$

If the air is assumed to have no effect upon the jet, the loss H_L will equal zero, and application of the physical laws of free fall,[5]

$$x = V_1 \cos \alpha\, t \quad \text{and} \quad z = V_1 \sin \alpha\, t - \tfrac{1}{2} g\, t^2 \dots\dots\dots\dots (7)$$

in which t is the time of fall, will yield for the nozzle inclination α and the efflux velocity V_1, a parabolic trajectory having the maximum height

$$z_{\max} = \frac{(V_1)^2}{2\,g} \sin^2 \alpha \dots\dots\dots\dots\dots\dots (8)$$

and the horizontal range at nozzle level

$$x_{\max} = 2\,\frac{(V_1)^2}{2\,g}\,\sin 2\,\alpha\dots\dots\dots\dots\dots\dots\dots(9)$$

Since, as will be shown in the following paragraphs, C_L for a free jet depends upon far more than simply the efflux velocity, the actual losses and hence the true form of the jet cannot be expressed in any simple manner.

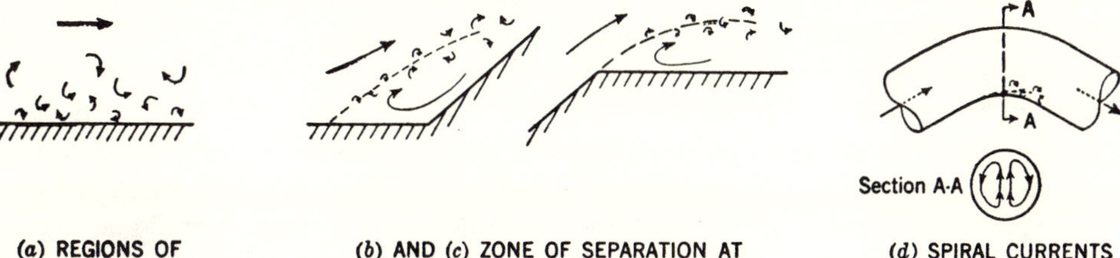

(a) REGIONS OF
SHEAR ALONG WALLS

(b) AND (c) ZONE OF SEPARATION AT
UNSTREAMLINED BOUNDARY CHANGE

(d) SPIRAL CURRENTS
FORMED IN BEND

FIG. 1.—SOURCES OF TURBULENT EDDIES

Effects of Turbulence.—Fluid turbulence[6] is generally defined as a haphazard secondary motion superposed on the mean translatory motion of a flowing fluid. It may thus be visualized as countless eddies of different shapes, sizes, orientations, and rotational speeds that are carried along with the flow. The sources of such eddies (see Fig. 1) are of several types: (1) They develop on a very small scale in regions of high shear along the walls of even very smooth conduits; (2) they are produced on a larger scale in zones of separation at boundary changes that are not fully streamlined; and (3) they result from major spiral currents formed in any conduit transition, such as a bend, that is not symmetrical about a linear axis. Practically all major losses of head, in fact, may be attributed to such eddy formation, and the magnitude of C_L for any portion of a conduit is a relative indication of the turbulence that will exist in the flow just downstream.

Turbulence in general is looked upon as an unwanted evil because of the power loss that it entails. In the case of fire streams, however, the loss of power accompanying the formation of turbulence within the monitor and nozzle is of small consequence compared with the subsequent loss that results from the presence of turbulence in the free jet. The latter is due to the fact that the secondary motion within each eddy of turbulent flow has transverse velocity components that, although usually small in comparison with the mean velocity, are nonetheless of finite magnitude. Within a conduit such eddies are laterally confined, as is the flow itself, by the conduit walls. A free jet is not confined, however, except by the rather slight effect of surface tension. Therefore, any lateral motion of the fluid due to an eddy will continue practically unabated past the original jet surface, with the result that only a short distance beyond the efflux section a turbulent jet will display a multitude of irregular surface eruptions.

---

[6] "The Mechanics of Turbulent Flow," by Boris A. Bakhmeteff, Princeton University Press, Princeton, N. J., 1936.

These liquid protuberances, readily visible in high-speed photographs (see Fig. 2), are not produced by the action of the air, for they would appear just as readily in a vacuum. In fact, the density of the air is so low (1/800 that of water) that its effect upon a smooth jet would be quite small in comparison to other factors. It is only after the surface of the jet has become sufficiently disrupted to produce an appreciable form resistance that the action of the air begins. Such resistance is roughly proportional to the square of the water velocity and to the cross-sectional area of the expanding jet. Thus, as ever more eddies carry water laterally out of the central stream, they are rapidly retarded in their longitudinal course by the surrounding air, but they continue to spread laterally at only slightly diminished speed. Although the outermost fringes of the jet at once form droplets that fall as a spray, the central portion appears simply to disintegrate in midair. In other words, since the equal and opposite reaction to the retardation of the water is the acceleration of the surrounding air, the originally intact but turbulent jet is transformed along its trajectory into an expanding mixture of dispersed water drops and air that travels at an ever-decreasing speed but brings an ever-increasing volume of air into motion.

Turbulence of the water emerging from a fire nozzle is hence seen to embody three distinct detriments to its efficiency as a fire-fighting tool. First, the original formation of the turbulence in monitor and nozzle represents a reduction in flow efficiency. Second, because of the lateral spread of the eddies in the free jet, air resistance reduces both the concentration and the carry of the stream. Third, entrainment of the surrounding atmosphere by the diffusing jet produces a powerful flow of air directly toward the fire.

Improvement of Fire Streams.—From the foregoing discussion it can be concluded that the only way to increase the efficiency of a fire stream is to decrease the turbulence of the flow before it leaves the nozzle. Although a

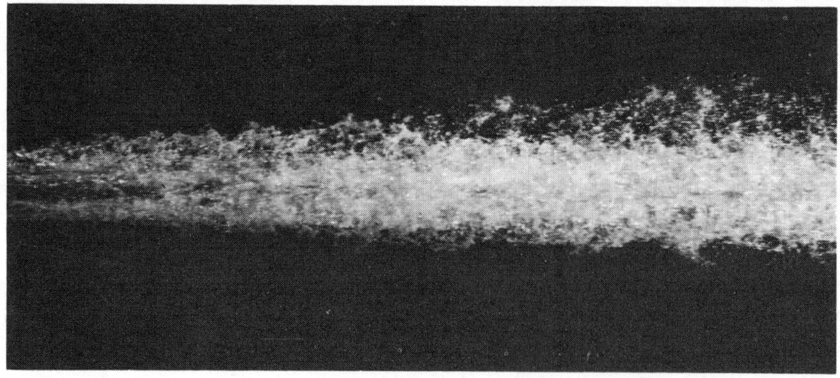

FIG. 2.—HIGH-SPEED PHOTOGRAPH OF TURBULENT JET, 1/20,000-SECOND EXPOSURE

wholly laminar (that is, nonturbulent) jet would be ideal, this is unfortunately out of the question for water jets having diameters and velocities of practical interest. Laminar flow seldom occurs at a Reynolds number greater than 2,000. With a water viscosity $\nu = 10^{-5}$ sq ft per sec, for example, flow in a 1-in. hose could not exceed a velocity of $\frac{1}{4}$ ft per sec without becoming turbulent. The fact remains, however, that the turbulence of a jet may be controlled to a considerable degree. Since the control of turbulence has not been the specific goal of monitor and nozzle designers in the past, it stands to reason that the

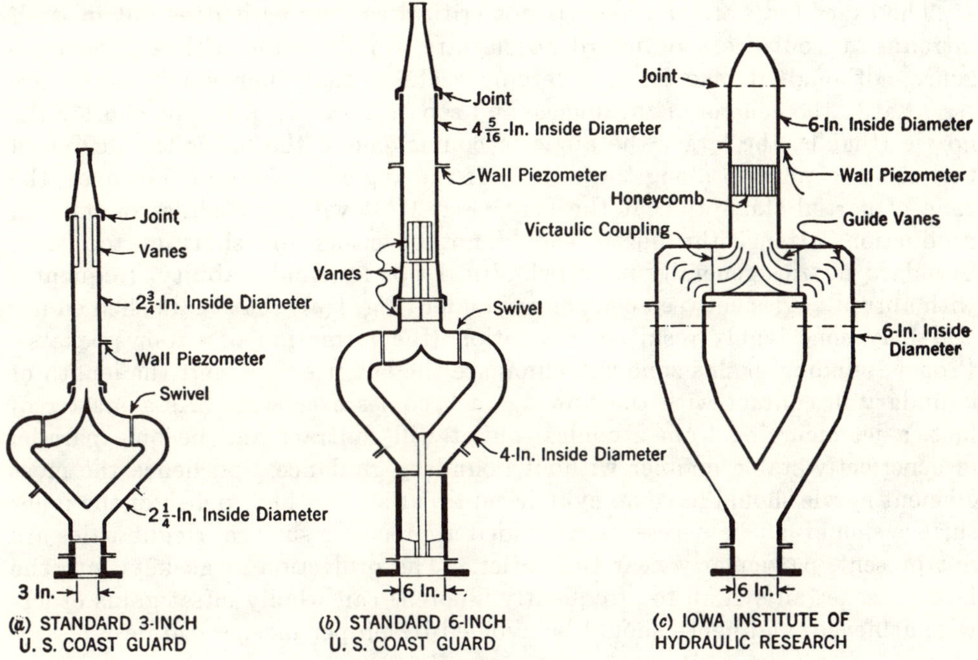

Fig. 3.—Fixed Monitors

incorporation of the principles of fluid mechanics in such design should effect a marked gain in fire-stream efficiency.

Wind-tunnel[7] and water-tunnel[8] requirements are closely allied with those of fire monitors and nozzles, for the flow at the test section of either type of tunnel must be of high velocity and yet as free from turbulence as possible. Nearly all such tunnels include: (a) Miter bends with curved vanes to guide the flow smoothly around every turn; (b) a very gradual expansion into a large approach section containing fine screens or honeycombs, in which the flow is finally allowed to "still" at low velocity; and (c) a well-formed nozzle, through which the fluid is accelerated continuously to the desired maximum speed.

In a fire monitor the passages should be free from cross-sectional discontinuities and should be of maximum dimensions commensurate with practicable over-all size—on the one hand to reduce the power loss and on the other hand to hold the energy of the turbulence to a minimum. The bends should not have the sweeping curves of the familiar rams-horn style, for these introduce secondary spiral currents (Fig. 1(d)) that persist far into the jet. Rather, miter bends with vanes should be used at all turns beyond the monitor base. In the approach section or barrel, a series of fine screens would be desirable to

<hr>

[7] "Experimental Methods—Wind Tunnels," by A. Toussaint, published in *Aerodynamic Theory*, Ed. by W. F. Durand, Julius Springer, Berlin, Germany, Vol. III, 1935, p. 252.

[8] "Cavitation and Pressure Distribution—Head Forms at Zero Angle of Yaw," by Hunter Rouse and John S. McNown, *Bulletin No. 32*, State Univ. of Iowa Studies in Engineering, Iowa City, Iowa, 1947.

break remaining large-scale eddies into small ones that rapidly decay; however, such screens would be turbulence generators rather than dissipators as they became clogged with debris, and a coarse honeycomb is the only stilling device that should be considered.

The exact form of the nozzle is not critical so long as it does not in itself introduce turbulence. Standard nozzles are usually made with a 7° convergence, although it is only in diverging sections that such small angles are essential in the control of turbulence and separation. In fact, the shorter the nozzle (that is, the larger the angle of convergence), the lower the degree of turbulence generated along the walls. If the angle is too great, however, the zone of partial stagnation at the base (Fig. 1(b)) will shed eddies of its own production. Hence the nozzle should not be either too short or too long. Standard nozzles generally have cylindrical tips for final stability, frequently with abrupt angles between cone and cylinder (Fig. 1(c)) that again shed eddies and may conceivably result in cavitation (the formation of vapor pockets). Proper rounding in this zone will eliminate the larger eddies, but the length of boundary in contact with the flow again becomes excessive. As a matter of fact, a jet emerging from a conical outlet will contract and become parallel in a perfectly stable manner without boundary guidance, and hence the most efficient nozzle should have no cylindrical tip at all. A high polish of the inner surface should not be necessary, provided appreciable surface irregularities are not present, particularly near the outlet. The projection of gaskets into the flow passage (an evil all too frequently ignored) can wholly offset gains otherwise achieved, and hence should be avoided by proper mechanical design.

Experimental Equipment and Procedure

Monitors and Nozzles.—Supplied by the Coast Guard as a basis for test comparison were standard 3-in. and 6-in. fixed monitors (Fig. 3), and

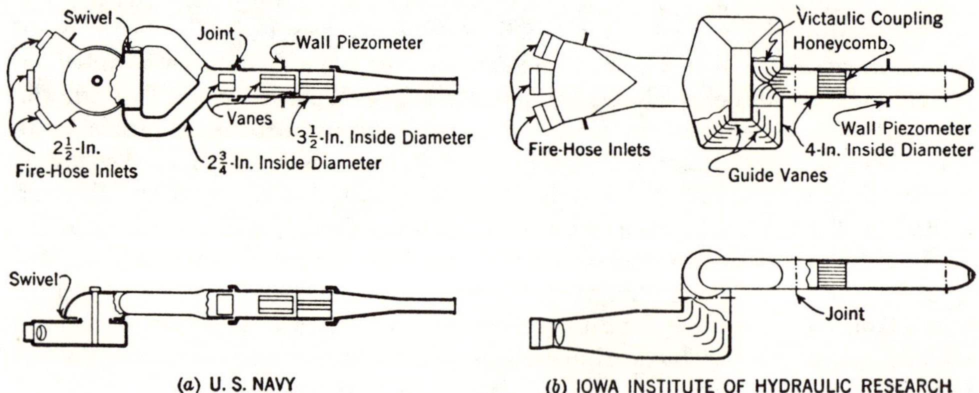

Fig. 4.—Portable Monitors

by the United States Navy a standard portable monitor (Fig. 4), each with a series of standard nozzles. The 3-in. monitor, originally proposed for improvement, had a 24-in. barrel with radial vanes, and could be directed manually through 360° horizontally and 70° vertically. Nozzles were of the usual 7° cone-

and-cylinder type with polished inner surfaces, abrupt boundary angles, and
inwardly protruding gaskets, and ranged from $1\frac{1}{2}$ to $2\frac{1}{2}$ in. in outlet diameter.
The 6-in. monitor shown in Fig. 3(b) had a 30-in. vaned barrel, and could be
directed by worm drives through 360° horizontally and 120° vertically. The
nozzles were quite similar to those of the smaller unit, and ranged from $1\frac{1}{2}$ in.
to 3 in. in outlet diameter. The portable monitor shown in Fig. 4(a) had three
hose inlets with check valves and a 15-in. barrel. It was designed for lashing in
position but could be swiveled manually through 360° horizontally and 90°
vertically. Nozzles ranged from $1\frac{1}{2}$ in. to 2 in. in outlet diameter and differed
from the others primarily in greater length of cylinder.

Only one fixed monitor and one portable monitor (also shown in Figs. 3 and
4) were built by the Institute as recommended forms. Other forms and sizes
were not investigated, since significant variation in design is not permitted by
the results to be accomplished and since essentially the same design may be
adapted to different monitor sizes by the principle of geometric similarity.
As indicated previously, the essential design criterion was the reduction of eddy
losses and secondary currents to a practicable minimum in order to attain flow
in the barrel with as low a degree of turbulence as possible. The flow passages
were, therefore, enlarged and straightened, and the sweeping curves of the
usual rams-horn were replaced with vaned miter bends. Since the purpose

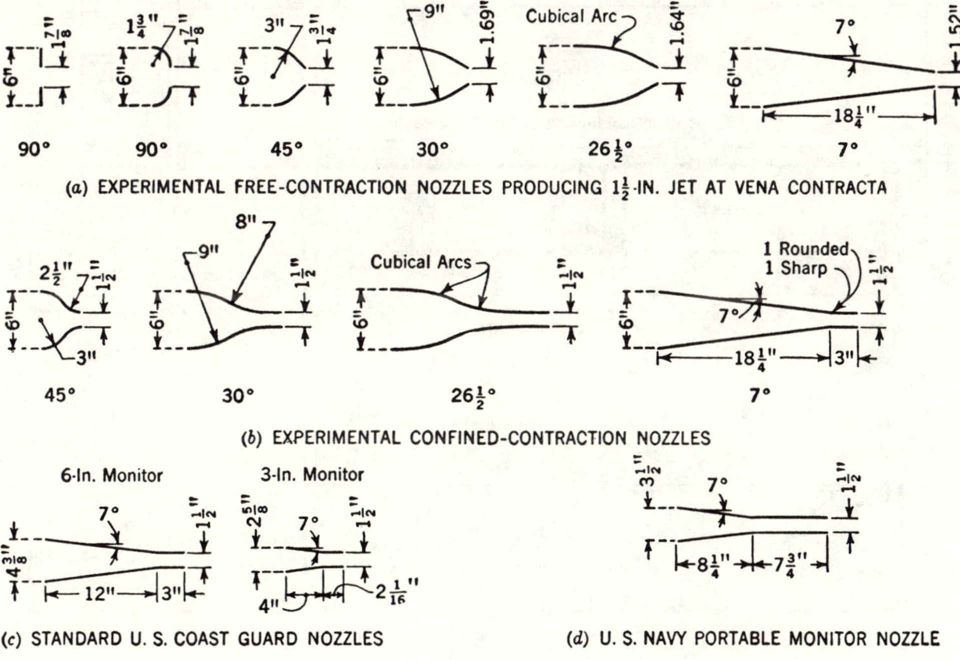

FIG. 5.—SCHEMATIC DRAWINGS OF 1½-IN. NOZZLE FORMS

of the project was the improvement of the flow characteristics only, no heed
was given either to external construction or to mechanization, and standard
pipe was used with welded, flanged, or Victaulic (swivel) joints, placed where
most convenient for test assembly and operation.

In order to verify the predicted criteria for nozzle design, a systematic series of contraction profiles, varying from a plate orifice to a long taper, both with and without cylindrical tips, were constructed of brass (see Fig. 5). The external form was again determined by expediency rather than ultimate use. In all cases, however, not only was the inner surface carefully machined, but a smooth, gasket-free, metal-to-metal joint was obtained between barrel and nozzle or nozzle parts. Supplementary nozzle forms included several with roughened surfaces and other internal obstructions, as well as one with a profile consisting of two cubical curves that electrical-analogy measurements had shown to yield a continuous pressure drop from base to tip. Each nozzle of this series was constructed and tested in the $1\frac{1}{2}$-in. size only, but the form finally recommended was repeated in diameters of $1\frac{1}{2}$, 2, $2\frac{1}{2}$, and 3 in. for the fixed monitor and $1\frac{1}{2}$, $1\frac{3}{4}$, and 2 in. for the portable monitor.

Pumping System.—Since the requirements of the monitor test program (maximum stream diameter of 3 in. and maximum base pressure of 200 lb per sq in.) were far in excess of the high-capacity, low-head pumps of the Institute laboratory, a battery of 8 standard Hale fire-boat pumps, driven by Chrysler gasoline engines, were supplied for the project by the Coast Guard. These

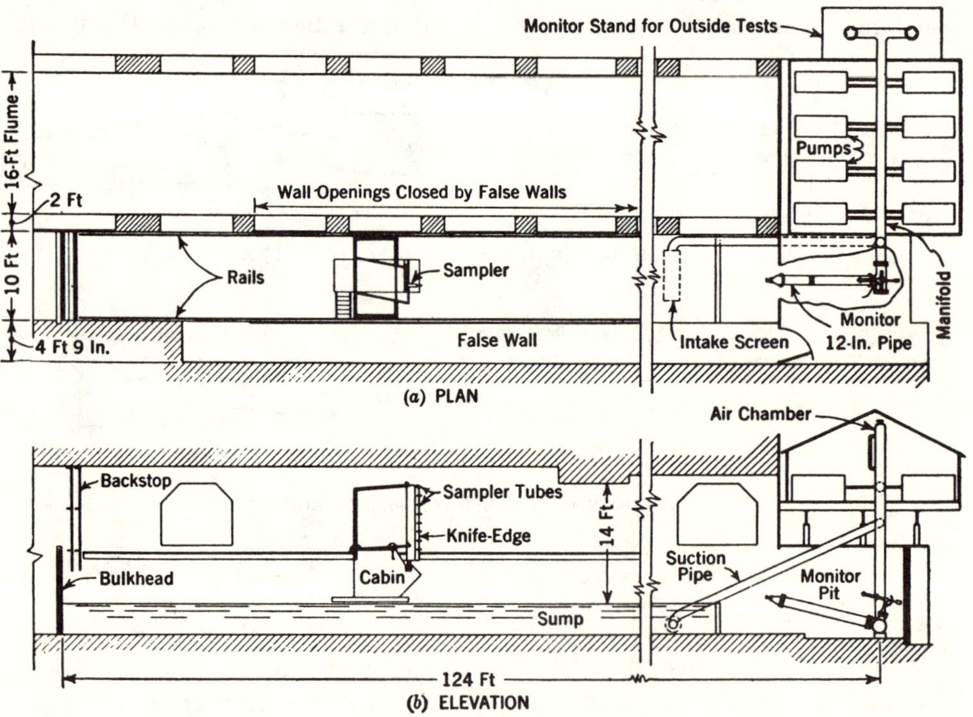

Fig. 6.—Layout of Test Gallery

pumps were housed in a temporary shelter at the south end of the laboratory (Fig. 6), and so arranged that they could either be connected in parallel to a 12-in. suction line leading from a sump in the 10-ft channel or drawn individually through standard 4-in. suction hoses from the 16-ft channel below. All pumps discharged into a common 12-in. pressure line leading both to the 10-ft channel and to a platform on the outside of the shelter.

Unless each of the centrifugal pumps in a parallel system of this type operates at such a speed as to yield the same change in head, one or more of

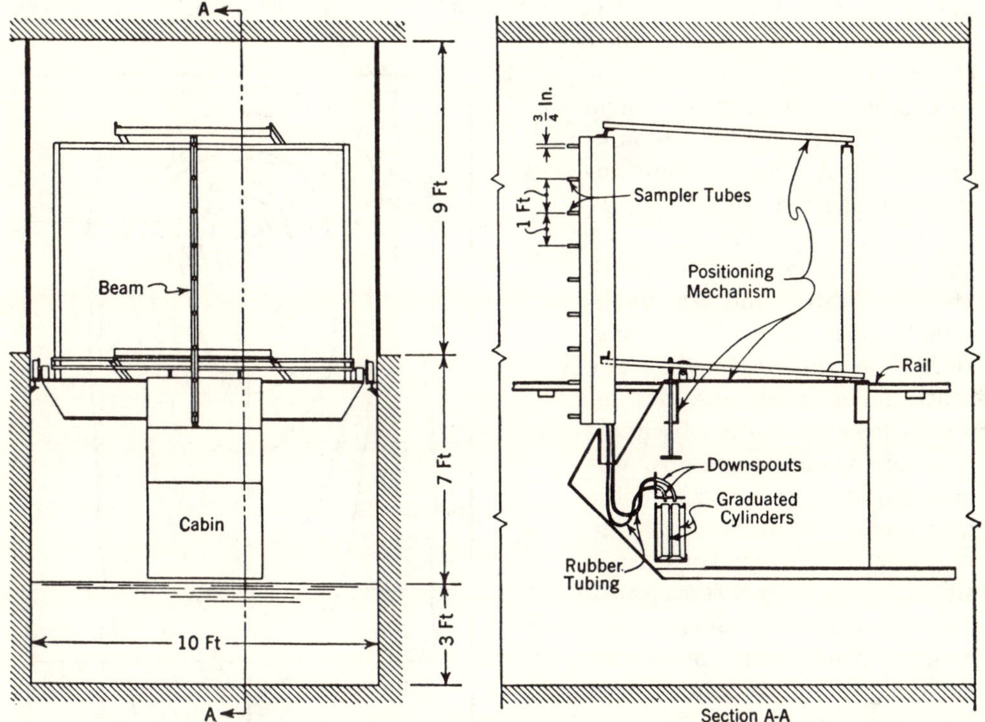

FIG. 7.—SAMPLER-TUBE BEAM AND OBSERVATION CAR

the pumps will be stalled by the others, with complete disruption of operation resulting. Therefore, it was necessary to have an individual operator for every two units in use. A 15-in. Bourdon gage mounted on the pressure manifold was under constant observation by a supervisor, who was responsible for maintaining the pressure within ± 1 lb per sq in. of the desired value during each run.

Pressure-Discharge Calibrations.—Secondary lines of fire hoses connected in parallel to a 6-in. pipe permitted each of the five monitors to be mounted above the two 18,000-lb scale tanks of the Institute laboratory for accurate determination of discharge coefficients. Since many of the nozzles tested produced contracted jets, the outlets were repeatedly machined until all jet diameters were within $\frac{1}{2}\%$ of $1\frac{1}{2}$ in. Jet profiles in the outlet vicinity were measured with a point gage that could be moved about 3 in. longitudinally and radially in any axial plane. A similar mounting permitted velocity traverses of the contracted jets to be made with a Pitot tube formed of a 0.05-in. hypodermic needle connected to a large Bourdon gage.

Wall piezometers were installed in each of the five monitors at key points (see Figs. 3 and 4), so that both the over-all losses and the relative losses at each transition could be determined. For comparative operation of the various monitors, the same pressure in the supply pipe at the monitor base was maintained in all cases.

Gallery Tests.—In order to compare both nozzle and monitor performance under laboratory conditions free from atmospheric disturbance, a test gallery was established in the 10-ft channel running under the Institute laboratory (Fig. 6). The ends of the channel were bulkheaded, and screens were installed along the sides, thus producing simultaneously a catch and storage basin for a 3-ft depth of clear water and an unobstructed test space 100 ft long, 10 ft wide, and 14 ft high. The south end of the gallery, containing the lower end of the discharge manifold from the pumps, was only partially screened, and the north end was fully vented to both the 16-ft channel on the one side and the 4-by-9-ft walkway on the other. Although this confined space could not be considered to simulate conditions in the open except as a first approximation, it nevertheless permitted operation of each monitor and nozzle under identical conditions, which would otherwise have been a practical impossibility.

Observations of the jets could be made in this gallery in three different ways: (1) Conditions of jet disruption could be studied by means of high-speed photography, a pair of flash lamps operating from a condenser permitting exposures of 1/20,000 sec; (2) the mean trajectory

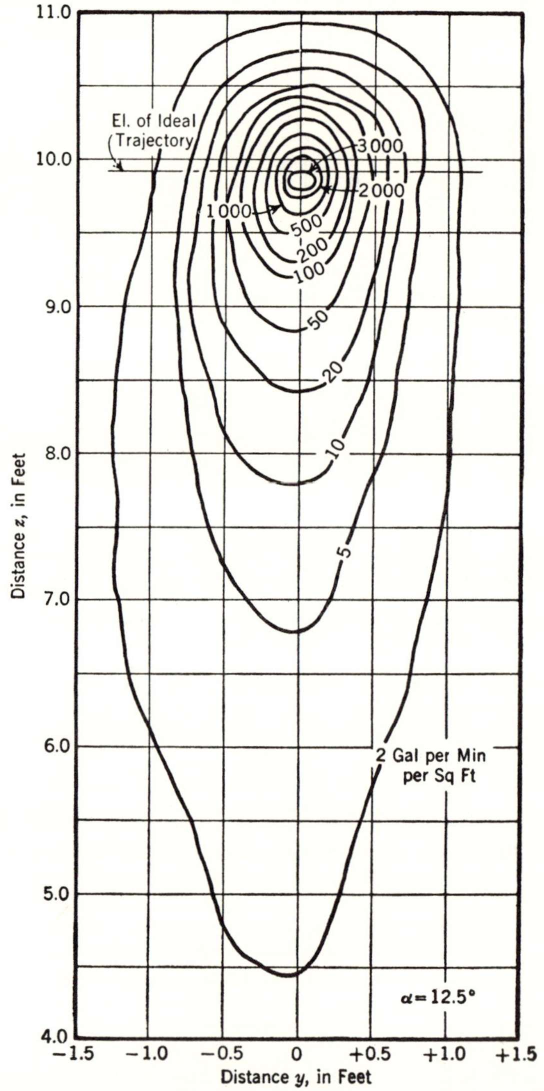

Fig. 8.—Typical Plot of Jet Concentrations 90 Ft from a 30° Nozzle

and jet thickness could be determined by means of a special profilometer operating by means of a horizontal beam of light; and (3) the discharge concentration of the jet at each and every point in any cross section up to 90 ft from the nozzle could be measured through use of a point sampler.

Since the sampler apparatus provided the basis for many performance ratings, it is described here in further detail. A small sampling car, with a cabin barely large enough for two observers, could be moved on the rails at either side of the channel to any section of the gallery. Mounted on the front of this car (see Fig. 7) was a vertical wedge-shaped beam that could be moved through 1 ft vertically and 4 ft either side of the centerline. From the leading edge of this beam projected a series of nine $\frac{3}{4}$-in. sampling tubes, 12 in. on center, that were connected by rubber tubing to a battery of downspouts within the cabin. Below these spouts was a series of graduated cylinders so framed that they could be abruptly alined with the downspouts, abruptly disalined, and rapidly inverted to empty. Concentration runs consisted of measuring the volume of inflow against time for each of the sampling tubes, the beam being shifted to section after section until a representative traverse (see Fig. 8) of the entire jet had been made. In zones of maximum concentration it was found

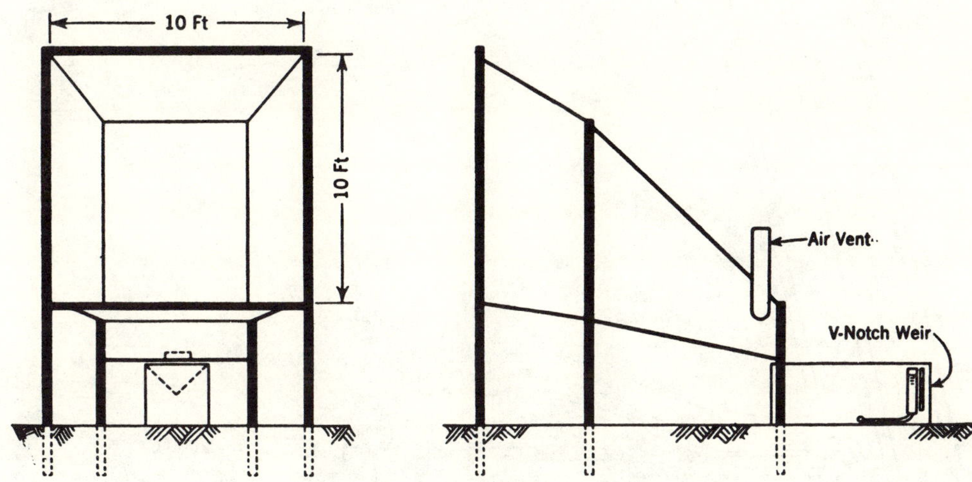

Fig. 9.—Outdoor Target

necessary to reduce the size of the sampling tube to $\frac{3}{8}$ in. by means of a streamlined insert and to reduce the distance between samples both horizontally and vertically. A typical run at the common distance of 90 ft from the point of efflux involved some 900 individual measurements.

Target Tests.—The test gallery had a maximum length that was far less than the throw of most of the streams, so further comparisons of performance were made in the open, along the west bank of the adjoining river, through use of a target mounted 200 ft from the monitor stand. The target (Fig. 9) had an opening 10 ft by 10 ft in vertical section, and the rate of flow entering the opening was determined by means of a suitably baffled V-notch weir directly behind and shielded by the target. Owing to the pronounced effect of wind upon the streams, tests could be made only on relatively calm days. Each stream was so directed that the maximum quantity entered the target, the same base pressure of 150 lb per sq in. and the same jet diameter of $1\frac{1}{2}$ in. being used in all cases. Since slight disturbances were almost immediately reflected in the head on the weir, repeated observations were made at intervals of $\frac{1}{4}$ min and the results were analyzed statistically to eliminate obviously erroneous values.

Trajectory Tests.—Initial comparison of the jet trajectories for the various monitors[9] was made by means of photographs of the nozzle streams taken from a point on the opposite side of the river, using a fixed-focus camera. The monitors were directed north at an inclination of 20°, with the laboratory building as a combined shield and background. This method could not be used for final performance tests, however, since at high pressures the spray from the streams interfered with traffic on the bridge 350 ft from the test stand. Moreover, at higher jet inclinations the stream could not be seen distinctly on the photographs against the background of sky above the laboratory.

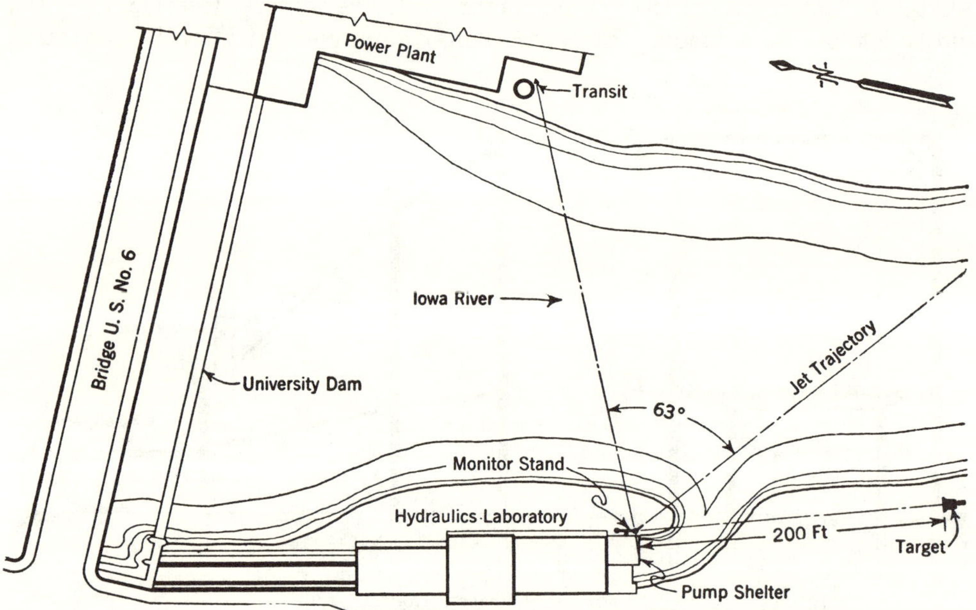

FIG. 10.—PLAN OF OUTDOOR TESTING AREA

The method of observation finally adopted involved the use of a surveyor's transit, on the telescope of which a nonmagnifying sight with cross wires and a large field of view had been mounted. The monitors were directed diagonally southeast across the river (see Fig. 10), and the absolute trajectories were calculated trigonometrically from readings of the horizontal and vertical circles of the transit and predetermined base-line characteristics. Measurements were carried as far along each trajectory as the stream retained a readily visible core. A very long period was required to complete these observations, because of the fact that satisfactory measurements could be made only under calm or nearly calm conditions. These conditions prevailed, unfortunately, only about 1% of the available time.

---

[9] "Characteristics of High-Velocity Jets," by Joseph W. Howe and Chesley J. Posey, *Proceedings, Third Hydraulics Conference; Bulletin No. 31*, State Univ. of Iowa Studies in Engineering, Iowa City, Iowa, 1947, p. 315.

DISCUSSION OF TEST RESULTS

Relative Performance of Nozzles.—Gallery tests on each of the nozzle forms shown in Fig. 5 resulted in contour plots of discharge concentration similar to those of Fig. 8 By integration of the successive contour areas, each of these plots was reduced to a single characteristic curve of percentage of total nozzle discharge against cross-sectional area.   A typical family of such curves, corresponding to the nozzle forms at the top of Fig. 5, can be seen in Fig. 11.   The logarithmic abscissa scale is used to permit convenient study of each portion of the field.   If the jet from the nozzle remained absolutely intact throughout the 90-ft distance to the sampler, the resulting curve would have the form indicated by the broken line labeled "intact jet" in Fig. 11.   The extent to which the actual curve falls below this ideal limit is an indication of the diffusion that the jet has actually undergone.   Unless a considerable amount of the spray from the jet has fallen into the reservoir before reaching the sampler, the curve should approach the 100% line as the cross-sectional area increases; this portion

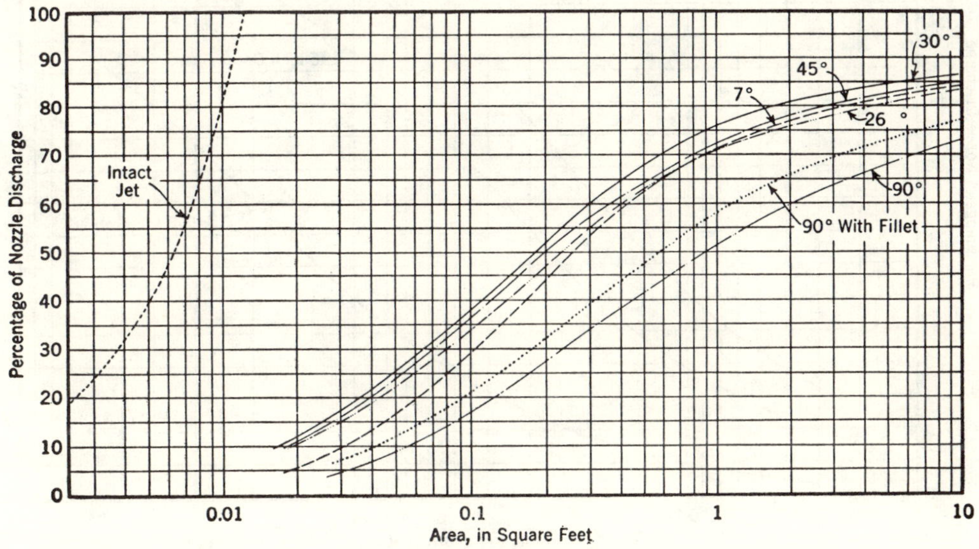

FIG. 11.—COMPARATIVE AREA-CONCENTRATION CURVES FOR FREE-CONTRACTION NOZZLE FORMS

of the curve is otherwise of little interest, since the jet at this point is dispersed over such a large area that a great quantity of air travels with the water at fairly high speed.   The initial portion of the curve, on the other hand, is of great significance, as it is a measure of the relative jet concentration at the axis.

With regard to the several curves shown in Fig. 11, it will be seen that the poorest performance (the lowest curve) corresponds to the 90° orifice plate. That this is due to disruption of the jet by eddies formed in the stagnation zone at the base of the plate is shown by the improvement in performance resulting from rounding the plate in this zone.   Decreasing the angle of convergence is likewise seen to improve the performance, largely because of the corresponding reduction in eddy formation in the steadily diminishing stagnation zone. Eventually, however, this effect is not only minimized, but at the same time the increased length of boundary begins in turn to augment the surface turbulence, so that optimum performance is found at an intermediate angle considerably greater than the standard 7°.

A similar family of curves for the 7° nozzles showed graphically the relative effects of the three types of nozzle tip. The standard type with abrupt intersection between cone and cylinder was very nearly comparable in performance to the 90° plate with fillet. Careful rounding of the intersection produced a marked improvement, owing to the elimination of the eddies (and possibly vapor bubbles) that are formed at such a discontinuity. The maximum concentration was produced by the cone without any cylindrical tip whatever,

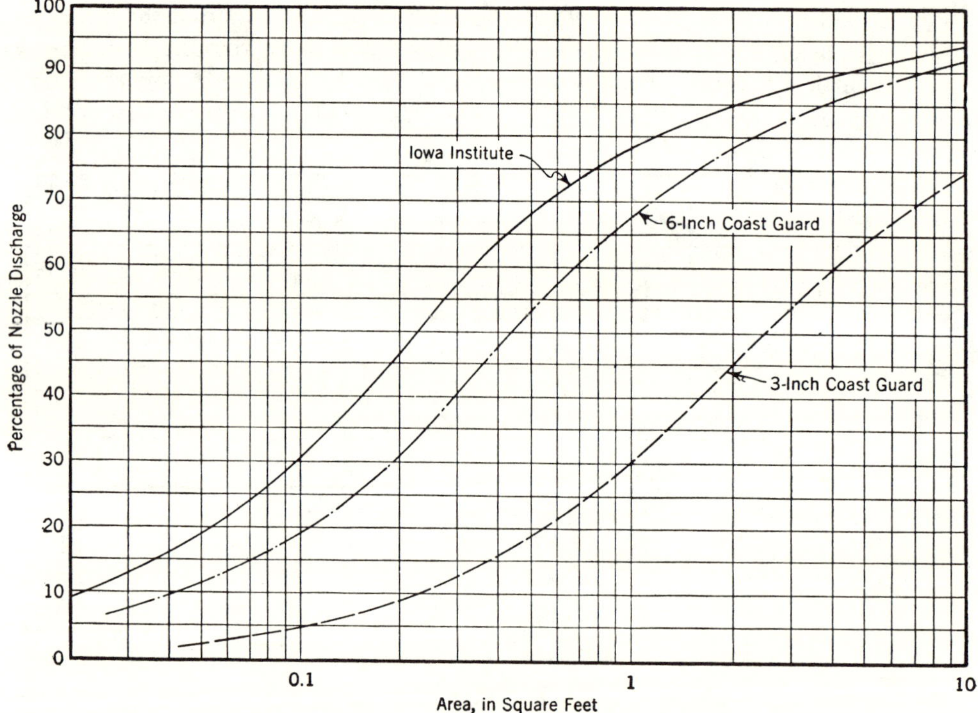

Fig. 12.—Comparative Area-Concentration Curves for Fixed Monitors

since omission of the cylinder shortened the zone of boundary contact with the high-velocity flow. Velocity traverses just beyond the efflux section revealed a significant reduction in the thickness of the boundary layer from one case to the next. Moreover, jets emerging from nozzles without tips invariably tended to appear crystal clear in comparison with the milky whiteness of jets from cylindrical tips with abrupt discontinuities.

Supplementary tests with the nozzles included artificial production of the three types of eddy formation leading to jet diffusion. Vanes and baffles in the approach barrel simulated the effects of upstream bends or obstructions, resulting in large-scale disruption of the stream. Boundary irregularities comparable to protruding gaskets or surface accretions caused dispersion to a smaller but still serious degree. Finally, increased length of cylindrical tip produced a marked tendency of the stream to surround itself with a fine mist that appeared to protect the central stream to some extent from the surrounding air but rapidly fell below the mean trajectory. It is noteworthy that minute nicks at the very tip of a nozzle yielded perceptible furrows in the jet which gradually expanded into regions of appreciable dispersion. Although the nozzles constructed by the Institute were all given a sharp 90° edge at the end of the contraction, it was subsequently found that a very short ($\frac{1}{8}$- to $\frac{3}{16}$-in.) cylindrical tip could be used for contraction angles as low as 26.5° without

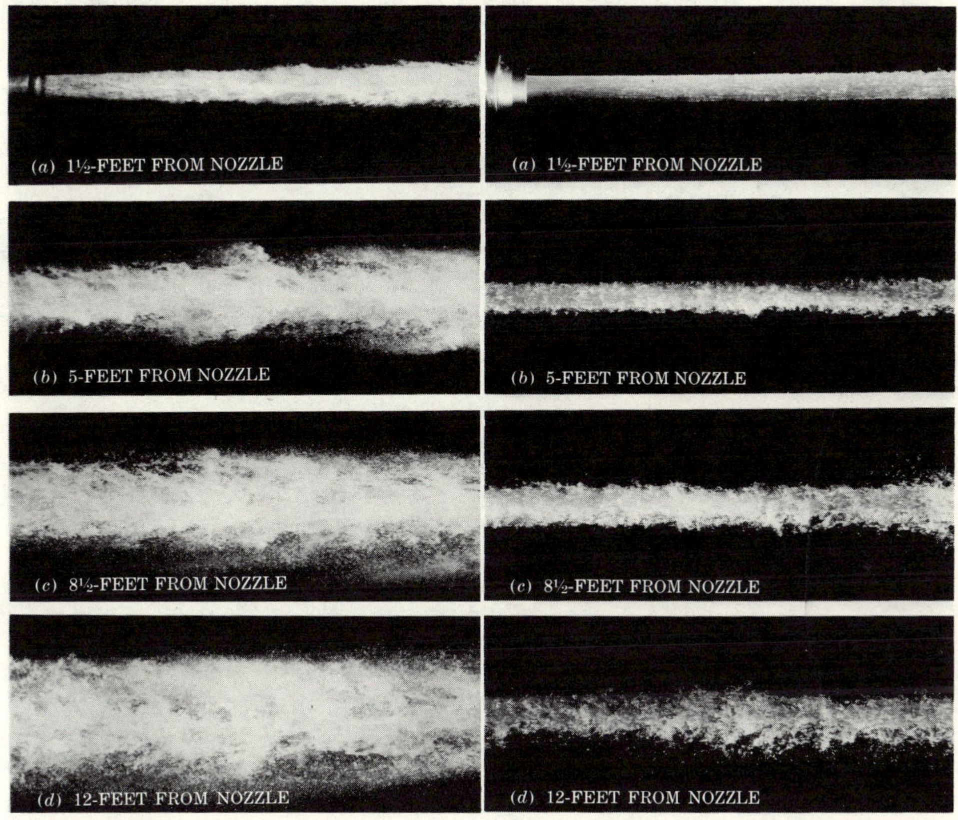

Fig. 13.—High-Speed Photographs of 1½-In. Jet from 3-In. Coast Guard Monitor

Fig. 14.—High-Speed Photographs of 1½-In. Jet from Institute Fixed Monitor

affecting the free jet. Not only is such an outlet easier to fabricate, but it also serves to protect the critical portion from damage.

The foregoing analysis of comparative tests on the nozzles themselves leads to the conclusion that, aside from the absence of boundary discontinuities, the best form of nozzle is one that results in a minimum length of contact between wall and fluid and yet produces a minimum eddy effect from the stagnation zone at the base. Although these factors tend in opposite directions, there is a considerable intermediate range of form that satisfies the requirements to an acceptable degree. The form chosen by the Institute was the 30° profile with well-rounded base and free contraction, shown in Fig. 5(a).

Relative Performance of Fixed and Portable Monitors.—Owing in part to the larger flow passages and in part to the reduced losses through the vaned bends, the flow efficiency of the Institute monitor with the recommended nozzle was appreciably higher than that of the 6-in. Coast Guard monitor (with a rounded rather than a standard nozzle) and considerably higher than that of the 3-in. Coast Guard monitor with a standard nozzle. If the rate of discharge is expressed in terms of the base pressure head h_b in the supply main and the area a of the contracted jet as

$$Q = C_d\, a\, \sqrt{2\,g\,h_b} \dots\dots\dots\dots\dots\dots(10)$$

the discharge coefficient C_d for the three monitors with $1\frac{1}{2}$-in. jets is as follows:

Monitor	C_d
3-in. Coast Guard	0.917
6-in. Coast Guard	0.955
Institute	0.982

In other words, the Institute monitor will discharge about 3% more water than the 6-in. Coast Guard monitor and 7% more than the 3-in. monitor at the same base pressure. Similarly, if the over-all loss in head H_L is expressed in terms of the jet velocity head as

$$H_L = C_L\, \frac{(V_i)^2}{2\,g} \dots\dots\dots\dots\dots\dots(11)$$

the following values result for the three monitors with $1\frac{1}{2}$-in. jets:

Monitor	C_L
3-in. Coast Guard	0.211
6-in. Coast Guard	0.104
Institute	0.049

Evidently, the power requirement of the Institute monitor is 5% below that of the 6-in. Coast Guard and 15% below that of the 3-in. one at the same discharge.

Gallery tests for jet concentration produced by the three monitors with $1\frac{1}{2}$-in. jets operating at the same base pressure of 100 lb per sq in. resulted in the three curves plotted in Fig. 12. The performance of the 3-in. standard monitor is obviously quite inferior. Although the Institute monitor does not at first glance seem much superior to the larger Coast Guard monitor, it should be noted that the peak concentration, 90 ft from the nozzle, was actually 60% greater than that for the larger and some 14 times as great as that for the smaller one. High-speed photographs of the jets from the 3-in. Coast Guard and the Institute monitors over the first 14 ft of their course (Figs. 13 and 14) show graphically the comparative degrees of turbulence producing the diffusion.

Essentially the same relative performance was indicated by the target tests over a 200-ft range with $1\frac{1}{2}$-in. jets at a base pressure of 150 lb per sq in. The 3-in. Coast Guard monitor fell short of this throw by some 15 ft and hence was omitted from further study. The Institute monitor showed an improvement of 26% over the 6-in. Coast Guard monitor, based on the proportion of discharged water striking the target, and an improvement of 32% based on the total amount of water caught in a given time while operating at the same base pressure.

Final comparisons involved the measured trajectories of streams from the three monitors with the same jet diameter of $1\frac{1}{2}$ in., for base pressures of 50 and 150 lb per sq in., and at 20° and 50° inclinations. The results are shown in Fig. 15, together with the ideal trajectories for the discharge of the Institute monitor under the four conditions. The Institute monitor obviously leaves much to be desired when compared with the ideal. On the other hand, its throw is some 20% better than that of the larger Coast Guard monitor and 40% better than that of the smaller. The relative performance is essentially independent of base pressure and inclination.

Although the portable Navy monitor was comparable in size and weight to the 3-in. Coast Guard unit, it was generally superior in performance. As a result, the improvement obtained with the portable model constructed by the Institute was considerably less than that with the fixed model. The discharge coefficients of the two portable monitors were as follows:

Monitor	C_d
Navy portable	0.900
Institute portable	0.950

At the same base pressure, the Institute monitor will thus discharge about 6% more water than the Navy monitor (under the same conditions, the Navy portable will deliver 2% less than the small Coast Guard unit). The loss coefficients were likewise as follows:

Monitor	C_L
Navy portable	0.288
Institute portable	0.118

Hence, at the same discharge, the Navy portable monitor will require 17% more power than the Institute monitor (it will, on the other hand, require 7% more than the small Coast Guard unit). Gallery measurements, the results of which are plotted in Fig. 16, showed the Institute portable to produce an appreciably lower rate of jet dispersion, with a peak concentration about $2\frac{1}{2}$ times that of the Navy portable monitor (which was, in turn, 4 times that of the small Coast Guard monitor). In the target tests, the Institute portable monitor showed improvements of 12% and 17% in proportion of discharged water and in total amount of water caught, respectively. Records of trajectory tests, reproduced in Fig. 17, indicate that the throw of the Institute model is about 15% greater than that of the Navy portable (the latter being some 10% greater than that of the small Coast Guard monitor). The performance of both portable units was essentially independent of the number of supply hoses used.

Mention should be made at this point of tests on honeycombs conducted with the larger Institute monitor. Since the passages of this monitor were all of the same diameter, the same honeycombs could be placed in either the barrel or the base riser. Two geometrically similar honeycombs, 4 in. and 6 in. long, were made of $\frac{1}{16}$-in. steel plate by slotting each set of plates over half

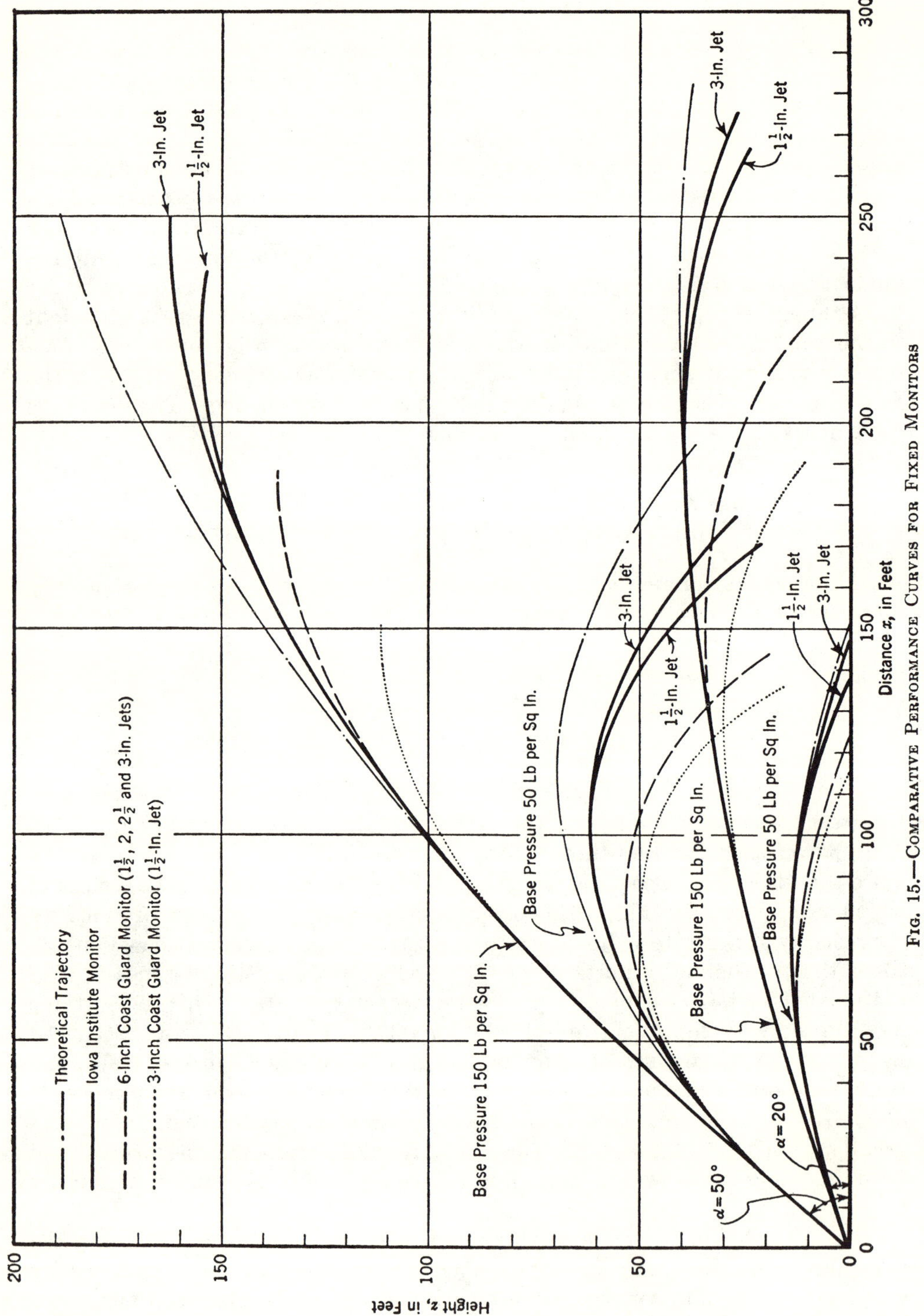

FIG. 15.—COMPARATIVE PERFORMANCE CURVES FOR FIXED MONITORS

their length and joining the respective sets in the familiar "egg-crate" fashion. The square passages of one unit were $\frac{1}{2}$ in. wide and those of the other were $\frac{3}{4}$ in. wide. Anti-swirl devices of this nature can be expected to reduce the scale (eddy size) of the turbulence only to a degree commensurate with the size of the passages. As in the case of the turning vanes, finer openings than those investigated were considered impracticable, since even the slightest fouling of the coarse units with debris produced a marked effect upon the jet dispersion. As a result, essentially the same jet characteristics were obtained either with or without a honeycomb in a barrel of the length shown in Fig. 3. Although shortening the barrel to half this length made a honeycomb necessary, doubling the length produced no noticeable change whether or not a honeycomb was

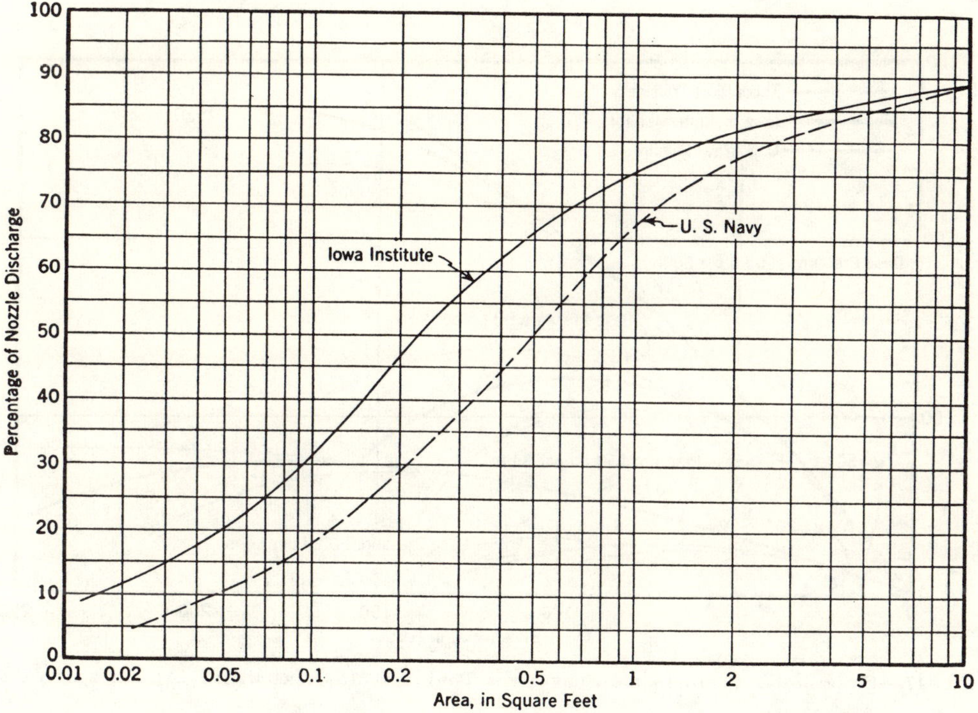

Fig. 16.—Comparative Area-Concentration Curves for Portable Monitors

used, indicating that the flow was already as good as honeycombs of this scale would ensure. That a relatively poor flow, on the contrary, would be considerably improved by even such a coarse honeycomb was graphically demonstrated by reversing the Institute monitor—that is, by connecting the barrel end to the supply line and placing the nozzle directly on the end of the inverted Y-shaped base. Without a honeycomb the jet was extremely poor, whereas with a honeycomb a short distance upstream from the nozzle the jet became fairly acceptable. Although nearly all measurements described herein were made with one or the other honeycomb in the barrel, the subsequent studies just described showed this to be neither necessary nor generally desirable with a barrel of normal length.

Equally illuminating was another supplementary test in which a nozzle was mounted at one end of the largest stilling chamber available—a pressure tank $2\frac{1}{2}$ ft in diameter and 6 ft long—in the hope of further reducing the turbulence. As is common practice in hydraulics, the discharge from the supply hoses at the upstream end was baffled by a 1-ft layer of $\frac{3}{4}$-in. crushed rock extending over the full cross section of the tank and held in place by wire mesh, thus leaving an additional 4 ft of unobstructed passage before the rounded entrance to the nozzle. The result was disappointingly poor, the jet being comparable to that from the small Coast Guard monitor. Later tests made by the Institute on the characteristics of flow through screens[10] pointed clearly to the source of the trouble: Unless the ratio of open to closed area is greater than unity, a baffle may do more harm than good. The crushed rock,

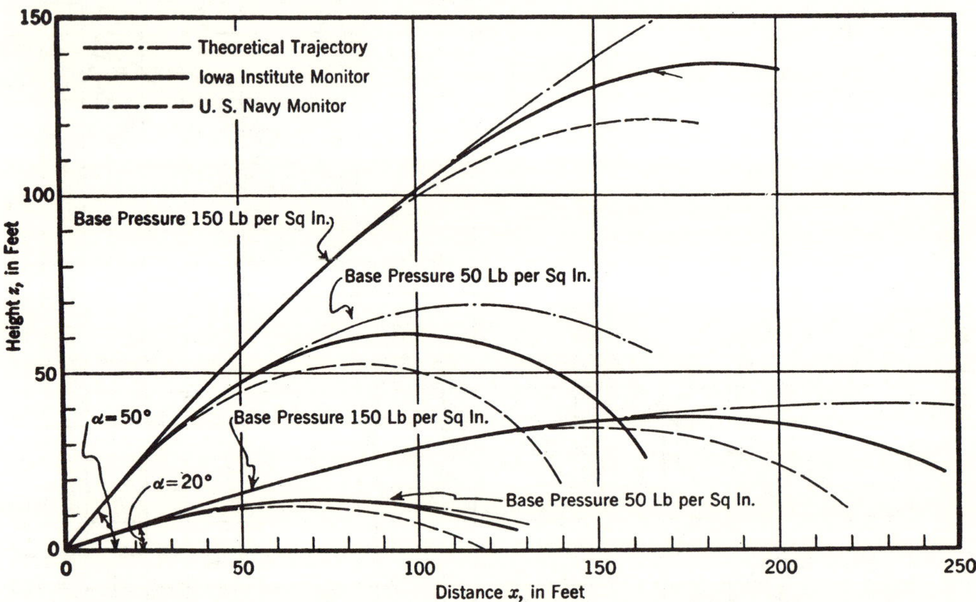

Fig. 17.—Comparative Performance Curves for Portable Monitors with $1\frac{1}{2}$, $1\frac{3}{4}$, and 2-In. Jets

in other words, very likely produced nearly as many disturbances as it eliminated, and the subsequent length of tank was not sufficient to damp the large-scale eddies remaining in the flow.

While the concentration and throw of a fire stream still can be vastly improved, it appears from the foregoing tests that this could be accomplished only at the expense of such other practical requirements of the monitor as lightness and maneuverability. If the approach passage were of the same relative diameter, length, and gradual taper as that of a stationary wind or water tunnel, the turbulence (and hence the rate of jet dispersion) could probably be reduced to a small percentage of that of the Institute monitors; such a system would, of course, be both prohibitively heavy and very difficult to

[10] "An Investigation of Flow Through Screens," by W. D. Baines and E. G. Peterson, *Transactions*, A.S.M.E., Vol. 73, July, 1951, p. 467.

mechanize.   Although a shorter approach might be fully as effective if it were possible to make the turning vanes as small and numerous as in a wind tunnel and to include a series of fine stilling screens of low solidity ratio, an arrangement of this nature would be practicable only with very clean water.   In view of these circumstances, the two monitors developed by the Institute were assumed to embody the optimum improvement possible without undue change in size or difficulty of manufacture.   Hence, these units were used as the basis of the final performance tests.

Performance Tests of Recommended Monitors.—Since three independent variables—jet diameter, base pressure, and inclination—were involved in the final tests of the Institute monitors, the effect of any one variable upon the stream could be studied only if the other two were at the same time held constant.   The series of tests, therefore, involved measurement of the stream coordinates at base pressures of 50, 100, 150, and 200 lb per sq in. and at angles of 20°, 30°, 40°, and 50° for all nozzles.   The latter included jet diameters of $1\frac{1}{2}$, 2, $2\frac{1}{2}$, and 3 in. for the fixed monitor and $1\frac{1}{2}$, $1\frac{3}{4}$, and 2 in. for the portable monitor.

To facilitate comparison, two different systems of plotting were followed. The first represents a series of superposed curves for different nozzles and pressures at constant inclination, successive plots of this nature corresponding to the different inclinations.   The second represents a similar series of superposed curves for different nozzles and inclinations at constant pressure, successive plots corresponding to the different pressures.   Since neither system contains information not readily available from the other, only the constant-pressure plots are presented (see Figs. 18 and 19).   In each instance the ideal trajectories for the corresponding discharges are shown.

Aside from the general value of these performance diagrams, careful inspection will indicate the following significant facts: First, the effect of jet diameter upon the departure from the ideal trajectory is relatively small for the fixed monitor and imperceptible for the portable monitor, at least over the range investigated.   Although this effect necessarily involves both the absolute size of the jet and its ratio to that of the monitor, it should be possible to utilize these curves with good approximation for monitor and nozzle dimensions a reasonable fraction larger or smaller than those investigated.   Second, since the degree of jet disruption at a given pressure depends primarily upon distance traveled, the horizontal length of throw of the actual jets tends to reach a maximum sooner and then to decrease more rapidly with increasing angle of inclination than does the ideal trajectory.   Except for the lowest pressure, the maximum horizontal throw is seen from Fig. 20 to occur at about 30° rather than at the ideal 45°.   Finally, whereas the coordinates of the ideal trajectory increase with the square of the efflux velocity (that is, with the first power of the base pressure), the dependence of the turbulence upon the mean velocity tends to make the angle of diffusion independent of the velocity.   However, the latter still controls the intensity of droplet dispersion by the air.   As a result, continued increase in the pressure does not produce a comparable increase in length of throw but eventually even leads to an appreciable reduction.   The summary plot of Fig. 20 indicates that a point of diminishing returns is reached

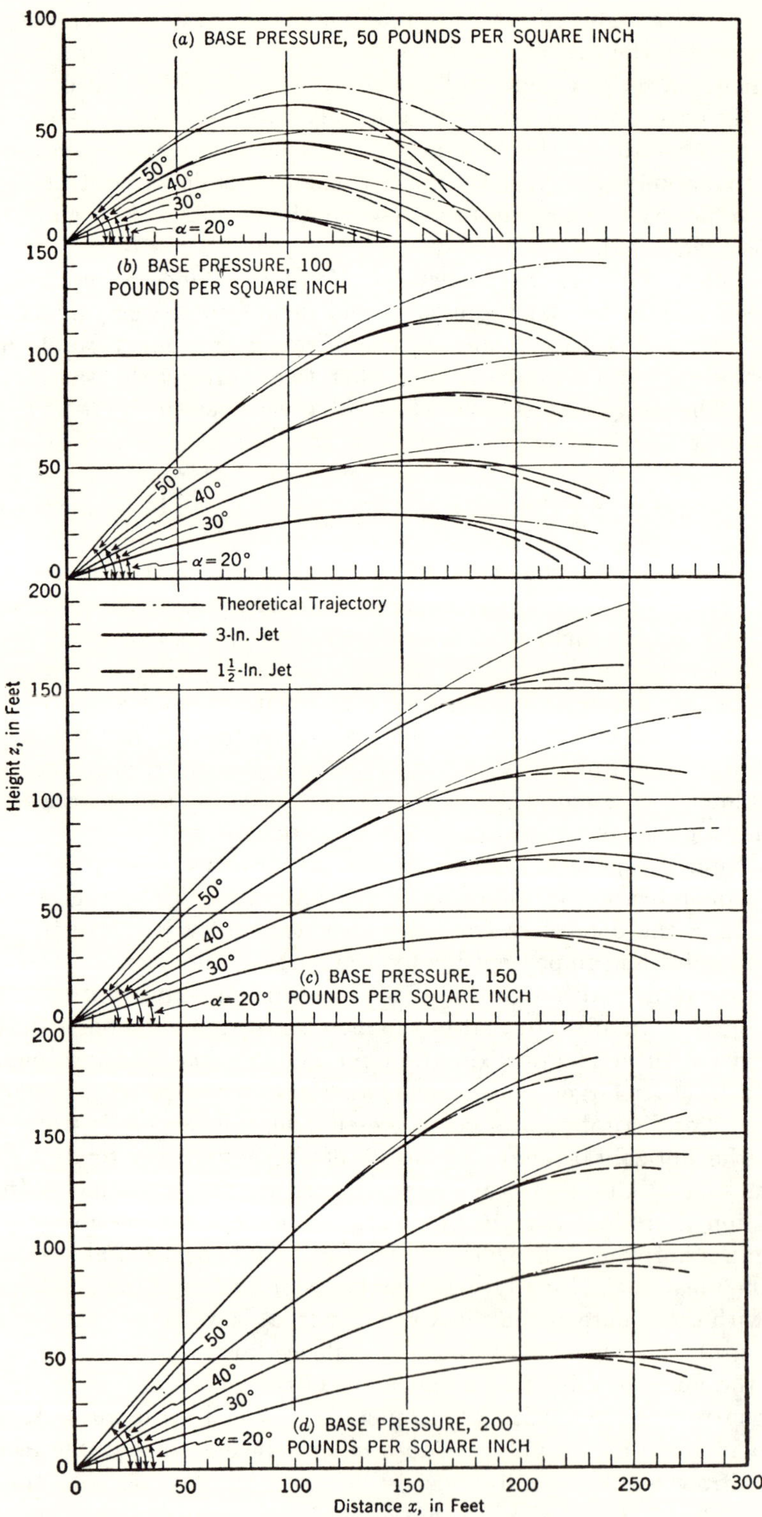

Fig. 18.—Performance Curves for Recommended Fixed Monitor with 1½ and 3-In. Jets

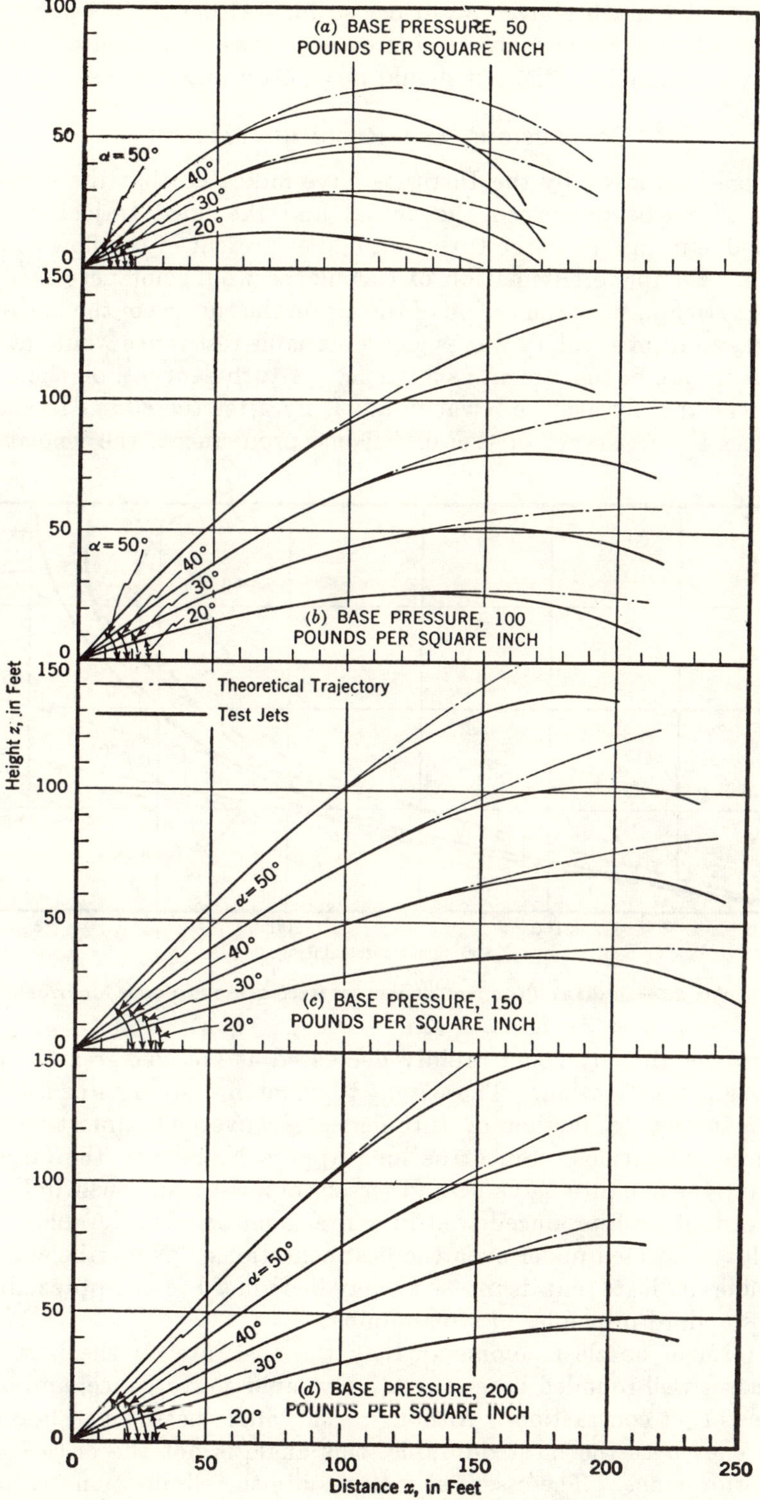

Fig. 19.—Performance Curves for Recommended Portable Monitor with 1½, 1¾, and 2-In. Jets

at a pressure not much above 150 lb per sq in. Although this point may be expected to advance somewhat with increasing nozzle and monitor size, higher pressures than those investigated should be of little practical use.

CONCLUSIONS AND RECOMMENDATIONS

Investigations made by the Institute have indicated that the considerable difference that exists between the actual and the ideal trajectories of fire streams is due primarily to the turbulent eddies present in the flow as it leaves the nozzle. Complete elimination of turbulence would not necessarily result in a perfect stream, since the action of the air on the surface of the high-velocity flow could still form capillary waves whose variable resistance would eventually lead to disruption of the stream as a whole. A turbulent jet, on the contrary, would spread rapidly even in a vacuum. Only after the eddies break the jet surface does the resistance of the air become pronounced, the velocity of the

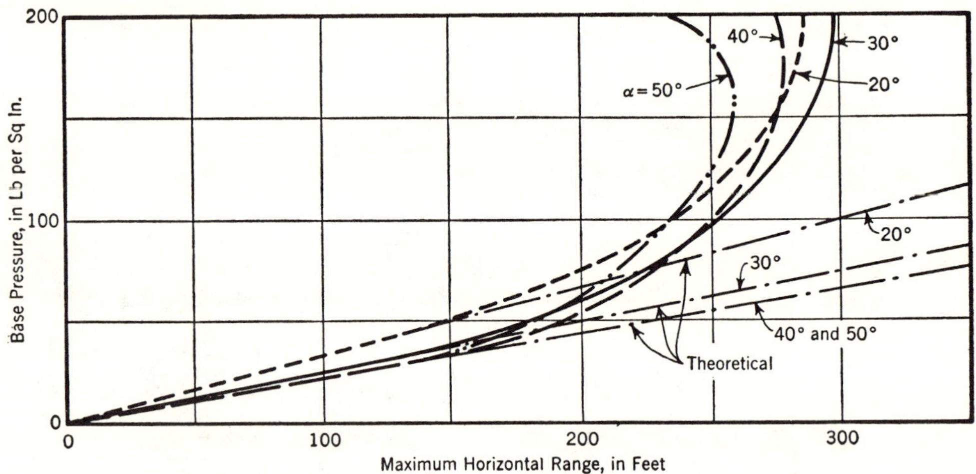

Fig. 20.—SUMMARY PLOT FOR THROW OF RECOMMENDED FIXED MONITOR

diffusing water thereby being rapidly decreased as an ever greater volume of air is brought into motion. The extent to which fire streams can actually be improved through reduction of turbulence is governed almost entirely by practical considerations. Since the long approach passage, the large stilling section, and the fine turning vanes and screens of low-turbulence wind and water tunnels cannot be reproduced well in a fire monitor, considerable turbulence must still be expected under even the best conditions. Nevertheless, the foregoing studies indicate that its intensity may be reduced to an appreciable degree through standard principles of streamlining.

The form of nozzle recommended by the Institute on the basis of these studies has a well-rounded base, a relatively rapid convergence, and an unconfined zone of jet contraction. Although a 30° angle between wall and centerline was chosen as the most desirable, this angle is not the critical factor in nozzle performance. The essential features are the elimination of: (a) A zone of eddy formation at the base that would exist in very short nozzles; (b) the excess surface resistance and formation of fine-scale turbulence in very long nozzles; and (c) the region of separation and possible cavitation at an abrupt juncture between the usual nozzle cone and cylinder. Of comparable importance but of somewhat different category is the elimination of a gasket that could (and frequently does) project into the flow at the nozzle base.

Table 1 shows schematically the general details of the recommended nozzle profile. The diameter of the barrel is taken as the reference dimension, and all other dimensions are given as ratios thereto. Although the relative jet diameters could conceivably be made either greater or smaller without undue change in performance, the range shown is judged to cover practical requirements. The table gives the relative diameter of outlet that will produce fully

TABLE 1.—CHARACTERISTICS OF RECOMMENDED FIRE NOZZLE

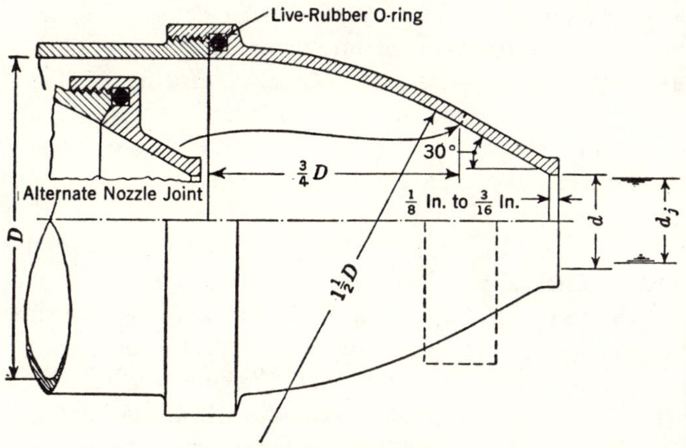

Schematic diagram	Ratio d_j/D	Ratio d/D	Coefficient of discharge C_d
	0.250	0.279	0.805
	0.300	0.335	0.806
	0.350	0.391	0.807
	0.400	0.446	0.811
	0.450	0.501	0.817
	0.500	0.557	0.825

contracted jets of the relative diameters indicated, and other sizes may readily be obtained through interpolation. Also indicated are the coefficients of nozzle discharge based upon the equation,

$$Q = C_d \, a \, \sqrt{2 g h_n} \dots\dots\dots\dots\dots\dots\dots\dots(12)$$

in which a is the area of the nozzle outlet and h_n is the piezometric head in the barrel approximately a diameter upstream from the nozzle.

Two alternative locations of the nozzle joint are shown in the figure—one at the base and one along the contraction cone. That at the base is preferable so far as the flow is concerned, but the resulting nozzles would be heavier and more costly. That along the cone will require greater precision of fabrication, but the nozzles would be far lighter and less expensive. In either instance the nozzle seats tightly against the barrel without a gasket between the surfaces; if a gasket is needed for an absolute seal, it should have the form of an external soft-rubber O-ring as shown. Although fire nozzles are usually turned painstakingly (but not always accurately) out of brass castings piece by piece, it would appear quite possible that far cheaper, lighter, and more accurate units could be formed of plastic from a single, carefully made die and mold. Under the latter circumstances, the base joint would be preferable and quite feasible, and the nozzles could be considered expendable if and as the tips became damaged. The latter tendency should be minimized by the very short tip cylinder shown in the drawing; this is merely for protection during handling since the stream springs completely free at the inner edge.

The monitor itself is subject to three general requirements: (1) The passages should be as long and large in cross section as possible; (2) the cross-sectional area should either be constant or increase very gradually in the direction of flow; and (3) section changes should be well rounded, and direction changes should contain guide vanes. Each of these requirements becomes the more important the more closely the nozzle is approached.

The fixed monitor and the portable monitor designed by the Institute were considered at the time of their design to meet these requirements to a generally satisfactory degree, and comparison of their operation with that of the corresponding standard monitors revealed a marked improvement in the concentration and throw of the resulting streams. The general forms of these monitors as shown in Figs. 3 and 4 are recommended as the basis for future specifications. Their performance characteristics as indicated by the several coefficients and jet trajectories already presented may be expected to apply to the extent that the essential features of the designs are followed. In this regard the fact should be emphasized that it is the requirements listed in the preceding paragraph, rather than incidental dimensions, that determine the performance. Moreover, it should be possible to change the absolute size of either unit by perhaps 30% without serious variation in any of the characteristics.

Like the design of the horizontal and vertical joints, the method by which the vaned units could be constructed in quantity is not within the scope of this paper. The welded construction utilized by the Institute was not expensive, and present-day casting techniques are capable of handling far more complicated units. The actual details of the vanes have been shown only schematically, since their number and fineness are governed by structural as well as hydraulic considerations. That is, the smaller and more numerous the vanes, the finer the scale of the eddies that they induce—but the costlier they are to fabricate and the greater their tendency to foul with debris. The number shown should be regarded as the minimum, the optimum being that commensurate with practicability. Although their arrangement is also somewhat flexible (two methods of subdivision being shown), the spacing of their leading and trailing edges must be such as to maintain constant proportions between the areal subdivisions at the beginning and end of each flow deflection.

One feature of the fixed monitor designed by the Institute that is subject to modification for several reasons is found in the initial Y-transition from the base pipe to two passages of the same cross-sectional area. This not only adds to the height of the unit but also results in an unguided deceleration of the flow with accompanying eddy formation. Although the distance from the Y to the nozzle appears to minimize the effect of the eddies upon the jet, it is probable that replacement of this **Y** by a vaned transition of essentially constant total cross-sectional area (similar to that of the portable unit) would improve the performance to at least an appreciable degree and would at the same time reduce the over-all height of the unit. A schematic indication of the resulting monitor form is shown in Fig. 21. The upper **T** containing the joints for vertical sweep remains the same as before, but the enlarged passage containing the joint for horizontal sweep ends in a vaned **T** with two tapered arms of noncircular section. These transitions terminate in vaned **L**'s with circular sections just prior to the original design. This improved design is obviously

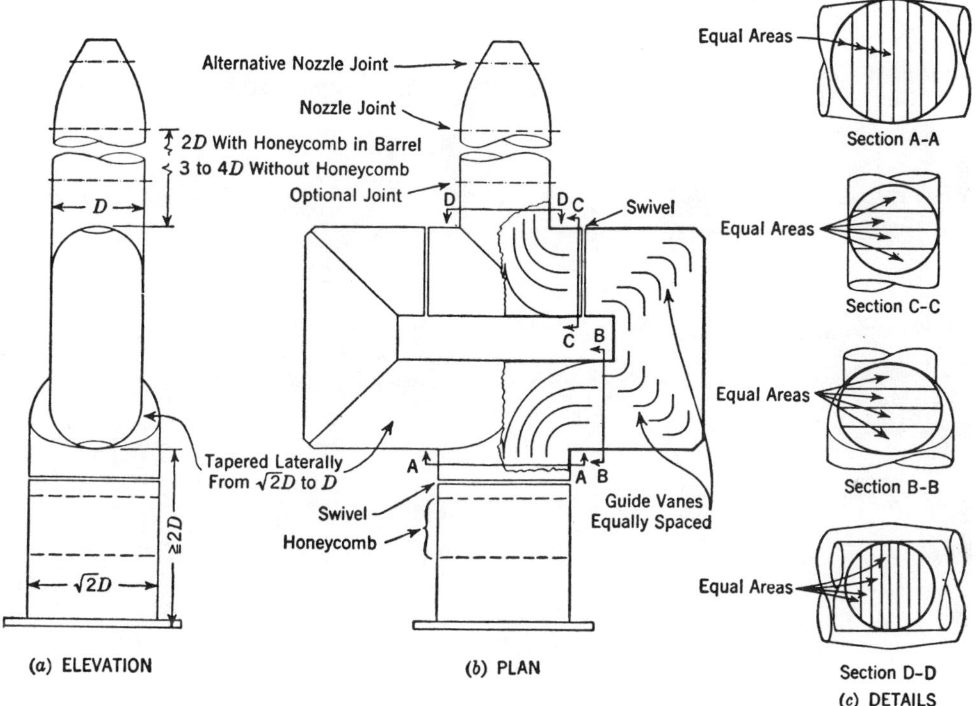

FIG. 21.—IMPROVED FORM OF RECOMMENDED FIXED MONITOR

more complex than that of the original recommendation, but casting should still be quite feasible. In this regard it is suggested that consideration be given to the use of stainless steel rather than brass, the vaned portions of the monitor being cast in symmetrical half sections that are ultimately welded together externally.

ACKNOWLEDGMENTS

The project described was conducted under Contract NObs-24084 with the Bureau of Ships, United States Navy, and Contracts N7onr-495 and N8onr-500 with the Office of Naval Research. The present paper summarizes the more detailed information presented under the same title in November, 1949, as Institute Report No. 3 to the Office of Naval Research. Additional copies of this report have been filed with the Engineering Societies Library and the Library of Congress for reference purposes.

Practically the entire staff of the Institute participated in the investigation in one manner or another. Particular credit is due Messrs. Chesley J. Posey, M. ASCE, who conducted many of the preliminary tests; E. M. Laursen, A. M. ASCE, who detailed the test monitors; and Dale C. Harris, who supervised the construction and operation of the experimental facilities. Lt. W. S. McConnell, of the Coast Guard, provided valuable assistance during his summer assignment to the project.

PRESENT-DAY TRENDS IN HYDRAULICS

HUNTER ROUSE

IOWA INSTITUTE OF HYDRAULIC RESEARCH, STATE UNIVERSITY of IOWA, IOWA CITY

UNTIL the beginning of the twentieth century, hydraulics was essentially the only engineering science dealing with fluid motion. For various reasons associated with its historical background, however, its underlying principles were not sufficiently general for application to the many new professional fields which have since developed. Instead, it became necessary to correlate the mathematical methods of classical hydrodynamics with systematic experimental investigations in both old and new fields of flow, thus establishing the more broadly useful science known today as the mechanics of fluids.

Like other old and conservative philosophies, hydraulics at first remained unaffected by the comparatively rapid progress being made around it, and only gradually did it begin to absorb such aspects of the broader principles as seemed directly useful. Within the past two decades this process of reorientation finally progressed to the point that the parent science is now not only taking good advantage of advancements made in the general subject of fluid motion but also is contributing its proper share. In fact, one can no longer delineate clearly the borders of hydraulics, since there is an almost continuous transition into allied sciences on every side. The following résumé of present-day tendencies is therefore restricted arbitrarily to those phases which have long been the principal lines of hydraulic endeavor.

Acceptance of the theory of potential flow well illustrates current trends. One of the original issues between classical hydrodynamics and empirical hydraulics arose from the fact that such mathematical theories were most nearly applicable to flow past streamlined boundaries, whereas hydraulic structures as a rule were poorly faired—and hence hardly susceptible to analytical design. However, requirements of high efficiency, absence of cavitation, and freedom from eddy-induced vibration in the costly structures of today are forcing the adoption of streamlining as an economic necessity. Potential theory is hence becoming an invaluable tool—as witness its application to the planning of conduit inlets, spillway crests, and other transitions, with the method of relaxation and the three-dimensional electrical analogy being used to a steadily increasing degree. While the same potential theories and methods had previously been applied with success to the flow of ground water, these have now been extended to flow in three dimensions and, at least two-dimensionally, to conditions of soil stratification.

Wave motion, long the only phase of free-surface hydraulics subjected to mathematical treatment, has maintained its lead in this manner of approach. Initial studies of the solitary wave and the deep-water wave have been enlarged to include the entire range between these limits. Such inherently unsteady phenomena as seiche and breaking waves, as well as the viscous dissipation of wave energy, are becoming steadily better understood. Interest in conditions of density stratification due to sediment suspension, temperature variation, or salt-water intrusion has gradually led to the investigation of subsurface waves and currents. The analogy between sound waves and gravity waves has permitted improvements in methods of designing high-velocity channels; in fact, use of the analogy has now gone full circle, in that wave flumes have become exploratory adjuncts to supersonic wind tunnels. Finally, the development of both mechanical and electrical analyzers has enabled great progress to be made in the study of flood-wave propagation. Water hammer—the closed-conduit counterpart of surface-wave hydraulics—was analyzed so effectively during the first quarter of the present century that efforts during the past two decades have been centered largely upon the application of the analysis to complex systems and conditions of wave generation; because of its great rapidity of solution, the electronic analog computer is proving to be of value in this regard.

In problems of conduit resistance, hydraulics has only begun to regain the position of leadership which it lost when the Reynolds-number concept, the boundary-layer theory, and the so-called universal laws of velocity distribution and resistance were originated in other scientific fields. As in the past, interest continues to be focused upon the resistance of rough surfaces, with particular attention to the natural roughness of new and aged commercial materials. Although average trends and values are now well defined, much remains to be accomplished in the statistical analysis of roughness and the prediction therefrom of its hydraulic behavior. Boundary-layer principles are slowly taking their place in the approach to such problems as cavitation at wall irregularities, establishment of flow in steep chutes, and fluid metering, with promise of considerable future success.

In matters of fluid turbulence, the tendency is still toward application of the simpler concepts developed in the not-too-recent past rather than contribution to current theoretical analysis. One of the principal obstacles to rapid progress is the fact that most turbulence in hydraulics is of the nonisotropic type, with wake and jet flow playing major roles, whereas existing theories of a practicable sort deal almost exclusively with conditions of isotropy. Problems of turbulence especially hydraulic in nature include eddy cavitation, the entrainment of air by high-velocity flow, the turbulent diffusion of material carried in suspension, and the mixing of stratified currents. In each of these problems there is now marked investigational activity.

Also peculiar to hydraulics in both historical background and present importance is the very complex problem of sediment transportation. Analysis of the settling of individual particles and particle groups through fluids has recently been extended over a considerable range of particle shape, concentration, and Reynolds number. The suspension phase of transport is understood in at least its basic aspects—up to the familiar point of distinction between momentum diffusion and mass diffusion, or the evaluation of the effect of the suspended material upon the flow itself. Bed-load movement remains in its empirical stage of evaluation, but gradual progress is being made. As in other phases of turbulent diffusion, the primary link in the chain of analysis which still remains to be forged is the necessary relation between the concentration of material in the flow just above the bed to that in the bed itself.

Because of its highly industrialized aspects, progress in the field of hydraulic machinery is somewhat difficult to segregate into its scientific and its engineering parts. Perhaps the greatest forward stride from the viewpoint of science as a whole has been the tendency to uniformalize the analysis of all turbomachinery, regardless of the fluid medium in question. Some progress—necessarily of a qualitative rather than a quantitative nature—has been made in the application of conformal methods to the control of blade design. Attention has also been given to the behavior of turbomachinery over a considerable Reynolds-number range. The commercial demand for high-pressure positive-displacement machinery has begun to stimulate analytical as well as purely developmental activity in both this field and the closely related one of hydraulic controls.

Hydraulics laboratories have played a considerable role during the present century in advancing the technique of scale-model investigations. An unsolved problem of long standing is the dependable simulation of river flow at reduced scale over movable beds of sediment, particularly under conditions of vertical (not to mention sedimentary) distortion. Of somewhat more recent importance is the simulation of the mixing produced by gravity underflows in reservoirs and tidal estuaries. The use of air for convenience in the study of many phenomena of hydraulics which do not involve a free surface is a noteworthy trend of the past decade. Quite recently the development of high-speed movie photography has permitted the detailed observation of such rapid occurrences as the growth and collapse of cavitation bubbles. Hydraulic instrumentation has also taken advantage of recent progress in electronics, but—despite good exploratory advances in the application of electrical methods to the measurement of liquid turbulence—a generally useful tool comparable to the hot-wire anemometer is yet to be perfected.

Hydraulic research as such is centered primarily in the university laboratories of the western countries, the strong position of leadership taken by the German institutions between World Wars I and II now being shared by various European countries and the United States. National laboratories, though restricted largely to the solution of specific problems, have contributed since the beginning of World War II to various aspects of instrumentation and fundamental knowledge as an outgrowth of their developmental projects, and governmental as well as private agencies are promoting university activity through provision of research funds.

Because of the great range of subject matter encompassed by this review, the selection of a few representative papers to highlight trends in the many different phases of hydraulics is hardly feasible. Instead, reference is made to Jaeger's "Technische Hydraulik" (AMR **3**, Rev. 712), and to the symposium volume, "Engineering Hydraulics" (AMR **4**, Rev. 1207), for résumés of present-day principles and their application and for the extensive bibliographies contained therein. Details of current investigations will be found in the annual summary, "Hydraulic Research in the United States," distributed by the National Bureau of Standards, and in a similar publication of the International Association for Hydraulic Research.

Gravitational Convection from a Boundary Source

By HUNTER ROUSE, C. S. YIH, and H. W. HUMPHREYS
Iowa Institute of Hydraulic Research, State University of Iowa, Iowa City

(Manuscript received 18 June, 1952)

Abstract

Elementary analyses of the mean patterns of free convection from a line source and a point source are presented without regard to the specific means by which the gravitational action is produced. The derived functional relationships are then verified and completed through use of velocity and temperature measurements above sources of heat, the generalized form of the results permitting characteristics of the mean flow to be determined over a considerable range of the primary variables. These results should enable meteorologists to evaluate the role of the basic convective process in the more complex movements of the atmosphere.

Introductory Remarks

Atmospheric disturbances are generally so complex in nature that the relative importance of the various factors which they involve can be appraised only by investigating the effect of each factor independently. In conducting an analysis of this nature, one often notes a close similarity between particular aspects of meterological and other flow occurrences, indicating that experience gained in related fields can be adapted to the problem in question. A case in point is the phenomenon of large-scale thermal updrafts in the atmosphere, one aspect of which is the comparatively simple process of gravitational or "free" convection from a boundary source, recently studied experimentally at the Iowa Institute of Hydraulic Research. The writers believe that the results of this investigation will permit meteorologists to evaluate the role of the basic convective mechanism in atmospheric phenomena.

Free convection due to a point source of heat is very simply illustrated by the plume of smoke which rises from a cigarette in otherwise stagnant air. Because of the buoyancy of the heated air in the immediate vicinity of the burning end, a continuous upward current is induced, with a corresponding radial inflow for reasons of continuity. While the initial steadiness of the smoke filament indicates purely laminar motion for a considerable distance above the heat source, the flow thereafter becomes unstable and the filament breaks up into the eddying clouds normally associated with turbulent motion. After the onset of turbulence, the effect of the convection upon the surrounding fluid becomes far more pronounced, the molecular shear and heat transfer of the initial laminar motion becoming dwarfed in scale by the macroscopic mixing action. Whereas the magnitude of the velocity along the vertical axis is then rapidly di-

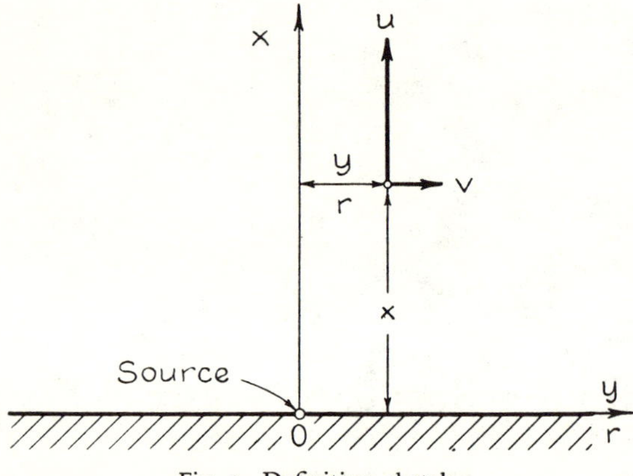

Fig. 1. Definition sketch.

minished, the angle of radial diffusion (and hence the rate of radial entrainment) is accordingly increased.

Although free convection of this nature is usually produced by a source of heat, it is the gravitational rather than the thermal aspects of the flow which are fundamental to the phenomenon. Indeed, essentially the same convection pattern can be produced in a liquid instead of a gas, and without any temperature change whatever. For example, the atmosphere may be replaced by a large body of water, and liquid having a different specific gravity may be introduced at a constant rate at a point on either horizontal boundary — a lighter liquid at the bottom or a heavier one at the top. In each event a similar pattern of convection will result, provided only that the local changes in density are relatively small in magnitude.

In the following pages an approximate analysis is made of the mean pattern of free convection from both a line source and a point source, without regard to the specific means of producing the gravitational action. Experimental evidence is then introduced to verify and complete the derived functional relationships, and generalized diagrams are presented to permit evaluation of the primary flow characteristics over a considerable range of the parameters involved.

Theoretical Analysis of the Mean Flow

If it is assumed that the scale of the initial laminar zone is small compared with that of the subsequent zone of turbulent convection,

notation for either two-dimensional flow over a line source or axially symmetric flow over a point source may be reduced to that shown schematically in Fig. 1. Herein the origin O represents the location of either source. The local mean component of velocity in the vertical direction x is denoted by u, and that in the lateral direction y or radial direction r is denoted by v.

In addition to the mean velocity of convection above the source, there will also be a mean local change $\Delta\gamma$ in the weight density of the fluid. This entails, to be sure, a comparable change in the mass density ϱ. However, the following analysis is based upon the assumption that, whereas the unit buoyant force $-\Delta\gamma$ is sufficiently great to produce vertical acceleration, the corresponding variation in the mass density of the fluid undergoing acceleration is sufficiently small, in comparison with the density itself, to be neglected.

The foregoing assumption as to constancy of the density makes it possible to proceed in close accordance with the elementary analysis of flow in boundary layers, jets, and wakes. Three additional approximations are thus involved: the pressure intensity is assumed to be hydrostatically distributed throughout the field of motion; transverse forces are ignored in comparison with those in the vertical direction; and turbulent mixing in the vertical direction is ignored in comparison with that in the horizontal. The fundamental equations of motion thus reduce to the equation of continuity and simplified equations of vertical acceleration and diffusion. The development of the relationships describing convection above a line source will be described in some detail, but for purposes of brevity only the results will be indicated for the case of axial symmetry.

Under the foregoing assumptions, the equation of vertical acceleration in two-dimensional flow reduces to

$$\varrho u \frac{\partial u}{\partial x} + \varrho v \frac{\partial u}{\partial y} = -\Delta\gamma + \frac{\partial \tau}{\partial y} \qquad (1)$$

in which the intensity of vertical shear τ is proportional to the mean product of the turbulent velocity components u' and v':

$$\tau = -\varrho \overline{u'v'}$$

The equation of continuity is simply

$$\frac{\partial u}{\partial x} + \frac{\partial v}{\partial y} = 0 \qquad (2)$$

Finally, the equation of diffusion of the quantity $\Delta\gamma$ has the form

$$u\frac{\partial(\Delta\gamma)}{\partial x} + v\frac{\partial(\Delta\gamma)}{\partial y} = -\frac{\partial w}{\partial y} \qquad (3)$$

in which the rate of lateral diffusion w due to turbulence is proportional to the mean product of the eddy scale l and the lateral velocity of fluctuation:

$$w = -\overline{v'l}\,\frac{\partial(\Delta\gamma)}{\partial y}$$

Since $u = 0$ at $y = \infty$, $v = 0$ at $y = 0$, and $\tau = 0$ at $y = 0$ and $y = \infty$, integration of Eq. (1), through use of Eq. (2), over a horizontal plane from the axis to infinity results in the following form of the momentum relationship:

$$\frac{d}{dx}\int_0^\infty \varrho u^2\, dy = -\int_0^\infty \Delta\gamma\, dy \qquad (4)$$

This states that the vertical gradient of the momentum flux is equal to the buoyancy of a horizontal stratum of unit thickness. As $\Delta\gamma = 0$ at $y = \infty$ and $w = 0$ at $y = 0$ and at $y = \infty$, a similar integration of Eq. (3) leads to the following expression for the constancy of the incremental weight flux past all successive planes:

$$\frac{d}{dx}\int_0^\infty u\,\Delta\gamma\, dy = 0 \qquad (5)$$

Finally, after multiplication of each term of Eq. (1) by u, integration of the resulting expression by parts, with the help of Eq. (1), yields the following form of the energy relationship:

$$\frac{d}{dx}\int_0^\infty \frac{\varrho u^3}{2}\, dy = -\int_0^\infty u\,\Delta\gamma\, dy - \int_0^\infty \tau\frac{\partial u}{\partial y}\, dy \qquad (6)$$

Because the quantity v^3 is negligible in comparison with u^3, this relationship states that the vertical gradient of the flux of kinetic energy is equal to the rate at which work is done by the buoyant force less the rate at which work is done by the turbulent shear. The last term at the same time represents the rate at which energy is lost to the mean motion through the generation of turbulence.

Although no indication is given as to the form of the functional relationships for u, $\Delta\gamma$, and τ which will satisfy these three necessary conditions, the customary hypothesis of dynamic similarity of the mean (and turbulent) motion at all elevations permits each of the relationships to be considered of the same form for every value of x. It is thus finally assumed that

$$\frac{u}{u_{\max}} = f(\eta) \qquad \frac{\Delta\gamma}{\Delta\gamma_{\max}} = g(\eta) \qquad \frac{\tau}{\varrho u_{\max}^2/2} = h(\eta)$$

Herein $\eta = y/\sigma$, the quantity σ representing some linear characteristic of the velocity profile $u = u(y)$, such as the value of y at which the ratio $u/u_{\max}$ has some arbitrary magnitude.

Inasmuch as the unknown functions f, g, and h are independent of elevation, any power or combination thereof must also be constant with x. It is hence convenient to formulate a series of pertinent integrals for substitution into the three primary relationships which the functions must satisfy:

$$I_1 = \int_0^\infty f\, d\eta \qquad I_2 = \int_0^\infty f^2\, d\eta \qquad I_3 = \int_0^\infty f^3\, d\eta$$

$$I_4 = \int_0^\infty g\, d\eta \qquad I_5 = \int_0^\infty fg\, d\eta \qquad I_6 = \int_0^\infty h\frac{df}{d\eta}\, d\eta$$

Introduction of these terms into Eqs. (4), (5), and (6) permits the latter to be rewritten as

$$\frac{d}{dx}\left(I_2\,\varrho\, u_{\max}^2\,\sigma\right) = -I_4\,\Delta\gamma_{\max}\,\sigma$$

$$I_5\, u_{\max}\,\Delta\gamma_{\max}\,\sigma = C_x$$

$$\frac{d}{dx}\left(I_3\,\frac{\varrho\, u_{\max}^3}{2}\,\sigma\right) = -C_x - I_6\,\frac{\varrho\, u_{\max}^3}{2}$$

Simultaneous solution of these relationships will yield the following significant results:

$$\sigma = \frac{I_4 I_6 x}{2 I_2 I_5 - I_3 I_4} \sim x$$

$$u_{max} = \left(\frac{C_x}{\varrho} \frac{I_3 I_4 - 2 I_2 I_5}{I_2 I_5 I_6}\right)^{1/3} \sim x^0$$

$$\Delta\gamma_{max} = -\left(C_x \varrho^{1/2} \frac{I_3 I_4 - 2 I_2 I_5}{I_2 I_5 I_6}\right)^{2/3} \frac{I_2}{I_4 x} \sim x^{-1}$$

Evidently, no matter what forms the distribution curves may take, the convection zone will expand linearly with elevation; the maximum velocity (and hence the velocity along any line of constant y/x) will be independent of elevation; and the maximum incremental weight density (and hence that along any line of constant y/x) will vary inversely with elevation.

With this information it is possible to reach corresponding conclusions as to the variation of the volume flux Q, the momentum flux M, the kinetic-energy flux E, the unit buoyant force F, and the flux W of the incremental weight above a source of length L. Thus, per unit length of source,

$$\frac{Q}{L} = 2 \int_0^\infty u\, dy \sim x \qquad \frac{M}{L} = 2 \int_0^\infty \varrho\, u^2\, dy \sim x$$

$$\frac{E}{L} = 2 \int_0^\infty \frac{\varrho\, u^3}{2}\, dy \sim x \qquad \frac{F}{L} = -2 \int_0^\infty \Delta\gamma\, dy \sim x^0$$

$$\frac{W}{L} = 2 \int_0^\infty u\, \Delta\gamma\, dy = 2 C_x$$

Herefrom it is seen that the flux of volume, the flux of momentum, and the flux of kinetic energy must increase continuously with elevation, for the force producing the convective motion is the same at all successive levels. As the flux of the incremental weight is also the same at every level, the quantity W must indicate the output of the source.

The corresponding analysis of the axially symmetric pattern of convection above a point source proceeds in a closely comparable manner. The resulting integral equations of momentum, diffusion, and energy have the forms

$$\frac{d}{dx} \int_0^\infty \varrho\, u^2\, r\, dr = - \int_0^\infty \Delta\gamma\, r\, dr \qquad (7)$$

$$\frac{d}{dx} \int_0^\infty u\, \Delta\gamma\, r\, dr = 0 \qquad (8)$$

$$\frac{d}{dx} \int_0^\infty \frac{\varrho u^3}{2}\, r\, dr = - \int_0^\infty u\, \Delta\gamma\, r\, dr - \int_0^\infty \tau \frac{\partial u}{\partial r}\, r\, dr \qquad (9)$$

Through the assumption of dynamic similarity of the mean flow at all elevations it is found that

$$\sigma \sim x \qquad u_{max} \sim x^{1/3} \qquad \Delta\gamma_{max} \sim x^{-5/3}$$

The following expressions are then obtained:

$$Q = 2\pi \int_0^\infty u\, r\, dr \sim x^{5/3} \qquad M = 2\pi \int_0^\infty \varrho u^2 r\, dr \sim x^{4/3}$$

$$E = 2\pi \int_0^\infty \frac{\varrho u^3}{2} r\, dr \sim x \qquad F = -2\pi \int_0^\infty \Delta\gamma r\, dr \sim x^{1/3}$$

$$W = 2\pi \int_0^\infty u\, \Delta\gamma\, r\, dr = 2\pi C_x'$$

Owing to the differences between these expressions and their counterparts for the two-dimensional case, basically different relationships are indicated for the flux of the several characteristic quantities past successive planes above a point source. Although the convection pattern again expands linearly with height above the source, the maximum velocity (and hence that for any constant value of r/x) now decreases continuously, and the incremental weight density decreases even more rapidly. On the other hand, the buoyant force on a stratum of unit thickness becomes steadily greater with increasing elevation, and Q, M, and E all increase at different rates. Once again, however, the flux of the incremental weight is the same at all elevations, so that W

must be equivalent to the output of the source.

Experimental Procedure and Results

In any experimental program to determine explicit relationships for either two-dimensional or axially symmetric convection from a boundary source, there are evidently four variables which can be controlled independently — two coordinate positions, the mass density of the fluid, and the strength of the source. These four quantities govern the magnitudes of such dependent variables as velocity, weight density, shear, turbulence, and their various combinations. While all of these factors would be of interest in a complete investigation of the phenomenon, attention is focused in the present study upon the major characteristics of the mean motion: the velocity component u and the change $\Delta \gamma$ in weight density. The resulting functional relationships between the dependent and independent variables for the two types of source can be indicated schematically as follows:

$$\frac{u}{\Delta \gamma} = f_{1-4}\left(x, \frac{y}{r}, \varrho, \frac{W/L}{W}\right)$$

Inasmuch as five variables involving three fundamental dimensions appear in each of these four functions, it is to be expected that the dimensionless form of each will contain only two terms. Thus, for two-dimensional conditions, the relationships may be expressed as

$$\frac{u}{(W/L\varrho)^{1/3}} = \varphi_1\left(\frac{y}{x}\right)$$

$$\frac{\Delta'\gamma}{(\varrho W^2/L^2 x^3)^{1/3}} = \varphi_2\left(\frac{y}{x}\right)$$

For conditions of axial symmetry, the counterparts of these expressions are

$$\frac{u}{(W/\varrho x)^{1/3}} = \varphi_3\left(\frac{r}{x}\right)$$

$$\frac{\Delta\gamma}{(\varrho W^2/x^5)^{1/3}} = \varphi_4\left(\frac{r}{x}\right)$$

In each instance the grouping of terms is seen to be in accordance with the general relationships obtained analytically.

Experiments with heated air to determine the forms of these functions were conducted in closed rooms, particular care being required in the study of axially symmetric conditions to eliminate all sources of disturbance. In this instance the room was approximately circular in plan and had a diameter of 25 feet and an overall height of 11 feet. The heat source was placed at the midpoint of a centrally located platform 8 feet in diameter and a few inches in height. For the two-dimensional study the flow was confined between two parallel walls 4 feet high, 8 feet long, and 4 feet apart, and the source was placed across the midsection of a low platform extending the length of the walls. The heat sources in both cases consisted of recessed gas burners yielding low, blue flames approaching as nearly as practicable the desired point and line concentrations.

Measurements of the temperature distribution were made by means of a copper-constantan thermocouple, with its cold junction placed well away from the heat source, in combination with a potentiometer reading to 0.002 millivolt. Velocity indications were obtained with a specially constructed vane anemometer $1^1/_4$ inch in diameter, the jewel bearings of which permitted the indication of velocities as low as 0.2 foot per second. These instruments were mounted alternately on remotely controlled traversing mechanisms.

The rates of heat output from both the line and the point source were evaluated from the measured distributions of velocity and temperature according to the following thermal counterparts of the equations for W/L and W:

$$\frac{H}{L} = 2c_p\varrho \int_0^\infty u\,\Delta T\,dy \tag{10}$$

$$H = 2\pi c_p\varrho \int_0^\infty u\,\Delta T\,r\,dr \tag{11}$$

for the closely approximate condition that $\Delta\gamma/\gamma = -\Delta T/T$, the corresponding value of W in either case was computed from the resulting conversion equation,

$$W = -\frac{gH}{c_p T} \tag{12}$$

Supplementary tests performed during the axially symmetric study were directed toward

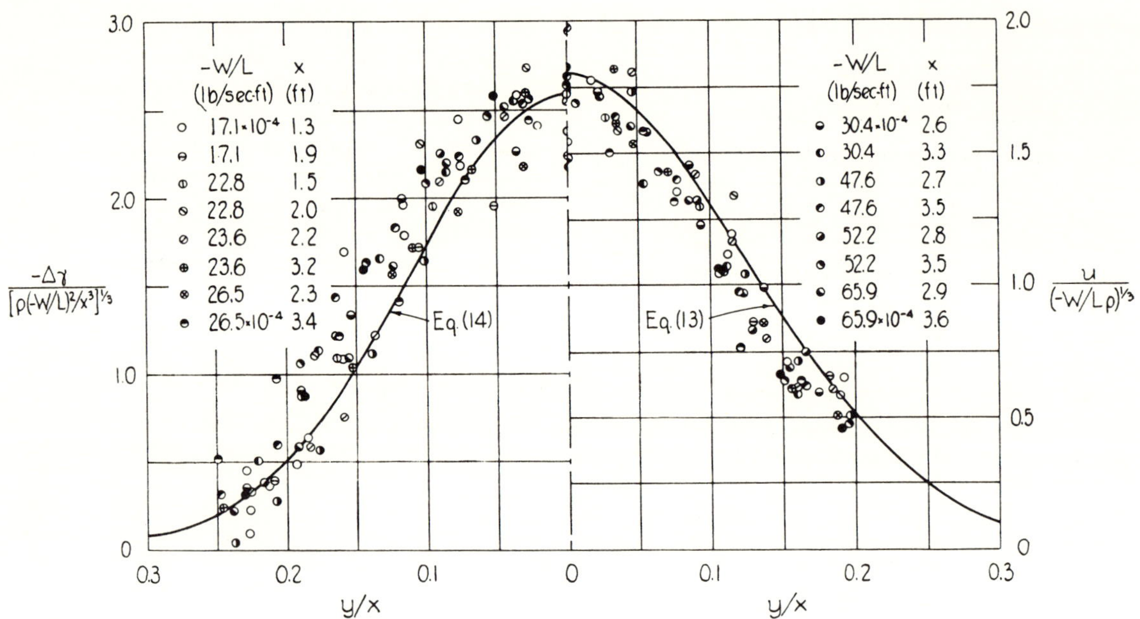

Fig. 2. Distribution functions for two-dimensional convection.

determination of the level at which, under ordinary circumstances, the initial laminar flow could be expected to become unstable. Because of the low rates of heat input required to yield readily measurable conditions of transition, it was necessary to use as the limiting heat source a burning cigarette (from a package carefully calibrated for rate of heat release), the smoke serving as a convenient tracing agent. The critical Reynolds number determined in this manner was of the form

$$\mathbf{R}_c = \frac{x_c}{\sqrt{\varrho \nu^3/(-W)}} \approx 10^5$$

in which ν is the kinematic viscosity. As can be determined herefrom, the critical height x_c was small in every case for which detailed measurements were made.

All velocity and temperature traverses were carried entirely across the convection zone, the plotted curves being essentially symmetrical about the maximum but often indicating a slight deviation of the flow from the vertical. In analyzing the results, the reference axis was adjusted to agree with the axis of symmetry. Upon reduction of data to the pertinent dimensionless forms, composite plots were made of velocity and weight distributions for both the two-dimensional

and the axially symmetric cases, to verify the assumed similarity of flow at different levels and with different strengths of source and to determine the explicit distribution functions. For ease in comparison as well as economy of space, these plots are shown in half section: in Fig. 2 for two dimensions and in Fig. 3 for axial symmetry.

At once apparent is the appreciable scatter of points which seems to typify experimental studies of this nature — in part because of the difficulty of precise measurement and in part because of the sensitivity of such flow to slight disturbances. The scatter is not appreciably systematic, however, nor is it difficult to construct mean curves to indicate the distribution functions. The curves shown are normal-probability functions which best fit the data and at the same time satisfy Eqs. (4) and (5), and Eqs. (7) and (8), respectively. At least as a first approximation, there seems little doubt that the plotted curves describe the basic free-convection phenomenon. To be remarked, however, is the fact that use of the Gaussian function — not to mention the simplified analysis itself — loses significance as y/x or r/x becomes large.

If the probability curves plotted in Fig. 2 are assumed to represent the true distributions of u and $\Delta\gamma$, the corresponding expressions

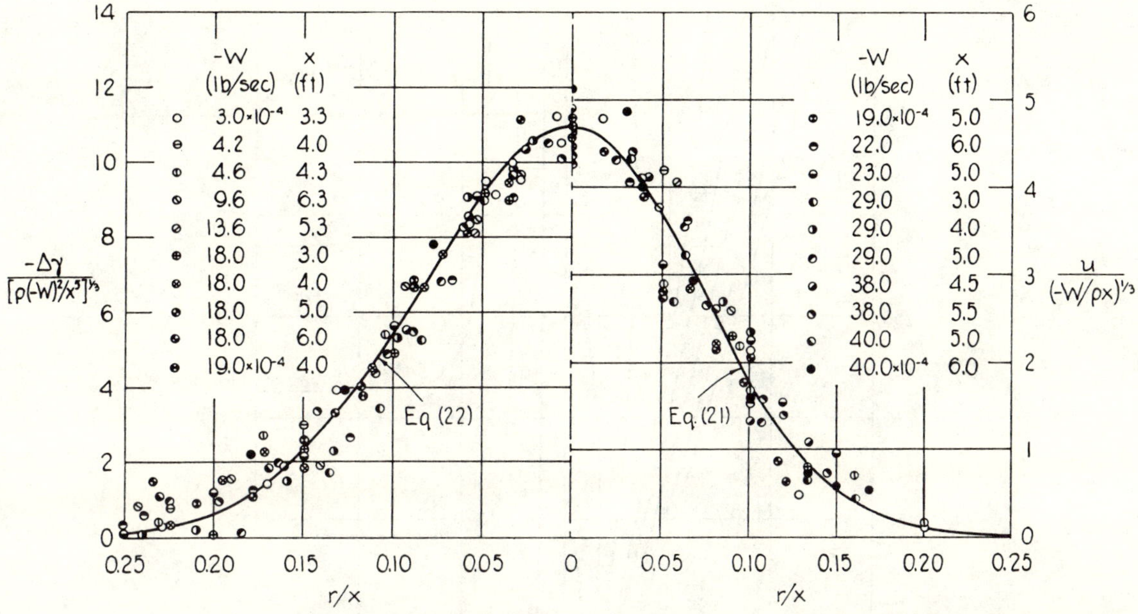

Fig. 3. Distribution functions for axially symmetric convection.

$$u = 1.80 \left(\frac{-W/L}{\varrho}\right)^{1/3} exp\left(-32\frac{y^2}{x^2}\right) \quad (13)$$

$$\varDelta\gamma = -2.6\left[\frac{\varrho(-W/L)^2}{x^3}\right]^{1/3} exp\left(-41\frac{y^2}{x^2}\right) (14)$$

can be used conveniently to evaluate other characteristics of the mean pattern of motion. Thus, since $\psi = \int u\,dy$, the stream function is expressible implicitly as

$$\psi = 1.80\left(\frac{-W/L}{\varrho}\right)^{1/3}\int_0^y exp\left(-32\frac{y^2}{x^2}\right)dy \quad (15)$$

in which values of the integral are obtainable from probability tables. The lateral velocity of entrainment just beyond the diffusion zone, otherwise available from the condition that $v = -\partial\psi/\partial x$, is instead computed from the expression $2v_e = -d(Q/L)/dx$ as

$$v_e = \mp 0.28\left(\frac{-W/L}{\varrho}\right)^{1/3} \quad (16)$$

Specific relationships for Q, M, E, and F per unit length of source are obtained by integration as follows:

$$\frac{Q}{L} = 0.57\left(\frac{-W/L}{\varrho}\right)^{1/3}x \quad (17)$$

$$\frac{M}{L} = 0.72\left(\frac{-W}{L}\right)^{2/3}\varrho^{1/3}x \quad (18)$$

$$\frac{E}{L} = 0.53\frac{-W}{L}x \quad (19)$$

$$\frac{F}{L} = 0.72\left(\frac{-W}{L}\right)^{2/3}\varrho^{1/3} \quad (20)$$

For conditions of axial symmetry these expressions take the following forms:

$$u = 4.7\left(\frac{-W}{\varrho x}\right)^{1/3} exp\left(-96\frac{r^2}{x^2}\right) \quad (21)$$

$$\varDelta\gamma = -11.0\left[\frac{\varrho(-W)^2}{x^5}\right]^{1/3} exp\left(-71\frac{r^2}{x^2}\right) (22)$$

$$\psi = 0.024\left(\frac{-Wx^5}{\varrho}\right)^{1/3}\left[1 - exp\left(-96\frac{r^2}{x^2}\right)\right] (23)$$

$$v_e = -0.041\left(\frac{-W}{\varrho}\right)^{1/3}\frac{x^{2/3}}{r} \quad (24)$$

$$Q = 0.153\left(\frac{-W}{\varrho}\right)^{1/3}x^{5/3} \quad (25)$$

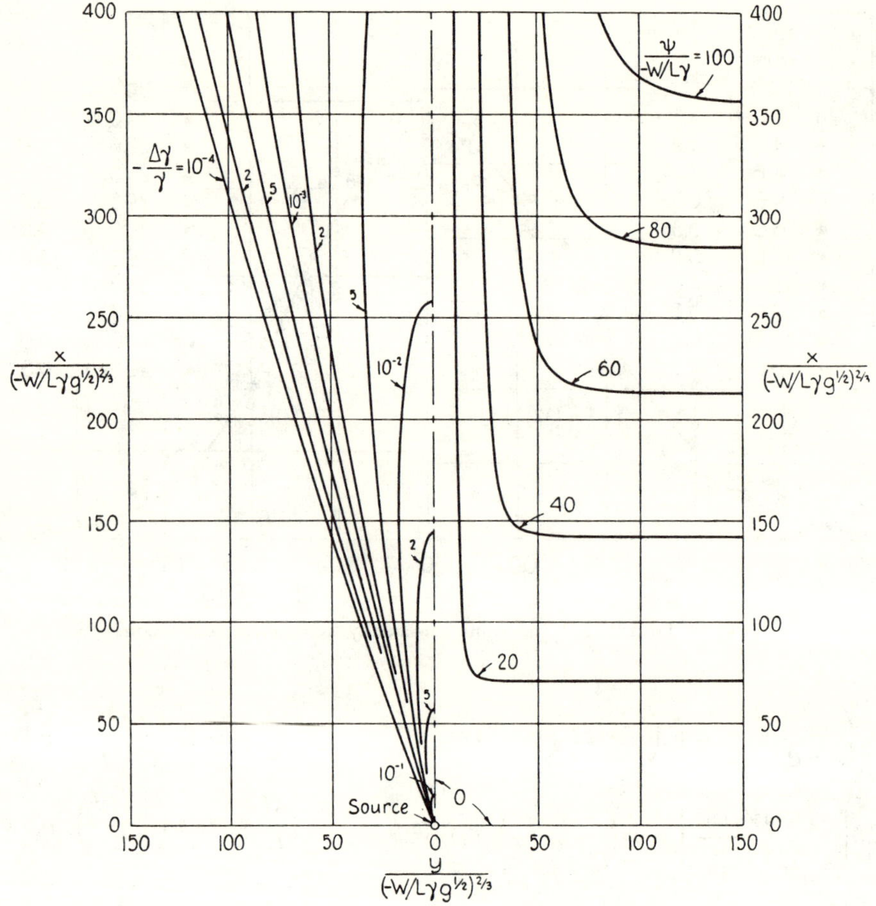

Fig. 4. Convection pattern over a line source.

$$M = 0.36\,(-W)^{2/3}\,\varrho^{1/3}\,x^{1/3} \tag{26}$$

$$E = 0.57\,(-W)\,x \tag{27}$$

$$F = 0.49\,(-W)^{2/3}\,\varrho^{1/3}\,x^{1/3} \tag{28}$$

So far as the mean flow is concerned, it remains only to indicate in generalized coordinates the pattern of stream lines and lines of constant incremental weight for the two cases. This, however, requires three rather than two dimensionless groups of variables in each instance — the additional variable which is needed being a characteristic length to serve as a reference coordinate. Such a length is found in the quantity $(W/L\gamma g^{1/2})^{2/3}$ for two dimensions, and in the corresponding quantity $(W/\gamma g^{1/2})^{2/5}$ for axial symmetry. Upon introduction of these quantities, the dimensionless coordinate relationships for ψ and $\varDelta\gamma$ become, for the line source,

$$\frac{\psi}{W/L\gamma} = \varphi_5\left[\frac{x}{(W/L\gamma g^{1/2})^{2/3}},\ \frac{y}{(W/L\gamma g^{1/2})^{2/3}}\right]$$

$$\frac{\varDelta\gamma}{\gamma} = \varphi_6\left[\frac{x}{(W/L\gamma g^{1/2})^{2/3}},\ \frac{y}{(W/L\gamma g^{1/2})^{2/3}}\right]$$

and, for the point source,

$$\frac{\psi}{W/\gamma} = \varphi_7\left[\frac{x}{(W/\gamma g^{1/2})^{2/5}},\ \frac{r}{(W/\gamma g^{1/2})^{2/5}}\right]$$

$$\frac{\varDelta\gamma}{\gamma} = \varphi_8\left[\frac{x}{(W/\gamma g^{1/2})^{2/5}},\ \frac{r}{(W/\gamma g^{1/2})^{2/5}}\right]$$

The corresponding plots of stream lines and lines of constant incremental weight are shown in half section in Figs. 4 and 5.

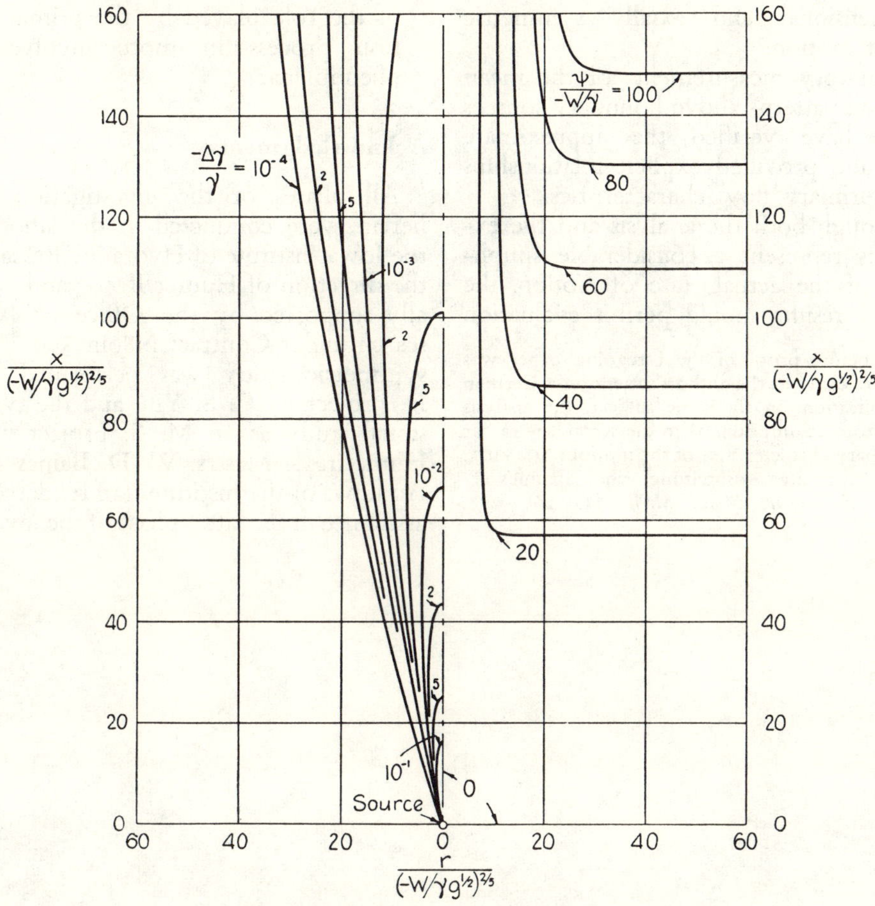

Fig. 5. Convection pattern over a point source.

As can be seen at once from these undistorted diagrams, the zones of pronounced vertical motion and change in weight density are concentrated about the area of symmetry and expand only slowly with elevation. Because of the fact that the measurements were restricted to these zones and that the analysis itself is not exact, the outlying portions of the stream-line patterns are purely schematic. Aside from this qualitative indication of the pattern of inflow, the diagrams are considered to represent with good approximation the basic characteristics of convection at any scale. They may readily be interpolated or extrapolated, as all curves in any diagram are geometrically similar by virtue of the dynamic similarity of the convection mechanism at successive elevations.

Conclusions

From the foregoing analytical and experimental studies of gravitational convection from both a line source and a point source, the following conclusions may be drawn:

(1) The motivating force in gravitational convection is the positive or negative buoyancy of the fluid, and the primary characteristics of the convection pattern can hence be analyzed without regard to the means whereby the buoyant effect is produced.

(2) Through use of an approximate method of analysis similar to that for boundary layers, wakes, and jets, it is possible to express as proportionalities the mean characteristics of gravitational convection for both

two-dimensional and axially symmetric turbulent motion.[1]

(3) Laboratory measurements of the mean convection patterns above boundary sources of heat have verified the approximate analysis and provided explicit relationships for the primary flow characteristics.

(4) Although both the analysis and the experiments represent a considerable simplification of the actual state of motion, the correlated results should permit evaluation of the role played by the primary convection process in more involved natural phenomena.

Acknowledgments

All phases of the investigation described herein were conducted in the laboratories of the Iowa Institute of Hydraulic Research under the direction of Hunter Rouse and were partially supported by the Office of Naval Research under Contract N8onr-500. The axially symmetric study was performed as a Ph. D. project by C. S. Yih, and the two-dimensional study as an M. S. project by H. W. Humphreys. Messrs. W. D. Baines and M. F. Andrews of the Institute staff rendered valuable assistance in the latter phase of the investigation.

---

[1] Shortly before proof of the foregoing paper was received, there came to the authors' attention a German wartime publication on the same subject, the analysis proceeding from assumptions as to the secondary rather than the primary characteristics of the motion: SCHMIDT, W., 1941, Turbulente Ausbreitung eines Stromes erhitzter Luft, *Z. angew. Math. Mech.* **21**, 265—278, 351—363.

Free Convection over Parallel Sources of Heat

By HUNTER ROUSE, W. D. BAINES and H. W. HUMPHREYS

Iowa Institute of Hydraulic Research, State University of Iowa, Iowa City, U.S.A.

MS. received 23rd October 1952

Abstract. One phase of a fundamental investigation of convection over point and line sources of heat, conducted through the past decade at the Iowa Institute of Hydraulic Research, simulated a method of fog dispersal used for the landing of aircraft in England during the war. This involved the release of heat from parallel lines of burners, the convection from which produced characteristic patterns of mean flow and of heat intensity above the burners. Experiments described in this paper consisted of the measurement of the velocity distribution and the temperature distribution at various model scales and rates of heat generation. The results have been reduced to dimensionless diagrams of the two distribution functions which satisfy the elementary continuity and impulse-momentum requirements. From these diagrams generalized families of stream lines and isotherms for the mean motion above the sources have been plotted. The general functions reveal the basic pattern of convection and permit the approximate evaluation of velocity and temperature rise for any desired combination of the independent variables.

---

§ 1. INTRODUCTION

DURING the past war, parallel lines of petrol burners were used in England as a means of dissipating fogs over aircraft landing strips, and several test installations were also built in America. The basic principle involved was that of free convection resulting from the buoyancy of the heated air, the elevated temperature in the convection zone over the runway causing the fog in this vicinity to evaporate.

Because atmospheric conditions in the regions of prevalent fog are quite different in England and America, to be effective the burner installations had to differ accordingly. English fogs generally occur with essentially no wind, whereas those in America and in the Aleutians are often accompanied by winds of appreciable strength. As a result, the English systems consisted of equivalent lines of burners on both sides of the runway, and the American modifications involved a primary line some distance in the prevailing upwind direction with secondary lines on the downwind side.

The Iowa Institute of Hydraulic Research, as a part of its wartime programme, undertook at model scale the empirical evaluation of burner requirements for cross-wind conditions, and the project was reported upon in 1944 to the Office of Scientific Research and Development. The experimental data were later analysed as a fundamental problem in free convection (Rouse 1947). As the field appeared to be a fruitful one for further study, investigations were then made of the basic convection patterns over point and line sources without cross flow (Rouse, Yih and Humphreys 1952). One phase of the latter project dealt with the induced motion and the accompanying distribution of temperature over parallel line sources similar to those used in the English system of fog dispersal (Rankine 1950). Although the need for such a method may now have been eliminated entirely by the development of blind-landing techniques, the results of the investigation are presented herewith for purposes of record.

§ 2. Physical and Dimensional Considerations

For conditions of two-dimensional convection over parallel sources of heat ocated at a horizontal boundary, the primary variables describing the mean flow are those indicated in fig. 1. Herein x and y are the vertical and horizontal coordinate positions with respect to a symmetrically located origin, y_0 is the distance from the origin to each of two sources of equal strength, u is the vertical and v is the lateral component of the mean velocity, and $-\Delta\rho$ is the local reduction in density corresponding to the temperature rise ΔT.

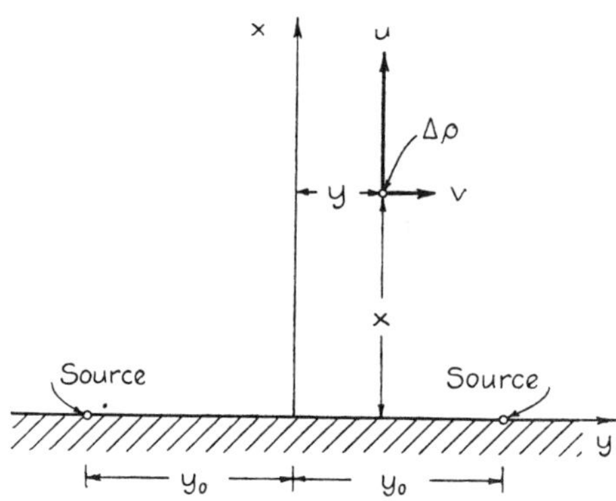

Fig. 1. Definition sketch.

If $\Delta\rho$ is assumed to be small in comparison with ρ, the density itself, the equation of continuity for constant-density conditions will apply with sufficiently close approximation:

$$\frac{\partial u}{\partial x} + \frac{\partial v}{\partial y} = 0.$$

If it is further assumed that the pressure intensity is the same at all points, the equation of acceleration in the vertical direction will reduce to a form comparable with the Prandtl boundary-layer equation,

$$\rho u\frac{\partial u}{\partial x} + \rho v\frac{\partial u}{\partial y} = -g\Delta\rho + \frac{\partial \tau}{\partial y},$$

in which $-g\Delta\rho$ represents the buoyant force per unit volume which produces the acceleration and τ is the intensity of turbulent shear. Because v and τ are zero at the axis of symmetry and u, $\Delta\rho$, and τ vanish at a great distance from the axis, integration of the equation of acceleration over a horizontal plane in accordance with the equation of continuity will result in the following form of the customary momentum equation

$$\frac{d}{dx}\int_0^\infty \rho u^2\,dy = -\int_0^\infty g\Delta\rho\,dy. \qquad\qquad \ldots\ldots(1)$$

A similar equation can be written for the diffusion of heat on the assumption that the vertical rate of mixing due to turbulence is small in comparison with the lateral rate q:

$$u\frac{\partial(\Delta T)}{\partial x} + v\frac{\partial(\Delta T)}{\partial y} = -\frac{1}{c_p\rho}\frac{dq}{dy}.$$

Because q likewise vanishes both at the axis and at a great distance from it, integration of this equation over a horizontal plane will lead to the following condition of heat flux:

$$\frac{d}{dx}\int_0^\infty c_p\rho u\,\Delta T\,dy = 0. \qquad\qquad \ldots\ldots(2)$$

The integral term itself, which is seen to be independent of elevation, thus represents the rate at which heat is generated per unit length of source.

The foregoing equations of momentum and heat flux not only specify conditions which must be approximately satisfied by the mean convection pattern, but they also define the variables which must be considered in an experimental investigation. The quantity $\Delta\rho$, first of all, can be related to the local temperature rise ΔT by the following simple expression for a perfect gas at constant pressure: $-\Delta\rho/\rho_0 = \Delta T/T_0$. If this expression is introduced into eqn. (1) and the difference between ρ and the ambient value ρ_0 is again ignored, the momentum equation can be rewritten in the form

$$\frac{d}{dx}\int_0^\infty u^2\,dy = \int_0^\infty \frac{\Delta T}{T_0/g}\,dy. \qquad\qquad \ldots\ldots(3)$$

By defining the rate at which heat is produced per unit length of source as H/L, one can then express the vertical flux of heat in the corresponding form

$$\frac{H/L}{c_p\rho_0 T_0/g} = \int_0^\infty u\frac{\Delta T}{T_0/g}\,dy. \qquad\qquad \ldots\ldots(4)$$

From the definition sketch of fig. 1 and the grouping of terms in eqns. (3) and (4), it will be seen that the simplest arrangements of the variables governing the velocity and temperature fields are as follows:

$$u = f_1\left(x, y, y_0, \frac{H/L}{c_p\rho_0 T_0/g}\right)$$

$$\frac{\Delta T}{T_0/g} = f_2\left(x, y, y_0, \frac{H/L}{c_p\rho_0 T_0/g}\right).$$

Inasmuch as only two-dimensional categories are represented by each series of five variables, the latter may be combined by means of the Π-theorem into functional relationships between three dimensionless parameters:

$$\frac{u}{A^{1/3}} = \phi_1\left(\frac{x}{y_0}, \frac{y}{y_0}\right) \qquad\qquad \ldots\ldots(5)$$

$$\frac{\Delta T}{A^{2/3}T_0/gy_0} = \phi_2\left(\frac{x}{y_0}, \frac{y}{y_0}\right) \qquad\qquad \ldots\ldots(6)$$

in which A is simply the group of terms with the dimension of (velocity)^3 appearing in the original arrangements of variables and indicating the strength of the source:

$$A = \frac{H/L}{c_p\rho_0 T_0/g}.$$

These, then, are the relationships to be determined experimentally.

§ 3. Experimental Investigation

Measurements of the variables involved in eqns. (5) and (6) were carried out as an extension of a closely related laboratory investigation of the mean pattern of convection above a single line source of heat. Indeed, the latter problem represents the limiting case of the one now under discussion, as may be seen by writing the terms at the right of eqns. (5) and (6) in the alternative forms of y/x and y_0/x and assuming y_0/x to approach zero. Because of this fact, it was possible to use the same experimental equipment for both phases of the project.

Two-dimensional conditions free from external disturbance were obtained by confining the flow between vertical walls parallel to the xy plane. These walls were 8 feet long and 4 feet high and were placed either side of a very low platform 8 feet long and 4 feet wide. The top and the two end sections of the flow passage were open, and the unit was centred in a room 25 ft. by 15 ft. by 10 ft. in size. The heat source consisted of a specially fabricated gas burner 4 feet in length with a single line of $\frac{1}{16}$ inch holes in the trough-shaped top; the burner was set into the platform at mid-section with its upper edges flush. Commercial bottle gas was used as fuel; with the proper admixture of oxygen, a blue flame with minimum radiant energy was obtained. To simulate by the method of images two parallel sources of variable spacing, a 4 ft. by 4 ft. vertical partition was inserted between the parallel walls at any desired location of the plane of symmetry.

Measurements of the vertical component of the mean velocity were made by means of a sensitive vane anemometer $1\frac{1}{4}$ inches in diameter, the jewel bearings of which permitted velocities as low as 0.25 feet per second to be observed. The anemometer was calibrated in a 7-inch duct just beyond the rounded inlet, against velocity measurements made in the parallel jet from a nozzle several diameters downstream. Measurements of the temperature distribution were made with a copper–constantan thermocouple in conjunction with a potentiometer reading to 0.002 millivolt; the cold junction was maintained at room temperature outside the working section. The anemometer and the thermocouple could be mounted interchangeably on a traversing mechanism supported by rails along the upper edges of the two parallel walls.

Data were obtained by these means for various values of x, y, y_0 and burner output, the combination usually being that which would yield measurable values of u over a horizontal section of significant width. Apart from preliminary surveys to ensure that the flow was essentially two-dimensional, the measurements were made midway between the parallel walls. The magnitude of H/L for each run was evaluated from the observed distributions of u and ΔT by means of eqn. (4).

§ 4. Discussion of Results

All measurements of u and ΔT were generalized in accordance with the parameters of eqns. (5) and (6) and plotted to yield the desired functional relationships. The results of six typical runs representing two distinct geometric conditions are shown in fig. 2; herein the velocity data are plotted to the right and the temperature data to the left of the plane of symmetry. Although the individual points display the appreciable scatter which characterizes measurements of this nature, values for the same geometric conditions show no consistent variation with different rates of heat release. On the other hand the smooth curves approximating the trend of the data, and at the same time satisfying the requirements of eqns. (3) and (4) for momentum and heat flux, are seen to fall somewhat above the velocity data and below the temperature data in each case. This discrepancy was probably due in part to effects of radiation on the thermocouple indication and in part to the fact that the temperature differences were not always small (the average maximum reading was about 60°F above room

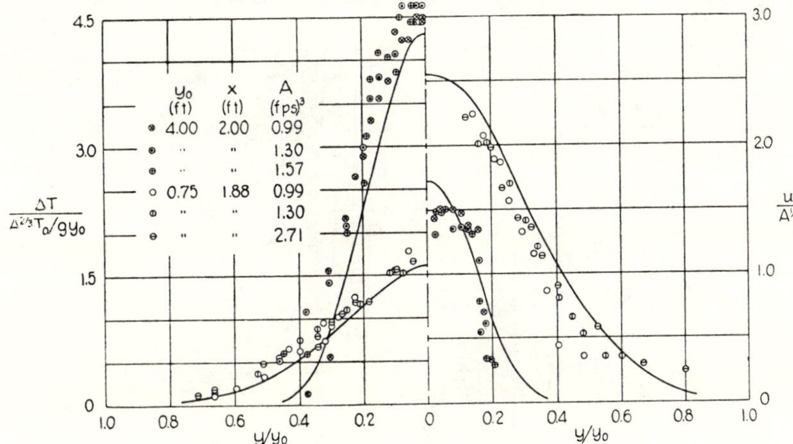

Fig. 2.   Typical experimental results.

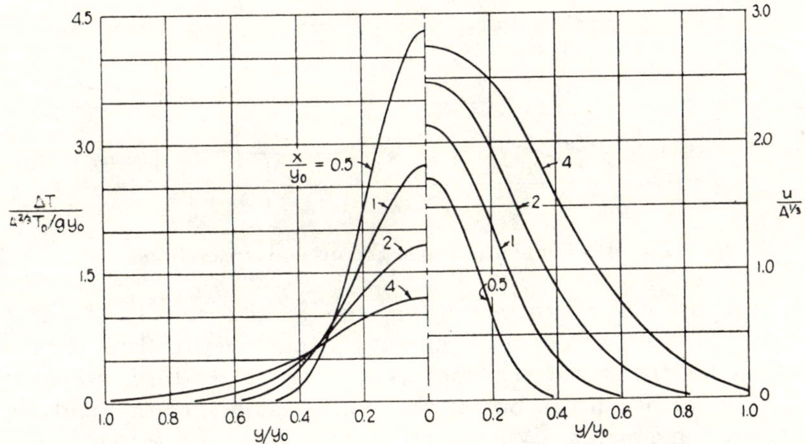

Fig. 3.   Generalized distribution curves.

temperature).   However, previous studies had not disclosed a satisfactory method of shielding, and a detailed evaluation of thermodynamic effects was not considered compatible with the elementary nature of the project. For these reasons the results were accepted without further correction, and by correlation of the data from all runs the systematic sequences of curves satisfying eqns. (3) and (4) and shown in fig. 3 were obtained.

The curves for the relative distribution of temperature at various relative elevations at once permitted construction of the dimensionless pattern of isotherms reproduced in the left half of fig. 4. Herein, it should be noted, the region below $x/y_0 = 0\cdot5$ represents pure extrapolation, although the fact that all isotherms must begin at the source provides an approximate control over their form.   The corresponding pattern of stream lines shown at the right of the figure was obtained from the systematized distribution curves of velocity by means of the definition equation for the stream function $\psi = \int u\, dy$.   The pattern in the region below $x/y_0 = 0\cdot5$, as well as that to the right of the dip in each line, was again obtained through extrapolation.   Inasmuch as the mean flow takes place in the direction of the stream lines and at a velocity which is

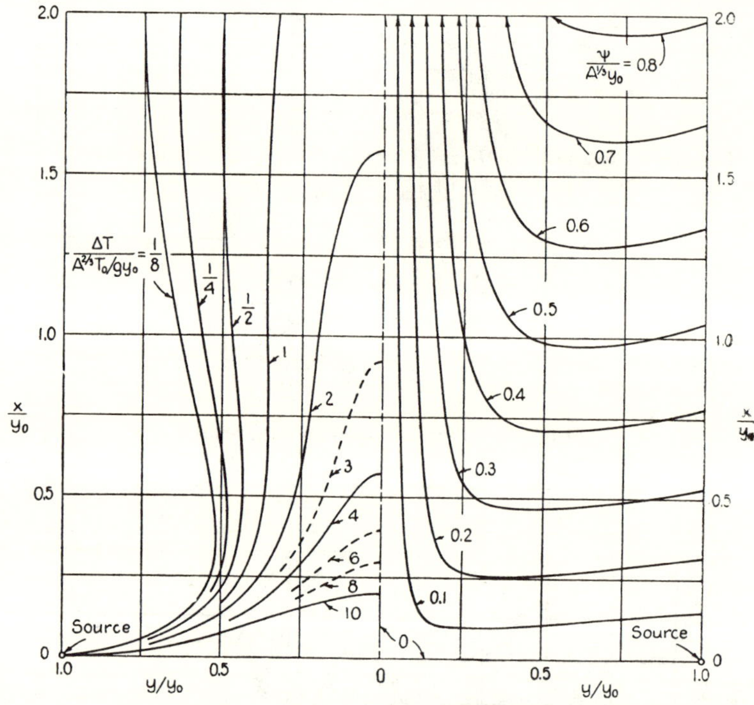

Fig. 4.   Dimensionless isotherms and stream lines.

inversely proportional to their spacing, the relative details of the velocity distribution can be seen from this figure at a glance.   Particularly noteworthy is the fact that the maximum velocities as well as temperatures are found—not directly above each source—but at the horizontal boundaries and along the vertical plane of symmetry.   Whereas the high boundary values would in practice be reduced to some degree by boundary-layer deceleration, the fact remains that the zone of primary convection is in the immediate vicinity of the plane of symmetry.

The applicability of these results to the prediction of field conditions depends upon two limitations of the experiments themselves: their small scale, and the artificial means by which symmetry was obtained.   The effect of scale is considered of less importance, for previous experiments with a point source had shown the Reynolds number of such motion to be far greater than that marking the transition from laminar to turbulent convection.   Considerably more important is the fact that the degree of asymmetry which might be expected above two independent sources—even presuming them to be of identical strength—is in no way indicated.   Indeed, slight winds in one direction or another (Rouse 1947) would probably offset any tendency towards symmetry which might otherwise exist.

Under the circumstances, the results presented must be regarded simply as a first approximation to the solution of a very complex problem. Surely, however, use of these functions to evaluate the order of magnitude of prototype conditions is preferable to the design of field installations with no generalized information at all. At the very least, the dimensionless parameters formulated in the analysis should permit field experience acquired at one scale to be utilized in determining the probable heat requirements at any other scale.

Acknowledgments

The investigation described herein was first undertaken by H. W. Humphreys as a graduate thesis project, the measurements thereafter being repeated and refined by W. D. Baines, with the assistance of M. F. Andrews. All phases of the investigation were conducted under the direction of Hunter Rouse and were partially supported by funds from Contract N8onr-500 between the Office of Naval Research and the Iowa Institute of Hydraulic Research.

References

Humphreys, H. W., 1950, *Thesis*, State University of Iowa.
Rankine, A. O., 1950, *Proc. Phys. Soc.* B, **63**, 225.
Rouse, Hunter, 1947, *J. Appl. Mech.*, **14**, No. 3, A–225.
Rouse, Hunter, Yih, C. S., and Humphreys, H.W., 1952, *Tellus*, **4**, No. 3, 201.

Cavitation in the mixing zone
of a submerged jet

Hunter ROUSE

IOWA INSTITUTE OF HYDRAULIC RESEARCH *

Whereas incipient cavitation at a streamlined boundary can be predicted by potential theory, that occurring at a surface of discontinuity depends upon correlation of the turbulent velocity and pressure fluctuations. Measurements of such fluctuations in a submerged jet are presented, together with direct observations of the cavitation phenomenon.

INTRODUCTION

Whenever the pressure intensity within a flowing liquid is reduced locally to the point of vaporization, vapor cavities form, persist momentarily, and then collapse as the pressure intensity in their vicinity increases. This flow phenomenon, known as cavitation, is generally to be avoided, because of the energy loss, boundary damage, and noise which it entails. Means are hence required of predicting its occurrence in terms of known flow characteristics.

There are two inherently different conditions of flow which can produce the local pressure drop upon which cavitation depends. The first of these is the acceleration of a liquid as it passes along a curved boundary, the point of minimum intensity then lying at the boundary itself. The second is the generation of vortices within a liquid as it enters a zone of velocity discontinuity, the points of minimum intensity then being more or less remote from the boundary. In the former case the pressure change depends upon the characteristics of the mean motion,

whereas in the latter it is the secondary motion which is directly involved.

For patterns of steady flow which are governed in form by the boundary geometry, use of the potential theory has been found to permit rather accurate prediction of the flow conditions at which the vapor pressure of the liquid will be attained at any boundary point. It is possible, in other words, to determine by analytical **means** the magnitude of a general cavitation **index at**

which the vaporization process can be expected to begin. This index is of the form

$$\sigma = \frac{p_0 - p_v}{\varrho\, u_0^2/2}\,,$$

in which p_0 and u_0 are the ambient values of the pressure intensity and the velocity, and p_v and ϱ are the vapor pressure and the density of the liquid. Experiments by many investigators have shown that a close agreement exists between analysis and measurement for flow without separation. Once separation occurs, however, cavitation is found to commence at magnitudes of σ which are considerably greater than those indicated by the pressure distribution along the boundary itself [1] **; because far lower intensities of pressure then occur within the eddies generated in the zone of velocity discontinuity. As yet no analytical means of evaluating these pressure intensities is at hand.

Efflux from a submerged orifice or nozzle represents an extreme case of separation, for the resulting processes of eddy generation and diffusion control not only the secondary motion but the mean motion as well. In other words, the boundary form has essentially no effect upon the pressure distribution, and the ambient pressure intensity p_0 also represents with very close approximation the mean intensity at all points. In order to predict the conditions under

---

* State University of Iowa, Iowa City (U.S.A.)

** *Numbers between brackets refer to the list of references at the end.*

which a submerged jet will begin to cavitate, it is essential that the pressure characteristics of the secondary motion be related in some manner to the mean velocity of the jet itself. Both the gross and the detailed aspects of this problem have been under investigation at the Iowa Institute of Hydraulic Research since 1949, and the present paper is a review of progress that has been made to date.

VELOCITY AND PRESSURE FLUCTUATIONS IN THE MIXING ZONE

From numerous analytical and experimental studies of the mean flow in a submerged jet, the velocity distribution is known to have the characteristics indicated schematically in fig. 1 [2]. Because the modification of the velocity profile is caused entirely by the turbulent mixing, comparison of the successive profiles will show that the mixing zone spreads inward as well as outward until the original constant-velocity core entirely disappears, whereafter the velocity of the central zone steadily diminishes as the outward spread continues. The sequence of velocity profiles (measured with a stagnation tube in a 6-inch jet of air) is presented in more complete detail in fig. 2.

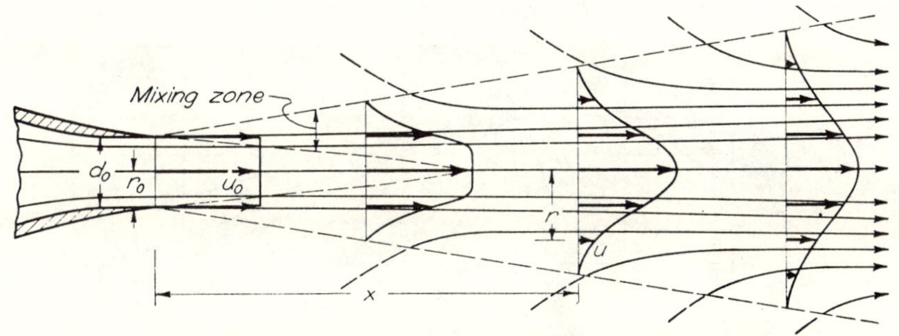

Fig. 1 — *Definition sketch for characteristics of mean flow in a submerged jet.*

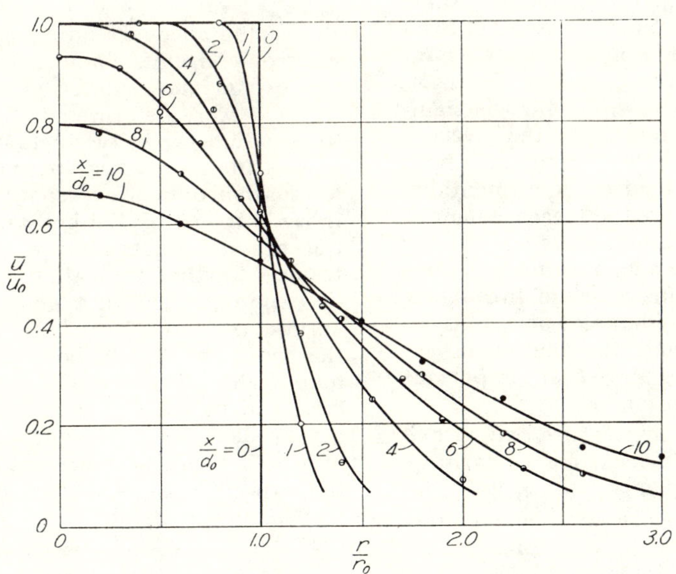

Fig. 2. — *Stagnation-tube measurements of mean longitudinal component of velocity.*

237

Both analytical and experimental studies of the secondary action have yielded at least partial information as to the velocity characteristics of the turbulence in the mixing zone. Typical root-mean-square values of the longitudinal component (measured in the 6-inch jet with a hot-wire anemometer) are indicated in fig. 3. Values of the other two components follow essentially similar profiles but have approximately two-thirds the magnitude. The maximum intensity of the turbulence is seen from the figure to coincide in radial location with the initial zone of discontinuity, to remain almost constant in magnitude over the length of the constant-velocity core, and then to diminish gradually as the mean velocity itself is reduced.

Since the velocity fluctuations at a given point of observation can be considered to result from the transit of successive eddies, and since the variation in velocity across any eddy entails a particular variation in the intensity of pressure, for every sequence of velocity fluctuations there must exist a specific — albeit unknown — sequence of pressure fluctuations. For the case of isotropic turbulence the correlation between the two has been expressed analytically, and Batchelor has found that the ratio of the root-mean-square pressure fluctuation to one-half the product of the density and the mean-square velocity fluctuation — i.e., $\sqrt{\overline{p'^2}}/(\varrho\,\overline{u'^2}\,2)$ — should have the value 1.17 [3].

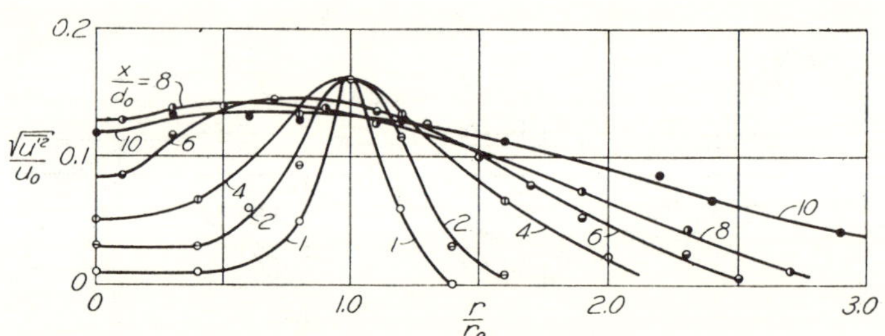

Fig. 3. — Hot-wire measurements of root-mean-square longitudinal velocity fluctuation.

Unfortunately, the point of incipient jet cavitation cannot at once be predicted from a value such as this for three pertinent reasons : First of all, the turbulence in a mixing zone is far from isotropic. Secondly, it is not the root-mean-square pressure fluctuation which is significant in cavitation so much as the average negative peak. Thirdly, the theory has been neither checked nor extended experimentally, because instrumentation for the measurement of pressure fluctuations within a mixing zone is still practically nonexistent. Further advancement in the analysis of this phase of turbulence must depend to a considerable degree upon experimental evidence, and the matter of instrumentation has hence played an important part in the Institute program.

Preliminary results of this nature, which had just become available at the time of writing, were obtained in the 6-inch air jet with a 1/4-inch tube enclosing an electro-mechanical pressure cell, at the end of which two hemispherical heads could be mounted interchangeably. The one was provided with a small tip opening, and the other with four openings at 90° intervals around a circle 47 1/2° from the tip. On the basis of tests made on a composite field instrument at much larger scale, the tip opening

was assumed to yield the instantaneous reading $1/2\,\varrho\,(\overline{u} + u')^2 + p' - p_0$ and the side openings the instantaneous reading $p' - p_0$. By suitable electronic circuits connected to the pressure cell, the two primary curves plotted dimensionlessly in fig. 4 were determined: the root-mean-square longitudinal velocity fluctuation, and the root-mean-square pressure fluctuation. The velocity-fluctuation data are seen to agree closely with those obtained by the hot-wire anemometer, and the pressure-fluctuation data follow a similar trend. Further indications as to the accuracy of the instrument are given by the other two curves, the data for which were obtained with the same two tips in conjunction with a liquid manometer. The curve of mean velocity is practically identical to that previously obtained with a very slender hypodermic needle; however, the indicated mean pressure intensity departs from the expected value of zero by a small amount which appears to depend upon the intensity of the velocity fluctuations.

Although these preliminary measurements are merely qualitative, they permit at least a rough evaluation of the pressure fluctuations accompanying jet diffusion. The maximum root-mean-square pressure deviation—like the maximum root-mean-square velocity deviation—was found

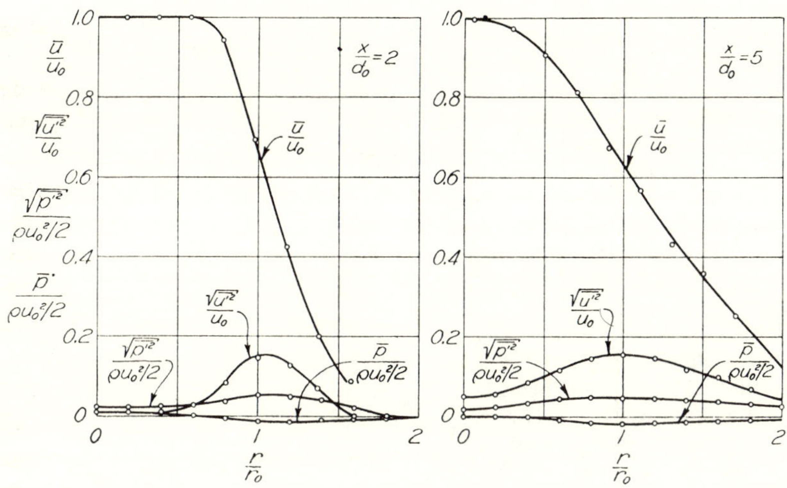

Fig. 4. — *Velocity and pressure measurements obtained with tube and hemispherical heads.*

from further exploratory tests to remain essentially constant through the initial zone of establishment. If the peak values of $\sqrt{\overline{u'^2}}/u_0$ and $\sqrt{\overline{p'^2}}/(\rho u_0^2/2)$ are taken from fig. 4 as representative, it will be found that the quantity $\sqrt{\overline{p'^2}}/(\overline{\rho u'^2}/2)$ then has a magnitude of about 2.15, in comparison with that of 1.17 predicted by Batchelor for isotropic turbulence. The primary value $\sqrt{\overline{p'^2}}/(\rho u_0^2/2) = 0.055$ is significant for the prediction of cavitation, however, only if multiplied by a numerical factor embodying the frequency of occurrence.

In arriving at a magnitude for this frequency factor, not only the temporal [4] but also the spatial distribution of the turbulence must be considered. The temporal distribution at any point may be assumed to agree with the normal-probability function, so that definition of incipience as that condition at which cavitation occurs during a specified percentage of time will permit a particular multiplying factor to be

evaluated. For example, if it is arbitrarily assumed that cavitation has begun when negative fluctuations reach the vapor magnitude 5 % of the time, this factor will have the magnitude $C_t = 1.6$. On the other hand, not only a single point but the mixing zone as a whole is involved. Thus, since combination of the temporal frequency with the mean velocity would yield a spatial distribution along only an elementary segment of the mixing zone, this must be multiplied by an additional spatial factor C_s representing the number of such segments included in the jet. Although this number is evidently controlled by either the relative scale of the eddy structure or that of the cavities themselves, there is at present no means of determining its magnitude except by observation of the cavitation phenomenon itself. For the present, therefore, the index for incipient cavitation must be written simply as

$$\sigma_t = C_s C_t \frac{\sqrt{\overline{p'^2}}}{\rho u_0^2/2}.$$

DIRECT OBSERVATIONS OF JET CAVITATION

For want of quantitative relationships between the fluctuations of velocity and pressure intensity in a mixing zone, preliminary observations were made in 1949 on the cavitation of a jet produced by a supplementary pumping system in the test section of the smaller Institute water tunnel [5]. Although pronounced cavitation could readily be obtained, the limited length of the test section was believed to influence the

pattern of diffusion to such an extent that measurements were at best merely qualitative. Moreover, unaccountable effects of temperature upon the incipient cavitation index were noted, even though every effort was made to hold the air content of the water to a minimum. For these reasons a special jet-cavitation tank was constructed the following year.

This tank was 5 feet in diameter and had an overall length of 10 feet, with convex ends. Nozzles could be mounted interchangeably on a 3-inch inlet pipe located on the axis at one end, and a 6-inch outlet pipe was connected at the center of the other end. A Chrysler-Hale fireboat pump was used to provide the required discharge at the necessarily high base pressure. Means of controlling the hydrostatic load on the tank, removing air, and either heating or cooling the water were similar to those on the Institute water tunnels [1]. Overall pressures and rates of flow were measured with a large Bourdon gage and a mercury manometer connected to suitable piezometer openings in the walls of the tank and the inlet pipe. The intensity of cavitation was observed through use of a crystal-type Massa hydrophone mounted below the jet on the floor of the tank, the hydrophone being connected through an amplifying circuit to a pen recorder. Background noise was largely eliminated by using a narrow frequency band centered at 25 kilocycles.

From previous cavitation investigations it was known that the apparent point of incipience is controlled to a variable degree by the amount of air carried in suspension as minute bubbles on particles of foreign matter. Although the presence of such nuclei in limited numbers is considered necessary to the onset of cavitation, when present to an excessive degree they tend to form cavities within the fluid before the pressure intensity falls to its true vapor magnitude. This corresponds, in effect, to replacing the term p_v in the cavitation index with a quantity p_c which depends to an unknown extent upon the amount and nature of the air carried in suspension. Evidently, if both p_0 and v_0 are low, this term can play a predominant role in controlling the apparent magnitude of σ. On the other hand, if both ambient values are high, the magnitude of p_c — as well as that of p_v — becomes relatively unimportant. For this reason, every series of cavitation measurements was preceded by 2 or 3 hours of de-aeration at a low rate of flow and a very low (i.e., a large negative) hydrostatic pressure, the air collecting at the top of the tank being removed continuously. The hydrostatic load was then increased by nearly two atmospheres, so that as much as possible of the remaining air would be dissolved. The hydrostatic pressure during the subsequent measurements was held at least 0.5 foot of mercury above that of the de-aeration run.

Observations with the hydrophone recorder indicated in general that the noise level remained constant with decreasing values of σ until the point of incipience was reached, whereupon the noise rapidly increased toward a maximum (fig. 5) and then gradually diminished. Use of a loud-speaker instead of the pen recorder showed the noise just below the point of incipience to be a very sharp crackling; however, at advanced stages it became progressively more mushy, indicating either that the collapse of the cavities was cushioned by air or that interference of the dispersed cavities affected the noise transmission.

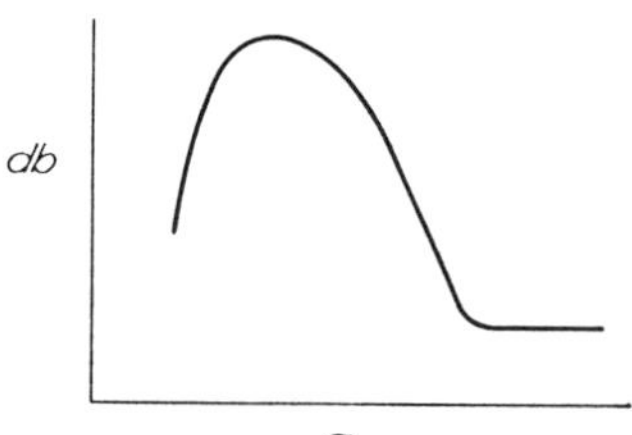

Fig. 5. — *Schematic indication of change in noise level with intensity of cavitation.*

Observations were hence restricted to the zone of incipience, for which mean levels of the cavitation noise were determined at various jet velocities, hydrostatic pressures, and water temperatures.

A typical plot of the observed mean noise levels against σ is shown in fig. 6 for the jet from a $1\frac{1}{2}$-inch nozzle with a short cylindrical tip.

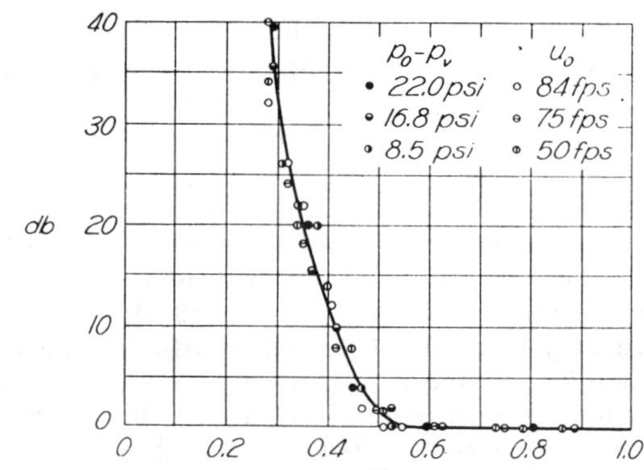

Fig. 6. — *Experimental data for determination of incipient cavitation index.*

Despite a certain amount of scatter, no systematic deviation with either pressure or velocity (the other being held constant) was noted, nor did supplementary runs in which the temperature was varied show thermal influences beyond that upon the vapor pressure. The line drawn through the plotted points indicates a value of 0.55 for the magnitude of σ_i, corresponding to a product of C_t and C_s of about 10. The fact should be emphasized that the plotted curve represents mean levels only, and that intermittent flashes of noise were observed at values of

σ as high as 0.7. Regarded in this light, designation of the point of incipience is just as arbitrary as the selection of a limiting frequency of pressure fluctuation.

Since it is the initial discontinuity between the jet and the surrounding liquid which leads to the generation of the low-pressure eddies, it is to be expected that any variation in approach conditions which would decrease the velocity gradient around the jet periphery would also decrease the intensity of the eddy formation and hence lower the magnitude of σ_i. This was demonstrated experimentally by extending the parallel portion of the nozzle tip about 8 diameters, the resulting boundary-layer growth reducing the magnitude of σ_i by 15 %. However, complete elimination of the parallel tip (the nozzle then producing a 60° converging jet) — also reduced the value of σ_i, but for reasons which

are not yet clear. The presence of air in suspension, on the other hand, gave rise to the pseudo-cavitation already discussed, the apparent magnitude of σ_i then varying with the magnitude of p_0 and with the air concentration over a wide range. Under these circumstances, a design value of 0.6 for σ_i seems a reasonable average.

Efforts to photograph the cavitation pockets in the diffusion zone further emphasized many of the points which have been discussed. In order to eliminate blurring of the image, high-speed (1/20,000 second) illumination was used. Just below the point of incipience, the chance of catching a readily visible vapor pocket with a single exposure was found to be negligible. At a sufficiently low value of σ, of course, representative cavities were always recorded, as shown in the upper photograph of fig. 7 obtained with rear illumination. In order to reveal the

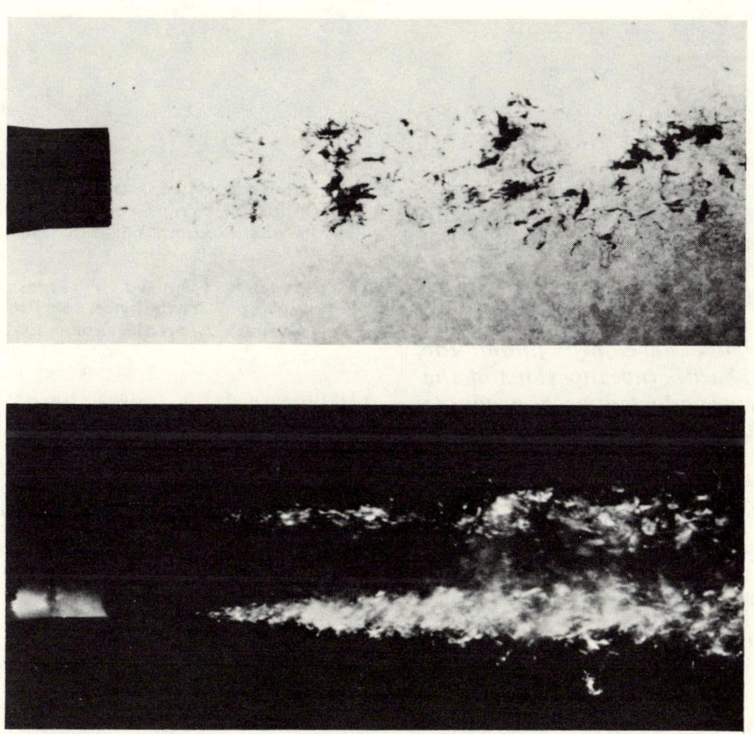

FIG. 7. — *Photographs of jet cavitation at* $\sigma = 0.2$. *Above, single exposure with rear illumination. Below, 25 exposures with only central plane illuminated.*

radial distribution of the cavities, a condensing lens and slot were used to produce a thin (1/4-inch) sheet of light at right angles to the camera axis. Then, however, multiple exposures were necessary to ensure the presence of cavities in sufficient density. The second photograph, obtained in this manner, shows clearly the outline of the cavitation zone. Noteworthy is the fact that — although the zone of maximum cavitation coincides in general with the zone of maximum turbulence — no vapor formation is evident over an initial length of about one diameter. The characteristics of the zone of collapse — beyond the range of these photographs — have yet to be investigated.

By way of summary, it must be noted that accomplishments to date have been largely of an exploratory nature, indicating directions for more refined observations rather than final values for immediate use. The pressure tube shows promise of permitting a quantitative study of the instantaneous fluctuations, provided that the tip can be made sufficiently small relative to the diffusion zone and free from appreciable bias; like the velocity tube, it has the particular advantage of being adaptable to use in both air and water. The matter of the spatial distribution of cavities will probably require an investigation of scale as well as intensity, by methods which are still far from obvious. And for an analysis of the complete history of a typical cavity, it will be necessary to carry both pressure measurements and photographic observations from the nozzle itself to the point of collapse. The only numerical result now at hand is that determined empirically for the point of incipient cavitation of an initially parallel jet with even velocity distribution and minimum air content; under these conditions, the design value $\sigma_i = 0.6$ is considered to represent a dependable figure.*

All experiments described herein were conducted at the Iowa Institute of Hydraulic Research under Contract N8onr-500 with the Office of Naval Research. Measurements of the velocity characteristics were made by H. C. Hsu and those of the pressure characteristics by S. C. Ling, with instruments developed under the direction of P. G. Hubbard. Investigations of the cavitation itself were conducted as a graduate thesis project by J. P. Whitehouse [6]. The paper was presented by the author at the Eighth International Congress on Theoretical and Applied Mechanics, Istanbul, 1952.

REFERENCES

[1] ROUSE, H., and McNOWN, J. S., « Cavitation and Pressure Distribution - Head Forms at Zero Angle of Yaw », Bulletin 32, Studies in Engineering, State University of Iowa, 1948.

[2] ALBERTSON, M. L., DAI, Y. B., JENSEN, R. A., and ROUSE, HUNTER, « Diffusion of Submerged Jets, » Trans. A.S.C.E., Vol. 115, 1950, p. 639, 1948.

[3] BATCHELOR, G. K., « Pressure Fluctuations in Isotropic Turbulence, » Proc. Camb. Phil. Soc. 47, part 2, p. 369, April 1951.

[4] ROESLER, F. C., « Ueber Kavitationsvorgänge in turbulenten Mischzonen, speziell an der Grenze schneller Flüssigkeitsstrahlen ». Oest. Akad. Wiss. math-nat. Kl. Anz. 87, 287-290, 1950.

[5] ROUSE, H., HUBBARD P. G., and SPENGO, A. C., « Cavitation of Submerged Jets, » Iowa Institute of Hydraulic Research, Interim Report to Office of Naval Research, March 1950.

[6] WHITEHOUSE, J. P., « An Investigation into the Point of Incipient Cavitation of Submerged Jets, » M. S. thesis, State University of Iowa, February 1952.

*For further discussion of this topic, the reader is referred to "Jet Diffusion and Cavitation," p. 529 in this volume.

MEASUREMENT OF VELOCITY
AND
PRESSURE FLUCTUATIONS IN THE TURBULENT
FLOW OF AIR AND WATER

by Hunter ROUSE

IOWA INSTITUTE OF HYDRAULIC RESEARCH
STATE UNIVERSITY OF IOWA

Although many phases of experimental hydraulics require the determination of various turbulence characteristics, the necessary instrumentation for their measurement in flowing water has lagged far behind that for air. Under some circumstances, to be sure, the similarity between air and water flow permits the results of hot-wire measurements in air to be applied to the analogous case for water. But there are at least two conditions for which such indirect means cannot suffice. On the one hand, free-surface phenomena in water are not subject to simulation with air; on the other hand, the hot-wire anemometer—whether used in air or water—gives no direct indication of pressure fluctuations, and these are sometimes fully as important in hydraulics as are the fluctuations in velocity. For some years, therefore, the Iowa Institute of Hydraulic Research has been seeking to develop an instrument which can be used in either fluid for the instantaneous measurement of both flow characteristics.

The essential qualities of such an instrument make its development extremely difficult. Accuracy of indication is a primary requisite, of course, but operating convenience and ease of evaluating results are almost as important. If it is to be used in the laboratory, the sensing element must be very small, although for field use this restriction is not so severe. To be most broadly applicable, moreover, the instrument should indicate simultaneously the pressure and the three components of the velocity at the point of measurement.

Because of the pressure requirements initially established, the Iowa Institute turned in 1949 to the adaptation of the familiar Pitot tube to the purpose in question. One part of the Institute's war-time research had involved the study of the pressure distribution around a systematic series of conical, rounded, and ellipsoidal head forms both for axial flow and at various angles of yaw. Thereafter certain of these studies were extended in the search for a head form with the most suitable pressure characteristics. Among all the forms tested, the simple hemispherical head at the end of a shaft of the same diameter proved to have the best characteristics for this purpose, and it was therefore retained as the basis of design.

The distribution of pressure around a tube of this nature at a zero angle of yaw follows the well known pattern shown schematically in

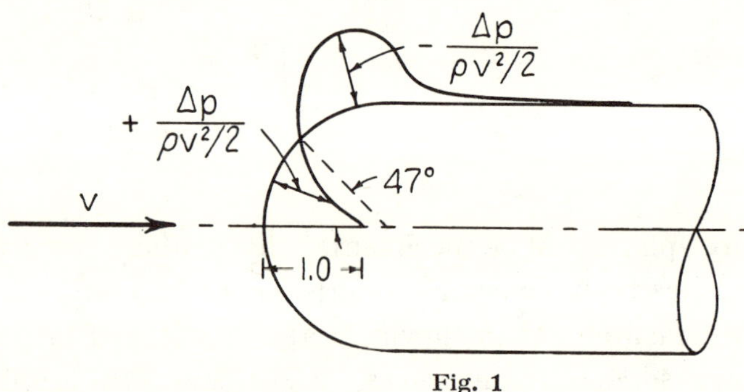

Fig. 1

figure 1, upon which the Prandtl type of Pitot tube was originally based. If now the tube is slightly yawed, the pressure at the tip and along the lee side of the profile will be reduced, whereas that on the opposite side will increase, as a function of the yaw angle. In fact, if a piezometer opening directly at the tip is made about $\frac{3}{16}$ the diameter of the tube, the pressure which it indicates will depart from the reference magnitude $\frac{\rho v^2}{2}$ in direct proportion to the square of the cosine of the angle of yaw; and if a piezometer opening of moderately small size $\left(\frac{d}{40}\right)$ is located at the point of zero pressure for zero yaw (approximately 47° from the axis in the plane of the angle), its indication will depart from zero in direct proportion to the product of the angle and the square of its cosine. Therefore, if both indications are plotted in the form $\dfrac{\Delta p}{\left(\dfrac{v^2 \cos^2 \alpha}{2}\right)}$ versus

angle of yaw, as shown in *figure* 2, the first will remain constant and the second will vary linearly—not only for slight angles but with good approximation up to about 15°.

If such a tube is placed in a moving fluid which changes continuously in speed and direction of flow, the differential pressure between the tip piezometer and two interconnected side piezometers $\pm$ 47° from the axis in the plane of yaw should remain proportional to the square of the axial component of the velocity. The differential pressure between the two side piezometers, moreover, should vary in direct proportion to the effective angle of yaw caused by the transverse velocity component in this plane. Two additional side openings in the normal longitudinal

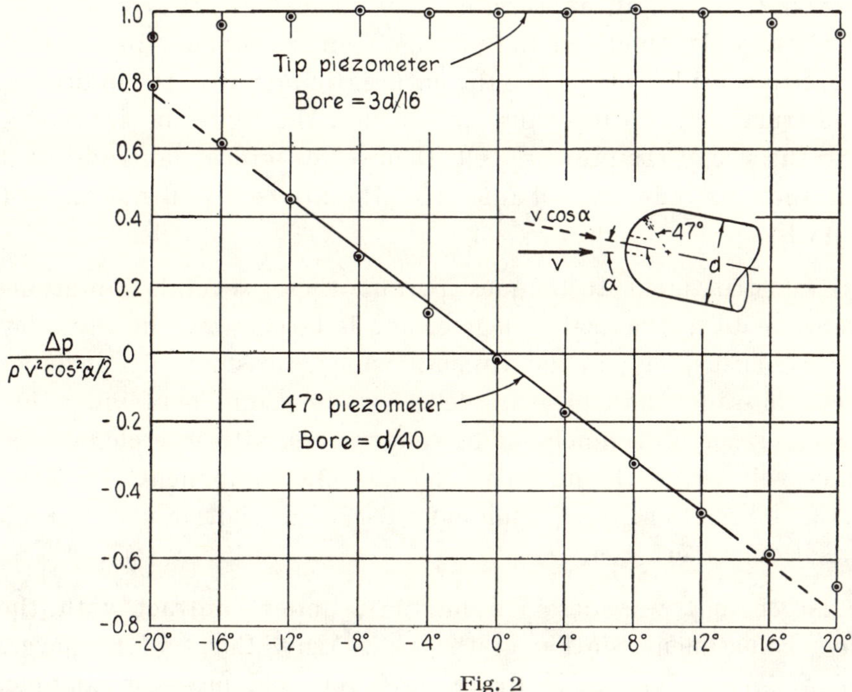

$$\frac{\Delta p}{\rho\, v^2 \cos^2\!\alpha/2}$$

Fig. 2

plane should then indicate the angularity due to the third velocity component. If there are pressure changes superposed upon the velocity changes, these should be indicated directly by the variation in the average of the pressures at the four side openings. There is thus available, at least on paper, a sensing element which should produce reactions that are proportional to the first or second powers of the four desired turbulence characteristics for turbulence intensities up to 25 %.

Despite the potential value of this form of tube for the measurement of turbulence in both water and air, certain practical factors still make it difficult to satisfy therewith all the requirements of a universal instrument. So far as the pressure field itself is concerned, accuracy comparable to that of the hot-wire anemometer should be readily obtainable, provided that proper scale relationships are maintained. Except for the matter of initial air removal, operational problems should be at a minimum; and electronic indication of fluctuating pressures and evaluation of statistical records are now standard practice. The primary obstacle at present is the matter of absolute size, and this arises not only from the requirement of accuracy in the construction of the tube with its various piezometers but also from the necessity of housing the pressure-sensitive devices sufficiently close to the piezometers to produce a satisfactory frequency response. Since the piezometers must be of the proper relative diameter and cleanly bored, even watchmaker's precision would require an appreciable tube diameter. Moreover, both the diameter and the length of the passage from piezometer to pressure cell control to a considerable degree the natural frequency of the indicator, so that constructional features thus further limit the degree of size reduction. And the size of the pressure cells themselves imposes an additional restriction upon the minimum diameter of the afterbody, if not that of the tube itself.

Until constructional techniques permit these several limitations to be overcome to a practicable degree, one is faced with the necessity of producing either (1) an instrument which satisfactorily includes piezometers, passages, and pressure cells for four-component indication regardless of size or (2) a much smaller instrument with interchangeable heads which will permit the indication of only one component at a time. To date the fabrication and preliminary use of a pilot model of each type have been undertaken.

The larger unit was designed in 1951, under contract with the Waterways Experiment Station of the U.S. Army Corps of Engineers by David W. Appel of the Institute staff, for field use—for example, in the zone of energy dissipation at the toe of a spillway. The diameter of head and shaft was 2 inches, and the head contained a tip piezometer and eight side piezometers 47º from the axis. Four pressure cells of the strain-gage type were housed in an afterbody $3\frac{5}{8}$ inches in diameter; suitable leads from the tip piezometer, from two pairs of opposite side piezometers, and from the four remaining side piezometers yielded indications of the longitudinal velocity component, the angularity in two normal planes, and the pressure. All connecting passages had a diameter of 0.5 inch and a maximum length of 12 inches, and calibration tests under a fluctuating pressure of controlled magnitude and frequency indicated a linear response up to a frequency of 100 cycles per second. Preliminary tests of the velocity elements of the instrument along the axis of a submerged water jet yielded results in fair agreement with hot-wire measurements in a jet of air.

The smaller unit was constructed by Sung Ching LING, also of the Institute staff, in the course of a study of jet cavitation under contract with the Office of Naval Research. Because of the size limitation, the tube diameter chosen was $\frac{1}{4}$ inch, and the two interchangeable heads contained (1) a tip piezometer and (2) four interconnected side piezometers 47° from the axis. Mechanical transmission of pressure from a thin diaphragm just behind the head to a cell of the transducer type yielded in air a linear response up to 1 000 cycles per second. Use of the pressure of the atmosphere as reference thus permitted measurement of either the longitudinal component of the velocity or the local pressure intensity throughout the diffusion zone of an air jet. As shown by diagrams included in the writer's paper "Cavitation in the Mixing Zone of a Submerged Jet" (*La Houille Blanche*, January-February, 1953), the transverse distribution of the root-mean-square as well as the mean velocity indicated by this instrument was in close accord with that obtained by the hot-wire anemometer. No such comparison was possible for the indicated pressure fluctuations, for these appear to have been the first measurements of their kind to be made. Although the distribution of the mean pressure pointed to a slight bias of the instrument, the distribution of the root-mean-square pressure fluctuation followed a logical trend, and its order of magnitude was the same as that obtained analytically by BATCHELOR for the case of isotropic turbulence.

The method of measurement under discussion is obviously still in its initial stage, and its future is by no means yet assured. At present the Iowa Institute is engaged with the refinement of the larger unit, in preparation for more extended use under field conditions, and of the smaller unit, to the end of further evaluation of the pressure characteristics of fluid turbulence. Despite its obvious limitations of size and frequency response, the fact remains that it is the only type of instrument available to date that has permitted the measurement of fluctuations in both velocity and pressure within both air and water.

Hunter Rouse
Tien-To Siao

FORM DRAG OF COMPOSITE SURFACES

Iowa Institute of Hydraulic Research, State University of Iowa

Following a free-streamline analysis of two-dimensional flow past multiple plates, experimental data are presented for the drag coefficient of axially symmetric and three-dimensional arrangements of sharp-edged disks, rings, and plates as a function of the number, form, and spacing of the elements. It is shown that the form drag of a composite surface of given solid area can be varied over nearly a twofold range by proper disposition of its parts.

STATEMENT OF THE PROBLEM

Knowledge of the effect of rearranging the parts of a composite body upon the drag of the body as a whole is often essential, because the net drag coefficient can be controlled to a considerable degree in this manner. A simple example is found in the case of disks arranged in tandem, their combined drag being far less when their spacing is small than when it is great. The drag of such devices as sea anchors and parachutes, on the contrary, can be greatly increased by dividing their surfaces into parts with small spaces between them. For lack of adequate theory even with respect to simple shapes, the drag coefficient of a complex system is determinable only by actual measurement. However, certain fundamental aspects of the problem can be assessed through the following considerations.

ANALYSIS OF FLOW PAST PARALLEL PLATES

The force F_D exerted upon an immersed body by the fluid moving past it is usually expressed as the product of the fluid density ρ, the normal projection of the solid area A, the square of the velocity V, and a coefficient C_D which depends primarily upon the form of the body:

$$F_D = C_D A \frac{\rho V^2}{2} \tag{1}$$

Since the drag of a body in the wake of another is known to be decreased because of the local reduction in velocity, one might reason that the drag of a body a short distance to the side of another would increase because of the local velocity rise. The latter problem thus reduces itself to the mutual effect of the velocity fields produced by two simple profiles, such as the parallel plates shown in the definition sketch of Fig. 1.

Now the free-streamline theory [1] (numbers in brackets refer to the bibliography at the end of the paper) does not in itself permit the actual drag of a single plate — not to mention that of several — to be determined even approximately, because this theory of necessity involves the assumption of free-stream velocity at the infinitely long surface of separation and hence of free-stream pressure within the wake. The computed drag coefficient of 0.88 [2] hence differs considerably from the measured value of 2.0, although this discrepancy can be greatly alleviated by arbitrarily assuming a higher velocity at the surface of separation (i.e., a wake of finite rather than infinite length) [3]. In either event, however, the theory should provide some measure of their interaction as the two plates of Fig. 1 are gradually separated.

A detailed analysis of flow past two such plates was performed for this paper by the second author on the as-

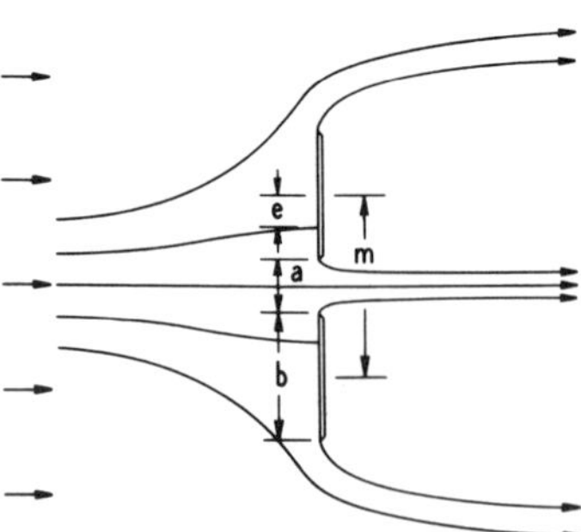

FIGURE 1. DEFINITION SKETCH FOR FLOW PAST PARALLEL PLATES.

sumption of free-stream pressure in the wake. Surprisingly enough, this indicated that the net force on each plate remained exactly the same as the spacing varied. Thus, although the location of the stagnation points and the contraction ratio of the inner jet changed continuously as indicated in Fig. 2, the drag coefficient based on the actual solid area maintained the constant magnitude $C_D = 2\pi/(\pi + 4) = 0.88$ for all values of the numerical factor $k = f(a/b)$.

One might argue that according to the free-streamline theory the velocity field in the vicinity of a single plate shows no local increase at any point — the longitudinal component, in fact, being everywhere less than the velocity of the free stream. In the actual case, on the contrary, the velocity attains its maximum value at the edge of the plate and diminishes laterally toward the free-stream limit. Nevertheless, pressure-distribution measurements made for this paper by D. W. Appel and Mikio Arie in an Institute air tunnel yielded results which were of the same nature as those of the free-sgreamline theory: no appreciable change in drag coefficient as the spacing of the two plates varied. Velocity measurements in the field

of a single plate, moreover, indicated only a small variation of the longitudinal component in the transverse plane except in the immediate vicinity of the edge.

If, now, instead of only two plates one considered a very large number, their drag could be evaluated once again in terms of the free-streamline theory, although the complexities of the analysis would be very great. However, at least in the centrally located region, for small to moderate spacing the flow pattern between lines of symmetry passing through neighboring plates should be approximately similar to that for flow through a slot. If written with respect to the free-stream velocity of the customary efflux equation [4]

$$v = \frac{1}{\sqrt{1 - C_c^2\, a^2/m^2}}\; \sqrt{2\, \Delta\, p/\rho} \tag{2}$$

the drag coefficient is found to increase as the ratio of width of opening to centerline spacing a/m (see Fig. 2) is increased:

$$C_D = \frac{1 - C_c^2\, a^2/m^2}{1 - a/m} \tag{3}$$

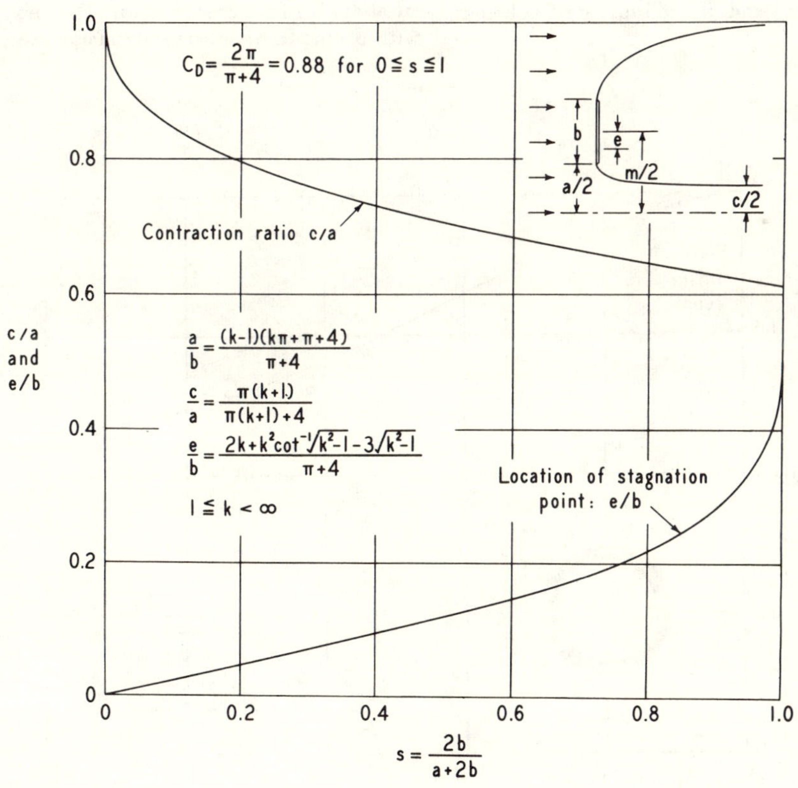

FIGURE 2. FREE-STREAMLINE SOLUTION FOR TWO PARALLEL PLATES.

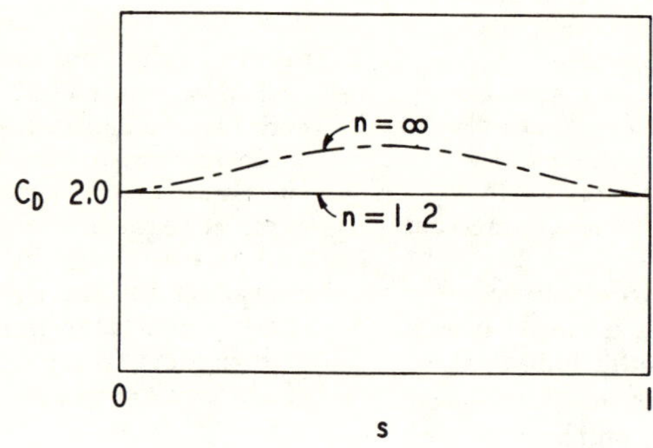

FIGUE 3. SCHEMATIC DRAG FUNCTION FOR MULTIPLE PLATES.

The effect of this central rise in drag should be to increase at least to some degree the net drag coefficient for a considerable number of plates with moderate spacing. Even for an infinite number of plates, of course, C_D would necessarily reduce to 2.0 as a/m approached 0 or 1 — i.e., as the elements attained a zero or infinite spacing. If the quantity s is used to indicate the solidity ratio of the system (the ratio of projected solid to total area) and n the number of elements, the drag coefficient may hence be expected to vary with s and n as indicated schematically in Fig. 3.

EXPERIMENTS ON COMPOSITE SURFACES

Although experimental evidence to support this analysis of the two-dimensional case is not yet at hand, a graduate-thesis investigation by the second author [5] involved the direct measurement of drag on a variety of three-dimensional surfaces which display the same general tendencies. The tests were conducted in an Institute air tunnel having a 4-by-6 foot cross section at an air speed of about 25 feet per second. The drag was determined with a simple beam-type dynamometer, corrections being

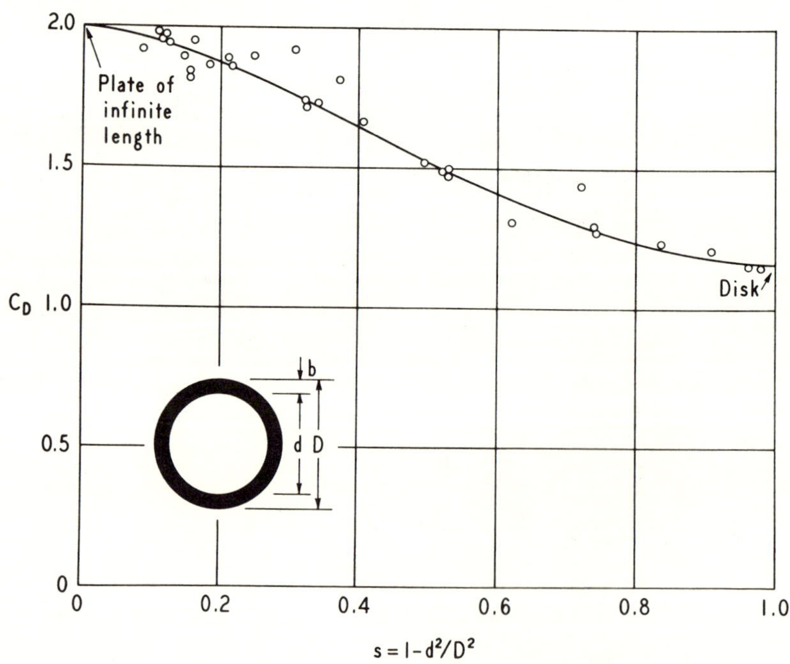

FIGURE 4. DRAG COEFFICIENT OF SINGLE RING AS A FUNCTION OF SOLIDITY RATIO.

made for the constriction effect through tests on a series
of similar forms at successively smaller scale. Further
details of the test procedure are presented in the thesis.

The axially symmetric counterpart of the two-dimensional plate was represented in these experiments by the circular disk, which has a drag coefficient of 1.1. The counterpart of two parallel plates with variable spacing was
then a ring of constant radial width b but variable diameter, the geometric limits of which were the plate (i.e., a
ring of infinite diameter), for $s = 0$, and the disk itself,
for $s = 1$. As seen from Fig. 4, the measured drag coefficient varies accordingly from the limit of 2.0 for the
plate to 1.1 for the disk; the continuous curve with zero
slope at each end evidently corresponds to the straight
line between the identical limits of 2.0 for the two parallel plates of Fig. 3.

Corresponding roughly to the inclusion of a third plate
in the two-dimensional case, the addition of a disk of
radius b within such a ring at once produces an appreciable increase in drag for the same solidity ratio, as shown
in Fig. 5. Moreover, as the number of rings around the
disk is increased, the solidity ratio remaining constant,
it is seen that the net drag based upon the solid area
continues to increase — except at the limits $s = 0$
and $s = 1$. In this figure the similar points represent
runs with constant ratio of ring width b to centerline

spacing m but with a successively greater number of
rings n (the broken curves are not continuous functions,
because of the necessary progression of n by integers).
The forms of these curves are defined by the plotted
points with reasonable approximation; furthermore, the
abscissa values of their upper ends are determined by the
limiting solidity ratio $s = b/m$ when $n = \infty$. From
this approximate location of their terminal points, the dot-
dash line for the drag of a finite body with an infinite
number of annular parts was drawn by eye. Although the
single ring does not, strictly speaking, represent the lower limit of this series of forms, its curve from Fig. 4 has
been superposed for purposes of comparison. The common
curve for $n = \infty$ evidently lies a considerable distance
above the curves for a single ring and for a single ring
and disk except near its two limits; its maximum point, in
fact, corresponds to a drag that is nearly double that of
the disk itself. To be noted, moreover, is the fact that
C_D changes only slowly when n is large, the major part
of the increase occurring in the range $1 < n < 4$.

Two further series of composite bodies were found to
display similar trends. The one was a lattice structure
composed of crossed strips enclosing open squares, and
the other its geometric complement: solid squares separated by open spaces. The results for the lattices, reproduced in Fig. 6, are comparable in nature to those for the

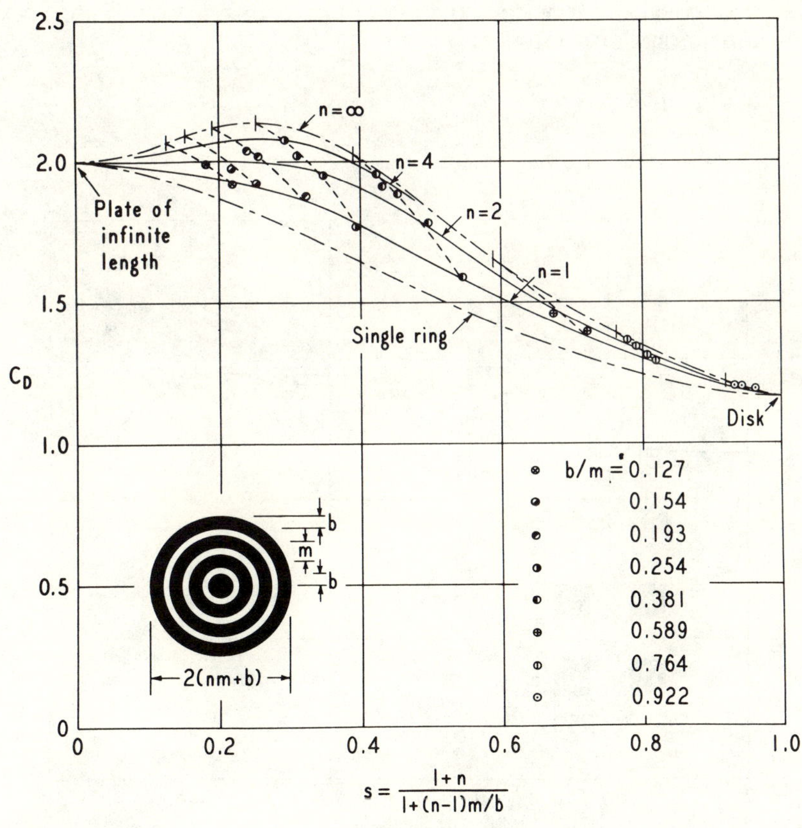

FIGURE 5. DRAG COEFFICIENT OF MULTIPLE RINGS AND DISK AS A FUNTION OF
SOLIDITY RATIO AND NUMBER OF RINGS.

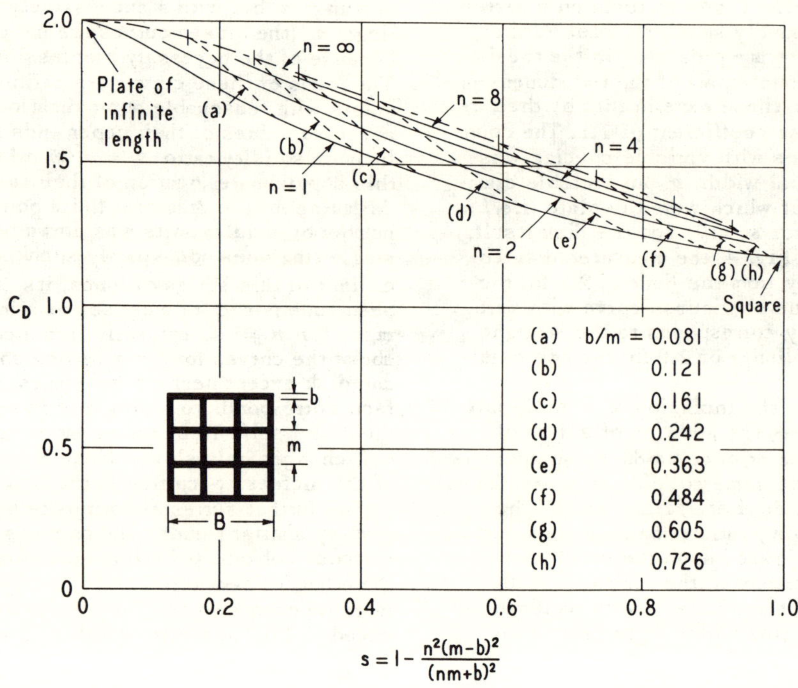

$$s = 1 - \frac{n^2(m-b)^2}{(nm+b)^2}$$

FIGURE 6. DRAG COEFFICIENT OF SQUARE LATTICES AS A FUNCTION OF SOLIDITY RATIO AND NUMBER OF ELEMENTS.

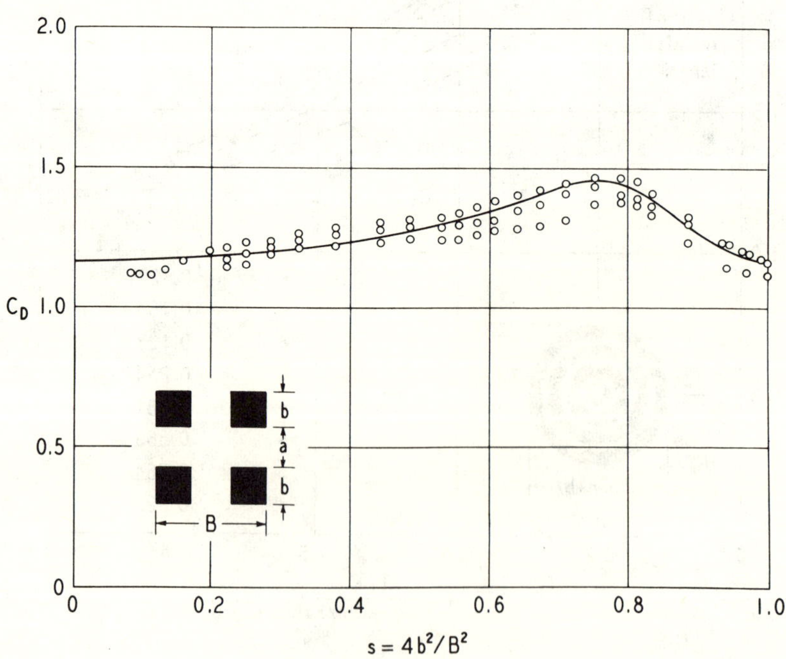

$$s = 4b^2/B^2$$

FIGURE 7. DRAG COEFFICIENT OF FOUR SQUARE PLATES AS A FUNCTION OF SOLIDITY RATIO.

annular plates, although an intermediate maximum is lacking. No points are shown in Fig. 6, because the nature of the independent variables used in the experiments made it necessary to derive these curves from those of C_D versus n with b/B or b/m as parameter. The multiple squares were investigated only in the symmetrical arrangement of four shown in Fig. 7, but even these display a pronounced rise in drag coefficient above that for the single square (or for the four with infinite spacing) in the range for which the spacing is moderately small. According to the free-streamline analysis of two parallel plates, no such rise should occur so long as the slots are narrow and the flow through them is essentially two-dimensional. However, the flow rapidly loses its two-dimensional aspects as the gap width a becomes appreciable, and the effect must hence be inherent to the three-dimensional motion. What further increase in drag would be produced by symmetrical combinations of greater numbers of squares can only be surmised. As for an even more significant geometric pattern — sectorial division of a disk — the writers can but presume that this would approach the maximum drag that a flat surface with proper subdivision can attain.

CONCLUSIONS

One is led to conclude from these observations that the form drag of a flat body with a given projected area can be varied over a considerable range by proper control of its solidity ratio and the number and form of its component parts. So far as elementary surfaces are concerned, the disk and the long plate have the minimum and maximum drag coefficients (1.1 and 2.0, respectively). Coefficients still greater than that of the plate (at least 2.1) can be obtained even for a surface of disk-like outline by dividing the surface into plate-like elements with small spaces between them. Although the drag coefficient increases continuously with the number of elements, its rate of increase rapidly diminishes; relatively few subdivisions, therefore, will produce almost the optimum effect.

BIBLIOGRAPHY

1. Lamb, H., *Hydrodynamics*, Cambridge, 1932, p. 94.
2. Kirchhoff, G. R., ''Zur Theorie freier Flüssigkeitsstrahlen,'' *Crelle*, 1869.
3. Plesset, M. S., and Shaffer, P. A. Jr., ''Cavity Drag in Three Dimensions,'' *Journal of Applied Physics*, Vol. 19, 1948.
4. Rouse, H., *Elementary Mechanics of Fluids*, New York, 1946, p. 87.
5. Siao, T. T., ''Drag Coefficients of Multiple Plates as a Function of Solidity Ratio,'' M. S. Thesis, State University of Iowa, 1950.

Turbulent diffusion across
a density discontinuity

H. ROUSE

IOWA INSTITUTE OF HYDRAULIC RESEARCH
STATE UNIVERSITY OF IOWA, IOWA CITY

J. DODU

LABORATOIRES DE MÉCANIQUE DES FLUIDES
ÉCOLE NATIONALE SUPÉRIEURE D'ÉLECTROTECHNIQUE ET D'HYDRAULIQUE DE GRENOBLE

To the end of understanding the influence of a given agitation on superposed layers of liquid of slightly different density — of which one encounters numerous examples in nature: clear and turbid water, pure and saline water, fluid layers differing in temperature — the authors undertook a laboratory study of the phenomenon.

Their experiments have permitted the definition of a velocity of diffusion for the liquid passing from an immobile layer into a layer that is artificially agitated. In conformance with the laws of dimensional analysis, the magnitude of that velocity enters into the formation of a diffusion parameter, wich in turn is a function of a Froude number (characteristic of gravitational effects) and of a Reynolds number (characteristic of viscous effects).

In the course of the following analysis, the authors present the results of the series of experimental measurements which they made to determine the foregoing function.

They conclude that, contrary to expectation, the experiments revealed no limit of stability; however, they disclosed one important fact: the agitation of a fluid layer has no effect upon a neighboring layer of slightly different density except in the immediate vicinity of the interface between them, that interface remaining continuously defined.

Diffusion of fluid between strata of slightly different densities plays a critical role in various phenomena of meteorology, hydraulics, and oceanography. These usually involve the motion of one stratum relative to another, such as that of warm or cold layers in the atmosphere, a sediment-laden current beneath the clear water of a reservoir, superposed fresh and salt water in estuaries, and thermal streams in the ocean. In each instance the rate of intermixing depends upon an as-yet-unknown function of Froude and Reynolds numbers characterizing the relative motion. At low values of these parameters, to be sure, the stabilizing influence of gravity and viscosity should be sufficient to prevent disruption of the interface, and from rather limited experimental data it has been assumed that the border line between stable and unstable regimes is such as shown in fig. 1 [1]. Since this is also a curve of constant (i.e., zero) mixing curves of constant but finite mixing at higher Froude and Reynolds numbers might be expected to be of similar functional form. Nevertheless, no general method has as yet been established for predicting even the onset of instability, much less the subsequent diffusion rate.

A closely related phenomenon—somewhat simpler, because relative motion between the layers is not involved— is found in the modification of the oceanic thermocline during a storm. Turbulence presumably generated by the breaking of surface waves is, under certain circumstances, known to produce diffusion between thermal layers some distance below. However, little information is at hand about either the requirements of stability or the effect of the diffusion process upon the thermocline, some contending that the local temperature gradient becomes more pronounced and others that an intermediate buffer layer is formed.

In the belief that a laboratory study of this simpler occurrence might also throw some light on that involving relative motion, the senior author initiated such experiments during his year as Fulbright Research Scholar at Grenoble, France. These were carried out by the junior

254

author in the Fluid Mechanics Laboratory of the Ecole Nationale Supérieure d'Electrotechnique et d'Hydraulique with equipment loaned for the purpose by the Etablissements Neyrpic. The apparatus consisted, in brief, of two closed plastic cylinders 20 cm in diameter and 50 cm high, above which was a motor-driven differential mechanism having a speed variation from 0 to 15 rps. For the present study, one of the cylinders was equipped with an agitator (see fig. 2) in the form of a single square-meshed lattice of 1-cm metal strips 5 cm on center in a horizontal plane; an eccentric device of variable amplitude permitted the grid to be oscillated vertically in simple harmonic motion through the available frequency range. The cylinder could thus be filled with fresh and saline water (one or both layers being colored with dye), the agitator set into motion some distance above the interface between them, and the resulting diffusion observed.

Preliminary tests revealed that the turbulence produced in the upper stratum by the motion of the grid, instead of penetrating far enough into the lower stratum to form a buffer zone as anticipated, did not penetrate at all—rather, it produced irregular interfacial cusps from which streamers were rapidly lifted and diffused through the upper stratum. Whenever the agitator mechanism was stopped, to be sure, the settling of the most recently developed streamers tended to form an intermediate layer of variable density similar to that sometimes observed in the ocean. But as long as the device continued to operate, the interface would gradually fall as the concentration of the upper layer increased,

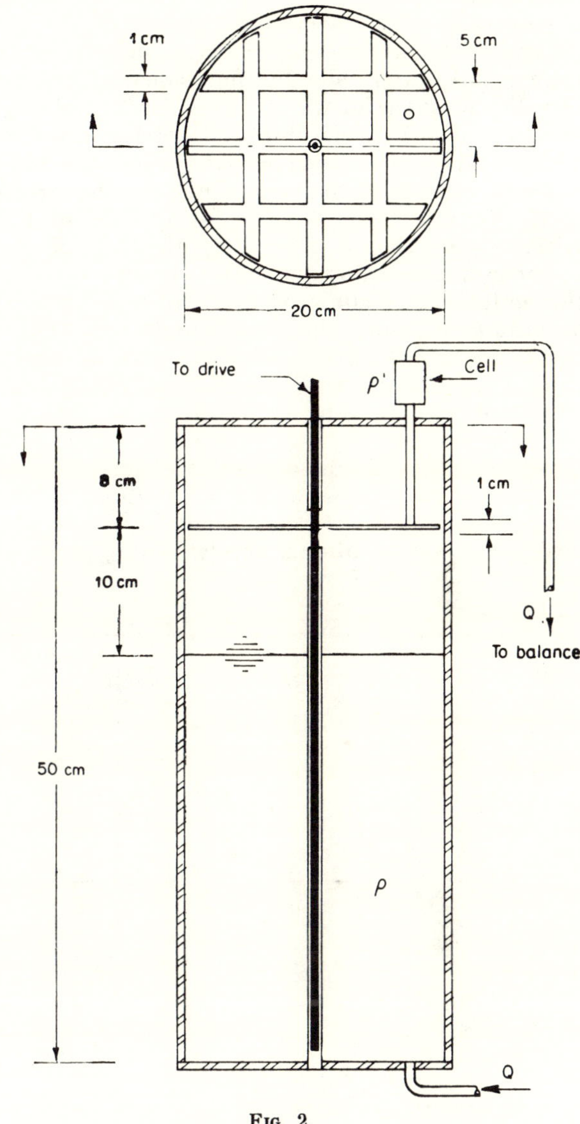

Fig. 2

Arrangement of the experimental equipment

the lower layer otherwise remaining undisturbed.

Since the intensity of the turbulence was presumed to diminish with distance from the agitator, it was decided that measurements of the diffusion rate would be significant only if the average distance between agitator and interface were held constant. Means were therefore provided of replenishing the lower fluid and removing the upper fluid at the rate necessary to maintain the interface at a predetermined level. The density of the increasingly saline effluent could then be measured continuously by passing it through a small resistance cell, and the rate of diffusion—indicated under these circumstances by the rate of efflux Q—could be

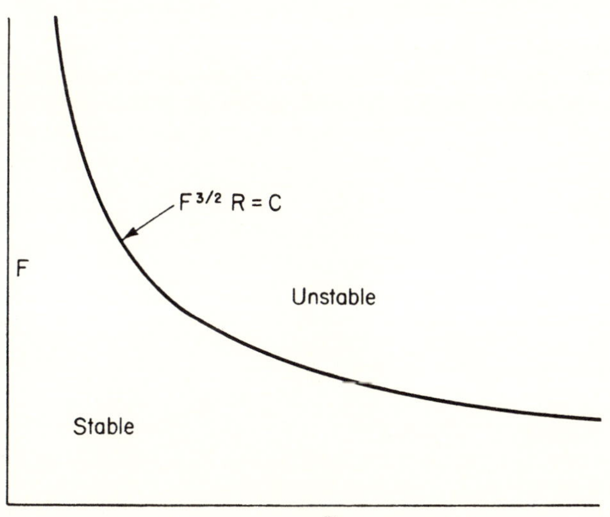

Fig. 1

Stability of the interface between moving strata of different densities.

determined by weighing the effluent collected during successive intervals of time.

Layers of any desired density and viscosity were readily preparable by the proper admixture of salt and sugar to water. The effect of changes in the position and amplitude of the agitator, which had only indirect bearing upon the phenomenon under study, was eliminated by arbitrarily placing the agitator (and the tip of the withdrawal tube) 8 cm below the top of the cylinder, maintaining the interface 10 cm below the agitator, and adhering to a constant agitator amplitude of 1 cm. The single length $h = 18$ cm, equal to the thickness of the upper stratum, thus sufficed to designate the scale of the system. The relationships to be determined were then (a) the difference $\Delta \varrho = \varrho - \varrho'$ between the measured densities of the fluids, and (b) the derived mixing rate $v = Q/A$, each as a function of the frequency n the reference density ϱ, the viscosity μ, the gravitational acceleration g, and the

time of observation t :

$$\Delta \varrho, v = f_{1,\,2}(h, n, \varrho, \mu, g, t)$$

Six carefully controlled runs with different densities, viscosities, and agitator frequencies were then made to determine these functions. For constant values of h, n, ϱ, μ, and g, $\Delta \varrho$ and v were fund to change with t as indicated in figs. 3 and 4. In every instance both v and the rate of change of $\Delta \varrho$ had finite magnitudes at the outset and thereafter increased steadily, the magnitude of v and the slope of the $\Delta \varrho$ curve eventually becoming indeterminate as $\Delta \varrho$ approached the limit zero. Control of the interfacial level, it should be remarked, thereby became increasingly difficult, for the surface disturbances finally seemed to grow without limit.

Since the tests thus indicated that each of the two diffusion functions changes continuously until this limit is reached, it appeared advisable

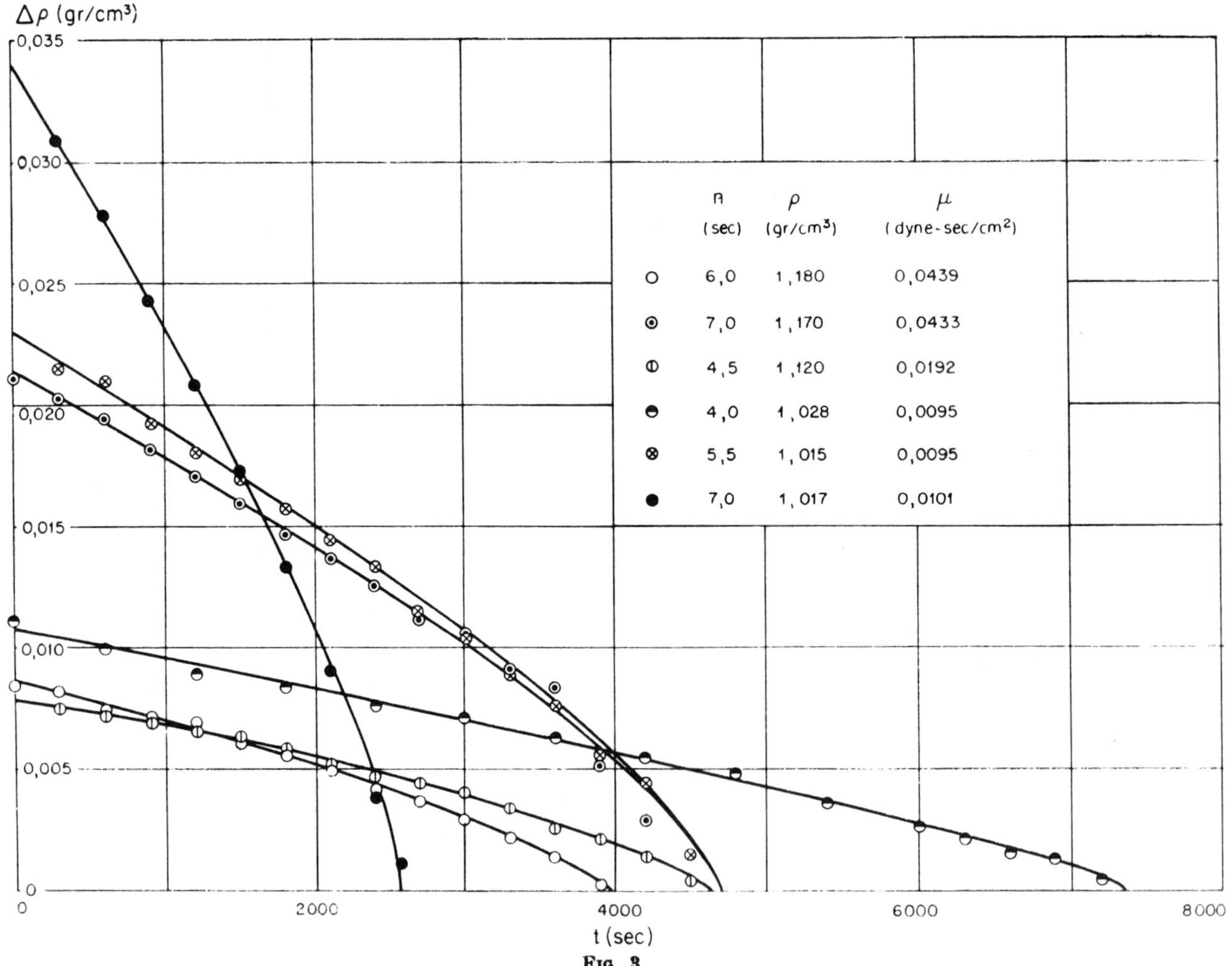

Fig. 3

Variation of the density difference with time.

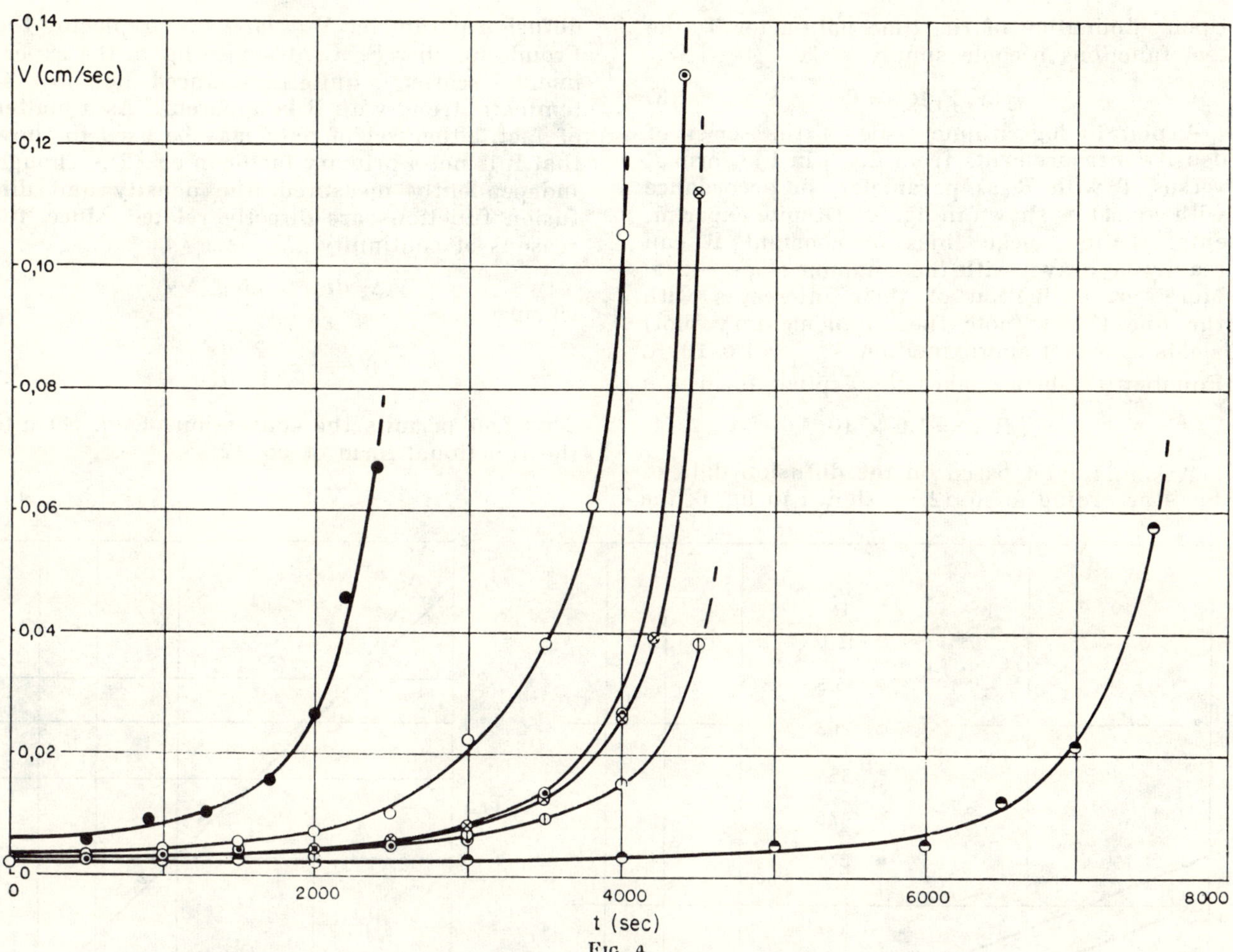

Fig. 4.

Variation of the diffusion rate with time

in the further analysis of the results to replace the absolute time t by the difference between it and its terminal value t_0, ill-defined as the latter often was. Moreover, because the experimental values of the two densities were such that $\Delta \rho$ was never greater than 3 % of ρ, inertial effects were essentially independent of changes in ρ'; $\Delta \rho$ and g could hence be combined into the single quantity $\Delta \gamma = g \, \Delta \rho$. Finally, to convert the agitator frequency and scale into a more general measure of the turbulence, the mixing coefficient ε in the vicinity of the interface was assumed proportional to the variable n and the square of the constant h — i.e., $\varepsilon = C \, n \, h^2$; for want of better information, the order of magnitude of the coefficient of proportionality

$$(C \approx 2.5 \times 10^{-4})$$

was estimated from data on the turbulence downstream from a lattice screen [2] for a velocity of flow equal to the mean relative speed of the agitator.

Through these modifications, the pertinent variables in the foregoing functions were reduced to the following :

$$\Delta \gamma, v = f_{3, 4} \, (h, \varepsilon, \rho, \mu, t_0 - t)$$

By means of the II-theorem of dimensional analysis, these variables may be combined into four dimensionless groups—a Froude number, a diffusion parameter, a Reynolds number, and a time parameter :

$$F = \frac{\varepsilon}{\sqrt{h^3 \, \Delta\gamma/\rho}} \qquad V = \frac{vh}{\varepsilon} \qquad R = \frac{\varepsilon}{\mu/\rho}$$

$$T = \frac{(t_0 - t) \, \varepsilon}{h^2}$$

These are related through the functions

$$\varphi_1 \, (F, R, T) = 0 \qquad\qquad (1)$$

$$\varphi_2 \, (V, R, T) = 0 \qquad\qquad (2)$$

Upon elimination of the time parameter T, the two functions become simply

$$\varphi_3 (V, F, R) = 0 \qquad (3)$$

A plot to logarithmic scale of the series of density measurements from fig. 3 in the form F versus T with R as parameter, in accordance with eq. (1), is shown in fig. 5. Despite experimental scatter, mean lines of constant R can readily be drawn with the common slope —2/5. Moreover, evaluation of their intercepts with the line $F = 1$ (note the supplementary plot) yields as a first approximation $T_{F=1} = 1.6/10^5 R$. Equation (1) hence takes the explicit form

$$F^{5/2} R T = 1.6 \times 10^{-5} \qquad (4)$$

A similar plot based on the diffusion data of fig. 4 according to eq. (2) is shown in fig. 6, the diffusion parameter $V = vh/\varepsilon$ now replacing the Froude number F. Contrary to fig. 5, the experimental scatter is quite pronounced, but no systemmatic trend with R is apparent. As a matter of fact, either set of data may be used to show that R is not a primary factor in eq. (2). Though independently measured, the density and diffusion functions are directly related, since, for reasons of continuity,

$$V \Delta \gamma \, dt = - h \, d \, (\Delta \gamma)$$

whence

$$V = - \frac{2}{F} \frac{dF}{dt}$$

This fact permits the conversion of eq. (4) into the functional form of eq. (2),

$$V T = 0.8 \qquad (5)$$

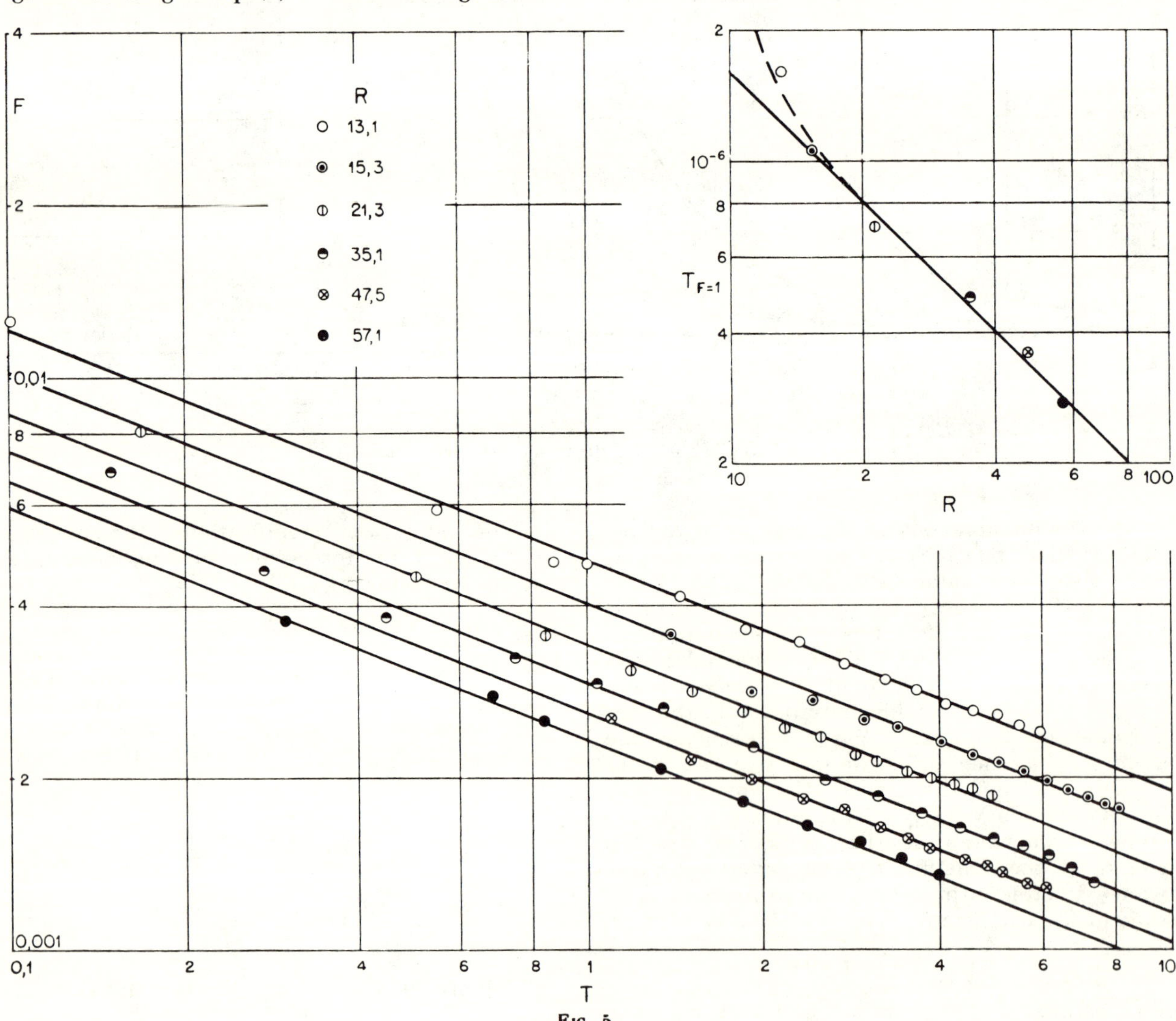

Fig. 5

The Froude number as a function of the time parameter and the Reynolds number.

which is represented by the single line plotted in fig. 6. The rather surprising simplicity of eq. (5) actually does not indicate a lack of dependence of V upon F and R, but merely the fact that any change in V produced by variation of F or R is counterbalanced by a reciprocal change in T.

Both the experimental scatter in fig. 6 and the incomplete agreement between the diffusion measurements (represented by the points) and the density measurements (represented by the line) warrant further comment. Despite the effective lag of the indicator, the density data are considered more dependable than the diffusion data, because the efflux rate was determined from incremental volumes collected in finite intervals of time and hence—unlike the density indication—was very sensitive to slight errors in the control of the interfacial level; this was particularly true toward the end of a run, and

the values of t_0 used in the diffusion computations were the more consistent ones of fig. 3 rather than those of fig. 4. However, another source of discrepancy lay in the density distribution itself. No instantaneous variation in density with elevation could be detected in the upper stratum with the rather sluggish resistance cell that was used for the general density indications, yet a gradient must have existed in order to maintain further diffusion of the fluid transported across the interface. Although the location of the outlet at the neutral level of the agitator tended to provide a first approximation to the mean density of the stratum, the fact that the point of true mean density was always at a somewhat lower elevation tended to make the diffusion rate computed from the density measurement slightly greater than that actually measured.

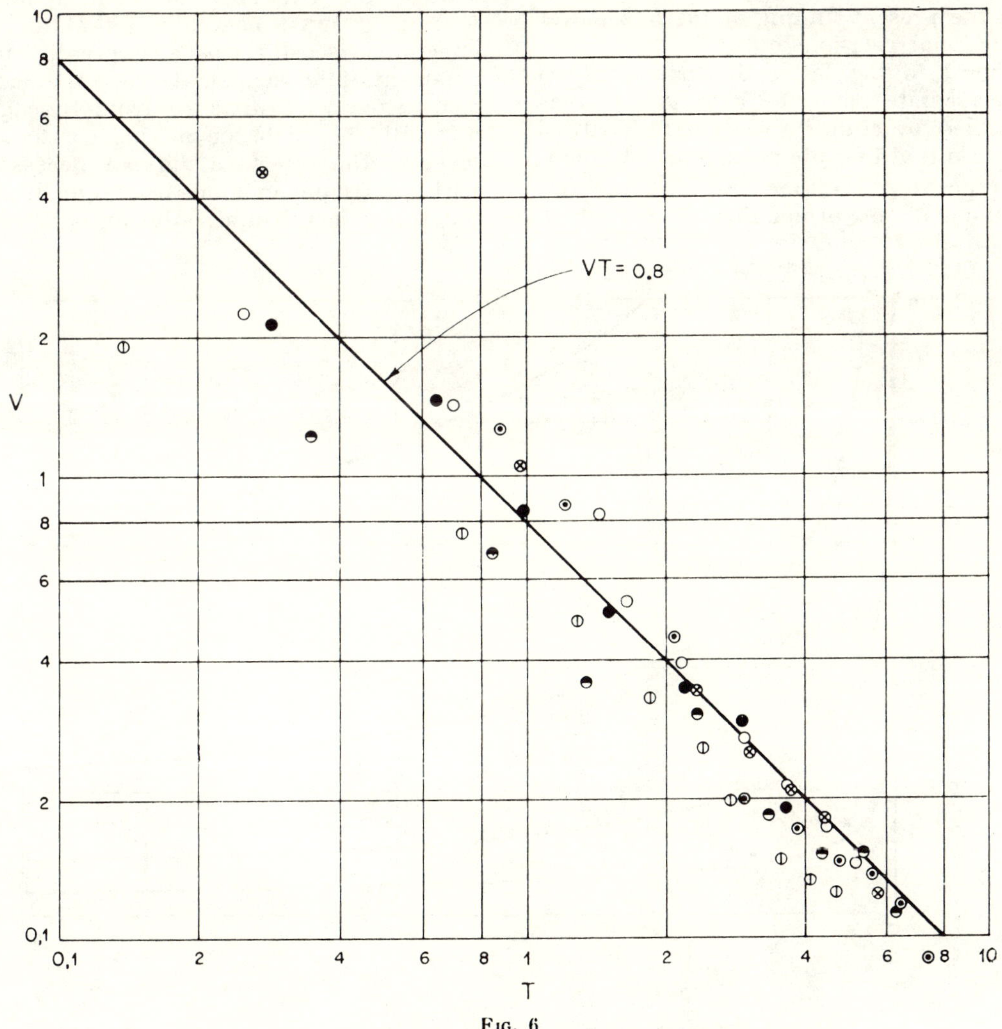

Fig. 6

The diffusion parameter as a function of the time parameter.

Under the circumstances, it may be concluded that eqs. (4) and (5) together describe the phenomenon under consideration with a reasonable degree of approximation. The more pertinent function represented by eq. (3) may hence be obtained from them by elimination of the quantity T, with the following result :

$$\frac{V}{F^{5/2}\,R} = 5 \times 10^4 \qquad (6)$$

For purposes of comparison with fig. 1, eq. (6) is plotted in fig. 7 in the form F versus R with V as parameter. At once evident is the fact that, as previously surmised, lines of constant relative diffusion are all similar functions of F and R. But whereas the only line of constant diffusion appearing in fig. 1 was that for the limiting value of zero, it is just this limit that is nonexistent in fig. 7; and there is an appreciable difference between previous indications of the 3/2 power and the present finding of the 5/2 power in the functional relationship.

With regard to the latter disparity, only a power intermediate between these two can as yet be justified by simple reasoning. That is, the assumption that the rate of increase of potential energy per unit volume, $v\,\Delta\gamma$, is proportional to the rate of production of turbulent energy by the agitator, $\rho\,n^3$, will lead to the form $V \sim F^2$. (The effect of viscosity can be introduced only by a still more arbitrary assumption.)

As for the former disparity, the indication that a finite rate of mixing may be expected to prevail regardless of the magnitudes of the Froude and Reynolds numbers would be contrary to previous indications as well as to reason itself were it not for one very essential distinction. In the present experiments the state of turbulence was independently imposed upon the interface, whereas in the case of relative motion it varies with the intensity of shear between the strata. There is, however, a slight restriction upon the generality of this conclusion. The displacement of the point for the lowest Reynolds number in the inset of fig. 5 indicates that the function may approach a limit according to the broken curve, rather than continue indefinitely according to the straight line used as the basis of eq. (4). Although the existence of such a limit seems logical, it must be realized that an unwanted effect of viscosity was also present in the experiments to a variable degree : the decay of the turbulence with distance from the agitator, as a result of which the mixing coefficient at the level of the interface did not necessarily vary with the frequency alone as assumed. Allowance for this effect would only strengthen the con-

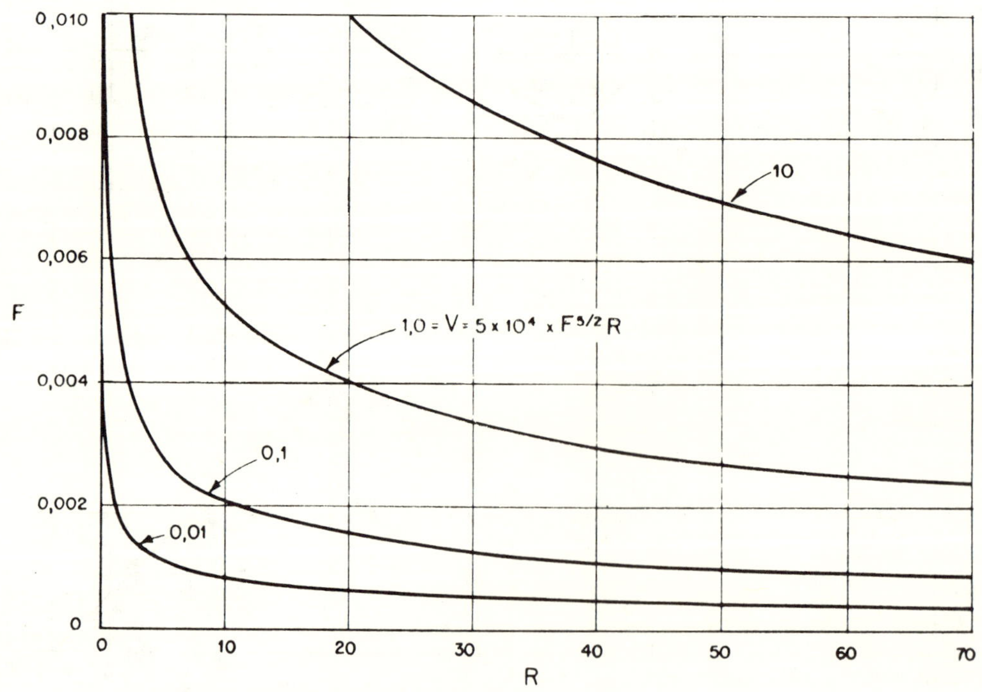

Fig. 7

Generalized plot of the diffusion function.

clusion that no limit of zero mixing was indicated by the experimental results.

Whatever the imperfections of the experiments herein described, they have disclosed two pertinent facts. First, turbulence generated in one of two fluid strata which are otherwise at rest neither penetrates the second stratum nor produces a buffer layer between the two; rather, fluid is entrained from the surface of the nonturbulent stratum and diffused throughout the turbulent zone. Second, so long as the turbulence is produced by some external mechanism, for small density differences a limit of stability comparable to that which characterizes interfacial shear is not to be expected; instead, there appears to be a finite rate of diffusion across the interface for any Froude and Reynolds numbers. Moreover, although the functional relationships derived from the experiments are linked quantitatively with the particular geometry of the apparatus, their form should nevertheless indicate the relative roles played by the Froude and Reynolds numbers no matter what the boundary proportions might be.

REFERENCES

[1] *Engineering Hydraulics*, edited by Rouse, New York, 1950, p. 762.
[2] *Engineering Hydraulics*, p. 92.

Discussion of "A Note on the Manning Formula"

VEN TE CHOW

[*Trans.*, **36**, 688, 1955]

Hunter Rouse (Iowa Institute of Hydraulic Research, State University of Iowa, Iowa City, Iowa, received Nov. 28, 1955)—Several statements with regard to the Manning formula that have appeared in this journal in recent months warrant further comment in order that the record may not be misunderstood.

To amplify Chow's admittedly second-hand account of the origin of the formula, the writer submits the following quotation [*Rouse* and *Ince*, 1954–5] based upon the paper itself and upon biographical information supplied by the Institution of Civil Engineers of Ireland:

"Quite a different tack [from those of Kutter and Bazin] had been followed in 1868 by Philippe-Gaspard Gauckler (1826–1905), Inspector General of the Ponts et Chaussées, who—in despair of finding a single formula which would describe the flow in all types of channels—proposed two for use in different slope ranges

$$V = \lambda_1 R^{4/3} S \quad \text{for} \quad S > 0.0007$$

$$V = \lambda_2 R^{2/3} S^{1/2} \quad \text{for} \quad S < 0.0007$$

An expression identical to the second of these was devised and investigated by the Irish engineer, Robert Manning (1816–1897), two decades later. Manning had been born in Normandy the year following the battle of Waterloo, in which his father had taken part. As chief engineer of the Office of Public Works, he was responsible for various drainage, inland-navigation, and harbor projects, and he also served a term as President of the Institution of Civil Engineers of Ireland. Although, judging from his writings, his primary interest was hydrology, in 1889 he presented to the Institution a paper 'On the Flow of Water in Open Channels and Pipes' in which he sought to remedy the various defects he believed to exist in previous resistance formulations. Apparently without knowledge of the Gauckler proposal, he showed that a relationship of the form

$$V = KR^{2/3}S^{1/2}$$

was in better agreement with available data than any in general use. However, not only did he consider the first fractional power inconvenient to use, but he clearly recognized the fact that '. . . if modern formulae are empirical with scarcely an exception, and are not homogeneous, or even dimensional, then it is obvious that the truth of any such equation must altogether depend on that of the observations themselves, and it cannot in strictness be applied to a single case outside them.'

"Manning therefore discarded formulas of the type now associated with his name, proposing instead the dimensionally homogeneous relationship

$$V = C \sqrt{gS} \left[R^{1/2} + \frac{0.22}{m^{1/2}} (R - 0.15m) \right]$$

Herein the factor C is not the Chezy coefficient but a pure number 'which varies with the nature of the surface,' and m (which he introduced for dimensional as well as physical reasons) 'the height of a column of mercury which balances the atmospheric pressure.' At no point did Manning suggest use of the Kutter n in either formula, the coefficient in question being obtained—much as Chezy had originally proposed—from known data for a channel of the type under consideration. What we now call the Manning formula was thus neither recommended nor even devised in full by Manning himself, whereas his actual recommendation received little further attention."

Although Manning did note that "the value of the reciprocal of $[K]$ . . . corresponds closely with that of n," who it was that first began to use the Kutter n in the Gauckler-Manning type of equation is not yet clear, nor are the circumstances known which led to the adoption of the same numerical values of this factor in both the metric and the English system. In any event, the Chezy C thereby became expressible in terms of what were apparently a constant number K, a variable number n, and the $\frac{1}{6}$ power of a length R

$$C = K \frac{R^{1/6}}{n}$$

When it eventually became understood that the Chezy and the Manning relationships could be dimensionally homogeneous only if each side of this equation had the dimension $L^{1/2}/T$, an argument began as to how this dimension should be distributed among the quantities at the right, opinions ranging all the way from entire absorption by K to entire absorption by n.

That the argument still continues is evident from *Kolupaila's* [1955] recent discussion of a paper on the Manning coefficient by *Boyer* [1954], in which he states that ". . . opposite to the widespread fallacious judgment, the Manning coefficient n has no dimension: the same values are used in both metric and English system and that can be easily checked by conversion factors." Now Boyer had originally followed the teaching of the writer

"

[*Rouse*, 1946, p. 218] that the Manning roughness factor must have the dimension $L^{1/6}$ in order that the ratio $n/R^{1/6}$ assume its proper role as relative-roughness parameter. In his closing discussion, however, *Boyer* [1955] seemed almost to welcome Kolupaila's claim to the contrary, with the words: "The author is interested in Kolupaila's statement that the Manning coefficient n has no dimension. It has long been his belief that even in the metric system, in which this formula was derived, the coefficient was dimensionless; if dimensional homogeneity is maintained."

Whether or not the factor n has the same set of numerical values in the two different systems is actually no indication of its dimensional characteristics, for the apparent values of any such series can be changed at will merely by introducing a compensating coefficient. In the present instance this coefficient has been absorbed numerically by the coefficient K. On the other hand, whereas the dimension $L^{1/6}$ of the roughness measure actually belongs to both the compensating coefficient and the remaining factor together, from the mathematical point of view it matters little whether it is placed entirely with the one or the other. Kolupaila is right insofar as he says that its arbitrary retention by the factor n, when the latter's numerical values are held the same in both systems, leads to apparent anomalies in values of $R^{1/6}/n$ for data in the two systems. Nevertheless, the physical inconsistency of treating n as a pure number is apparently even more dangerous, because it leads to such errors [*Kolupaila*, 1955] as the plotting of the velocity ratio $V_{0.2}/V_{0.8}$ versus n alone; since the velocity ratio is a function solely of the resistance coefficient C [*Rouse*, 1946, pp. 197–200], this is tantamount to saying that the resistance coefficient depends upon the absolute rather than the relative roughness (that is, that C is independent of the depth of flow).

It is obvious that the source of the misunderstanding originally lay in the use of the same Kutter values in both the English and metric forms of the relationship, and that these values are now so much a part of engineering usage throughout the world that their proper conversion would no longer be acceptable. But it should also be apparent that, regardless of the actual numerical relationship between the roughness factor and the measurement in feet or in meters of the roughness that it characterizes, n must continue to be recognized in either system as an absolute (that is, dimensional) rather than a relative (that is, non-dimensional) roughness measure. Under these conditions it must be conceded to have the same dimension as the quantity to which it is related: length to the one-sixth power. The magnitude of the coefficient K will continue to depend upon the conversion of hydraulic radius and velocity from one system to another and the apparent lack of conversion of n; but the dimension of K must remain $L^{1/2}/T$ (that is, the same as that of $g^{1/2}$), however troublesome this may make the comparison of values of the non-dimensional ratio $R^{1/6}/n$ between systems.

References

Boyer, M. C., Estimating the Manning coefficient from an average bed roughness in open channels, *Trans. Amer. Geophys. Union*, **35**, 957–961, 1954.

Boyer, M. C., Closing discussion of *Boyer* [1954], *Trans. Amer. Geophys. Union*, **36**, 916, 1955.

Kolupaila, S., discussion of *Boyer* [1954], *Trans. Amer. Geophys. Union*, **36**, 912–916, 1955.

Rouse, H., *Elementary mechanics of fluids*, John Wiley and Sons, 376 pp., 1946.

Rouse, H., and Ince, S., History of hydraulics, supplement to *La Houille Blanche*, 9–10, 250 pp., 1954–5.

Experiments on two-dimensional flow over a normal wall

By MIKIO ARIE

Faculty of Engineering, Hokkaido University

and HUNTER ROUSE

Iowa Institute of Hydraulic Research, State University of Iowa

(*Received 3 January* 1956)

SUMMARY

Measurements of the velocity, pressure, and turbulence behind a series of normal plates in the uniform test section of an air tunnel are described, the oscillation of the wake being prevented in all but one test through use of symmetrically located tail plates. By a combination of experimental and computational techniques, details of the pattern of flow over a wall on a plane boundary in an infinite fluid are closely approximated. A significant difference is indicated between the characteristics of such a flow and those of the flow past an isolated plate with oscillating wake.

INTRODUCTION

For want of experimental information about the characteristics of flow past a sufficiently great variety of boundary shapes, one is forced either to perform specific tests upon each body design that comes into question, or to estimate the expected characteristics from what is known about flow past the individual elements of which the body consists. Under some conditions the latter method is quite effective; often the purely qualitative indication that it provides is far better than none at all, and sometimes the combination of available data with mathematical analysis permits the indication to be of quantitative value as well. Under other conditions the information at hand is merely sufficient to throw serious doubt upon even the qualitative value of such a procedure.

A case in point is the utilization of experimental data for two-dimensional flow past a cylindrical body to predict the conditions of motion that would prevail if a boundary were to exist at the plane of symmetry. Assume, for example, that it was desired to estimate the force exerted by the wind upon a wall, or the effectiveness of such a wall in either shielding something behind it or in promoting the diffusion of some substance in its wake. Drag coefficients and eddy patterns are well known for rectangular plates (Fage & Johansen 1927, Goldstein 1938) but it is also known (Roshko 1955) that interference with an oscillating wake is likely to change the primary nature of the flow.   Just how great this change might be in any particular case can be determined as yet only by experiment.

Whereas the wake behind an immersed body is of basic importance in such professions as aeronautics, ballistics, and marine engineering, it is the wake behind a boundary irregularity, such as the wall, that is of primary interest in hydraulics.   Whether the wall typifies roughness elements or part of an orifice or valve makes little difference; and the standing eddy in its wake also has much in common with those which form at any abrupt changes in conduit section.   Yet so little is known about their geometrical proportions, about the dynamics of the mean flow within them, and about the generation of turbulence along their borders, that the experimental analysis of such standing eddies has received the attention of the Iowa Institute of Hydraulic Research for nearly a decade.   The most recent in this series of investigations has been a study of the wake behind a simple wall (Arie 1955), the purpose of the present paper being to make generally available the most essential of the results.

Flow past a plate that is fully surrounded by fluid differs in several ways from flow past one that is in contact with a continuous boundary.   First of all, the former may be considered to lie in the path of essentially irrotational flow, whereas the latter will be immersed in a boundary layer of indefinite thickness.   Secondly, as has already been noted, the flow in the wake of an isolated plate is free to oscillate about the plane of symmetry, whereas such oscillation is wholly prevented in the vicinity of a rigid longitudinal boundary.   Now the lateral freedom or restraint of the wake marks a dynamically basic distinction between the two cases, whereas the boundary layer is a secondary factor that is not determined by local conditions. Hence, in the experiments under discussion, boundary layer development both upstream and downstream was obviated by placing the plate at the axis, rather than at the floor, of an air tunnel, the lateral boundary restraint at the rear being effected by means of a tail plate which was only slightly longer than the standing eddy.

Whether a plate fully surrounded by the fluid or one in contact with a boundary is simulated in a tunnel, the presence of the test section floor and ceiling will introduce an additional threefold effect.   First, these surfaces will correspond geometrically to planes of symmetry between the given plate and standing eddy, on the one hand, and their first images in a series having a finite lateral spacing, on the other; the velocity of the surrounding field being increased thereby.   Second, there will be an additional augmentation of velocity as a result of the development of the boundary layer along the tunnel surfaces.   Finally, despite the use of a foreshortened downstream boundary, the conversion of energy in the wake of the test plate will produce something akin to a boundary layer effect, which will cause the velocity of the flow outside the wake to become still higher than it would be in an infinite fluid.

Three courses were thus open to the investigator.   The pattern of flow could be determined as a function of the relative spacing of a series of plates, with the double aim of permitting extrapolation to the infinite case and of providing supplementary information about flow between opposite walls in a passage of finite size.   Secondly, by means of various experimental artifices, an effort could be made to simulate the infinite case itself without recourse to extrapolation.   Finally, the data obtained for any spacing could be recomputed for infinite spacing by means of an analytical approximation.   In actuality, these three courses were followed in sequence, the first serving as a guide to the partial refinements of the second, and the final correction proceeding analytically.

EXPERIMENTAL PROCEDURE

The air tunnel used for the experiments in question was of the closed circuit type, with a uniform test section 3 feet high, 3 feet wide, and 10 feet long.   Air speeds could be varied at will from 10 to 50 feet per second, but for ease in measurement all tests were made in the upper part oi this range.   The main air speed was indicated by the change in pressure at the bell entrance to the test section; the corresponding pressure-velocity relationship had been determined through test-section traverses with a Prandtl-type Pitot tube, which in turn had been calibrated in the irrotational core of a jet from a 6-inch rounded orifice.   No effort was made to reduce the turbulence of the flow, beyond the use of vanes at the turns and a fourfold reduction in cross-sectional area between the plenum chamber and the test section; the root-mean-square velocity fluctuation was about $1\frac{1}{2}\%$ of the mean velocity of flow.

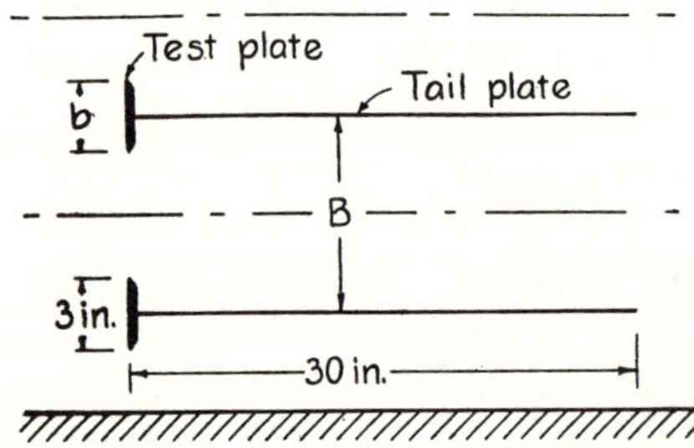

Figure 1. Definition sketch of plate arrangement.

All test plates were introduced into the tunnel approximately midway between the ends of the test section, completely spanning its 3-foot width. They were invariably 3 inches in effective height, 1/4 inch in thickness, and bevelled at an angle of 60° along the edges on the rear side.   All but one were of tempered hardboard, the one being of carefully machined brass. Each was guyed with piano wire, at its one-third points, to the floor and ceiling of the tunnel to maintain its alignment and eliminate all possibility of vibration.   The tail plates (see figure 1), which were of 1/16-inch sheet aluminium, were attached to the test plates at mid-height, and likewise guyed in place.   They extended a distance of 30 inches to the rear, which had been found by observation to be longer than the eddy by a sufficient amount for the size of the latter to be unaffected by a further lengthening of of the plate. By successively introducing dummy plate assemblies at planes of symmetry on either side of the centrally located test plate, the ratio B/b of spacing to width could be varied through the consecutive values of 12, 6, 3, etc.

Because of the two-dimensional nature of the flow, it was possible to make measurements of the magnitude and direction of the mean velocity at any point by means of one of two alternative Pitot cylinders. Each was constructed of 1/8-inch brass tubing mounted horizontally on a fork-shaped support with bearings 6 inches on centre. One cylinder contained a 0·015-inch hole on each side of a central plug (see figure 2), the two portions

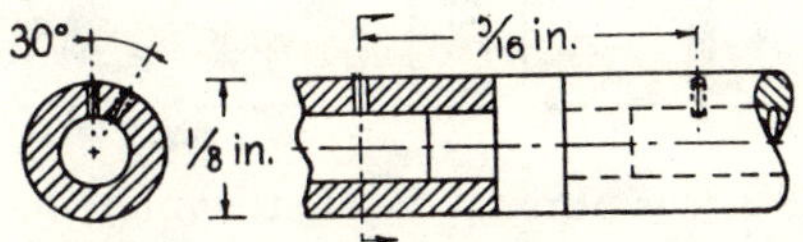

Figure 2. Pitot-cylinder details.

of the tube having been rotated differentially about their common axis till the holes lay at an angle of 30° to each other in the normal plane. A protractor at either bearing permitted the common angular position of the holes to be read to within about 0·2°. The other cylinder differed from the first only in the angle—now 180°—between the holes. After careful calibration of each cylinder, the magnitude of the velocity could be computed from the differential pressure between the respective holes, the 180° cylinder being oriented to yield the maximum reading, and the 30° cylinder being turned so that first one and then the other hole lay directly upstream (in the latter instance the direction of the flow was first determined by noting the angle at which the differential reading was exactly zero). The 30° cylinder was used so long as the velocity gradient was not excessive; since, in zones of high gradient, this cylinder was subject to a pronounced error in both direction and magnitude of the velocity, the 180° cylinder was then employed, and the direction determined by successive steps in plotting the streamlines.

In the zones in which both the inclination and the magnitude of the velocity vector were directly measurable, it was a simple matter to determine and plot, from vertical traverses with the 30° cylinder, the distribution of the longitudinal component of velocity over a series of normal sections. Graphical calculations of the areas $\psi = \int u\,dy$ enclosed by these curves (figure 3), as a function of distance normal to the tail plate, then permitted successive streamlines $\psi = C$ to be located for any selected increment $\Delta\psi$. In those zones for which only the magnitude of the velocity could be measured, this was first assumed to be indicative of the longitudinal component. Evaluation of the streamline location according to the foregoing method yielded the inclination of the velocity vectors as a first approximation, and a second approximation was usually sufficient to determine the flow pattern with acceptable precision. Figure 4 shows the pattern obtained from the single run made without tail plate for purposes of reference.

This method of determining the flow pattern was followed for B/b equal to 12 and to 6 without undue experimental difficulty. All attempts to repeat the measurements for B/b equal to 3 met with no success, however,

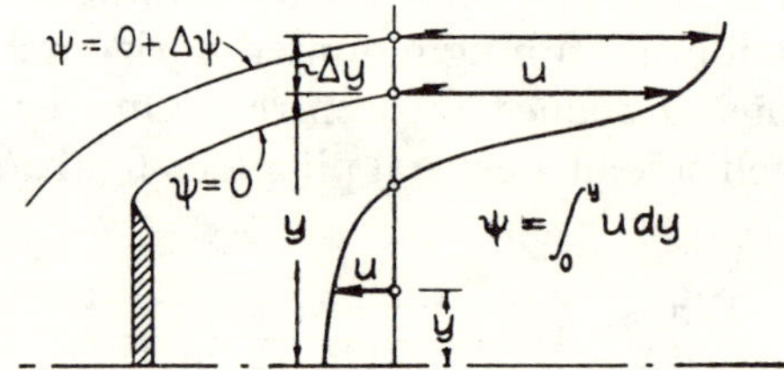

Figure 3. Determination of streamline location.

as the pattern was then not only unsymmetrical about the axis of any plate but unstable as well. This situation is in approximate accord with previous experience, flow through lattice screens being known to become unstable as the solidity ratio approaches the magnitude 0·5. For this reason, no attempt was made to carry the study any farther in the direction of decreasing spacing, which meant that a sufficient number of cases to

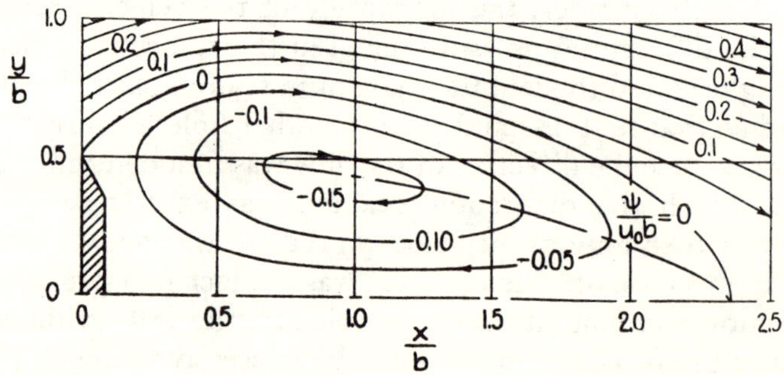

Figure 4. Approximate pattern of mean streamlines for flow without tail plate
$B/b = 12$.

permit extrapolation to greater spacing could not thus be realized. A different tack was hence followed in approximating the desired conditions of infinite spacing.

Since the profile of the standing eddy behind the test plate was roughly elliptical in form (figure 5), it was assumed that the streamlines at a considerable distance would approximate those of potential flow past either an ellipse or a Rankine oval of comparable dimensions. The latter being somewhat easier to describe mathematically, the location and strength of the source and sink were determined so as to yield a profile coinciding most closely with the measured eddy profile for $B/b = 12$. The resulting stream function had the form

$$\psi = -yu_0 + 0.56\,bu_0\left[\tan^{-1}\frac{y}{x - 3.4\,b} - \tan^{-1}\frac{y}{x + 3.4\,b}\right],$$

the profile of the oval corresponding to $\psi = 0$. From this relationship it was possible to evaluate the coordinates of any desired streamline in the vicinity of the test-section wall; the maximum deviation of the streamline at the approximate lateral distance of 1·5 feet was found to be about 1·5 inches. Because of the boundary layer development along these walls, however, a still further change in their shape would be necessary to eliminate this additional wall effect. The displacement thickness of the boundary layer as a function of distance along this tunnel was hence evaluated by a combination of velocity measurement with extrapolation, in accordance with known boundary layer relationships; its entire variation was found to be slightly less than 1/4 inch over the 10-foot length of test section. False boundaries of hardboard were then introduced along the ceiling and floor of the test section; they were curved according to the stream function of the eddy profile, flared by an amount equal to the displacement thickness of the boundary layer, and smoothly joined to the bell entrance of the test section.

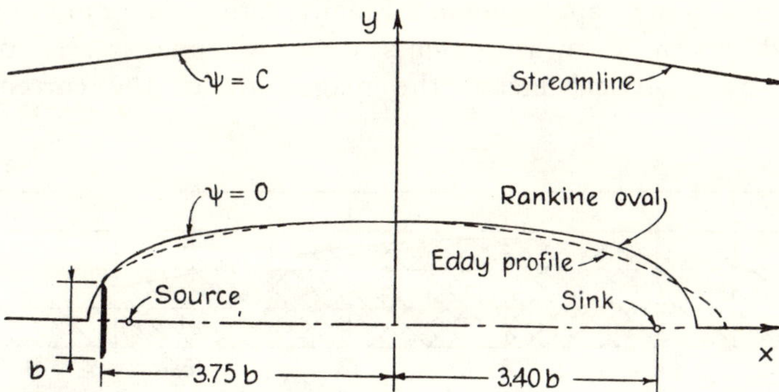

Figure 5. Derivation of stream function for Rankine oval having proportions of standing eddy.

With the falsework in position, the velocity distribution over successive vertical sections throughout the eddy region was again measured with the Pitot cylinder. In addition to such velocity traverses, measurements were also made of the pressure distribution at the same sections, and of the pressure over the faces of the test plate and tail plate. Pressure measurements at the plates themselves required simply the introduction of piezometer orifices along the one side, the velocity measurements being restricted to the same side in order to avoid the disturbing effect of the piezometer connections on the other. The pressure traverses in a plane normal to the two plates involved the use of a third plate, 1/8 inch thick, 5 inches wide (in the longitudinal direction), and about 18 inches high, sharply bevelled on the leading edge, and having a series of piezometer orifices on one side in a vertical row, which could be placed at any section. Finally, measurements were made with the constant-temperature type of hot-wire anemometer (Hubbard 1954) at the same vertical sections, the wires and circuit being so arranged as to yield the root-mean-square values of the three components of fluctuation and the cross product of the components in the plane normal to the two test plates.

ANALYSIS OF MEASURED DATA

Had the revision in tunnel profile actually been sufficient to simulate infinite-fluid conditions, the experimental results would have yielded directly the patterns of velocity, pressure, and turbulence produced by flow over a normal wall.   However, it will be recalled that the boundary profile was determined on the assumption of irrotational flow around a Rankine oval, whereas flow over a wall—though it may be essentially irrotational at the outset—departs from this state more and more with distance downstream, as a result of the pronounced shear and generation of turbulence at the border of the standing eddy.   In other words, although a fairly successful effort had been made to compensate for the presence of the eddy and the development of the boundary layer on the tunnel walls, there remained uncompensated the effect of the wake of the plate itself, the importance of which was not at once appreciated.   Finally, rather than modify the tunnel profile and repeat the measurements a second time, it was decided to determine by analytical means the magnitude of the corrections still required.

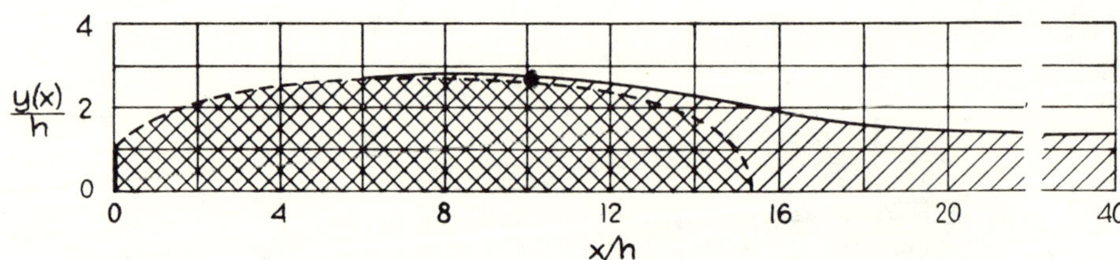

Figure 6. Displacement effect of standing eddy and turbulent wake.

With reference to figure 6 (in which the wall height h is introduced as reference length), the total effect of the wall upon the surrounding flow could be considered equivalent to that of the combination of the Rankine body (shown by double cross-hatching) with an additional tail section (shown by single cross-hatching) corresponding to the displacement thickness of the flow in the wake.   The total thickness of the combination was assumed to be given to a first approximation by

$$\delta* = \int_0^\delta \frac{u - u_{\max}}{u_{\max}} \, dy,$$

the quantity δ representing the value of y, just outside the wake, at which the velocity becomes a maximum.

Just as the profile of the eddy was simulated by the combination of a source and a sink with a uniform flow, the composite profile of the eddy and the tail section (or the profile of either one alone) could be realized by a distribution of sources and sinks, or, more conveniently, by a distribution of doublets (a method analogous to that of Landweber 1951), in place of the single source and sink of the Rankine body. For the case of an infinite fluid the stream function would then have the form

$$\psi = -yu_0 - \int_a^\infty \frac{m(t)\,y}{(x-t)^2+y^2}\,dt,$$

in which $m(t)$ is the doublet strength per unit length as a function of distance t along the line of symmetry and a is the abscissa of the initial point of the distribution which will yield the desired profile for the condition $\psi = 0$. For the case of flow that is confined between parallel walls (i.e., for a series of similar bodies) with the spacing B, on the other hand, the stream function would have to be written as

$$\psi = -yu_0 + \frac{\pi}{B}\int_a^\infty \frac{m(t)\sin 2\pi y/B}{\cosh 2\pi(x-t)/B - \cos 2\pi y/B}\,dt,$$

the doublets being distributed symmetrically along a series of parallel axes midway between the walls.

Because the tunnel boundaries in the present experiments were formed according to the approximate stream function of the eddy alone, the differences between the velocity distributions indicated by the proper space derivatives of these two equations must be due almost entirely to the tail section or wake. Evaluation of the difference, in the form of an approximate correction factor, hence depended upon determination of the doublet distribution $m(t)$ for the tail section alone. Fortunately, the elements shown in figure 6 were sufficiently slender for a considerable simplification (again by analogy with the method of Landweber (1951)) to be made. Thus, upon replacing $m(t)$ by $m(x)$ with $\psi = 0$ in the stream function for the infinite fluid, it could be shown that, with acceptable accuracy,

$$m(x) = -u_0 y(x)\left[\tfrac{1}{2}\pi + \tan^{-1}\frac{x-a}{y}\right]^{-1}.$$

The computed distribution of $m(x)$ for the tail section (and as well for the composite profile, with a value of a that would yield a satisfactory approximation in the vicinity of the plate) was that shown in figure 7. Use of the values indicated by the full curve thus permitted calculation of the factor by which the measured velocity outside the wake should be reduced to correspond with that of the infinite case. (Use of the composite values, it should be noted, will lead to an approximate stream function for the infinite case.)

The magnitude of the resulting corrections may be judged from figure 8 *a,* in which are superposed the longitudinal components of the measured velocities (the points themselves) and the corrected values (the distribution

curves) at every other measured section. Also shown are the streamlines
determined from the corrected distribution curves in the region of the eddy.
From these it will be seen that it was actually of little moment whether or
not the correction (based upon the assumption of irrotational flow) was made
in the rotational region, where the velocity was already quite low. As a
matter of fact, the velocity correction was so small in all zones as to have
almost no effect upon the flow pattern itself. Its effect upon the pressure
distribution, on the other hand, was considerable.

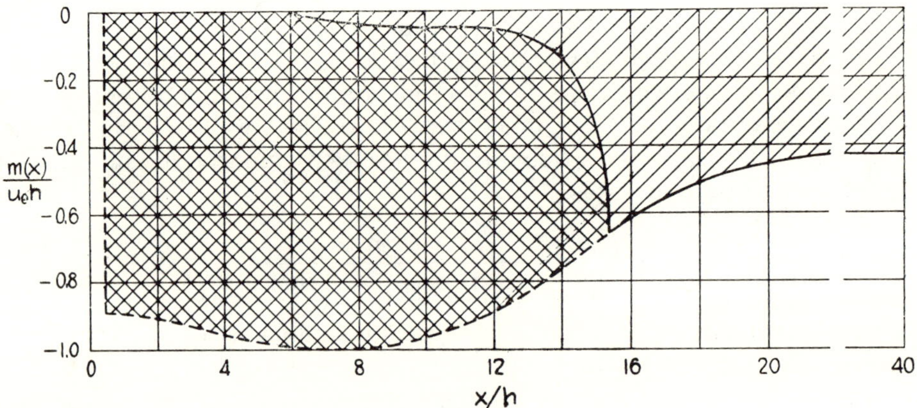

Figure 7. Distribution of doublet strength for eddy and wake displacement
of figure 6.

Figure 8 *b* reproduces the actual measurements of the pressure intensity
as points and the corrected values as continuous curves plotted to the left
of the respective sections (i.e., negatively). Once again the correction
was strictly applicable only outside the wake, for it was based upon the
assumption of irrotational flow and constancy of the sum $\frac{1}{2}\rho(u^2+v^2)+p$.
Unlike the velocity correction, however, that of the pressure could by no
means be ignored within the wake, since there it was evidently a maximum.
For want of a more precise method of estimating it, the customary boundary
layer assumption of negligible pressure change over the distance δ had to be
made; in other words, the correction at any point in the eddy was assumed
to be the same as that just outside at the same cross-section. That the
probable error in this assumption was small may be concluded from the fact
that the measurements themselves varied only slightly across the eddy.

With the corrected magnitudes of the velocity and pressure at hand
throughout the zone in question, it was a simple matter to compute and
plot the distribution of total pressure (i.e., as would be read at the stagnation opening of a Pitot tube) relative to that in the ambient fluid, which is
also shown in figure 8 *b*. Comparison of this sequence of curves with the
locus of points of maximum velocity will indicate that the latter corresponds
rather closely to the approximate limit of the zone of irrotational flow.

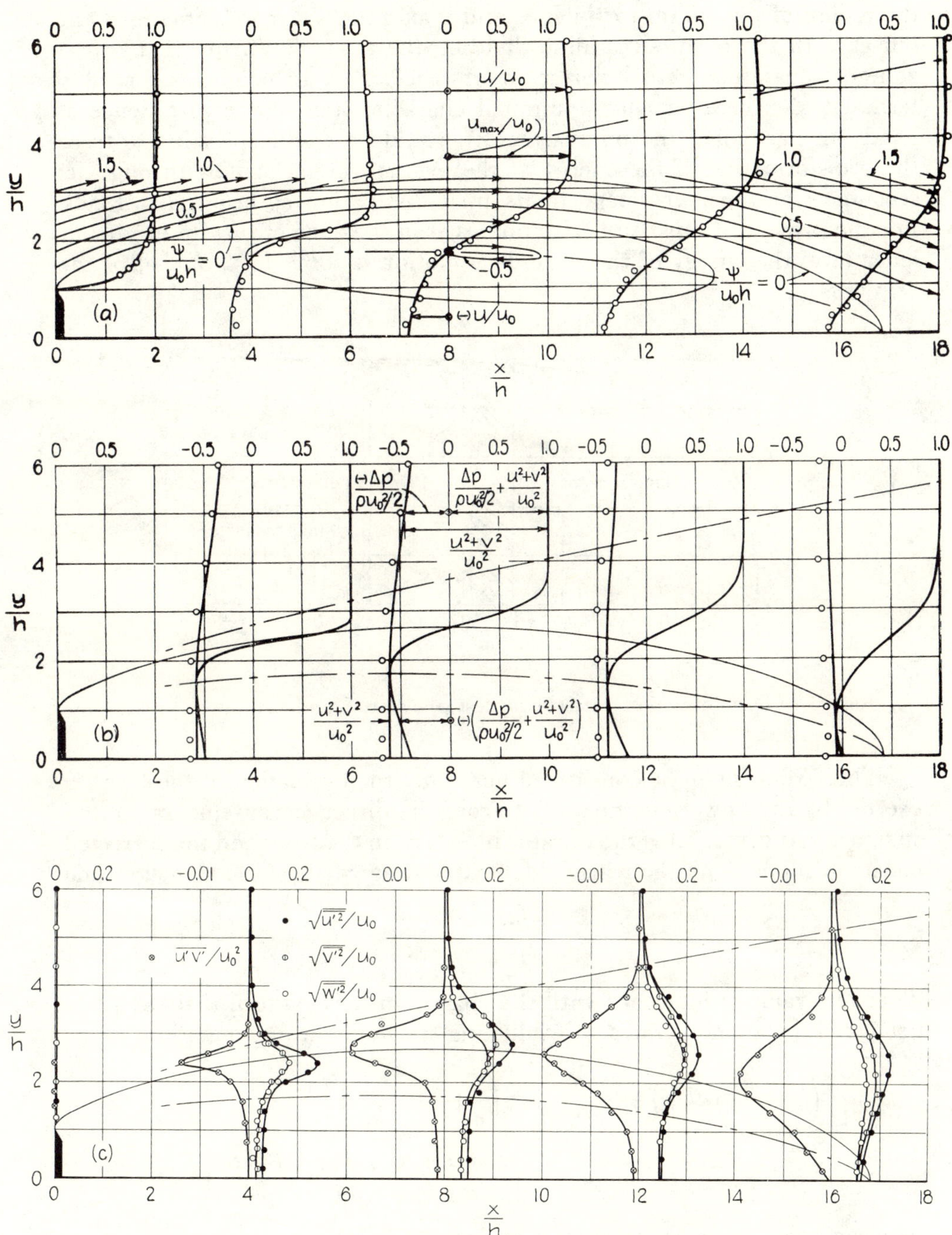

Figure 8. Experimental data and corrected distributions of (*a*) velocity,
(*b*) pressure, and (*c*) turbulence, behind test plate.

The summary plots in figure 8*c* of the three components of the turbulent velocity and of the cross product of the two components in the plane
of motion (obviously uncorrected for the wake effect) quite logically show

the region of maximum turbulence and maximum shear to coincide at the outset with the mean streamline dividing the main flow from that in the zone of separation, remaining at about twice the wall height as the eddy terminates. Were data for additional characteristics of the turbulence at hand—in particular, the microscale—it would be readily possible to trace the transformation of the energy of the primary or mean flow through its secondary or turbulent stage to its final dissipation as heat (Hsu 1950). For the present, it must suffice to note that in the zone of maximum turbulence level the energy of the secondary motion is some 15% of that of the primary

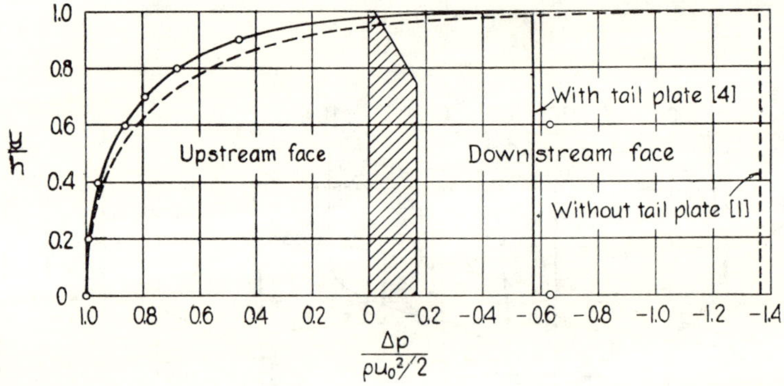

Figure 9. Distribution of pressure on test plate, with and without tail plate.

There remains to be considered here only the evaluation of the force F exerted by the flow upon the wall.   From the direct measurements plotted in figure 9, the integral of the variable intensity on the front and the corrected constant intensity on the rear yields for the drag coefficient the magnitude

$$C_D = \frac{F/hL}{\tfrac{1}{2}\rho u_0^2} = 1\cdot 38.$$

That this value is in accord with the change in flow pattern that the plate produced was demonstrated by evaluating the momentum equation

$$\int_0^\infty \left(1 - \frac{u^2}{u_0^2}\right) d\left(\frac{y}{h}\right) + \int_0^\infty \frac{p_0 - p}{\rho u_0^2} d\left(\frac{y}{h}\right) - \int_0^\infty \frac{\overline{u'^2}}{u_0^2} d\left(\frac{y}{h}\right) +$$

$$+ \frac{\nu}{u_0 h} \int_0^\infty \frac{\partial(u/u_0)}{\partial(x/h)} d\left(\frac{y}{h}\right) = \frac{F/hL}{\rho u_0^2} = \tfrac{1}{2} C_D$$

between a section far upstream (i.e. where $u/u_0 = 1$) and an arbitrary section downstream.   For the latter the section $x/h = 4$ was chosen, and the integration was carried out to—in lieu of infinity— the limit $y/h = 100$ (extrapolation beyond the zone of measurement being facilitated by use of the stream function based upon the distribution of $m(t)$ for the composite profile plotted in figure 7).   The successive integrals were found to have the values $-2{\cdot}29$, $3{\cdot}04$, $0{\cdot}05$, and $0{\cdot}00$, respectively, with the result that $C_D = 1{\cdot}40$, which, as a residual value, is in unexpectedly good agreement with $1{\cdot}38$.

Interpretation of results

One must conclude from this investigation that the use of measurements made on isolated cylindrical bodies to approximate conditions in which the wake is not free to oscillate can result in rather large errors. The actual drag coefficient of the wall ($C_D = 1{\cdot}4$), for example, is only two-thirds that of the isolated plate ($C_D = 2{\cdot}1$), and the relative pressure change $\Delta p/\frac{1}{2}\rho_0^2$ behind the wall ($-0{\cdot}57$) is less than half the mean value ($-1{\cdot}36$) behind the plate. That this corresponds to a considerable difference in the length of the mean eddy zone (in accordance with the Riabouchinsky and related theories (Birkhoff 1950) that the free-streamline curvature must increase to produce a pressure decrease in a wake or cavity) is verified by the comparative measurements made in the course of the present study, the standing eddy behind the wall being more than three times as long ($l/h = 17$) as the homologous portion of the temporal mean pattern of flow past an isolated plate ($l/b = 2{\cdot}3$).

Such reduction in drag and pressure drop, and such lengthening of the average eddy zone, point to the fact that a major difference between the two cases lies in the considerable amount of mean-flow energy that is required to produce the vortices that are shed alternately into the oscillating wake. At low Reynolds numbers these vortices are known to persist for a considerable distance prior to their disintegration into turbulence; in such circumstances, the wake behind the isolated plate should be narrower than that behind the wall. At high Reynolds numbers, on the other hand, the vortices tend to disintegrate quite rapidly; the rate of lateral spread of the turbulent zone behind the plate then probably exceeds that behind the wall. Since at all Reynolds numbers beyond a few thousand the standing eddy behind the wall appears to maintain an essentially constant form, the same can reasonably be expected of its wake as a whole; and since both the scale and the energy of the turbulence generated at the periphery of the standing eddy are smaller than the corresponding quantities in the oscillating wake, the disturbance caused by the wall should invariably dissipate more rapidly.

How great an effect the existence of boundary shear upstream from the wall (deliberately set aside at the outset because of its undefined magnitude) would actually have upon these characteristics can only be surmised. The complexity of its experimental evaluation is seen from the fact that under even the simplest probable conditions of velocity variation—for example, the semi-logarithmic distribution $u = A + B \log y$—at least two variables must be considered: in brief, a magnitude and a rate of change. Fortunately, a phenomenon analogous to that under discussion is found in flow through a pipe orifice, for which it is both logical and experimentally demonstrable that maintaining a constant mean velocity and varying the transverse distribution by roughening the pipe surface will reduce both the pressure change and the jet contraction. There is little basic difference between the axisymmetric orifice and the two-dimensional slot, and the slot differs even less from the boundary arrangement investigated in the present study. One may therefore conclude that an increase in the velocity variation over a given vertical distance (say $2h$, the velocity at level h thereby remaining constant) could be expected to cause a decrease in the drag of a wall, in the pressure reduction behind it, and in the size of the standing eddy which it produces. Until the extent of this effect has been determined experimentally, however, the prediction of such related phenomena as cavitation in the eddies formed by boundary projections must continue according to the rule of thumb that the significant velocity is that at a boundary distance equal to the height of the projection.

This investigation was undertaken by the first author under the direction of the second, in part as a thesis project for the degree of Master of Science at the State University of Iowa and in part as a sponsored project under Contract N8onr–500 between the Iowa Institute of Hydraulic Research and the Office of Naval Research. In the course of the hydrodynamic analysis of the flow pattern, considerable assistance was provided by Dr Louis Landweber of the Institute staff.

REFERENCES

ARIE, M. 1955 Flow past a normal plate in contact with a boundary, M. S. thesis, State University of Iowa.

BIRKHOFF, G. 1950 *Hydrodynamics*. Princeton University Press.

FAGE, A. & JOHANSEN, F. C. 1927 On the flow of air behind an inclined flat plate of infinite plan, *Proc. Roy. Soc.* A, **116**, 170.

GOLDSTEIN, S. (Ed.) 1938 *Modern Developments in Fluid Dynamics*, vol. 2. Oxford University Press.

HSU, H. C. 1950 Characteristics of mean flow and turbulence at an abrupt two-dimensional expansion, Ph.D. dissertation, State University of Iowa.

HUBBARD, P. G. 1954 Constant-temperature hot-wire anemometry with application to measurement in water, Ph.D. dissertation, State University of Iowa.

LANDWEBER, L. 1951 The axially symmetric potential flow about elongated bodies of revolution, *David Taylor Model Basin, Rep.* no. 761.

ROSHKO, A. 1955 On the wake and drag of bluff bodies, *J. Aero. Sci.* **22**, 124.

DIFFUSION IN THE LEE OF A TWO-DIMENSIONAL JET

by

Hunter Rouse

Iowa Institute of Hyraulic Research
State University of Iowa

INTRODUCTION

At the Sixth International Congress for Applied Mechanics, held in
Paris in 1946, the writer described an investigation of the diffusion of
heat released upwind from an airplane landing strip for purposes of fog
dispersal [1] . Because of the very low efficiency that this procedure was
found to have, an effort was next made to produce a standing eddy of roughly the same proportions as the zone to be cleared, so that the heat could
be released and diffused primarly within this zone. Two alternative means
of eddy formation were studied : a wall parallel to and upwind from the
landing strip, and a vertical jet of air from a continuous slot located in
essentially the same position [2] . The eddies produced by these two methods
were strikingly similar in shape (see *Fig. 1*). The scale of the one behind the wall, of course, was determined by the scale of the wall itself.
The eddy behind the jet, on the contrary, was found to vary in length L
with the slot width B , the velocity of efflux V , and the velocity U
of the oncoming flow, according to the empirical relationship

$$\frac{L}{B} = 20 \left(\frac{V}{U}\right)^{\frac{3}{2}}$$

In the intervening period a detailed study of the diffusion behind the
wall has been made [3] , but until two years prior to the time of writing
no further attention had been given to the question of the analogous jet
phenomenon, except for the publication of a theoretical analysis that had
been made a decade earlier [4] . Correspondance on the matter then raised a
question as to the validity of the $\frac{3}{2}$ power in the foregoing relationship,
because the customary dimensional, momentum, and similarity considerations
seemed to require that the power be 2. For this reason the question has
recently been subjected to further examination from both the analytical
and the experimental points of view.

PHYSICAL CONSIDERATIONS

However similar the eddy patterns produced by the wall and the jet
may be, there are two basic distinctions between the phenomena themselves.
On the one hand, the very presence of the jet between the oncoming flow
and the eddy entails both a geometric difference (note the displacement
of the streamline shown as a broken line in *Figs. 1 and 2*) and a dynamic
difference (note in *Fig. 2* the change in direction of the shear upon the
surrounding flow). On the other hand, not only is the slot geometrically
the reverse of the wall, but it involves one more variable ; that is, the
variation in the width as well as the velocity of the jet could only be
matched by variation in the shape of the wall in addition to its heigth.

Both conditions of flow are, needless to say, subject to the identical equations of motion as embodied in the momentum integrals ; it is the
conditions themselves that are different. To be sure, the pressure distributions on the horizontal boundaries are roughly comparable (see *Fig.2*),
there being a gradual increase to the stagnation limit just upstream from
the eddies and then a negative magnitude throughout the eddy zones. However, in the case of the wall the pressure integral over the horizontal
boundary must be equal to zero, because there is no overall change in the
vertical flux of momentum, whereas in the case of the jet the integral
must be equal and opposite to the momentum flux from the slot :

$$\int p \, dx = - \rho \, BV^2$$

Similarly, in the horizontal direction there must be a balance between
the force exerted upon the flow by the vertical boundary, the difference
between the pressures over vertical sections well upstream and downstream
from the boundary discontinuity, and (ignoring the relatively minor effects
of viscous and turbulent shear [3]) the increase in momentum flux from the
one section to the other :

$$- \int p_b \, dy + \int p_1 \, dy - \int p_2 \, dy = \rho \int u_2^2 \, dy - \rho \int u_1^2 \, dy$$

Now the force exerted upon the flow by any wall-like projection is directed upstream. If such flow is confined, this force is offset primarily
by the drop in pressure from one section to the next ; if the flow is not
confined, the force is counterbalanced primarily by the reduction in momentum flux in the wake. The force exerted by the faces of a slot, on the
contrary, will be directed downstream, as may be seen from the limiting
pressures (compare *Figs. 2 and 3*) that must prevail on them at the efflux
section (this is, of course, tantamount to an influx of momentum with a
horizontal as well as a vertical component). If such flow is confined,
the pressure will again drop in the direction of flow, as may be seen from
one-dimensional energy considerations ; if the flow is not confined, one
can only conclude from analogy with the case of a source superposed upon a
uniform stream that a pressure drop in the direction of the increased flow
is still possible. Finally, the momentum flux itself must now increase in

the direction of flow, as will be seen not only from the signs of the other
terms in the equation but from the very fact that the volume flux is augmented.

A detailed examination of the flow pattern in the slot vicinity will
clarify not only the distinction that has been made between flows past wall
and jet but also the reason for the departure of the latter from the anticipated conditions of similarity. If U and V are regarded as the reference velocities of the oncoming flow and of the jet (i.e., some distance
upstream from the efflux section in both distances), then only for the condition $U = V$ can no slipping be considered to occur beyond the point of
juncture. As shown in $Fig.$ 3 , the dividing streamline will necessarily
leave the common point of stagnation at an inclination of $45°$ and - except
for the region of separation downstream from the slot - the combined flow
can be treated as irrotational. If $U \neq V$, however, slipping will take
place beyond the point of juncture, and the fact that the pressure must be
the same in the two fluids at every point of contact will then permit only
the more slowly moving fluid to attain the stagnation limit at the juncture
point (i.e., in accordance with the Bernoulli principle, the same stagnation
pressures and pressure distributions would entail the same velocity distributions as well). In other words, if $V > U$, the dividing streamline will
leave the corner of the slot vertically, whereas with $V < U$ it will leave horizontally. In either event the radius of curvature of the dividing
streamline will increase with increasing deviation of the ratio V/U from
unity. Obviously, the greater the magnitude of V/U , the greater the relative size of the eddy L/B.

$Still$ further insight into the matter may be obtained by adopting the
eddy length rather than the slot width as reference. For very large values
of V/U (i.e., very small values of B/L) the foregoing discussion would
indicate a minimum initial deflection of the jet at the efflux section. As
V/U is decreased, the ratio between the horizontal and vertical components
of momentum flux should increase continuously, the eddy hence growing ever
lower in proportion to its length. Much as the Riabouchinsky and related
theories [5] show that increased free-streamline curvature produces a decrease in pressure in a cavity or wake, the pressure reduction within the eddy
should be alleviated as the eddy becomes relatively lower. By these very
facts it is evident that the power n in the expression $L/B = C(V/U)^n$
must be less than 2, for the latter value would correspond to constancy
of the non-dimensional geometric and dynamic characteristics - i.e., similarity - of the overall pattern. That the power must be greater than 1
follows from the simple observation that otherwise no inertial effect whatever could be involved.

EXPERIMENTAL PROCEDURE

To verify and supplement the foregoing considerations, a systematic
program of experimental measurements was undertaken in the same low-velo-

city air tunnel in which the studies of flow past a wall had been performed. The *3-foot* height of the test section was reduced *6 inches* by the insertion of a *3x12-foot* false floor to accomodate the supply duct for the jet. Five interchangeable slots having fixed widths of *0.001, 0.0032, 0.01, 0.032,* and *0.1 foot* were used. The jet velocity could be varied from *25* to *500 feet* per second with decreasing slot width ; its magnitude was determined from centerline traverses with a stagnation tube [6] at zero tunnel velocity and checked during operation against an orifice meter in the supply duct. Nominal tunnel velocities varying from *10* to *50 feet* per second were measured with a vaned anemometer in the test section and checked during operation against the drop in pressure across the entrance bell. Eddy lengths from *1* to *4 feet* could thus be obtained for each of the slot widths.

Now an investigation of this nature conducted in a tunnel of limited size must proceed through three successive stages : first, the determination of the mean pattern of flow for the restricted section ; second, the reshaping of the test section, in accordance with the preliminary measurements, to eliminate the unwanted boundary effects ; third, the final evaluation of the mean and fluctuating characteristics of the simulated semi-infinite flow. In the current jet investigation, only the first stage has been completed. However, the difference between the first and last is of secondary rather than primary magnitude. Therefore, in order to illustrate the present paper, the preliminary velocity measurements have been subjected to first-order corrections and supplemented with exploratory measurements of pressure and turbulence for the measured eddy length producing minimum constriction : $L = 1$ *foot*.

The distribution of mean velocity throughout the field of motion was determined with a hot-wire anemometer having a rather coarse filament compensated for local variations in ambiant temperature. The first-order correction consisted of using the velocity above the eddy rather than that of the approaching flow as the reference value U . The end of the eddy was established by balancing the indications of a fine Pitot cylinder mounted on the floor parallel to and 1 foot downstream from the slot, with piezometers 180° apart on upstream and downstream sides. Pressure distributions were obtained from a series of piezometers mounted in the floor, in the ceiling, and in one of two *6x18-inch* vanes that had been installed upright 1 foot from each side of the test section to minimize the effect of secondary flows in the eddy zone ; the pressures were read with a modified Wahlen-type gage sensitive to *0.001 inch* of alcohol. The first-order correction involved the subtraction of the ceiling pressure (which decreased longitudinally because of boundary-layer and constriction effects) from the vane and floor readings at the same section. Values of the longitudinal component of the turbulence intensity and of the cross product of the lontudinal and vertical components were measured with a Hubbard constant-temperature hot-wire anemometer [7] ; there were not corrected.

DISCUSSION OF RESULTS

The composite plot of eddy profiles and boundary pressures shown in
Fig. 4 is a graphic verification of the physical reasoning presented in
the foregoing pages. Somewhat unexpected, to be sure, is the fact that
departure from regularity of spacing occurs at low values of V/U rather
than high ; in other words, if similarity of the flow pattern is ever approached, this should take place as V/U becomes very great. Nevertheless,
the graph of L/B as a function of V/U seen in *Fig. 5* gives no indication of departure from the $\frac{3}{2}$-power function in the upper zone, whereas
the power changes more and more perceptibly as V/U approaches unity.
Supplementary graphs of the relative eddy height H/L and limiting jet
thickness T/L (nominally, $T = BV/U)$ are shown in the same figure.

Limitations of space make it necessary to restrict the detailed patterns of flow to a single example, that for the intermediate length-width
ratio $L/B = 100$ being reproduced in *Fig. 6*. Typical velocity distribution curves, from which the streamline intercepts were determined by evaluation of the integral $\psi = \int u\, dy$, are shown as dashed lines in the upper diagram, together with pressure-distribution curves as dotted lines.
The corresponding distributions of turbulence intensity and shear are shown
in the same manner in the lower diagram. So far as general magnitudes are
concerned, the results for this value of L/B are essentially the same as
those for the case of the plate. However, whereas in each case the maximum level of turbulence occurs in the zone of prenounced shear, in the present instance the shear necessarily reverses direction somewhat above the
upper streamline of the jet ; that is, the jet is decelerated by the relatively stagnant fluid within the eddy, but in turn it accelerates the surrounding flow by which it has been deflected. In the wake of the eddy,
nevertheless, the velocity near the boundary is lower than that of the undisturbed flow. Because the momentum flux must be greater than that upstream from the slot, it follows that there must continue to be a positive
velocity increment well above the nominal limit of the zone of diffusion
indicated by the dot-dash line.

ACKNOWLEDGMENTS

Throughout the course of this investigation the writer has profited
greatly by discussions with Dr. C.S. Yih, Research Engineer of the Iowa
Institute. The actual measurements were carried out by O.M. Erickson, with
the assistance of S. Nagaratnam, both graduate-student members of the Institute staff. The project was supported by the Office of Naval Research under Contract N8onr-500.

———

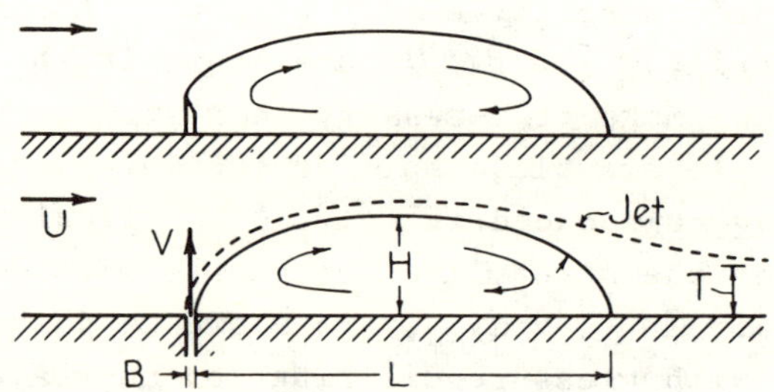

Fig. 1 Profiles of eddies behind wall and jet.

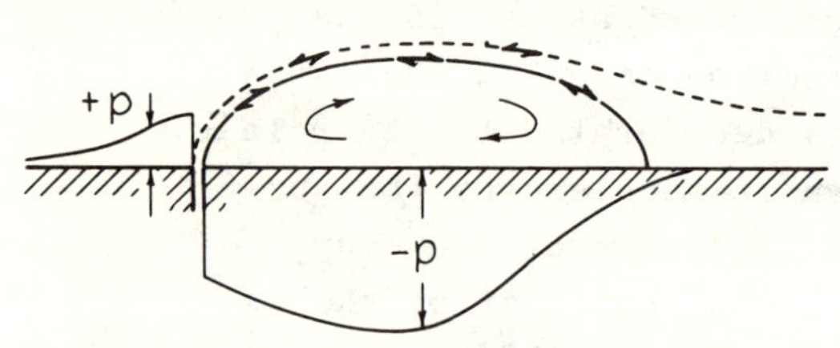

Fig. 2 Distribution of pressure and shear.

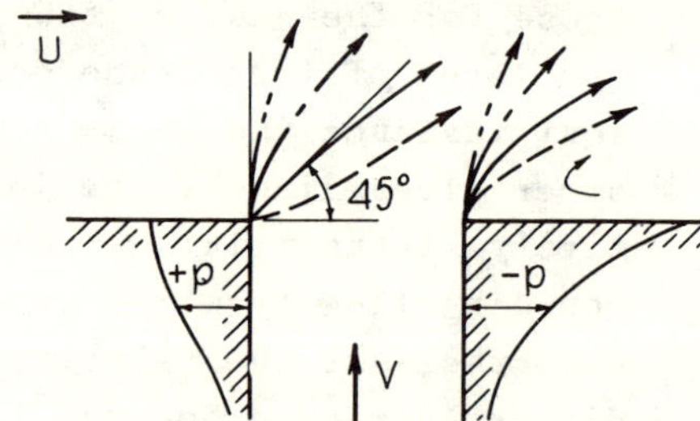

Fig. 3 Streamline configuration in the slot vicinity.

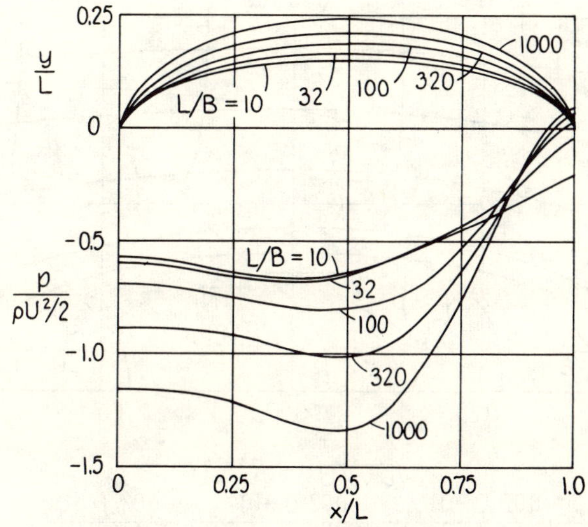

Fig. 4 Plots of eddy profiles and boundary pressures.

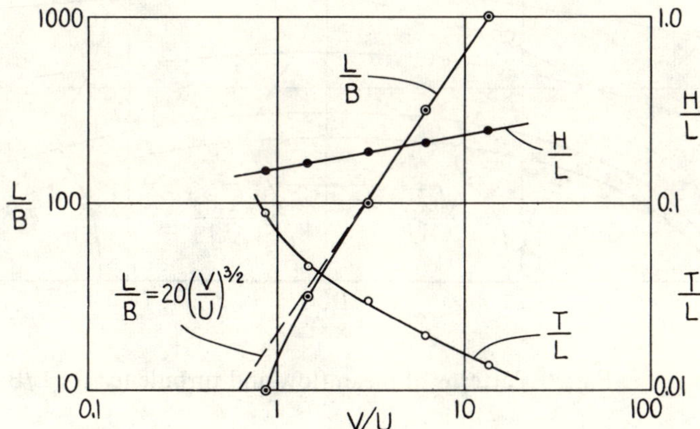

Fig. 5 L/B, H/L and T/L as functions of V/U.

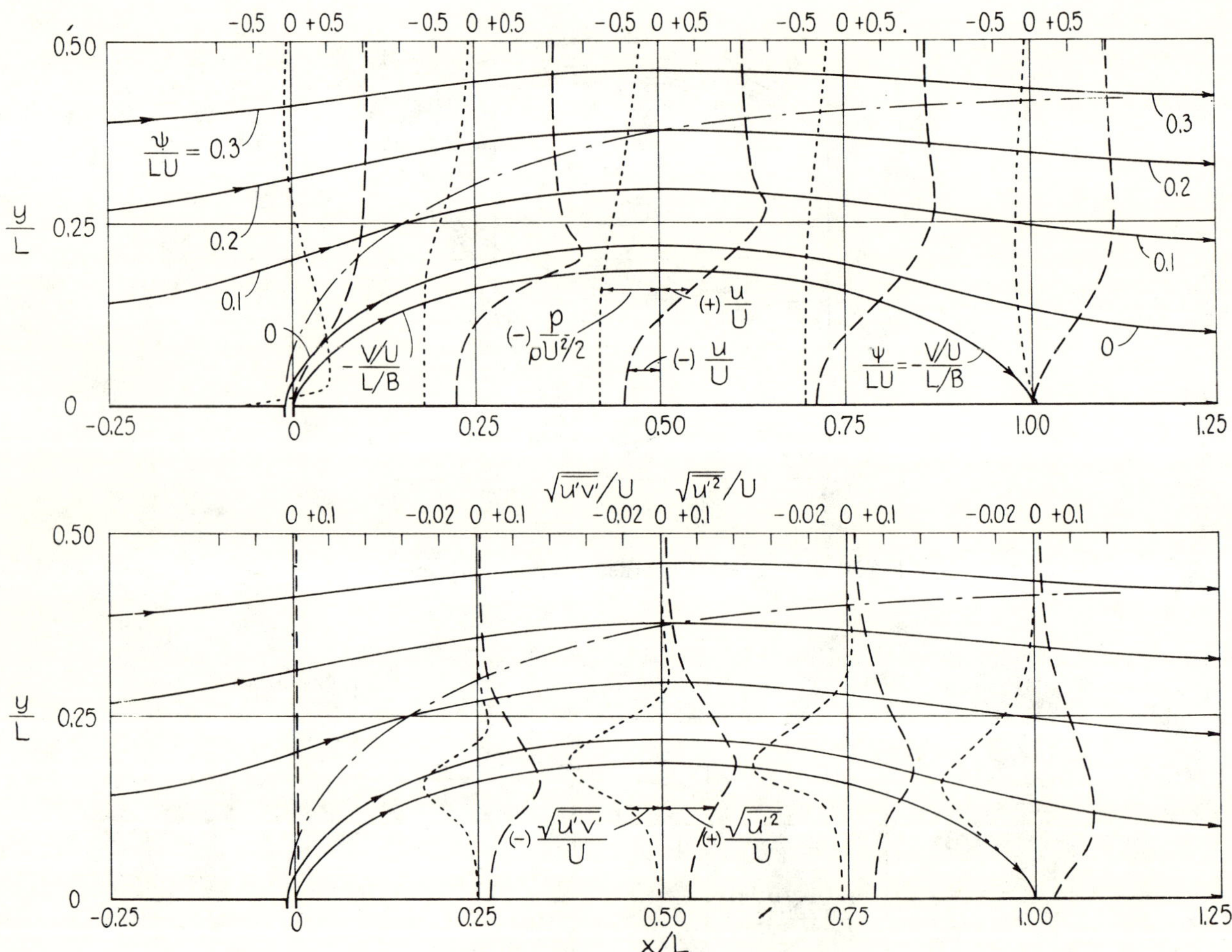

Fig. 6 Patterns of mean flow and turbulence for L/B = 100

REFERENCES

1. ROUSE, H.
"Gravitational Diffusion From a Boundary Source in Two-Dimensional Flow"
Journal of Applied Mechanics, September 1947.

2. ROUSE, H., ALBERTSON, M.L., JENSEN, R.A., and SCHADT, C.F.
"Wind-Tunnel Studies of the Diffusion of Heat by Single Wind Curtains and Baffles"
report to the National Defense Research Committee, Contract OEMsr-1243 July 1944.

3. ARIE, M., and ROUSE, H.
"Experiments on Two-Dimensional Flow Over a Normal Wall"
Journal of Fluid Mechanics, vol 1, n° 2, 1956.

4. TAYLOR, G.I.
"The Use of a Vertical Air Jet as a Wind Screen"
Mémoires sur la Mécanique des Fluides, Ministère de l'Air, 1954.

5. BIRKHOFF, G.
Hydrodynamics – *Princeton, 1950.*

6. ALBERTSON, M.L., DAI, Y.B., JENSON, R.A., and ROUSE, H.
"Diffusion of Submerged Jets"
Transactions A.S.C.E. – vol, 115, 1950.

7. HUBBARD, P.G.
"Constant-Temperature Hot-Wire Anemometry with Application to Measurement in Water"
Ph. D. dissertation, State University of Iowa, 1954.

Journal of the

HYDRAULICS DIVISION

Proceedings of the American Society of Civil Engineers

CHARACTERISTICS OF FLOW OVER TERMINAL WEIRS AND SILLS

P. K. Kandaswamy,[1] A.M. ASCE and Hunter Rouse,[2] M. ASCE
(Proc. Paper 1345)

ABSTRACT

Generalized experimental results are presented to show the variation in both the discharge coefficient and the nappe profile for two-dimensional flow over a vertical sharp-crested weir at the end of a horizontal channel as the ratio of head to depth of flow changes continuously from zero to unity.

SYNOPSIS

Whereas most existing weir formulas lose significance as the ratio of head to height of weir becomes great, it is known that the rate of flow over a sill at the end of a channel can be estimated from the basic weir equation by assuming the velocity of approach to have its critical magnitude. This gives rise to a discharge function for sills which meets that for weirs in a sharp peak when the head is about ten times the weir height. There are related indications that the nappe attains a maximum elevation and thickness at the same head-height ratio. Laboratory tests to determine quantitatively the variation in discharge coefficient and nappe profile over the entire range are described, and the experimental results are presented in generalized form for design purposes.

Historical Background

Poleni,[1] in 1717, was the first to publish an equation for weir discharge obtained by integrating over the outlet section an expression for the velocity in terms of the square root of distance below the free surface. Although he

---

Note: Discussion open until January 1, 1958. Paper 1345 is part of the copyrighted Journal of the Hydraulic Engineering Division of the American Society of Civil Engineers, Vol. 83, No. HY 4, August, 1957.
1. Iowa Inst. of Hydr. Research, State Univ. of Iowa.
2. Prof., Fluid Mechanics & Div., Inst. of Hydr. Research, State Univ. of Iowa.

neither took account of the nappe contraction nor explicitly introduced the gravitational acceleration g, Poleni's name continues to be associated with the elementary discharge equation

$$q = \frac{2}{3} C_c \sqrt{2g} \; h^{3/2} \tag{1}$$

in which q is discharge per unit width of flow section, C_c the coefficient of contraction, and h the head on the crest. Contraction of the nappe was taken indirectly into consideration by Du Buat[2] sixty years later in setting as arbitrary limits of integration the crest level and a level the distance h/2 above the crest; it was also Du Buat who introduced present-day methods of estimating the effect of submergence. Bidone,[3] better known for his study of the hydraulic jump, appears to have been the first to show an interest in expressing algebraically the geometry of the weir nappe.

Not until about the middle of the 19th century was attention focused upon the velocity-of-approach "correction" for weir discharge, which was to dominate certain branches of the literature for another century. In 1845 Weisbach[4] published the familiar integral form

$$q = \frac{2}{3} C_c \sqrt{2g} \left[\left(h + \frac{v_0^2}{2g} \right)^{3/2} - \left(\frac{v_0^2}{2g} \right)^{3/2} \right] \tag{2}$$

Since the solution of this equation has to proceed by successive approximation because of the implicit presence of the approach velocity v_0 on the left side of the equation, the impression has persisted that this velocity is independently variable. Weisbach himself appreciated the fact that a unique solution existed for any ratio of head h to depth of approach h + w, as he proposed an empirical formula for its evaluation:

$$q = \frac{2}{3} \left(C_c \right)_0 \left[1.041 + 0.3693 \left(\frac{h}{h+w} \right)^2 \right] \sqrt{2gh^3} \tag{3}$$

Bazin[5] in 1888 adopted the same form of function, with specific evaluation of the limiting coefficient of contraction $(C_c)_0$

$$q = \left(0.405 + \frac{0.003}{h(m)} \right) \left[1 + 0.55 \left(\frac{h}{h+w} \right)^2 \right] \sqrt{2g} \; h^{3/2} \tag{4}$$

the dimensional term 0.003/h being introduced to compensate for low-head effects now attributed to surface tension and viscosity. In a series of investigations extending over the next ten years, Bazin also went deeply into such varied aspects of weir flow as the distribution of velocity and pressure through the nappe, and the effects of incomplete ventilation, submergence, weir inclination, and geometry of the crest; his nappe-profile measurements are still used as the basis of spillway design.

Bazin's formulation of the discharge coefficient soon led others to attempt further improvements by slight changes in the type of algebraic expression used. From 1911 to 1928 Rehbock[6] proposed a number of variants, the most pertinent of which was

$$q = \left(0.605 + 0.08\,\frac{h}{w} + \frac{1}{305\,h(ft)}\right)\frac{2}{3}\sqrt{2g}\;h^{3/2} \tag{5}$$

Comparison of Eq. (5) with Eq. (4) will show that the primary distinction lies in the use of the ratio h/w in place of h/(h+w) as the geometric parameter. The remaining term of Eq. (5)—which, like the corresponding one of Bazin, is not dimensionally homogeneous with the others—again reflects capillary and viscous effects at low heads. If the latter factors are ignored, and if the basic weir equation is written in terms of a discharge coefficient C_d,

$$q = \frac{2}{3}\,C_d\,\sqrt{2g}\;h^{3/2} \tag{6}$$

it will be seen that the essential aspects of the Bazin and Rehbock formulas may be reduced for comparison to

$$C_d = 0.608 + 0.334\left(\frac{h}{h+w}\right)^2 \tag{7}$$

and

$$C_d = 0.605 + 0.08\,\frac{h}{w} \tag{8}$$

which are plotted against h/(h+w) in Fig. 1. The two curves evidently differ by little more than 1% for heads as great as twice the height of weir.

When Mises,[7] in 1917, determined the coefficient of contraction analytically for various two-dimensional forms of conduit outlet, he noted that the values for a symmetrical slot in a normal plate at the end of a conduit would—in combination with Eq. (2)—duplicate the trend of the Rehbock formula very closely over a considerable range if the ratio of slot dimension b to conduit dimension B were used to represent the ratio h/(h+w). His curves for C_c and C_d are also shown in Fig. 1. To be noted in particular is the limiting theoretical value $\pi/(\pi+2) = 0.611$ in comparison with the empirical values 0.608 and 0.605 chosen by Bazin and Rehbock. One of Mises' students, Lauck,[8] nearly a decade later approximated not only the discharge coefficient but also the entire nappe profile for the weir of infinite relative height, by means of the Cauchy integral theorem.

Inspection of the Bazin and the Rehbock formulas will indicate that the latter must approach the limit $C_d = \infty$ as the relative height of the weir approaches the limit zero (i.e., the free overfall), a trend that is necessarily shared by the orifice function of Mises, whereas the Bazin function has the finite limit $C_d = 0.942$. Now it is obvious that discharge over the free overfall is finite rather than infinite, since the effective head is the critical depth. However, it was noted by Böss[9] that the critical stage of flow is already established upstream from terminal sills well before they attain zero height, which would yield a trend contrary to that of Eq. (6). That is, following the procedure of Rouse,[10] the discharge coefficient for such conditions can be determined by introducing into Eq. (6) the critical-depth relationship

$$q = \sqrt{gy_c^3} = \sqrt{g(h+w)^3} \tag{9}$$

Thus,

$$C_d = 1.06\left(1 + \frac{w}{h}\right)^{3/2} \tag{10}$$

which is plotted at the upper right of Fig. 1. Evidently, the discharge characteristics of terminal sills follow quite a different functional relationship from that of even moderately low weirs, with an intermediate maximum value of the discharge coefficient. Now the coefficient of contraction C_c can be determined from simultaneous solution of Eqs. (10), (6), and (2) and plotted as at the lower right of Fig. 1. If, after Rouse, this coefficient is assumed to approximate the relative thickness of the nappe at its highest point, then this plot indicates that the flow profile will also attain a maximum stage at the borderline between weirs and sills.

In order to investigate these aspects of the problem, the second author undertook in 1931-33 in the River Hydraulic Laboratory of the Department of Civil and Sanitary Engineering at M.I.T. a systematic series of profile (as well as velocity and pressure) measurements on the nappe of a weir that was successively reduced in height until the overfall limit was reached. The results of the profile tests were reflected in a subsequent book,[11] but they were found to cover the transition range inadequately and hence were never published in their entirety. Similar measurements over a less extensive range of the parameter h/(h+w) were undertaken later by the Bureau of Reclamation and published[12] for purposes of spillway design. Because it was believed that certain characteristics of the weir-sill range were nevertheless of technical interest, generalized plots of the M.I.T. data showing a parallel between the discharge-coefficient and nappe-profile functions were included in the Proceedings of the Fourth Hydraulics Conference[13] of the Iowa Institute of Hydraulic Research.

Continued efforts have since been made at the Iowa Institute to complete both the analytical and the experimental phases of the weir and sill problem. The relaxation method has already been applied to the improvement of Lauck's approximation of the nappe profile for the infinitely high weir[14] and at the time of writing the study is being extended to other values of h/(h+w). However, since numerical integration is not as readily applicable to low weirs and sills as to high, the gaps left in the M.I.T. study were in 1956 filled in experimentally by the first author[15] under the direction of the second, to the end of providing a basis for both design and further theoretical analysis. It is to the presentation of the composite results that this paper is devoted.

Experimental Procedure

The M.I.T. experiments were conducted in a 50-cm glass-walled flume, the rate of flow through which was measured with a ventilated sharp-crested weir 40 cm high in accordance with Eq. (5). The experimental weir was initially of identical characteristics, but its effective height was changed by the addition of successive false floors on the upstream side. These were constructed of troweled cement mortar over sand, and each consisted of a 6-ft. level section preceded by a very gradual rise from the original floor of the flume. Three weir heights—40, 20, and 10 cm—were produced in this way. The 40-cm weir was then removed, and still another false floor was added at a height of 37.5 cm above the flume floor, with a brass-angle end to simulate the free overfall. A supplementary weir crest could be attached to the end with wingscrews to produce additional weir heights of 5, 2 1/2, and 1 1/4 cm. Full ventilation was provided below the nappe in all instances.

Profile measurements were made at 1-cm sections upstream and downstream from the crest with point and hook gages reading by vernier to 0.1 mm. For each of the seven weir heights the same nine rates of discharge were investigated; these corresponded to heads on the measuring weir ranging by 2-cm increments from 2 to 18 cm. Supplementary measurements of pressure distribution along the floor, up the weir plate, and through the nappe, as well as checks upon the constancy of the total head at various points, were also made, but neither the experimental techniques nor the results are pertinent to the present discussion.

The tests conducted at the Iowa Institute employed the same type of free-overfall structure as the one finally constructed at M.I.T., except that it was built of brass plate and angles and mounted in a 1-ft. flume. The level portion was 4 ft. in length, its height above the floor of the flume was 1 ft., and the approach curve had the form of a quarter ellipse with an axis ratio of 3.5:1. The adjustable weir plate was machined to a sharp upstream edge, the top surface having a width of about 1/32 in. To minimize capillary distortion of the lower surface of the nappe at small weir heights, the crest surfaces were rubbed with paraffine. For the very smallest height (0.005 ft.) the crest was remachined to a knife edge. The test procedure was essentially the same as that at M.I.T., although the range of head was limited to $0.1 \text{ ft.} < h < 0.5 \text{ ft.}$ and the range of weir height to $0.005 \text{ ft.} < w < 0.05 \text{ ft.}$ Discharge measurement was effected by means of a 90° elbow meter in the 6-in. supply line that had been previously calibrated against a 12-in. sharp-crested weir mounted in the flume itself, the Rehbock formula being used as reference just as in the M.I.T. experiments.

Discussion of Results

As has long been realized, the head on a weir must be measured a considerable distance upstream from the crest, where the effect of the curvilinearity of flow upon the surface elevation is still negligible. A distance equal to three or four times the head is generally considered to suffice. Unfortunately, with relatively low weirs the effect of channel resistance becomes appreciable, so that the nappe profile is asymptotic to a sloping rather than a horizontal line. In other words, the head measurement then is subject to two types of error, the one increasing with proximity to the crest and the other with distance from it.

In order to determine that section at which the net error would be a minimum, so that the heads for different weir heights could be evaluated consistently, a composite dimensionless plot was made of all profile measurements upstream from the crest for the most extreme case—the free overfall—

as shown in Fig. 2. Since the computed critical depth was used as the reference length, and since the critical section has been shown to be the last at which hydrostatic distribution of pressure (i.e., absence of curvilinearity) can be assumed to prevail,[10] the abscissa of the desired section should be that having an ordinate value of unity, or $x/h = 4$. Inasmuch as this value for the free overfall agrees with that commonly used for weirs of great relative height, it was adopted for all subsequent evaluations in the intermediate range as well.

From the experimental determinations of discharge and surface profile for the various heights of weir and sill, the coefficient of discharge was evaluated according to Eq. (5) and plotted as in Fig. 3. Herein the Rehbock parameter h/w is used as abscissa in the weir range, because of the linearity of the resulting function. Its reciprocal w/h then logically serves as the abscissa in the sill range, the relative scales of the two being so chosen as to yield approximate agreement in the intermediate range. All of the Iowa data are shown, together with a comparable number from M.I.T. (i.e., all at the higher heads). Inspection of the diagram will indicate that Eq. (8) for weirs applies with good approximation over the range $0 < h/w < 6$, and Eq. (9) for sills over the range $0 < w/h < 0.06$. Although some discrepancy between the two sets in the zone of overlap is apparent, a smooth transition between the two functions can readily be drawn.

In Fig. 4 are superposed a systematic series of nappe profiles extending through the weir range to the case of maximum coefficient, $h/w = 10$. Figure 5 shows the corresponding family of profiles for sills. Like the coefficient plot of Fig. 3, the profiles (except for the lower surface of the weir nappe) are seen to be characterized by similar trends, the case $h/w = 10$ or $w/h = 0.1$ evidently representing a maximum for both the discharge coefficient and the nappe deflection.

As had previously been noted by the second author and used as the basis of the generalized profile diagrams already mentioned,[13] plots of z/h as functions of h/w and w/h with x/h as parameter bear a striking resemblance to the coefficient function shown in Fig. 3. Revised in accordance with the sill data now at hand, these diagrams are presented as Figs. 6 and 7, together with measured points for two representative values of x/h. The confocal rectilinear portions of the curves below $h/w = 5$ are identical with those previously published, but the remainder will be found to differ appreciably. Similarity with the coefficient curve is maintained well into the sill range, however, except in the very neighborhood of the overfall limit.

Although the foregoing experimental data were obtained with heads of relatively small magnitude and hence can be expected to apply only approximately to conditions at larger scale, it is believed that the degree of approximation will not be appreciably lower than the experimental precision evident in the figures. The Rehbock equation, for example, indicates that capillary and viscous effects change the discharge by about 1% at a head of 0.3 ft. Since this is representative of the heads that were investigated, there is no reason to expect scale effects of a higher order to be present in the results.

CONCLUSIONS

Through the combination of experimental measurements made at considerably different times and places, information is now at hand for the coefficient of discharge and the form of the nappe profile as the relative height of a ventilated, sharp-crested weir at the end of a channel varies over its complete range from infinity to zero. In the range $0 < h/w < 6$ the coefficient of weir discharge is given approximately by the formula

$$c_d = 0.61 + 0.08 \, \frac{h}{w}$$

In the range $0 < w/h < 0.06$, the coefficient of discharge is given approximately by

$$c_d = 1.06 \left(1 + \frac{w}{h}\right)^{3/2}$$

In the intermediate zone there is a continuous transition between the two functions with a maximum at approximately $h/w = 10$; the latter might hence be regarded as the borderline between weir flow and sill flow. This value of h/w also marks the condition of maximum nappe deflection, the nappe for both weirs and sills gradually rising and thickening as this common value is approached. Nappe profiles for design purposes may be determined either by interpolation of the profile curves or by direct reading of the generalized profile coordinates presented in the paper.

REFERENCES

1. Poleni, G., De Motu Aquae Mixto, Padua, 1717.

2. Du Buat, P.L.G., Principes d'Hydraulique, Paris, 1779.

3. Bidone, G., "Expériences sur la dépense des reversoirs," Memorie della Reale Accademia delle Scienze di Torino, Vol. 28, 1824.

4. Weisbach, J., Lehrbuch der Ingenieur- und Maschinen-Mechanik, Brunswick, 1845.

5. Bazin, H. E., Expériences nouvelles sur l'écoulement en déversoir, Paris, 1898.

6. Rehbock, T., discussion of E. W. Schoder and K. B. Turner's "Precise Weir Measurements," Transactions A.S.C.E., Vol. 93, 1929.

7. Mises, R. von, "Berechnung von Ausfluss- und Ueberfallzahlen," Zeitschrift des V.D.I., 1917.

8. Lauck, A., "Ueberfall über ein Wehr," Zeitschrift für angewandte Mathematik und Mechanik, Vol. 5, 1925.

9. Böss, P., "Berechnung der Abflussmengen und der Wasserspiegellage bei Abstürzen und Schwellen unter besonderer Berücksichtigung der dabei auftretenden Zusatzspannungen," Wasserkraft und Wasserwirtschaft, No. 2-3, 1929.

10. Rouse, H., "Discharge Characteristics of the Free Overfall," Civil Engineering, Vol. 6, No. 4, 1936.

11. Rouse, H., _Fluid Mechanics for Hydraulic Engineers_, New York, 1938.

12. "Studies of Crests for Overfall Dams," Bulletin 3, Part IV, Boulder Canyon Project Final Reports, Bureau of Reclamation, 1948.

13. Ippen, A. T., "Channel Transitions and Controls," _Engineering Hydraulics_, New York, 1950.

14. McNown, J. S., Hsu, E. Y., and Yih, C. S., "Applications of the Relaxation Technique in Fluid Mechanics," _Transactions A.S.C.E._, Vol. 120, 1955.

15. Kandaswamy, P. K., "Discharge Characteristics of Low Weirs and Sills," M. S. Thesis, State University of Iowa, 1957.

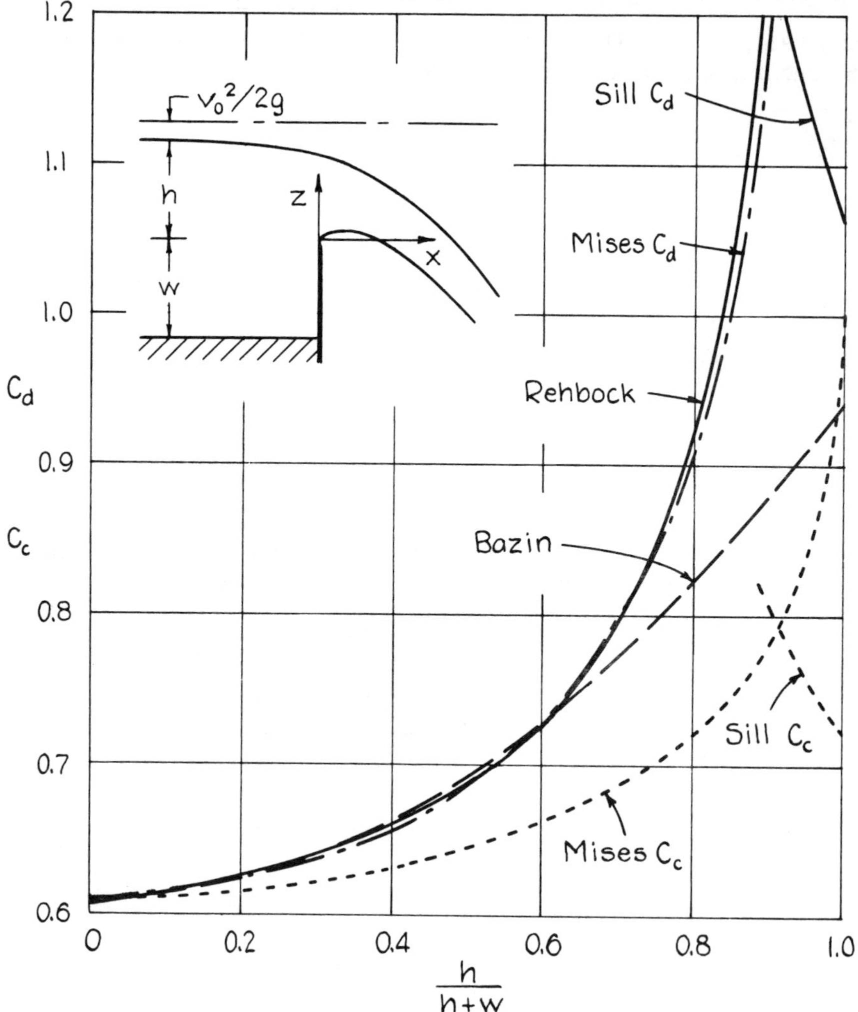

Fig. 1. Comparison of Discharge Functions for Weirs and Sills.

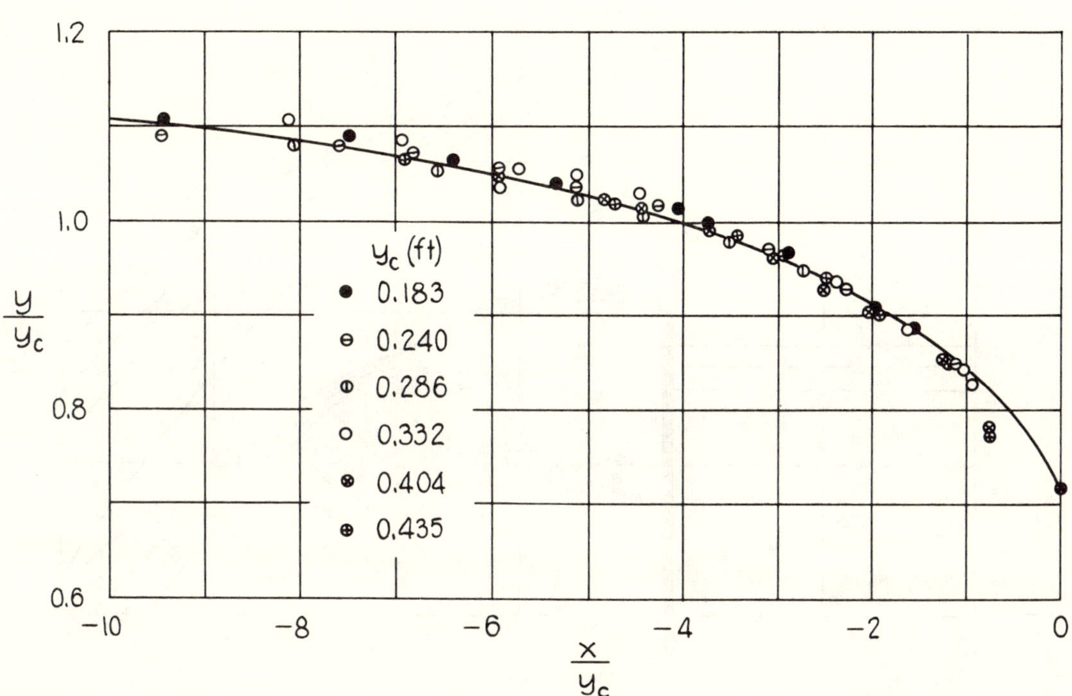

Fig. 2. Location of Critical Section Upstream from Free Overfall.

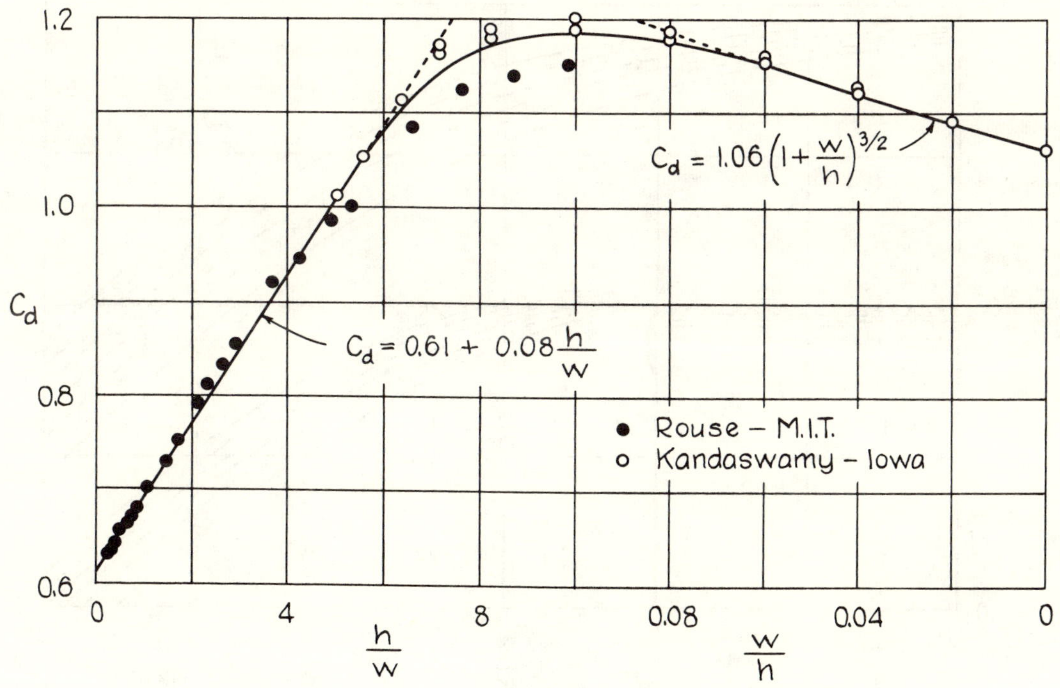

Fig. 3. Experimental Evaluation of Discharge Coefficient for Entire
Weir-Sill Range.

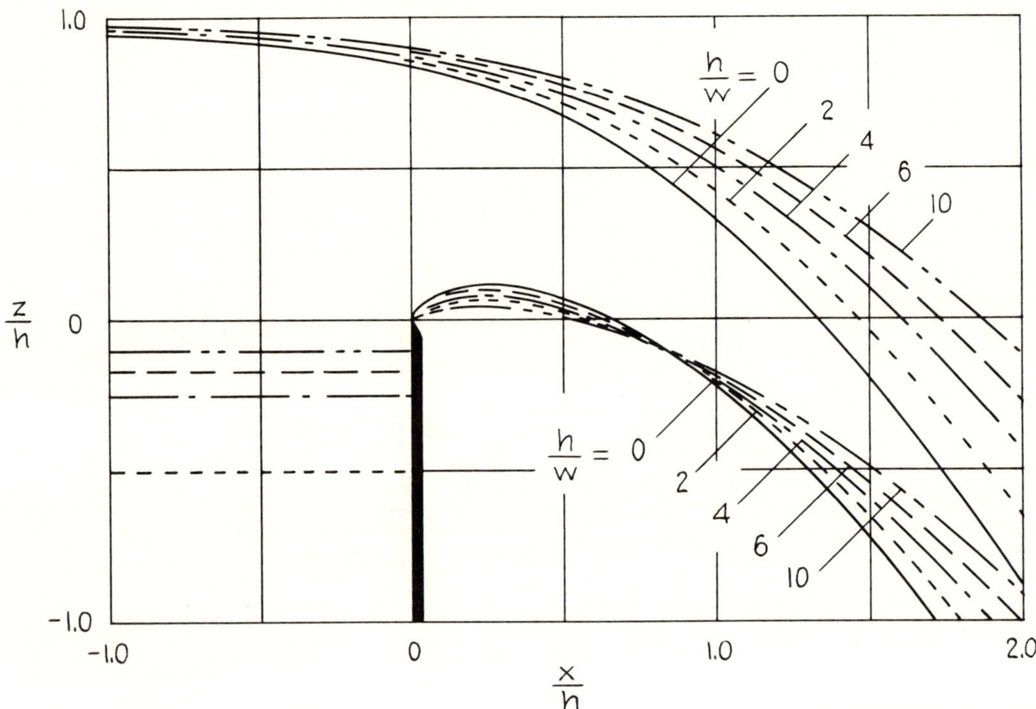

Fig. 4. Variation of Nappe Profile with Relative Height of Weir.

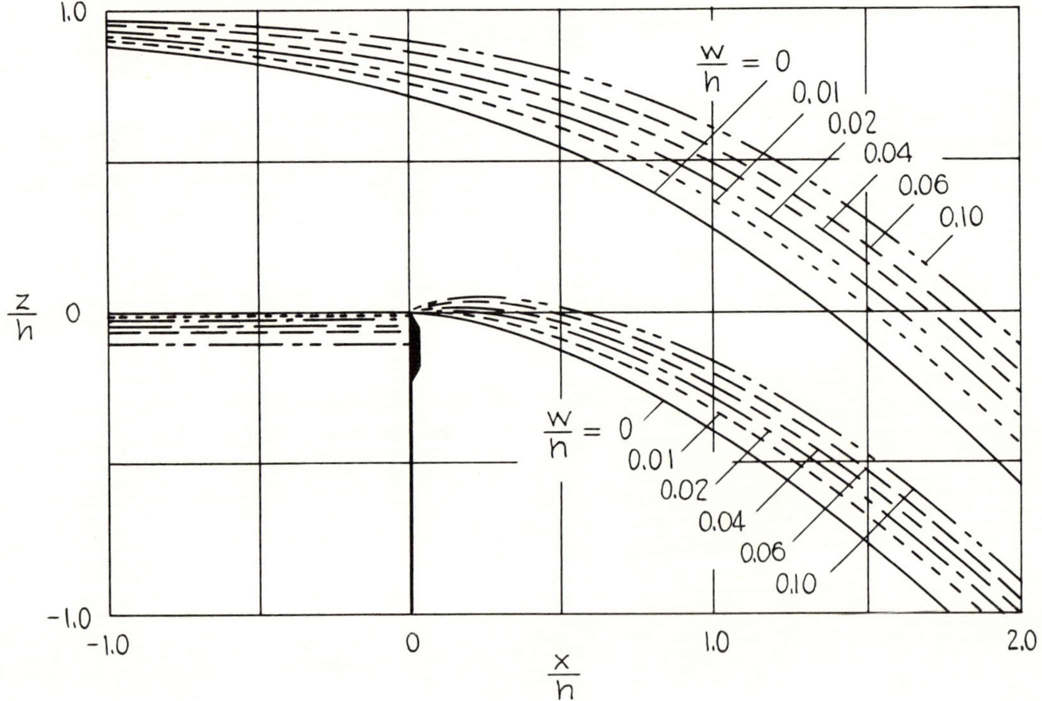

Fig. 5. Variation of Nappe Profile with Relative Height of Sill.

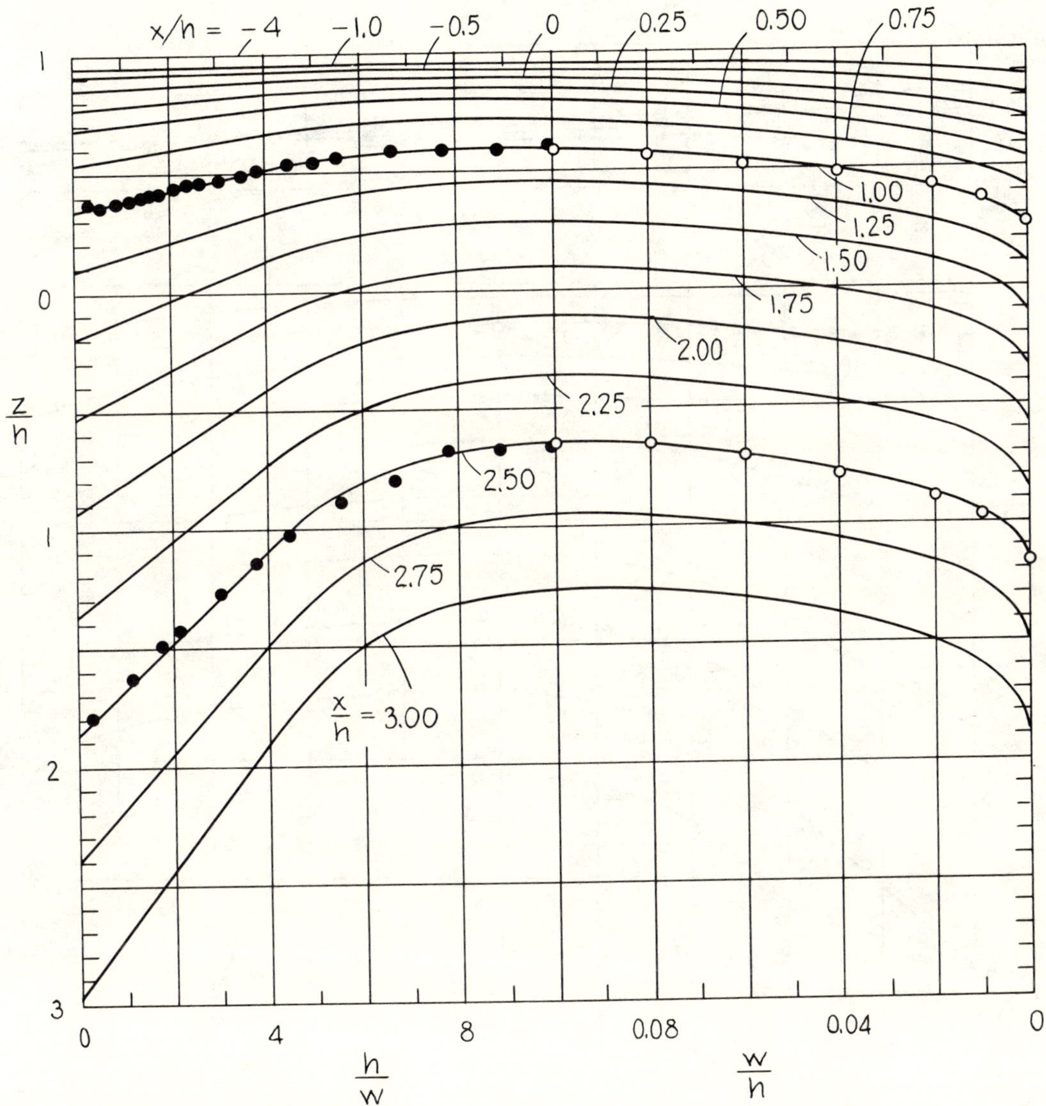

Fig. 6. Generalized Coordinates of Upper Nappe Surface for
Terminal Weirs and Sills.

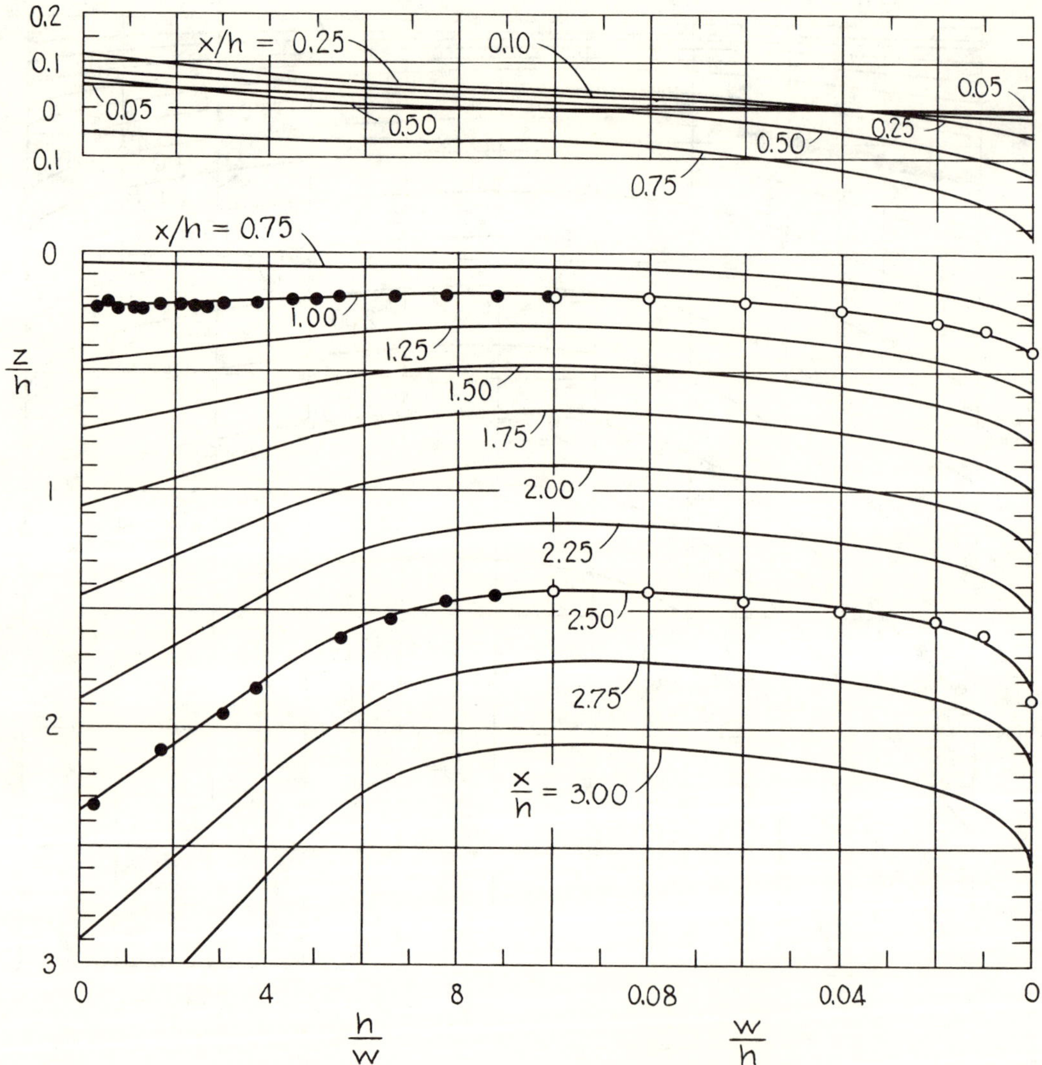

Fig. 7. Generalized Coordinates of Upper Nappe Surface for Terminal Weirs and Sills.

Paper No. 3006

TURBULENCE CHARACTERISTICS OF THE HYDRAULIC JUMP

By Hunter Rouse,[1] M. ASCE, Tien To Siao,[2] and S. Nagaratnam[3]

With Discussion by Messrs. Edward Silberman; Alvin J. Peterka and Joseph N. Bradley; James M. Robertson; Donald R. F. Harleman; Philip G. Hubbard; and Hunter Rouse, Tien To Siao, and S. Nagaratnam

Synopsis

Because the fluid discontinuities produced by entrapped air interfere with the measurement of turbulence in the hydraulic jump itself, the flow pattern of the jump has been simulated in an air duct and the hot-wire anemometer used to determine the corresponding pattern of turbulence for Froude numbers of 2, 4, and 6. In the analytical section of the paper the modeling method is justified and the differential and integral forms of the pertinent momentum and energy equations are explained. The experimental apparatus and techniques are described next, and the basic data are presented in diagrams. The measured characteristics of the mean flow and of the turbulence are then correlated in accordance with the momentum and energy equations, and the process of energy conversion from the mean flow through turbulence to heat is followed in detail. Various aspects of the hydraulic jump, long subject to conjecture or misunderstanding, are thereby clarified.

Introduction

Three sorts of information pertinent to the present investigation of the hydraulic jump are contained in discussions of an earlier paper[4] by Boris A.

Note.—Published, essentially as printed here, in February, 1958, in the Journal of the Hydraulics Division, as *Proceedings Paper 1528*. Positions and titles given are those in effect when the paper or discussion was approved for publication in *Transactions*.

[1] Director, Iowa Inst. of Hydr. Research, State Univ. of Iowa, Iowa City, Iowa.

[2] Hydr. Engr., Inst. of Hydr. Research, Academia Sinica, Peking, China.

[3] Engr., Harza Eng. Co., Chicago, Ill.

[4] "The Hydraulic Jump in Terms of Dynamic Similarity," by Boris A. Bakhmeteff and Arthur E. Matzke, *Transactions*, ASCE, Vol. 101, 1936.

Bakhmeteff and Arthur E. Matzke, A. M. ASCE: First, a well-annotated account of the gradual development of knowledge on this subject; second, a somewhat contradictory mixture of factual matter, hypothesis, and error; third, and by inference only, a guide to further research. In 1936, when the paper and discussions were published, the tools for such research had only recently come into use in other fields, and it is hardly to be expected that they would already have been applied in hydraulics. In fact, the investigation herein described had the purpose of demonstrating the efficacy of the new approach as well as providing greater insight into the phenomenon under consideration.

As was evident from the aforementioned paper and its discussions, past advancements had been concerned with the vertical elements (that is, depths and heads) of the mean-flow pattern as correlated by the one-dimensional forms of the continuity, momentum, and energy relationships. Some attention had been given to the empirical evaluation of the longitudinal elements as well, and to the expression of the functional relationships in terms of what is now called the Froude number. However, concern with the accompanying pattern of turbulence was almost entirely speculative. For lack of even an approximate evaluation of the turbulence characteristics, their importance to the phenomenon was both overestimated and underestimated. For example, the senior writer claimed that the computed loss in head at the jump represented merely a transfer of energy from the mean motion to the turbulence, and, hence, that the total head would be found to remain practically constant through the jump if the kinetic energy of the turbulence could be measured and included as velocity head. Others, on the contrary, did not sense the importance of eddy generation along the surface of discontinuity between the secondary flow of the roller and the primary flow beneath, some even going so far as to claim that the roller played no essential part in the phenomenon.

The difficulty lay, as is now apparent, in the general lack of clarity that prevailed in nearly all matters involving fluid turbulence. Not only did existing turbulence theories seem to have little connection with hydraulic reality, but the only practicable means of measuring turbulence—the hot-wire anemometer—was restricted to use in air. During the intervening years, progress has been made in the interpretation, if not in the quantitative prediction, of the transformation from energy of the mean motion through energy of turbulence to that of heat, and the hot-wire technique itself has been extended to use in water. Means would therefore seem to be at hand for the measurement of the primary and secondary patterns of flow in the hydraulic jump, and at least the evaluation of the various energy forms at representative sections. Unfortunately, the jump will probably be one of the last hydraulic phenomena to prove susceptible to exploration with the hot-wire instrument, because of the presence of countless fluid discontinuities (bubbles of entrained air) in the region of greatest interest.

To overcome this rather formidable obstacle (yet without having recourse to the extremely tedious process of statistically analyzing motion pictures of suspended particles), it was resolved at the Iowa Institute of Hydraulic Research, at Iowa City, Iowa, to simulate the flow pattern of the hydraulic jump in an air duct, and then use the hot-wire anemometer for turbulence surveys, just as in previous Institute studies of other diffusion phenomena. The analog will seem, at first glance, a rather artificial one, to say the least. However, it must be realized that few model flows portray perfect dynamic similarity with their prototypes, and that, for all practical purposes, only the pertinent part of the phenomenon need be closely approximated. In the present instance it is not the gravitational action that is to be evaluated but the viscous action. If the mean-flow patterns are geometrically similar, and if the changes in energy are comparable, then it would seem safe to assume that the patterns of turbulence are also similar, and it is specifically these with which the present investigation is concerned.

Analytical Bases of Experimentation

Notation.—The letter symbols adopted for use in this paper are defined where they first appear, in the illustrations or in the text, and are assembled, for convenience of reference, in the Appendix.

Air-Water Correlation.—Application of the simplified continuity, momentum, and energy relationships for one-dimensional flow to the conditions shown in Fig. 1 leads to the following well-known equations for the hydraulic jump in

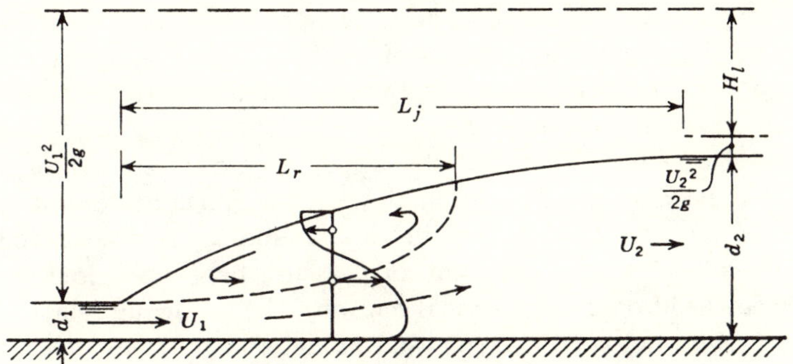

Fig. 1.—Definition Sketch of Hydraulic Jump

homogeneous liquid flowing through a horizontal, resistance-free channel of rectangular cross section:

$$\frac{d_2}{d_1} = \frac{1}{2}\left(\sqrt{1 + 8\frac{U_1^2}{g\,d_1}} - 1 \right) \dots\dots\dots\dots\dots (1)$$

and

$$\frac{H_l}{d_1} = \frac{1}{2}\frac{U_1^2}{g\,d_1}\left(1 - \frac{d_1^2}{d_2^2} \right) - \frac{d_2}{d_1} + 1 \dots\dots\dots\dots (2)$$

in which d is the depth of flow; U denotes the average across a vertical section of the instantaneous longitudinal component of velocity at a point (u); g is the acceleration of gravity; and H_l represents the head loss in the jump. In Fig. 1, L_j is the length of the jump, and L_r denotes the length of the roller. Both the depth ratio, d_2/d_1, and the relative loss of total head, H_l/d_1, are explicit functions of the Froude number, $\mathbf{F}_1$, of the approaching flow,

$$\mathbf{F}_1 = \frac{U_1}{\sqrt{g\,d_1}} \dots\dots\dots\dots\dots\dots\dots (3)$$

as are also such longitudinal characteristics as the relative length of the jump, L_j/d_1, and the shape of the flow profile.

It is not too difficult to visualize the application to the flow of an upper boundary having the same profile as the mean free surface. Like the spillway profile that is shaped according to the lower surface of the corresponding weir nappe, the presence of the solid boundary may be expected to have some effect on the flow. In the case of the spillway the predominant effect is the addition of shear. In the case of the jump it is rather the elimination of both the freedom of the surface to fluctuate and the presence of entrained air. However, measurements of the pressure head along the floor of a flume under a jump yield values that differ from the local depths by no more than the experimental error, indicating at once that the pressure distribution can be assumed to be hydrostatic and that the quantity of air in suspension is actually insufficient to change appreciably either the density or the specific weight. As for the surface fluctuations, measurement of the resulting wave amplitudes just downstream and approximation therefrom of the resulting wave energy have shown this to be less than 1% of the energy of the flow itself. It would thus seem reasonable to assume that the presence of the upper boundary would have, at most, only a secondary effect on the flow pattern.

If the foregoing premise is accepted, then it must also be granted (in accordance with the general mechanics of confined flow) that the same pattern would obtain between the upper and lower boundaries at other rates of flow, if only the Reynolds numbers did not differ greatly. However, no longer would the mean pressure intensity along the upper boundary be atmospheric at all points, for it must change in accordance with the Bernoulli equation. On the other hand, if the linear quantities in Eqs. 1 and 2 are considered to be the piezometric heads, h (that is, the sum of pressure head and elevation, which is presumably constant over any vertical section), the following relationships must continue to prevail between the enclosed model and its free-surface prototype:

$$\frac{h_2 - h_1}{\dfrac{U_1^2}{2g}} = \frac{d_2 - d_1}{\dfrac{U_1^2}{2g}} \dots\dots\dots\dots\dots (4)$$

and

$$1 - \frac{h_2 - h_1}{\dfrac{U_1^2}{2g}} - \frac{d_1^2}{d_2^2} = \frac{H_l}{\dfrac{U_1^2}{2g}} \dots\dots\dots\dots\dots (5)$$

Again in accordance with the principles of fluid mechanics, it makes no difference whether the fluid is a liquid or a gas as long as the Reynolds numbers are still comparable and the velocities of the gas are not so high (that is, 200 or 300 ft per sec) as to cause compressibility effects. Hence, the same patterns of mean flow and turbulence should be obtainable through use of air of density ρ as the experimental fluid. Moreover, since the elevation term then could be neglected in the Bernoulli and related equations, the foregoing model-prototype expressions would reduce to the following pressure forms:

$$\frac{p_2 - p_1}{\dfrac{\rho U_1^2}{2}} = \frac{d_2 - d_1}{\dfrac{U_1^2}{2g}} \dots\dots\dots\dots\dots (6)$$

and

$$1 - \frac{p_2 - p_1}{\frac{\rho\, U_1{}^2}{2}} - \frac{d_1{}^2}{d_2{}^2} = \frac{H_l}{\frac{U_1{}^2}{2\,g}} \quad \ldots \ldots \ldots \ldots \ldots \ldots \ldots (7)$$

in which p is the pressure intensity.

Although the Froude number cannot be expressed by replacing d_1 with either h_1 or p_1/γ (the term γ is the specific weight of the fluid), since the pressure load on the system is no longer determined by free-surface conditions, it will be seen that the right-hand sides of Eqs. 4 and 6 still are governed completely by the prototype magnitude of $\mathbf{F}_1$, as are hence the left-hand sides. Moreover, for any Froude number the pressure-distribution terms $(p - p_1)/(p_2 - p_1)$ and $(h - h_1)/(h_2 - h_1)$ should be numerically equal to the profile dimension $(d - d_1)/(d_2 - d_1)$ at homologous sections. To the extent that these equalities are actually realized in prototype and model, the flow patterns themselves may be considered comparable.

Momentum Relationship.—To obtain a more precise form of the momentum equation than that customarily used in one-dimensional hydraulics, recourse must be had to the differential equations of motion as adapted to turbulent flow by Osborne Reynolds from the form derived for laminar flow by G. G. Stokes, and as presented in standard texts on hydrodynamics.[5] Space limitations prevent the inclusion in full of the derivations outlined herewith. However, they are readily available in the literature, and they can probably be checked by those with advanced training even without reference to the papers cited. What is important is the implicit demonstration that (a) the customary jump relationships are one-dimensional approximations, (b) a more refined analysis of the role of turbulence must stem from the basic equations of motion, (c) the derived expressions must in turn be simplified, but only in accordance with experimental evidence, and (d) the final results are broadly significant.) Reynolds' adaptation involved, in brief, the replacement of every instantaneous velocity magnitude by the sum of its mean value and its deviation therefrom, and then the retention of only those terms having a temporal average other than zero. Thus, in general, the local vector magnitude of the velocity, V, is considered to have the mean velocity components, $\bar{u}$, $\bar{v}$, and $\bar{w}$, in the directions x, y, and z, respectively, with the corresponding instantaneous deviations, u', v', and w'. For the direction x, the Reynolds equation has the typical form (bars denoting temporal means):

$$\bar{u}\frac{\partial \bar{u}}{\partial x} + \bar{v}\frac{\partial \bar{u}}{\partial y} + \bar{w}\frac{\partial \bar{u}}{\partial z} + \overline{u'\frac{\partial u'}{\partial x}} + \overline{v'\frac{\partial u'}{\partial y}} + \overline{w'\frac{\partial u'}{\partial z}}$$

$$= -\frac{1}{\rho}\frac{\partial \bar{p}}{\partial x} + X + \frac{\mu}{\rho}\left(\frac{\partial^2 \bar{u}}{\partial x^2} + \frac{\partial^2 \bar{u}}{\partial y^2} + \frac{\partial^2 \bar{u}}{\partial z^2} \right) \ldots (8)$$

in which μ is the dynamic viscosity and X the x-component of the body force (such as the attraction of gravity) per unit mass.

Eq. 8 is fundamentally an equation of acceleration, for the terms to the left of the equality sign represent the rates of change of the velocity components

---

[5] "Hydrodynamics," by H. Lamb, Dover Publications, New York, N. Y., 6th Ed., 1946.

of the mean motion and of the turbulence, and the terms to the right of the equality sign represent the pressure, body, and viscous forces per unit mass by which the velocity changes are produced. Through introduction of the two equations of continuity,

$$\frac{\partial \bar{u}}{\partial x} + \frac{\partial \bar{v}}{\partial y} + \frac{\partial \bar{w}}{\partial z} = 0 \dots\dots\dots (9)$$

and

$$\frac{\partial u'}{\partial x} + \frac{\partial v'}{\partial y} + \frac{\partial w'}{\partial z} = 0 \dots\dots\dots (10)$$

the acceleration terms can be written in the alternative form,

$$\frac{\partial (\bar{u}^2)}{\partial x} + \frac{\partial (\bar{u}\,\bar{v})}{\partial y} + \frac{\partial (\bar{u}\,\bar{w})}{\partial z} + \frac{\partial \overline{u'^2}}{\partial x} + \frac{\partial \overline{u'\,v'}}{\partial y} + \frac{\partial \overline{u'\,w'}}{\partial z}$$

which will be used in a subsequent derivation.

Because there is an equation of acceleration for each of the three coordinate directions, a considerable saving in space, without sacrifice of clarity, is gained by the use of tensor notation, in which the quantities u, v, and w are represented by u_i or u_j, and x, y, and z by x_i or x_j. In shorthand form the indices i and j denote, in turn, each of the three coordinate directions. Thus, Eq. 8 and its counterparts for the other two directions are represented at one and the same time by

$$\frac{\partial (\bar{u}_i\,\bar{u}_j)}{\partial x_j} + \frac{\partial \overline{u_i'\,u_j'}}{\partial x_j} = -\frac{1}{\rho}\frac{\partial \bar{p}}{\partial x_i} + X_i + \frac{\mu}{\rho}\frac{\partial^2 \bar{u}_i}{\partial x_j\,\partial x_j} \dots\dots\dots (11)$$

in which the tensor subscript i refers to the direction characterizing each equation, and j denotes the direction of each independent term within it. Therefore, repetition of indices represents a summation of terms.

Any equation of acceleration (like the Newtonian equation itself) may also be regarded as an equation of momentum. It will be noted that multiplication of both sides of Eq. 11 by the mass density will give each term the dimension of rate of change of momentum per unit volume or impulse per unit volume per unit time. The physical significance of the resulting expression becomes greater as it is integrated over a given space through which the fluid moves. By the Gaussian rule of the calculus relating volume integrals and surface integrals,

$$\int_v \frac{\partial (-)}{\partial x_j}\, dV = \int_s (-)\frac{\partial x_j}{\partial n}\, dS \dots\dots\dots (12)$$

the result can be put in the following general form:

$$\int_s \rho\,\bar{u}_i\,\bar{u}_j\,\frac{\partial x_j}{\partial n}\, dS + \int_s \rho\,\overline{u_i'\,u_j'}\,\frac{\partial x_j}{\partial n}\, dS$$

$$= -\int_s \bar{p}\,\frac{\partial x_i}{\partial n}\, dS + \int_v \rho\,X_i\, dV + \int_s \mu\,\frac{\partial \bar{u}_i}{\partial x_j}\,\frac{\partial x_j}{\partial n}\, dS \dots (13)$$

Herein S denotes the surface of the region over which the integration is performed, n the outward normal to the surface, and **V** the enclosed volume. The terms at the left of the equality sign (each representing nine different quantities) embody the net flux of momentum (three components in each of three directions) of the mean flow and of the turbulence out of the region in question—that is, the difference between the rates of efflux and influx. The first term at the right denotes the three components of the mean normal force exerted externally on the surface of the region; the second term, the three components of the weight of the fluid contained within the region; and the third term, the three components of the mean tangential force exerted on the surface.

For the case in which the hydraulic jump is formed on a horizontal bed of great width, the pertinent momentum equation is that for the direction x. If it is assumed that the turbulence is negligible at the initial section, that the pressure distribution is hydrostatic at any arbitrary section, and that both the viscous stress and the turbulent stress are negligible over the free surface, Eq. 13 reduces to

$$\int_0^d \rho\,\bar{u}^2\,dy - \int_0^{d_1} \rho\,\bar{u}^2\,dy + \int_0^d \rho\,\overline{u'^2}\,dy$$

$$= \frac{\gamma\,d_1^2}{2} - \frac{\gamma\,d^2}{2} - \int_0^x \mu\left(\frac{\partial \bar{u}}{\partial y}\right)_{y=0} dx \cdot\cdot (14)$$

For the limiting case in which $x = L_j$ and $d = d_2$, Eq. 14 will be seen to differ in the following ways from the simpler version normally used in studying the jump. First, the change in the mean momentum flux, which is expressed as the difference between the first two integrals, perforce includes the effect of local departure from the average velocity over the vertical section. Second, the change in the momentum flux of the turbulence is represented by the next integral, which embodies the sole effect of turbulence in the force balance. Finally, the effect of variable bed shear,

$$\tau_0 = \mu\left(\frac{\partial \bar{u}}{\partial y}\right)_{y=0} \dots\dots\dots\dots\dots\dots\dots\dots (15)$$

over the length of the jump is included in the integral to the right. Obviously, in applying Eq. 14 each integral must be determined from known distributions of the variables in question, an evident reason for the more customary use of the simplified relationship. However, only through the quantitative evaluation of the integral terms can the magnitude of the error arising from the neglect of the distribution details be appraised.

Equations of Energy.—An essential characteristic of the Reynolds equations of motion, and hence of the momentum equations derived from them, is the absence of all viscous stresses except those involving the mean velocities. However, although the viscous terms containing the velocity fluctuations are invariably eliminated by the averaging process, there remain certain inertial terms consisting of fluctuation products which can conveniently be considered to have the nature of a stress and hence to take their place. These are the momentum-transport terms of the secondary motion, which may be combined with the terms for mean viscous stress to yield

$$\bar{\tau}_{ij} = \mu \left(\frac{\partial \bar{u}_i}{\partial x_j} + \frac{\partial \bar{u}_j}{\partial x_i} \right) - \rho \, \overline{u_i' \, u_j'} \ldots \ldots \ldots \ldots \ldots (16)$$

The momentum-transport terms are known as Reynolds stresses.

If, before averaging, each of the equations of motion is multiplied by the corresponding component of the instantaneous velocity, and the three equations are then added, there will result a differential equation of work and energy[6] of the form

$$\bar{u}_i \frac{\partial}{\partial x_j} \left(\frac{\rho \, \bar{V}^2}{2} + \rho \, \frac{\overline{V'^2}}{2} \right) + \overline{u_j' \frac{\partial (\rho \, V'^2/2)}{\partial x_j}} + \frac{\partial (\bar{u}_i \, \rho \, \overline{u_i' \, u_j'})}{\partial x_j}$$

$$= - \bar{u}_i \frac{\partial \bar{p}}{\partial x_i} - \overline{u_i' \frac{\partial p'}{\partial x_i}} + \rho \, \bar{u}_i \, X_i + \mu \, \bar{u}_i \frac{\partial^2 \bar{u}_i}{\partial x_j \, \partial x_j} + \mu \, \overline{u_i' \frac{\partial^2 u_i'}{\partial x_j \, \partial x_j}} \cdot \cdot (17)$$

Inspection of Eq. 17 will reveal not only that many of the fluctuating stresses which disappeared from the momentum equations during the averaging process are now retained, but also that there is a fairly consistent parallel between the terms for the mean flow and those for the turbulence. In fact, it is instructive to segregate the terms of the two categories in the following work-energy relationships for the mean flow and for the turbulence (which are also directly derivable by combining the pertinent equations and velocity components):

$$\bar{u}_j \frac{\partial (\rho \, \bar{V}^2/2)}{\partial x_j} + \rho \, \bar{u}_i \frac{\overline{\partial u_i' \, u_i'}}{\partial x_j} = - \bar{u}_i \frac{\partial \bar{p}}{\partial x_i} + \rho \, \bar{u}_i \, X_i + \mu \, \bar{u}_i \frac{\partial^2 \bar{u}_i}{\partial x_j \, \partial x_j} \cdot \cdot (18)$$

and

$$\bar{u}_j \frac{\partial (\rho \, \overline{V'^2}/2)}{\partial x_j} + \overline{u_j' \frac{\partial (\rho \, V'^2/2)}{\partial x_j}} + \rho \, \overline{u_i' \, u_j'} \frac{\partial \bar{u}_i}{\partial x_j} = - \overline{u_i' \frac{\partial p'}{\partial x_i}} + \mu \, \overline{u_i' \frac{\partial^2 u_i'}{\partial x_j \, \partial x_j}} \cdot (19)$$

As was done with the momentum form of the equations of motion, it is now in order to integrate the differential work-energy equations over a given region of space. After the appropriate volume integrals have been changed by the Gaussian rule to surface integrals, the work-energy equation for the mean flow assumes the form,

$$\int_S \frac{\rho \, \bar{V}^2}{2} \, \bar{u}_j \frac{\partial x_j}{\partial n} \, dS + \int_S \rho \, \overline{u_i' \, u_j'} \frac{\partial x_j}{\partial n} \, dS - \int_V \rho \, \overline{u_i' \, u_j'} \frac{\partial \bar{u}_i}{\partial x_j} \, dV$$

$$= - \int_S \bar{p} \, \bar{u}_i \frac{\partial x_i}{\partial n} \, dS + \int_V \rho \, X_i \, \bar{u}_i \, dV + \int_S \mu \left(\frac{\partial \bar{u}_i}{\partial x_j} + \frac{\partial \bar{u}_j}{\partial x_i} \right) \bar{u}_i \frac{\partial x_j}{\partial n} \, dS$$

$$- \int_V \mu \left(\frac{\partial \bar{u}_i}{\partial x_j} + \frac{\partial \bar{u}_j}{\partial x_i} \right) \frac{\partial \bar{u}_i}{\partial x_j} \, dV \cdot \cdot (20)$$

The successive terms have the following significance: The term at the extreme left represents the net flux of kinetic energy of the mean motion out of the

<hr>

[6] "The Structure of Turbulent Shear Flow," by A. A. Townsend, Cambridge University Press, Cambridge, 1956.

region in question (that is, the rate of increase in kinetic energy as the fluid passes through the region).   The second and third terms, while also representing energy transport, can best be explained as the rates at which work is done by the Reynolds stresses over the surface and throughout the interior of the region, respectively.   The first and second terms on the right of the equality sign are the rates at which work is done by the external pressures and the body forces.   The third and fourth terms are evidently the rates at which work is done by the viscous stresses of the mean flow over the surface and throughout the interior of the region, respectively.

Now the work done externally by the viscous stresses (the third term at the right) is wholly conservative (since, as is evident from the Gaussian transformation, it is derivable from a potential or space derivative), whereas that done internally (the last term) is wholly dissipative.   Because of the analogy between the turbulent stresses and viscous stresses, one might seek a further parallel between the types of work done in each instance.   In fact, whereas the second term at the left (like the third term at the right) is conservative, the third term at the left can be considered dissipative like the last at the right if one assumes that energy transferred from the mean flow to the turbulence can never be recovered.   The third term at the left, in other words, represents the rate at which turbulence is produced at the expense of the mean flow.   (The integral itself is inherently negative, so that the negative sign before it denotes a positive rate of production.)

The corresponding work-energy equation for the secondary motion is

$$\int_S \frac{\rho \, \overline{V'^2}}{2} \, \bar{u}_j \frac{\partial x_j}{\partial n} \, dS + \int_S \frac{\rho \, \overline{V'^2}}{2} \, u_j' \frac{\partial x_j}{\partial n} \, dS + \int_v \rho \, \overline{u_i' \, u_j'} \, \frac{\partial \bar{u}_i}{\partial x_j} d\mathbf{V}$$

$$= -\int_S \overline{p' \, u_i'} \frac{\partial x_i}{\partial n} \, dS + \int_S \mu \, \overline{\left(\frac{\partial u_i'}{\partial x_j} + \frac{\partial u_j'}{\partial x_i} \right) u_i' } \frac{\partial x_j}{\partial n} \, dS$$

$$- \int_v \mu \, \overline{\left(\frac{\partial u_i'}{\partial x_j} + \frac{\partial u_j'}{\partial x_i} \right) \frac{\partial u_i'}{\partial x_j}} \, d\mathbf{V} \quad .. (21)$$

The first two terms at the left, by analogy to the first term in Eq. 20, represent the net flux of kinetic energy of turbulence out of the region by convection and diffusion (that is, by the mean flow and by the turbulence), respectively. The third term is again the rate of production of turbulence, now a negative quantity in so far as the work accomplished by the turbulence is concerned. The first term on the right of the equality sign is the rate at which work is done externally on the surface of the region by the fluctuating pressures.   The last two terms represent the rates at which work is done by the viscous stresses of the turbulence over the surface of the region and throughout the interior, respectively, the former being conservative and the latter dissipative.

On the previous assumptions that the initial section is free from turbulence, that both the mean stresses and turbulent stresses can be neglected over the free surface, and that the pressure distribution is hydrostatic throughout, reduction of the work-energy equation for the mean motion to the case of the hydraulic jump on a level bed of great width yields

$$\int_0^d \frac{\bar{V}^2}{2\,g}\,\bar{u}\,dy \;-\; \int_0^{d_1} \frac{\bar{V}^3}{2\,g}\,dy \;+\; \int_0^d \frac{\bar{u}\,\overline{u'^2} + \bar{v}\,\overline{u'v'}}{g}\,dy$$

$$-\int_0^d \int_0^x \left[\frac{\overline{u'\,v'}}{g}\left(\frac{\partial \bar{u}}{\partial y} + \frac{\partial \bar{v}}{\partial x} \right) + \frac{\overline{u'^2} - \overline{v'^2}}{g}\,\frac{\partial \bar{u}}{\partial x} \right] dy\,dx$$

$$= q\,d_1 - q\,d + \int_0^d \frac{\mu}{\gamma}\left[2\,\bar{u}\,\frac{\partial \bar{u}}{\partial x} + \bar{v}\left(\frac{\partial \bar{u}}{\partial y} + \frac{\partial \bar{v}}{\partial x} \right) \right] dy$$

$$-\int_0^d \int_0^x \frac{\mu}{\gamma}\left[4\left(\frac{\partial \bar{u}}{\partial x} \right)^2 + \left(\frac{\partial \bar{u}}{\partial y} + \frac{\partial \bar{v}}{\partial x} \right)^2 \right] dy\,dx \quad .. (22)$$

For the limiting case that $x = L_j$ and $d = d_2$, the first two terms on each side of Eq. 22 correspond to those of the usual one-dimensional open-channel relationship, with due account being taken of the variation in velocity over the cross section. The third terms represent work done on the end section by the turbulent and viscous stresses. The last terms combine to form the quantity $H_l\,q$, since they indicate the rate at which turbulence is produced by the mean motion and the rate at which energy is dissipated by the mean viscous stresses.

Reduction of the energy equation for the turbulent motion to the same conditions yields

$$\int_0^d \frac{\overline{V'^2}}{2\,g}\,\bar{u}\,dy \;+\; \int_0^d \frac{\overline{V'^2}}{2\,g}\,u'\,dy$$

$$+\int_0^d \int_0^x \left[\frac{\overline{u'\,v'}}{g}\left(\frac{\partial \bar{u}}{\partial y} + \frac{\partial \bar{v}}{\partial x} \right) + \frac{\overline{u'^2} - \overline{v'^2}}{g}\,\frac{\partial \bar{u}}{\partial x} \right] dy\,dx$$

$$= -\int_0^d \frac{\overline{p'}}{\gamma}\,u'\,dy \;+\; \int_0^d \frac{\mu}{\gamma}\left[\frac{\partial}{\partial x}\left(\frac{\overline{V'^2}}{2} + \overline{u'^2} \right) + \frac{\partial \overline{u'\,v'}}{\partial y} \right] dy$$

$$-\int_0^d \int_0^x K\,\frac{\mu}{\gamma}\overline{\left(\frac{\partial u'}{\partial x} \right)^2}\,dy\,dx \quad .. (23)$$

In Eq. 23 terms appear only for the arbitrary section, as it was assumed that no turbulence existed at the initial section. The proportionality factor, K, permits what are actually nine terms to be written as one. Its use is justified herein by the fact that (a) the dissipation is most intense in the smallest eddies, (b) the smallest eddies tend to be isotropic, and (c) in isotropic turbulence this factor has the constant magnitude[7] of 15. Only if the cumulative production (the third term on the left in Eq. 23) and the cumulative dissipation (the third term on the right) are equal will the turbulence at the final section (that is, for $d = d_2$ at $x = L_j$) be negligible. The extent of the actual inequality of these terms evidently can be ascertained only through detailed measurement and analysis of the mean and fluctuating patterns of flow.

---

[7] "Modern Developments in Fluid Dynamics," by S. Goldstein, Oxford University Press, 1938.

Experimental Equipment and Procedure

To simulate the mean and fluctuating patterns of the hydraulic jump in air, a special duct having a length of 9 ft, a width of 1 ft, and a maximum depth of 1 ft was built of plywood and plexiglass. Both the front wall and the curved surface were made transparent to simplify the visual setting of instruments, and, for added convenience in supporting them, the plane surface corresponding to the bed of the channel was placed on top (Fig. 2). The form of the curved surface was controlled by successive pairs of adjusting screws at 6-in. intervals over the entire length of the duct. Both the plane surface and the curved surface were provided with piezometer holes along the center line, and a grid with a 0.02-ft mesh was ruled on the rear wall as a guide in controlling the surface profile. At the downstream end the duct underwent a smooth

Fig. 2.—Air-Flow Model

transition to a 12-in.-diameter circular section connected to the intake of a 5-hp centrifugal blower, which discharged into the surrounding space through a baffled diffuser. The intake end was provided with a rectangular bell having carefully rounded sides. The bottom side of the bell was adjustable in position, and the curved surface of the test section was hinged to it after a 2.5-in. tangent.

Because the velocity of the flow varied greatly in magnitude and direction, even undergoing full reversal in the roller, it was considered desirable to utilize a single instrument that would respond in the same manner to flow in either direction. This precluded the ordinary type of pitot tube, but either a pitot cylinder with opposing holes or a simple hot-wire anemometer would have sufficed. The instrument actually used was a double pitot, symmetrical upstream and down, formed of two L's of 0.04-in. tubing placed back to back with stagnation openings 0.6 in. apart. The tube was calibrated in an air stream for both magnitude and inclination of the velocity vector, differential pressures being indicated on a precision manometer reading to 0.001 in. of alcohol. In operation the tube was inserted through each of a series of center-line openings at 6-in. intervals in the plane top of the duct, and a velocity traverse was made with the tube in its normal position. On the assumption that the velocity indication corresponded to a zero inclination, the location of the streamlines for equal increments of the stream function, ψ, was evaluated from plots of the velocity distribution (Fig. 3) according to the relationship,

$$\psi = \int_0^y \bar{u}\, dy \dots\dots\dots\dots\dots\dots\dots(24)$$

With the approximate angularity thus indicated by the streamline pattern, the velocity readings were corrected by use of the angularity calibration, and the

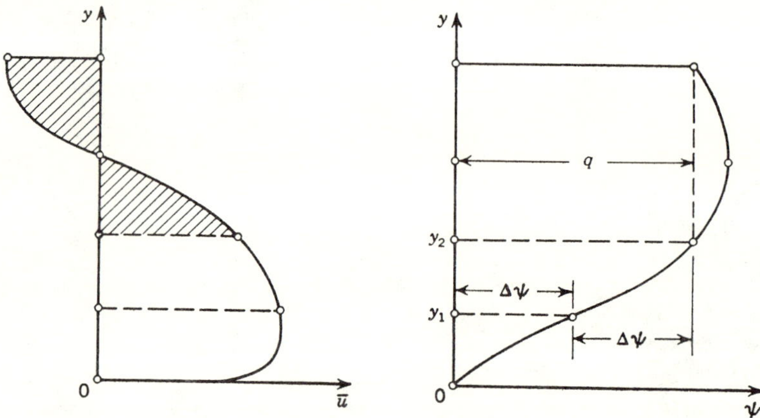

FIG. 3.—LOCATION OF STREAMLINES FROM VELOCITY DISTRIBUTION

determination of streamline location was repeated. A single correction of this nature usually proved sufficient.

Supplementary measurements for the determination of the local intensity of boundary shear were made with a 0.04-in. stagnation tube in contact with the plane boundary. The tube had been calibrated in connection with other boundary-layer measurements, and J. H. Preston's method of shear calculation[8] was followed in both instances. The various characteristics of turbulence ($\sqrt{\overline{u'^2}}$, $\sqrt{\overline{v'^2}}$, $\sqrt{\overline{w'^2}}$, $\overline{u'v'}$, and $\overline{(\partial u'/\partial t)}$) were measured with a constant-temperature hot-wire anemometer,[9] the operation of which is also too specialized for further comment here. It should be remarked, however, that—although designed for a precision of $\pm 2\%$ under ideal conditions—the hot wire (like any method of turbulence measurement) must be expected to have a probable error ranging from $\pm 5\%$ to $\pm 10\%$ in practice and to as much as $\pm 20\%$ in zones of abnormally great intensity of fluctuation.

[8] "The Determination of Turbulent Skin Friction by Means of Pitot Tubes," by J. H. Preston, *Journal of the Royal Aeronautical Society*, London, Vol. 58, 1954.

[9] "Operating Manual for the IIHR Hot-Wire and Hot-Film Anemometers," by P. G. Hubbard, *Bulletin 37*, State Univ. of Iowa Studies in Eng., Iowa City, 1957.

Because the efficacy of the air-model method depends entirely on the degree of similarity between the model and its water prototype, rather extensive preliminary tests were conducted to determine both the mean-flow characteristics of the jump itself and the accuracy with which they could be simulated in the air duct. The jump was produced at various Froude numbers in the 1-ft glass-walled flume of the Institute at approximately the same scale as the air duct (that is, $d_2 = 1$ ft). The form of the free surface was determined by point gage, the pressure distribution on the bottom by water manometer, and the end of the roller by surface float. The several length ratios thus obtained are plotted in Fig. 4. For the Froude number $\mathbf{F}_1 = 4$, the velocity distribution was also measured by the double pitot tube (using the technique of flushing before each reading to eliminate errors due to entrapped air). The streamlines are plotted in Fig. 5 at a distorted scale to exaggerate the vertical displacement.

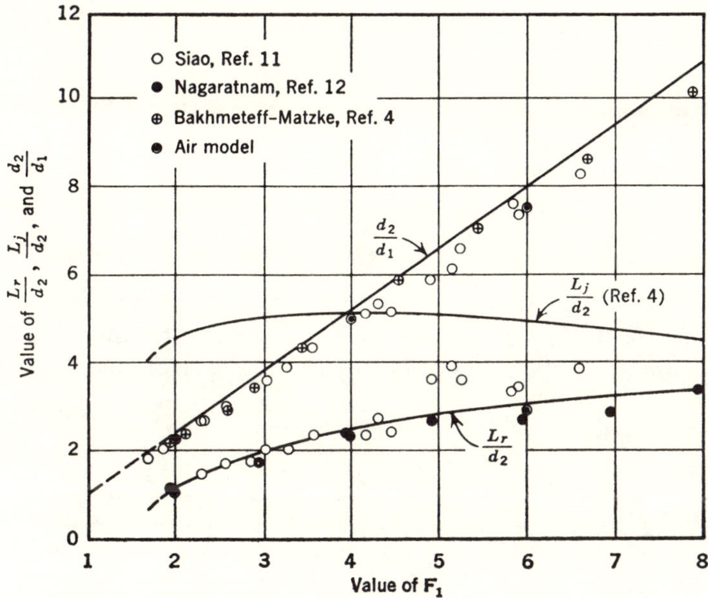

FIG. 4.—LINEAR CHARACTERISTICS OF THE JUMP

Superposed on Fig. 5 are the corresponding streamlines for flow in the air duct at a comparable depth ratio, surface profile, and Reynolds number, together with the pressure distribution on plane and curved surfaces. The agreement between the model and prototype results is seen to be quite satisfactory.

Fig. 4 shows the usual slight deviation of the measured depth ratio from that computed by means of the simplified momentum relationship of Eq. 1. The variation is attributable primarily to the neglect of boundary shear in the course of the derivation. In fact, it was the measured depth ratio rather than the computed ratio that had to be used in setting the duct profile. Moreover, in order to reduce wall shear (and, hence, its influence on the flow pattern) to a minimum, after introductory measurements the original duct was rebuilt to a width of 2.5 ft. All additional data presented herein were obtained with the wider flow section. However, even then, wall effects on the center-line flow

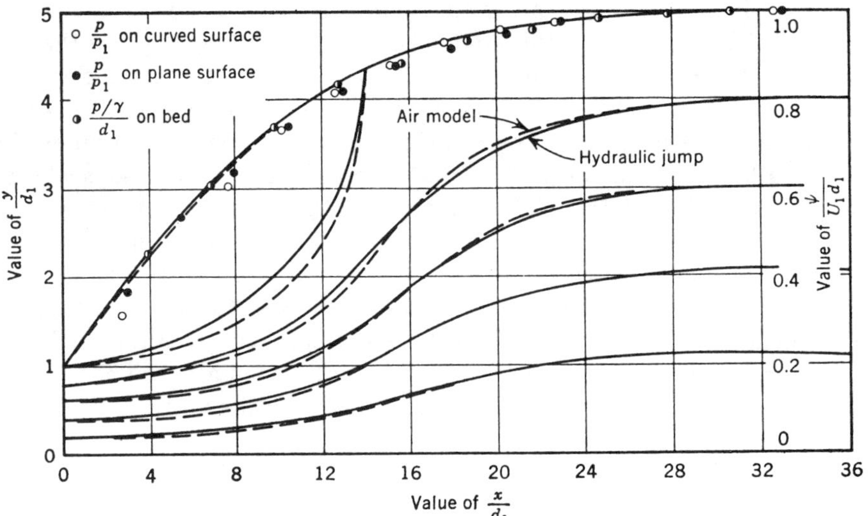

FIG. 5.—COMPARISON OF MEAN PATTERNS OF FLOW IN MODEL AND PROTOTYPE FOR $F_1 = 4$

were still noticeable, because the unit discharge,

$$q = \int \bar{u}\, dy \dots\dots\dots\dots\dots\dots\dots (25)$$

gradually increased with distance downstream. As a first-order correction for this variation, the local unit discharge rather than the initial unit discharge was used in computing the reference velocity. In no case did the change exceed

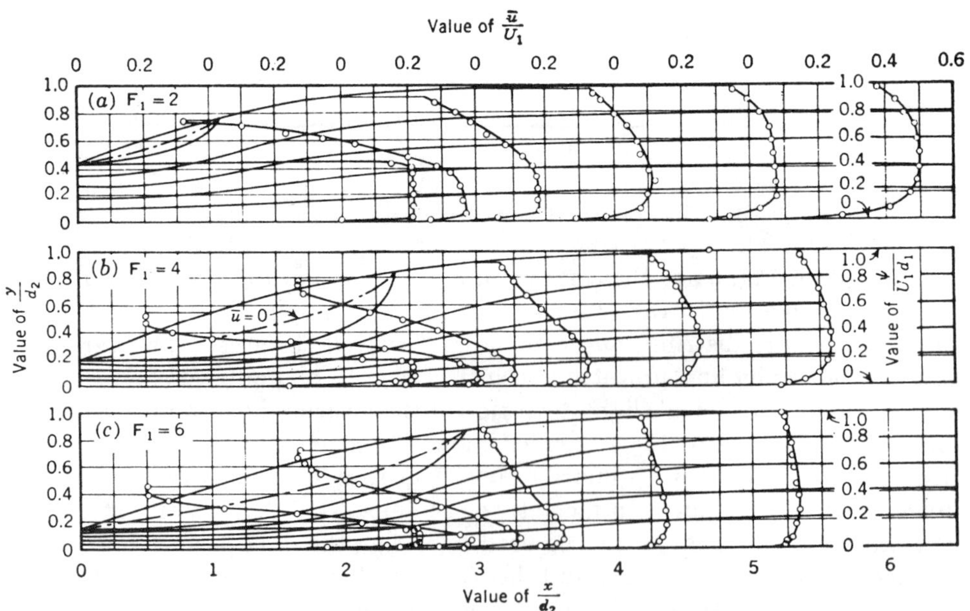

FIG. 6.—DISTRIBUTION OF MEAN LONGITUDINAL VELOCITY AND
CORRESPONDING PATTERNS OF FLOW

6%.　Unfortunately, correction for the volume flux is not the same as that for either the momentum or energy flux, the only method of handling all together being to flare the duct laterally in proportion to the boundary-layer growth.

Following the preliminary tests, systematic measurements were made of the primary and secondary flow patterns for Froude numbers of 2, 4, 6, and 8, though without attaining a satisfactory degree of approximation at the highest of these.　In each instance the downstream dimension was maintained constant at 1 ft, and the upstream dimension and the surface profile were varied in accordance with the prototype requirements.　Reproduction of the necessary conditions was found to become increasingly difficult as the Froude number increased.　In no event, of course, was it possible to obtain exact agreement of

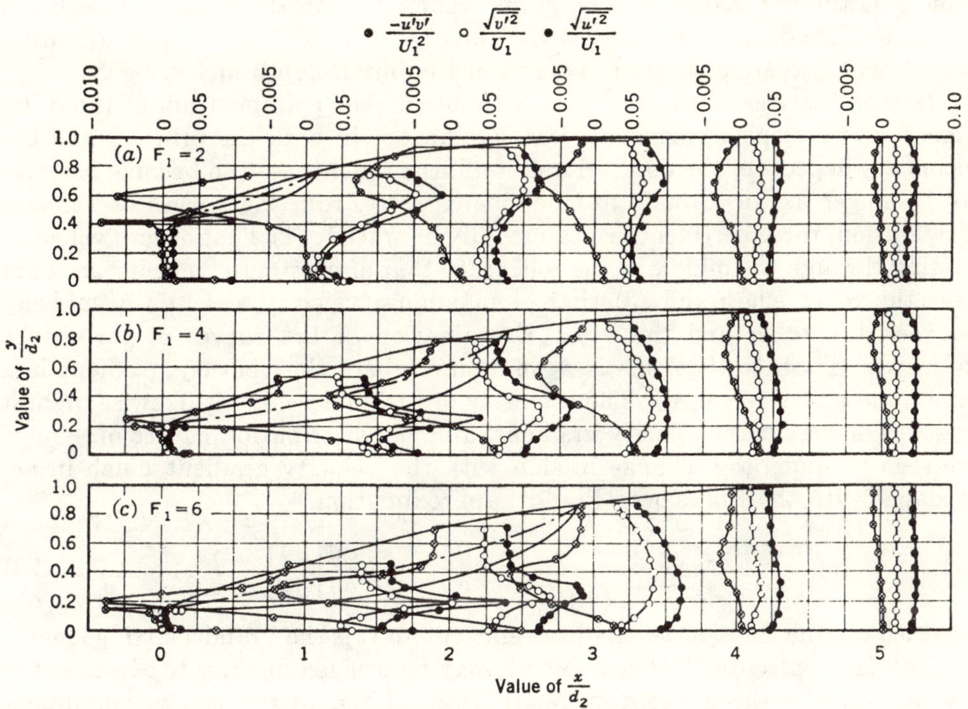

FIG. 7.—DISTRIBUTION OF TURBULENCE INTENSITIES AND MEAN PRODUCT OF COMPONENTS

surface profile, pressure distribution, and flow pattern (as indicated by the length of roller), and, hence, each case involved slight inaccuracies in all three factors to the end of obtaining the best over-all duplication.　A barely perceptible inclination of the short tangent section of the bell was of considerable use in this regard.　However, no such artifice sufficed to produce a state even roughly approximating similarity at $F_1 = 8$, for despite the use of intermediate splitter vanes the rapidly divergent flow could not be stabilized either laterally or longitudinally.

The basic results for the other three Froude numbers are plotted in Figs. 6 and 7.　Fig. 6 embodies the mean-flow data and Fig. 7 the turbulence data. (For reasons of simplicity the w'-component of turbulence is not shown.　It did not differ greatly from v', and, moreover, the measurements indicated that

$$\int_0^d \overline{u'^2}\, dy \approx \int_0^d (\overline{v'^2} + \overline{w'^2})\, dy \dots\dots\dots\dots\dots (26)$$

that is,

$$\int_0^d \overline{V'^2}\, dy \approx 2 \int_0^d \overline{u'^2}\, dy \dots\dots\dots\dots\dots (27)$$

Values of the time derivative of u' will be introduced subsequently.) Whereas the initial depth is commonly used as the length parameter in dealing with the jump as such, in comparing the results for various Froude numbers it proved to be more convenient to use the final depth, as this maintains essentially the same longitudinal scale. Hence, purely geometric ratios are given in terms of d_2. The Froude number necessarily involves d_1, of course, and all other parameters also are expressed in terms of the initial depth and velocity.

Interpretation of Results.—Even without further manipulation of the data, Figs. 6 and 7 display trends of great significance in both the primary and the secondary aspects of the flow. First of all, the roller is seen to become progressively longer as the Froude number increases. From the curves of velocity distribution, three pertinent loci can readily be traced: (1) That of zero velocity, passing through the middle of the roller; (2) that of maximum velocity gradient near the roller edge; and (3) that of maximum velocity, essentially midway between the roller and the bed. Examination of the curves of turbulence intensity will show that this is a minimum where the velocity gradient is a minimum, and vice versa. The intensity of turbulent shear is likewise highest in zones of maximum velocity gradient, and the distribution curves of $-\overline{u'v'}$ are seen to undergo a change in sign with the velocity gradient much in accordance with the Boussinesq-Prandtl approximation,[10]

$$\bar{\tau} = -\rho\,\overline{u'v'} \approx \rho\,\epsilon\,\frac{\partial \bar{u}}{\partial y} \approx \rho\,l\,\frac{\partial \bar{u}}{\partial y}\left|\frac{\partial \bar{u}}{\partial y}\right| \dots\dots\dots\dots (28)$$

in which ϵ is the Boussinesq mixing coefficient and l is the Prandtl mixing length.

The data presented in Figs. 6 and 7 may be utilized directly to evaluate the several terms in the momentum relationship of Eq. 14 reduced to nondimensional form through division by

$$\rho\, U_1{}^2\, d_1 = \rho\, U_1\, q \dots\dots\dots\dots\dots (29)$$

The relative magnitudes of the terms can be shown most conveniently by computing and plotting the sum

$$\frac{1}{U_1{}^2\, d_1}\int_0^d \bar{u}^2\, dy + \frac{1}{U_1{}^2\, d_1}\int_0^d \overline{u'^2}\, dy + \frac{1}{2\,\mathbf{F}_1{}^2}\frac{d^2}{d_1{}^2} + \frac{1}{\mathbf{R}_1\, U_1}\int_0^x \left(\frac{\partial \bar{u}}{\partial y}\right)_{y=0} dx \,. \,(30)$$

as a function of longitudinal distance, as shown in Fig. 8. In Eq. 30, the first term represents the mean momentum flux, and in Fig. 8 this is represented by MM; the second term is pressure, represented by P; the third term denotes

<hr>

[10] "Modern Conceptions of the Mechanics of Fluid Turbulence," by H. Rouse, *Transactions*, ASCE, Vol. 102, 1937.

turbulent momentum flux—TM in Fig. 8; and the fourth term is shear, denoted by S.   Because the sum should be a constant at all successive sections, departure from its original value is a measure of the experimental error.   Of principal significance in connection with these results is the fact that the contribution of the turbulence to the momentum flux, although it is of considerable importance at intermediate sections, is small at each final section.   The contribution of the bed shear, although small also, necessarily increases with both

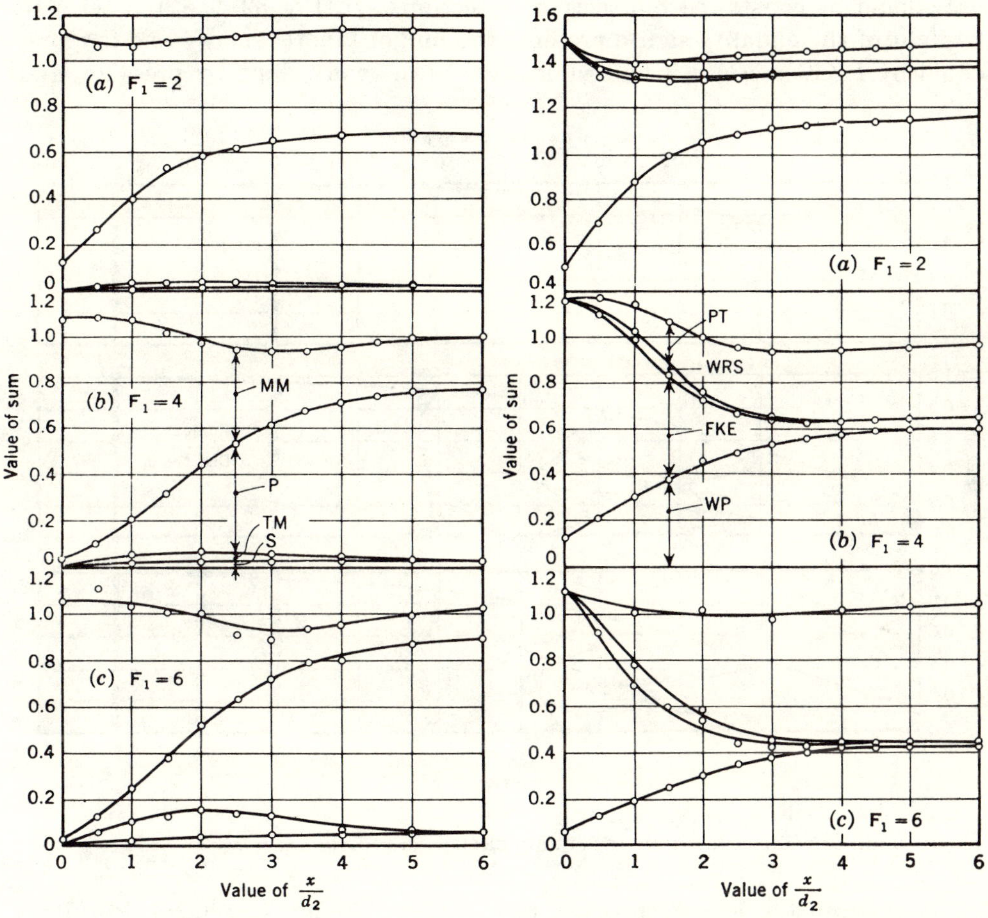

FIG. 8.—DIMENSIONLESS PLOTS OF MOMENTUM BALANCE

FIG. 9.—DIMENSIONLESS PLOTS OF MEAN-ENERGY BALANCE

distance and Froude number (and, it should be noted, with the viscous and roughness effects as well).

A comparable analysis of the energy relationship for the mean flow (Eq. 22) proceeds in much the same fashion.   On the assumption (verified experimentally) that the terms involving the mean viscous stresses are of much smaller order than the others, and with

$$d_1 \frac{U_1^3}{2g} = q \frac{U_1^2}{2g} \dots\dots\dots\dots\dots\dots\dots\dots\dots (31)$$

as the common denominator, now it is the nondimensional sum

$$\frac{1}{U_1{}^3 d_1}\int_0^d \bar{V}^2\,\bar{u}\,dy + \frac{2}{\mathbf{F}_1{}^2}\frac{d}{d_1} + \frac{2}{U_1{}^3 d_1}\int_0^d (\bar{u}\,\overline{u'^2} + \bar{v}\,\overline{u'\,v'})\,dy$$

$$- \frac{2}{U_1{}^3 d_1}\int_0^d \int_0^x \left[\overline{u'\,v'}\left(\frac{\partial\bar{u}}{\partial y} + \frac{\partial\bar{v}}{\partial x}\right) + (\overline{u'^2} - \overline{v'^2})\frac{\partial\bar{u}}{\partial x}\right] dy\,dx \dots (32)$$

that should be constant from section to section. In Eq. 32, the first term to the right of the equality sign represents the flux of kinetic energy and is represented by FKE in Fig. 9; the second term is the work done by pressure, and

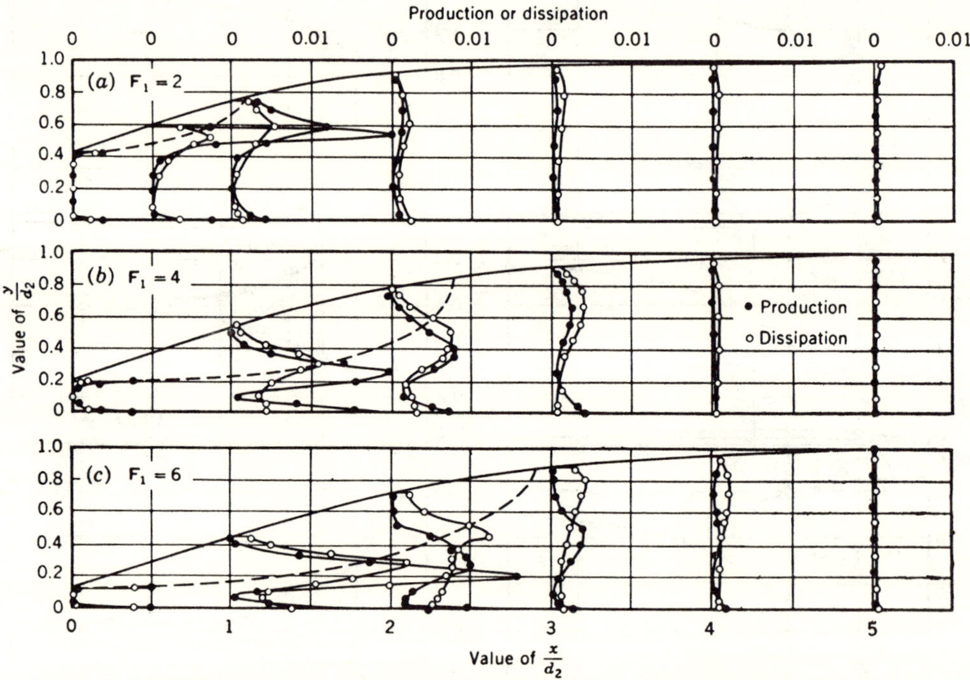

FIG. 10.—DISTRIBUTION OF TURBULENCE PRODUCTION AND DISSIPATION

WP is its symbol in Fig. 9; the third term denotes the work performed by Reynolds stresses—WRS in Fig. 9; and the last term is the production of turbulence and is denoted by PT. Calculation of the various quantities from the data presented in Figs. 6 and 7 yields the curves shown. The gradual change that characterized the corresponding sum in the momentum calculations is even more pronounced (particularly for $\mathbf{F}_1 = 4$). No explanation is evident beyond experimental error, including the effect of wall retardation on the central flow.

Since it reappears in the energy equation of the turbulence, it is the turbulence-production term (the last member of Eq. 32) that remains of primary interest. In fact, its variation in comparison with that of the turbulence-dissipation term is particularly significant, not only from section to section but over each section as well. For this reason the nondimensional form of the

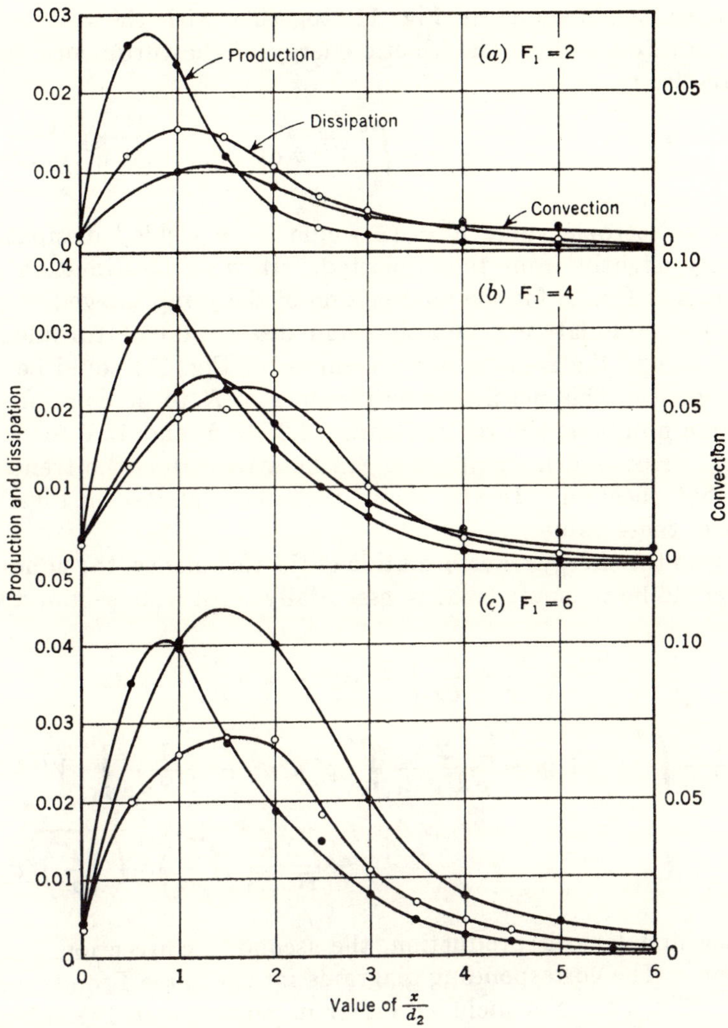

FIG. 11.—LONGITUDINAL VARIATION IN RATES OF PRODUCTION, CONVECTION, AND DISSIPATION OF TURBULENCE

local rate of production,

$$- 2 \frac{d_1}{U_1{}^3} \left[\overline{u'v'} \left(\frac{\partial \bar{u}}{\partial y} + \frac{\partial \bar{v}}{\partial x} \right) + (\overline{u'^2} - \overline{v'^2}) \frac{\partial \bar{u}}{\partial x} \right] \ldots\ldots\ldots (33)$$

(the double integral of which was used in preparing Fig. 9) is shown in detail in Fig. 10. Also shown in Fig. 10 is the corresponding distribution of the local rate of dissipation,

$$2 \frac{K}{\mathbf{R}_1} \frac{d_1{}^2}{U_1{}^2} \overline{\left(\frac{\partial u'}{\partial x} \right)^2} \ldots\ldots\ldots\ldots\ldots\ldots\ldots (34)$$

The measured mean-square temporal gradient $\overline{(\partial u'/\partial t)^2}$ was used to approximate the required mean-square spatial gradient through division by $\bar{u}^2$. The integrals of these distribution curves over the vertical sections are plotted against longitudinal distance in Fig. 11, together with the corresponding integral for the rate at which the kinetic energy of the turbulence is convected past each section:

$$\frac{1}{U_1{}^3 d_1} \int_0^d \overline{V'^2}\, \bar{u}\, dy \dots\dots\dots\dots\dots\dots (35)$$

Use of the isotropic value $K = 15$ would have yielded dissipation curves differing only slightly from those plotted. However, estimation of the remaining terms of Eq. 23 for the final section of the jump showed them to be so small that the cumulative production and dissipation of turbulence energy through the jump (the areas under the curves of Fig. 11) could be considered essentially equal. This permitted bulk values of K to be computed for each of the Froude numbers, the results being 12.2, 13.5, and 15.0 for 2, 4, and 6, respectively. How much significance can be attached to the trend in magnitude is a moot question. In any event, the curves plotted in Figs. 10 and 11 correspond to these values.

In studying the energy transformation of the turbulence, the nondimensional sum that should be constant (that is, essentially zero) from section to section is

$$-\frac{2}{U_1{}^3 d_1} \int_0^d \int_0^x \left[\overline{u'v'}\left(\frac{\partial \bar{u}}{\partial y} + \frac{\partial \bar{v}}{\partial x}\right) + (\overline{u'^2} - \overline{v'^2})\frac{\partial \bar{u}}{\partial x} \right] dy\, dx$$

$$-\frac{1}{U_1{}^3 d_1} \int_0^d \overline{V'^2}\, \bar{u}\, dy - \frac{2}{\rho\, U_1{}^3 d_1} \int_0^d \overline{p'\, u'}\, dy - \frac{1}{U_1{}^3 d_1} \int_0^d \overline{V'^2\, u'}\, dy,$$

$$-\frac{2\, K}{R_1\, U_1{}^2} \int_0^d \int_0^x \overline{\left(\frac{\partial u'}{\partial x}\right)^2}\, dy\, dx \dots (36)$$

in which the first term is production, the second is convection, and the fifth is dissipation. The corresponding diagrams for the three Froude numbers are presented in Fig. 12. Not included either in Eq. 36 or in Fig. 12 itself is the rate at which work is done externally by the viscous stresses, for calculation of this term (the fifth) in Eq. 23 showed it to be of much lower order than the others. On the other hand, it was impossible to evaluate either the rate at which work was done by the fluctuating pressures or the rate of diffusion of the turbulent energy, since neither the velocity-pressure correlation involved in the former nor the triple velocity correlation in the latter could be measured with available equipment. The small difference between the computed values of the other terms might hence be assumed to represent—except for experimental error—the sum of these two.

Several related characteristics of the turbulence should be noted in Figs. 10, 11, and 12. In Fig. 10, the locus of points of maximum production coincides closely with the border of the roller, whereas that for maximum dissipation lies appreciably higher, and the vertical spread of the dissipation curves is likewise greater. This emphasizes the fact that the turbulence is not wholly dissipated at its point of formation, but is convected by the mean motion (including the secondary flow of the eddy) and diffused by its own mixing action. Much

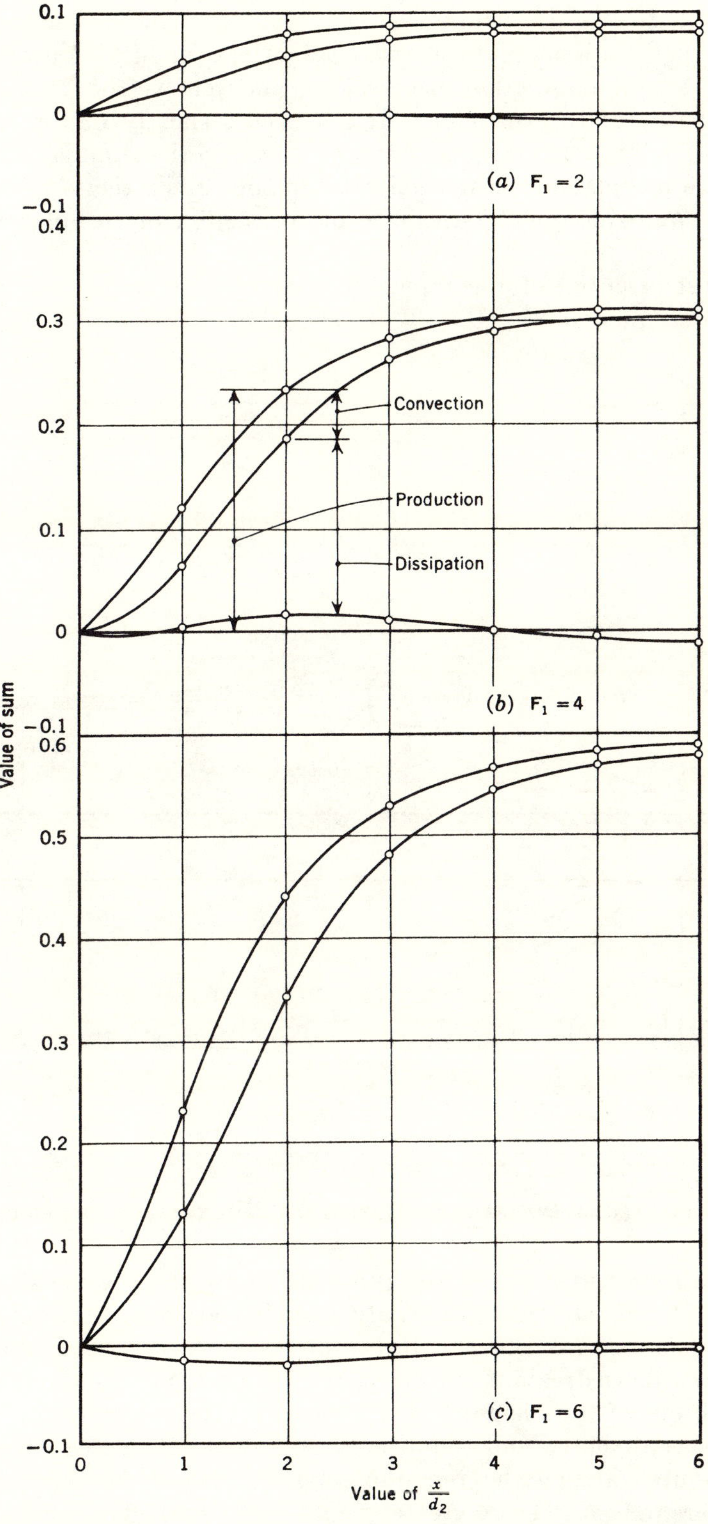

Fig. 12.—Dimensionless Plots of Turbulent-Energy Balance

the same situation is seen from Fig. 11. Even though the maximum rate of production per section occurs even ahead of the middle of each roller, the maximum rate of dissipation per section (as well as the maximum cross-sectional energy of the turbulence itself) occurs shortly before its end. As shown graphically by Fig. 12, the difference between the cumulative production rate and the cumulative dissipation rate up to a given section is represented with close approximation by the rate of convection of turbulence past that section.

The most essential of the foregoing results are summarized in the three geometric diagrams of Fig. 13. Whereas these correspond to the usual one-

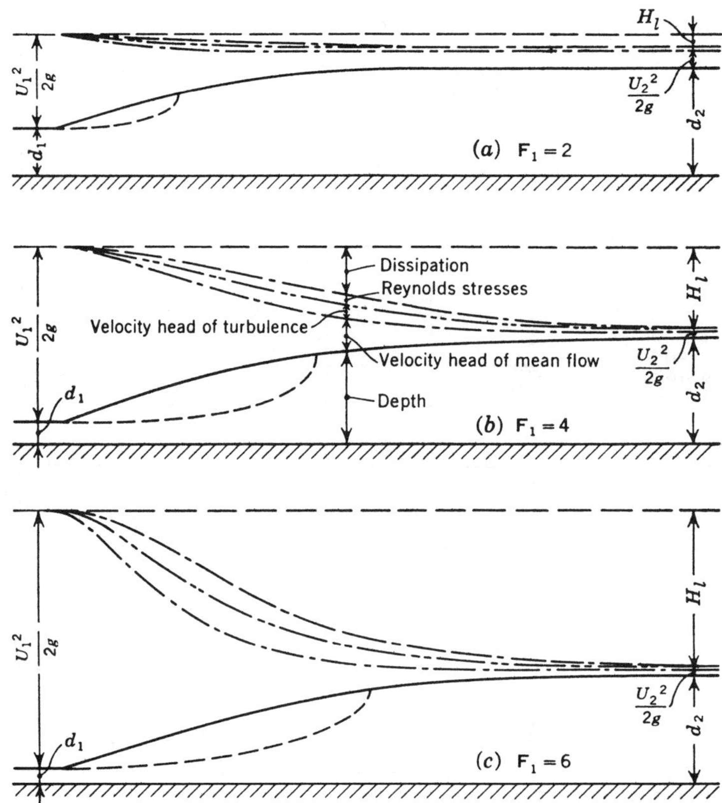

FIG. 13.—GENERALIZED DIAGRAMS OF ENERGY TRANSFORMATION THROUGH THE JUMP

dimensional representations of the jump, not only are they drawn to proper scale longitudinally as well as vertically, but they are quantitatively significant within the nonuniform zone of transition as well as in the uniform zones upstream and downstream. They are, to be sure, approximations rather than exact renditions of the phenomenon, because the least important of the variables have been ignored and the discrepancies evident in Figs. 9 and 12 have been distributed among the remaining terms. All in all, however, the inaccuracies are judged to be no greater than the variable effects of bed roughness and slope encountered in the field.

Lowermost among the curves plotted in Fig. 13 are those showing the variation of depth (that is, the profile of the free surface) for each of the Froude numbers. Next are those representing the addition of the velocity head of the mean flow, $\dfrac{1}{q}\displaystyle\int_0^d \dfrac{\bar V^2}{2\,g}\,\bar u\,dy$, to the depth, much as in the usual manner. However, although the correction factor

$$\alpha = \frac{1}{U^2\,q}\int_0^d \bar V^2\,\bar u\,dy \dots\dots\dots\dots\dots\dots\dots\dots (37)$$

compensating for the use of the mean velocity $U = q/d$ is usually assumed to be unity, its magnitude within the nonuniform zone was found to rise to as high as 4; on the other hand, at the end of the transition it was within 5% of unity in every case. It should be noted that the two sets of curves just cited are the only two included in the customary diagrams.

Third in the sequence illustrated are the curves showing the increase in head resulting from the inclusion of the velocity head of the turbulence, $\dfrac{1}{q}\displaystyle\int_0^d \dfrac{\overline{V'^2}}{2\,g}\,\bar u\,dy$. For the effect that the turbulence has on the flow the relative magnitude of this term is surprisingly small, even at the sections where it is a maximum; at the beginning of the jump, of course, it was purposely made insignificant; at the end it is small compared to the maximum value that it attained, although still an appreciable fraction of the final mean-velocity head. Next in order are the curves indicating the addition of the terms for the work done by the Reynolds stresses, $\dfrac{1}{g\,q}\displaystyle\int_0^d (\bar u\,\overline{u'^2} + \bar v\,\overline{u'\,v'})\,dy$—quantities that are very small in zones of normal turbulence but too great to be ignored completely in zones of pronounced mixing. The only other quantity with a magnitude comparable to those already plotted is the head lost through viscous dissipation, $\dfrac{K\,\nu}{g\,q}\displaystyle\int_0^d\int_0^x \overline{\left(\dfrac{\partial u'}{\partial x}\right)^2}\,dy\,dx$, and this is represented by the drop in the sum of all the other terms from section to section. As has already been emphasized, this drop is essentially complete at the end of the jump.

Conclusions

Through the analysis of turbulence measurements conducted in an air-flow model of the hydraulic jump under conditions simulating three representative Froude numbers, significant information has been obtained on the energy transformation associated with the phenomenon. In particular, the part played by the roller in producing the turbulence by which the transformation is effected has taken on new clarity. Although the measurements were necessarily lacking in both precision and completeness, their analysis proceeded in accordance with the basic equations of motion, and conclusions based thereon can be considered to have both qualitative and quantitative worth.

As in any other type of flow past a zone of discontinuity, at the juncture between the oncoming stream and the return flow or roller of the hydraulic jump there is a pronounced velocity gradient, and the resulting shear gives rise to the rapid generation of turbulence. The effect of the turbulence, as usual, is twofold. On the one hand, the mixing process diffuses all characteristics of the flow—momentum, energy, and even the turbulence itself—both laterally and longitudinally. On the other hand, the increased viscous shear produces a rapid conversion of mechanical energy to heat. Were it not for the roller, the formation of the turbulence—and hence the dissipation of energy —would be minimal, as can be demonstrated in a closed duct so formed that the roller is eliminated. Except for Froude numbers well below 2, however, the hydraulic jump without a breaking front is physically impossible. The roller is thus an inseparable part of the phenomenon.

In view of the customary practice of considering the length of the jump to be the distance between the limiting sections of uniform flow, it should be noted that the roller extends about one-half this distance, and that one-half the energy of the turbulence is produced within the first half length of the roller. Because of both the convective effect of the mean flow and the diffusive effect of the turbulence itself, this energy is not wholly dissipated at the point that it is produced, but rather some distance away both vertically and longitudinally. In other words, there is a distinct lag between the loss of energy to the mean flow in the form of turbulence and its ultimate dissipation in the form of heat.

Of interest in this connection is the surprisingly small kinetic energy of the turbulence at any section, even in the zone of maximum production, convection, and dissipation at the midsection of the roller. At the final section of the jump, moreover, it is already small in comparison with the maximum value that it attained. Hence the belief that the energy lost to the mean flow persists for a considerable distance in the form of turbulence is definitely in error. In fact, the hydraulic jump is not only an effective means of reducing mean-flow energy but a very efficient mechanism for restoring conditions of uniformity to the flow.

Acknowledgments

Development of the experimental technique and method of analysis, as well as the complete investigation for the intermediate Froude number in the original duct, formed the dissertation project[11] of the second writer and one of his assignments as Research Associate of the Iowa Institute. Measurement and evaluation of data for all three Froude numbers in the enlarged duct served as the thesis project[12] of the third writer and his sole assignment as Research Assistant. Both were frequently assisted by Philip G. Hubbard, A. M. ASCE, in the operation of the hot-wire anemometer. The senior writer conceived and directed the investigation and prepared the final manuscript. Partial support was received from the Office of Naval Research, Washington, D. C., under Contract N8onr-500(03) with the Iowa Institute.

---

[11] "Characteristics of Turbulence in an Air-Flow Model of the Hydraulic Jump," by T. T. Siao, thesis presented to the State University of Iowa, at Iowa City, in 1954, in partial fulfilment of the requirements for the degree of Doctor of Philosophy.

[12] "The Mechanism of Energy Dissipation in the Hydraulic Jump," by S. Nagaratnam, thesis presented to the State University of Iowa, at Iowa City, in 1957, in partial fulfilment of the requirements for the degree of Master of Science.

APPENDIX. NOTATION

The following symbols, adopted for use in the paper, conform essentially with "American Standard Letter Symbols for Hydraulics" (ASA Y10.2—1958), prepared by a committee of the American Standards Association with Society representation, and approved by the Association in 1958:

d = depth of flow;

$\mathbf{F}$ = Froude number;

g = acceleration of gravity;

H = total head (sum of velocity and piezometric heads);

H_l = head lost in jump;

h = piezometric head (sum of pressure head and elevation);

i = tensor subscript, corresponding to coordinate direction of equation of motion;

j = tensor subscript, corresponding to each coordinate direction in succession;

K = dissipation coefficient;

L_j = length of jump;

L_r = length of roller;

l = Prandtl mixing length;

n = the outward normal to the surface;

p = pressure intensity;

q = rate of flow per unit width;

$\mathbf{R}$ = Reynolds number;

S = surface area;

t = time;

U = average of u across vertical section;

u = instantaneous longitudinal component of velocity at a point;

V = local magnitude of velocity vector;

$\mathbf{V}$ = volume;

v = instantaneous vertical component of velocity at a point;

w = instantaneous lateral component of velocity at a point;

X = longitudinal component of body force per unit mass;

x = longitudinal coordinate;

y = vertical coordinate;

z = lateral coordinate;

α = velocity-head coefficient for energy equation;

γ = specific weight;

ϵ = Boussinesq mixing coefficient;

μ = dynamic viscosity;

ν = kinematic viscosity;

ρ = mass density;

ψ = stream function; and

τ = bed shear.

A bar ($\overline{}$) denotes the temporal mean value of the quantity beneath it, and a prime ($'$) denotes an instantaneous deviation from the temporal mean.

THÈSES

PRÉSENTÉES A LA

FACULTÉ DES SCIENCES DE L'UNIVERSITÉ DE PARIS

POUR OBTENIR LE GRADE DE DOCTEUR ÈS SCIENCES PHYSIQUES

PAR

Hunter ROUSE

Répartition de l'énergie
dans des zones de décollement

par Hunter ROUSE

S.B., S.M., M.I.T.; DR.-ING., KARLSRUHE

INTRODUCTION

Dans le domaine de l'hydraulique, le décollement local de l'écoulement principal d'avec la paroi est une caractéristique très connue des phénomènes. Soit pour raison d'économie, par indifférence du projeteur, ou pour cause d'impraticabilité technique, l'usage de profils aérodynamiques, pour éviter ce phénomène, est toujours l'exception plutôt que la règle. Des éléments de plomberie, des raccordements de canaux, des entrées de conduites et des piles de ponts ont ainsi souvent un profil coudé plutôt qu'arrondi, où le décollement est habituellement la cause de vibrations, de pertes de charge, ou d'érosion. Mais il est vrai que, parfois, le décollement et la formation de tourbillons sont des aspects inévitables du mode d'écoulement, comme dans le cas des vannes, des chicanes et des bassins d'amortissement. En outre, les profils de parois coudés sont en certains cas préférables à ceux insuffisamment arrondis, parce que ceux-ci peuvent provoquer des intermittences dans le processus de décollement, lequel n'a presque rien en sa faveur et beaucoup contre lui. En tout cas, la mécanique de l'écoulement dans une zone de décollement semblerait en beaucoup de cas être plus importante pour l'hydraulicien que pour tout autre savant ou ingénieur. C'est peut-être pourquoi de tels écoulements ont trouvé jusqu'à maintenant aussi peu d'attention sur le plan théorique. Cependant, il y a là une bonne raison pour l'hydraulicien de prendre une part importante dans son étude.

L'hydraulique s'est depuis longtemps intéressée presque exclusivement aux analyses de l'écoulement moyen, et en conséquence, elle a négligé le mouvement secondaire ou turbulent qui se superpose au mouvement primaire ou moyen. Dans les dernières années, d'autres professions ont apporté beaucoup d'attention à la turbulence même, en limitant le phénomène à tel point que l'écoulement moyen pouvait être négligé. Au moyen de telles simplifications, on a beaucoup appris sur les écoulements moyens et sur les bases de la turbulence. Dans la plupart des états du mouvement fluide réel, cependant, ni l'un ni l'autre de ces aspects ne peut être étudié tout seul d'une manière satisfaisante, car chacun a un rôle déterminant des caractéristiques de l'autre. Et cela est spécialement vrai pour le phénomène de décollement vu sous ses aspects les plus étendus.

Des progrès considérables ont été faits dans l'aérodynamique en prévoyant le point de décollement sur des parois progressivement courbées en fonction de la géométrie des parois et du nombre de Reynolds — pourvu que le décollement n'ait aucun effet appréciable sur l'écoulement voisin, ou l'affecte simplement d'une manière qui peut être évaluée par l'analyse des lignes de jet. Cependant, en cas de parois simples, comme celles que montre la figure 1, dans

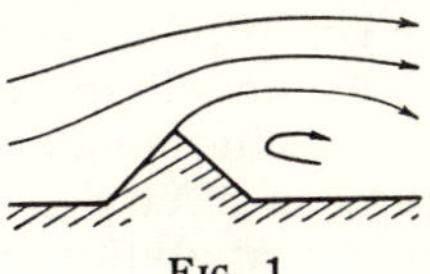

Fig. 1

Décollement à partir d'une irrégularité de paroi

laquelle le point de décollement est fixé, non seulement il est impossible de prévoir avec exactitude la forme moyenne de la zone de décollement, mais encore, même avec cette connaissance, on ne pourrait estimer qu'approximativement la répartition de la vitesse et de la pression dans le tourbillon et la perte d'énergie qui en résulte; et d'autres caractéristiques de l'écoulement secondaire, telles que l'intensité de la turbulence dans et derrière le tourbillon et la fréquence des variations à son extrémité postérieure, pourraient seulement faire l'objet d'hypothèses.

Le rôle de tels tourbillons dans la transformation de l'énergie a été depuis longtemps sousestimé par les uns, et surestimé par les autres. Nécessairement, le schéma des lignes de courant de l'écoulement moyen montre un vortex stable, avec un système de lignes aussi bien défini que celui du courant extérieur, et on est tenté d'attribuer au tourbillon la même quantité d'énergie par unité de volume que celle de l'écoulement propre. C'est une grave erreur, car la ligne de décollement n'est pas une ligne de courant libre, mais une ligne au long de laquelle l'énergie décroît sans cesse. Cependant, puisque le tourbillon est en effet une source variable de tourbillons plus petits perdus dans son sillage, il est fréquemment représenté comme une sorte de « moulin à tourbillons » dans lequel il se produit de la turbulence et par suite, une perte d'énergie. En réalité, la zone primaire de production de la turbulence se trouve entre le tourbillon et l'écoulement même, comme dans tous les cas de mélange à travers une surface de vitesse discontinue, et la turbulence est convoyée et diffusée avant que son énergie soit tout à fait dissipée. C'est l'écoulement inversé qui distingue ce phénomène des cas plus simples de mélange : d'une part, le fluide entraîné du côté du tourbillon est déjà fortement turbulent; d'autre part, l'extrémité aval du tourbillon change de position pendant que d'appréciables quantités se déchargent de façon intermittente dans l'écoulement propre.

Le problème est évidemment bien loin des deux états du mouvement avec lequel on a déjà obtenu des succès analytiques. Il diffère du cas isotrope, représenté par des sections normales de l'écoulement en aval d'une grille, parce qu'il constitue un problème d'écoulement fortement anisotrope avec des tensions tangentielles; et il diffère du cas de similitude, symbolisé par l'écoulement établi dans un sillage ou une conduite, parce qu'il est spécifiquement un phénomène localisé, bien que son effet s'étende loin en aval. Faute d'instruments analytiques sûrs, l'étude du phénomène doit procéder d'une manière quasiempirique : il faut examiner des cas caractéristiques avec autant de détails expérimentaux

que possible; apprécier les valeurs des quantités non encore mesurables en comparant les mesures possibles et les équations du mouvement; et essayer, lorsque le mécanisme du tourbillon devient mieux connu, d'en donner l'expression sous une forme mathématique approchée.

Sous notre direction, plusieurs études d'exploration de ce genre ont déjà été faites dans les Laboratoires de l'Institut des Recherches Hydrauliques de Iowa. La première fut un projet de doctorat, dressé par H.C. Hsu [1] sur le décollement en aval d'un élargissement brusque dans une conduite à deux dimensions; les caractéristiques de l'écoulement moyen et de la turbulence furent mesurées et rattachées aux équations du mouvement, de telle sorte que le passage de l'énergie de l'écoulement primaire à l'écoulement secondaire et jusqu'à sa dissipation thermique, put être représentée en gros. Celle-ci fut suivie par une étude de M. Arie [2] sur l'écoulement derrière une paroi perpendiculaire [3], dont les résultats furent analysés sous forme énergétique [4]. Des études semblables, quoique moins dépendantes de la géométrie des parois, furent celles conduites par O. Erickson et nous-même sur un jet normal à un écoulement transversal [5], par T.T. Siao [6] et S. Nagaratnam [7] sur le ressaut hydraulique [8], et par T.K. Rao [9] sur un jet dirigé axialement contre un écoulement. Chacun de ces cas permet de constater l'existence commune d'une zone stable d'écoulement inversé — un tourbillon stationnaire — dont la turbulence était engendrée le long des bords; la convection, la diffusion et la dissipation consécutives à cette turbulence furent des facteurs essentiels de détermination de la forme de l'écoulement entier.

La thèse suivante s'avance davantage vers la compréhension finale de ce phénomène; elle décrit une étude de l'écoulement en symétrie axiale auprès de deux configurations comparables de parois, d'intérêt fondamental en hydraulique et dans les domaines voisins : une entrée de conduite et un cylindre à bout trempé. Les mouvements primaire et secondaire sont étudiés, en s'efforçant de déterminer le processus du transfert d'énergie depuis l'écoulement moyen, par la turbulence jusqu'à sa forme finale, la chaleur. En prenant ces écoulements comme exemples, la signification des relations entre les formes d'énergie différentes est discutée en détail dans le but de faciliter à l'avenir de semblables études.

EXAMEN PRÉLIMINAIRE DES ÉCOULEMENTS A ÉTUDIER

Une entrée brusque dans une conduite située dans la paroi d'un grand réservoir et un cylindre de section normale sont des parois qui semblent au premier coup d'œil bien différentes l'une de l'autre, car l'une représente les voies par lesquelles un fluide s'écoule et l'autre les corps qui sont transportés généralement par les fluides. Bien entendu, le principe du mouvement relatif élimine facilement l'une des deux différences apparentes. Pourtant, quant à l'autre, la véritable similitude entre les deux espèces de parois ne devient apparente qu'après les avoir étudiées en fonction de leurs homologues à deux dimensions. En fait, une telle représentation a un double but, puisque seules les images planes sont, même approximativement, susceptibles d'être soumises à l'analyse mathématique, et cette approximation est souhaitable au moins pour comprendre qualitativement les principes de physique qui sont à la base des écoulements respectifs.

En examinant la figure 2, on voit que les élé-

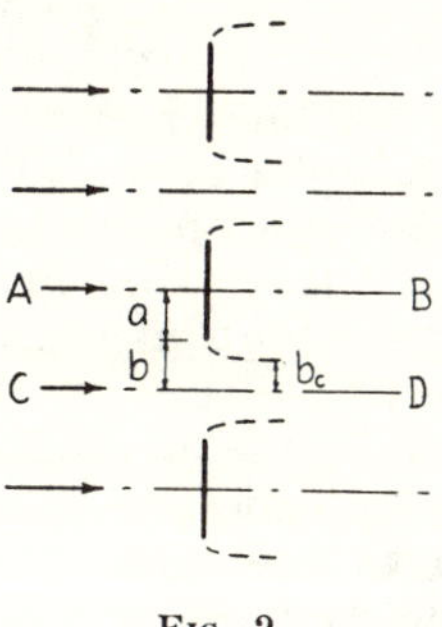

Fig. 2

Ecoulement par une série de plaques normales

ments multiples de l'écoulement autour d'une série de plaques à arêtes vives peuvent être considérés de deux manières différentes. Ainsi l'élément ABCD se combine au-dessus avec son image pour former un passage symétrique en relation avec la ligne AB représentant alors l'écoulement renfermé autour d'une plaque; la ligne CD et son image marquent la position des parois solides. De même, on peut considérer CD comme axe de symétrie, avec des parois solides le long de AB et de son image, ainsi que les deux demi-plaques, tout en simulant un passage avec un étranglement local de chaque côté. Si, pour un instant, on néglige le mélange des sillages voisins en aval, les lignes de décollement sembleraient être les mêmes dans chaque cas. La véritable distinction entre les deux

points de vue ne devient évidente que lorsque la demi-épaisseur a et la demi-distance b entre les plats deviennent de plus en plus différentes. Lorsque le rapport a/b devient très inférieur à l'unité, par exemple, c'est l'aspect de corps immergé de cet arrangement qui est accentué; et quand le rapport devient très grand, c'est l'aspect de l'écoulement qui prédomine. Naturellement, la limite $a/b=0$ du premier cas est la plaque de dimension finie dans un fluide infini, alors que la limite $a/b=\infty$ de l'autre cas est une fente de dimension finie dans une paroi infiniment grande.

En utilisant la méthode de l'analyse conforme, avec l'hypothèse de lignes de courant libres, von Mises [10] put évaluer le rapport b_c/b de contraction du jet pour toutes les valeurs du rapport a/b des parois, y compris, bien entendu, la limite de Kirchhoff :

$$\pi/(\pi+2)=0{,}61$$

pour la valeur $a/b=\infty$. Ces résultats ont été bien approchés par des expériences avec des jets libres aussi bien qu'immergés. Cependant, l'application de la même méthode d'analyse à l'écoulement autour d'une plaque donne des résultats qui ne sont vérifiés que pour une interface entre un liquide et un gaz (c'est-à-dire une poche de cavitation) de grande étendue, car c'est seulement en ce cas que la pression dans la poche est proche de celle de l'écoulement ambiant. Cependant, comme l'a démontré Riabouchinsky [11], l'hypothèse soit d'une pression plus basse soit d'une courbure plus rapide permettra à des méthodes d'analyse de donner des résultats plus conformes aux mesures.

Bien sûr, une telle concordance est seulement approchée, car elle dépend des forces ou des variations de pression plus que des particularités locales de l'écoulement. Celles-ci, en effet, sont tout à fait contraires à l'hypothèse de la ligne de courant libre, sauf au voisinage immédiat du point de décollement. Par suite des tensions tangentielles élevées le long de la surface libre du décollement, sauf aux faibles nombres de Reynolds, c'est ici le siège d'une intense production de turbulence, laquelle diffuse transversalement pendant qu'elle se transporte en aval avec l'écoulement et en même temps modifie les particularités d'une zone devenant toujours plus large. Dans les cas limites $a/b=0$ et $a/b=\infty$, les conditions sont celles, respectivement, du sillage croissant et du jet croissant, dont les solutions approchées sont connues depuis plusieurs

327

décennies [12]. La base de telles solutions est la similitude, observée expérimentalement, de la répartition de la vitesse en sections successives bien en aval des irrégularités initiales. Bien entendu, dans leur voisinage immédiat, les conditions de similitude n'existent ni dans un cas ni dans l'autre.

Des considérations semblables s'appliquent à l'écoulement à travers la cascade de barres, montrée dans la figure 3, car, dans l'une des

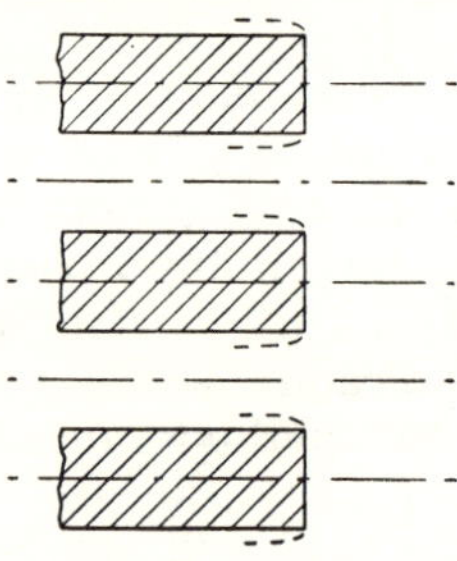

Fig. 3

Ecoulement par une série de barres parallèles

limites, on peut supposer qu'elle représente l'écoulement d'un réservoir dans une conduite uniforme, et à l'autre limite, l'écoulement non confiné autour d'un mur ou d'une pile isolés. Finalement, elles aussi mènent à des états de mouvement qui ont été analysés avec succès en fonction des répartitions de similitude dont l'écoulement plat et uniforme est l'une des limites, et l'autre, le développement d'une couche limite le long d'une surface plane. Au voisinage immédiat des sections initiales, cependant, la similitude n'existe pas, et les méthodes conformes n'ont pas d'autre utilité qu'un sens qualitatif. Par contre, il est évident que les deux états de mouvement ont des traits communs, car, dans chacun, il faut que la poche de décol-

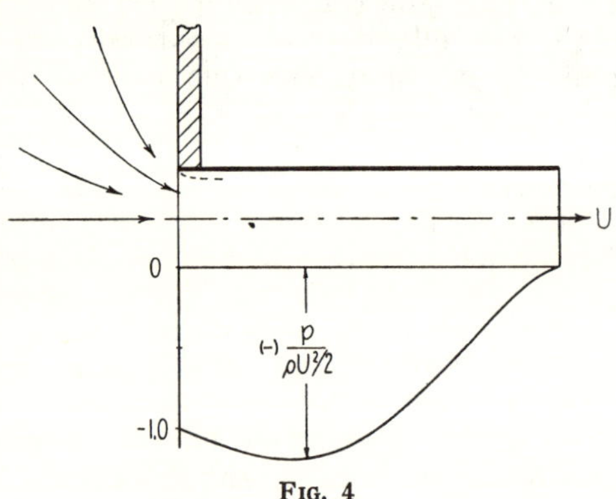

Fig. 4

Répartition de la pression
pour l'écoulement hors d'un tube court

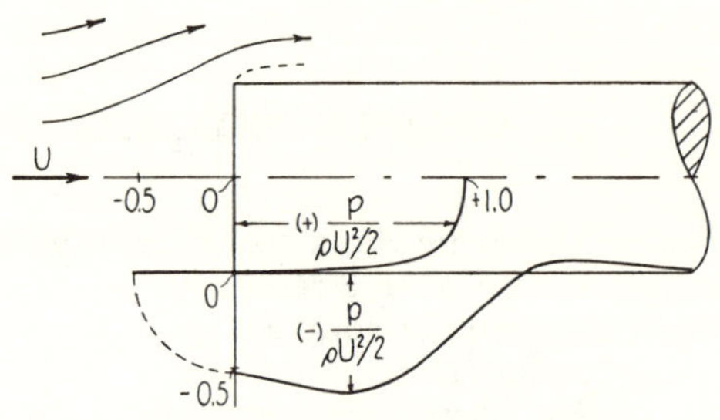

Fig. 5

Répartition de la pression
pour l'écoulement autour d'un cylindre tronqué

lement ait une longueur finie, et que les conditions entre la paroi et le bord de la poche soient essentiellement pareilles. En effet, c'est en dehors de la zone de décollement que la différence fondamentale se trouve, car au-delà de l'une il y a une grande étendue de fluide relativement non perturbé, tandis qu'en face de l'autre existe la même zone de décollement en image réfléchie.

Si l'on examine ensuite les homologues de ces schémas en symétrie axiale (fig. 4 et 5), on verra que les similarités pertinentes dans les cas bidimensionnels sont toutes retenues, tandis que les dissimilarités deviennent encore plus accentuées. Evidemment, les cas de symétrie axiale ne peuvent plus être considérés comme les limites d'un cas paramétrique; au contraire, chacun est en effet la version inverse de l'autre. Bien que ces cas soient encore moins sujets à l'analyse rigoureuse que leurs homologues en deux dimensions, les mêmes observations générales leur sont applicables, et on verra que d'autres encore pourront être trouvées.

Si l'on néglige l'effet de la viscosité le long de la paroi et la déviation croissante des lignes de courant d'un cours purement linéaire, un débit Q à travers une ouverture circulaire dans la paroi d'un réservoir peut être défini au moyen de la fonction de courant d'une source négative,

$$\psi = \frac{Q}{2\pi} \frac{x}{\sqrt{x^2 + r^2}} \qquad (1)$$

pour laquelle les composantes de la vitesse sont, comme d'habitude, $u = \partial\psi/r\partial r$ dans la direction axiale x, et $v = -\partial\psi/r\,\partial x$ dans la direction radiale r. Puisque l'écoulement est rapidement accéléré, l'effet de la viscosité peut être, en fait, négligé sans erreur appréciable, mais la déviation des lignes de courant devient assez importante dans le voisinage de la zone d'émission. Ce désaccord pourrait être éliminé en partie au moyen de la fonction de courant pour une répartition de sources négatives sur le plan de l'ouverture; pourtant, la connaissance expérimentale du schéma de l'écoulement (ou au moins de la surface du décollement) est nécessaire pour déterminer la répartition cherchée, et le sujet est par conséquent d'un intérêt plutôt académique que pratique. En revanche, il est connu, grâce à des études numériques [13], que les coefficients de contraction pour des orifices circulaires, considérés comme rapports d'aire, sont à peu près, sinon même exactement, égaux à ceux des fentes ayant les mêmes proportions linéaires; à la limite, par exemple, $(d_c/d)^2 = 0{,}61$. De plus, l'expérience [14] a montré que cette valeur est presque aussi exacte pour le jet immergé que pour le jet libre. En supposant que le même degré de contraction soit accusé par l'écoulement à l'entrée d'une conduite, la perte de charge à la Borda résultante :

$$\Delta H = \frac{(U_1 - U_2)^2}{2\,g} \qquad (2)$$

a été, pendant presque un siècle, combinée avec le coefficient de contraction limite :

$$\left(\frac{d_c}{d}\right)^2 = 0{,}61 = \frac{U}{U_c} \approx \frac{U_2}{U_1} \qquad (3)$$

pour donner comme perte de charge à l'entrée la valeur :

$$\frac{\Delta H}{U^2/2\,g} = 1 - 0{,}61 \approx 0{,}5 \qquad (4)$$

qui est utilisée d'habitude dans les projets d'hydraulique. Sauf pour plusieurs mesures de la répartition de pression dans un tuyau court, reproduites dans la figure 4 [15], et d'innombrables mesures de pression et de vitesse dans les conduites, loin de l'entrée [16], il semble qu'on en connaît un peu plus sur l'écoulement dont il s'agit.

On connaît encore moins l'écoulement axial autour d'un cylindre à section droite, au moins du point de vue historique, car c'est seulement à propos de la cavitation des projectiles qu'ont été faites des recherches, et cela seulement dans les dernières décennies. Si, pourtant, on conduit la discussion de la même manière, on peut remarquer [17] ceci : les mesures ont montré que la zone de décollement en aval d'un disque a un diamètre environ 1,7 fois celui du disque, le coefficient de traînée a une valeur :

$$\frac{F/(\pi d^2/4)}{\varrho U^2/2} = 1{,}20 \qquad (5)$$

et le coefficient de pression du sillage immédiat est :

$$\frac{p}{\varrho U^2/2} = -0{,}45 \qquad (6)$$

Les mesures de la répartition de pression le long du cylindre [18] sont comme il est indiqué dans la figure 5 où le diamètre de la poche est $1{,}5\,d$, et le coefficient de pression minimum $-0{,}62$. Si la tension tangentielle à la surface du cylindre est négligée, le coefficient de traînée du cylindre doit être au plus 0,78 résultant seulement de la pression sur la face en amont. Pour les sections du sillage sur lesquelles la décroissance de la pression est négligeable, le coefficient de traînée doit correspondre à l'intégrale de la quantité de mouvement [19] :

$$F = 2\,\pi\varrho \int_{d/2}^{\infty} u\,(U - u)\,r\,dr \qquad (7)$$

Si la forme de la zone de décollement et de la répartition de vitesse subséquente était connue même approximativement, de telle sorte que l'épaisseur du déplacement de la perturbation puisse être calculée, la configuration de l'écoulement irrotationnel environnant (sauf peut-être au voisinage immédiat de la face tronquée) pourrait être exprimée par la fonction de courant pour une répartition de doublets [3, 19] de puissance $\Delta\,(t)$ par unité de longueur le long de l'axe de symétrie depuis $x = t = a$ jusqu'à b :

$$\psi = r^2\,U\left\{\, 1/2 - \int_a^b \frac{\Delta(t)}{[(t-x)^2 + r^2]^{3/2}}\,dt \,\right\} \qquad (8)$$

Evidemment, cela exigerait, de nouveau, beaucoup de données expérimentales, quoique la partie limite soit connue pour la couche limite le long d'une paroi cylindrique [20].

Les deux cas d'écoulement dont il s'agit sont évidemment assujettis à l'analyse mathématique dans les régions irrotationnelles en amont, à la condition que certaines valeurs déterminantes soient fournies par des expériences; de plus, une connaissance considérable — analytique autant qu'empirique — existe déjà pour les zones rotationnelles de similitude bien en aval. Jusqu'à ce jour, néanmoins, on n'a pu obtenir qu'une idée qualitative de la transition entre les zones initiale et finale de l'écoulement, sans la mesurer en détails. Les phénomènes de production et de dissipation de la turbulence sont bien connus pour être extrêmement complexes et sauf si l'on emploie à tout le moins les équations hydrodynamiques de base, une telle mesure serait certainement un procédé hasardeux. C'est pourquoi, beaucoup d'attention sera apportée dans la suite à la validité de ces équations.

FORMULATION ET INTERPRÉTATION DES ÉQUATIONS D'ÉNERGIE

Equations de Stokes et de Reynolds

Les équations fondamentales du mouvement d'un fluide en coordonnées cylindriques x, r et θ se dérivent en égalant à la variation totale des composantes de la vitesse u, v et w, les composantes respectives de la force par unité de masse agissant sur un élément fluide tel que celui montré

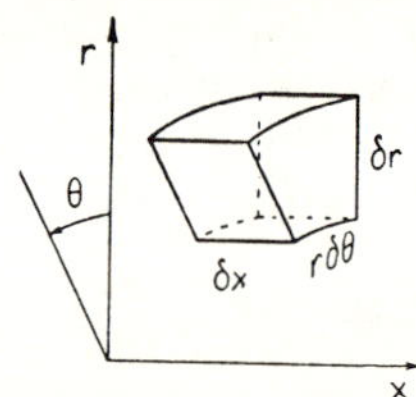

FIG. 6
Notation des coordonnées cylindriques

par la figure 6. Lorsque le volume de l'élément approche de la limite zéro, ces équations se réduisent à la forme [21] :

$$\left.\begin{aligned}
\frac{\partial u}{\partial t}+u\,\frac{\partial u}{\partial x}+v\,\frac{\partial u}{\partial r}+w\,\frac{\partial u}{r\partial\theta} &=-\frac{\partial\Omega}{\partial x}+\frac{1}{r}\left[\frac{\partial(r\sigma_x)}{\partial x}+\frac{\partial(r\tau_{rx})}{\partial r}+\frac{\partial(r\tau_{\theta x})}{r\partial\theta}\right] \\
\frac{\partial v}{\partial t}+u\,\frac{\partial v}{\partial x}+v\,\frac{\partial v}{\partial r}+w\,\frac{\partial v}{r\partial\theta}-\frac{w^2}{r} &=-\frac{\partial\Omega}{\partial r}+\frac{1}{r}\left[\frac{\partial(r\tau_{xr})}{\partial x}+\frac{\partial(r\sigma_r)}{\partial r}+\frac{\partial(r\tau_{\theta r})}{r\partial\theta}-\sigma_\theta\right] \\
\frac{\partial w}{\partial t}+u\,\frac{\partial w}{\partial x}+v\,\frac{\partial w}{\partial r}+w\,\frac{\partial w}{r\partial\theta}+\frac{vw}{r} &=-\frac{\partial\Omega}{r\partial\theta}+\frac{1}{r}\left[\frac{\partial(r\tau_{x\theta})}{\partial x}+\frac{\partial(r\tau_{r\theta})}{\partial r}+\frac{\partial(r\sigma_\theta)}{r\partial\theta}+\tau_{r\theta}\right]
\end{aligned}\right\} \quad (9)$$

Ici Ω représente la fonction « potentiel dû à la pesanteur » (hauteur multipliée par le poids effectif du fluide par unité de masse); σ_x est la tension normale par unité de superficie sur une surface perpendiculaire à l'axe des x, nécessairement dans la direction x; τ_{rx} est la tension tangentielle par unité de superficie sur une surface perpendiculaire à l'axe des r dans la direction x, etc. Pour des raisons d'équilibre, $\tau_{rx}=\tau_{xr}$, etc. Aux équations de mouvement doit être ajoutée celle de continuité,

$$\frac{\partial(ru)}{\partial x}+\frac{\partial(rv)}{\partial r}+\frac{\partial(rw)}{r\partial\theta}=0 \qquad (10)$$

qui se dérive en exprimant ou bien le fait que le volume d'un élément fluide de densité ρ constante ne doit pas changer pendant qu'il se déplace d'un point à l'autre, ou bien, ce qui équivaut, que l'écoulement instantané qui sort d'une zone élémentaire de l'espace est simultanément égal à l'écoulement qui y entre.

Il y a à peu près un siècle que Stokes [22] exprima les tensions précédentes en fonction de la viscosité dynamique μ, comme suit :

$$\left.\begin{aligned}
\sigma_x &=-p+2\,\mu a & \sigma_r &=-p+2\,\mu b & \sigma_\theta &=-p+2\,\mu c \\
\tau_{r\theta} &=\mu f & \tau_{\theta x} &=\mu g & \tau_{xr} &=\mu h
\end{aligned}\right\} \qquad (11)$$

Ici, p représente l'intensité moyenne de la pression :

$$p=-\frac{\sigma_x+\sigma_r+\sigma_\theta}{3} \qquad (12)$$

et les lettres a à h représentent les déformations linéaires et angulaires :

$$\left.\begin{aligned}
a &=\frac{\partial u}{\partial x} & b &=\frac{\partial v}{\partial r} & c &=\frac{\partial w}{r\partial\theta}+\frac{v}{r} \\
f &=\frac{\partial v}{r\partial\theta}+\frac{\partial w}{\partial r}-\frac{w}{r} & g &=\frac{\partial w}{\partial x}+\frac{\partial u}{r\partial\theta} & h &=\frac{\partial u}{\partial r}+\frac{\partial v}{\partial x}
\end{aligned}\right\} \qquad (13)$$

L'introduction de ces tensions dans les équations du mouvement eut pour résultat celles qu'on appelle les équations de Navier-Stokes :

$$\begin{aligned}
\varrho\left(\frac{\partial u}{\partial t}+u\,\frac{\partial u}{\partial x}+v\,\frac{\partial u}{\partial r}+w\,\frac{\partial u}{r\partial\theta}\right) &=-\,\frac{\partial}{\partial x}\,(p+\varrho\Omega)+\mu\left(\frac{\partial^2 u}{\partial x^2}+\frac{\partial^2 u}{\partial r^2}+\frac{1}{r}\,\frac{\partial u}{\partial r}+\frac{\partial^2 u}{r^2\partial\theta^2}\right)\\[2mm]
\varrho\left(\frac{\partial v}{\partial t}+u\,\frac{\partial v}{\partial x}+v\,\frac{\partial v}{\partial r}+w\,\frac{\partial v}{r\partial\theta}-\frac{w^2}{r}\right) &\\[2mm]
&=-\,\frac{\partial}{\partial r}\,(p+\varrho\Omega)+\mu\left(\frac{\partial^2 v}{\partial x^2}+\frac{\partial^2 v}{\partial r^2}+\frac{1}{r}\,\frac{\partial v}{\partial r}-\frac{v}{r^2}+\frac{\partial^2 v}{r^2\partial\theta^2}-\frac{2}{r}\,\frac{\partial w}{r\partial\theta}\right)\\[2mm]
\varrho\left(\frac{\partial w}{\partial t}+u\,\frac{\partial w}{\partial x}+v\,\frac{\partial w}{\partial r}+w\,\frac{\partial w}{r\partial\theta}+\frac{vw}{r}\right) &\\[2mm]
&=-\,\frac{\partial}{r\,\partial\theta}\,(p+\varrho\Omega)+\mu\left(\frac{\partial^2 w}{\partial x^2}+\frac{\partial^2 w}{\partial r^2}+\frac{1}{r}\,\frac{\partial w}{\partial r}-\frac{w}{r^2}+\frac{\partial^2 w}{r^2\partial\theta^2}+\frac{2}{r}\,\frac{\partial v}{r\partial\theta}\right)
\end{aligned}\tag{14}$$

Si chacune de ces équations était multipliée par la composante respective de la vitesse et si les trois étaient additionnées, le membre de gauche de l'équation résultante représenterait, par unité de volume, la variation temporelle de l'énergie cinétique du fluide, et le membre de droite le travail correspondant. L'énergie est alors conservée. Le travail total, en revanche, se calcule en multipliant la tension sur chaque face d'un élément typique par la composante correspondante de la vitesse, en additionnant les produits algébriquement, et en divisant par le volume de l'élément. Si le travail conservatif est alors soustrait du travail total, le résultat sera la dissipation de l'énergie sous l'action de la viscosité [23] :

$$\Phi=\mu\,(2\,a^2+2\,b^2+2\,c^2+f^2+g^2+h^2)$$

$$=\mu\left[\left(\frac{\partial u}{\partial x}\right)^2+\left(\frac{\partial v}{\partial x}\right)^2+\left(\frac{\partial w}{\partial x}\right)^2+\left(\frac{\partial u}{\partial r}\right)^2+\left(\frac{\partial v}{\partial r}\right)^2+\left(\frac{\partial w}{\partial r}\right)^2+\left(\frac{\partial u}{r\partial\theta}\right)^2+\left(\frac{\partial v}{r\partial\theta}\right)^2+\left(\frac{\partial w}{r\partial\theta}\right)^2\right]\tag{15}$$

Reynolds adapta les équations de Stokes à l'étude de l'écoulement turbulent en introduisant une valeur moyenne et la déviation instantanée de la moyenne, pour les trois composantes de la vitesse, et pour la pression, et en éliminant dans la suite tous les termes dont la valeur temporelle moyenne était nécessairement égale à zéro. Si on indique comme d'habitude les valeurs moyennes par des tirets supérieurs et les déviations par des « primes », et si, pour raison de simplification, on réduit les équations à celles qui s'appliquent aux problèmes de l'écoulement moyen stationnaire avec $w=0$ dans la direction θ, les équations de Reynolds deviendront :

$$\begin{aligned}
\varrho\left(\bar u\,\frac{\partial\bar u}{\partial x}+\bar v\,\frac{\partial\bar u}{\partial r}+\overline{u'\,\frac{\partial u'}{\partial x}}+\overline{v'\,\frac{\partial u'}{\partial r}}+\overline{w'\,\frac{\partial u'}{r\partial\theta}}\right) &=-\,\frac{\partial}{\partial x}\,(\bar p+\varrho\Omega)+\mu\left(\frac{\partial^2\bar u}{\partial x^2}+\frac{\partial^2\bar u}{\partial r^2}+\frac{1}{r}\,\frac{\partial\bar u}{\partial r}\right)\\[2mm]
\varrho\left(\bar u\,\frac{\partial\bar v}{\partial x}+\bar v\,\frac{\partial\bar v}{\partial r}+\overline{u'\,\frac{\partial v'}{\partial x}}+\overline{v'\,\frac{\partial v'}{\partial r}}+\overline{w'\,\frac{\partial v'}{r\partial\theta}}-\frac{\overline{w'^2}}{r}\right) &=-\,\frac{\partial}{\partial r}\,(\bar p+\varrho\Omega)+\mu\left(\frac{\partial^2\bar v}{\partial x^2}+\frac{\partial^2\bar v}{\partial r^2}+\frac{1}{r}\,\frac{\partial\bar v}{\partial r}-\frac{\bar v}{r^2}\right)
\end{aligned}\tag{16}$$

L'équation de continuité correspondante est simplement :

$$\frac{\partial(r\bar u)}{\partial x}+\frac{\partial(r\bar v)}{\partial r}=0\tag{17}$$

Au premier coup d'œil, les équations sembleraient être changées seulement par le remplacement des valeurs instantanées par les valeurs moyennes, et par l'addition de quelques termes renfermant des combinaisons moyennes des variations. Les membres de droite de ces équations ont ainsi été simplifiés par l'élimination des tensions fluctuantes visqueuses, qui — si compliquées qu'elles soient — ont nécessairement une valeur temporelle moyenne de zéro. En même temps, pourtant, les termes « accélérateurs » variables, qui n'ont pas nécessairement une valeur moyenne de zéro, ont rendu les membres de gauche d'autant plus compliqués. En effet, ces termes accélérateurs peuvent se construire comme transferts moyens de quantités de mouvement, et par conséquent (par analogie avec la théorie de la viscosité moléculaire) comme tensions dues au mélange d'une échelle molaire au lieu de moléculaire. Cela s'éclaircit au cours de la manipulation suivante des quantités en question :

$$\left.\begin{aligned}
u'\,\frac{\partial u'}{\partial x} + v'\,\frac{\partial u'}{\partial r} + w'\,\frac{\partial u'}{r\partial\theta} &= \frac{\partial\,(u'u')}{\partial x} + \frac{\partial\,(u'v')}{\partial r} + \frac{\partial\,(u'w')}{r\partial\theta} - u'\left(\frac{\partial u'}{\partial x} + \frac{\partial v'}{\partial r} + \frac{v'}{r} + \frac{\partial w'}{r\partial\theta}\right) + u'\,\frac{v'}{r} \\[2mm]
u'\,\frac{\partial v'}{\partial x} + v'\,\frac{\partial v'}{\partial r} + w'\,\frac{\partial v'}{r\partial\theta} - \frac{w'^2}{r} &= \frac{\partial\,(u'v')}{\partial x} + \frac{\partial\,(v'v')}{\partial r} + \frac{\partial\,(v'w')}{r\partial\theta} \\[2mm]
& \qquad - v'\left(\frac{\partial u'}{\partial x} + \frac{\partial v'}{\partial r} + \frac{v'}{r} + \frac{\partial w'}{r\partial\theta}\right) + v'\,\frac{v'}{r} + \frac{w'w'}{r}
\end{aligned}\right\} \quad (18)$$

Les termes à l'intérieur des grandes parenthèses sont égaux à zéro, en vertu de l'équation de continuité, et les produits de ϱ par le restant des termes — logiquement nommés tensions de Reynolds — seront tous trouvés analogues aux tensions de Stokes exprimées en fonction de la vitesse moyenne. En d'autres mots :

$$\left.\begin{aligned}
\overline{\sigma}_x &= -\,p + 2\,\mu\,\frac{\partial\overline{u}}{\partial x} - \varrho\overline{u'^2}; \quad \overline{\sigma}_r = -\,p + 2\,\mu\,\frac{\partial\overline{v}}{\partial r} - \varrho\overline{v'^2}; \quad \overline{\sigma}_\theta = -\,p + 2\,\mu\,\frac{\overline{v}}{r} - \varrho\overline{w'^2} \\[2mm]
\overline{\tau}_{r\theta} &= 0; \quad \overline{\tau}_{\theta x} = 0; \quad \overline{\tau}_{xr} = \overline{\tau}_{rx} = \overline{\tau} = \mu\left(\frac{\partial\overline{u}}{\partial r} + \frac{\partial\overline{v}}{\partial x}\right) - \varrho\overline{u'v'}
\end{aligned}\right\} \quad (19)$$

L'introduction de ces tensions dans les équations de Reynolds, avec une manipulation des termes moyens de la gauche, se ramène à la forme assez symétrique :

$$\left.\begin{aligned}
\frac{\varrho}{r}\left[\frac{\partial\,(r\overline{u}^2)}{\partial x} + \frac{\partial(r\overline{uv})}{\partial r}\right] &= -\,\varrho\,\frac{\partial\Omega}{\partial x} + \frac{1}{r}\left[\frac{\partial\,(r\overline{\sigma}_x)}{\partial x} + \frac{\partial\,(r\overline{\tau})}{\partial r}\right] \\[2mm]
\frac{\varrho}{r}\left[\frac{\partial(r\overline{uv})}{\partial x} + \frac{\partial\,(r\overline{v}^2)}{\partial r}\right] &= -\,\varrho\,\frac{\partial\Omega}{\partial r} + \frac{1}{r}\left[\frac{\partial\,(r\overline{\tau})}{\partial x} + \frac{\partial\,(r\overline{\sigma}_r)}{\partial r} - \overline{\sigma}_\theta\right]
\end{aligned}\right\} \quad (20)$$

Des équations d'accélération comme celles-ci peuvent aussi être interprétées comme des relations entre la variation de quantité du mouvement, à gauche, et l'impulsion correspondante, à droite. N'importe laquelle d'entre elles peut ainsi être intégrée sur une zone donnée de l'espace pour produire une équation d'impulsion du type fréquemment rencontré en hydraulique. L'intégrale volumétrique de l'équation de Reynolds pour la direction axiale sera évidemment :

$$\varrho\int_{\textit{v}}\frac{1}{r}\left[\frac{\partial\,(r\overline{u}^2)}{\partial x} + \frac{\partial(r\overline{uv})}{\partial r}\right]dV = -\,\varrho\int_{\textit{v}}\frac{\partial\Omega}{\partial x}\,dV + \int_{\textit{v}}\frac{1}{r}\left[\frac{\partial\,(r\overline{\sigma}_x)}{\partial x} + \frac{\partial\,(r\overline{\tau})}{\partial r}\right]dV \quad (21)$$

En appliquant la relation de Gauss :

$$\int_{\textit{v}}\frac{1}{r}\left[\frac{\partial\,(rP)}{\partial x} + \frac{\partial\,(rQ)}{\partial r}\right]dV = \int_{s}\left(P\,\frac{\partial x}{\partial n} + Q\,\frac{\partial r}{\partial n}\right)dS \quad (22)$$

dans laquelle n représente la distance perpendiculaire hors de la surface, les intégrales volumétriques convenables peuvent être transformées en intégrales superficielles, avec le résultat significatif que :

$$\varrho\int_{s}\left(\overline{u}^2\,\frac{\partial x}{\partial n} + \overline{uv}\,\frac{\partial r}{\partial n}\right)dS = -\,\varrho\int_{\textit{v}}\frac{\partial\Omega}{\partial x}\,dV + \int_{s}\left(\overline{\sigma}_x\,\frac{\partial x}{\partial n} + \overline{\tau}\,\frac{\partial r}{\partial n}\right)dS \quad (23)$$

Le membre de gauche représente alors la variation de la composante x de la quantité de mouvement par unité de volume pendant que le fluide passe par la zone en question — c'est-à-dire la différence entre la quantité sortante et la quantité entrante par unité de temps. En général, le fluide entre ou sort de la zone par tous les points de la surface environnante; cependant, en hydraulique, la zone considérée est ordinairement limitée sur les côtés par des parois solides, au travers desquelles aucun fluide ne s'écoule, et, aux extrémités, par des sections normales, auxquelles, seule, une des composantes de la vitesse s'applique. Le terme gravitationnel est simplement la composante exacte de la pesanteur du fluide contenu dans la zone. Toutes les tensions internes sont évidemment sans effet (i.e. contrebalancées). Pour le reste des tensions, la partie moyenne visqueuse est limitée en général à l'action tangentielle le long des parois, dont les tensions moyennes turbulentes disparaissent; de même, la partie moyenne turbulente est limitée en général aux sections normales de l'écoulement, où les tensions visqueuses sont, en comparaison, négligeables.

Equations d'énergie

Puisque les relations comme celles de l'accélération ou de la quantité de mouvement s'écrivent nécessairement en fonction de composantes vectorielles, elles ne peuvent pas comprendre une mesure directe de la perte d'énergie, qui est une quantité scalaire. De même que pour les équations de l'écoulement visqueux, afin d'obtenir des expressions pour l'énergie de l'écoulement turbulent [25], il faut multiplier chaque équation par la composante respective de la vitesse (écrite partout comme la somme des parties moyennes et variables) et après, additionner les résultats. (On doit remarquer que, pour retenir tous les termes importants, il faut effectuer la multiplication avant de faire les moyennes temporelles et d'omettre des termes.) L'issue finale moyenne est la variation temporelle totale de l'énergie par unité de volume :

$$\frac{\rho}{2}\left(\bar{u}\,\frac{\partial \overline{V^2}}{\partial x} +\bar{v}\,\frac{\partial \overline{V^2}}{\partial r} +\bar{u}\,\frac{\partial \overline{V'^2}}{\partial x} +\bar{v}\,\frac{\partial \overline{V'^2}}{\partial r} +\overline{u'\,\frac{\partial V'^2}{\partial x}} +\overline{v'\,\frac{\partial V'^2}{\partial r}} +\overline{w'\,\frac{\partial V'^2}{r\partial \theta}} \right)$$

$$=-\,\bar{u}\,\frac{\partial}{\partial x}\,(\bar{p}+\rho\Omega) -\bar{v}\,\frac{\partial}{\partial r}\,(\bar{p}+\rho\Omega) -\overline{u'\,\frac{\partial p'}{\partial x}} -\overline{v'\,\frac{\partial p'}{\partial r}} -\overline{w'\,\frac{\partial p'}{r\partial \theta}}$$

$$+\mu\left(\bar{u}\,\frac{\partial^2 \bar{u}}{\partial x^2} +\bar{u}\,\frac{\partial^2 \bar{u}}{\partial r^2} +\frac{\bar{u}}{r}\,\frac{\partial \bar{u}}{\partial r} +\bar{v}\,\frac{\partial^2 \bar{v}}{\partial x^2} +\bar{v}\,\frac{\partial^2 \bar{v}}{\partial r^2} +\frac{\bar{v}}{r}\,\frac{\partial \bar{v}}{\partial r} -\frac{\bar{v}^2}{r^2} \right.$$

$$+\overline{u'\,\frac{\partial^2 u'}{\partial x^2}} +\overline{u'\,\frac{\partial^2 u'}{\partial r^2}} +\overline{\frac{u'}{r}\,\frac{\partial u'}{\partial r}} +\overline{u'\,\frac{\partial^2 u'}{r^2\,\partial \theta^2}}$$

$$+\overline{v'\,\frac{\partial^2 v'}{\partial x^2}} +\overline{v'\,\frac{\partial^2 v'}{\partial r^2}} +\overline{\frac{v'}{r}\,\frac{\partial v'}{\partial r}} +\overline{v'\,\frac{\partial^2 v'}{r^2\,\partial \theta^2}} -2\,\overline{v'\,\frac{\partial w'}{r\partial \theta}} -\overline{\frac{v'^2}{r^2}}$$

$$+\overline{w'\,\frac{\partial^2 w'}{\partial x^2}} +\overline{w'\,\frac{\partial^2 w'}{\partial r^2}} +\overline{\frac{w'}{r}\,\frac{\partial w'}{\partial r}} +\overline{w'\,\frac{\partial^2 w'}{r^2\,\partial \theta^2}} +2\,\overline{w'\,\frac{\partial v'}{r\partial \theta}} -\left. \overline{\frac{w'^2}{r^2}} \right)$$

$$-\rho\left(\bar{u}\,\frac{\partial \overline{u'^2}}{\partial x} +\bar{u}\,\frac{\partial \overline{u'v'}}{\partial r} +\bar{u}\,\frac{\overline{u'v'}}{r} +\bar{v}\,\frac{\partial \overline{u'v'}}{\partial x} +\bar{v}\,\frac{\partial \overline{v'^2}}{\partial r} +\bar{v}\,\frac{\overline{v'^2}}{r} \right.$$

$$+\overline{u'^2}\,\frac{\partial \bar{u}}{\partial x} +\overline{u'v'}\,\frac{\partial \bar{u}}{\partial r} +\overline{u'v'}\,\frac{\partial \bar{v}}{\partial x} +\left. \overline{v'^2}\,\frac{\partial \bar{v}}{\partial r} \right) \tag{24}$$

Si les équations de Reynolds avaient été multipliées par les composantes respectives de la vitesse et ensuite additionnées, les termes pourraient être arrangés pour obtenir la forme suivante de l'équation d'énergie pour le mouvement moyen tout seul :

$$-\rho\bar{u}\,\frac{\partial \Omega}{\partial x} -\rho\bar{v}\,\frac{\partial \Omega}{\partial r} +\frac{1}{r}\left[\frac{\partial}{\partial x}\,r\,(\overline{u}\bar{\sigma}_x+\overline{v}\bar{\tau}) +\frac{\partial}{\partial r}\,r\,(\overline{u}\bar{\tau}+\overline{v}\bar{\sigma}_r) \right]$$

$$=\frac{\rho}{2}\left(\bar{u}\,\frac{\partial \overline{V^2}}{\partial x} +\bar{v}\,\frac{\partial \overline{V^2}}{\partial r} \right) +\bar{\sigma}_x\,\frac{\partial \bar{u}}{\partial x} +\bar{\sigma}_r\,\frac{\partial \bar{v}}{\partial r} +\bar{\sigma}_\theta\,\frac{\bar{v}}{r} +\bar{\tau}\left(\frac{\partial \bar{u}}{\partial r} +\frac{\partial \bar{v}}{\partial x} \right) \tag{25}$$

En fonction des tensions individuelles, ceci devient :

$$-\bar{u}\,\frac{\partial}{\partial x}\,(\bar{p}+\rho\Omega) -\bar{v}\,\frac{\partial}{\partial r}\,(\bar{p}+\rho\Omega)$$

$$+\frac{\mu}{r}\left\{ \frac{\partial}{\partial x}\,r\left[2\,\bar{u}\,\frac{\partial \bar{u}}{\partial x} +\bar{v}\left(\frac{\partial \bar{u}}{\partial r} +\frac{\partial \bar{v}}{\partial x} \right) \right] +\frac{\partial}{\partial r}\,r\left[\bar{u}\left(\frac{\partial \bar{u}}{\partial r} +\frac{\partial \bar{v}}{\partial x} \right) +2\,\bar{v}\,\frac{\partial \bar{v}}{\partial r} \right] \right\}$$

$$-\frac{\rho}{r}\left[\frac{\partial}{\partial x}\,r\,(\bar{u}\,\overline{u'^2}+\bar{v}\,\overline{u'v'}) +\frac{\partial}{\partial r}\,r\,(\bar{u}\,\overline{u'v'}+\bar{v}\,\overline{v'^2}) \right]$$

$$=\frac{\rho}{2}\left(\bar{u}\,\frac{\partial \overline{V^2}}{\partial x} +\bar{v}\,\frac{\partial \overline{V^2}}{\partial r} \right) +\mu\left[2\left(\frac{\partial \bar{u}}{\partial x} \right)^2 +2\left(\frac{\partial \bar{v}}{\partial r} \right)^2 +2\left(\frac{\bar{v}}{r} \right)^2 +\left(\frac{\partial \bar{u}}{\partial r} +\frac{\partial \bar{v}}{\partial x} \right)^2 \right]$$

$$-\rho\left[\overline{u'^2}\,\frac{\partial \bar{u}}{\partial x} +\overline{v'^2}\,\frac{\partial \bar{v}}{\partial r} +\overline{w'^2}\,\frac{\bar{v}}{r} +\overline{u'v'}\left(\frac{\partial \bar{u}}{\partial r} +\frac{\partial \bar{v}}{\partial x} \right) \right] \tag{26}$$

dont la soustraction de l'équation pour la variation totale donne l'équation de l'énergie pour le mouvement turbulent tout seul :

$$
-\overline{u'\frac{\partial p'}{\partial x}} -\overline{v'\frac{\partial p'}{\partial r}} -\overline{w'\frac{\partial p'}{r\partial\theta}} -\varrho\left[\overline{u'^2}\frac{\partial\overline{u}}{\partial x} +\overline{v'^2}\frac{\partial\overline{v}}{\partial r} +\overline{w'^2}\frac{\overline{v}}{r} +\overline{u'v'}\left(\frac{\partial\overline{u}}{\partial r} +\frac{\partial\overline{v}}{\partial x}\right)\right]
$$

$$
-\frac{\mu}{r}\left\{\frac{\partial}{\partial x}\,r\,\overline{\left[u'\frac{\partial u'}{\partial x} +v'\left(\frac{\partial u'}{\partial r} +\frac{\partial v'}{\partial x}\right) +w'\left(\frac{\partial w'}{\partial x} +\frac{\partial u'}{r\partial\theta}\right)\right]}\right.
$$

$$
+\frac{\partial}{\partial r}\,r\,\overline{\left[u'\left(\frac{\partial u'}{\partial r} +\frac{\partial v'}{\partial x}\right) +v'\frac{\partial v'}{\partial r} +w'\left(\frac{\partial v'}{r\partial\theta} +\frac{\partial w'}{\partial r} -\frac{w'}{r}\right)\right]}
$$

$$
\left.+\frac{\partial}{r\partial\theta}\,r\,\overline{\left[u'\left(\frac{\partial w'}{\partial x} +\frac{\partial u'}{r\partial\theta}\right) +v'\left(\frac{\partial v'}{r\partial\theta} +\frac{\partial w'}{\partial r} -\frac{w'}{r}\right) +w'\left(\frac{\partial w'}{r\partial\theta} +\frac{v'}{r}\right)\right]}\right\}
$$

$$
=\frac{\varrho}{2}\left(\overline{u}\,\frac{\partial\overline{V'^2}}{\partial x} +\overline{v}\,\frac{\partial\overline{V'^2}}{\partial r} +\overline{u'\frac{\partial V'^2}{\partial x}} +\overline{v'\frac{\partial V'^2}{\partial r}} +\overline{w'\frac{\partial V'^2}{r\partial\theta}}\right)
$$

$$
+\mu\left[\left(\overline{\frac{\partial u'}{\partial x}}\right)^2 +\left(\overline{\frac{\partial v'}{\partial x}}\right)^2 +\left(\overline{\frac{\partial w'}{\partial x}}\right)^2 +\left(\overline{\frac{\partial u'}{\partial r}}\right)^2 +\left(\overline{\frac{\partial v'}{\partial r}}\right)^2 +\left(\overline{\frac{\partial w'}{\partial r}}\right)^2 +\left(\overline{\frac{\partial u'}{r\partial\theta}}\right)^2 +\left(\overline{\frac{\partial v'}{r\partial\theta}}\right)^2 +\left(\overline{\frac{\partial w'}{r\partial\theta}}\right)^2\right]
$$

$$\tag{27}$$

L'interprétation réfléchie des divers termes de ces équations est d'une très grande importance. Les trois dernières ont été arrangées de telle sorte que les membres de gauche représentent le travail et les membres de droite la variation d'énergie correspondante. Les termes qui indiquent le travail fait par la pression et par la gravité sont évidents, de même que les variations de l'énergie cinétique moyenne et turbulente $\varrho\,\overline{V}^2$ et $\varrho\,\overline{V'^2}/2$ par suite de la convection par le mouvement moyen et de la diffusion par la turbulence. Exactement comme dans le cas de l'écoulement purement visqueux déjà discuté brièvement, le travail fait par les tensions dans l'écoulement turbulent est en partie conservateur et en partie dissipateur. En analogie avec l'analyse précédente de l'écoulement visqueux, les termes conservateurs venant des équations de Reynolds ont été remplacés par la différence entre les termes totaux et les termes dissipateurs et mis à part comme dans la première équation pour l'écoulement moyen — totaux à gauche et dissipateurs à droite. Mais les tensions consistent maintenant dans les composantes visqueuses et turbulentes, comme décrites dans l'équation alternative de l'écoulement moyen. La signification de la partie dissipatrice visqueuse est telle qu'auparavant. Pourtant, la partie comprenant ce qui est réellement des termes « accélérateurs » dus à la variation turbulente, ne peut pas être vraiment dissipatrice, parce qu'elle ne dépend pas de la viscosité; néanmoins elle est décidément non conservatrice en ce qui concerne le mouvement moyen. En d'autres termes, la quantité :

$$
-\varrho\left[\overline{u'^2}\frac{\partial\overline{u}}{\partial x} +\overline{v'^2}\frac{\partial\overline{v}}{\partial r} +\overline{w'^2}\frac{\overline{v}}{r} +\overline{u'v'}\left(\frac{\partial\overline{u}}{\partial r} +\frac{\partial\overline{v}}{\partial x}\right)\right]\tag{28}
$$

représente la rapidité avec laquelle l'énergie est perdue irrécouvrablement par le mouvement moyen et gagnée par la turbulence.

Ces termes pour la production de turbulence paraissent, encore une fois, dans l'équation de l'énergie pour le mouvement turbulent, mais à gauche, car ils représentent maintenant une fourniture extérieure d'énergie, quelque peu comme le travail fait par des forces externes. Du travail est fait encore par les tensions visqueuses aussi bien que par celles du mouvement secondaire. Enfin, tout à droite, se trouvent les termes représentant la dissipation visqueuse dans la turbulence, laquelle est de beaucoup la partie majeure de la dissipation.

L'équation d'énergie, comme les trois composantes de l'équation de la quantité de mouvement, devient pratiquement utile lorsqu'elle est intégrée sur la zone considérée. Pour le moment, cette opération sera indiquée schématiquement sous la forme :

$$
-\varrho\int_{V}\left(\overline{u}\frac{\partial\Omega}{\partial x} +\overline{v}\,\frac{\partial\Omega}{\partial r}\right)dV +\int_{V}\frac{1}{r}\left[\frac{\partial}{\partial x}\,r\,(\overline{u\sigma_x}+\overline{v\tau_{xr}}+\overline{w\tau_{x\theta}})\right.
$$

$$
\left.+\frac{\partial}{\partial r}\,r\,(\overline{u\tau_{rx}}+\overline{v\sigma_r}+\overline{w\tau_{r\theta}}) +\frac{\partial}{r\partial\theta}\,r\,(\overline{u\tau_{\theta x}}+\overline{v\tau_{\theta r}}+\overline{w\sigma_\theta})\right]dV
$$

$$
=\frac{\varrho}{2}\int_{V}\left(\overline{u\,\frac{\partial V^2}{\partial x}} +\overline{v\,\frac{\partial V^2}{\partial r}} +\overline{w\,\frac{\partial V^2}{r\partial\theta}}\right)dV +\int_{V}\left[\overline{\sigma_x\frac{\partial u}{\partial x}} +\overline{\sigma_r\frac{\partial v}{\partial r}} +\overline{\sigma_\theta\frac{\partial w}{r\partial\theta}}\right.
$$

$$
\left.+\overline{\tau_{r\theta}\left(\frac{\partial v}{r\partial\theta} +\frac{\partial w}{\partial r} -\frac{w}{r}\right)} +\overline{\tau_{\theta x}\left(\frac{\partial w}{\partial x} +\frac{\partial u}{r\partial\theta}\right)} +\overline{\tau_{xr}\left(\frac{\partial u}{\partial r} +\frac{\partial v}{\partial x}\right)}\right]dV\tag{29}
$$

qui, en appliquant la transformation de Gauss entre les intégrales spaciales et superficielles, s'écrit dans une manière plus significative telle que :

$$-\varrho \int_{\mathsf{V}} \left(\bar{u}\,\frac{\partial\Omega}{\partial x} + \bar{v}\,\frac{\partial\Omega}{\partial r} \right) d\mathsf{V} + \int_{s} \left[(\overline{u\sigma_x} + \overline{v\tau_{xr}} + \overline{w\tau_{x\theta}})\,\frac{\partial x}{\partial n} + (\overline{u\tau_{rx}} + \bar{\sigma}_r + \overline{w\tau_{r\theta}})\,\frac{\partial r}{\partial n} \right.$$

$$+ (\overline{u\tau_{\theta x}} + \overline{v\tau_{\theta r}} + \overline{w\sigma_\theta})\,\frac{r\partial\theta}{\partial n} \left. \right] dS = \frac{\varrho}{2} \int_{s} \left(\overline{uV^2}\,\frac{\partial x}{\partial n} + \overline{vV^2}\,\frac{\partial r}{\partial n} + \overline{wV^2}\,\frac{r\partial\theta}{\partial n} \right) dS$$

$$+ \int_{\mathsf{V}} \left[\overline{\sigma_x\,\frac{\partial u}{\partial x}} + \overline{\sigma_r\,\frac{\partial v}{\partial r}} + \overline{\sigma_\theta\,\frac{\partial w}{r\partial\theta}} + \overline{\tau_{r\theta}\left(\frac{\partial v}{r\partial\theta} + \frac{\partial w}{\partial r} - \frac{w}{r}\right)} + \overline{\tau_{\theta x}\left(\frac{\partial w}{\partial x} + \frac{\partial u}{r\partial\theta}\right)} + \overline{\tau_{xr}\left(\frac{\partial u}{\partial r} + \frac{\partial v}{\partial x}\right)} \right] d\mathsf{V}$$

$$\tag{30}$$

Ici, les termes successifs ont à peu près la même signification qu'auparavant, travail à gauche et transformation d'énergie à droite. Pour évaluer les termes, il faut se souvenir que chaque composante de la vitesse consiste en une partie moyenne et une partie variable, que chaque tension consiste en une partie visqueuse et une partie due à l'inertie, et que tous les produits qui ne sont pas nécessairement égaux à zéro ont leur place dans l'expression qui en résulte.

Interprétation du point de vue de l'analyse isotrope

Les relations précédentes indiquent évidemment un état d'équilibre entre la production, le transfert, et la dissipation de l'énergie turbulente : production aux dépens de l'écoulement moyen, transfert par convection et diffusion, et dissipation au moyen de la viscosité. Dans chaque série d'action, les composantes moyennes et variables de la vitesse jouent manifestement un rôle primaire, mais la seule indication de l'échelle de l'action se trouve dans les dérivées, en fonction de l'espace, des composantes et de leurs produits. Une évaluation quantitative de cette échelle n'est possible qu'au moyen d'une analyse statistique des composantes de la vitesse comme fonction de l'espace ou du temps.

Chaque composante de la vitesse, à un point quelconque d'un écoulement turbulent, varie généralement avec le temps dans une manière continue mais qui ne peut pas être autrement pré-

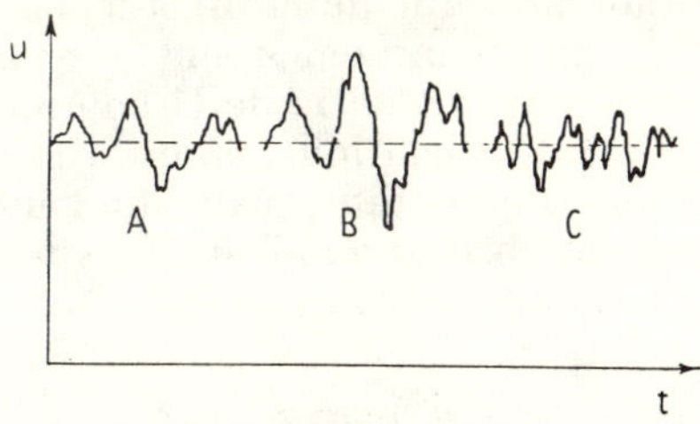

Fig. 7

Enregistrements temporels des fluctuations de vitesse

dite, comme l'indique la figure 7. Si $f(u)$ représente la fréquence avec laquelle la composante u se trouve entre les valeurs u et $u+du$, de telle sorte que :

$$\int_0^\infty f(u)\,du = 1 \tag{31}$$

alors la répartition de la fréquence de la variation aura en gros une forme de la fonction de probabilité montrée par la figure 8. La valeur

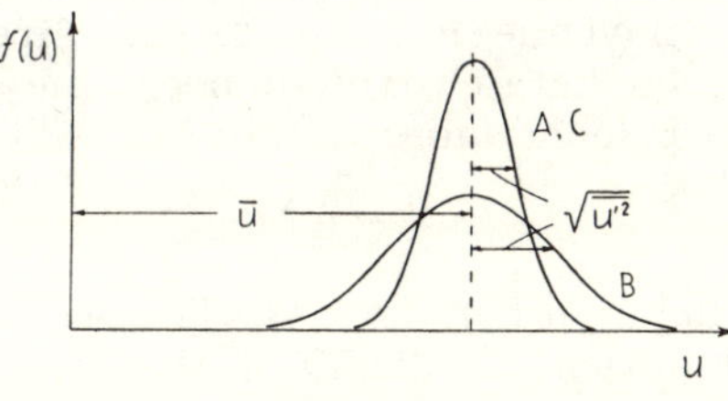

Fig. 8

Diagrammes de fréquence des fluctuations de vitesse

moyenne de la composante correspond à l'abscisse du centroïde de la surface délimitée par la courbe de fréquence :

$$\bar{u} = \int_0^\infty u f(u)\,du \tag{32}$$

et la racine du carré moyen de la déviation à partir de cette moyenne est représentée par le rayon de giration de la même surface de fréquence autour de son axe centroïdal :

$$\sqrt{\overline{u'^2}} = \left[\int_0^\infty (u - \bar{u})^2\,f(u)\,du \right]^{1/2} \tag{33}$$

De cette façon, tandis que les traces A et B de la figure 7 ont la même valeur moyenne, les racines des carrés moyens de leurs déviations diffèrent considérablement.

La trace C a évidemment la même moyenne que A et B et la même racine du moyen carré de la déviation que A, mais elle diffère de tous les deux d'une manière qu'il faut caractériser par références aux abscisses plutôt qu'aux ordonnées, c'est-à-dire, en déterminant la corrélation moyenne R_t entre deux valeurs de la distance Δt

variable, séparées sur l'enregistrement :

$$R_t = \frac{\overline{u'(t)\,u'(t+\Delta t)}}{\overline{u'^2}} \qquad (34)$$

Pendant que Δt tend vers zéro, R_t tend nécessairement vers l'unité (fig. 9), mais en revan-

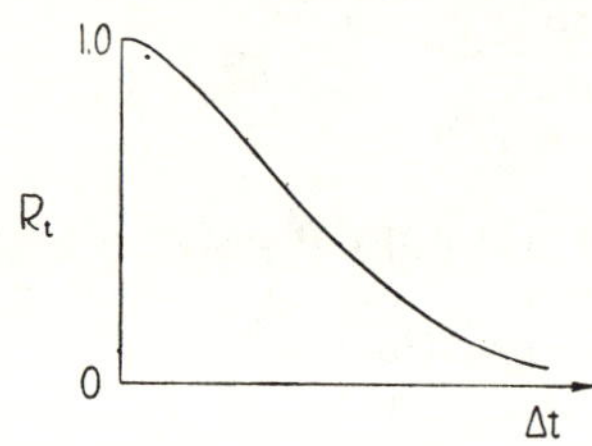

FIG. 9

Corrélation temporelle des fluctuations

che R_t tend vers zéro lorsque Δt excède une certaine valeur. Si la déviation est petite par rapport à la moyenne, le temps Δt peut être changé avec une bonne approximation en une distance axiale correspondante :

$$\Delta x \approx \overline{u}\,\Delta t \qquad (35)$$

Avec cela, le diagramme de la corrélation R_x (fig. 10) devient l'indicatif de l'échelle linéaire

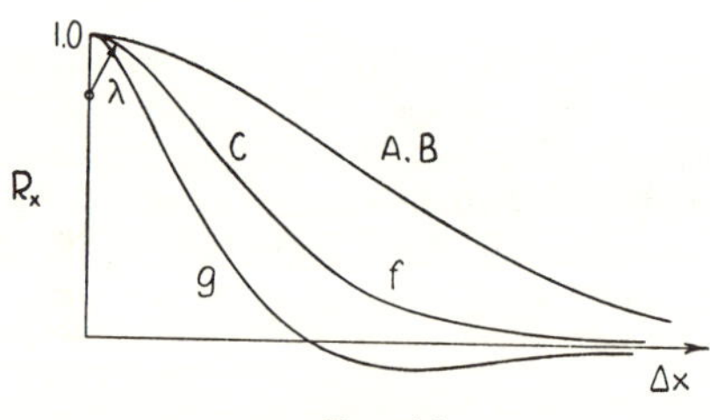

FIG. 10

Corrélations spatiales des fluctuations

du mouvement secondaire. Soit que le point où la corrélation s'approche de zéro, soit que, préférablement, l'aire de la surface sous le diagramme, s'utilisent comme mesure de cette longueur, on voit que les traces A et B sont caractérisées par une échelle qui est sensiblement plus grande que celle de C.

Si l'on imagine les variations de la vitesse dans l'écoulement turbulent comme étant le résultat d'un système hétérogène de tourbillons transportés par le courant moyen, l'échelle linéaire ensuite obtenue de la fonction de corrélation, peut être considérée comme représentant la grandeur moyenne des tourbillons. Que la notion de tourbillon ne soit pas déraisonnable est indiqué par la forme de la fonction de corrélation pour les

composantes de la vitesse normale à la ligne entre les points considérés, ce qui est illustré par la courbe g de la figure 10; c'est-à-dire que la fonction devient négative pour des distances un peu plus petites que celles auxquelles la courbe f s'approche de zéro, parce que, en effet, le fluide sur l'un des côtés du tourbillon a tendance à se déplacer en amont pendant que celui de l'autre côté se déplace en aval.

Semblable aux composantes de la vitesse, la fonction de corrélation doit varier avec la direction, parce qu'elle dépend de l'orientation de chacune des deux composantes et de celle de la ligne entre les deux — c'est-à-dire le vecteur de séparation s. Une telle double corrélation est, par le fait, une fonction tensorielle avec neuf composantes différentes, dont toutes sont liées à l'échelle des tourbillons. (On doit noter que les tensions de Reynolds sont les valeurs limites de certaines de ces composantes, sous la condition d'une séparation égale à zéro, de telle sorte qu'elles ont perdu leur signification linéaire.) Chaque composante est nécessairement une quantité qui par elle-même ne donne aucune indication de la véritable répartition de grandeur des tourbillons.

Une telle connaissance s'obtient de la même manière que pour résoudre un enregistrement de vitesse tel que celui de la figure 7, en une fonction spectrale qui montre la répartition de l'énergie attribuable à la composante de la vitesse donnée, pour toutes les fréquences de variation ou longueurs d'ondes présentes dans l'enregistrement.

Une telle fonction est, de fait, une transformée de Fourier de la fonction de corrélation [26]. Toutefois, le spectre général a la même nature d'un tenseur que la corrélation générale, et c'est le tenseur, au lieu d'une de ses composantes, qui est d'intérêt actuel. Si l'énergie cinétique moyenne de la turbulence par unité de volume était reconnue comme étant le produit de la densité par la moitié de la somme des carrés moyens des variations, alors celle-ci pourrait s'écrire comme l'intégrale de l'énergie par l'unité de nombre d'onde $E(k)$ sur toute la gamme de nombres d'onde :

$$\frac{\rho}{2}\left(\overline{u'^2}+\overline{v'^2}+\overline{w'^2}\right) = \int_0^\infty E(k)\,dk \qquad (36)$$

Ici k est la grandeur du vecteur du nombre d'ondes [27], c'est-à-dire un paramètre tensoriel égal à 2π divisé par une sorte de longueur d'onde spatiale. Le point essentiel est que k peut être regardé comme la valeur réciproque d'une longueur qui caractérise l'échelle des tourbillons. Le diagramme d'un tel spectre (fig. 11) représente par conséquent la répartition de l'énergie entre les différentes grandeurs de tourbillons, lesquels

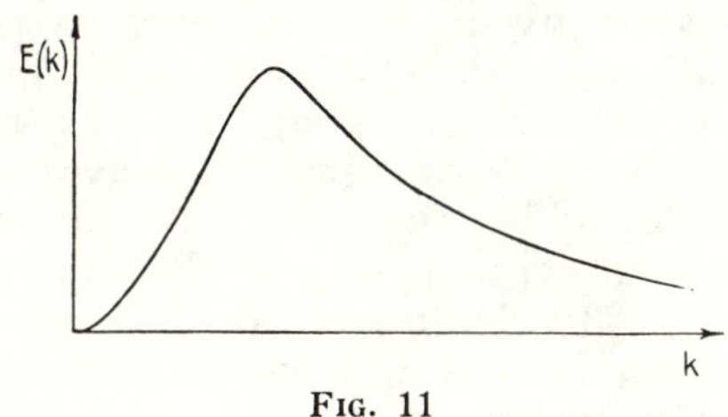

FIG. 11

Répartition spectrale d'énergie turbulente

varient d'une très grande dimension près de l'origine jusqu'à une très petite à l'extrême-droite.

Pour le cas spécial, introduit par Taylor [28], dans lequel la turbulence est non seulement homogène (indépendante de la position) mais aussi isotrope (indépendante de la direction), les conditions sont simplifiées à tel point que les carrés moyens de la fluctuation sont tous identiques :

$$\overline{u'^2}=\overline{v'^2}=\overline{w'^2} \tag{37}$$

Les carrés moyens de leurs dérivées sont alors deux fois plus petits dans les directions des composantes que dans les directions perpendiculaires :

$$\overline{\left(\frac{\partial u'}{\partial x}\right)^2} = \overline{\left(\frac{\partial v'}{\partial r}\right)^2} = \overline{\left(\frac{\partial w'}{r\partial\theta}\right)^2}$$

$$= \frac{1}{2}\overline{\left(\frac{\partial u'}{\partial r}\right)^2} = \frac{1}{2}\overline{\left(\frac{\partial v'}{r\partial\theta}\right)^2} \text{ etc.} \tag{38}$$

et le tenseur pour la double corrélation se réduit à deux fonctions indépendantes [29] :

$$g=f+\frac{s}{2}\frac{\partial f}{\partial s} \tag{39}$$

La dissipation de l'énergie est ensuite exprimable en fonction d'une seule dérivée de la vitesse :

$$\varepsilon=15\,\mu\,\overline{\left(\frac{\partial u'}{\partial x}\right)^2} \tag{40}$$

et aussi en fonction du rayon de courbure λ de la fonction de corrélation f de la figure 10 à $R=1$:

$$\varepsilon=15\,\mu\,\frac{\overline{u'^2}}{\lambda^2} \tag{41}$$

La fonction spectrale générale, de plus, peut être réduite pour fournir l'expression [25] :

$$\frac{\partial E(k)}{\partial t} = T(k) - 2\,\mu k^2\,E(k) \tag{42}$$

laquelle énonce que la variation de l'énergie d'un nombre d'onde donné est égale au gain net d'énergie au moyen du transfert des plus petits

aux plus grands nombres d'onde, moins la dissipation à ce nombre d'onde.

Le transfert d'énergie d'un nombre d'onde à un autre est un effet inertiel, pendant que celui de la dissipation est visqueux, et le rapport entre les deux devrait donc varier avec le nombre de Reynolds de l'écoulement et avec la position relative dans le spectre du nombre d'onde considéré. Néanmoins Kolmogoroff [30] a formulé l'hypothèse qu'aux nombres de Reynolds assez élevés les effets visqueux seraient concentrés dans les très petits tourbillons, et que la structure turbulente dans cette partie du spectre atteindrait une condition de « similitude universelle » ou d'équilibre statistique qui ne dépend que de la fourniture cinématique d'énergie S et la viscosité cinématique ν, sur quoi $S=\varepsilon/\varrho$. Cette hypothèse a reçu une sérieuse confirmation expérimentale. C'est d'un intérêt spécial de noter que la portion du spectre contenant la majeure partie de l'énergie (celle montrée en fig. 11) est bien distincte de celle dans laquelle la dissipation a lieu (qui est réellement bien à droite de la partie montrée dans la figure). On ne peut s'attendre qu'à cette situation, ainsi que cela découle d'une simple comparaison des quantités $E(k)$ et $k^2 E(k)$ qui caractérisent les deux zones spectrales.

De telles conclusions, au sujet de la turbulence isotrope homogène, auraient peu de signification pour le phénomène anisotrope non homogène dont il s'agit, s'il n'y avait pas deux facteurs influant sur le problème : en premier lieu, il existe, en général, assez de similitude qualitative entre les deux sortes de turbulence, pour pouvoir faire une interprétation utile; en second lieu, cette similitude semble prendre des aspects quantitatifs lorsque le nombre d'onde devient très grand. Dans un écoulement quelconque, par exemple, il est raisonnable de conclure [21], qu'à tous les nombres de Reynolds, sauf les petits, la structure des grands tourbillons dépend presque entièrement de la forme de l'écoulement moyen lui-même. En outre, plus grand est le nombre de Reynolds, plus petite doit être la taille des tourbillons à laquelle la viscosité commence à jouer un rôle appréciable. Les tensions aux échelles plus grandes que celles-ci sont par conséquent dues à l'inertie, et leur effet est de produire une rupture continue du mouvement tourbillonnaire dans la direction de nombres d'onde plus élevés. Alors, de même que la source des tourbillons dans une zone de rapide variation de vitesse est l'écoulement primaire, on peut s'imaginer que les tourbillons voisins produisent eux-mêmes des zones de tension qui engendrent des tourbillons encore plus petits. Plus grande est la dispersion sur laquelle ce transfert d'énergie a agi, et moins évidentes doivent être les préférences di-

337

rectionnelles de la source originaire. En d'autres termes, dans un champ turbulent donné, plus petite est l'échelle de certains tourbillons, plus proche d'un état homogène et isotrope doit être le mouvement dans ces tourbillons. En même temps, l'effet dissipateur de la viscosité devrait croître pendant que l'échelle décroît (en fait presque comme le carré du nombre d'onde), et aux nombres de Reynolds suffisamment grands, la dissipation doit être à peu près toute concentrée dans la portion du spectre dans laquelle la turbulence est essentiellement isotrope.

La conclusion significative à laquelle on est ainsi arrivé est que même dans les zones de turbulence anisotrope non homogène, il peut être possible d'évaluer la dissipation en fonction d'une seule dérivée de vitesse au lieu des neuf qui se trouvent dans la fonction générale de dissipation. Il est déjà évident [31] que cette conclusion est valable pour des cas relativement simples de turbulence libre avec les tensions de Reynolds, bien que cela ne soit pas nécessairement vrai pour l'écoulement près des parois [32, 33]. Malheureusement, des études même plus poussées que celles-ci ne peuvent être utilisées facilement pour vérifier la théorie dans les conditions plus complexes qu'elles représentent, car des limitations pratiques rendent impossibles à mesurer d'autres termes encore, qu'on ne peut que supposer négligeable. L'hypothèse selon laquelle la théorie est valable est donc essentielle pour conduire l'analyse dans son état actuel d'approximation.

LES MESURES EXPÉRIMENTALES

Soufflerie, cylindre et conduite

Puisque les deux phénomènes dont il s'agit sont entièrement indépendants de l'action de la gravité, on a jugé souhaitable d'employer l'air au lieu de l'eau comme fluide expérimental [34]. Les mesures autour du cylindre se firent alors dans une soufflerie, et, pour les mesures à l'entrée de la conduite, un simple tuyau de la forme désirée fut monté. A cet égard, les appareils furent très semblables à ceux déjà employés pour des études précédentes de cette nature [1 à 9].

La soufflerie était du type fermé, d'une section droite allant de 2×3 mètres immédiatement au-delà du ventilateur centrifuge jusqu'à $0,9 \times 0,9$ mètre dans la zone d'essai. Celle-ci avait une longueur sans obstacle de 3 mètres, depuis la trompe d'entrée jusqu'aux redresseurs d'écoulement dans le coude en aval. Elle était munie de parois vitrées et de deux lignes de piézomètres en barres de laiton disposés axialement dans le bois dur du plancher et du plafond. La combinaison de vannes et d'un moteur à vitesse variable permettait le réglage de la vitesse de l'air, dans la section d'essais, entre 3 et 15 mètres par seconde. La soufflerie n'était certainement pas du type à basse-turbulence, car l'intensité dans l'air ambiant dans la zone de mesures était à peu près de 1 % de la vitesse moyenne de l'air. Alors que cela aurait été excessif pour des expériences poussées, pour une telle étude exploratrice, cela était considéré comme acceptable.

Le cylindre fut formé d'un tube bouché en laiton tourné au diamètre de 7,5 cm sur une longueur de 120 cm. Au-delà, il était relié à un tuyau presque du même diamètre qui s'étendait jusqu'aux redresseurs d'écoulement au bout de la section des essais. Le cylindre entier était suspendu par des fils fins à l'axe de la section avec sa face tronquée en amont, à mi-chemin entre l'entrée et le coude. Des ouvertures piézométriques étaient pratiquées à travers la face normale et le long de la portion tournée du cylindre, aux endroits convenables.

La conduite était en matière plastique commerciale, d'un diamètre de 15 cm et d'une épaisseur de parois de 5 mm. Son extrémité amont fut coupée au tour et cimentée dans un orifice percé dans un morceau carré de matière plastique de 30 cm de long et 6 mm d'épaisseur; lequel fut alors placé dans un ouvrage en bois de 2×2 m pour simuler le mur d'un réservoir. La conduite était d'un diamètre constant sur une longueur de 2,5 m. Son extrémité en aval était reliée à l'entrée d'un ventilateur centrifuge, dont une vanne-papillon permettait de régler la vitesse entre 3 et 30 m par seconde. Des ouvertures piézométriques latérales se trouvaient placées à des intervalles appropriés sur toute sa longueur. De plus, environ deux douzaines d'ouverture d'accès étaient pratiquées par couples opposés au-dessus et au-dessous, avec des bouchons étanches faits de telle sorte que la surface intérieure était continue quand les ouvertures n'étaient pas en service.

Instruments

Dans les deux cas, trois sortes de mesures se firent : un contrôle continu sur la constance de l'écoulement ambiant, la répartition de la vitesse et de la pression, et les caractéristiques de la turbulence, dont la seconde seule différait dans les deux essais. La première dépendait de la variation de pression due au changement brusque de la section — à travers la trompe d'entrée de la soufflerie et à travers l'entrée de la conduite même — conformément aux mesures faites au moyen d'un tube de Pitot, des vitesses ambiantes pendant les essais d'étalonnage. Un simple manomètre différentiel à eau, sensible à 0,3 mm, assurait le contrôle en chaque cas. La troisième sorte de mesure — celle de la turbulence — s'effectua toujours au moyen de l'anémomètre à fil chaud et des circuits connexes, dont la description exigerait déjà un petit livre : heureusement un tel livret a déjà été préparé à cet effet [35]. Il suffira de dire ici que les sondes se montèrent sur le même agencement traversant que les indicateurs de vitesse, encore à décrire, et que les quantités mesurées étaient l'intensité axiale, l'intensité radiale, leur produit, et la dérivée de l'intensité axiale par rapport au temps. Les circuits étaient de telle sorte qu'ils indiquaient directement la racine du carré moyen ou le carré moyen de chaque valeur et son rapport à la vitesse moyenne locale ou à son carré. Bien que des erreurs de 5 à 10 % puissent être attendues au stade actuel, pour des intensités relativement faibles, il est évident que dans des régions de décollement de faible vitesse moyenne, une assez grande inexactitude de mesure peut en résulter.

Les indications de la vitesse moyenne autour du cylindre s'obtinrent au moyen d'un tube de 3 mm de diamètre pourvu d'orifices piézométriques, à 30° l'un de l'autre, à mi-chemin de sa longueur. Par rotation du tube autour d'un axe perpendiculaire au plan d'écoulement, il était possible de déterminer la position de la pression différentielle maximum et ensuite la direction de la vitesse vectorielle; la lecture du manomètre permettait de plus de trouver la valeur de la vitesse sur une courbe d'étalonnage. Le tube avait une longueur de 15 cm entre les pivots, sur l'un desquels était monté un petit rapporteur, et se plaçait symétriquement sur l'appareil traversant qui était bien en aval de la section étudiée. La position radiale du tube était lue sur une graduation sensible à 0,3 mm d'eau

et la pression observée au moyen d'un indicateur Wahlen modifié [36] sensible à 0,03 mm d'alcool. La répartition de la pression dans l'écoulement se déterminait au moyen d'une plaque de laiton de 2,5×40×0,3 cm de grandeur, à arêtes vives, au milieu de laquelle étaient douze orifices piézométriques sur une ligne droite. Cette plaque pouvait se monter comme un aileron radial parallèle au courant pour permettre la détermination de la répartition radiale de pression.

Une plaque piézométrique (voir fig. 12 *a*), semblable à celle du cylindre, était utilisée dans la conduite. Cependant, elle ne comportait que deux orifices, et la répartition de la pression était mesurée pendant que la plaque se déplaçait par

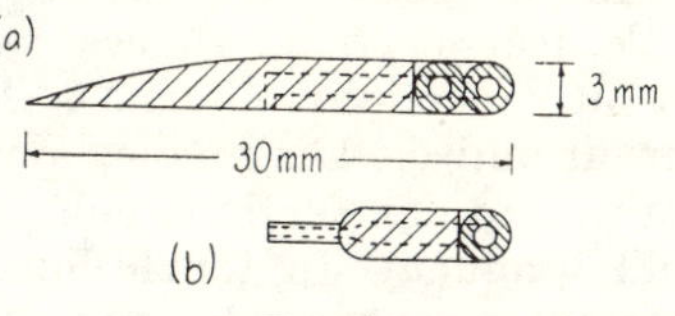

FIG. 12

Plaque piézométrique et prise de pression

étapes dans la section. Pendant les mesures, elle franchissait complètement le conduit. Au lieu du cylindre tournant, dont n'importe quelle forme aurait causé trop d'étranglement dans la conduite, on employait une simple prise de pression au moyen d'une aiguille hypodermique (fig. 12 *b*). Celle-ci pouvait se monter de manière interchangeable avec la plaque piézométrique sur l'appareil traversant de telle sorte qu'elle se déplaçait, sa pointe étant sur la ligne suivie auparavant par les orifices piézométriques. La direction du tube pouvait être inversée sans déplacer la pointe pour les mesures postérieures. Car une telle prise de pression indique correctement la pression totale moyenne même pour des angles d'inclinaison relativement grande [37], la soustraction de la pression statique mesurée, de celle-là, donnait la hauteur moyenne due à la vitesse au point considéré. A cause de leurs valeurs relativement grandes, ces pressions se mesuraient au moyen d'un manomètre différentiel à eau, sensible à 0,3 mm.

Bien que n'étant pas du même genre, la méthode de détermination de la position de l'extrémité aval du tourbillon, dans chaque cas, doit être aussi citée. Car cette zone était toujours l'endroit d'une variation intense, et comme les indications de vitesse devenaient indéterminées près de la limite zéro, une faible longueur de fil à coudre fut montée de façon à flotter librement au bout d'un fil métallique et introduite dans l'écoulement près de la paroi. La section où il flottait en amont aussi souvent qu'en aval était considérée comme l'extrémité du tourbillon.

Présentation des mesures

Dans les figures 13 à 18 sont reproduites les données essentielles obtenues pendant les expériences avec l'entrée, et dans les figures 19 à 24 les mesures correspondantes pour le cylindre. Toutes les valeurs ont été réduites à une forme sans dimension au moyen d'une division par les grandeurs de référence convenables. Dans les deux cas, les mesures avaient été faites dans des sections identiques — c'est-à-dire pour des valeurs de x/R égales à 0,5; 1,0; 1,5; 2,0; 3,0; 4,0; 6,0 et 8,0. Toutefois, vu que la détermination indépendante de la longueur du tourbillon avait montré que celle-ci était presque exactement deux fois plus grande pour le cylindre ($x/R=3,1$) que pour l'entrée ($x/R=1,5$), les valeurs correspondant aux sections alternées le long du cylindre furent omises. Les mêmes symboles sont utilisés alors pour des sections comparables par rapport à la longueur du tourbillon.

La vitesse fut mesurée à la section initiale du cylindre et aussi à des valeurs de r/R supérieures à 2,0, mais à cause d'anomalies évidentes, ces mesures n'ont pas été reproduites. Toujours dans un but d'homogénéité, les vitesses vectorielles sont indiquées dans les deux cas, bien que les directions et par conséquent les composantes aient été obtenues pour le cylindre. Puisque les conditions de continuité rendent possible d'évaluer les composantes au moyen d'une seule approximation pour toutes les sections sauf la première (où aucune mesure acceptable n'eut lieu), les composantes sont ainsi représentées implicitement sur les diagrammes. Les valeurs de pression se rapportent à celles de l'écoulement ambiant dans les deux cas, à la suite desquelles les mesures pour l'entrée apparaissent comme très négatives; elles peuvent naturellement se rapporter à n'importe quelle valeur de base sans changer de forme. Toutes les mesures pour l'entrée ont été corrigées à cause de l'effet général (mais non local) d'étranglement de la plaque piézométrique, comme indiqué par le piézomètre dans la paroi, avec et sans plaque en place; et toutes les mesures pour le cylindre ont été corrigées pour le gradient de pression le long de la soufflerie, comme indiqué par les piézomètres dans la paroi. Seules deux composantes de la turbulence furent mesurées; la composante axiale se détermina séparément par fil unique et, en même temps que la composante radiale, par fil croisé; et seules ces mesures-ci sont représentées. Enfin, toutes les valeurs de la dérivée, par rapport au temps, de la vitesse axiale ont été converties en celles de la dérivée spatiale correspondante en divisant par la composante moyenne axiale de la vitesse au point considéré.

Un simple coup d'œil sur les points mesurés révèle une dispersion expérimentale considérable, et un examen plus attentif ferait découvrir beaucoup de contradictions, dont l'élimination partielle sera discutée dans les pages suivantes. Avant de terminer ce chapitre, il importe de noter que la façon de procéder nécessairement adoptée dans le cas présent — expérimentation et analyse à des temps et emplacements différents — n'est pas la manière idéale d'étudier de tels écoulements, parce que les deux phases de l'étude auraient pu être conduites beaucoup plus efficacement sous forme de collaboration. Pourtant, peu d'études d'écoulement aussi complexes sont vraiment satisfaisantes dans leur forme initiale, et c'est seulement lorsqu'elles sont refaites à la lumière des découvertes préliminaires que des connaissances de valeur permanente sont acquises. Donc, pour plus d'une raison, il faut regarder l'étude présente comme tout à fait provisoire, avec, pour objet, celui de dégager, par n'importe quel moyen, les aspects approximatifs de la répartition de l'énergie.

ÉVALUATION ET DISCUSSION DES RÉSULTATS

Détermination des schémas des écoulements moyens

Bien que les données présentées dans le chapitre précédent comprennent en général la connaissance expérimentale qui est nécessaire à l'analyse, on y trouve des discordances de trois types différents, qu'il faut tout d'abord éliminer. A savoir : la dispersion des points d'une seule série, de même que des différences entre les séries successives dans le même système; le déplacement systématique des points dû à l'infidélité bien connue des instruments; et des discordances entre les données expérimentales et les exigences des équations du mouvement. Il a fallu tout réduire à des proportions admissibles par un moyen ou par un autre, ainsi qu'il est décrit dans les pages suivantes.

La dispersion des points d'une série se compense d'habitude en traçant une courbe moyenne entre eux, soit à vue d'œil, soit par la méthode des moindres carrés. Un semblable procédé peut être employé pour raccorder plusieurs courbes successives, pour lesquelles un bon « lissage »

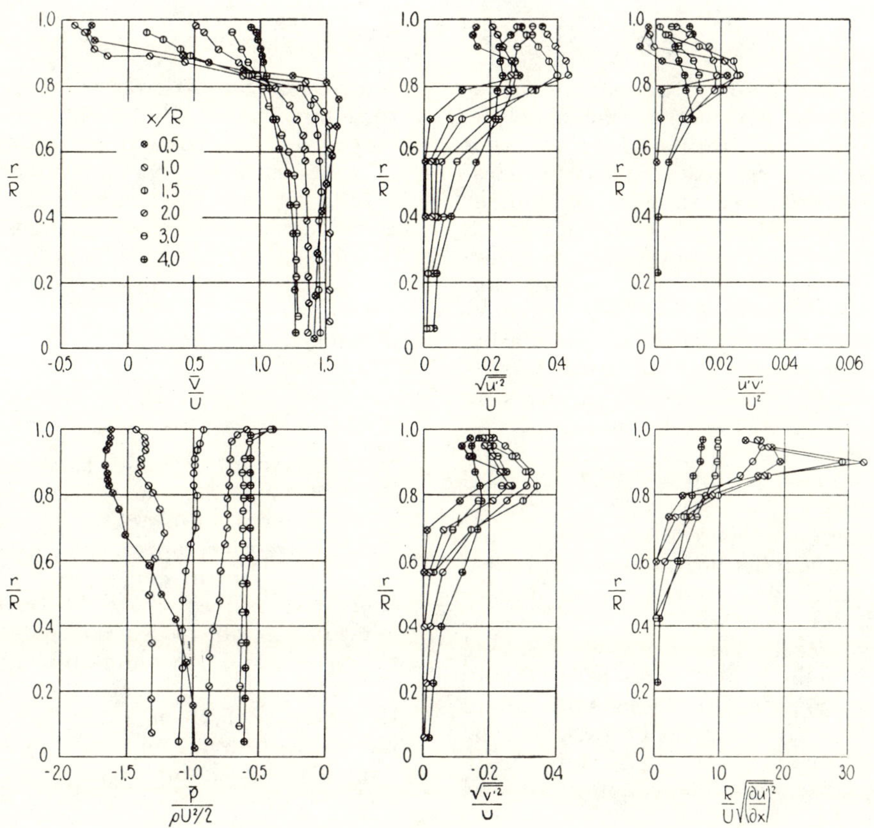

Fɪɢ. 13 - 18

Mesures de vitesse, de pression et de turbulence pour l'entrée

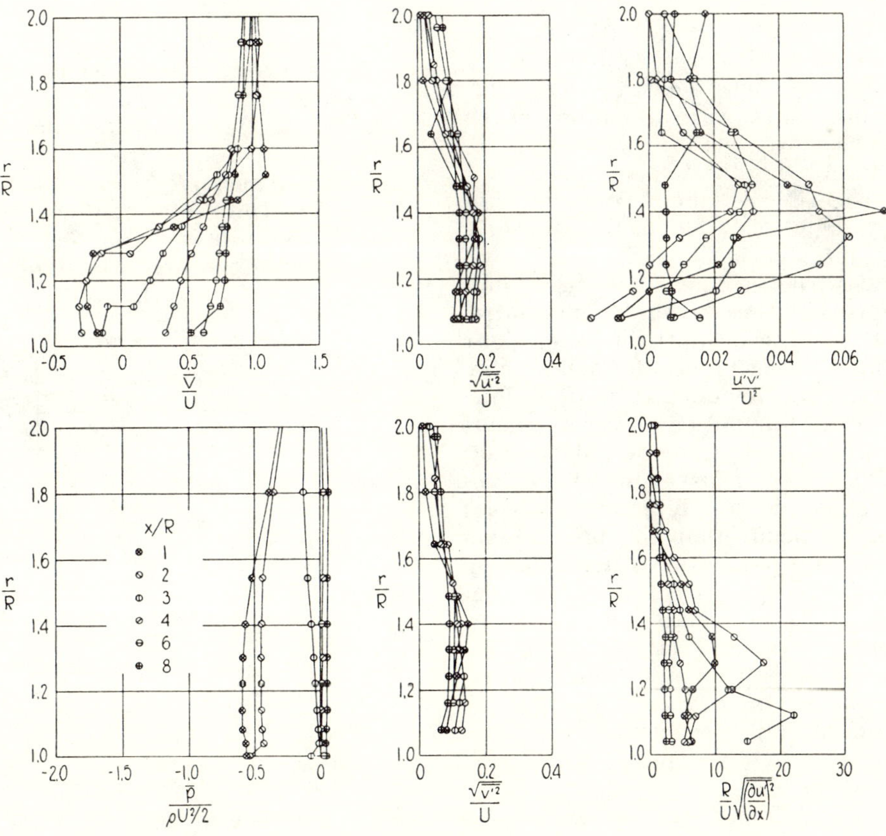

Fɪɢ. 19 - 24

Mesures de vitesse, de pression et de turbulence pour le cylindre

341

doit toujours exister même si une autre variable intervient comme paramètre. Par exemple,
une série de courbes montrant u/U en fonction
de r/R pour des valeurs succcessives du paramètre x/R peut être tracée en vue de montrer
u/U en fonction de x/R avec r/R comme paramètre. L'ajustement des deux systèmes de courbes, de telle sorte que tous les deux soient con-
tinus et rationnels, s'opère donc avec des données comportant trois variables, ce que l'ajustement d'une seule courbe fait seulement dans
le cas de deux variables. Toutes les mesures en
question ont été soumises à cette procédure, et
les valeurs utilisées pour les calculs subséquents
sont celles des courbes rectifiées, plutôt que les
mesures elles-mêmes.

Certaines corrections à cause des erreurs ins-
trumentales — en particulier celles sur la répartition de la pression — ont été mentionnées
comme étant déjà comprises dans les mesures
présentées. Une correction du même genre reste
encore à noter : celle pour l'effet de la turbulence sur l'indication de la vitesse. On entend
généralement que, par suite de la variation de
vitesse, un instrument capable d'enregistrer la
grandeur instantanée du vecteur donnerait une
indication moyenne égale à :

$$\Delta p = \frac{\rho}{2}\ (\overline{V^2} + \overline{V'^2}) \tag{43}$$

Divers instruments [37] sont sensibles à de
telles variations, à des degrés différents, et
l'exactitude de leur réponse dépend aussi beaucoup du rapport entre la fluctuation et la
moyenne locale. Par exemple, une prise de pression formée d'une aiguille hypodermique est
pratiquement insensible à l'inclinaison du courant jusqu'à un angle voisin de 15° de telle sorte
que les composantes perpendiculaires de la vitesse jusqu'à 25 % de la valeur moyenne, doivent être enregistrées sans erreurs. Par contre,
le tube de Pitot du type Prandtl est conçu pour
indiquer sûrement la composante axiale elle-
même dans la même étendue, et il doit donc
être insensible à des fluctuations normales, sauf
si elles sont grandes ou si la vitesse moyenne locale est faible. Le tube transversal est évidemment sensible à des degrés différents dans les
deux directions perpendiculaires. Après réduction à la composante axiale du vecteur moyen,
les mesures dont il s'agit furent corrigées seulement selon la variation axiale. A l'égard d'autres ajustements, la valeur réelle de la correction était probablement peu importante, spécialement parce que la turbulence était presque
toujours maximum dans les zones de vitesse
minimum, où l'erreur était la plus grande dans
la correction comme dans la mesure.

Pour le cas de l'écoulement autour du cy-

lindre, les composantes axiales des mesures vectorielles étaient disponibles tout de suite au
moyen des inclinaisons mesurées. Pour le cas de
l'écoulement dans la conduite, au contraire, il
a fallu évaluer les composantes par approximations successives. Les vecteurs mesurés furent
d'abord supposés représenter les composantes;
celles-ci furent intégrées, une section après l'autre, pour donner la fonction de courant non
dimensionnelle :

$$\frac{\psi}{R^2 U} = \int \frac{\overline{u}}{U}\ \frac{r}{R}\ \frac{dr}{R} \tag{44}$$

en fonction de r/R, et la sélection des valeurs
de r/R pour des accroissements constants de
$\psi/R^2 U$ dans la fonction tracée donna ensuite
comme première approximation les positions des
lignes de courant correspondantes dans les sections successives. Avec le tracé des lignes de
courant, on obtint les inclinaisons des vecteurs,
et par elles une seconde approximation des composantes et de la fonction de courant. Bien entendu, la répartition pour la section de l'entrée
elle-même n'avait pas été mesurée et il fallait
l'obtenir autrement. Elle fut prise comme première approximation des valeurs théoriques pour
la répartition dans le plan d'un orifice circulaire [13] et corrigées selon la différence entre
les pressions théoriques et mesurées, respectivement à l'arête de l'orifice et de l'entrée. Toutes les courbes de vitesse pour la conduite furent alors subordonnées à des ajustements mi-
neurs aux environs de la paroi pour obtenir la
grandeur constante de 0,5 exigée par l'inté-
grale $\int (\overline{u}/U)\ (r/R)\ dr/R$ dans les conditions de
continuité, c'est-à-dire en fonction du débit,
$Q = 2\pi\psi$, d'où $\psi/R_2 U = \frac{1}{2}\ Q/\pi R^2 U = \frac{1}{2}$). Le système de profils résultant est reproduit dans la
figure 25 et le schéma des lignes de courant
correspondantes, dans la figure 26.

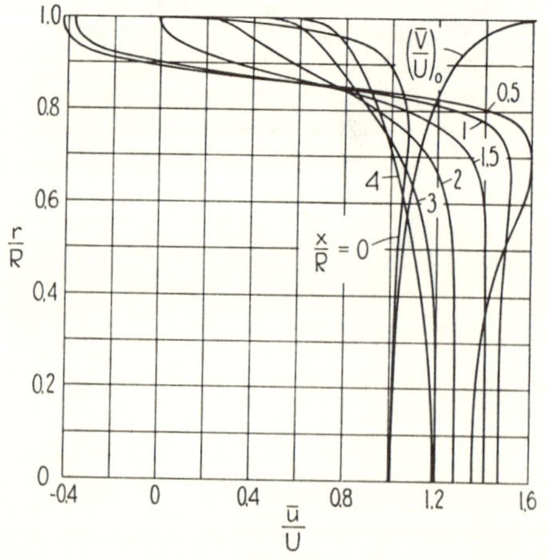

Fig. 25

Courbes rectifiées de la vitesse pour l'entrée

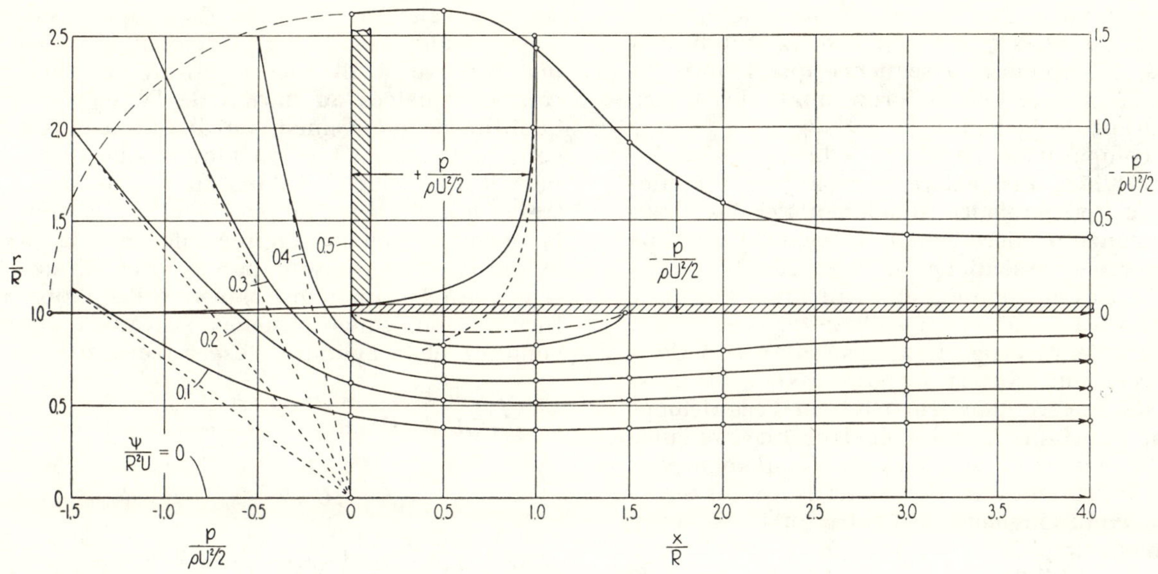

Fig. 26

Schéma des lignes moyennes du courant pour l'entrée

Pour le cylindre, les constructions comparables étaient un peu plus complexes. La répartition de la grandeur vectorielle de la vitesse dans la section initiale fut d'abord déduite approximativement de celle de l'écoulement potentiel à la section centrale d'une sphère, et corrigée selon la différence entre la pression minimum correspondante et celle mesurée au bout du cylindre. Le schéma des lignes de courant au voisinage du cylindre fut alors déterminé de la même manière que pour le tuyau. Or, cependant, une condition de continuité semblable à celle de la conduite uniforme n'existait pas, aux fins de contrôle, et, néanmoins il apparaissait que l'augmentation de vitesse mesurée juste en dehors du tourbillon manquait de grandeur. Ce qui correspond à l'épaisseur du déplacement [19] de la couche limite :

$$\frac{\delta}{R} = \int \left(1 - \frac{\bar{u}}{U} \right) \frac{r}{R} \frac{dr}{R} \qquad (45)$$

qui représente la distance à laquelle l'écoulement est déplacé radialement par le dérangement local, fut donc calculé d'après des courbes de vitesse. En évaluant l'intégrale, on doit remarquer que sa limite était prise nécessairement au point où la vitesse atteint la première fois la valeur U. Un diagramme de S/R en fonction de x/R est superposé aux lignes brisées sur le schéma des lignes de courant de la figure 28. Il représente évidemment la force d'un corps imaginaire qui déplacerait un écoulement irrotationnel environnant dans la même mesure que

le ferait la combinaison du cylindre, du tourbillon et du sillage.

Au voisinage du tourbillon même, le profil du déplacement se trouve en bon accord avec celui d'un ovoïde de Rankine (figuré en ligne pointillée) ayant un demi-diamètre de 1,5 et une demi-longueur de 2,2. La fonction de courant pour un tel corps :

$$\frac{\psi}{R^2 U} = \frac{1}{2}\left(\frac{r}{R}\right)^2 + 0{,}87\left[\frac{(x/R) - 1{,}3}{\sqrt{[r/R]^2 + [(x/R) - 1{,}3]^2}} \right.$$
$$\left. - \frac{(x/R) + 1{,}3}{\sqrt{[r/R]^2 + [(x/R) + 1{,}3]^2}} \right] \qquad (46)$$

doit donc donner avec une belle approximation la forme des lignes de courant autour de l'extrémité du cylindre — mais bien entendu, ni assez près pour qu'il y ait un effet de face plane plutôt qu'arrondie, ni assez loin radialement pour qu'il y ait un effet de profil étendu plutôt que symétrique. Le premier effet ne s'élimine pas facilement par l'analyse. Mais, en revanche, le second se réduit sans difficulté en appliquant à la fonction du courant donné par l'équation (8) l'approximation de Munk, $\Delta(t) \approx \frac{1}{4}\,\delta$ pour de minces profils, avec ce résultat que :

$$\frac{\psi}{R^2 U} = \left(\frac{r}{R}\right)^2 \left\{ \frac{1}{2} \right.$$
$$- \int \frac{(\delta/R)^2}{4\left[\left(\frac{t}{R} - \frac{x}{r}\right)^2 + \left(\frac{r}{R}\right)^2\right]^{3/2}} \frac{dt}{R} \left. \right\} \qquad (47)$$

Par ajustement des parties irrotationelles des courbes de vitesse entre ces deux conditions limites, fut obtenue la séquence que montre la figure 27 et avec elle la forme finale des lignes de courant de la figure 28.

La comparaison des deux schémas d'écoulement est bien instructive, car on y voit facilement la ressemblance fondamentale des deux phénomènes de décollement et en même temps la différence essentielle de leur étendue dans l'écoulement en conduite et à l'air libre. Les deux tourbillons ne diffèrent pas d'une façon marquée soit dans leurs proportions générales, soit dans leurs relations avec la variation axiale de pression, soit encore dans leur effet sur l'écoulement voisin. Pourtant, à cause de leur dissemblance essentielle dans l'interaction de l'écoulement voisin, le tourbillon dans le cas de la conduite n'a que la demi-longueur de l'autre, et il est aussi sensiblement plus mince, malgré l'identité de la dimension radiale de la paroi. D'autres ressemblances et différences deviendront apparentes dans les relations d'énergie du chapitre suivant.

Avant que l'analyse d'énergie fut commencée, une épreuve se fit encore sur la précision des courbes ajustées au moyen de l'équation de la quantité de mouvement en direction axiale. En supposant que la tension tangentielle à la paroi ne soit appréciable ni dans un cas, ni dans l'autre, et que l'intégration soit portée plus loin que la région de tension appréciable dans le cas du cylindre, et en divisant par la quantité de référence $\pi\rho R^2 U^2$, on a pu ramener l'équation aux formes non dimensionnelles suivantes pour la conduite et le cylindre, respectivement :

$$2\int\left(\frac{\bar{u}}{U}\right)^2\frac{r}{R}\frac{dr}{R}+2\int\frac{\overline{u'^2}}{U^2}\frac{r}{R}\frac{dr}{R}$$

$$+\int\left(\frac{\bar{p}}{\rho U^2/2}-1\right)\frac{r}{R}\frac{dr}{R}=0 \qquad (48)$$

$$2\int\left[\left(\frac{\bar{u}}{U}\right)^2-1\right]\frac{r}{R}\frac{dr}{R}+2\int\frac{\overline{u'^2}}{U^2}\frac{r}{R}\frac{dr}{R}$$

$$+2\left(\frac{\bar{u}}{U}\right)_m\int\left(1-\frac{\bar{u}}{U}\right)\frac{r}{R}\frac{dr}{R}$$

$$+\int\frac{\bar{p}}{\rho U^2/2}\frac{r}{R}\frac{dr}{R}=0 \qquad (49)$$

Dans chaque équation, le premier terme représente la différence dans la quantité de mouvement moyenne (égale au facteur β de Boussinesq, pour l'entrée) et le second celle dans la quantité de mouvement turbulente, entre la section considérée et une très lointaine en amont. Le dernier terme représente l'effet de la différence de pression entre les deux sections. Les équations diffèrent en forme, pour trois raisons. La première est simplement que la vitesse relative en amont est nulle dans le cas de l'entrée de conduite et égale à l'unité dans le cas du cylindre. La seconde réside dans le fait que la pression est reliée à celle où la vitesse relative est l'unité dans les deux cas — c'est-à-dire, loin du

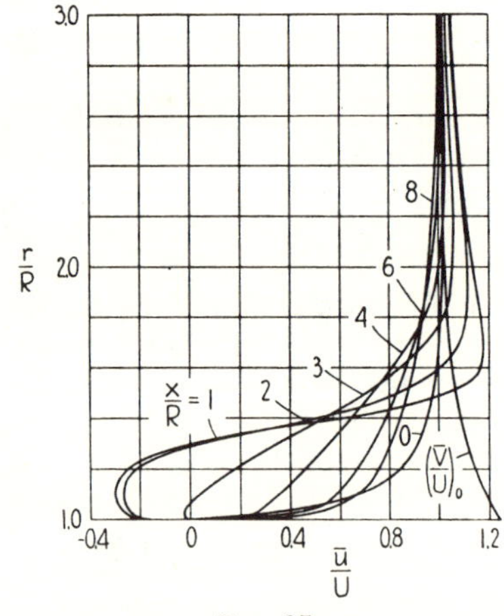

Fig. 27

Courbes rectifiées de la vitesse pour l'entrée

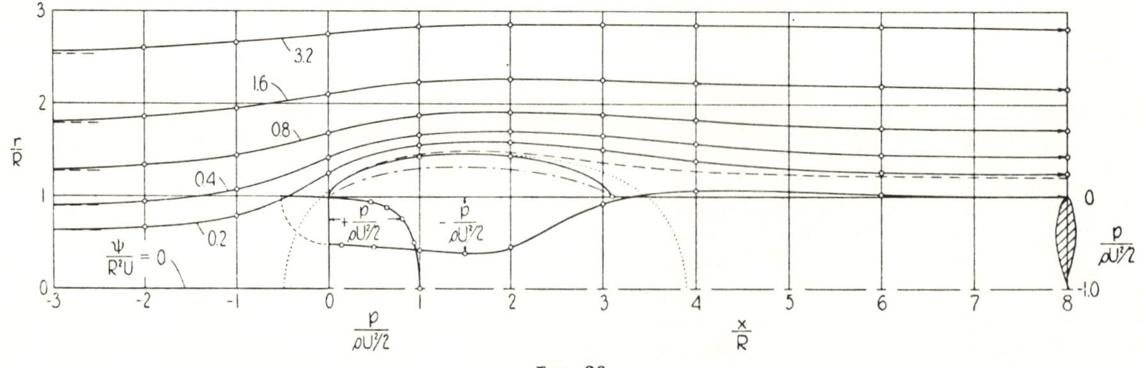

Fig. 28

Schéma des lignes moyennes du courant pour le cylindre

cylindre, et ce qui existerait loin en aval de l'entrée si la tension tangentielle à la paroi n'existait pas. La troisième se trouve dans la différence entre l'écoulement en conduite et à l'air libre, par suite de laquelle un terme additionnel est nécessaire dans l'équation du cylindre pour tenir compte du flux de la quantité de mouvement à travers la périphérie de la région considérée, où la vitesse axiale a la valeur moyenne $(\bar{u}/U)_m$; cette valeur diffère peu de l'unité à la limite utilisée de l'intégration $r/R=3$.

Ces équations contiennent évidemment deux caractéristiques non encore discutées en détail : la pression moyenne et l'intensité axiale de la turbulence, dont les courbes ajustées se trouvent dans les figures 29 à 32. Tandis que les courbes spécifiques diffèrent souvent considérablement des mesures originales, elles concordent

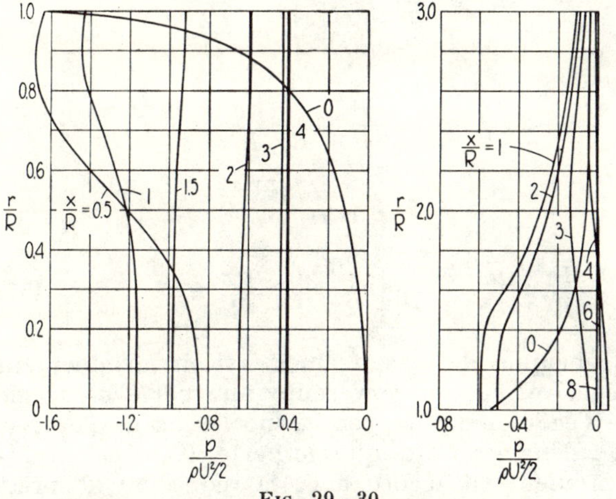

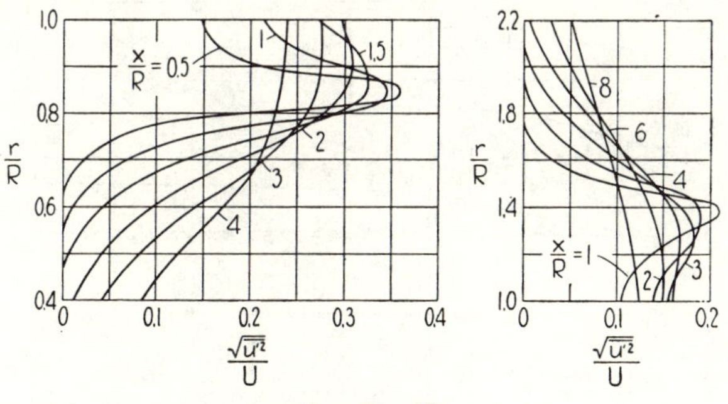

Fig. 29 - 30

Courbes rectifiées de la pression pour l'entrée
et le cylindre

Fig. 31 - 32

Courbes rectifiées des fluctuations axiales
de la vitesse

avec elles-mêmes, avec la tendance générale des mesures, et avec les indications trouvées pour d'autres parois semblables [3, 8]. De plus, lors-

que les divers termes de la relation de quantité de mouvement se calculent section par section, les résultats (voir fig. 33 et 34) ne dévient pas

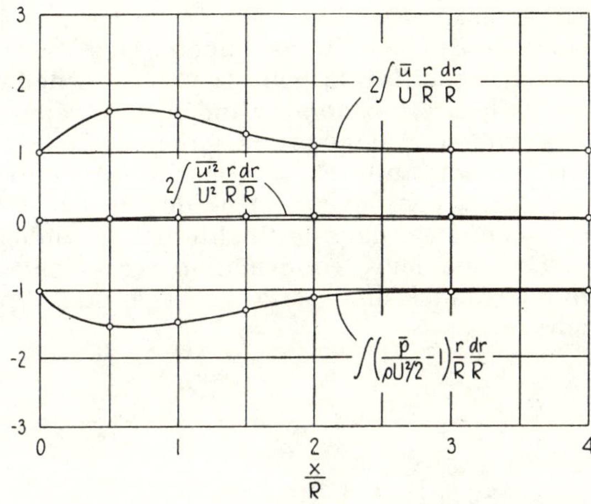

Fig. 33

Analyse de la quantité de mouvement pour l'entrée

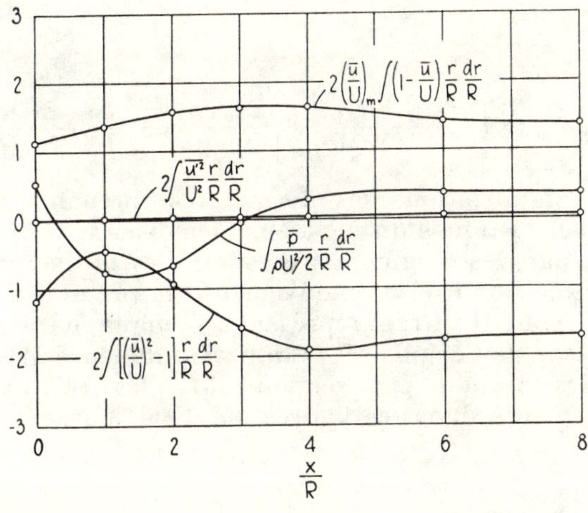

Fig. 34

Analyse de la quantité de mouvement pour le cylindre

des conditions requises par l'équation, plus que de 7 % de la variation de la quantité de mouvement. En ce qui concerne les deux systèmes de courbes, on doit remarquer qu'à l'entrée la réduction de la pression sur la face du réservoir (0,32) est comprise dans la valeur totale pour chaque section, mais jamais explicitement, pendant que la valeur finale du cylindre (0,37) représente exactement l'augmentation de pression sur sa face tronquée. Tout aussi digne de remarque est le fait que l'effet des composantes de la turbulence sur l'équation est petit.

Avec ces contrôles sur les mesures ajustées, on tend à accepter les courbes correspondantes comme représentatives du schéma d'écoulement réel avec un degré d'approximation satisfaisant. On pourrait aussi supposer qu'avec plus d'ajustement la précision deviendrait encore meilleure. Pourtant, cela ne fut pas essayé, parce que la satisfaction des conditions imposées est nécessaire, mais insuffisante pour la précision définitive (c'est-à-dire, un nombre indéfini de répartitions de vitesse et de pression satisferait les conditions de continuité et de quantité du mouvement), et par conséquent on ne doit pas mettre trop de confiance dans la finalité de ces indications. Une meilleure confirmation sera trouvée, cependant, dans l'analyse de l'énergie au chapitre suivant.

Production, transfert, et dissipation de la turbulence

A cause des conditions simplifiées sous lesquelles la relation de la quantité de mouvement fut évaluée, et avec la supposition également raisonnable d'un nombre de Reynolds assez grand pour que les tensions visqueuses moyennes soient partout négligeables, l'équation de l'énergie de l'écoulement moyen peut être ramenée aux formes convenables suivantes pour l'entrée et pour le cylindre, rendues maintenant non dimensionnelles en divisant par la quantité de base $\frac{1}{2}\,\varrho\pi\,R^2U^3$:

$$2\int \frac{\bar{u}}{U}\left(\frac{\bar{V}}{U}\right)^2 \frac{r}{R}\frac{dr}{R} +2\int \frac{\bar{u}}{U}\left(\frac{\bar{p}}{\varrho U^2/2}-1\right)\frac{r}{R}\frac{dr}{R} + 4\int\left(\frac{\bar{u}}{U}\frac{\overline{u'^2}}{U^2}+\frac{\bar{v}}{U}\frac{\overline{u'v'}}{U^2}\right)\frac{r}{R}\frac{dr}{R}$$

$$-4\iint\left[\frac{\overline{u'v'}}{U^2}\left(\frac{\partial u/U}{\partial r/R}+\frac{\partial\bar{v}/U}{\partial x/R}\right)+\frac{\overline{u'^2}}{U^2}\frac{\partial\bar{u}/U}{\partial x/R}+\frac{\overline{v'^2}}{U^2}\frac{\partial\bar{v}/U}{\partial r/R}+\frac{\overline{w'^2}}{U^2}\frac{\bar{v}/U}{r/R}\right]\frac{r}{R}\frac{dr}{R}\frac{dx}{dr}=0 \qquad (50)$$

$$2\int\left[\frac{\bar{u}}{U}\left(\frac{\bar{V}}{U}\right)^2-1\right]\frac{r}{R}\frac{dr}{R} +2\int\frac{\bar{u}}{U}\frac{\bar{p}}{\varrho U^2/2}\frac{r}{R}\frac{dr}{R} +2\int\left(1-\frac{\bar{u}}{U}\right)\frac{r}{R}\frac{dr}{R}$$

$$+4\int\left(\frac{\bar{u}}{U}\frac{\overline{u'^2}}{U^2}+\frac{\bar{v}}{U}\frac{\overline{u'v'}}{U^2}\right)\frac{r}{R}\frac{dr}{R}$$

$$-4\iint\left[\frac{\overline{u'v'}}{U^2}\left(\frac{\partial\bar{u}/U}{\partial r/R}+\frac{\partial\bar{v}/U}{\partial x/R}\right)+\frac{\overline{u'^2}}{U^2}\frac{\partial\bar{u}/U}{\partial x/R}+\frac{\overline{v'^2}}{U^2}\frac{\partial\bar{v}/U}{\partial r/R}+\frac{\overline{w'^2}}{U^2}\frac{\bar{v}/U}{r/R}\right]\frac{r}{R}\frac{dr}{R}\frac{dx}{R}=0 \qquad (51)$$

La signification des termes comprenant les caractéristiques du mouvement moyen sera donnée par des commentaires sur celles de la relation de quantité de mouvement, (le premier terme pour l'entrée représentant maintenant le facteur de Coriolis) et l'évaluation de leurs grandeurs section par section aura les résultats reproduits dans les figures 35 et 36. La somme algébrique de ces termes est invariablement négative, et elle représente une perte d'énergie de l'écoulement moyen. Dans le cas de l'entrée, la valeur asymptotique de perte (0,36) est nécessairement en accord avec la réduction de pression et l'augmentation de vitesse satisfaisant déjà à l'équation de la quantité de mouvement. Pour le cylindre, la valeur asymptotique (0,74)

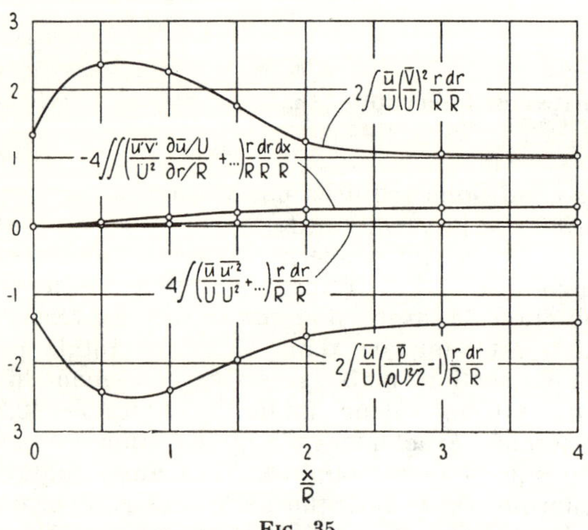

Fig. 35

Analyse de l'énergie moyenne pour l'entrée

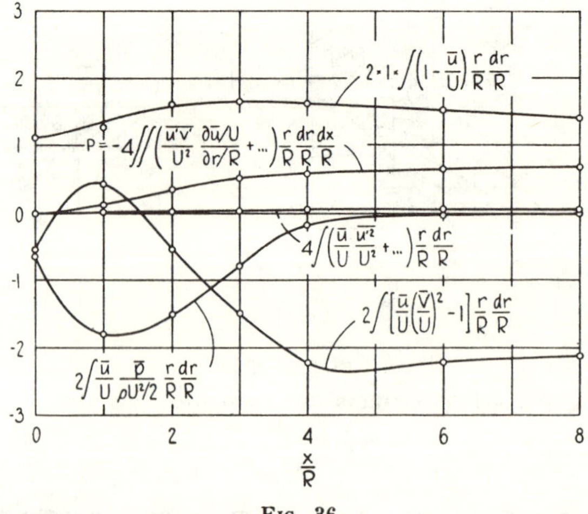

Fig. 36

Analyse de l'énergie moyenne pour le cylindre

est le produit de la vitesse ambiante et de la pression sur la face du cylindre.

Dans chaque cas, une telle perte équivaut à un accroissement de l'énergie du mouvement secondaire, comme indiqué par le terme final pour la production de la turbulence. Ce terme est le même dans les deux cas et il contient trois quantités non encore introduites : l'intensité radiale et tangentielle de la turbulence, et le produit moyen des fluctuations axiales et radiales, qui est proportionnel à la tension tangentielle turbulente. La première fut supposée reliée à la composante axiale de la manière $\overline{u'^2} \approx 2\,\overline{v'^2} \approx 2\,\overline{w'^2}$, déjà trouvée applicable à un degré d'approximation suffisant pour un tel besoin [8] en effet, nous avons trouvé que seul le produit de la tension tangentielle et du gradient radial de la vitesse axiale contribue de façon appréciable à la production de la turbulence. Les courbes ajustées pour les produits moyens des variations sont montrées dans les figures 37 et 38, et les intégrales doubles résultantes dans les figures 35 et 36. La somme des termes tracés dans ces dernières sera trouvée satisfaisant les équations d'énergie moyenne à moins de 5 % du flux différentiel d'énergie. Il est à remarquer que, pour

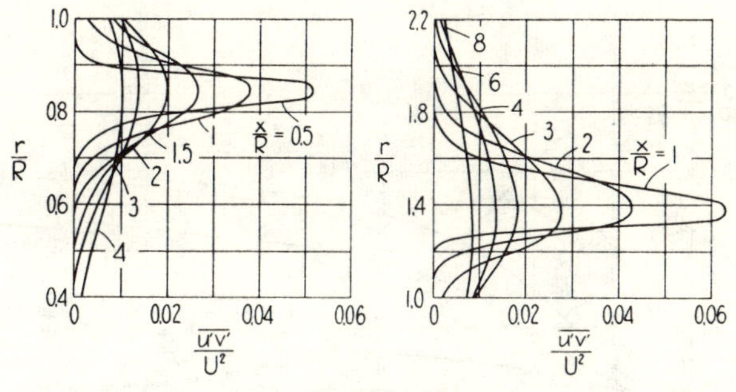

Fig. 37 - 38

Courbes rectifiées du produit des fluctuations

obtenir un tel accord, une augmentation appréciable des produits moyens à l'entrée est nécessaire — ceci en dépit du fait que les composants individuels aient paru, peut-être, trop grands. Quand l'équation de l'énergie pour le mouvement turbulent (y compris l'approximation de Kolmogoroff) s'écrit en utilisant les mêmes restrictions que celles introduites pour le mouvement moyen, elle aura la force identique dans chacun des deux cas :

$$4 \iint \left[\frac{\overline{u'v'}}{U^2}\left(\frac{\partial \overline{u}/U}{\partial r/R} + \frac{\partial \overline{v}/U}{\partial x/R} \right) + \frac{1}{2}\,\frac{\overline{u'^2}}{U^2}\,\frac{\partial \overline{u}/U}{\partial x/R} \right] \frac{r}{R}\,\frac{dr}{R}\,\frac{dx}{R} + 2\int \frac{\overline{u}}{U}\,\frac{\overline{V'^2}}{U^2}\,\frac{r}{R}\,\frac{dr}{R}$$

$$+ 2\int \frac{\overline{u'V'^2}}{U^3}\,\frac{r}{R}\,\frac{dr}{R} + 4\int \frac{\overline{u'p'}}{\varrho U^3}\,\frac{r}{R}\,\frac{dr}{R} + \frac{60}{\mathcal{R}_R}\iint \left(\frac{\partial u'/U}{\partial x/R}\right)^2 \frac{r}{R}\,\frac{dr}{R}\,\frac{dx}{R} = 0 \quad (52)$$

En plus du nombre de Reynolds, $\mathcal{R}_R = RU/\nu$, trois termes nouveaux apparaissent ici. L'un est une corrélation triple de variation de vitesse qui ne fut pas mesurée et peut seulement être supposée négligeable (avec peu de chances d'erreur appréciable). Un autre consiste en une corrélation double de variation de vitesse et de pression, qu'il faut supposer négligeable de la même manière. Le troisième est le gradient axial de la variation de vitesse axiale; les courbes ajustées sont reproduites dans les figures 39 et 40. Par

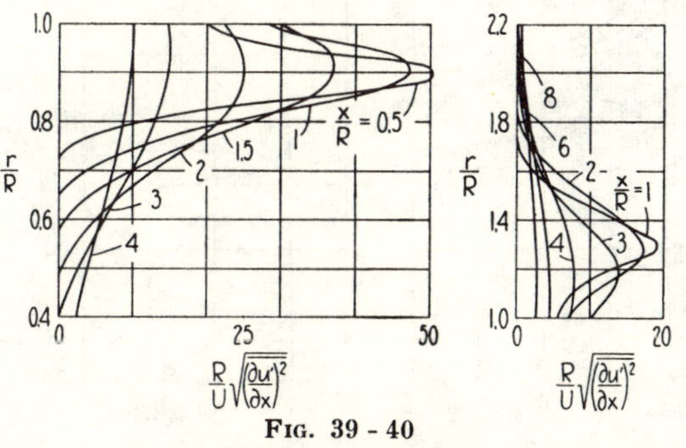

Fig. 39 - 40

Courbes rectifiées
du gradient des fluctuations axiales

l'élimination des deux termes non mesurés, l'équation de l'énergie turbulente se réduit à l'égalité entre la production de la somme de la convection et de la dissipation de la turbulence — autrement dit, la turbulence produite dans un endroit donné est en partie dissipée localement et en partie emportée par l'écoulement pour être dissipée autre part. L'évaluation des valeurs de la convection et de la dissipation pour les comparer avec celles de la production déjà calculées aboutit aux courbes des figures 41 et 42. Bien que l'accord entre les points calculés et les courbes satisfaisant aux conditions physiques de l'équation soit évidemment excellent, il faut remarquer que les valeurs du gradient (en particulier celles de l'entrée, de même que pour les valeurs du produit) furent arbitrairement ajustées pour atteindre ce but. Les figures montrent, alors, des tendances probables des fonctions énergétiques plutôt que des variations précisément mesurées. Il est quand même significatif de noter que la dissipation cumulative reste en arrière de la production cumulative, d'un montant égal à la convection locale; de plus, toutes deux ne s'approchent de leur limite commune que si la turbulence même — toujours

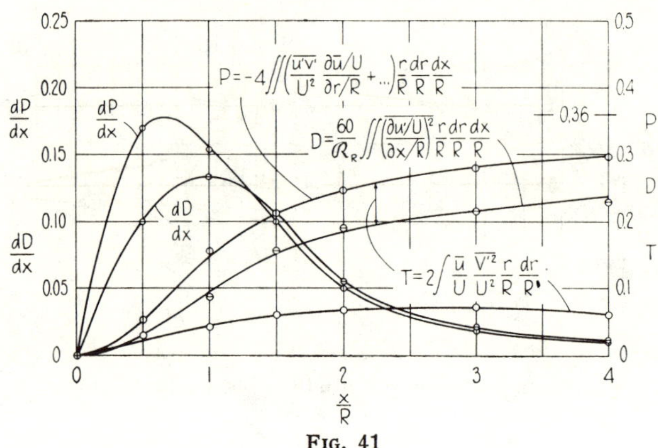

FIG. 41

Analyse de l'énergie turbulente pour l'entrée

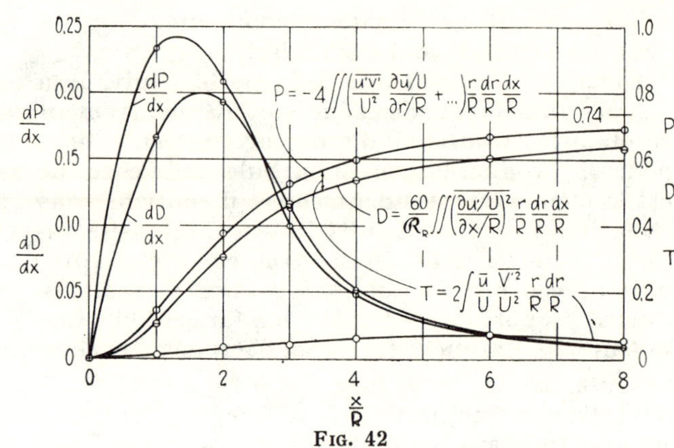

FIG. 42

Analyse de l'énergie turbulente pour le cylindre

une petite quantité — tend graduellement vers zéro.

Parce que les équations précédentes d'énergie s'exprimaient réellement en fonction de la puissance — c'est-à-dire de l'énergie multipliée par le débit — il y a intérêt à examiner leur forme en fonction de l'énergie par unité de volume, mieux connue et appelée équation de Bernoulli. Celle-ci s'obtient, en effet, en intégrant l'équation de base, divisée par le débit différentiel dQ, le long d'un tube de courant. Dans le cas d'un

écoulement stationnaire irrotationel, la somme de Bernoulli qui en résulte, $B = \varrho V^2/2 + p$, est nécessairement constante le long du tube. Cependant, pour un écoulement rotationel dont seulement le champ de vitesse moyenne est stationnaire, trois termes additionnels seront inclus :

$$B = \frac{1}{2}\, \varrho\, \overline{V}^2 + \overline{p} + \frac{1}{2}\, \varrho\, \overline{V'^2} + \varrho\overline{u'^2} + \Delta \quad (53)$$

Le terme pour l'énergie cinétique de l'écoule-

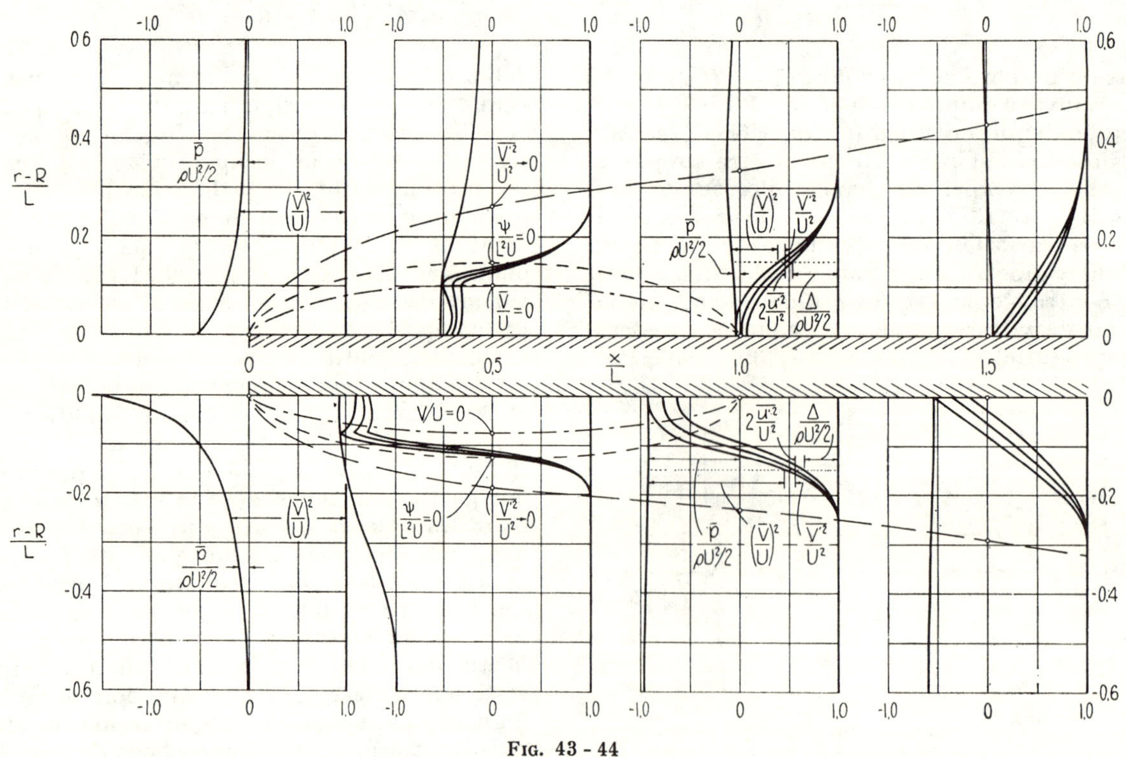

FIG. 43 - 44

Répartition des termes de Bernoulli pour l'entrée (*en bas*) et pour le cylindre (*en haut*)

ment moyen trouve ainsi sa contrepartie dans le terme pour l'énergie cinétique de la turbulence, de même que la pression moyenne a sa contrepartie dans une pression de Reynolds. On incline à regarder Δ comme un terme de dissipation, parce que $B—\Delta$ représente l'énergie nette par unité de volume en un point quelconque. Pourtant, Δ n'a pas seulement sa source dans le travail dissipateur fait par les tensions visqueuses internes, mais aussi dans le travail conservateur fait par les tensions externes de Reynolds. En d'autres termes, Δ peut même décroître le long d'une ligne de courant pendant qu'il croît le long d'une autre, par suite d'un transfert latéral d'énergie au moyen du mélange turbulent. L'intégrale $D=\int \Delta dQ$ sur la section d'écoulement, bien entendu, est équivalente à la dissipation déjà calculée, parce que les éléments individuels du processus de transfert s'équilibrent à l'intérieur.

Les figures 43 et 44 montrent les variations des cinq termes successifs de l'équation étendue de Bernoulli après division par la pression dynamique, $\varrho U^2/2$; la somme des termes devient alors égale à l'unité. En utilisant la longueur L du tourbillon comme référence linéaire pour l'entrée et pour le cylindre, on voit que la répartition de l'énergie offre une ressemblance surprenante dans les deux phénomènes.

CONCLUSION

Par une combinaison de moyens analytiques, expérimentaux et graphiques quelque peu arbitraires, on a obtenu des représentations approximatives de l'écoulement primaire et secondaire dans une entrée brusque de conduite et autour d'un cylindre à face tronquée. Celles-ci ne peuvent pas être considérées comme fournissant une connaissance quantitative sur les détails internes du mécanisme de la turbulence, parce que les mesures n'étaient pas assez complètes ou précises, et que les ajustements visaient trop catégoriquement à équilibrer les relations générales. Par contre, la méthode employée devrait servir le double but de révéler les caractéristiques fondamentales des deux écoulements et d'illustrer les relations qui existent entre les termes principaux des équations de la quantité de mouvement et de l'énergie. Puisqu'il faut comprendre celles-ci avant que les détails puissent être proprement étudiés, cette étude peut être regardée comme le prologue indispensable d'un travail plus complet et plus précis.

Aux observations générales qui sont d'une grande importance, il faut ajouter les suivantes.

Le décollement qui est provoqué par les caractéristiques dues à l'inertie de l'écoulement moyen est arrêté par les caractéristiques dues à l'inertie de la turbulence, chacune des deux sans qu'il soit besoin d'une action visqueuse. La zone de décollement dépend donc en forme et en grandeur de la géométrie des parois, et elle est plus limitée en étendue dans les écoulements en conduites que dans les écoulements à l'air libre. Le domaine des vitesses autour d'une zone de décollement est semblable à celui d'un écoulement irrotationnel autour d'un corps solide de même forme, et la répartition de la pression à l'intérieur ainsi qu'alentour varie pareillement. De l'énergie est perdue, à l'écoulement moyen, presque uniquement par la production de la turbulence. La quantité de turbulence qui existe à un point quelconque dans une zone de décollement est relativement petite. Pourtant, elle a le double effet de ramener le domaine des vitesses vers son état uniforme originaire, par l'effet de mélange de tourbillons plus grands, et de causer sa propre disparition, par les tensions visqueuses dans les tourbillons les plus petits.

REFERENCES

[1] Hsu (H. C.), "Characteristics of Mean Flow and Turbulence at an Abrupt Two-Dimensional Expansion," Ph. D. Dissertation, State University of Iowa, 1950.

[2] Arie (M.), "Flow Past a Normal Plate in Contact with a Boundary," M. S. Thesis, State University of Iowa, 1955.

[3] Arie (M.), and Rouse (H.), "Experiments on Two-Dimensional Flow over a Normal Wall," *Journal of Fluid Mechanics*, Vol. 1, Part 2, 1956.

[4] Arie (M.), "Characteristics of Two-Dimensional Flow behind a Normal Plate in Contact with a Boundary on Half Plane," *Memoirs, Faculty of Engineering*, Hokkaido University, Vol. 10, No. 2, 1956.

[5] Rouse (H.), "Diffusion in the Lee of a Two-Dimensional Jet," *Proceedings, Ninth International Congress of Applied Mechanics*, Brussels, 1957.

[6] Siao (T.-T.), "Characteristics of Turbulence in an Air-Flow Model of the Hydraulic Jump," Ph. D. Dissertation, State University of Iowa, 1954.

[7] Nagaratnam (S.), "The Mechanism of Energy Dissipation in the Hydraulic Jump," M. S. Thesis, State University of Iowa, 1957.

[8] Rouse (H.), Siao (T.-T.), and Nagaratnam (S.), "Turbulence Characteristics of the Hydraulic Jump," *Proceedings, American Society of Civil Engineers*, February, 1958.

[9] Rao (T. R. K.), "Investigation of the Penetration of a Jet into a Counterflow," M. S. Thesis, State University of Iowa, 1958.

[10] Mises (R. von), "Berechnung von Ausfluss- und Ueberfallzahlen," *Zeitschrift VDI*, 1917.

[11] Birkhoff (G.), "Hydrodynamics," Princeton, 1950.

[12] Schlichting (H.), "Grenzschichttheorie," Karlsruhe, 1951.

[13] Rouse (H.), and Abul-Fetouh (A.-H.), "Characteristics of Irrotational Flow Through Axially Symmetric Orifices," *Journal of Applied Mechanics*, Vol. 17, No. 4, 1950.

[14] Rouse (H.), "Seven Exploratory Studies in Hydraulics," *Proceedings, American Society of Civil Engineers*, August, 1956.

[15] Angus (R. W.), "Hydraulics for Engineers," London, 1937.

[16] Nikuradse (J.), "Gesetzmässigkeiten der turbulenten Strömung in glatten Rohren," *VDI Forschungsheft 356*, 1932; "Strömungsgesetze in rauhen Rohren," *VDI Forschungsheft 361*, 1933.

[17] Rouse (H.), "Elementary Mechanics of Fluids," New York, 1945.

[18] Rouse (H.), and McNown (J. S.), "Cavitation and Pressure Distribution," *State University of Iowa, Studies in Engineering*, Bulletin 32, 1948.

[19] Rouse (H.), "Advanced Mechanics of Fluids," New York, 1959.

[20] Yu (Y.-S.), "Effect of Transverse Curvature on Turbulent-Boundary-Layer Characteristics," *Journal of Ship Research*, December, 1958.

[21] Pai (S. I.), "Viscous Flow Theory, Vol. 1, Laminar Flow," New York, 1956.

[22] Stokes (G. G.), "On the Theories of the Internal Friction of Fluids in Motion, and of the Equilibrium and Motion of Elastic Solids," *Transactions, Cambridge Philosophical Society*, Vol. 8, 1845.

[23] Lamb (H.), "Hydrodynamics," 6th edition, New York, 1946.

[24] Reynolds (O.), "On the Dynamical Theory of Incompressible Fluids and the Determination of the Criterion," *Philosophical Transactions, Royal Society*, Vol. 186, 1894.

[25] Townsend (A. A.), "The Structure of Turbulent Shear Flow," Cambridge, 1956.

[26] Taylor (G. I.), "The Spectrum of Turbulence," *Proc. Royal Society A*, Vol. 164, 1938.

[27] Batchelor (G. K.), "The Theory of Homogeneous Turbulence," Cambridge, 1953.

[28] Taylor (G. I.), "Statistical Theory of Turbulence," *Proc. Royal Society A*, Vol. 151, 1935.

[29] Karman (T. von), and Howarth (L.), "On the Statistical Theory of Isotropic Turbulence," *Proc. Royal Society A*, Vol. 164, 1938.

[30] Batchelor (G. K.), "Kolmogoroff's Theory of Locally Isotropic Turbulence," *Proc. Cambridge Philosophical Society*, Vol. 43, 1947.

[31] Corrsin (S.), "An Experimental Verification of Local Isotropy," *Journal of the Aeronautical Sciences*, Vol. 16, 1949.

[32] Klebanoff (P. S.), "Characteristics of Turbulence in a Boundary Layer with Zero Pressure Gradient," *Technical Note No. 3178, National Advisory Committee for Aeronautics*, 1954.

[33] Laufer (J.), "The Structure of Turbulence in Fully Developed Pipe Flow," *Report 1174, National Advisory Committee for Aeronautics*, 1954.

[34] Rouse (H.), "Use of the Low-Velocity Air Tunnel in Hydraulic Research," *Proceedings of the Third Hydraulics Conference, State University of Iowa Studies in Engineering*, Bulletin 31, 1947.

[35] Hubbard (P. G.), "Operating Manual for the IIHR Hot-Wire and Hot-Film Anemometers," State University of Iowa, Studies in Engineering, Bulletin 37, 1957.

[36] Howe (J. W.), "Flow Measurement," Chapter III, *Engineering Hydraulics*, New York, 1950.

[37] Peters (H.), "Druckmessung," Vol. IV, *Handbook der Experimentalphysik*, Leipzig, 1932.

On the Bernoulli theorem for turbulent flow

By Hunter Rouse

(Iowa Institute of Hydraulic Research, State University of Iowa, Iowa City)

Because its several terms represent directly measurable quantities, the so-called Bernoulli equation (actually first drived by Euler) for an incompressible, inviscid fluid is frequently adapted to viscous flow by adding a term representing cumulative losses between successive sections. Moreover, if such flow is turbulent, it is sometimes assumed that terms should be introduced as well for the kinetic energy of the turbulence. In order to be correct, of course, the resulting expression must be derivable from the basic equations of motion, an integral form of which is generally known as the equation of energy. Since the Bernoulli equation itself is also called an energy relationship, it is significant to note that neither is written in terms of energy as such — the former being an equation of power (i. e., temporal rate of energy change) and the latter involving energy per unit volume, mass, or weight. This distinction, as well as other ponts essential to the following discussion, is illustrated by the customary derivation of the two relationships for inviscid flow.

The Eulerian equation of acceleration for steady incompressible flow in the direction s tangent to a streamline is simply

$$\frac{\partial}{\partial s}\left(\frac{\varrho\, U^2}{2}\right) = -\frac{\partial\,(p + \gamma\, z)}{\partial s} \tag{1}$$

in which ϱ is the density, U the velocity, p the pressure, γ the specific weight, and z the elevation. Integration between points 1 and 2 on the same streamline immediately yields the Bernoulli equation

$$\frac{\varrho\, U_1^2}{2} + p_1 + \gamma\, z_1 = \frac{\varrho\, U_2^2}{2} + p_2 + \gamma\, z_2\,. \tag{2}$$

However, if each term of Eq. (1) is first multiplied by the velocity, as in the usual energy analysis, integration with respect to s can proceed in a comparable fashion only by virtue of the fact that in the differential expression

$$U\,\frac{\partial}{\partial s}\left(\frac{\varrho\, U^2}{2}\right) ds\, dA = -\,U\,\frac{\partial\,(p + \gamma\, z)}{\partial s}\, ds\, dA \tag{3}$$

the product $U\, dA$ (i. e., the rate of flow past any section of a streamtube of cross-sectional area dA) is independent of s. Usually a second integration — now with respect to A — is performed thereafter to yield a power relationship for the gross flow section. If, instead, the result of the first integration is divided by $U\, dA$, Eq. (2) will again result. This procedure would hence be of trivial importance were it not for the fact that it is the most significant way by which a true dissipation term can eventually be obtained in the BERNOULLI equation for a viscid fluid.

The relationship for turbulent flow [1] that corresponds to the foregoing product of the velocity and the equation of EULER is derived from the sum of the REYNOLDS equations for the tangential, normal, and binormal directions of the mean motion, each multiplied by the corresponding component of the mean velocity:

$$\overline{U}\,\frac{\partial}{\partial s}\left(\frac{\varrho\,\overline{U}^2}{2}\right) + \varrho\,\overline{u}_i\,\frac{\partial \overline{u_i'\,u_j'}}{\partial x_j} = -\,\overline{U}\,\frac{\partial\,(\overline{p}+\gamma\,z)}{\partial s} + \mu\,u_i\,\frac{\partial^2 \overline{u}_i}{\partial x_j\,\partial x_j}\,. \tag{4}$$

Herein, as usual, the bars denote temporal mean values and the primes the instantaneous deviations; the subscripts $i = 1, 2, 3$ and $j = 1, 2, 3$ each indicate the tangential, normal, and binormal directions, respectively, and the use of a particular subscript twice in any term implies summation over all values that the subscript can take. Since the mean velocity $\overline{U}$ is wholly tangential, it follows that $\overline{u}_1 = \overline{U},\ \overline{u}_2 = 0,\ \overline{u}_3 = 0$.

If the notation of Eq. (4) for the mean flow is simplified by the usual designation for viscous and turbulent stresses

$$\overline{\tau}_{ij} = \mu\left(\frac{\partial \overline{u}_i}{\partial x_j} + \frac{\partial \overline{u}_j}{\partial x_i}\right) - \varrho\,\overline{u_i'\,u_j'} \tag{5}$$

and if the result is multiplied by the elementary volume $ds\, dA$, it will have the form corresponding to Eq. (3),

$$\overline{U}\,\frac{\partial}{\partial s}\left(\frac{\varrho\,\overline{U}^2}{2}\right) ds\, dA = -\,\overline{U}\,\frac{\partial\,(\overline{p}+\gamma\,z)}{\partial s}\,ds\, dA + \overline{u}_i\,\frac{\partial \overline{\tau}_{ij}}{\partial x_j}\,ds\, dA\,. \tag{6}$$

This can be integrated along the streamtube to yield, after division by the elementary rate of flow,

$$\frac{\varrho\,\overline{U}_1^{\,2}}{2} + \overline{p}_1 + \gamma\,z_1 = \frac{\varrho\,\overline{U}_2^{\,2}}{2} + \overline{p}_2 + \gamma\,z_2 - \int_{s_1}^{s_2} \frac{\overline{u}_i}{\overline{U}}\,\frac{\partial \tau_i}{\partial x_{ij}}\,ds \tag{7}$$

which presumably corresponds to the expression that is sought when the BERNOULLI equation is arbitrarily modified for the case of turbulent flow. As will be seen at once, however, no explicit terms for the kinetic energy of the turbulence are present. Moreover, the fact that the term at the right is not restricted as to sign indicates that the viscous and turbulent stresses can cause the unit energy along a particular streamtube to increase as well as decrease. Evidently, a reduction in the BERNOULLI sum — i. e., a negative value of the integral — can be produced not only by energy dissipation but also by energy transfer to neighboring streamtubes; accordingly, a positive value will represent a local gain in excess of the local dissipation at the expense of the surrounding flow. Thus it is a mistake to assume that the customary additive term (even so far as the mean viscous effects are concerned) represents purely an irrecoverable loss.

By means of the rule of the calculus for the derivative of a product, the rate at which the stresses do work in changing the Bernoulli sum can be written as the difference between the total rate at which work is done and what is usually considered the rate at which energy is dissipated:

$$\bar{u}_i \frac{\partial \bar{\tau}_{ij}}{\partial x_j} = \frac{\partial(\bar{u}_i\,\bar{\tau}_{ij})}{\partial x_j} - \bar{\tau}_{ij}\frac{\partial \bar{u}_i}{\partial x_j}\,. \tag{8}$$

Incorporation of the latter rates into Eq. (7) then yields the more explicit form of the equation for the mean flow:

$$\frac{\varrho\,\bar{U}_1{}^2}{2} + \bar{p}_1 + \gamma\,z_1 = \frac{\varrho\,\bar{U}_2{}^2}{2} + \bar{p}_2 + \gamma\,z_2 - \int_{s_1}^{s_2}\frac{1}{\bar{U}}\frac{\partial(\bar{u}_i\,\bar{\tau}_{ij})}{\partial x_j}\,ds + \int_{s_1}^{s_2}\frac{\bar{\tau}_{ij}}{\bar{U}}\frac{\partial \bar{u}_i}{\partial x_j}\,ds\,. \tag{9}$$

This statement of the Bernoulli theorem is complete and rigorous. In many cases, however, the viscous stresses expressed in terms of mean velocity are small in comparison with the Reynolds stresses of the turbulence and can be neglected without appreciable error. The Bernoulli equation then reduces to the form

$$\frac{\varrho\,\bar{U}_1{}^2}{2} + \bar{p}_1 + \gamma\,z_1 = \frac{\varrho\,\bar{U}_2{}^2}{2} + \bar{p}_2 + \gamma z_2 + \varrho\int_{s_1}^{s_2}\frac{1}{\bar{U}}\frac{\partial(\overline{u_i'\,u_j'})}{\partial x_j}\,ds - \varrho\int_{s_1}^{s_2}\frac{\overline{u_i'\,u_j'}}{\bar{U}}\frac{\partial \bar{u}_i}{\partial x_j}\,ds\,. \tag{10}$$

The last or dissipative term now warrants the following consideration: Including the negative sign before it, it is inherently positive; but whereas it thus represents a loss to the mean flow, it is not a true measure of dissipation in that it does not contain the viscosity. In other words, it indicates solely the extent to which the energy of the turbulence has been increased at the expense of the mean flow.

To illustrate the physical interrelationship of the several terms of Eq. (10) under rather extreme conditions, use is made of published data for the mean and turbulent characteristics of flow over a simple wall [2], in which the zone of separation behind the wall is of particular significance. The mean pattern of streamlines for such flow is reproduced in the upper half of the accompanying figure (Fig. 1); the two supplemental lines indicate the approximate upper limit of the turbulent diffusion and the locus of points of zero horizontal velocity. Since the approaching flow was essentially irrotational, the usual three terms of the Bernoulli equation should have a constant sum up to the diffusion zone. In fact, if the values of the piezometric sum $p + \gamma z$ are referred to that of the ambient flow and if all terms are then divided by the ambient value $\varrho\,U_0^2/2$, the sum of the nondimensional kinetic and piezometric terms K and P will be unity over the entire wall section and at all other points not affected by the turbulence. Even within the diffusion zone, moreover, the difference between these terms and the change Δ produced by the Reynolds stresses should remain constant along a streamline, in accordance with Eq. (7). Thus, outside the zone of separation the quantity $K + P - \Delta$ is seen to retain at all cross sections the magnitude of unity determined by conditions at the initial section. The streamlines of

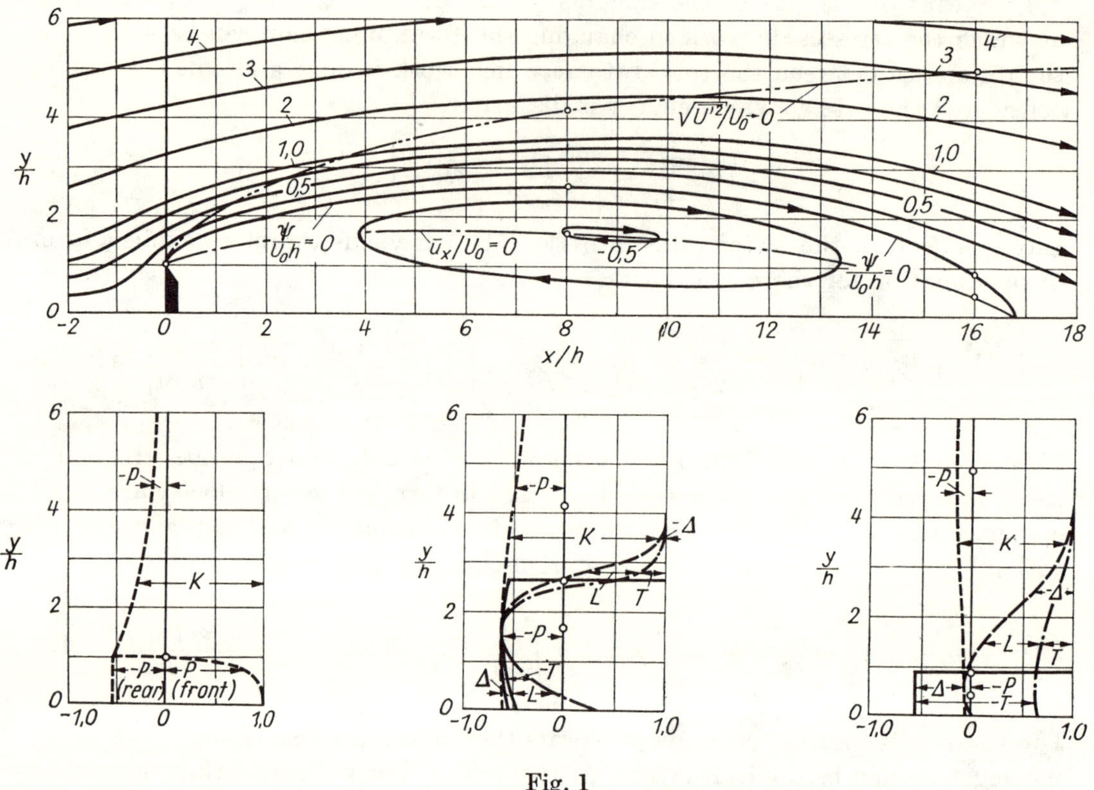

Fig. 1

the eddy, on the other hand, are closed curves, and their "initial" section must be chosen arbitrarily — the curve of zero forward velocity being the most logical choice. The initial energy level of the eddy, given as before by the local sum of K and P, is evidently well below that of the primary flow. Comparison of the two zones at successive sections, on the other hand, will show that the sum $K + P$ for the primary flow decreases (that is, Δ is negative) as energy is transferred through the REYNOLDS stresses to the fluid below, the sum of $K + P$ for the eddy thereby increasing (and Δ accordingly becoming positive).

Since the quantity $K + P - \Delta$ must remain constant along any streamline, the loss term can be included only by simultaneously adding and subtracting it. The right side of Eq. (10), in other words, is of the form $K + P + T + L$, in which T and L denote energy transfer and loss. As indicated qualitatively in the figure (the shear measurements being too inexact to permit quantitative analysis), the loss necessarily becomes continuously greater in the direction of flow. This produces an apparent anomaly within the eddy, since each streamline closes upon itself; evidently, the circuit is simply repeated indefinitely.

If, as in the usual energy (i. e., power) analysis, instead of division by $\overline{U}\,dA$ the relationship is integrated laterally to a limit beyond the zone of diffusion (or from wall to wall of a closed conduit), the internal effects of lateral transfer will disappear and the term for total work will reduce to the difference between the integrals of the velocities times the REYNOLDS stresses over the two end sections. The "dissipation" term will then correspond to the net loss per unit volume which occurs over the reach under consideration, usually evaluated as a residual quantity in the one-dimensional application of the BERNOULLI theorem.

Whether the analysis is one-, two-, or three-dimensional, the fact remains that the loss term under discussion does not represent true dissipation, nor has the expected term for the kinetic energy of the turbulence as yet appeared. Only through introduction of the energy equation for the turbulence itself [1] can terms of the types sought be obtained. This equation, in contradistinction to Eq. (4) for the mean flow, is derived by multiplying each of the NAVIER-STOKES equations by the corresponding velocity component before substituting the REYNOLDS notation and averaging. Subtraction of Eq. (4) therefrom then yields

$$\overline{U} \frac{\partial}{\partial s}\left(\frac{\varrho\,\overline{U'^2}}{2}\right) + \overline{u'_j \frac{\partial}{\partial x_j}\left(\frac{\varrho\,U'^2}{2}\right)} + \varrho\,\overline{u'_i\,u'_j \frac{\partial u_i}{\partial x_j}} = -\,\overline{u'_i \frac{\partial p'}{\partial x_i}} + \mu\,\overline{u'_i \frac{\partial^2 u'_i}{\partial x_j\,\partial x_j}} \qquad (11)$$

wherein $\overline{U'^2}$ is the sum of the mean squares of the turbulence components u'_1, u'_2, u'_3. Like the term for the work done in transferring energy by the mean stresses in Eq. (4), the term at the end of Eq. (11) can be replaced by the difference between the total rate at which work is done by the instantaneous stresses and the rate of dissipation which they produce:

$$\mu\,\overline{u'_i \frac{\partial u'_i}{\partial x_j\,\partial x_j}} = \mu\,\frac{\partial}{\partial x_j}\overline{\left(u'_i \frac{\partial u'_i}{\partial x_j}\right)} - \mu\,\overline{\frac{\partial u'_i}{\partial x_j} \frac{\partial u'_i}{\partial x_j}}\,. \qquad (12)$$

Multiplication of the resulting equation by $ds\,dA$, integration along the stream tube, division by $\overline{U}\,dA$, and solution for the term representing rate of turbulence production will then yield the desired equality:

$$-\varrho \int_{s_1}^{s_2} \frac{\overline{u'_i\,u'_j}}{\overline{U}} \frac{\partial \overline{u}_i}{\partial x_j}\,ds = \frac{\varrho\,\overline{U_2'}^2}{2} - \frac{\varrho\,\overline{U_1'}^2}{2} + \int_{s_1}^{s_2} \frac{\overline{u'_j \frac{\partial}{\partial x_j}\left(\frac{\varrho\,U'^2}{2}\right)}}{\overline{U}}\,ds$$

$$+ \int_{s_1}^{s_2} \frac{\overline{u'_i \frac{\partial p'}{\partial x_i}}}{\overline{U}}\,ds - \mu \int_{s_1}^{s_2} \frac{1}{\overline{U}} \frac{\partial}{\partial x_j}\overline{\left(u'_i \frac{\partial u'_i}{\partial x_j}\right)}\,ds + \mu \int_{s_1}^{s_2} \frac{1}{\overline{U}} \overline{\frac{\partial u'_i}{\partial x_j} \frac{\partial u'_i}{\partial x_j}}\,ds\,. \qquad (13)$$

Herein the term at the left is obviously the loss integral already discussed. At the right are, first of all, the anticipated terms for the kinetic energy of the turbulence at the two sections; these represent, it should be remarked, the effect of longitudinal mean-flow convection on the turbulence distribution. The next integral, accordingly, represents the corresponding effect of turbulent diffusion — now lateral as well as longitudinal. Thereafter comes a term for the fluctuating pressure, corresponding to those already at hand for the mean pressure, and then a term for the total work done by the instantaneous viscous stresses of the turbulence, corresponding to that already at hand for the total work done by the REYNOLDS stresses. The final term — at last one that is truly dissipative — represents the rate at which mechanical energy is transformed into heat by the turbulence.

Substitution of these six additional terms into the equation of BERNOULLI would obviously introduce such complexities as to render the equation for the present quite unusable. As a matter of fact, not all of the terms have as yet been measured satisfactorily for even the simplest of conditions — uniform pipe flow [3]. There is, to be sure, evidence [4] that under conditions of separation, such as in the example illustrated, the equation of turbulent energy can be reduced with fair approximation to terms for production, convection, and dissipation — provided that the relationship is integrated laterally beyond the limits of diffusion. For the analysis of flow-pattern details, however, one is evidently restricted to the BERNOULLI relationship of Eq. (9) or (10) until such time as the terms of Eq. (13) become either more readily measurable or better understood.

REFERENCES

[1] A. A. TOWNSEND: The Structure of Turbulent Shear Flow, Cambridge (1956).
[2] M. ARIE and H. ROUSE: Experiments on Two-dimensional Flow over a Normal Wall, Journal of Fluid Mechanics 1 (1956) Part 2.
[3] J. LAUFER: The Structure of Turbulence in Fully Developed Pipe Flow, N.A.C.A. Report 1174 (1954).
[4] H. ROUSE, T. T. SIAO and S. NAGARATNAM: Turbulence Characteristics of the Hydraulic Jump, Trans. A.S.C.E. 124 (1959).

Current Trends in American Hydraulics

Hunter Rouse

During the past decade two of us at the State University of Iowa wrote and published a general history of world hydraulics. Much of the final material was gathered in France, and the first version of the completed work appeared as a fold-in supplement of a French engineering journal. The last chapter of this history contained an appraisal of the science at midcentury. I now propose to begin at the point at which the previous work left off and to assess American contributions since that time, following the same style and spirit. In particular, emphasis will again be laid upon the fundamentals of the science, as opposed to their engineering application, in keeping with the theme of this symposium.

Our National Bureau of Standards has published annually since 1939 a review of American hydraulic-laboratory research comparable to that published by the International Association for Hydraulic Research for its non-American members. I should like to make considerable, though purposely unspecific, reference to current and back numbers of this publication throughout much of my presentation. In the 1960 issue of the Bureau review, for example, research activity is described in five private American laboratories; in 28 laboratories of six different federal departments; and in 67 laboratories of 47 private, municipal, and state universities, for a grand total of nearly a thousand distinct experimental projects. Even if one grants at the outset that many of these are applied rather than fundamental and that not a few are redundant or otherwise ill-conceived, one must still conclude that considerable attention is currently being given in America to the scientific aspects of the subject.

I intend to survey the most pertinent of these projects, retaining as far as is feasible the main divisions of subject matter of the earlier history. Once again, for the same reasons, specific reference will be made neither to the laboratories in which the work is being, or has been, performed nor to the investigators who are responsible. Such information can readily be obtained by reference to the Bureau reviews. Rather, full attention will be given to the present status of the science itself in the country as a whole.

Two somewhat related influences may be cited as factors contributory to the rather flourishing situation in which American hydraulics now finds itself. On the one hand, the various federal agencies, some of which are prohibited by policy or precedent from undertaking basic investigations in their own establishments, find it fruitful in many ways to underwrite the fundamental research of universities and even of private firms. Projects supported by each of the executive departments are usually relevant to that department's interests. Organizations like the National Science Foundation, on the other hand, are very broad and general in scope and their influence is accordingly of vast extent. It is estimated that some two thirds of all university research is now supported by the government, and hydraulics is probably as typical a case as any. On the other hand, for the past 25 years American hydraulics has been taking its rightful place as one of the branches of that newer and more inclusive science, the mechanics of fluids. As such, it both borrows from and contributes to the vast pool of knowledge which the broader field encompasses. The Bureau of Standards lists hydraulics projects not only of civil and mechanical engineering departments but also of some that are primarily chemical, marine, agricultural, and meteorological. Hydraulics has gained greatly in recent years through use of material produced by people as varied as mathematical physicists, aerodynamicists, and astronomers. But it has also contributed in its turn to such apparently unrelated matters as atomic power, papermaking, and experimental medicine. In fact, it is this very breadth of viewpoint which is the present strength of American hydraulics at, so to speak, the leading edge of its endeavors.

A case in point is found at once in the field of classical hydro-dynamics, long considered foreign territory by and for hydrauli-cians. The recent generalization of the mathematical concept of the stream function to three-dimensional flow was a con-tribution by a civil engineer working in an American hydraulics laboratory—though not illogically it was first published by him in the French technical press. Other staff members of hydraulics institutes have contributed in no small measure to problems in such related fields as naval architecture, and mathematicians in their turn have recently helped to advance such purely hydraulic matters as free-streamline flow at weirs and gates. Application of the circle theorem to the investigation of the pressure distributions about two-dimensional profiles of arbi-trary form and of the Lagally theorem to the forces and mo-ments on three-dimensional bodies even in unsteady motion has been found of value not only in naval architecture but in other branches of fluid mechanics dealing with streamlined boundaries—as those in hydraulics are gradually becoming. Because of the mathematical complexities that can result, more and more frequent use is made of the various high-speed com-puters, complete programs now being available for pressure-distribution solutions. Currently, at least two attempts are being made to bring the problem of free-streamline flow in a gravita-tional field into the realm of solvability, initially with the re-laxation procedure or its electrical-network analogue, and more recently with high-speed computers. Related studies are under way in nonuniform or unsteady flows with density stratification, whether of continuous or discontinuous density variation. To come full circle, the irrotational flow of a stratified (and even compressible) fluid has just received a generalized treatment—again at the hands of a hydraulician.

Gravity-wave analysis, perhaps the first aspect of mathematical hydrodynamics to be accepted by hydraulicians, now provides the basis for an approximate solution of many complex engi-neering problems. Among instances of recent investigations are the following: the generation or deformation of waves by the movement or presence of bodies near or at the free surface; the resonance of surface oscillation—seiche—in harbors; wave refraction, diffraction, and reflection by breakwater systems; statistical analysis of wave spectra, and the study of the so-called "confused" sea; interaction of wind and wave at coastlines and in inland seas; the damping of solitary, oscillatory, and standing waves by boundary shear; forces on shore and offshore structures produced by breaking as well as stable waves; and interfacial waves and surges in flow with density stratification. Of related import is the sound-wave analogy to gravity waves; with the completion of parallel studies of water-channel bends, contractions, and expansions for hydraulic design, attention has been turned to the simulation of supersonic occurrences in the water channel for exploratory studies related to aerodynamic design. The close relationship among the goals of aerodynamicists, oceanographers, naval physicists, and hydraulicians in these pursuits is conducive to joint effort, and the present availability of high-speed computers has greatly extended the range of problems subject to analytical treatment.

Except for the specialty of Professor Escande—the evaluation of surges and water hammer in complex systems—other cases of unsteady flow had previously been avoided by the American hydraulician. Now, however, there are numerous examples of work in this rather difficult field. Among these are problems of added (or so-called "virtual") mass involved in the pitching, rolling, or vibration of immersed and floating bodies. Somewhat related is the behavior of an immersed body (such as a sediment particle) in an oscillating fluid, as is the more general case of any body undergoing relative acceleration or deceleration. The drag of an oscillating plate is a noteworthy example. This has its counterpart, of course, in oscillatory or uniformly accelerated flow through conduit transitions, which is proving to be far from simple even experimentally. Analysis of the well-known phenomenon of oscillating cross thrust due to the formation of a regular Kármán vortex trail has been extended experimentally to include on the one hand the vibration of very thin blades and on the other the harmonic evaluation of the frequency of cross-thrust reversal once the vortex trail has lost its regularity at high Reynolds numbers. Finally, the study of temporal variation in flow through porous media is well under way.

Cavitation research has progressed on so many fronts as to warrant separate comment in its own right. Much attention has been given to the thermodynamics of bubble formation and collapse, with particular bearing on the problem of scaling. Previous photographic studies of individual bubble behavior have been supplemented by related experiments on the very start of boundary damage. Cavitation has been produced ex-

perimentally not only by the vibratory method but also sonically, and not only in flow along curved boundaries but also in
separation zones of turbulence generation. Conditions of
"super" cavitation, unsteady as well as steady, have proved
amenable to free-streamline analysis. Of considerable bearing
upon the following section has been a very recent study of
cavitation at small boundary irregularities.

Since the mechanics of surface resistance is intimately related
to the theory of the boundary layer, it might be well to note
at this point the extensions to and uses of boundary-layer principles that are being made by hydraulicians and by other scientists working in hydraulics laboratories. The development of
the mean-velocity profile characterizing the boundary layer has
been studied experimentally on flat plates in both zero and
finite pressure gradients, in two-dimensional and conical
diffusers, along circular cyclinders, and very recently around
an ellipsoid having three unequal axes. The theory has been
applied to the calculation of the discharge coefficient for the
broadcrested weir; to the prediction of the point of onset of
aeration of a spillway nappe; to the analysis of energy dissipation of gravity waves; to the evaluation of resistance to rolling
of cylindrical bodies; and to the design of water tunnels. Because of the frequent use of the Preston method—involving
a calibrated stagnation tube in contact with the boundary—for
the determination of boundary shear, attention is currently being given to a re-examination of the equivalence of the so-called
"inner" laws of velocity distribution for pipe flow (in which
the Preston tubes are usually calibrated) and boundary-layer
flow. A similar investigation of the relationship between the
resistance functions for rough plates and rough conduits was
made previously in considerable detail, with the result that
experimental values obtained for the one type of motion are
now directly applicable to the other. Studies have been made
with either plates or conduits roughened by applications of
emery cloth, wire screening, transverse bars, and regularly
spaced cubes, and considerable clarity has been realized in our
understanding of the effect of areal concentration, as well as
size, of roughness elements. Currently some success is being had
in the analytical synthesis (by high-speed computer) of smooth-
surface resistance with the drag of individual elements of arbitrary concentration and size distribution. Finally, results have
been obtained pointing to the variation of open-channel re-

sistance with the Froude number once the limit of free-surface stability has been surpassed; this in turn has instigated a re-examination of the stability criterion, which involves the Reynolds number and the relative roughness, as well as the Froude number, to a degree that is still imperfectly understood.

In the field of turbulence, the endeavors of hydraulicians in America have continued to be directed toward the application to the flow of water of techniques of measurement and analysis already perfected for the flow of air. Activity of this sort has been notable in such problems as river control, cavitation, diffusion, energy dissipation, sedimentation, sediment suspension, air entrainment, jet dispersion, and the swimming of fish. At the same time many a hydraulics laboratory resorts to air as the test fluid so that standard techniques of turbulence measurement can be applied. The distribution of pressure fluctuations within a free turbulent stream has now been studied experimentally. Though not all characteristics of the turbulence have as yet been measured even in air, approximate energy analyses for both the mean and the secondary flows have been conducted for such instances of shear flow as mixing streams, wakes, jets, quasi-stable eddies, and the hydraulic jump. Closely related to the problem of turbulence is the analysis of stability of liquid films, high-velocity streams, and stratified flow already mentioned; whereas the first has been performed with good approximation and the second with lesser degrees, the third still defies successful attack.

Perhaps the greatest advance in the field of sediment transportation has been in the evaluation of the total load of streams —if not as yet through the ultimately essential correlation of the suspended and bed loads, at least as an empirical function of stream and sediment characteristics. Investigations of the mechanics of sediment suspension have concerned themselves largely with the effect of sediment concentration on the fluid turbulence. Likewise, investigations of bed load have dealt to a considerable extent with the effect of modes of movement on bed resistance, and vice versa, with particular attention to ripple and dune formation. In the field of scour, it has been shown that erosion by clear water and by sediment-laden water differs greatly in limiting depth and that local scour in the bed of a stream undergoing transport tends toward geometric similarity at all scales, including model and prototype. Considerable study has been given to the mechanics of particle movement with respect to both shape and concentration, and mention has already been made of the question of the effect of turbulence upon the velocity of fall. Standard deviations from mean values, as well as the means themselves, are at last receiving attention. The historical review mentioned at the outset noted the close relationship between certain aspects of sediment transportation, air entrainment, and evaporation. A new version of particle transport in suspension has recently been brought into the hydraulic laboratory on behalf of the paper industry: the flow of paper stock, a very dilute suspension of fine fibers, the behavior of which is markedly thixotropic. The tendency of such paper stock to flocculate is also reminiscent of that of river-borne clays when they meet the saline tidal wedge in estuaries —likewise a current problem of hydraulicians. Many of the aspects of sediment transportation in rivers are encountered as well along coasts, so much so that transport and sorting by waves and littoral drift receive continuous attention from a number of federal and university laboratories; this, to be sure, is a world-wide situation.

Under the heading of modeling, the frequent use of air to simulate all but the gravitational and cavitation aspects of water continues to be characteristic of American laboratory practice. In fact, the structural simplicity of air-flow models is such that the model can be, when this is desirable, several times larger than the prototype: for example, a portion of the "slice" or sluicelike outlet section of a papermaking machine enlarged to permit the study of turbulence characteristics with the hot-wire anemometer. Other models used in coastal, oceanographic, and even meteorological studies have been so greatly reduced in scale that the molecular viscosity of the model water or air is assumed to simulate the eddy viscosity of the prototype. For detailed study of the turbulence distribution, of course, this is of no use whatever; for the simulation of over-all effects, on the other hand, it may well provide the only possible approach, just as the assumption of a homogeneous mixing process once permitted considerable advancement to be made in the gross analysis of sediment transport. The modeling of such involved phenomena as air entrainment, cavitation, and cavitation damage has progressed only very slowly beyond original accomplishments. The use of hydraulic model techniques to stimulate at least the initial spurts in the development of oceanographic and meteorological model techniques was an occurrence of the past decade; meteorological modeling, in fact, experienced several of its formative years in American hydraulics laboratories. Many of the resulting laboratory devices are, in actuality, analogues rather than dynamically similar models, for they undergo distortion in many ways to achieve similarity in one or two ways only. The distorted models of vast reaches of American rivers that are now in operation are of this nature; indeed, they differ primarily in detail of solution from the analogue computers that are assembled from electronic components to yield the same gross results of flood prediction. Mention should also be made at this point of the electrical analogy to irrotational flow, which has proved its worth in problems of three-dimensional boundary configuration, and of the resistance-network counterpart of both the electrolytic bath and the numerical process of relaxation which has been devised for two-dimensional and axisymmetric flow.

Models and analogues verge rather closely on laboratory instrumentation as a whole. Devices for the programming and recording of tidal and other models are in various degrees of development in America, as they are throughout the model world; a typical case in point is the complex system of controls and gages for the flood-prediction models just mentioned. The same is true of the use of radioactive tracers in the study of snow, water, and sediment distribution: for example, diffusion studies conducted in representative streams. Perhaps more typically American are such instruments as sonic and ultrasonic devices for sounding (i.e., the measurement of water depth); for recording fluctuating free-surface elevation, as in wave studies; and for the instantaneous measurement of the spatial mean velocity of flow in penstocks and similar passages. Equally characteristic are instruments using electromagnetic principles for both mean and fluctuating velocities in water; also, thermistors have been applied to the indication of fluctuations in water velocity, depth, and density; and not only has the hot-wire anemometer received its counterpart in the hot-film anemometer but both have been employed successfully in water. At the same time considerable use has been made of crystal, strain-gage, and transducer pickups for the indication of small displacements or force intensities. Intimately associated with the evaluation of the continuously fluctuating signals from such instruments is the necessity of performing one or another type of statistical analysis, and electronic circuits have been devised for obtaining means, root-mean-squares, derivatives, integrals, and autocorrelations from magnetic-tape records played back at perhaps higher or lower speed. In a quite different category is the recording of unsteady occurrences (for example, cavitation) by high-speed photography—particularly with motion-picture film—and cameras and lights therefore have reached a high state of development.

In discussing the future of world hydraulics 10 years ago, I made the following remarks:

Even a limited extrapolation beyond the present would be sheer folly, despite the availability of several centuries of record as a guide, for the record has been seen to display an appreciable change in course in the very recent past. However, one can surely hope, if not predict, that hydraulics will continue to widen its break with pure empiricism, which long since reached a point of diminishing returns, and to strengthen its tie with physical analysis, for the latter is still unlimited in its possibilties. On the other hand, one could avoid a direct forecast by merely citing . . . what seem to be the major gaps in hydraulic knowledge, on the assumption that at least some of them will be filled in the near future. . . . Let us therefore list as problems for the future just four that are typically hydraulic in nature yet little subject to solution by measurement alone. One is the incorporation of gravitational action in the mathematical analysis of flow with respect to fixed boundaries such as gates and spillways. Another follows from the presumption that hydraulic structures will never be fully streamlined: the understanding of the mean and fluctuating patterns of flow in a zone of separation. A third is the statistical evaluation of surface roughness and its effect upon boundary resistance. The last is a problem that is broadly inclusive: the mechanism of turbulent mixing at any boundary, whether the surface of air-entraining water, the bed of sediment-entraining water or air; or the interface between two moving fluids of only slightly different density.

Perhaps in the selection of these problems I was subconsciously influenced by American needs or even consciously by my own interests. At any rate, all four have been mentioned again in the foregoing pages describing current American activities. However, degrees of accomplishment were noted that varied over a very great range. Free-streamline flow in a gravitational field is being pursued ever more vigorously, but success remains just around the corner. Separation zones are becoming slowly more familiar in their complexities, despite a tendency for the horizon to recede as the point of observation advances.

Good progress has been made with artificial roughness and synthesis of the corresponding resistance coefficient, though a comparable evaluation of natural roughness remains well in the future. But so far as the boundary condition of turbulent diffusion is concerned, we in America, like our colleagues in the rest of the world, appear to have made no noticeable progress in the past decade.

If I were to attempt a comparable selection of essential topics of research for the coming decade, surely the same four would have a prominent part therein. The matter of the characteristics of turbulence at the juncture between the actual flow and a boundary across which material mixing is occurring remains one of the most important in general fluid mechanics, as in hydraulics itself, and on its eventual understanding will depend our further progress in at least the three specified fields of sediment, air entrainment, and evaporation. Free-streamline flow in a gravitational field, surface roughness, and phenomena of separation are equally essential matters in hydraulics, but apparently more amenable to attack. With the attention that each is receiving in this country, there is good hope for continued progress.

As far as additional topics that should receive particular attention are concerned, at least a dozen were mentioned in the foregoing pages. One that is important—though perhaps beyond the mathematical ability of even our better hydraulicians—is the formulation of general stability criteria for the interface between fluids of different density—whether the air-water interface of a river or canal or the fresh-water/salt-water interface in an estuary. The one condition portends the formation of roll waves, to be sure, and the other the onset of turbulent mixing, but there are reasons to believe that both phenomena are part of the same general occurrence. Nearly as difficult and quite as important is the question of unsteady flow with respect to nonuniform boundaries. Such problems have proved subject to approximate solution so long as the assumption of irrotationality can be made. When separation occurs, however, there is introduced not only the vagary of quasi-steady flow with eddy formation but the additional question of the variation of the separation pattern with the degree of acceleration. For the present, at least, progress in this problem will undoubtedly remain experimental.

As a check upon the adequacy of coverage of this review, I would now like to refer to a list of suggested research topics which recently appeared in a news leaflet of the Hydraulics Division of the American Society of Civil Engineers. Of the 50-odd items listed, the majority naturally were of the specific or applied type to be anticipated from hydraulic engineers. Even several of those included under the heading of hydrodynamics were somewhat of this nature—spillway characteristics, forces on baffles—but it is significant that nearly a dozen were of the fundamental nature discussed herein. In fact, four warrant more than passing mention, if only for their confirmation of the trend in American hydraulics depicted in these pages.

One topic, specified as the spreading of jets on sloping floors, is actually a combination, in restricted form, of two distinct divisions of the subject that were previously discussed: irrotational flow in a gravitational field, and the sonic analogy to gravity waves. Just a decade ago American hydraulicians published a symposium on high-velocity flow in open channels based upon the analogy, including the spreading of jets. Channel slopes were perforce limited to the so-called "friction" slope, simply because the gravitational effect could not yet be incorporated in the existing method of analysis. The matter of slope is still handled only experimentally, and this is clearly a field for theoretical development.

A second topic is the hydraulic evaluation of natural roughness. Whereas the use of artificial roughness permits functional variations to be studied individually, the types of roughness employed often bear little resemblance to natural surfaces. Several devices have already been produced to record actual surface profiles either mechanically or electronically, and various means are at hand for their statistical analysis. The correlation of the resulting parameters with the hydraulic behavior of the surface, however, is still a matter for the future.

Third among the topics for which further knowledge is desired was listed as shear-flow turbulence—a field so broad as to include boundary layers, pipes and channels, wakes, jets, and zones of separation. Every one of these is vital to some aspect of hydraulics, yet none of them falls within the turbulence category that is presently subject to mathematical analysis. Little wonder that each was mentioned at least once in the foregoing pages—and how significant that their common basis is recognized by hydraulic engineers as requiring the researcher's attention!

The fourth topic—hardly a scientific one in itself—is likely to require the contributions of more than one other science before it is adequately handled. It is the provision of a small device for the indication of very low velocities. Other countries have had considerable success in the construction of midget current meters, of the order of a centimeter in size. In many laboratory models this is still large, however, and such a slowly rotating device is not well suited to the indication of fluctuations as well as mean values, say, in the mixing zone of a stratified flow. Unfortunately, hot-wire instruments become increasingly unreliable as velocities decrease, and outwardly attractive methods such as the electrolytic seem to have unpredictable qualities of the same order as the desired indications. Simple as it is, the problem remains a major one of the science.

To end this discussion of current trends in American hydraulics on a note that is both positive and pertinent, let me return to the extensive Bureau of Standards list of hydraulic-research projects mentioned at the outset. A third and less evident reason for the great activity of America in this field lies in the close dependence which it indicates between university research and postgraduate education. There is, to be sure, nothing purely American in the employment of the advanced student to assist the professor in laboratory experimentation. But American hydraulics has had the advantage during the past two or more decades of a most fortunate sequence of occurrences: first, the gradual transition from art to science; second, the ever-expanding federal support of university research; third, the employment of postgraduate students on a part-time basis to conduct this research, and the resultant provision of both facilities and subject matter for thesis projects. This salutary situation has now been in existence for a sufficient length of time to have produced a new generation of well-trained hydraulicians who are represented on the staffs of nearly every one of the hundred and fifty accredited American engineering colleges (not to mention a comparable number in other countries of the world), and they in turn are producing a still better quality of hydraulician in even greater quantity. This is perhaps the most notable trend in present-day American hydraulics.

AMERICAN SOCIETY OF CIVIL ENGINEERS

Founded November 5, 1852

TRANSACTIONS

Paper No. 3289
(Vol. 127, 1962, Part I)

INTERFACIAL MIXING IN STRATIFIED FLOW

By Enzo Oscar Macagno,[1] Aff. ASCE, and Hunter Rouse,[2] F. ASCE

SYNOPSIS

In order to effect a state of stratified flow in which the source of interfacial turbulence would be the instability of the interface itself, apparatus was devised which would permit the counterflow of two superposed streams of different densities to be maintained for a considerable period of time over an appreciable length of uniform passage. By means of chronophotography of injected droplets and analysis of point samples of fluid, measurements were made of velocity and density distributions across the interface for various flow regimes. In accordance with indications of dimensional and stability analyses, results were presented in the form of dimensionless ratios of interfacial shear and of mass transport as functions of the Froude and Reynolds numbers.

INTRODUCTION

Density stratification occurs in such natural fluid systems as the oceans and the atmosphere as well as in those that are man-made, and this can have an important influence upon the flows that take place. In many instances two or more layers of fluid of different densities flow in such a way that they preserve their identity even when the fluids are miscible and appreciable diffusion

Note.—Published essentially as printed here, in October, 1961, in the Journal of the Engineering Mechanics Division, as Proceedings Paper 2964. Positions and titles given are those in effect when the paper was approved for publication in Transactions.

[1] Asst. Prof. of Fluid Mechanics, State Univ. of Iowa, and Research Engr., Iowa Inst. of Hydr. Research, Iowa City, Iowa.
[2] Prof. of Fluid Mechanics, State Univ. of Iowa, and Dir., Iowa Inst. of Hydr. Research, Iowa City, Iowa.

occurs across the transition zones between them. As a first approximation such layers can be considered to be separated by an interface, rather than by an interlayer of finite thickness, along which shear occurs and across which mass transfer takes place.

The analysis of the flow of strata of different densities requires a knowledge of conditions at solid boundaries as well as along the internal boundaries between the individual layers. The conditions at solid boundaries are fairly well known, although the problem of boundary roughness is far from solved. The conditions at inner boundaries, on the contrary, have been only scantily explored. To be sure, it is possible to state them in the case of laminar flow, and some laminar-flow problems have been solved mathematically. However, as soon as the flow becomes unstable and the interface is deformed by waves, it is no longer possible to evaluate even the mean shear stresses. When turbulence develops, still another unknown is introduced: the rate of mixing of the neighboring fluids and the resulting change in the density distribution.

As in many other problems, when mathematical analysis is not yet feasible, one can resort to studies at model scale. However, no model study can make ultimate sense without knowledge of fundamental relationships on which to build either exact or approximate criteria of similitude. In the present instance, approximate criteria for the general case of stratified flow must be based upon the parameters derived for the case of discrete strata separated by interlayers of negligible thickness.

The goal of this investigation was an experimental study of interfacial instability and subsequent mixing - due, however, to conditions prevailing directly at the interface as contrasted to previous studies involving rather arbitrary or specialized boundary conditions. Only to mention among those studies the ones that were the most useful during the present research, reference is made to the articles by G. I. Taylor[3] on the stability of superposed streams of different densities, by G. H. Keulegan[4] on interfacial mixing and instability, by A. T. Ippen, F. ASCE, and D. R. F. Harleman,[5] M. ASCE, on subsurface flow, and by J. P. Raynaud[6] on underflow of silt-laden water. All involved the flow of one layer under or over another that was stagnant except for the motion induced by shear.

Following the previously cited investigations, various papers have been published in the 1950's presenting new findings either in the theoretical or in the experimental field. The series on the hydrodynamical discussion of stability was culminated by the contributions of S. Feldman[7] and W. P. Graebel.[8] Both papers contain solutions of the Orr-Sommerfeld equation for two-layered stratified flows, taking into account the effect of surface tension in addition to that of

---

3 "An Experiment on the Stability of Superposed Streams of Fluid," by G. I. Taylor, Proceedings, Cambridge Phil. Soc., Vol. 23, 1927, pp. 730-731.

4 "Interfacial Instability and Mixing in Stratified Flow," by G. H. Keulegan, Journal of Research of NBS, Vol. 43, 1949, pp. 487-500.

5 "Steady-State Characteristics of Subsurface Flow," by A. T. Ippen and D. R. F. Harleman, Proceedings, Semicentennial Symposium on Gravity Waves, NBS Circular 521, 1952, pp. 79-93.

6 "Etudes des Courants d' Eau Boueuse dans les Retenues," by J. P. Raynaud, 4th Congress of Large Dams, New Delhi, Question 14, 1951, pp. 137-161.

7 "On the Hydrodynamic Stability of Two Viscous Incompressible Fluids in Parallel Uniform Shearing Motion," by S. Feldman, Journal of Fluid Mechanics, Vol. 2, Part 4, 1957, pp. 343-370.

8 "The Stability of a Stratified Flow," by W. P. Graebel, Journal of Fluid Mechanics, Vol. 8, Part 3, 1960, pp. 321-336.

density and of viscosity. The case of flow considered by Graebel corresponds to the one in the present experiments, but his solution was confined to essentially vertical flow, while that considered herein is nearly horizontal. Graebel's results indicate that rather low critical Reynolds numbers can be expected. On the experimental side, the decade has seen frequent contributions by Keulegan and his associates. Particular reference is made to the paper by K. Lofquist,[9] which presents results similar to those for shear stress contained herein, although for a dissimilar case of flow.

A different and also somewhat abstract approach to the problem of interfacial mixing was followed by H. Rouse, F. ASCE, and J. Dodu[10] in an experimental study of the turbulent diffusion between miscible fluids of different densities induced by oscillation of a grid in one of the layers. The present investigation, in contradistinction, was intended to cover a process of diffusion across an interface due to self-generated turbulence rather than turbulence imposed by external means. Knowledge existing at the time the present investigation was started, in 1957, indicated both the feasibility and the desirability of this approach. The idea of making the interface the plane of maximum shear had already been under investigation at the Iowa Institute of Hydraulic Research by Rouse and C. S. Yih. A considerable amount of work was still necessary to reach the stage of successfully controlling the flow, and to develop a technique of observation which would reveal the mechanism of interfacial mixing and allow the researcher to obtain quantitatively useful data. That the investigation was to be carried on at a small scale was justified partly by economical reasons but mainly by previous experience showing that important relationships can often best be revealed at a small scale in which much more effective control of the phenomena can be exercised. A subsequent investigation should then continue the exploration of the interfacial behavior at a much larger scale, either through controlled experiments in the laboratory or through the analysis of field data obtained under natural conditions.

Notation.—The letter symbols adopted for use in this paper are defined where they first appear, in the illustrations or in the text, and are arranged alphabetically, for convenience of reference, in the Appendix.

STABLE LAMINAR STRATIFIED FLOW

The following analysis of two-layered laminar stratified flow has the purpose of defining the conditions under which a nearly antisymmetrical flow is possible. In the case of the uniform two-dimensional motion indicated in Fig. 1, the Navier-Stokes equations for the upper layer reduce to

$$0 = \rho_1 \, g \, \sin \theta - \frac{\partial p_1}{\partial x} + \mu_1 \, \frac{\partial^2 u_1}{\partial y^2} \quad \ldots \ldots \ldots \ldots \quad (1)$$

$$0 = - \rho_1 \, g \, \cos \theta - \frac{\partial p_1}{\partial y} \quad \ldots \ldots \ldots \ldots \ldots \quad (2)$$

---

[9] "Flow and Stress Near an Interface Between Stratified Liquids," by K. Lofquist, The Physics of Fluids, Vol. 3, No. 2, 1960, pp. 158-175.

[10] "Diffusion Turbulente à Travers une Discontinuité de Densité," by H. Rouse and J. Dodu, La Houille Blanche, Vol. 10, No. 4, 1955, pp. 522-532.

and for the lower layer to

$$0 = \rho_2 \, g \, \sin \theta - \frac{\partial p_2}{\partial x} + \mu_2 \, \frac{\partial^2 u_2}{\partial y^2} \quad \ldots \ldots \ldots \quad (3)$$

$$0 = - \rho_2 \, g \, \cos \theta - \frac{\partial p_2}{\partial y} \quad \ldots \ldots \ldots \ldots \quad (4)$$

The boundary conditions at the top are

$$u_1 = 0 \text{ for } y = h_1 \quad \ldots \ldots \ldots \ldots \quad (5)$$

and at the bottom

$$u_2 = 0 \text{ for } y = - h_2 \quad \ldots \ldots \ldots \ldots \quad (6)$$

while at the interface

$$u_1 = u_2 \text{ for } y = 0 \quad \ldots \ldots \ldots \ldots \quad (7)$$

$$\mu_1 \, \frac{\partial u_1}{\partial y} = \mu_2 \, \frac{\partial u_2}{\partial y} \text{ for } y = 0 \quad \ldots \ldots \ldots \ldots \quad (8)$$

and

$$p_1 = p_2 \text{ for } y = 0 \quad \ldots \ldots \ldots \ldots \quad (9)$$

As u_1 and u_2 are functions only of y, Eqs. 1 to 4 together with Conditions 9 give

$$p_1 = - \rho_1 \, g \, y \, \cos \theta + K \, x + C \quad \ldots \ldots \ldots \ldots \quad (10)$$

$$p_2 = - \rho_2 \, g \, y \, \cos \theta + K \, x + C \quad \ldots \ldots \ldots \ldots \quad (11)$$

Upon integration of Eqs. 1 and 3 and use of the conditions at the interface, there results

$$u_1 = \frac{K_1}{2 \, \mu_1} \, y^2 + \frac{A}{\mu_1} \, y + B \, \ldots \ldots \ldots \ldots \quad (12)$$

and

$$u_2 = \frac{K_2}{2 \, \mu_2} \, y^2 + \frac{A}{\mu_2} \, y + B \, \ldots \ldots \ldots \ldots \quad (13)$$

where

$$K_1 = K - \rho_1 \, g \, \sin \theta \quad \ldots \ldots \ldots \ldots \quad (14)$$

$$K_2 = K - \rho_2 \, g \, \sin \theta \quad \ldots \ldots \ldots \ldots \quad (15)$$

The coefficients A and B are constants that can be determined through the use of Conditions 5 and 6:

$$A = \frac{\mu_1 \, K_2 \, h_2{}^2 - \mu_2 \, K_1 \, h_1{}^2}{2 \left(\mu_1 \, h_2 + \mu_2 \, h_1 \right)} \quad \ldots \ldots \ldots \ldots \quad (16)$$

and

$$B = \frac{K_1 \, h_1{}^2 \, h_2 + K_2 \, h_2{}^2 \, h_1}{2 \left(\mu_1 \, h_2 + \mu_2 \, h_1 \right)} \quad \ldots \ldots \ldots \ldots \quad (17)$$

Upon substitution of A and B in Eqs. 10 and 11, it follows that

$$u_1 = \frac{K_1}{2 \, \mu_1} \, y^2 + \frac{\left(K_2 \, h_2{}^2 - K_1 \, h_1{}^2 \frac{\mu_2}{\mu_1} \right) y - \left(K_1 \, h_1{}^2 \, h_2 + K_2 \, h_2{}^2 \, h_1 \right)}{2 \left(\mu_2 \, h_1 + \mu_1 \, h_2 \right)} \quad . \, .(18)$$

$$u_2 = \frac{K_2}{2 \, \mu_2} \, y^2 + \frac{\left(K_2 \, h_2{}^2 \frac{\mu_1}{\mu_2} - K_1 \, h_1{}^2 \right) y - \left(K_1 \, h_1{}^2 \, h_2 + K_2 \, h_2{}^2 \, h_1 \right)}{2 \left(\mu_2 \, h_1 + \mu_1 \, h_2 \right)} \quad . \, .(19)$$

If the zero-velocity line and the interface coincide, $u_1 = u_2 = 0$ at $y = 0$, which results in

$$K_1 \, h_1 = - \, K_2 \, h_2 = M \quad \ldots \ldots \ldots \ldots \ldots \quad (20)$$

and it then follows that

$$u_1 = \frac{M \, y^2}{2 \, \mu_1 \, h_1} - \frac{M \, y}{2 \, \mu_1} = \frac{M}{2 \, \mu_1} \left(\frac{y^2}{h_1} - y \right) \quad \ldots \ldots \ldots \ldots \quad (21)$$

and

$$u_2 = - \, \frac{M}{2 \, \mu_2} \left(\frac{y^2}{h_2} + y \right) \quad \ldots \ldots \ldots \ldots \quad (22)$$

If the rates of flow of the two fluids are made equal,

$$\int_0^{h_1} u_1 \, dy = \frac{M}{2 \, \mu_1} \frac{h_1{}^2}{3} - \frac{M \, h_1{}^2}{4 \, \mu_1} = - \, \frac{M}{12 \, \mu_1} h_1{}^2 \quad \ldots \ldots \ldots \quad (23)$$

and

$$\int_{-h_2}^{0} u_2 \, dy = - \, \frac{M}{2 \, \mu_2} \frac{h_2{}^2}{3} + \frac{M \, h_2{}^2}{4 \, \mu_2} = \frac{M}{12 \, \mu_2} h_2{}^2 \quad \ldots \ldots \ldots \quad (24)$$

which yields, finally,

$$\frac{h_1}{h_2} = \sqrt{\frac{\mu_1}{\mu_2}} \quad \ldots \ldots \ldots \ldots \ldots \ldots \quad (25)$$

Evidently, if the ratio μ_1/μ_2 is kept nearly equal to unity, the ratio h_1/h_2 will deviate still less from unity, and for all practical purposes the stratified flow can then be considered to consist of two layers of equal depth separated by an interface coincident with the plane of zero velocity.

Once it is possible to produce the laminar flow of two layers with a velocity distribution which is like a sine curve, the stability of the motion can be studied. With the inflection point at the interface, it is there that disturbances can be expected to grow rather than at any other level. It is in this sense that antisymmetrical flow is considered to have inherent conditions of instability, as opposed to other cases in which disturbances develop elsewhere and are propagated to the interfacial region.

DISCUSSION OF INSTABILITY

In 1945, Rouse[11] concluded that, other things being equal, the relative stability of a given flow should depend upon the velocity gradient, the wall distance,

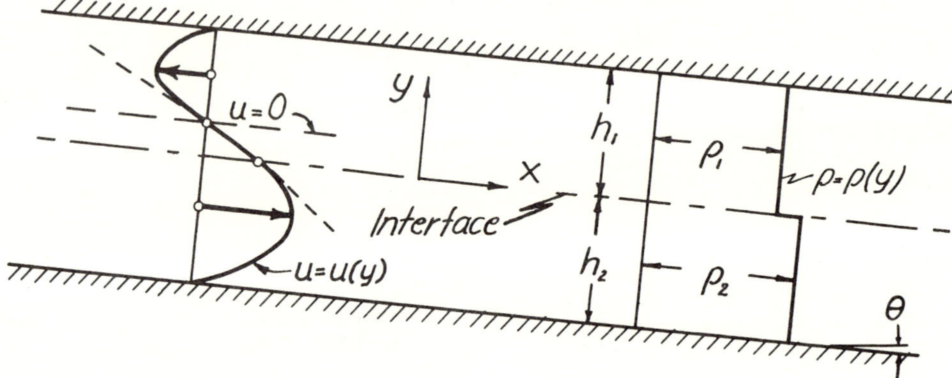

FIG. 1.—DEFINITION SKETCH FOR STRATIFIED FLOW WITHOUT MIXING

and the kinematic viscosity, three variables which combine to yield the nondimensional parameter

$$\chi = \frac{y^2 \dfrac{du}{dy}}{\nu} \qquad \dots\dots\dots\dots\dots \ (26)$$

Evaluation of this parameter for various points in laminar flow near a boundary, as shown in Fig. 2, results in a curve with a definite maximum, the location of which should correspond to the zone of minimum stability. Accordingly, comparison of maximum values for different flows - other things again being equal - should permit their relative stability to be assessed. Rouse also found that for conditions corresponding to the lower critical Reynolds number for pipes (that at which turbulence will no longer persist) the maximum value of

<hr>

11 "A General Stability Index for Flow near Plane Boundaries," by H. Rouse, Journal of the Aeronautical Sciences, Vol. 12, No. 4, 1945, pp. 329-332.

χ was about 500 - not only for pipes and wide channels but also for boundary layers and plane Couette flows.

Evaluation of the same parameter for the antisymmetric condition diagrammed in Fig. 3 indicates that two maxima exist: one in the wall vicinity, as before, but another and much larger one at the center line, where $\chi = 6\,u\,h/\nu$. The latter, according to this criterion, is by far the more susceptible to disturbances of a comparable nature. If, moreover, the value of 500 is assumed to represent the critical limit, this would be found to correspond to a Reynolds number of only $u\,h/\nu = 83$, which is evidently a full order below that for either a wide channel or the boundary region of the antisymmetric case, for which $u\,h/\nu = 2{,}250$. Apart from the stabilizing effect of gravity, which vanishes as the density ratio approaches unity, the zone of interfacial shear in a stratified fluid appears to be relatively susceptible to instability of the Reynolds type.

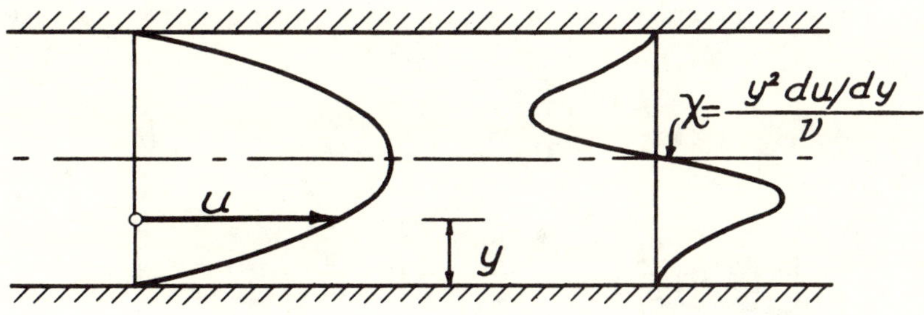

FIG. 2.—VARIATION OF χ IN PLANE POISEUILLE FLOW

To introduce the effect of gravity on stability, the behavior of small disturbances can be examined through a one-dimensional analysis[12] of interfacial waves. For each layer of the stratified flow the following equations apply:[13]

$$\frac{\partial u}{\partial x} + \frac{\partial v}{\partial y} = 0 \quad \ldots\ldots\ldots\ldots\ldots\ldots\ldots (27)$$

$$\frac{Du}{Dt} = \rho\,g\,\sin\theta - \frac{\partial p}{\partial x} + \frac{\partial \sigma_x'}{\partial x} + \frac{\partial \tau}{\partial y} \quad \ldots\ldots\ldots\ldots (28)$$

and

$$\frac{Dv}{Dt} = -\rho\,g\,\cos\theta - \frac{\partial p}{\partial y} + \frac{\partial \sigma_y'}{\partial y} + \frac{\partial \tau}{\partial x} \quad \ldots\ldots\ldots (29)$$

The two-dimensional case is considered, because it has been shown that there is no need to take three-dimensional disturbances into account for the reason that they lead less rapidly to instability (H. B. Squire,[14] Yih[15]). For long

---

12 "Effect of Turbulence and Channel Slope on Translation Waves," by G. H. Keulegan and G. W. Patterson, Journal of Research of NBS, 1953, pp. 461–512.

13 "Advanced Mechanics of Fluids," by H. Rouse, John Wiley and Sons, Inc., New York, 1959, p. 59.

14 "On the Stability of Three-Dimensional Disturbances of Viscous Flow between Parallel Walls," by H. B. Squire, Proceedings, Royal Soc., A. Vol. 142, 1933, pp. 621–628.

15 "Stability of Two-Dimensional Parallel Flows for Three-Dimensional Disturbances," by C. S. Yih, Quarterly of Applied Mathematics, Vol. 12, No. 4, 1955, pp. 434–435.

waves the acceleration terms in the y-direction can be neglected. Additional assumptions will involve the neglect of $\partial \sigma'_x/\partial x$, $\partial \sigma'_y/\partial y$, and $\partial \tau/\partial x$, and the hypothesis of similar velocity profiles. After integration over y across the strata, for each layer a pair of equations similar to the following is obtained:

$$h \frac{\partial U}{\partial x} + U \frac{\partial h}{\partial x} + \frac{\partial h}{\partial t} = 0 \quad \dots\dots\dots\dots\dots \quad (30)$$

and

$$\rho \left[\frac{\partial U}{\partial t} + (1 - \alpha) \frac{U}{h} \frac{\partial h}{\partial t} + \alpha U \frac{\partial U}{\partial x}\right] = \rho g \sin\theta - \frac{\partial p_i}{\partial x} - \rho \frac{\partial h}{\partial t} g \cos\theta - \frac{\rho f U^2}{8 h} \quad .. (31)$$

Herein τ has been expressed in terms of the resistance coefficient, α is the ratio $\int_0^h u^2 \, dy / U^2 h$, and p_i is the pressure at the interface. If the linearization of the equation is performed and p_i is eliminated, one obtains the differential equation

$$\left(1 + \frac{\rho_2}{\rho_1}\right) \frac{1}{U_1^2} \frac{\partial^2 h'}{\partial t^2} + 2\left(\alpha_1 + \frac{\rho_2}{\rho_1} \frac{U_2}{U_1} \alpha_2\right) \frac{1}{U_1} \frac{\partial^2 h'}{\partial x \, \partial t} + \frac{h_1 \Delta\gamma}{\rho_1 U_1^2} \left(1 - \frac{\rho_2}{\rho_1} \frac{h_2}{h_1}\right) \frac{\partial^2 h'}{\partial x^2}$$

$$+ \left[\alpha_1 + \alpha_2 \frac{\rho_2}{\rho_1} \left(\frac{U_2}{U_1}\right)^2 \frac{\partial^2 h'}{\partial x^2} + R' \frac{\partial h'}{\partial t} + R'' \frac{\partial h'}{\partial x}\right] = 0 \quad \dots \quad (32)$$

where the subscripts 1 and 2 correspond to the upper and lower layer, respectively, and R' and R'' indicate terms containing the resistance coefficients f_1 and f_2:

$$R' = \frac{\rho_2}{\rho_1} \left(\frac{U_2^2}{8 U_1^2 h_2} \frac{\partial f_2}{\partial U_2} + \frac{f_2 U_2}{4 h_2 U_1^2}\right) - \frac{1}{8 h_1} \frac{\partial f_1}{\partial U_1} + \frac{f_1}{4 U_1 h_1} \quad .. \quad (33)$$

and

$$R'' = \frac{\rho_2}{\rho_1} \left(\frac{U_2^3}{8 U_1^2 h_2} \frac{\partial f_2}{\partial U_2} - \frac{U_2^2}{8 U_1^2} \frac{\partial f_2}{\partial h_2} + \frac{3 f_2 U_2^2}{8 h_2 U_1^2}\right)$$

$$- \left(\frac{U_1}{8 h_1} \frac{\partial f_1}{\partial U_1} - \frac{1}{8} \frac{\partial f_1}{\partial h_1} + \frac{3 f_1}{8 h_1}\right) \quad \dots\dots\dots\dots \quad (34)$$

Eq. 32 is a linearized form of an approximate equation for the interfacial waves. However, all the essential factors have been preserved in it, so that the pertinent dimensionless parameters involved can be recognized: ρ_2/ρ_1, the ratio of densities; h_2/h_1, the ratio of depths of the two layers; $U_1/\sqrt{h_1 \Delta\gamma/\rho}$, the Froude number; R_1, R_2, the Reynolds numbers of each layer, included in the corresponding resistance coefficients; α_1, α_2, the velocity distribution coefficients. It should be noted that two Reynolds numbers are equivalent to one

Reynolds number and the viscosity ratio. The ratio of mean velocities U_1/U_2 is not listed among the dimensionless parameters, because of the imposed requirement that $h_1 U_1 = h_2 U_2$. Eq. 32 can be written completely in nondimensional form, but no new nondimensional variables will be added.

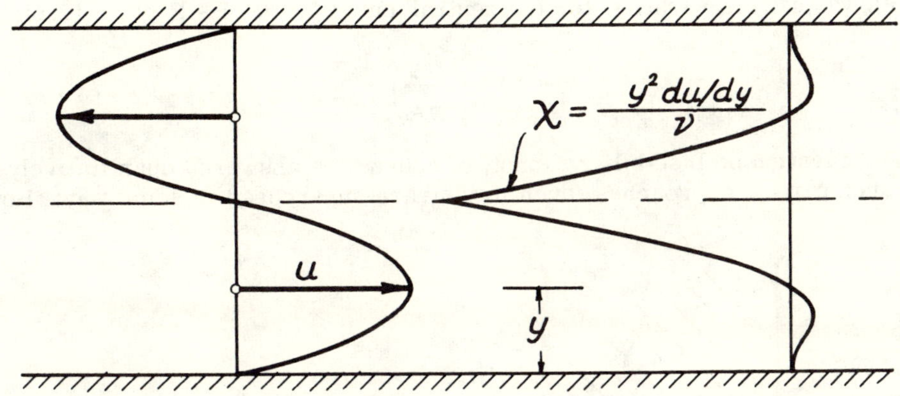

FIG. 3.—VARIATION OF χ IN ANTISYMMETRIC FLOW

When the two layers are made of equal thickness, the equation for the disturbance h' becomes simpler, and under the assumption of equal resistance in both layers with practically the same viscosity it follows that

$$\frac{\partial^2 h'}{\partial t^2} + 2\,\alpha\,U\,\frac{\partial^2 h'}{\partial x\,\partial t} + \frac{1}{2}\frac{h\,\Delta\gamma}{\rho\,U^2}\cos\theta\,\frac{\partial^2 h'}{\partial x^2}$$

$$+ \alpha\,U^2\,\frac{\partial^2 h'}{\partial x^2} + r'\,\frac{\partial h'}{\partial t} + r''\,\frac{\partial h'}{\partial x} = 0 \quad \ldots\ldots\ldots (35)$$

in which r' and r'' are simpler expressions of the terms R' and R''. The relationship

$$h' = C\,e^{i(\xi t + \eta x)} \quad \ldots\ldots\ldots\ldots\ldots (36)$$

which satisfies this equation, represents a disturbance which can be made as small as desired by assigning convenient values to C. After substitution of h' into Eq. 35 an equation for ξ is obtained, from which the condition that will make $\xi = 0$ can be derived - that is, the condition for neutral disturbances which neither grow nor decay. Because in this approximation the terms involving resistance cancel, there results, for laminar flow and for small differences in density, the rather simple condition

$$F^2 = \frac{\cos\theta}{3\sqrt{2}} \quad \ldots\ldots\ldots\ldots\ldots (37)$$

Because the Froude and Reynolds numbers are related in the case of laminar flow by the equation

$$F^2 = \frac{1}{12}\, R \sin \theta \quad \ldots\ldots\ldots\ldots\ldots\ldots (38)$$

it follows that the equation for the neutral curve in the F‑R plane is simply

$$F = \left(18 + \frac{144}{R^2} \right)^{-1/4} \quad \ldots\ldots\ldots\ldots (39)$$

The question of instability cannot, of course, be answered quantitatively by two such partial approaches, even if they are superposed in some way. How-

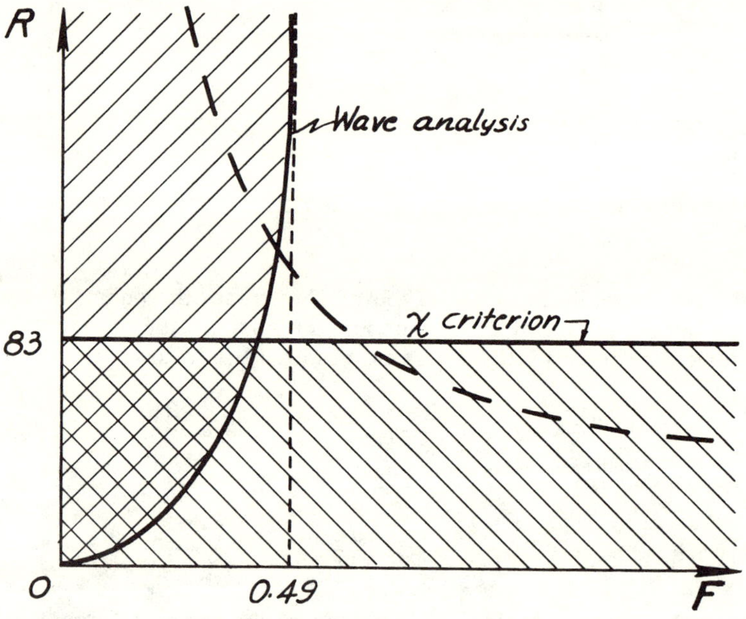

FIG. 4.—SCHEMATIC REPRESENTATION OF THE STABILITY
FUNCTION

ever, one can expect to acquire at least a qualitative indication from such a combination. If the lines represented by the two analyses are plotted, a zone in the Froude-Reynolds plane is determined (doubly hatched in the Fig. 4) which must certainly belong to the stable region. At least a portion of the neighboring zones in the singly hatched regions should also belong to the stable portion of the plane. In all probability the boundary between the two zones is of the form indicated schematically by the dashed curve. It is the experimental determination of the indicated function that is the goal of this investigation.

EXPERIMENTAL APPROACH

There is a simple and interesting experiment attributed to Helmholtz in which two fluids of different density are made to flow one over the other by carefully tilting the tube that contains them. The interface is first seen to be very smooth, then to become regularly wavy, and finally to disrupt with the onset of turbulent motion. The experiment can be repeated indefinitely if the two fluids are not miscible; it can also be performed very well with miscible fluids, although each repetition becomes more difficult.

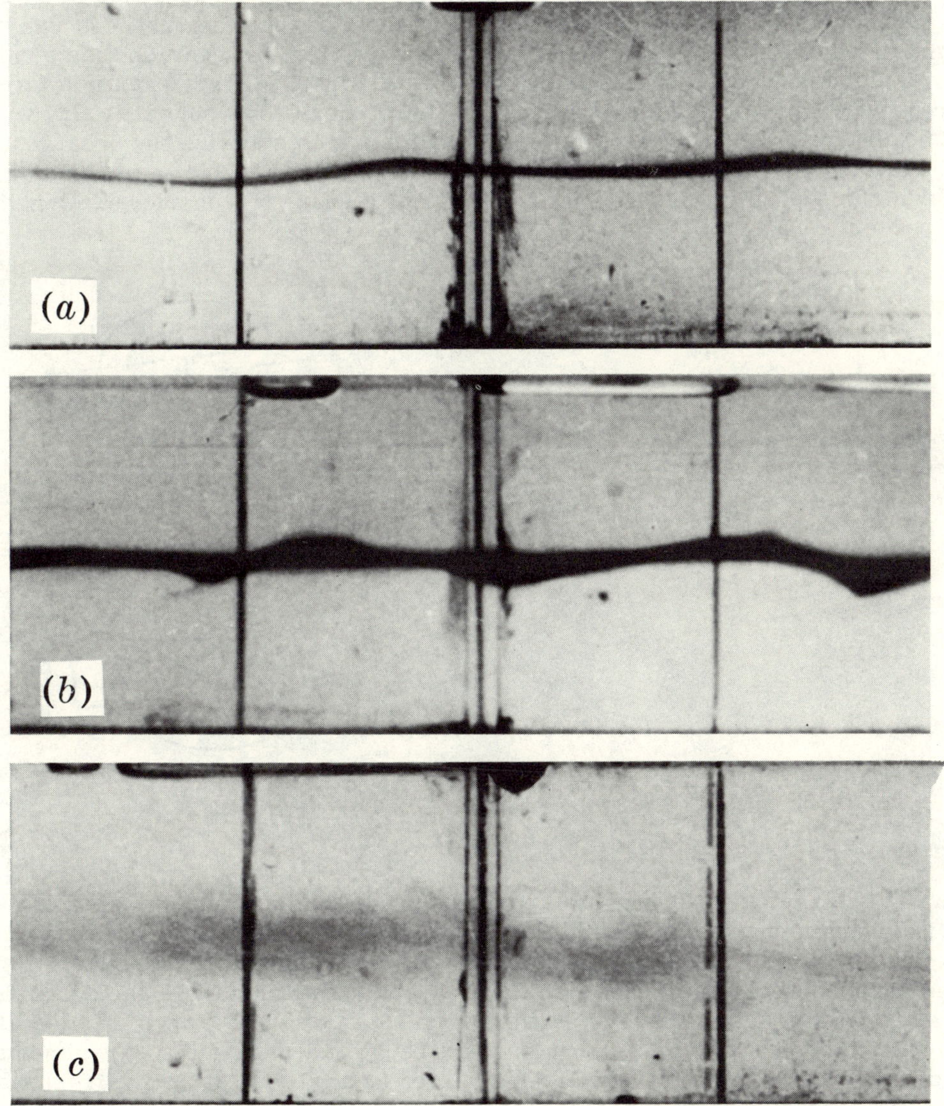

FIG. 5.—SUCCESSIVE STAGES OF INTERFACIAL DISTURBANCE

For the experiment with miscible fluids, care has to be exercised to have an initially neat interface. To make this region clearly visible, dye can be injected between the two fluids. The process of mixing is very well defined if two chemical indicators (such as phenolphthalein and sodium hydroxide) are used in such a way that color will appear where mixing takes place. Figs. 5(a), (b), and (c) show interfacial conditions with regular waves, with waves on the verge of breaking, and with turbulent mixing, respectively. The experiments were performed with fresh and salt water. Dye was used in the first two, and chemical indicators were introduced in the third.

In the Helmholtz experiment, which has been studied in some of its aspects by G. H. Mittendorf,[16] A. M. ASCE, when the difference in densities is small the symmetry of the flow is almost perfect and external disturbances can be minimized; the phenomenon has a transient character, however, and lasts only for a minute or two unless a rather large scale is used. Yet even if a large scale were to be adopted, every aspect of the phenomenon would still vary with time. Therefore, to keep both layers in motion and to stabilize the occurrence at different stages, an apparatus was devised that would bring two streams flowing in opposite directions into contact for a reach of considerable length.

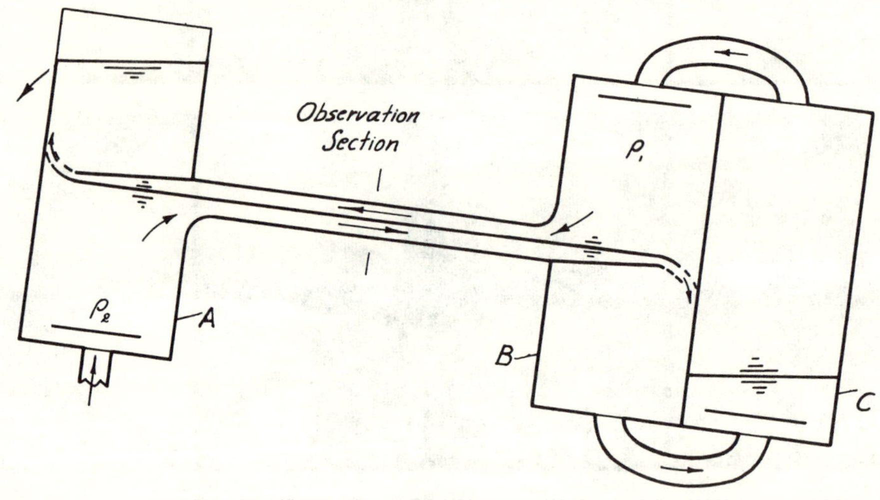

FIG. 6.—SKETCH OF EXPERIMENTAL APPARATUS

In Fig. 6 a sketch is given of the apparatus used, which consisted mainly of two tanks connected by a 36-in. conduit of 2-in. by 6-in. rectangular cross section. A third tank provided the necessary capacity for running the experiment during the time required to establish the flow and to perform the measurements. The apparatus was mounted on a tiltable frame in order to give it an adequate slope. Tank B had a lid so that it could be closed during the experiments. Tank A was open, and one of its sides served as an overflow for the fresh water leaving the apparatus.

To perform an experiment, the unit was first completely filled with fresh water and then an inflow of salt water was started very slowly at the bottom of tank A. There, salt water was diverted radially and parallel to the bottom by

---

¹⁶ "The Instability of Stratified Flow," by G. H. Mittendorf, thesis presented to the State Univ. of Iowa, at Iowa City, Iowa, in 1961, in partial fulfilment of the requirements for the degree of Master of Science.

a plate and thence upward through wire screens. In this way a neat interface would be formed, and the inflow could be gradually increased as the interface became more remote. When the interface reached the inlet of the rectangular duct, the salt water would start to flow down the slope toward tank B and then to tank C. As the salt water entered tank C, it displaced fresh water, which moved through the apparatus in the opposite direction and wasted over the weir of tank A. The duration of the experiments depended on the rate of inflow of the salt water and the capacity of tank C.

During the experiment the discharge and the slope were varied in such a way that two layers of equal thickness would form. To aid in this operation the two tailplates indicated in Fig. 6 had to be adjusted somewhat. One of the main experimental difficulties was to do this in a short time and in such a way that no appreciable disturbances could originate where the fresh and the salt water entered the working section and thus affect the flow within. A very persistant disturbance of another type was the intrusion of the fluid of intermediate density resulting from mixing of the two fluids; this, instead of flowing away, would accumulate in the end tanks, and sometimes had to be bled off in order to keep the interface clean. The great change in the thickness ratio of the two strata in the tailplate sections tended to stabilize the interface in the approach section even while considerable turbulence was developing in the zone under study.

TECHNIQUES OF OBSERVATION AND MEASUREMENT

In addition to measurement of the discharge, slope, temperature, and geometrical dimensions involved, the velocity and density distributions at the center line of the central section of the rectangular duct were subjected to careful and detailed investigation. The choice of this one section for the distribution studies was determined for reasons not only of convenience but of achieving a state of symmetry in the two strata. It was further assumed that the wall effects were of small magnitude, so that the central portion of the flow could be treated as two-dimensional. Preliminary measurements as well as calculations were made to verify this assumption.

The velocity measurement was based on the technique of photographing the motion of small particles suspended in the fluid, because the velocities were in the low range and disturbances of the flow were to be avoided. The well-known procedure of chronophotography was adapted to the case by using as timer a Strobolume activated by an electrical clock combined with a photocell. After considering and experimenting with several different tracers, the choice was made in favor of a mixture of two different fluids in the proper proportion to yield the desired density. For each experiment three mixtures of n-butyl phthalate and xylene with a bit of white paint were prepared to match the densities of fresh water, salt water, and the intermediate mixture. The mixture was injected by means of hypodermic needles in such a way as to produce small clusters of droplets essentially instantaneously but with practically no disturbance, as was verified under conditions of laminar flow. To take pictures of the droplets, a narrow strip was illuminated with both continuous and intermittent light. The channel was made of black plastic, so that the droplets moved against a dark background and left continuous traces on the film with bright dots corresponding to the flashes emitted by the Strobolume. Usually some 30 to 50 pictures were taken in each run in order to ensure a good average, especially in the case of turbulent motion. The traces left on the film generally appeared to have a head and a tail of unequal length, and this helped to determine the direction of the motion. No effort was made to synchronize the camera shutter and the flashing light in such a way as to make different heads and tails, because it was found that mere chance was enough to obtain the desired result in most of the pictures.

The frequency of the flashing light was controlled first by means of a mechanical contactor and then by a combination of an electric clock operating a notched aluminum disk and an electric circuit with a photocell. A beam of light passing through the notches excited the photocell, which in turn operated

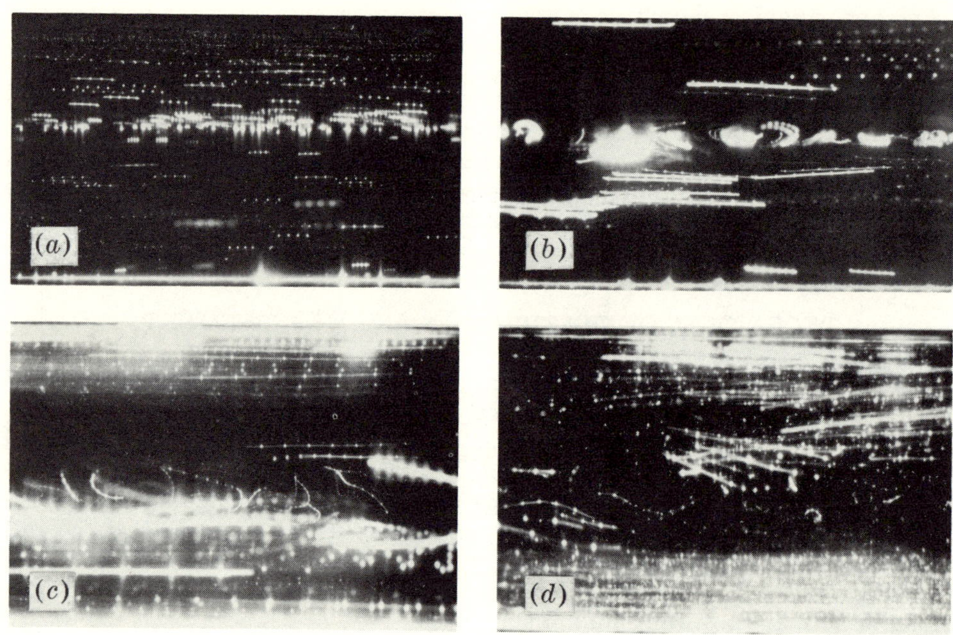

FIG. 7.—CHRONOPHOTOGRAPHS OF ONSET OF MIXING

the Strobolume. Both timers were built at the Iowa Institute using standard equipment and instruments.

To determine the density distribution, samples were taken at different levels in sufficiently small amounts to minimize disturbance of the flow and to withdraw fluid only from the desired elevation. The samples were diluted afterward to a larger volume and their electrical conductivity was measured to determine the salinity and therefrom the density. Sampling at and near the lateral walls was carried on in a few experiments to explore the conditions of mixing in that region. It was found that the density distribution there was somewhat distorted with respect to that of the center and the results much less consistent. This vindicated an earlier rejection of optical methods based on transverse beams of light for determining the density distribution, because they would include wall effects. It was concluded that the use of very fine hypodermic needles would yield a truer profile even if they slightly disturbed the flow.

DIRECTLY OBSERVED INTERFACIAL CONDITIONS

Qualitative as well as quantitative observations can be of great value in the study of a phenomenon which has modes of occurrence yet to be disclosed. To the naked eye and to the photographic camera interesting features of the stratified flow under consideration were made observable by refraction due to the density stratification or through the addition of dyes or the injection of particles. The accompanying photographs (Fig. 7) illustrate the principal features thus observed.

In Fig. 7(a) the motion is laminar with straight streamlines; the interface coincides with the line of zero velocity, which is indicated by droplets of intermediate density appearing as dots at the center line. Fig. 7(b) corresponds to an interface with regular waves, and Fig. 7(c) to an interface with waves which break and start to show irregularity and randomness. Fig. 7(d) depicts the motion after turbulence has developed and active mixing occurs across the interface.

Based on the qualitative conditions observed both visually and photographically, each experiment was labeled with the symbol L, W, I, or T, corresponding, respectively, to uniform laminar motion, laminar motion with regular waves, irregular motion with breaking waves and incipient turbulence, and, finally, pronounced turbulence in the interfacial region. These symbols were then plotted as points in the **F - R** plane. As can be seen in Fig. 8, the four corresponding regions are clearly distinguishable. To express the correlation in quantitative form, however, some measurable quantity had to be considered. The resistance coefficient, the shear stress at the interface, and the rate of mixing across it were therefore evaluated as described in the following sections.

THE RESISTANCE COEFFICIENT

The stratified flow under investigation consists of two streams flowing in opposite directions and separated by a plane of zero velocity. Considering the forces acting on two fluid prisms of length Δx, as indicated in Fig. 9, one can write

$$m_1\, g \sin \theta + (\tau_1)_m\, P\, \Delta x - A\, \Delta p = 0 \quad \ldots\ldots\ldots (40)$$

and

$$m_2\, g \sin \theta - (\tau_2)_m\, P\, \Delta x - A\, \Delta p = 0 \quad \ldots\ldots\ldots (41)$$

where P is the wetted perimeter, A is the cross-sectional area of each stream, and $(\tau_1)_m$ and $(\tau_2)_m$ represent the corresponding mean shear stresses over floor, walls, and interface. From the two preceding equations, after elimination of Δp one obtains

$$\tau_m = \frac{(\tau_1)_m + (\tau_2)_m}{2} = \frac{1}{2}\frac{m_2 - m_1}{P\, \Delta x}\, g \sin \theta \quad \ldots\ldots (42)$$

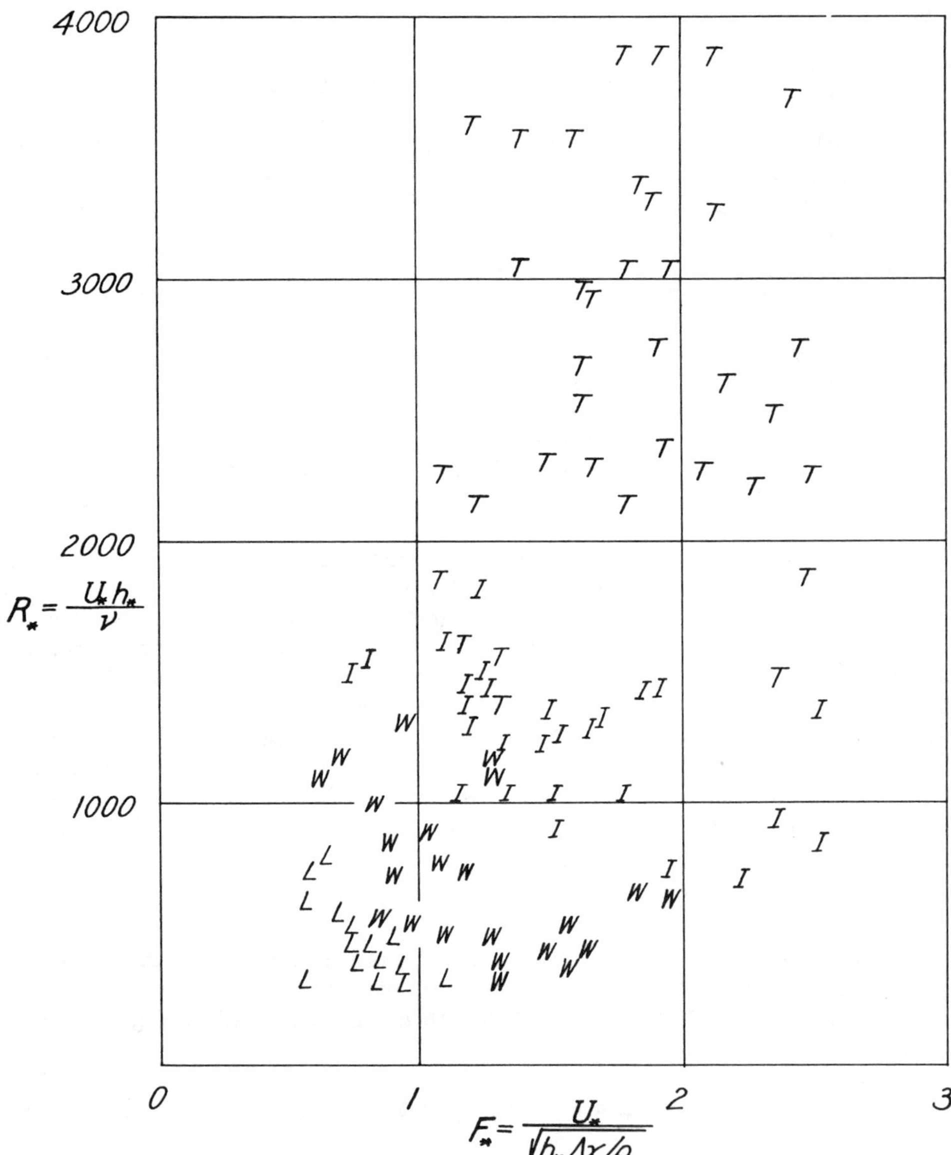

FIG. 8.—OBSERVED ZONES OF LAMINAR FLOW, WAVES, INSTABILITY, AND TURBULENCE IN FROUDE-REYNOLDS PLANE

The masses m_2 and m_1 depend on the density distribution and are to be calculated through integrals of $\rho(y)$ over the corresponding volumes. Approximate values of m_1 and m_2 can be obtained if a linear distribution of density is as-

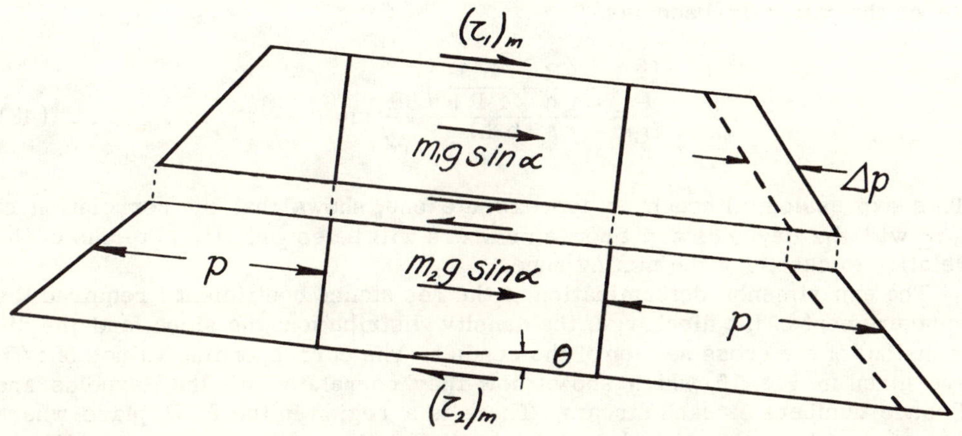

FIG. 9.—FREE–BODY DIAGRAM OF FLUID ELEMENTS

sumed over the region where it changes from ρ_1 in the upper layer to ρ_2 in the lower layer. In this way it is seen that

$$m_1 \approx \rho\left(h - \frac{\delta}{2}\right)\Delta x + \int_0^{\delta/2}\left[\rho_1 + \left(\frac{1}{2} - \frac{y}{\delta}\right)\Delta\rho\right]\Delta x\, dy = \rho_1 h\, \Delta x + \frac{\Delta\rho}{8}\delta\, \Delta x \quad . . (43)$$

and, similarly, that

$$m_2 \approx \rho_2 h\, \Delta x - \frac{\Delta\rho}{8}\delta\, \Delta x \quad . . . . (44)$$

Substituting m_1 and m_2 in the expression for τ_m yields

$$\tau_m \approx \frac{1}{2}\Delta\rho\left(R - \frac{b\,\delta}{4\,P}\right)g\sin\theta \quad . . . . (45)$$

where $R = A/P$ is the hydraulic radius. The resistance coefficient is then given by

$$f \approx \frac{4\,g\,\Delta\rho\left(R - \dfrac{b\,\delta}{4\,P}\right)}{\rho\,U^2}\sin\theta \quad . . . . (46)$$

Since J. Boussinesq[17] has treated the general problem, the solution for laminar flow in a rectangular pipe can be evaluated therefrom. The corre-

---

17 "Mémoire sur l' Influence des Frottements dans les Mouvements Réguliers des Fluides," Journal de Mathématiques Pures et Appliquées, 2nd Series, Vol. 13, 1868, pp. 377–424.

sponding resistance coefficient is given by

$$f_L = \frac{\phi(b/h)}{R} \quad \dots\dots\dots\dots\dots\dots \quad (47)$$

Hence the ratio f/f_L becomes

$$\frac{f}{f_L} \approx \frac{4\left(\dfrac{R}{h} - \dfrac{b\,\delta}{4\,P\,h}\right)}{\phi\,(b/h)} \frac{R}{F^2} \sin\theta \quad \dots\dots\dots\dots \quad (48)$$

This expression, although an approximate one, shows that the correlation of f/f_L with the Reynolds and Froude numbers will be essentially in terms of the relative thickness of the mixing zone.

The experimental determination of the resistance coefficient f required the measurement of the discharge, the density distribution, the slope, and the dimensions of the cross section of the conduit. The corresponding values of f/f_L are given in Fig. 10, which shows how they correlate with the Reynolds and Froude numbers of each stream. There is a region in the **F - R** plane where f/f_L approaches unity, though in the mean it remains somewhat greater. Where the flow is established and completely stable, the value should be exactly unity; this leads to the conclusion that the lack of establishment and the presence of initial disturbances are responsible for f/f_L being slightly greater than unity even in the zone where stable laminar flow is to be expected. That the magnitude of the ratio otherwise steadily increases with **F** and **R**, as anticipated, is quite evident from the diagram.

DIMENSIONAL ANALYSIS FOR SHEAR STRESS AND RATE OF MIXING

To study the interfacial shear stress and the rate of transfer of mass across the interface, these were next correlated with the independent variables of the problem. The quantities U_* and h_* (Fig. 11) were selected to represent the geometric and kinematic variables of the interfacial layer, in order to divorce the analysis so far as possible from the dimensions of the test section itself. Because the variations in density and viscosity were small, the mean values ρ and μ were considered accurate enough for the expression of inertial and viscous forces. On the other hand, gravitational effects being essentially dependent on the difference between the two specific weights rather than upon their absolute values, $\Delta\gamma$ was introduced as the corresponding independent variable. The dimensionless variables found through the approximate analysis of an earlier section are thus reduced to a Froude number and a Reynolds number, the others being eliminated by the conditions h_1/h_2, $\rho_1 \approx \rho_2$, and $\mu_1 \approx \mu_2$.

Instead of considering directly the interfacial shear stress τ_i, its ratio to the viscous stress $(du/dy)_0$, which also depends on the above-listed variables, will be analyzed. For turbulent flow the shear stress has been represented by the sum of two terms $\mu(du/dy) - \rho\,u'\,v'$. In the present case values of the shear stress greater than $\mu(du/dy)$ can be expected as soon as the flow ceases to correspond to the steady-state solution of the parallel laminar flow, though the initial increase is not due to turbulent transfer of momentum so much as to a difference in regime. If desired, the ratio can be considered as a "total"

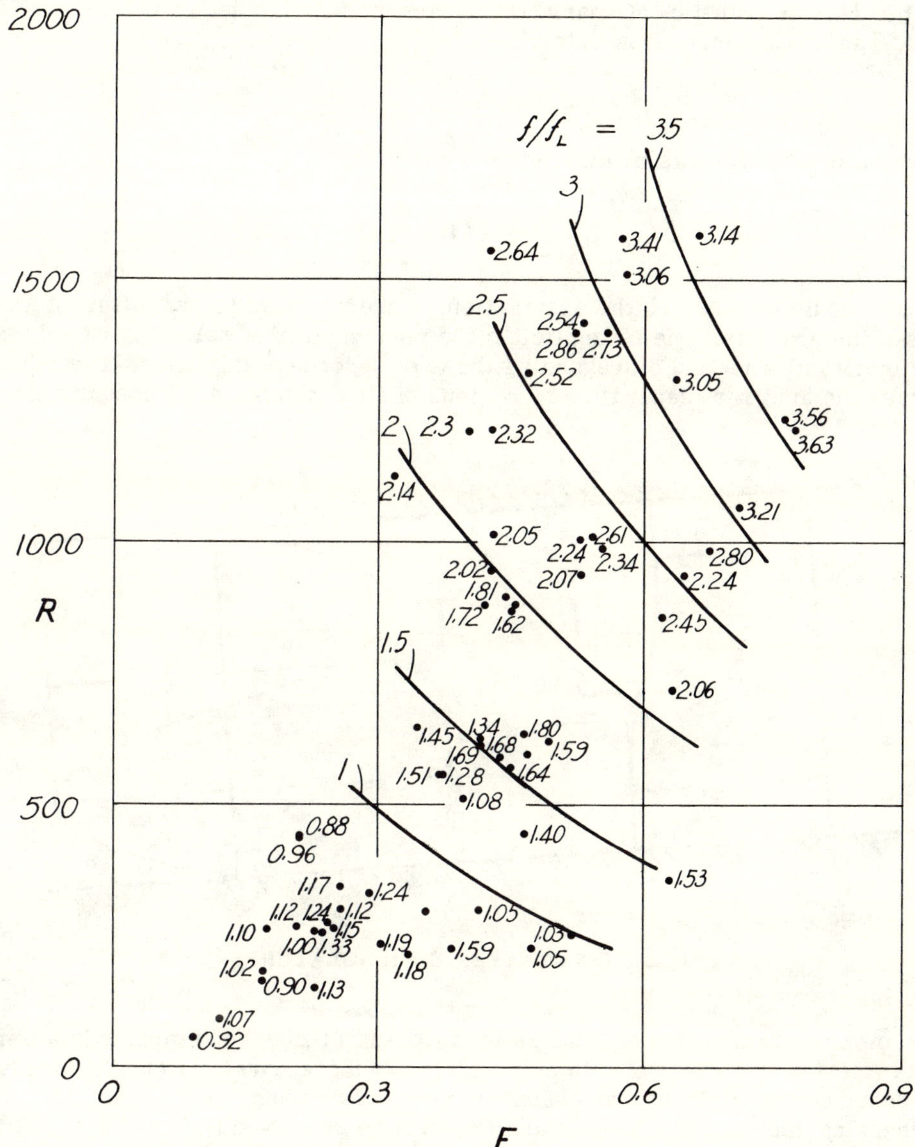

FIG. 10.—VARIATION OF THE MEAN RESISTANCE COEFFICIENT WITH **F** AND **R**

viscosity coefficient. If, then

$$\tau_i = f(U_*, h_*, \rho, \mu, \Delta\gamma) \quad \dots\dots\dots\dots (49)$$

dimensional analysis will immediately yield

$$S = \frac{\tau_i}{\mu(du/dy)_0} = f(F_*, R_*) \quad \dots\dots\dots\dots (50)$$

For the rate of transfer of mass as represented by the mean velocity V of transverse motion across the interface,

$$V = f(U_*, h_*, \rho, \mu, \Delta\gamma) \quad \ldots\ldots\ldots\ldots (51)$$

for which it again follows by dimensional analysis that

$$T = \frac{V}{U_*} = f(F_*, R_*) \quad \ldots\ldots\ldots\ldots\ldots (52)$$

Note should be taken of the assumptions made relative to two different aspects of the analysis. One is implicit in the usual method of selecting variables of primary influence and disregarding those of secondary influence. The other involves the arbitrary separation of regions of flow which are supposed to have

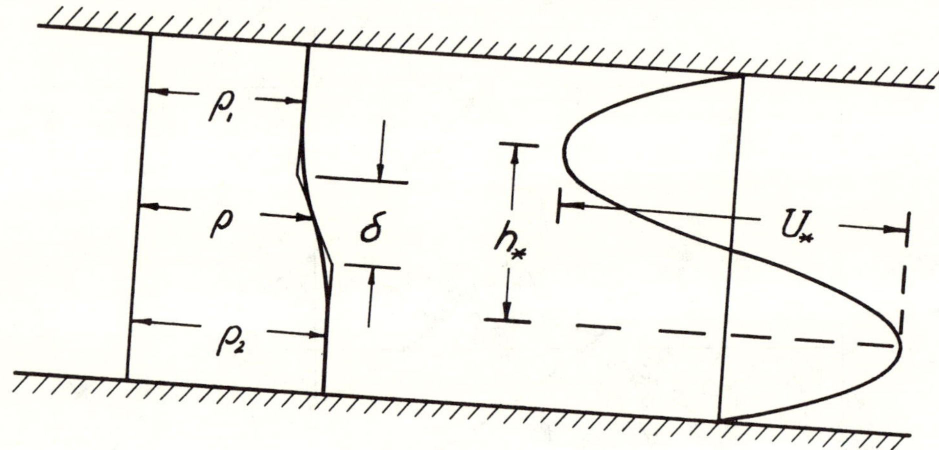

FIG. 11.—INTERLAYER NOMENCLATURE

their own mechanical inner laws at least to a first degree of approximation. This is certainly a questionable step which should be carefully checked under different conditions. In the present experiments it was subjected to verification by changing the variable h* in the ratio of 1 to 2, with satisfactory results.

SHEAR STRESS AT THE INTERFACE

The resistance coefficient f gives only an indirect measure of the interfacial conditions, because it averages effects over the whole perimeter of each stream, including the walls. It is thus a variable which corresponds to a given rectangular section, whereas in order to obtain from these experiments a picture that is focused on the interface itself, it is desirable to interpret them in a manner that minimizes boundary effects.

The calculation of the shear stress at the interface for two-dimensional conditions is similar to that of the resistance coefficient except that it requires consideration of the local value of the shear stress at the top and bottom boundaries of the conduit. This would involve considerable difficulty were it not that the rather low values of the Reynolds number ensured the existence of laminar motion at those boundaries (as could be seen from the photographic records) even when turbulence occurred at the interface. If the equations for two elementary prisms in each layer are written and any change of momentum is neglected, it is found that

$$\tau_1 \, \Delta x + \tau_i \, \Delta x + \left(\rho_1 \, h \, \Delta x + \frac{\Delta \rho}{8} \, \delta \, \Delta x \right) g \, \sin \theta - h \, \Delta p = 0 \quad \ldots \quad (53)$$

$$- \tau_2 \, \Delta x - \tau_i \, \Delta x \quad \left(\rho_2 \, h \, \Delta x - \frac{\Delta \rho}{8} \, \delta \, \Delta x \right) g \, \sin \theta + h \, \Delta p = 0 \quad \ldots \quad (54)$$

where τ_i is the shear stress at the interface. A linear variation of the density has been assumed over the thickness δ of the mixing zone. After elimination of Δp the expression for τ_i becomes

$$\tau_i = \frac{1}{2} \, \Delta \rho \left(h - \frac{\delta}{4} \right) \sin \theta - \frac{1}{2} \, (\tau_1 + \tau_2) \quad \ldots \ldots \ldots \quad (55)$$

The flow in the wall boundary layers being laminar, it follows that

$$\tau_1 + \tau_2 = \mu_1 \left(\frac{du}{dy} \right)_{y=h} + \mu_2 \left(\frac{du}{dy} \right)_{y=-h} \quad \ldots \ldots \ldots \quad (56)$$

and hence

$$\tau_i = \frac{1}{2} \, \Delta \rho \left(h - \frac{\delta}{4} \right) g \, \sin \theta - \frac{1}{2} \left[\mu_1 \left(\frac{du}{dy} \right)_n + \mu_2 \left(\frac{du}{dy} \right)_{-h} \right] \quad \ldots \ldots \quad (57)$$

In accordance with the foregoing section, the ratio $S = \tau_i/(du/dy)_0$ has been chosen to represent the interfacial shear in a dimensionless form. Like the ratio f/f_L, S must be unity for parallel laminar flow and should become greater than unity as soon as the flow becomes either nonuniform or unsteady. In order to correlate the ratio S with the Froude and Reynolds numbers, the latter parameters were so evaluated that only the region of flow in the vicinity of interface need be considered. The representative length chosen was h_*, the distance between the two maxima of the velocity distribution curve, and the representative velocity was selected to be U_*, the maximum relative velocity between the two layers. (This amounts, it should be remarked, to treating the interfacial region as a layer with its own inner laws, an approach which must be tested further to verify to what extent it may hold with different velocity distributions in the rest of the fluid.)

The correlation of S with the variables

$$R_* = \frac{U_* \, h_*}{\nu} \quad \ldots \ldots \ldots \ldots \ldots \ldots \quad (58)$$

and

$$F_* = \frac{U_*}{\sqrt{h_* \, \Delta \gamma / \rho}} \quad \ldots \ldots \ldots \ldots \ldots \quad (59)$$

is shown in Fig. 12. The scattering is in this case greater than in the case of the resistance coefficient, which is probably due to the difficulties involved in determining exact values of du/dy and δ. Nevertheless, there is certainly a definite trend, and tentative curves have been fitted which show the approximate dependence of the shear-stress ratio upon the special Froude and Reynolds numbers. These results are rather similar to those reported by Lofquist for another case of flow.[9]

TRANSFER OF MASS

The onset of instability brings about a more active interfacial diffusion, which up to that point was due only to molecular motion. The more turbulent the motion becomes, the more intense must be the transfer of mass from the denser to the less dense fluid. The mean value of the velocity of mass transfer was calculated in the following manner: On the basis of symmetry it was assumed that, whereas across such an interface the volumetric flux must be zero, the mass flux per unit width is given by ρ_2 V L - ρ_1 V L, where V represents the mean velocity of transfer. The longitudinal flux of mass at the entrance section of each layer is given simply by the product of the rate of flow and the corresponding density. This flux is increased with longitudinal distance for the fresh-water stream and decreased for the salt-water stream by diffusion across the interface. The amounts of increase and decrease can be evaluated with respect to the mass flux at the central section through integrals of the form $\int \rho(y)\, u(y)\, dy$ over the corresponding section. An approximate expression of these integrals is obtained by assuming that for the mixing zone

$$\rho = \frac{1}{2}\left(\rho_1 + \rho_2\right) - \alpha\, y = \rho - \alpha\, y \quad \ldots \ldots \ldots \ldots \quad (60)$$

where

$$u = \beta\, y \quad \ldots \ldots \ldots \ldots \ldots \ldots \quad (61)$$

which consists in substituting for the density and velocity curves the tangents at their inflection points. It then follows that

$$V\, \Delta\rho\, \frac{L}{2} = \int_0^h \rho(y)\, u(y)\, dy - \rho_1\, q \quad \ldots \ldots \ldots \ldots \quad (62)$$

$$V\, \Delta\rho\, \frac{L}{2} = \rho_2\, q - \int_{-h}^0 \rho(y)\, u(y)\, dy \quad \ldots \ldots \ldots \ldots \quad (63)$$

or

$$V\, \Delta\rho\, \frac{L}{2} = \int_0^{\delta/2} (\rho - \alpha\, y)\, \beta\, y\, dy + \rho_1 \int_{\delta/2}^h u\, dy - \rho_1\, q \quad \ldots \ldots \quad (64)$$

$$V\, \Delta\rho\, \frac{L}{2} = \rho_2\, q - \int_0^{\delta/2} (\rho + \alpha\, y)\, \beta\, y\, dy - \rho_2 \int_{\delta/2}^h u\, dy \quad \ldots \ldots \quad (65)$$

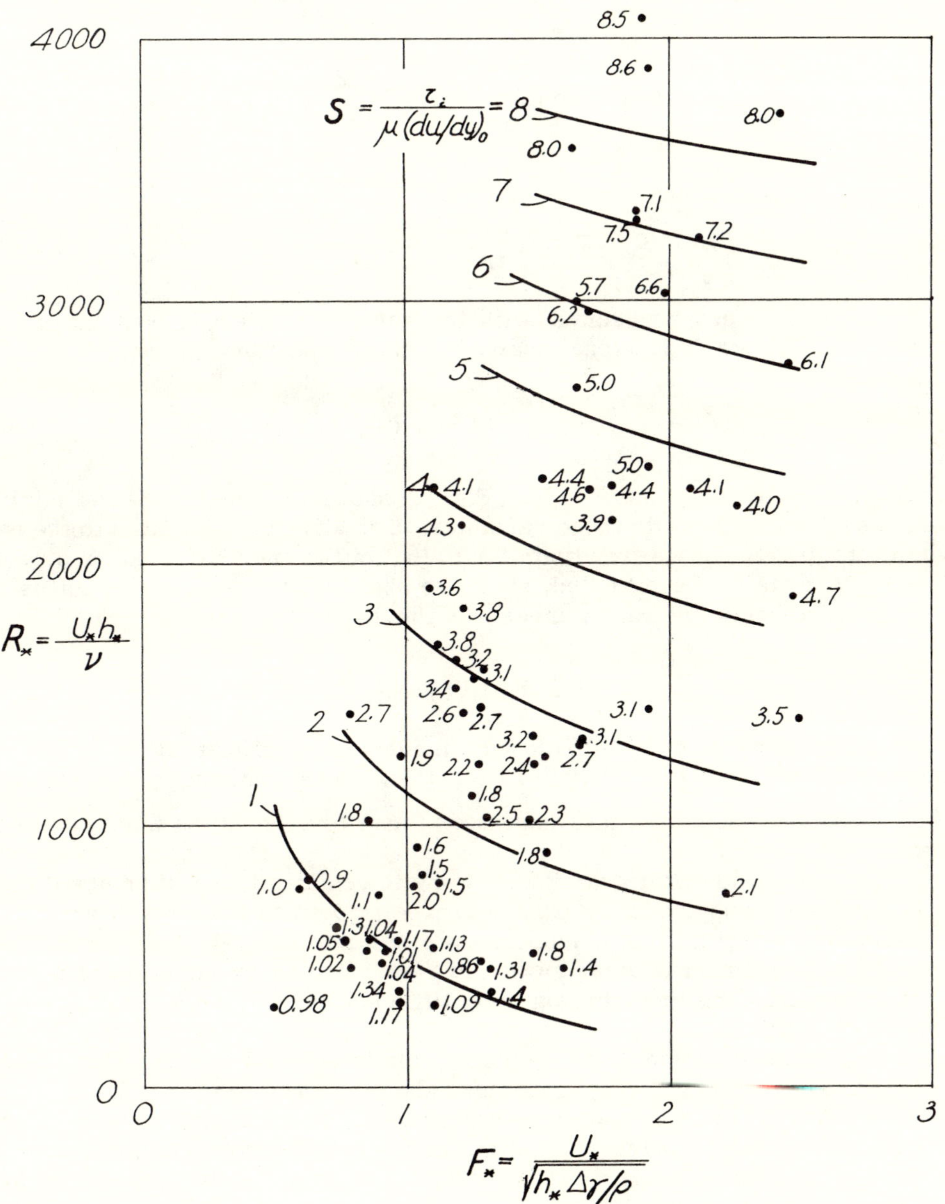

FIG. 12.—VARIATION OF INTERFACIAL SHEAR WITH $\mathbf{F}_*$ AND $\mathbf{R}_*$

Combination of these two expressions yields

$$V \, \Delta\rho \, L = - 2 \, \alpha \, \beta \int_0^{\delta/2} y^2 \, dy - \Delta\rho \int_{\delta/2}^h u \, dy + q \, \Delta\rho$$

$$= - \alpha \, \beta \frac{\delta^3}{12} + \Delta\rho \int_0^{\delta/2} u \, dy = \beta \frac{\delta^2}{8} \Delta\rho - \alpha \, \beta \frac{\delta^3}{12} \quad \ldots \ldots (66)$$

and finally,

$$V \approx \frac{\beta \, \delta^2}{8 \, L} - \frac{\alpha \, \beta \, \delta^3}{12 \, L \, \Delta\rho} \quad \ldots \ldots \ldots \ldots \ldots (67)$$

This velocity was correlated with the same variables as was the shear stress, and through dimensional analysis it was found that

$$\frac{V}{U_*} = f(F_*, R_*) \quad \ldots \ldots \ldots \ldots \ldots (68)$$

A plot of V/U_* on the plane $F_* - R_*$ (Fig. 13) shows that the interfacial mixing varies systematically with these variables. Tentative curves fitted to the experimental points show a trend similar to that of the curves for shear stress. This could certainly be expected, since the shear stress depends on diffusion of momentum, while the rate V measures the diffusion of matter.

CONCLUSIONS

This laboratory study has disclosed a number of significant facts regarding interfacial conditions in a stratified flow; these can be summarized as follows:

1. Four successive regimes have been found to exist in the interfacial region:

a. The initial uniform laminar regime is confined to a rather small zone of the $F - R$ plane, the latter parameters being defined in terms of variables describing a layer in the neighborhood of the interface.

b. Such uniform laminar motion is first affected by the onset of motion of a non-turbulent type with relatively little secondary flow and a very small amount of mixing.

c. In a third region of the $F - R$ plane the interfacial conditions are characterized by breaking waves, which produce a more effective mixing.

d. Finally, the waves become of complex spectrum with random characteristics, and turbulence develops in the neighborhood of the interface.

2. The interfacial shear stress, expressed in terms of the ratio of the actual stress to the viscous stress for the mean flow, shows a definite correlation with the Froude and Reynolds numbers.

3. A similar correlation exists for the rate of mixing across the interface. This was to be anticipated, since - though determined independently - the shear is a measure of momentum transport and the mixing a measure of mass transport.

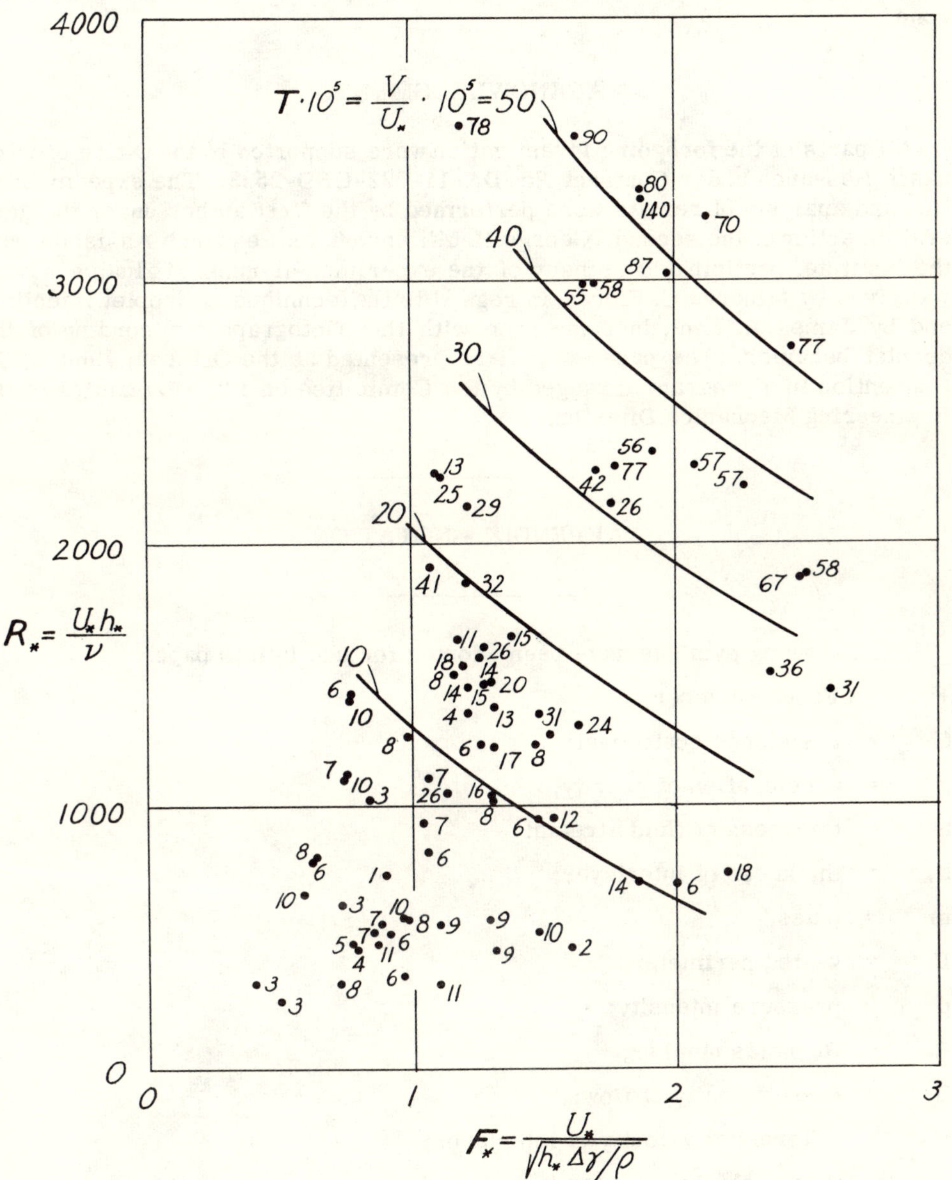

FIG. 13.—VARIATION OF THE MIXING COEFFICIENT WITH F_* AND R_*

4. To be of general significance, these studies must be now extended to large-scale conditions in which the boundaries are remote from the interfacial zone.

ACKNOWLEDGMENTS

All parts of the foregoing investigation were supported by the Office of Ordnance Research under Contract No. DA-11-022-ORD-2523. The experimentation and analysis of results were performed by the first author under the general direction of the second. George H. Mittendorf, as Research Assistant with the Institute, participated in many of the experimental runs. Valuable advice was given by Minerva C. Flores, in regard to the technique of droplet injection, and by James A. Kent, in connection with the photographic recording of the droplet behavior. The paper was first presented at the October, 1960 ASCE Convention in a program arranged by the Committee on Fluid Dynamics of the Engineering Mechanics Division.

APPENDIX.—NOTATION

The following symbols have been adopted for use in this paper:

F = Froude number;

f = resistance coefficient;

g = acceleration of gravity;

h = thickness of fluid stream;

h_* = thickness of interlayer;

m = mass;

P = wetted perimeter;

p = pressure intensity;

R = Reynolds number;

U = mean velocity of flow;

U_* = differential velocity of interlayer;

V = mean velocity of mass transfer;

u, v = velocity components;

x, y = Cartesian coordinates;

α = $\int_0^h u^2 \, dy / U^2 h$;

γ = specific weight;

δ = interlayer thickness;

μ =   dynamic viscosity;

ν =   kinematic viscosity;

ρ =   mass density;

σ =   normal stress;

τ =   shear stress;

τ_m =   mean shear stress along wetted perimeter; and

χ =   stability parameter.

ENERGY TRANSFORMATION IN ZONES OF SEPARATION

HUNTER ROUSE

IOWA INSTITUTE OF HYDRAULIC RESEARCH
STATE UNIVERSITY OF IOWA,
IOWA CITY — USA

Because of the part that separation plays in the performance of flow structures, hydraulics is at least as deeply involved with the phenomenon of eddy motion as any other engineering science. Analysis of the problem by the assumption of irrotational flow is unrealistic, whereas flow with a reasonable distribution of vorticity is very difficult to handle. The most effective progress in understanding the eddy mechanism can probably be realized through evaluation of the various terms in the equations of motion from detailed measurements of the mean-flow and turbulence characteristics. Such an investigation is presented for the pertinent case of two-dimensional flow over a normal wall. Not only are the integrated equations of momentum and of mean energy examined, but particular attention is given to variation of the Bernoulli sum along a mean streamline in either the primary flow or the zone of separation.

A cause du rôle joué par le décollement dans les différentes formes prises par l'écoulement, l'Hydraulique est au moins aussi intéressée par le phénomène du mouvement tourbillonnaire que n'importe quelle autre science. L'analyse du problème, supposant que l'écoulement est irrotationnel, n'est pas réaliste, tandis qu'un écoulement supposant une répartition raisonnable des tourbillons est très difficile à manipuler. Le progrès le plus effectif quant à la compréhension du méchanisme des tourbillons peut probablement être réalisé en évaluant les différents termes des équations du mouvement en utilisant des mesures détaillées des caractéristiques de l'écoulement moyen ainsi que de la turbulence. Une telle investigation est présentée pour le cas de l'écoulement à deux dimensions au-dessus d'une paroi normale à l'écoulement. Les équations de la conservation de l'énergie et de la quantité de mouvement sont non seulement examinées après integration, mais une considération toute particulière est donnée à la variation de la somme des termes de l'équation de Bernoulli le long d'une ligne de courant, soit dans la région de l'écoulement principal, soit dans la zone de décollement.

INTRODUCTION

One of several persistent paradoxes in hydraulics is the fact that the flow phenomena that seem to be the least well understood are those that have been the most familiar for the longest time. Fluid motion in zones of separation is a case in point. Leonardo da Vinci's notebooks, which have now been in existence almost five hundred years, contain numerous and perceptive sketches of eddy patterns in the wake of boundary irregularities. Nearly two hundred years ago Dubuat showed that the drag exerted on an immersed body is due largely to the reduction in pressure within the zone of separation at the rear. Venturi, shortly thereafter, described not only the pressure reduction accompanying a local contraction of the flow but also the diffusive effect of the eddies downstream. Within our time, IAHR founder Theodor Rehbock was among the first to recognize the overall part played by standing eddies or rollers in energy dissipation, and his papers were illustrated by careful depiction of the flow patterns in such regions. In no instance, however, has the mechanism of eddy formation been more than vaguely appreciated, nor is it yet possible after all these years to predict even the basic geometry of a separation zone.

One might ask, to be sure, why separation phenomena are still of interest in hydraulics, when today every effort is being made to eliminate them by careful streamlining. At least three reasons can be cited in reply. First, the hydraulic engineer is a notorious avoider of streamlining except where it is absolutely necessary. Second, perfect streamlining is generally impracticable if not impossible. Third, there are devices (for example, energy dissipators) which streamlining would render wholly ineffective. It thus behooves the hydraulician to seek a better understanding of the phenomenon, for he will probably continue to encounter it the rest of his professional life. Even the fundamental matter of improving the operating efficiency of a flow structure requires an awareness of separation dangers, while such recurrent problems as those of gate vibration, cavitation damage, and sediment scour can usually be traced to one phase or another of eddy formation.

A zone of separation of the sort that Rehbock and his successors have depicted actually presents to the hydraulician just about every aspect of fundamental hydraulics except the gravitational. Such a zone embodies, first of all, non-uniform and unsteady motion (though the unsteadiness is, to be sure, restricted to the sort that can be resolved into a mean and a fluctuating part). Secondly, it consists of two distinct regions of flow — the main or primary stream, and a standing eddy or secondary stream that the primary induces. Thirdly, between the two regions is a transition zone of intense turbulence, and it is the generation, convection, diffusion, and dissipation of the turbulence that represents the most essential feature of the separation phenomenon. Not only is the flow mechanism itself very incompletely understood, but even its measurement remains an unfinished process. As a matter of fact, such simple operations as the determination of the velocity and pressure distribution of the mean motion are still replete with minor difficulties, and the precise evaluation of the turbulence characteristics continues to await its turn. It is the partial clarification of this very complex yet very pressing situation that is the subject of the present lecture.

SIMPLIFIED MODELS OF A SEPARATION ZONE

Separation patterns that are well known in hydraulics are those associated with abrupt expansions, contractions, and bends; with submerged bodies of angular form; and with variants and combinations of the two classes, even including the hydraulic jump. Although the geometric proportions of eddy zones vary from case to case, all have the same underlying characteristics, and hence only one will be used as the basis of the present discussion. This is a thin wall of great length, fixed normal to both the boundary and the flow to yield essentially two-dimensional motion.

In Fig. 1 is shown a photograph of the mean pattern of flow past such a wall, obtained from a time exposure of aluminum particles floating on the surface of a tank along which were towed the wall and a short section of boundary (actually a symmetrical plate

Fig. 1 — Photograph of flow pattern downstream from a wall

with trailing splitter vane was used, to avoid
the secondary effect of upstream separation,
as in Reference [1]; the photograph has been
divided along the splitter-vane center line).
Clearly visible are the trends of the primary
and secondary motion, as well as an approximate line of demarcation between the two —
that is, the profile of the standing eddy. Evidently, the eddy shape plays a fundamental

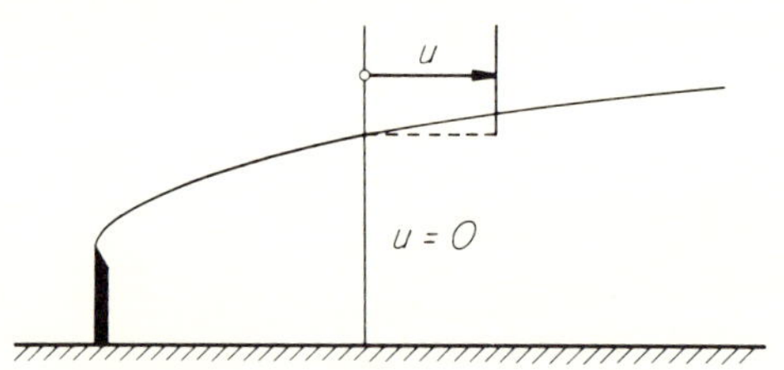

Fig. 2 — Separation zone with free-streamline
border

part in the phenomenon, for the fluid outside
it behaves much as though the eddy boundary
were that of a fixed body around which the
primary flow passes. The eddy, in turn, is
not only driven by the surrounding flow but
subjected to the same dynamic conditions
along the entire surface of demarcation. As a
first approximation, in other words, the two
zones are mutually dependent, and the combined pattern is thus at least one step removed
from the simpler state of flow around a void
or cavitation pocket.

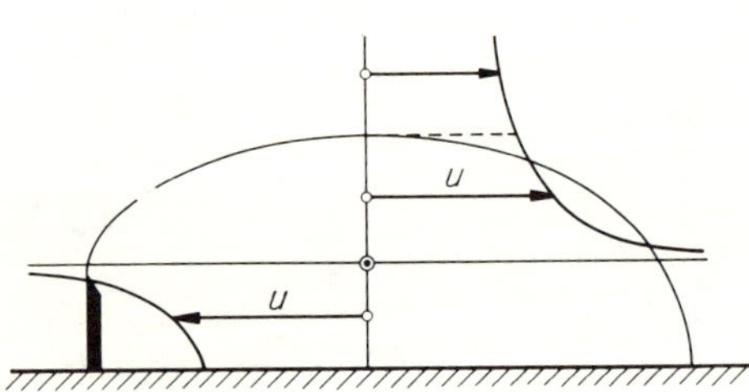

Fig. 3 — Separation zone with irrotational vortex

The temptation to analyze the separation
pattern in the Helmholtz free-streamline
manner, as is done for cavitation pockets, is
nonetheless enticing. However, not only can
such a free streamline end only at infinity
unless certain conditions of symmetry or
reentrant flow are imposed upon it (Fig. 2),
but the eddy profile that one is seeking is far
from that of a constant-pressure void. These
objections would be eliminated by the con-

formal superposition of an irrotational or free
vortex upon the flow in the separation zone,
both the primary and the secondary parts of
the flow being included by the transformation function (Fig. 3). Unfortunately, such a
solution would require the velocity to increase
toward infinity at the eddy center, whereas
the true separation eddy is rather of the rotational or forced-vortex type.

At the Iowa Institute attention has been
directed recently to approximating the mean
flow pattern in a zone of separation by means
of an arbitrary distribution of vorticity. The
simplest condition, of course, is one of zero
vorticity outside the eddy and a constant
magnitude within, such that there is no discontinuity of velocity (only of velocity gradient) at the streamline dividing the secondary from the primary flow, as in Fig. 4.
Were slip to occur along this streamline, flow
in the vicinity of the ends of the eddy, which
are necessarily stagnation points in both the
primary and the secondary flows, would be
physically impossible. Slip is permitted at the

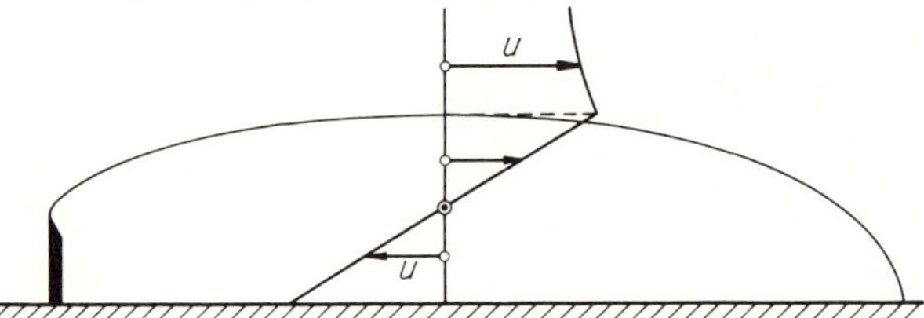

Fig. 4 — Separation zone with rotational vortex

solid boundary in both zones, however, as
boundary shear is not considered essential to
the phenomenon. The flow pattern corresponding to this assumed condition was first
sought by means of the relaxation process,
but it soon became apparent that too many
variables were involved to render such a solution practicable, and use was then made of a
digital computer. To date an approximate
solution has been obtained for a single boundary geometry: a simple lateral displacement
corresponding to one side of an abrupt expansion. Though the eddy profile differs appreciably from that indicated by measurement,
the difference is in the direction to be expected by virtue of the simplifications that were
made, and the result is considered to be a
significant first step in the direction of eddy-
pattern synthesis.

The principal way in which the foregoing
approximation differs from the true mean
pattern of separation lies in the assumed

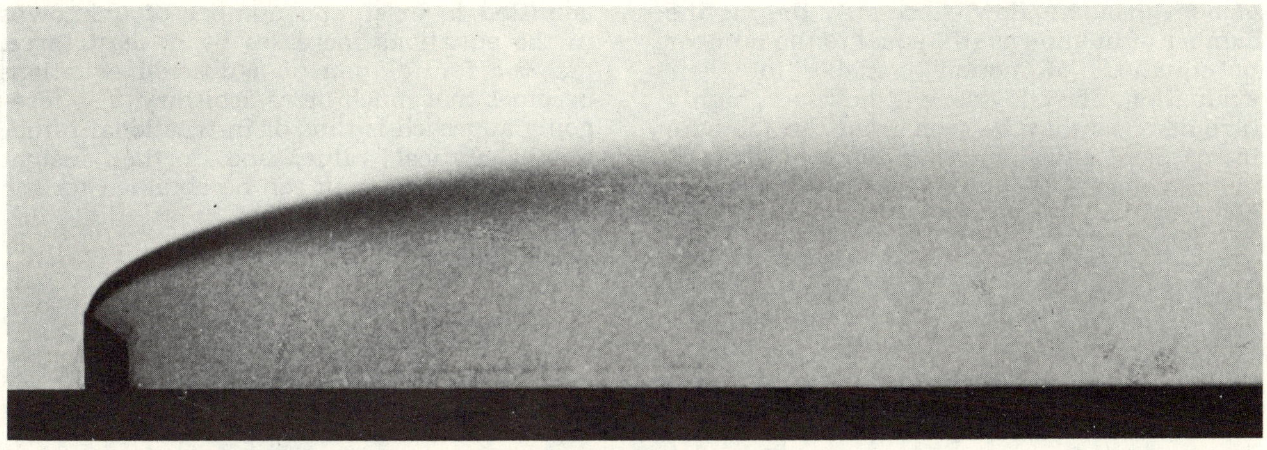

Fig. 5 — Photograph of dyed shear layer bordering separation zone

constancy of vorticity throughout the secondary flow. In actuality the vorticity in the lower part of the eddy is much less than assumed, whereas the point of separation is the origin of an initially thin zone of far higher vorticity, which expands laterally as the vorticity level decreases with distance downstream. At the final stagnation point the fluid within this zone divides, part of it forming the low-vorticity return flow of the eddy and part passing off into the otherwise irrotational wake. The profile of the expanding zone is clearly seen from the diffusion of dye introduced at the edge of the wall (i. e. symmetrical plate with trailing splitter vane) in the time exposure of Fig. 5.

Presumably a similar sort of numerical evaluation could be performed for such a. model, once a reasonable distribution of vorticity were assumed, with a comparable chance of success. However, not only would the complexity of the evaluation be greatly increased, but the assumed distribution would be much more arbitrary, with the relative magnitude and the rate of diffusion of vorticity being open to selection. One could, perhaps, adopt the vorticity and the rate of spread of a submerged jet as a first approximation, but a logical way of handling the distribution at the end of the eddy is not yet apparent. Even were the synthesis a possible one, in fact, the solution would be significant only for the case

Fig. 6 — Photograph of scour of a sand bed (containing dye crystals) in the lee of a wall

of non-turbulent flow, since only then is the number of unknowns still equal to the number of equations of motion available for their evaluation. The eddy zone is, however, highly turbulent, as may be seen from the increase in sediment entrainment in zones of lowered mean velocity (Fig. 6). Once turbulence is admitted to exist, the number of unknowns in the equations increases by at least three, and the formulation of additional equations becomes that much more arbitrary. The foregoing approach is thus of instructional rather than analytical value, and further insight into the phenomenon can be obtained for the

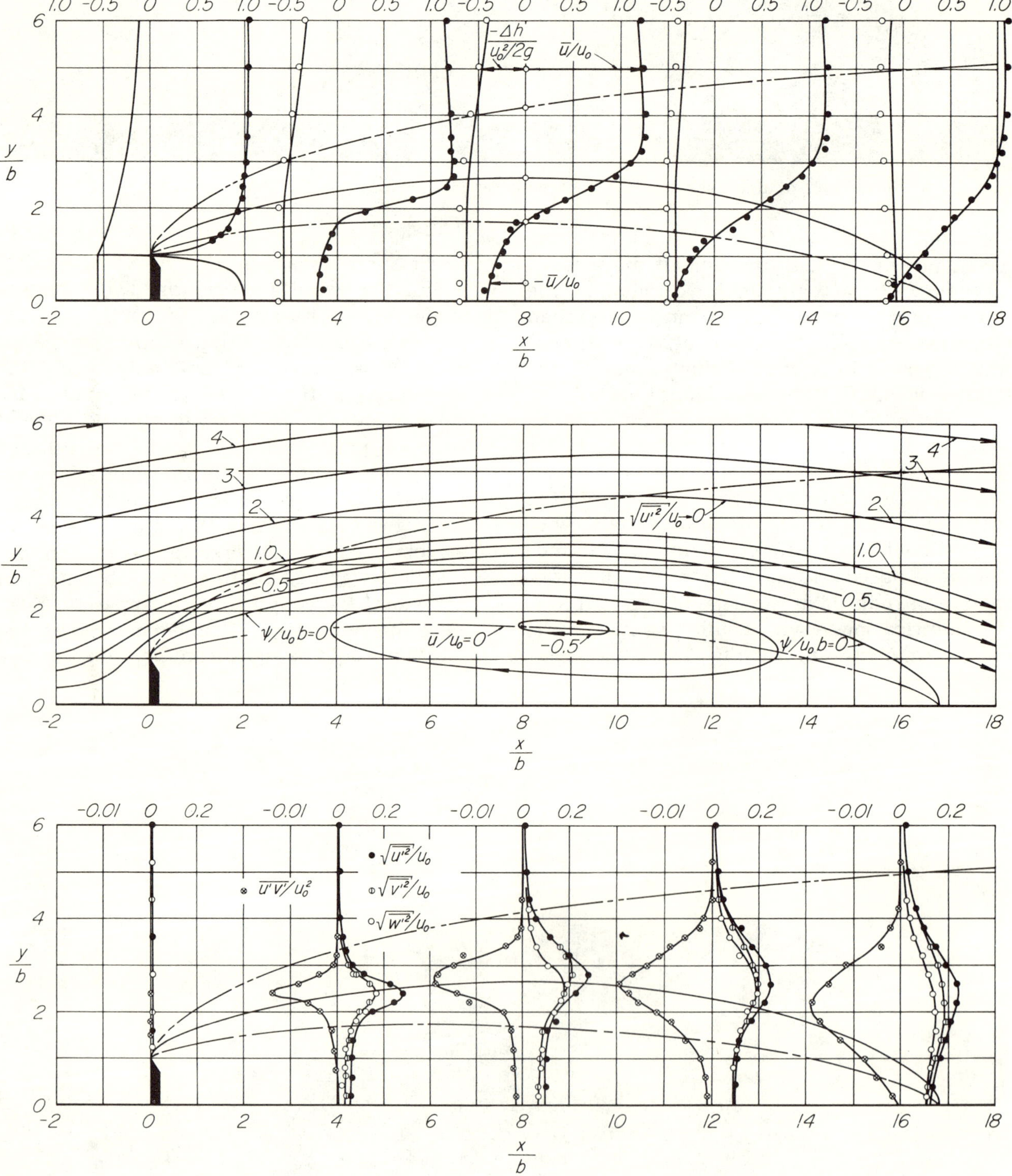

Fig. 7 — Mean-flow and turbulence measurements [1] for flow past a wall

present only through an interpretation of mean-flow and turbulence measurements in the light of the fundamental equations.

EXPERIMENTAL MEASUREMENTS AND THE EQUATIONS OF MOTION

In addition to its elemental significance in more complex boundary configurations, the wall was chosen to illustrate this presentation because of the availability in the literature of measured data for the mean and turbulent flow in its vicinity [1]. These are reproduced in Fig. 7. Since the writer was partly responsible for their determination and publication, he can with propriety point out that they are somewhat incorrect and decidedly incomplete; in fact, their inconsistencies are as pertinent to this discussion as their more direct implications.

From the data for mean velocity and pressure, it was shown that the primary flow could indeed be analyzed as a state of irrotational motion — not around a cavity, as first suggested, but around a solid body having the shape of the standing eddy, augmented in lateral dimension by a fraction of the width of the diffusion zone, calculated much like the displacement thickness of the boundary layer. The plotted streamlines and the plotted velocity and piezometric curves are thus in accord with the simple continuity and Bernoulli equations — outside the broken line that indicates the limit of the diffusion zone. Within the standing eddy the velocity is seen to decrease toward zero at the neutral point and then to reverse in direction on the lower side, the streamlines being shaped in such manner that the continuity requirement is again fulfilled at every section. The simple Bernoulli equation is no longer valid from one streamline to the next, however, for the piezometric pressure within the eddy is seen to be practically constant over each section, as in the usual boundary-layer approximation.

In order to relate the velocity and pressure to the other measured characteristics of the eddy zone, use must be made of the Reynolds equations of motion simplified as follows:

$$\bar{u}\frac{\partial \bar{u}}{\partial x} + \bar{v}\frac{\partial \bar{u}}{\partial y} = -g\frac{\partial h}{\partial x} - \frac{\partial \overline{u'^2}}{\partial x} - \frac{\partial \overline{u'v'}}{\partial y} \qquad (1)$$

$$\bar{u}\frac{\partial \bar{v}}{\partial x} + \bar{v}\frac{\partial \bar{v}}{\partial y} = -g\frac{\partial h}{\partial y} - \frac{\partial \overline{u'v'}}{\partial x} - \frac{\partial \overline{v'^2}}{\partial y} \qquad (2)$$

Herein u and v are velocities in the directions x and y, the bars and primes represent mean values and instantaneous deviations therefrom, g is the gravitational acceleration, h is the piezometric head, and such viscous stresses as $\mu\, \partial u/\partial y$ are assumed to be neglible in comparison with the apparent — i. e., Reynolds — stresses $\bar{\tau}_{xx} = -\rho\overline{u'^2}$, $\bar{\tau}_{yy} = -\rho\overline{v'^2}$, and $\bar{\tau}_{xy} = -\rho\overline{u'v'}$ of the turbulence.

A correct and complete set of measured data for a given zone of separation must satisfy these equations at every point of the field. So far as is known to the writer, no such detailed test has ever been applied to experimental results except in the rather simple case of a submerged jet now under study at the Iowa Institute. As a matter of fact, only for the jet have the characteristics of even the mean flow been measured correctly enough to satisfy the requirements of the equations.

The distribution of mean velocity in such a field is usually determined by measuring the total head and the piezometric head separately and assuming the difference to represent the velocity head. This is correct if the turbulence is of negligible magnitude, but in zones of appreciable fluctuation the difference is really the sum of a mean value squared and a mean-square fluctuation. If the total-head tube is properly rounded and the bore is small, its directional sensitivity will approach the cosine function and the fluctuation effect (quite apart from fluctuation errors due to manometric or piezometric bias) will be restricted to the longitudinal component of the turbulence. If, on the contrary, the tube is blunt-ended and very thin-walled, it will be almost insensitive to direction over a very large angle, and the measurement will then tend to include all components of the turbulence. A piezometric-head tube or disk with side openings, on the other hand, cannot be made insensitive to direction; not only must it be aligned with the mean flow to yield a correct indication without turbulence, but the effective presence of angularity due to turbulence will cause a reduction in reading for eddy scales of order comparable to (or conceivably an increase in reading for scales of order much lower than) that of the instrument diameter. Evidently, errors that arise from this cause will increase in relative magnitude as the relative intensity of turbulence increases — the discrepancies hence approaching a maximum value within the eddy where the turbulence is high and the mean velocity is low (note, for example, the misalignment of velocity data near the eddy center in Fig. 7).

Added experimental care should thus be exercised in measuring the mean flow in a separation zone, and in addition an understanding of conditions indicated by the equations of motion should guide the interpretation of results — particularly in the face of assumptions of long standing. In the simpli-

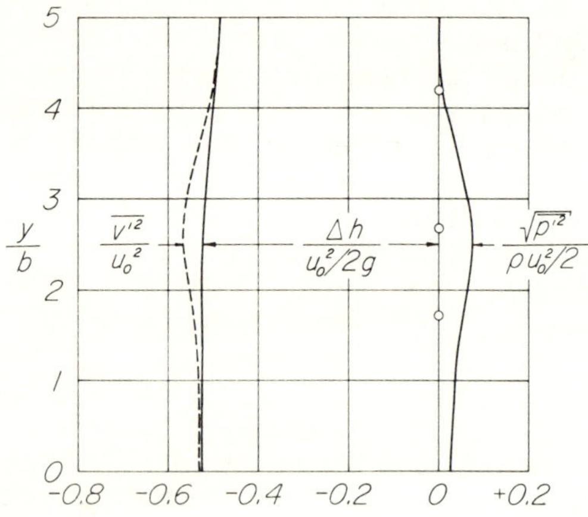

Fig. 8 — Distribution of pressure at midsection of eddy behind wall

fied analysis of wakes and jets, for instance, it has long been assumed that the pressure was hydrostatically distributed throughout the zone of diffusion. As predicted by Townsend [2] and recently demonstrated experimentally by Miller [3], however, for jet flow in the x-direction all velocity terms in the Reynolds equation for the y-direction will be small in comparison with $\overline{v'^2}$, and there must hence be a local decrement in piezometric head comparable in magnitude to $1/g$

times this component of the turbulence. With rough approximation it can therefore be anticipated that the zone of maximum turbulence along the edge of a separation eddy is of even lower mean pressure than the eddy itself (a condition illustrated in Fig. 8 but not reflected in the data of Fig. 7). Furthermore, the pressure must be expected to fluctuate about this mean at any point as the result of the transit of small edies with low-pressure cores, the root-mean-square fluctuation (which is exceeded *15%* of the time) being estimated in the figure from measurements in a jet [4, 5]. Evidently, a triple reason exists for cavitation to develop initially around the periphery of separation zones, as shown in the multiple high-speed photograph of Fig. 9.

An overall check upon the general accuracy of such measurements can be obtained by integrating the first equation of Reynolds with respect to both x and y over a region bounded by a reference section well upstream from the wall where $\bar{u} \approx u_0$, a final vertical section the distance x beyond the wall, the lower boundary, and a limiting section at an elevation y equal to many (say B) times the wall height b. The result may be written for convenience in the following nondimensional form of an equation of momentum:

$$2\int_0^B \left[\left(\frac{\bar{u}}{u_0}\right)^2 - 1\right]\frac{dy}{b} + \int_0^B \frac{\Delta h}{u_0^2/2g}\frac{dy}{b}$$

$$+ 2\int_0^B \frac{\overline{u'^2}}{u_0^2}\frac{dy}{b} - \frac{F/\ell b}{\rho u_0^2/2} = 0 \tag{3}$$

Herein the first term represents the change in momentum flux between the initial and final sections; the second, the differential

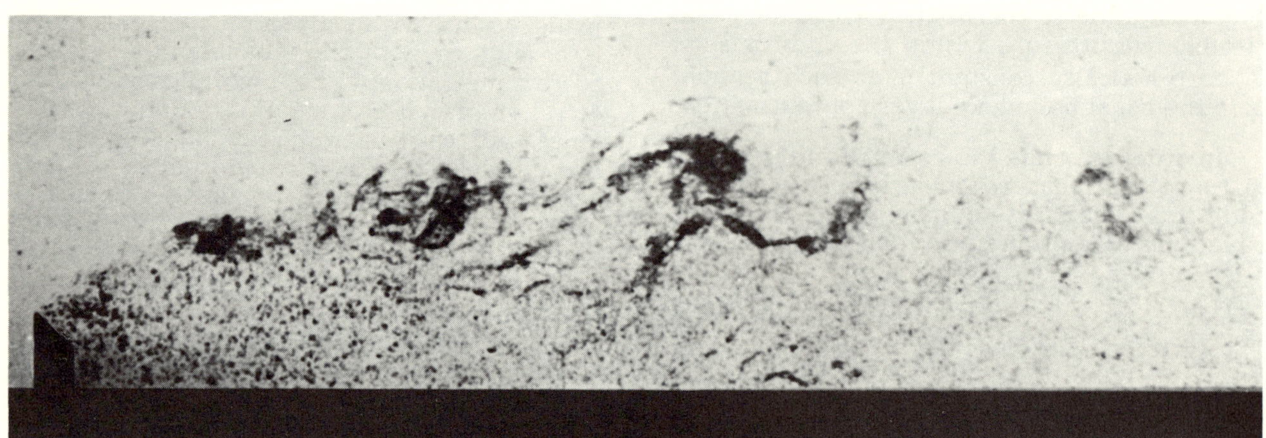

Fig. 9 — Photograph of cavitating eddies along edge of separation zone (exposure, 1/20,000 second)

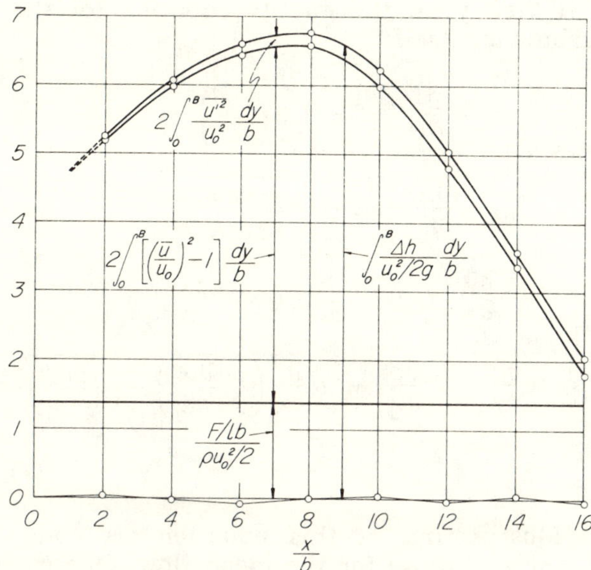

Fig. 10 — Evaluation of experimental data [6] according to momentum integrals of Eq. (3)

piezometric head between the same sections; the third, the normal force of the turbulence over the downstream section; and the fourth, the resultant force per unit length exerted by the wall (taken as positive in the direction of flow). The sum of the terms should evidently be zero, whatever the location of the final section. Each has been evaluated by Arie [6] for the data previously cited, and his results are plotted in Fig. 10. The excellence of the agreement is evident from the proximity of the final points to the zero line. It must be recalled, however, that in the integration process essentially all internal detail was lost.

EQUATIONS OF ENERGY FOR THE MEAN FLOW AND THE TURBULENCE

If each of the equations of Reynolds is multiplied by the corresponding component of mean velocity and the two are added, a slight amount of manipulation will permit the result to be written in the following form:

$$\bar u\frac{\partial}{\partial x}\left(\frac{\rho\overline{V}^2}{2}\right)+\bar v\frac{\partial}{\partial y}\left(\frac{\rho\overline{V}^2}{2}\right)=-\bar u\frac{\partial(\gamma h)}{\partial x}$$

$$-\bar v\frac{\partial(\gamma h)}{\partial y}\quad-\bar u\left[\frac{\partial(\rho\overline{u'^2})}{\partial x}+\frac{\partial(\rho\overline{u'v'})}{\partial y}\right]$$

$$-\bar v\left[\frac{\partial(\rho\overline{u'v'})}{\partial x}+\frac{\partial(\rho\overline{v'^2})}{\partial y}\right]\tag{4}$$

This is the equation of work and energy for the mean flow. The left side (with

$\overline{V}^2=\overline{u}^2+\overline{v}^2)$ indicates the rate of increase of the kinetic energy of the fluid per unit volume as it passes through an elementary portion of space, and the right side the rate at which work is done by the piezometric pressre and the Reynolds stresses. By analogy with the work done by the mean viscous stresses (here ignored as negligible), and in accord with the elementary rule of differentiation $d(xy)=x\,dy+y\,dx$, the terms for work done by the Reynolds stresses can be shown to be the difference between the total work which they do and that which they do "dissipatively" — that is, in producing turbulence, the energy of which is irrecoverably lost to the mean flow:

$$-\bar u\left[\frac{\partial(\rho\overline{u'^2})}{\partial x}+\frac{\partial(\rho\overline{u'v'})}{\partial y}\right]$$

$$-\bar v\left[\frac{\partial(\rho\overline{u'v'})}{\partial x}+\frac{\partial(\rho\overline{v'^2})}{\partial y}\right]$$

$$=-\frac{\partial(\bar u\rho\overline{u'^2})}{\partial x}-\frac{\partial(\bar u\rho\overline{u'v'})}{\partial y}$$

$$-\frac{\partial(\bar v\rho\overline{u'v'})}{\partial x}-\frac{\partial(\bar v\rho\overline{v'^2})}{\partial y}$$

$$+\rho\overline{u'^2}\frac{\partial\bar u}{\partial x}+\rho\overline{v'^2}\frac{\partial\bar v}{\partial y}+\rho\overline{u'v'}\left(\frac{\partial\bar u}{\partial y}+\frac{\partial\bar v}{\partial x}\right)\tag{5}$$

Whether with or without this substitution, the equation of energy must be satisfied by properly measured characteristics of the mean flow and turbulence at all points of a separation zone. Except for the jet studies now in progress at the Iowa Institute, however, it is not known that such a check has ever been made. Arie, on the other hand, has evaluated for the case in question the various terms of the following integral relationship corresponding to that of the previous momentum analysis:

$$\int_0^B\left(\frac{\bar u}{u_0}\frac{\overline{V}^2}{u_0^2}-1\right)\frac{dy}{b}+\int_0^B\frac{\bar u}{u_0}\frac{\Delta h}{u_0^2/2g}\frac{dy}{b}$$

$$+2\int_0^B\left(\frac{\bar u}{u_0}\frac{\overline{u'^2}}{u_0^2}+\frac{\bar v}{u_0}\frac{\overline{u'v'}}{u_0^2}\right)\frac{dy}{b}$$

$$-2\int_0^x\!\!\int_0^B\left[\frac{\overline{u'^2}-\overline{v'^2}}{u_0^2}\frac{\partial\bar u/u_0}{\partial x/b}\right.$$

$$\left.+\frac{\overline{u'v'}}{u_0^2}\left(\frac{\partial\bar v/u_0}{\partial x/b}+\frac{\partial\bar u/u_0}{\partial y/b}\right)\right]\frac{dx}{b}\frac{dy}{b}=0\tag{6}$$

The first term of this equation represents the increase in energy flux between the initial and final sections; the second, the rate at which work is done by the piezometric pressure difference; the third, the total rate at which work is done by the Reynolds stresses; the fourth, the rate at which work is done by these stresses in producing turbulence. Once again, the sum of the terms must be zero at all sections. The extent to which this condition is satisfied by Arie's calculations can be seen from Fig. 11.

Either the differential or the integral form of the energy equation for the mean flow is evidently complete in itself, yet obviously neither one tells the entire story about the transformation of energy to its ultimate form of heat. In other words, although the differential form was manipulated to embody the rate at which energy is irrecoverably lost to the mean flow through the production of

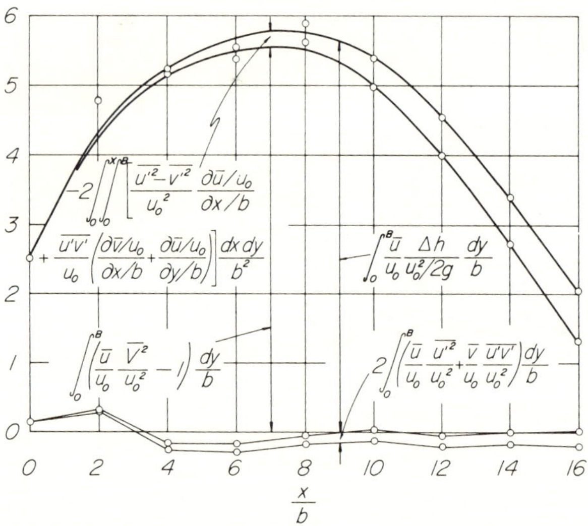

Fig. 11 — Evaluation of experimental data [6] according to mean-energy integrals of Eq. (6)

turbulence, what happens to this turbulence thereafter is in no way indicated. To obtain such an indication, much the same procedure must be followed as in deriving the energy equation for the mean flow, except that the temporal averages are not taken till after the equations of acceleration have been multiplied by the instantaneous velocity components [7]. Subtraction of the equation for the mean motion from this more general relationship

will then yield the energy equation for the turbulence itself:

$$\bar{u}\frac{\partial}{\partial x}\left(\frac{\rho\overline{V'^2}}{2}\right) + \bar{v}\frac{\partial}{\partial y}\left(\frac{\rho\overline{V'^2}}{2}\right)$$

$$+ \overline{u'\frac{\partial}{\partial x}\left(\frac{\rho V'^2}{2}\right)} + \overline{v'\frac{\partial}{\partial y}\left(\frac{\rho V'^2}{2}\right)}$$

$$+ \rho\overline{u'^2}\frac{\partial\bar{u}}{\partial x} + \rho\overline{v'^2}\frac{\partial\bar{v}}{\partial y} + \rho\overline{u'v'}\left(\frac{\partial\bar{u}}{\partial y} + \frac{\partial\bar{v}}{\partial x}\right)$$

$$= -\overline{u'\frac{\partial p'}{\partial x}} - \overline{v'\frac{\partial p'}{\partial y}} + \mu\overline{\frac{\partial}{\partial x_j}\left(u_i'\frac{\partial u_i'}{\partial x_j}\right)} - \mu\overline{\left(\frac{\partial u_i'}{\partial x_j}\right)^2}$$

$$(7)$$

Most terms of this equation are counterparts of those for the mean flow. One can thus distinguish terms for the convection and diffusion of turbulence energy by the mean flow and by the turbulence, respectively; for the transfer of energy from the mean flow by turbulence production; for work done by the pressure fluctuations; and for total and dissipative work done by the viscous stresses within the tiny eddies. (In each of the latter terms u_i' and x_j' are to be given the successive values u', v', w' and x, y, z, respectively, and the resulting nine mean values added.) Correct and complete turbulence measurements must of necessity satisfy this equation. However, not only has no such detailed analysis ever been conducted for a separation zone, but measurements that are extensive enough to permit all of the terms in the equation to be evaluated have never been made. Not even the integral form of the equation (from which only ·the next-to-last term would disappear) can be investigated, for much the same reason.

A rough approximation to the integral relation can be obtained if the rate of con-

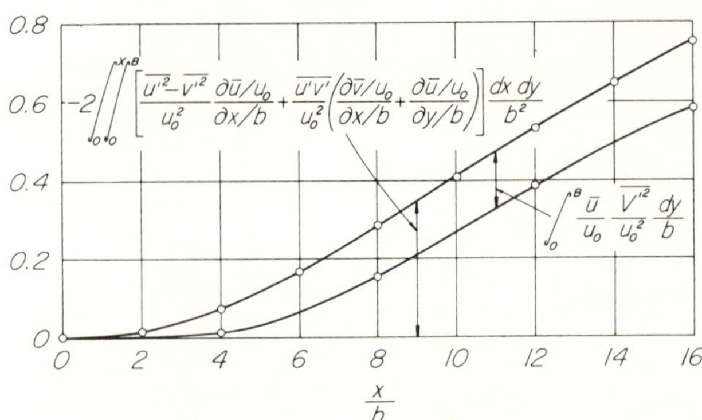

Fig. 12 — Approximation of dissipation function by difference between cumulative turbulence production and convection

vection of turbulence energy past successive sections is computed and subtracted from the cumulative rate of turbulence production, as shown in Fig. 12. The major part of the difference should represent the cumulative rate of dissipation over the zone under consideration, from the trend of which at least general conclusions can be drawn. For example, three significant facts are to be noted in this diagram: first, the rate at which turbulence is convected by the flow past any section becomes increasingly smaller in comparison to the cumulative rate at which it is produced; second, the turbulence is carried an appreciable distance downstream from the point of production during the dissipation process; third, the cumulative rates of turbulence production and subsequent dissipation must approach as a limit the product of the force on the wall and the mean velocity of the ambient flow.

THE BERNOULLI THEOREM FOR TURBULENT FLOW

Though also an energy relationship, the Bernoulli equation differs from the other equations of energy primarily in dimension. Each term in the equations just discussed had the dimension of work or energy per unit time — i. e., power. The terms in the usual Bernoulli equation, on the other hand, have the dimension of work or energy per unit volume (or, optionally, per unit mass or weight). When written in the familiar hydraulics form

$$\frac{V_1^2}{2g} + \frac{p_1}{\gamma} + z_1 = \frac{V_2^2}{2g} + \frac{p_2}{\gamma} + z_2 \qquad (8)$$

each of the terms is dimensionally equivalent to energy per unit weight, or simply length, and the subscripts refer to any two points in the field of motion, to two points on the same streamline, or to two successive sections of a gross flow passage, depending upon the method of derivation. Noteworthy is the absence of terms for either energy of turbulence or energy dissipation. These are sometimes introduced arbitrarily — but, as will be seen, as likely as not erroneously.

It is the streamline version of the Bernoulli equation that is pertinent to the present situation. This can be rigorously derived by writing the Reynolds equations in terms of the Bernoulli sum and components of vorticity [7] before multiplying by the mean velocity components in their ratio to V. For

present purposes, the foregoing differential form of the energy equation for the mean flow is merely rewritten by letting the direction x become the tangential direction s and y the normal direction n, under which circumstances $\overline{u} = \overline{V}$ and $\overline{v} = O$, and dividing by $\overline{V}$. If, for simplicity, tensor notation is again introduced in connection with the Reynolds stresses (the result then being valid as well for three-dimensional flow), the equation will take the form

$$\frac{\partial}{\partial s}\left(\frac{\rho \overline{V}^2}{2}\right) = -\frac{\partial(\gamma h)}{\partial s} - \frac{\overline{u}_i}{\overline{V}}\frac{\partial \overline{\tau}_{ij}}{\partial x_j} \qquad (9)$$

Integration between points on the same streamline will then yield the simplest form of the Bernoulli equation for turbulent flow:

$$\frac{\rho \overline{V}_1^2}{2} + \gamma h_1 = \frac{\rho \overline{V}_2^2}{2} + \gamma h_2 - \int_{s_1}^{s_2} \frac{\overline{u}_i}{\overline{V}}\frac{\partial \overline{\tau}_{ij}}{\partial x_j}\,ds \qquad (10)$$

However, not only does this equation still contain no term for the kinetic energy of the turbulence, but the added integral — far from representing energy losses — is not restricted as to sign and can even represent a gain (is it must in any zone of induced flow like a standing eddy). As before, a term for effective loss can nevertheless be introduced by replacing the integral term by the equivalent difference between the total work of the Reynolds stresses and the work done by them in producing turbulence:

$$\frac{\rho \overline{V}_1^2}{2} + \gamma h_1 = \frac{\rho \overline{V}_2^2}{2} + \gamma h_2$$

$$-\int_{s_1}^{s_2} \frac{1}{\overline{V}}\frac{\partial(\overline{u}_i \overline{\tau}_{ij})}{\partial x_j}\,ds + \int_{s_1}^{s_2} \frac{\overline{\tau}_{ij}}{\overline{V}}\frac{\partial \overline{u}_i}{\partial x_j}\,ds \qquad (11)$$

The interrelationship of the several terms in these two equations can be studied more conveniently by assigning to the nondimensional form of each ($\rho \overline{u}_0^2/2$ being used as common denominator) a characteristic symbol: K for the kinetic energy, P for the piezometric pressure, Δ for the change represented by the integral of Eq. (10), W for the total-work integral of Eq. (11), and L for the loss due to turbulence production (the equality $W = \Delta + L$ prevailing by definition). Equations (10) and (11) then state, in effect, that the following sums must have the same magnitude at all successive points on the same streamline:

$$K + P - \Delta = K + P - W + L = C_s \qquad (12)$$

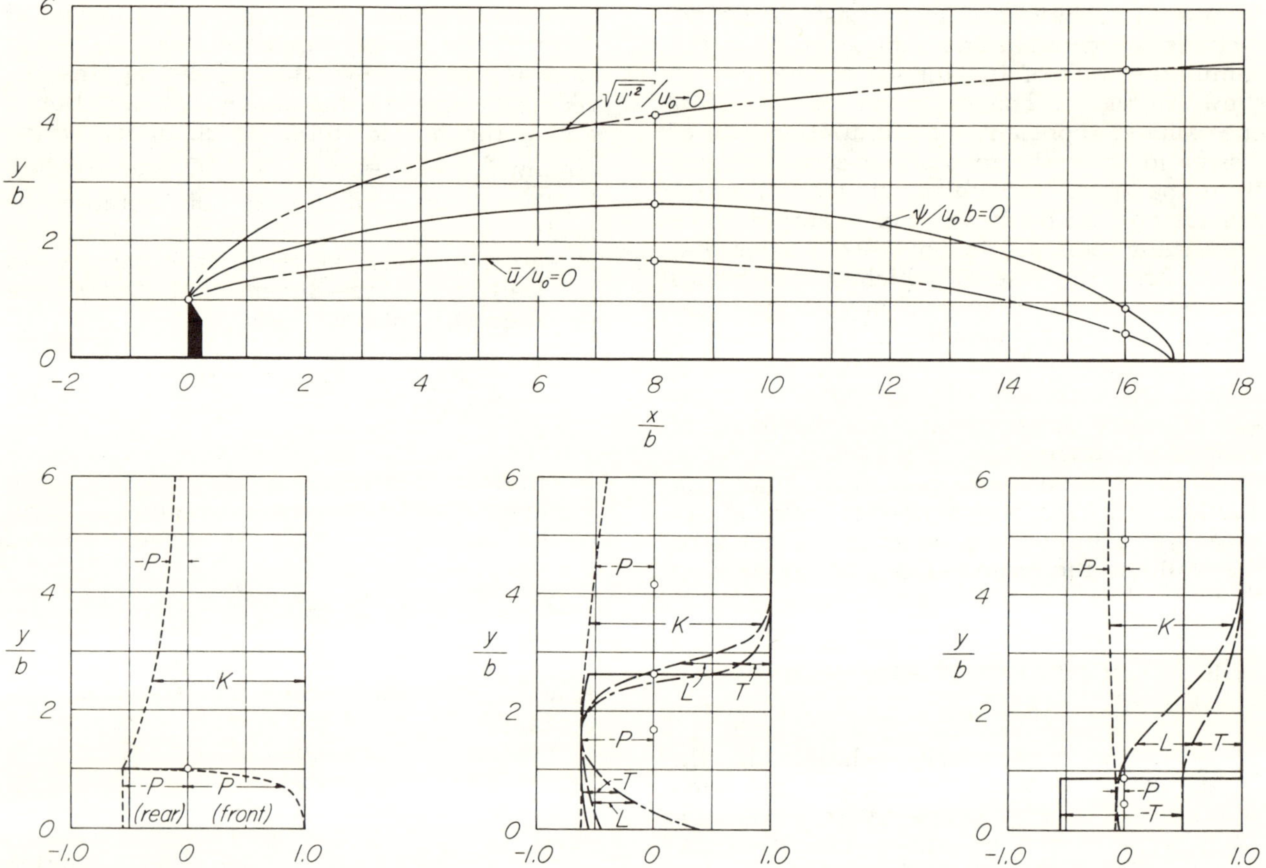

Fig. 13 — Distribution of Bernoulli terms — kinetic energy, piezometric pressure, energy transfer, and loss — in separation zone

Now the change Δ in the Bernoulli sum $K + P$ (negative if the sum decreases) must involve either a local transformation (turbulence production) or a transfer to or from the surroundings — or, in general, a combination of the two. As a matter of fact, the total work W done upon an elementary filament over the reach under consideration is simply the negative of the work done by that particular length of filament in increasing the energy of the flow in its vicinity, and $-W$ therefore measures the outward transfer T.

To the end of yielding further clarity, the distribution of the four quantities involved in the sum

$$K + P + T + L = C_s \qquad (13)$$

is shown schematically in Fig. 13 for the zone of separation behind the wall [8]. In the irrotational region outside the diffusion zone, the sum $K + P$ is seen to retain the constant magnitude of unity of the approaching flow, T and L both remaining negligibly small. Within the diffusion zone but outside the eddy, L and T are both positive, indicating

that the Bernoulli sum is decreasing along each streamline due to lateral transfer and local production of turbulence, but the sum C_s still remains equal to unity at all points. Within the eddy, on the contrary, C_s is even less than zero. Not only is T everywhere negative, indicating a gain through transfer, but in the upper half this gain even exceeds the loss through turbulence production. In the lower half of the eddy, however, the latter situation is reversed, because all streamlines within the eddy are closed circuits; in other words, by the end of a circuit all terms must be restored to the values that they had as the arbitrary initial section (the line of zero forward velocity) was first passed.

As was true of the general relationship for the energy of the mean flow, this form of the Bernoulli equation evidently contains neither the kinetic energy of the turbulence nor a true measure of the energy that has been lost through viscous transformation to heat within the turbulent eddies. While the former can readily be evaluated from available measurements, the latter still defies rigorous determination; moreover, without other still unmeasurable terms belonging in a completely detailed version of the Bernoulli equation, neither one is of present import in more than a preparatory way. The status of the Bernoulli theorem for turbulent flow, in other words, is rather similar to that of the problem of energy distribution in zones of separation which it is intended to describe: qualitatively familiar and significant, but still quantitatively obscure and incomplete.

CONCLUSION

A considerable amount of knowledge about zones of separation has evidently been gained since their existence was first recognized, though it must be admitted that at the same time the apparent scope of the subject has expanded many fold. The characteristics of standing eddies are becoming more and more familiar as they are studied experimentally, so far as both their mean and their turbulent components are concerned. In a number of instances laboratory measurements have been shown to satisfy the integral forms of the momentum and the mean energy equations. Not yet, however, has either the differential form of these equations or the differential or integral form of that for the energy of the turbulence been similarly utilized, primarily because of the scarcity of accurate and complete experimental data. The analytical prediction of even the mean characteristics of a separation zone remains still more remote.

Hydraulicians are surely as deeply concerned with advancements in this field as any professional group, because of the recurrent part that eddies play in the performance of flow structures. Yet their immediate concern is not only the mechanism of the energy-transformation process, but also the effect of this process upon the use of their daily tools: measuring instruments, and methods of calculation. In regions of separation, Pitot tubes (not to mention current meters and other devices) cannot be expected to yield precise results unless the effects of the turbulence are properly taken into account. And for such regions the familiar Bernoulli theorem can no longer presume a simple constancy of velocity head plus pressure head plus elevation — nor can discrepancies be counterbalanced by the arbitrary addition of terms for turbulence and energy losses; instead, an expression must be properly derived to include the effects that predominate in zones of separation: energy transformation and transfer by the Reynolds stresses.

ACKNOWLEDGMENTS

Relaxation and computer studies of a separation zone described herein were conducted by Dr. and Mrs. E. O. Macagno and Mr. S. P. Garg. Certain of the supplementary measurements were made by Mr. T. J. Carmody, computations by Mr. M. C. Chaturvedi, and photographs by Messrs. P. J. Collins and J. A. Kent. Dr. Macagno and Dr. L. M. Brush, Jr. critically reviewed the manuscript during its preparation. The project as a whole represents one of a series partially supported by the Office of Naval Research under Contract Nonr 1509 (03) with the Iowa Institute of Hydraulic Research.

REFERENCES

[1]. **Arie, M., and Rouse, H.:** "EXPERIMENTS ON TWO-DIMENSIONAL FLOW OVER A NORMAL WALL," Journal of Fluid Mechanics, Vol. 1, Part 2, 1956.

[2]. **Townsend, A. A.:** THE STRUCTURE OF TURBULENT SHEAR FLOW, Cambridge, 1956.

[3]. **Miller, D. R., and Comings, E. W.:** "STATIC PRESSURE DISTRIBUTION IN THE FREE TURBULENT JET", Journal of Fluid Mechanics, Vol. 3, Part 1, 1957.

[4]. **Rouse, H.:** "CAVITATION IN THE MIXING ZONE OF A SUBMERGED JET," La Houille Blanche, Vol. 8, No. 1, 1953.

[5]. **Hinze, J. O.:** TURBULENCE, New York, 1959.

[6]. **Arie, M.:** "CHARACTERISTICS OF TWO-DIMENSIONAL FLOW BEHIND A NORMAL PLATE IN CONTACT WITH A BOUNDARY ON HALF PLANE," Memoirs of Faculty of Engineering, Hokkaido University, Vol. 10, No. 2, 1956.

[7]. ADVANCED MECHANICS OF FLUIDS, edited by Rouse, New York, 1959.

[8]. **Rouse, H.:** "ON THE BERNOULLI THEOREM FOR TURBULENT FLOW", Tollmien-Festschrift, Berlin, 1962.

THE ROLE OF THE FROUDE NUMBER IN OPEN-CHANNEL RESISTANCE

by Hunter Rouse
Iowa Institute of Hydraulic Research
and
H. J. Koloseus and Jacob Davidian
U.S. Geological Survey

A frequent assumption regarding channel resistance is that the mean intensity of boundary shear is dependent upon the mean depth and velocity of flow, the shape of the cross section, a linear measure of the boundary roughness, and the mass density, specific weight, and dynamic viscosity of the water:

$$\tau = \varphi(d,\ V,\ \text{shape},\ r,\ \varrho,\ \gamma,\ \mu)$$

These quantities may be combined to yield the following nondimensional form of functional relationship:

$$\frac{\tau}{\varrho V^2} = \varphi\left(\frac{Vd}{\mu/\varrho},\ \frac{r}{d},\ \text{shape},\ \frac{V}{\sqrt{d\gamma/\varrho}}\right)$$

Through introduction of the equilibrium condition involving the hydraulic radius and channel slope, $\tau = \gamma RS$, the dependent parameter can also be written in terms of the familiar Chezy C or Weisbach f:

$$\frac{\tau}{\varrho V^2} = \frac{g}{C^2} = \frac{f}{8}$$

Though the form of the general relationship for open channels has yet to be evaluated, there is ample evidence that the resistance coefficient is truly a function of the Reynolds number, the relative roughness, and the shape of the flow section, for open as well as closed conduits. The presence of the Froude number, on the other hand, is less satisfactorily justified by experience. The resistance of open-channel transitions, like that of ships, is assuredly dependent upon gravitational effects, because of the formation of standing waves. However, such phenomena involve boundary configurations which are inherently nonuniform, whereas the resistance problem is normally restricted to conditions approaching uniformity. (More than one experimental investigation of channel "roughness", nevertheless, have utilized boundary irregularities which were so large as to represent changes in cross section rather than surface texture.)

Since the few indications of gravitational influence that are to be found in the literature [see References 1–3] were all obtained at Froude numbers well in excess of unity, it was eventually surmised that the phenomenon was in some way connected with the surface instability involved in the formation of roll waves. Now the stability of open-channel flow is analyzed in much the same manner as that of closed-conduit flow, by assuming a periodic disturbance of small amplitude to occur and determining from the equations of motion whether the perturbation will be damped or amplified. In the case of confined flow, augmentation of the disturbance leads to the onset of turbulence, whereas in free-surface flow it corresponds to the development of travelling waves, the larger of which then continue to grow as they overtake and assimilate the smaller. Although the problem of internal stability has been solved with a considerable degree of rigor, that of surface stability is so complex that all of the analyses to date [see, for example, References 4–7] have involved varying degrees of simplification. The most elementary result [4] for the limit of stability is a constant magnitude of the Froude number equal to 2.

A considerably more extensive solution would yield a magnitude which varied with the Reynolds number, relative roughness, and cross-sectional shape; this is indicated for a wide, hydraulically rough channel in Fig. 1 [7].

If the gravitational influence upon open-channel resistance actually is associated with the development of roll waves, then the unsteadiness as well as the nonuniformity of such motion would seem to place it even farther

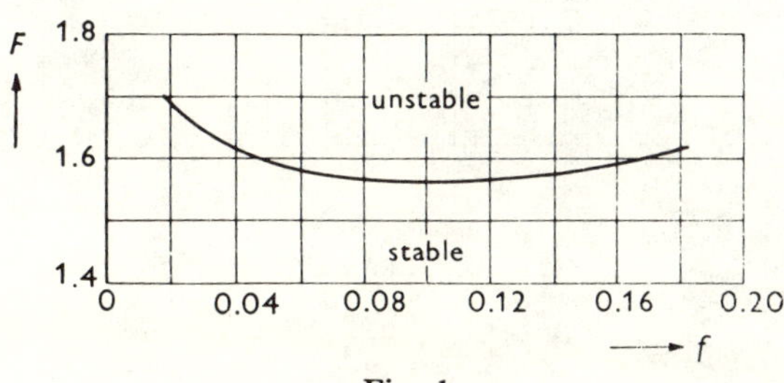

Fig. 1.

from the realm of surface resistance than is mere cross-sectional irregularity. On the other hand, to become perceptible, roll waves normally must develop over reaches that are several orders greater than the depth, and within such reaches the influence of gravity might well be treated in terms of the usual resistance function for steady, uniform flow.

In the course of experiments at Iowa on the role of roughness concentration in surface resistance, an excellent opportunity was recently afforded to extend the range of Froude numbers well beyond the computed stability limit. The roughness elements were $3/16$-inch brass cubes cemented in a diamond-shaped pattern to the bottom of a tilting flume 2 feet wide and 30 feet long. So significant were the results obtained with areal concentrations of $1/128$ and $1/32$ that additional concentrations of $1/512$ and $1/8$ were subsequently studied in the same flume, and the $1/32$ concentration was also reproduced in a flume 2.5 feet wide and 85 feet long. As described elsewhere [8], all experimental results obtained for the stable regime of flow could be expressed as a unique function of the relative height k/d and concentration λ of the roughness elements. For the unstable regime, on the other hand, the results were found to deviate therefrom in proportion to the growth of the Froude number F beyond its limiting value F_s for stable flow.

Fig. 2 contains a plot of all data obtained for a typical roughness concentration $(1/32)$ in the 85-foot flume, from which one can clearly discern the deviation of values for particular rates of flow (*i.e.*, constant Reynolds number R) as the Froude number increases beyond the stability limit. Similar plots were obtained with the same and other concentrations in the shorter flume. In every case, division of the abscissa values by the two-thirds power of the ratio F/F_s was found to reduce the data for unstable flow to the functional trend of those for stable flow, as shown in Fig. 3 for the unstable-flow data of Fig. 2. Of particular import is the fact that although the smaller flume was too short to permit the formation of perceptible roll waves, they could usually be observed near the end of the longer flume for runs in which $F/F_s > 1$. Thus, not only is the role of gravity in open-channel resistance clearly related to the inherent instability of the free surface at high Froude numbers, but in the early stages of roll-wave formation the loss of head that this entails can be estimated through a simple modification of the usual resistance relationship for steady, uniform, openchannel flow.

References

1. Jegorow, S. A., Turbulente Überwellenströmung (Schiessen) im offenen Gerinne mit glatten Wänden, Wasserkraft und Wasserwirtschaft, Vol. 35, 1940, p. 55–57.
2. Powell, R. W., Flow in a Channel of Definite Roughness, Transactions American Society of Civil Engineers, Vol. 111, 1946, p. 531–66.
3. Homma, M., Fluid Resistance in Water Flow of High Froude Number, Proceedings 2nd Japanese National Congress for Applied Mechanics, 1952, p. 251–254.
4. Jeffreys, H., The Flow of Water in an Inclined Channel of Rectangular Section, Philosophical Magazine and Journal of Science, Vol. 49, May, 1925, p. 793–807.
5. Vedernikov, V. V., Characteristic Features of a Liquid Flow in an Open Channel, Transactions USSR Academy of Sciences, Vol. 52, 1946, p. 207–10.
6. Iwasa, Y., The Criterion for Instability of Steady Uniform Flows in Open Channels, Memoirs, Faculty of Engineering, Kyoto University, Vol. 16, No. VI, 1954, p. 264–75.
7. Koloseus, H. J., The Effect of Free-Surface Instability on Channel Resistance, Ph. D. Dissertation, State University of Iowa, 1958.
8. Koloseus, H. J., and J. Davidian, Flow in an Artificially Roughened Channel, Vol. B, U.S. Geological Survey Annual Review for 1961.

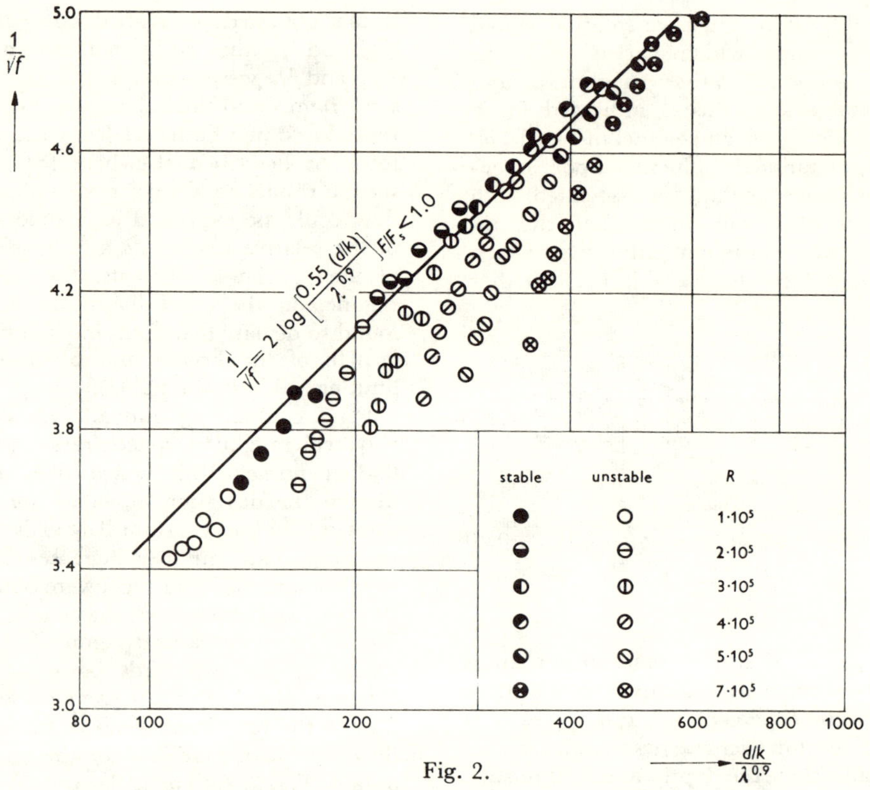

Fig. 2.

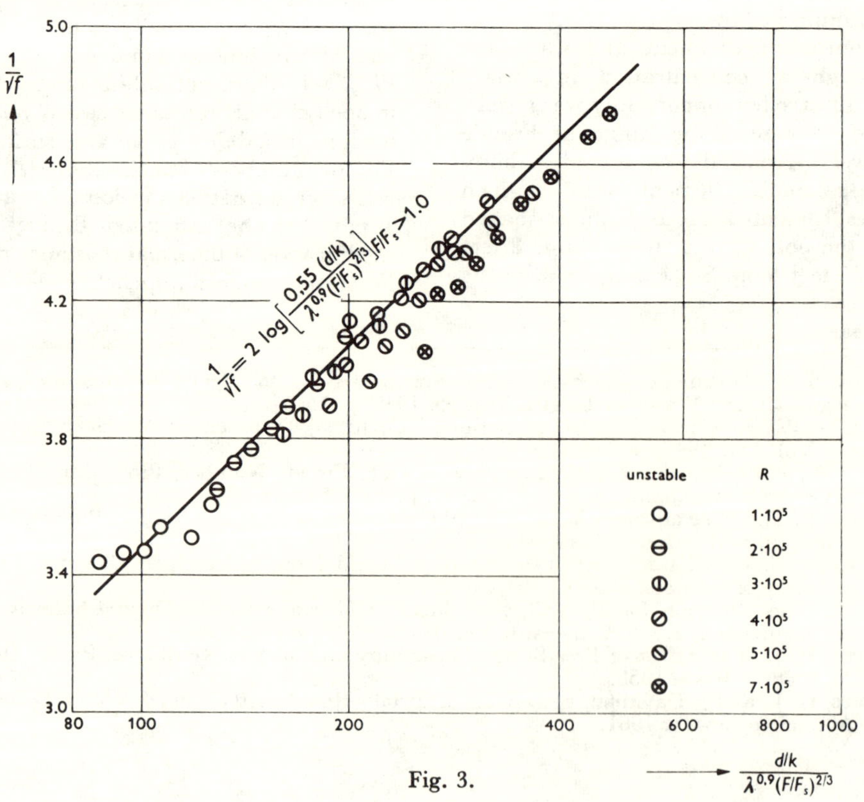

Fig. 3.

410

ON THE ROLE OF EDDIES IN FLUID MOTION

By Hunter Rouse
University of Iowa

Some thirty years ago the writer published in the M.I.T. *Technology Review* an article entitled "Whirling Enemies of Motion," wherein eddies were treated primarily as the "root of the resistance evil," which one could eliminate—or at least control—by the relatively new process of streamlining. Actually, there is very little inherently eddy-free flow in the world today, if one exclude the motion of lubricants in bearings, blood in capillaries, and groundwater, gas, and oil in the pervious strata of the earth. Nor would one wish that the situation were basically different, for the lack of eddy-induced resistance to the motion of a vehicle through a fluid (or of a fluid through a conduit) would be of little worth to mankind in surroundings that were not made physiologically endurable by processes of eddy diffusion on a great variety of scales.

What an eddy looks like is perhaps most popularly exemplified by the various spiral galaxies in photographs of the heavens (Fig. 78), and indeed classical hydrodynamics is one of the tools of astronomical analysis. Somewhat closer to earth is the large-scale motion of the atmosphere, and its similarity of pattern has become the more apparent through

Fig. 78. "Whirlpool" galaxy in Canes Venatici,
photographed at Palomar Observatory.

Fig. 79. Extratropical cyclone in the South Atlantic Ocean
as recorded by U.S. Weather Bureau *Tiros I*.

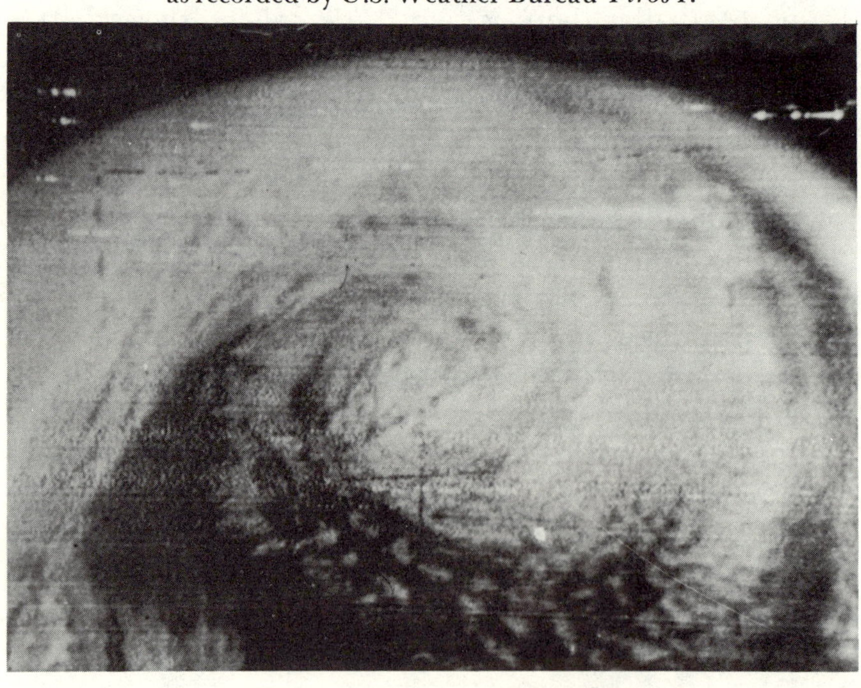

recent satellite photographs (Fig. 79) of storm formations. We are all familiar, at least through newspapers, with such phenomena as cyclonic twisters and waterspouts (Fig. 80), and who has not produced his own visible eddy trails with a canoe paddle—or even with a spoon in a coffee cup? By far the most plentiful eddies, however—the majority of those involved in the heterogeneous motion known as turbulence—are usually so small as to escape notice completely, except for their bulk action in phenomena of diffusion and energy dissipation.

Complex as they all are, each of the occurrences just mentioned has been subjected to mathematical investigation, sometimes of a very extensive and profound nature. In fact, several of these investigations have been carried so far that their experimental verification perforce lags far behind. By

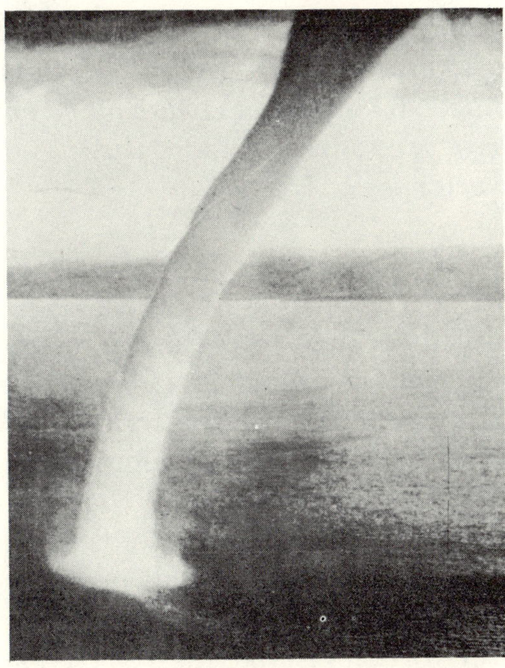

Fig. 80. Waterspout off the coast of Italy (courtesy of *Weatherwise*).

such an approach, moreover, the physical aspects of the
analysis are often subordinated to the mathematical, with
the result that much contact with reality appears to be lost
—at least by the onlookers, if not by the investigators themselves. The present discussion will place first emphasis on
the physical, in the effort to correlate the many isolated facts
that are known about eddy behavior, whether these have
been learned through analysis or observation. To clarify
the underlying aspects of this behavior, use will still be made
of various simplified mathematical models. Through this
procedure some degree of realism may momentarily be sacrificed; however, the gain in clarity that is accomplished
will invariably be of a higher order, and the picture that is
ultimately assembled should be both clear and sound.

Vorticity

Classical hydrodynamics [1] has long embodied the mathematical concept known as vorticity, a vector quantity ω having three orthogonal components which are expressible in
terms of transverse gradients of the three components of
the velocity vector v:

$$\omega_x = \frac{\partial v_z}{\partial y} - \frac{\partial v_y}{\partial z} \; ; \; \omega_y = \frac{\partial v_x}{\partial z} - \frac{\partial v_z}{\partial x}; \quad \omega_z = \frac{\partial v_y}{\partial x} - \frac{\partial v_x}{\partial y} \quad (1)$$

Vorticity is directly related to the apparent tendency of a
cubical fluid element to rotate as a result of the velocity
distribution, since each pair of terms represents—except for
the missing factor $\frac{1}{2}$— the average angular velocity of two
orthogonal axes of the element in their coordinate plane. In
a fluid mass actually rotating about an axis like a solid body,
therefore, the vorticity is equal at all points to just twice
the angular velocity.

In general the vorticity is, like the velocity, a function of
both space and time. Just as streamlines are tangent to the

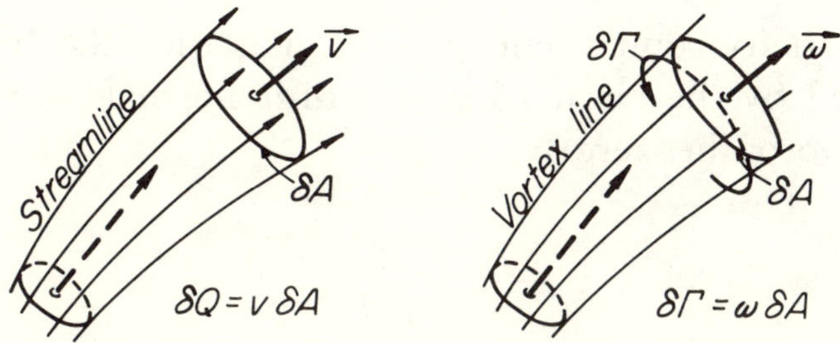

FIG. 81. Comparative sketches of stream and vortex filaments.

velocity vector at each point through which they pass, vortex lines can be imagined tangent at every point to the vector of vorticity. Similarly (see Fig. 81), in much the same way as a stream filament is bounded by a surface containing streamlines, a vortex filament can be visualized as enclosed by a surface controlled in shape by vortex lines. For a fluid that is incompressible, the fact that no flow can occur through the walls of a stream filament makes the rate of flow Q necessarily the same at successive cross sections at the same instant—i.e. $v_1\delta A_1 = v_2\delta A_2\ldots$; the velocity is thus inversely proportional to the cross-sectional area of the filament. The analogy is complete in that the product of vorticity and cross-sectional area is also constant along a vortex filament: $\omega_1\delta A_1 = \omega_2\delta A_2\ldots$; the vorticity thus also varies inversely with the cross-sectional area of the corresponding filament.

The counterpart of the rate of flow in this analogy—the product of the cross-sectional area and the vorticity—is what is known in hydrodynamics as circulation. The circulation Γ around a closed curve of arbitrary shape is customarily defined as the line integral of the tangential component of the velocity:

$$\Gamma = \oint V_L \, dL \qquad\qquad (2)$$

The vorticity component at a point is thus the limit approached by the circulation per unit normal area as the latter approaches zero:

$$\omega\perp_A = \lim_{A \to 0} \frac{\Gamma}{A} \qquad\qquad (3)$$

Helmholtz, who concerned himself with the mathematics of vorticity to a considerable degree about a century ago,[2] established three theorems (later refined by Kelvin[3]) which are pertinent to the subsequent discussion. The first of these was the condition already noted that the circulation must be the same at any instant at all sections of a vortex filament; in addition to the corollary that the vorticity varies inversely with the cross-sectional area, there is another to the effect that a vortex filament must either extend to a flow boundary or form a closed circuit like a smoke ring. The second theorem states that a particular vortex filament will always consist of the same fluid, regardless of how it is convected or deformed by the flow. The third embodies the principle that the degree of rotationality of a fluid—i.e. the circulation around a closed curve connecting the same fluid particles—can be changed only through the action of viscous shear.

Because viscous shear occurs to some degree wherever there is relative motion between neighboring fluid elements, conditions of negligible vorticity are generally limited in actuality to loci of symmetry or cases in which the entire fluid mass is either at rest, rapidly accelerated from rest, or moving as a unit. Nevertheless, for convenience of analysis it is customary in hydrodynamics to assume rather idealized situations that can be handled mathematically. The most common is the state of complete irrotationality, in which the vorticity at every point is zero. Another is a discontinuity in velocity at the juncture between two irrotational zones, the resulting surface of infinite vorticity being known as a vortex sheet. A third is a line vortex—a filament of infinitesimal diameter and infinite vorticity (though finite circulation)—in an otherwise irrotational fluid. Each of these is obviously a physically impossible condition, but its assumption permits the pertinent effect to be analyzed without complication of the picture by extraneous influences.

The Irrotational Vortex

In normal circumstances the circulation around a closed curve in an irrotational fluid is equal to zero. If such a curve surrounds a line vortex, on the other hand, the magnitude of the circulation around it will necessarily be the same as that of the vortex itself. For the case of plane motion, the flow pattern that satisfies the requirement of constant circulation evidently consists of a series of concentric streamlines, from one to another of which the velocity varies inversely with the radius:

$$v = \frac{\Gamma}{2\pi r} \tag{4}$$

For the steady, irrotational flow of an inviscid, weightless, incompressible fluid, the Bernoulli theorem [1] states that the sum B of what are sometimes called dynamic and static pressure

$$B = \frac{\rho v^2}{2} + p \tag{5}$$

has a constant magnitude. The static or actual pressure p would hence vary through an irrotational vortex from a maximum equal to its magnitude $p\infty$ at infinite radius to a minimum of negative infinity at zero radius:

$$p_\infty - p = \frac{\rho v^2}{2} = \frac{\rho \Gamma^2}{8\pi^2 r^2} \tag{6}$$

Since the quantity $\rho v^2/2$ represents the kinetic energy per unit volume (ρ being the mass density of the fluid), its integral over the field of motion should yield the total kinetic energy (per unit distance normal to the plane of motion) of the irrotational vortex:

$$E = \pi\rho \int_0^\infty v^2 r\, dr = \frac{\rho \Gamma^2}{4\pi} \ln r \Big|_0^\infty = \infty \tag{7}$$

The infinite result will be seen to stem from both limits of integration.

Rankine's Combined Vortex

Conditions at zero radius (if not at infinite radius as well) are obviously an extreme that must be eliminated from consideration if the flow pattern as a whole is to have physical significance. This was done by Rankine [4] a century ago by considering, instead of the line vortex, a filament of finite radius a and constant vorticity ω, surrounded by an irrotational field of the same circulation $\Gamma = \pi\omega a^2$ as the periphery of the filament. Within the filament the velocity obviously varies directly with the radius—i.e. $v = r/2$. The pressure is hence found from integration of the equation of centripetal acceleration

$$\frac{\partial p}{\partial r} = \frac{\rho v^2}{r} \tag{8}$$

to rise above its value at the axis in direct proportion to the square of the velocity:

$$p - p_{min} = \frac{\rho v^2}{2} = \frac{\rho \omega^2 r^2}{8} \tag{9}$$

The kinetic energy of such a filament is the finite value

$$E = \pi\rho \int_0^a v^2 r \, dr = \frac{\pi\rho\omega^2 a^2}{16} \tag{10}$$

With the velocity distribution through the combined vortex as shown in Figure 82, the pressure evidently decreases according to a second-degree hyperbola from its value at infinity to that at the periphery of the filament, and then parabolically to a minimum at the axis; exactly half the total variation will be found to occur in each zone, its magnitude depending upon the interrelated characteristics of the two zones together. As a result of the assumed identity of velocity (and pressure) at the border between the zones, however, any two characteristics of either zone are sufficient to define all characteristics of the combination—for example, the vorticity and diameter of the filament, or the diameter and limiting circulation.

A very graphic depiction of the Rankine vortex is obtained if the fluid is not an essentially weightless gas but a liquid with a free surface. The Bernoulli sum B must accordingly be increased by a term representing the product of the elevation z and the weight density γ. If each term is then divided by γ, all will acquire the dimension of length or so-called head,[5]

$$H = \frac{v^2}{2g} + \frac{p}{\gamma} + z \tag{11}$$

the sum of which is constant in the irrotational zone. Since along the free surface the pressure can be considered zero, the surface elevation will be found to vary in the same man-

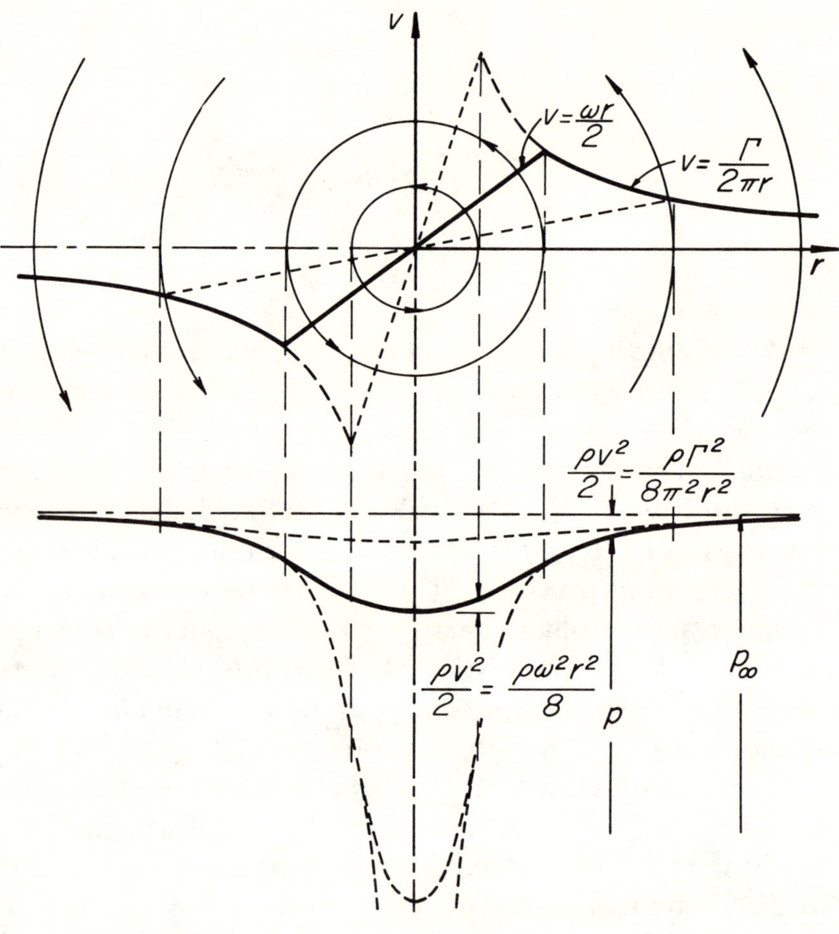

Fig. 82. Characteristics of the Rankine combined vortex
in plan and profile.

ner as did the pressure for the case of a gas, and the distribution curve of Figure 82 will represent the surface profile.
The depth of the depression at the axis evidently depends
upon any two characteristics of the combined flow. Thus,
for a given circulation—and hence a given profile of the
irrotational flow—the depression will increase in depth as

the diameter of the filament decreases, the vorticity increasing accordingly.

The kinetic energy of the Rankine combined vortex, per unit length of filament, has the form

$$E = \frac{\rho\Gamma^2}{16\pi} + \frac{\rho\Gamma^2}{4\pi} \ln r \Big|_a^\infty \tag{12}$$

for a filament of circulation Γ at radius a. Two significant facts should be noted about the several parts of this equation. Evidently, the energy of the filament (the left-hand term) is independent of its size once the circulation is prescribed. The filament size thus affects the total only as the lower limit of the irrotational field (the right-hand term), and this becomes of little importance in comparison with the infinite remainder, for the major part of the energy lies in the continuous succession of annular elements, since the incremental contribution of this type of distribution approaches zero only as the radius becomes very great—obviously a rather unrealistic situation.

Cyclonic Flow Toward a Sink

Let us now consider some actual approximations to the Rankine idealization. One is found in the hurricane, cyclone, or tornado. Such motion, however, displays a radial as well as a tangential velocity component. Indeed, it is the radial inflow toward the center of the storm which begets the circulation through Coriolis effects; that is, by virtue of the earth's rotation, air moving toward either pole tends to develop an easterly velocity component and air moving toward the equator a westerly component; a counterclockwise circulation is hence induced in the northern hemisphere and a clockwise circulation in the southern as the fluid approaches the storm center. So far as the outer regions of flow are concerned, the composite spiral motion can be simulated analytically [6] by superposing a line sink (the velocity of flow toward which also varies inversely with the radius) upon the axis of the irrotational vortex. However, the initial motivation for the cyclonic development results from the upward displacement of the central fluid through thermal convection, and near the axis this vertical component of velocity is no longer negligible in comparison with the other two. Although a point sink rather than a line sink would permit a somewhat better approximation to the irrotational aspects of this flow, the very central zone is decidedly rotational. As a vortex filament it is far from the Rankine ideal, moreover, because of secondary currents that result from the viscous drag of the passing quasi-irrotational flow.

Even the liquid simulation of cyclonic motion by the water above a drain, which exhibits in inverted form the most basic of the storm characteristics, is in itself far more complex than it would seem from the familiar photographs

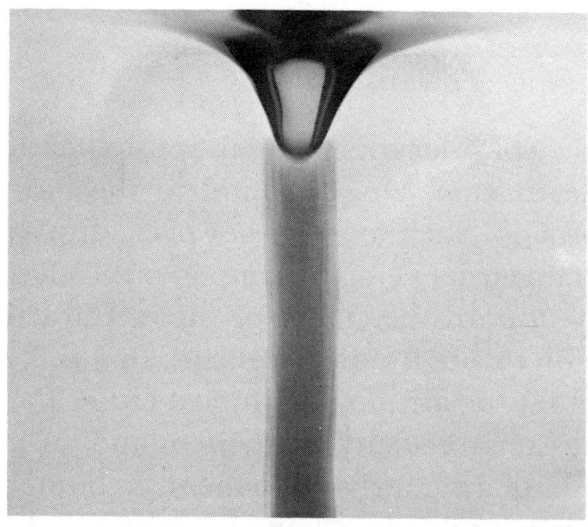

Fig. 83. Silhouette of water surface above a drain, with vortex filament made visible by dye.

of its profile, such as that in Figure 83. By no means restricted to a household scale, vortices large enough to engulf small motorboats have been formed in reservoirs at hydroelectric intake towers, and the Maelstrom remains as strong as ever in fable if not in fact. (At this point customary inquiries might well be anticipated by the comment that the rotation of the earth has an almost negligibly small influence upon the direction of rotation of waterspouts, dust devils, or the vortex at even a sizable drain, and that only if the many extraneous disturbances normally in evidence are kept sufficiently small will the Coriolis effect actually prevail.)

From these brief comments it is apparent that cyclonic motion resembles the conditions of the Rankine idealization most closely rather far from the central core, where present interest in it is minimal, and that the zone which is of the greatest import eddywise is that which departs most markedly from Rankine's simplification because of neglected viscous influences. The kinetic energy of such a system is limited only by the size of the field of motion. It results from work done by either unbalanced pressure or gravity, for the sink is invariably a region of reduced pressure or elevation.

Vortex Generation and Dissipation by Viscous Shear

Almost diametrically opposite conditions result from the artifice of rotating a cylinder in a viscous fluid; although this departs from the Rankine-Helmholtz idealization by the introduction of shear, and though there are various obstacles (for example, limitations of hydrodynamic stability [6]) to the experimental realization of the effect, pursuing the analysis mathematically again yields indications that are of significance in real flow.

Rotation of a cylinder at constant speed in a fluid of viscosity μ (assuming the resulting flow to remain stable despite the pronounced centrifugal effects) will cause the circulation at any radial distance to increase continuously through transmission of shear according to the Newtonian equation [6]

$$\tau \;=\; \mu \left(\frac{\partial v}{\partial r} - \frac{v}{r} \right) \tag{13}$$

The ultimate limit of such development, paradoxically, is the irrotational case of constant circulation—reached after an infinite time and an infinite amount of mechanical work. Such work, it should be noted, is involved not only in increasing the kinetic energy of the fluid but also in converting mechanical energy into heat through viscous shear. The vortex filament is only crudely simulated by the generating cylinder, however, and a more significant result is obtained by the following sequence of assumptions.[7] First, the generator is imagined to approach the zero radius of a line vortex, simultaneously increasing in rotational speed to maintain a constant peripheral circulation Γ_g; second, it is allowed to accelerate the surrounding fluid by the transmission of shear over a finite generation time t_g instead of an infinite time, the circulation hence still diminishing radially rather than attaining a constant magnitude; third, the generator is abruptly brought to rest, and the surrounding fluid is then permitted in turn to be decelerated by the resulting negative shear. At any time t_d during the latter phase the velocity will be distributed according to the twofold effect embodied in the expression

$$v \;=\; \frac{\Gamma_g}{2\pi r} \left(\exp \frac{-r^2}{4v\,(t_g + t_d)} - \exp \frac{-r^2}{4v t_d} \right) \tag{14}$$

(v being the kinematic viscosity μ/ρ), typical plots of which for different values of t_d/t_g are shown in Figure 84. If any one of such curves is assumed to represent the instantaneous

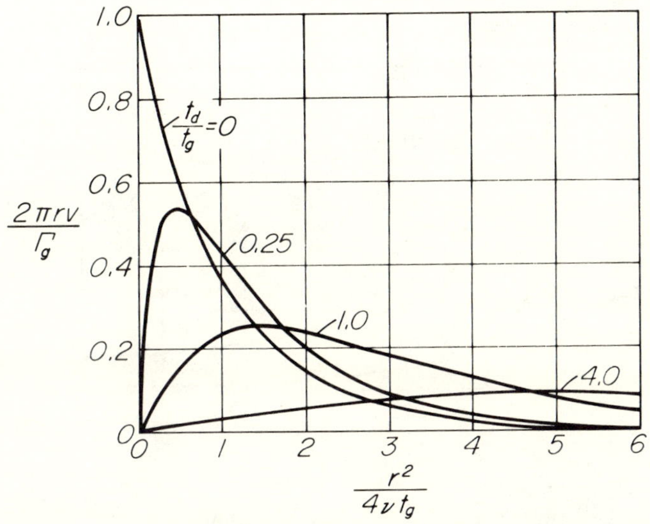

FIG. 84. Variation of circulation pattern with time through viscous shear.

velocity distribution through an actual vortex, subsequent curves will indicate its modification with time through viscous shear. Designation of the nominal filament border as the radius of maximum circulation will then permit the trend of filament size, minimum pressure, and kinetic energy to be evaluated, as plotted nondimensionally in Figure 85. The mathematical model will, in other words, permit a very significant analysis to be made of the gradual viscous decay of a simple vortex once it has been formed.

Impulsive Generation and Interaction of Vortex Pairs

There is, unfortunately, a considerable experimental limitation concealed in the words "formation of a simple vor-

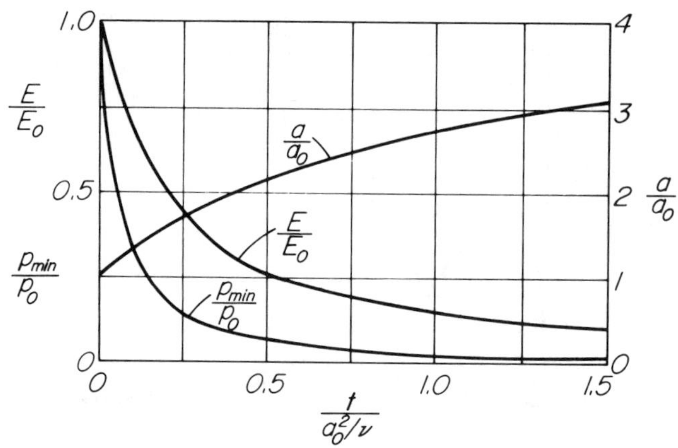

Fig. 85. Temporal change in core size, differential pressure, and kinetic energy.

tex," and this stems directly from the third of Helmholtz's theorems, which indicates that a vortex can be produced only through viscous shear. Now it is true that separation from a curved boundary and the formation of a vortex in the separation zone can be traced to viscous action in the boundary layer; nevertheless, as the radius of boundary curvature is decreased toward zero, the absolute role played by viscosity must approach the same limit. Consider, for example, motion at the edge of a thin plate that is accelerated from rest at a constant rate. At the initial instant of movement, the flow tends to be symmetrical about the plane of the plate. If the edge is sufficiently sharp, however, the very low pressure required there to maintain symmetry would entail either cavitation, if the fluid were a liquid, or severe rarefaction, if a gas, and separation hence begins almost at once in the form of a tiny vortex just behind the edge. As the velocity of the plate increases, its distance from the vortex center also increases, as does the circulation of the surrounding field. In fact, so long as the displacement of the plate is still small compared with its width, both the form of the streamlines (Fig. 86) and the velocity distribution of the flow appear to be in close accord with the Rankine model.

While one might quibble about the part placed by viscosity in causing the separation—arguing that its relative (as

Fig. 86. Impulsive generation of vortex by acceleration of plate.

opposed to its absolute) importance remains the same as the radius of boundary curvature is reduced—its action thereafter is clearly one of resistance rather than motivation, and the circulation of the nearly irrotational field around the core continues to grow by virtue of the difference in pressure between the fore and after sides of the plate. Even the initial velocity distribution within the vortex core is an inertial rather than a viscous effect, for its diameter and vorticity vary directly with time, and the velocity differential across the surface of separation continues to have the same small magnitude relative to the increasing velocity of the plate.

If at any time the plate is carefully withdrawn, the streamlines of the vortex will close and the motion will tend to follow that described in the foregoing section on viscous action—to the extent that the third Helmholtz theorem will permit. In other words, if the cumulative action of viscous shear is restricted to the central filament, the circulation around a curve enclosing the paddle must tend to retain its original magnitude of zero. This is possible only if another vortex of equal and opposite strength is formed at the other paddle edge—which is just what invariably happens. (In fact, if a plate is brought into motion and then brought again to rest rather than withdrawn, a pair of vortices will be formed at each edge.) It might be noted, on the other hand, that dexterous handling of the plate before it is withdrawn will permit one filament to be made with small diameter and high vorticity and the other with large diameter and low vorticity; the relative proportions of the two are probably unlimited in range.

Because emphasis has been laid upon the infinite energy involved in the single Rankine vortex, note should be made of the reason for the limited energy expenditure that appears to be involved in the impulsive generation of a vortex pair even in an infinite expanse of fluid. If one pictures the veloc-

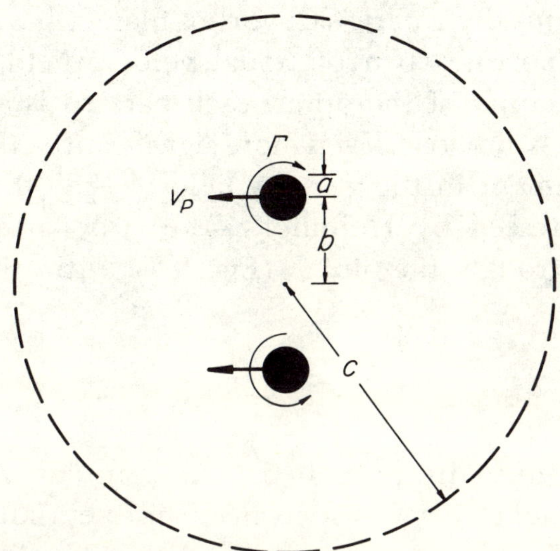

FIG. 87. Definition sketch for self-propulsion of vortex pair.

ity distribution of the two superposed irrotational fields (Fig. 87), one can readily see that, as the radial distance relative to the spacing of the two filaments increases, the induced velocities tend more and more nearly to cancel one another. The approximate relationship for the kinetic energy per unit length—compare with equation (12)—

$$E = \frac{\rho \Gamma^2}{2\pi} \left[\frac{1}{4} + \ln \left(\sqrt{1 + \frac{b}{a}} + \frac{b}{a} \right) - \frac{b^2}{c^2} \right] \qquad (15)$$

becomes more and more accurate as the ratio of filament size to spacing, a/b, and the ratio of spacing to expanse of fluid, b/c, become smaller. Obviously, the energy is not only finite —contrary to that of the single vortex—but is contributed to most heavily by the central zone.

We would do well at this point to observe the further behavior of the two (or more) vortices that are invariably produced with a paddle. Each has been seen to take on the more essential characteristics of the Rankine combination: the central, primarily rotational, vortex filament; and the surrounding, more nearly irrotational, velocity field. Since each is in the vicinity of the other, each pattern is of necessity somewhat asymmetric. Even more significantly, the presence of each filament in the velocity field of the other requires that—as foreseen by Helmholtz—the two vortex patterns advance through the fluid (Fig. 87) with the velocity

$$v = \frac{\Gamma}{4\pi b} \tag{16}$$

as though mutually propelled. (An annular vortex—the common smoke ring—propels itself in essentially the same manner.) If the vortex pair (or the vortex ring) approaches a solid wall in its motion, the parts will then separate (or the ring will increase in diameter) as though propelled by their mirror images. Because, according to one of Helmholtz's theorems, the fluid within a vortex filament cannot change, it is evident that such motion involves a forward displacement of a portion of the fluid as well as the velocity field.

While such phenomena provide amusing material for experimentation over a cup of coffee, their bearing upon the present discussion is much farther reaching. Picture, for example, the interaction of two vortices which are not mutually parallel (or the elements of a single one that is not straight [8]). The velocity field surrounding each one will necessarily affect every part of the other in a different manner (Fig. 88), with the result that each individual filament will eventually be bowed and stretched beyond recognition. (It is inconceivable that the reverse type of evolution—i.e. in the direction of symmetry or simplicity—could ever occur on more than a very local scale.) During the stretching, since the filament invariably consists of the same fluid, it will

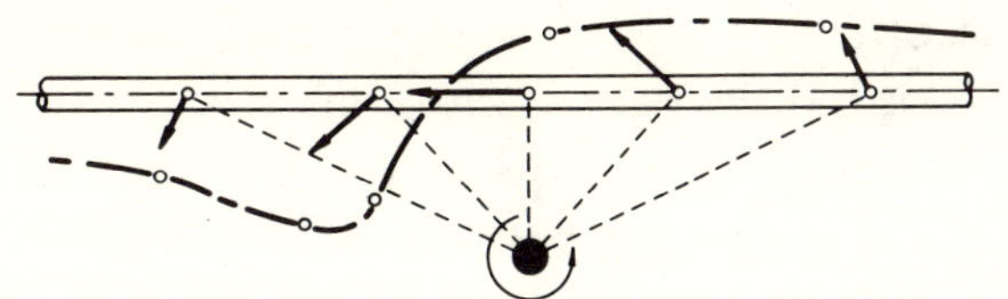

FIG. 88. Interaction of nonparallel vortex filaments.

necessarily become thinner, and—as prescribed by another of Helmholtz's theorems—its rotational speed will increase accordingly. One need only recall that the motion within and around each filament is further complicated by viscous shear to secure a preliminary glimpse of the extreme complexity involved in a moving fluid containing a series of vortices, whatever their origin. Little wonder that the rather ideal term of "vortex" soon becomes replaced by the less pretentious word "eddy"!

The Kármán Vortex Trail

As a plate or any other bluff cylindrical body is brought from rest to a constant speed at right angles to its axis, the vortex pair formed through separation at either side will first grow symmetrically at an ever-reducing rate. Eventually one vortex will tend to overbalance the other and pass off into the wake. As the second continues to become larger, a third will form in the place of the first, and a process of alternate formation and detachment is thus ultimately established. Because of the periodicity of the occcurrence, each growing vortex displays much the same nature as those formed impulsively by a constant acceleration of the body. Now, however, the flow around the body itself is involved in the circulation balance—to the extent of just half that of the detaching vortex and in the opposite sense.[9] Such oscillation of the flow as a whole gives rise to the fluctuating lateral thrust that causes cylindrical bodies (telephone wires, periscopes, struts) to vibrate in cross flows. As was shown by Strouhal,[10] the phenomenon is an inertial rather than a viscous one. The frequency of oscillation relative to the size and speed of the body (i.e. the Strouhal number fD/V) thus varies with geometry and elastic characteristics of the body rather than with the relative magnitude of the inertial and viscous characteristics of the flow (represented by the Reynolds number VD/μ.[5] For example, the Strouhal number for a rigid cylinder maintains a magnitude of about 0.2 over a very wide Reynolds-number range.

Kármán showed a half century ago [11] that a trail of ir-rotational vortices will move stably through a fluid (Fig. 89) only if the ratio of lateral to longitudinal spacing is 0.28, in which circumstances the self-induced group velocity will be

$$v_G = \frac{\cosh^{-1}\sqrt{2}}{4\pi\sqrt{2}}\frac{\Gamma}{b} \tag{17}$$

i.e. $0.05\Gamma/b$, as compared with $0.08\Gamma/b$ for the simple vortex pair. Because of Kármán's analytical contributions to the phenomenon, both the mathematical model and the phenomenon itself commonly bear his name. However, several characteristics of the actual flow [12] which could not be foreseen from the irrotationality of the mathematical model warrant at least mention, if not emphasis, in the present dis-

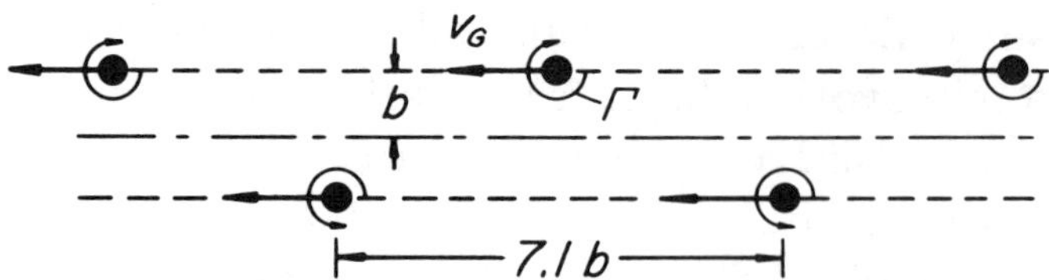

FIG. 89. Characteristics of the Kármán vortex trail.

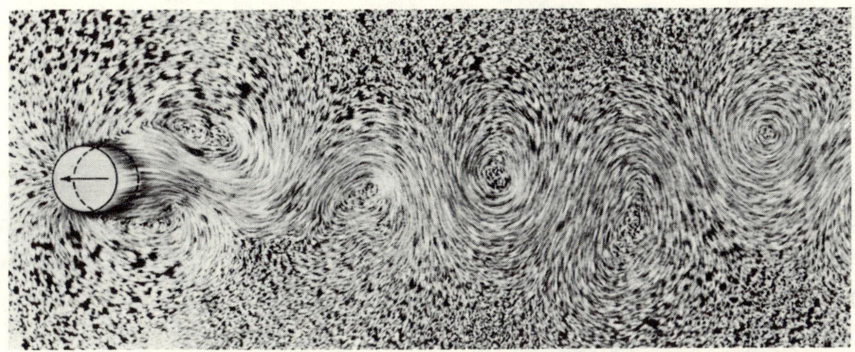

FIG. 90. Formation of eddies in the wake of a cylinder.

cussion. First, only over an intermediate range of the Reynolds number does anything like an orderly vortex trail appear in the wake of a cylindrical body. At very low Reynolds numbers the viscous action is so great in comparison with the inertia of the flow that separation does not even occur. At higher Reynolds numbers (of the order of 50) the vortex pair forms but remains symmetrical, and only at still higher values do asymmetry and alternate detachment occur. In their early stages of formation (Fig. 90), vortex trails display the geometric characteristics of the stable irrotational pattern analyzed by Kármán. As they progress down-

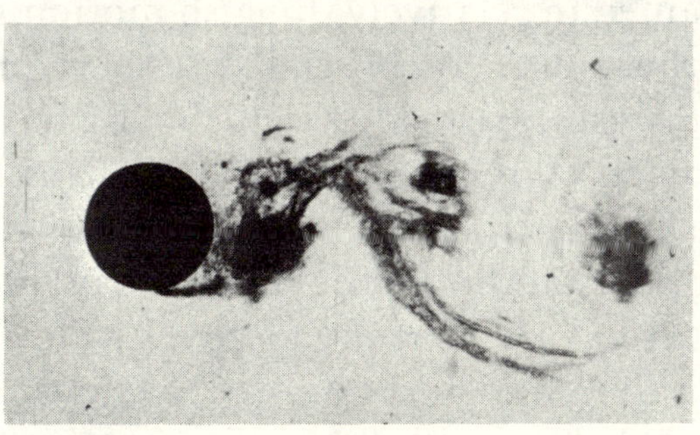

FIG. 91. Eddy trail made visible by cavitation.

stream, however, the individual vortices are subject to two distinct effects. On the one hand, the filaments increase in size and the circulation diminishes through viscous action. On the other hand, the slight disturbances invariably present in any flow gradually produce an additional departure from the original geometry of the pattern (Fig. 91). Since the Reynolds number indicates the relative importance of the inertial and viscous effects, one might expect the stable portion of the trail to become shorter as the Reynolds number increases. As a matter of fact, at a magnitude of 2,500 a trail as such can no longer be detected. Although the generating body continues to undergo an alternating side thrust to Reynolds numbers as high as 10^6, each vortex becomes progressively more unstable during its formation, and there is left in the wake of the body only a heterogeneous series of eddies which act upon each other as they decay—in general accordance with the foregoing simple principles, to be sure, but in a manner so complex as to defy detailed study.

Quasi-Stable Eddies

Between the impulsive generation of the first vortex pair and the alternate formation and detachment of vortices in the Kármán trail, there is a transition period in wake development in which a rather symmetrical zone of reverse flow is maintained in relatively sluggish motion by the fluid passing on either side. Closely related to such zones in their basic characteristics are eddies which—by virtue of the boundary geometry—remain permanently fixed in axial position and are kept in circulatory motion by the neighboring flow. The familiar standing eddies in riverbank indentations are of this nature, as are the more violent sort formed in gate slots, behind valves, and at conduit expansions. In fact, any separation eddy behind a bluff body in contact with a longitudinal boundary is of this nature, and one (or, rather, two) can be produced instead of the vortex trail simply by introducing a longitudinal fin behind the generating cylinder to prevent communication between the two alternate zones of separation. If the cylinder considered is a thin plate, so that the initial separation will be wholly a geometric—i.e. inertial—rather than partly a viscous matter, either half of the resulting pattern will, except for the absent surface drag upstream, be similar to that of flow over a wall in contact with the ground. (There is, of course, an axisymmetric—if not even three-dimensional—counterpart to every two-dimensional phenomenon discussed except lateral oscillation of the wake.)

The flow pattern behind such a wall, like that of any quasi-stable eddy, possesses a series of significant characteristics which can best be discussed under three different categorie. The first is the geometry of the mean streamlines (Fig. 92). The line of separation will divide the passing flow

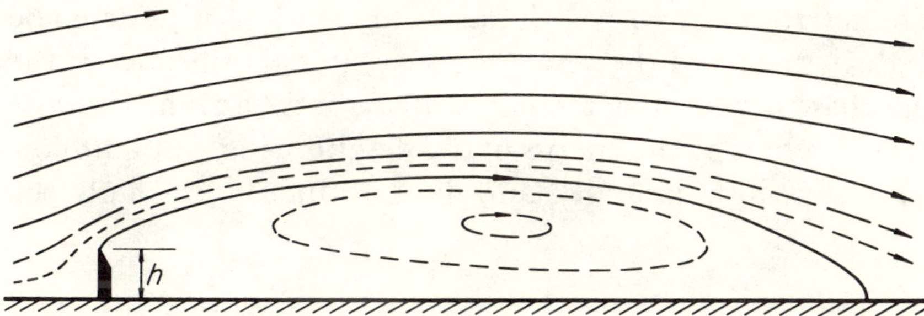

FIG. 92. Pattern of mean streamlines in the lee of a wall.

of the main stream from the circulatory flow of the eddy, much as in the case of a line vortex or any other singularity superposed mathematically upon an irrotational flow in a manner also devised by Rankine.[13] Now, however, the circulatory flow is rotational and hence cannot be treated by the classical velocity-potential methods of hydrodynamics. Nevertheless, some progress has been made in evaluating the eddy profile by the relaxation method on the assumption of constant vorticity throughout the eddy zone.[14] This is somewhat akin to the synthesis of flow patterns around the so-called Rankine bodies, for the ambient fluid is in effect assumed to follow the irrotational pattern around a body having the same configuration as the enclosed eddy.

The second approximation to reality will result from an analysis like that of the boundary layer introduced by Prandtl. Therein the zone of deceleration due to viscous shear between a body and the ambient stream displaces the stream outward much as would an increase in the thickness of the body profile. There is, in fact, an expanding zone of pronounced shear between a standing eddy and the passing flow (Fig. 93). A better approximation of the distribution of vorticity would hence be a constant magnitude through the eddy proper, and, in the intermediate zone, a variable magnitude that decreased toward the constant limit of the eddy in inverse proportion to the width of the shear zone. The surrounding irrotational field would, of course, be slightly displaced thereby. Of far greater significance is the fact that there is now a fairly realistic mechanism—intense shear—whereby the energy of the neighboring ambient flow can vary both laterally and longitudinally through the

FIG. 93. Shear zone delineated by diffusion of dye.

processes of transfer and dissipation. The energy of the eddy, in other words, stems from the work done upon it by the ambient stream—in the course of which energy is steadily lost to the main flow, and to the eddy as well by viscous dissipation.

Such loss of energy can be a purely viscous effect, if the flow is wholly laminar. However, flow of this nature becomes unstable to small disturbances at relatively low Reynolds number and breaks down into a succession of expanding vortices (Fig. 94), the generation of which then gradually diminishes in regularity in all but a statistical sense as the Reynolds number continues to increase. As in the case of flow that is free to oscillate about a cylindrical body, the frequency of vortex formation along a surface of separation is governed by the geometry of the boundary and the velocity of the flow, unless extraneous effects like elastic vibration of the boundary control. Thus, the Strouhal number for the dominant pressure fluctuation a short distance behind a wall of height h is about 0.1.[15] Unless the oncoming flow is unusually disturbance-free, however, there will be a considerable variation in frequency about its mean value. Even by moderately large Reynolds numbers, moreover, the shear zone will have broken down into an almost

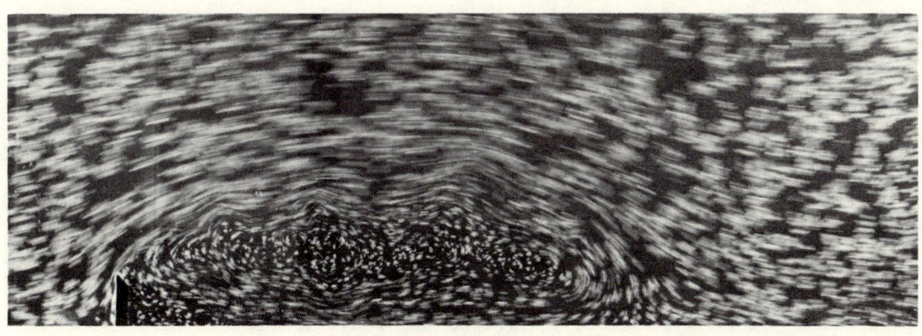

FIG. 94. Generation of eddies in zone of instability.

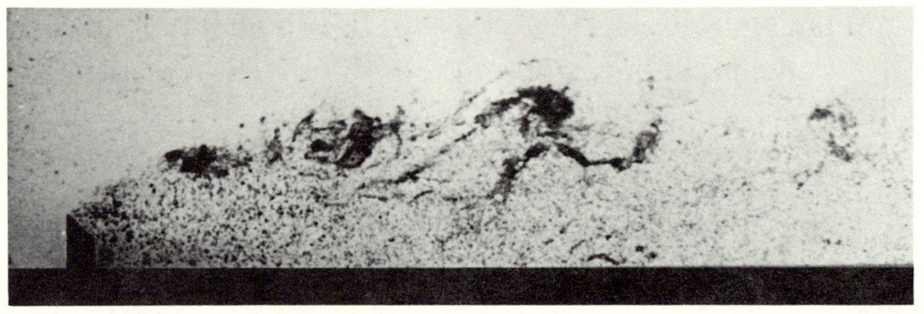

Fig. 95. Successive eddies made visible by cavitation.

haphazard mixture of eddies of changing size and orientation, each of which influences and is influenced by its neighbors as it moves along its course. Some further idea of the irregularity in eddy size, form, intensity, and spacing can be gained from Figure 95, in which the low-pressure cores are made visible by vapor bubbles formed under conditions of cavitation.

Vortex Generation in Zones of High Velocity Gradient

Because the "rolling-up" of a shear zone is encountered in many more instances than that just described, further attention to the mechanism involved is now in order. Since the time of Helmholtz it has been understood that a vortex sheet is inherently unstable and will degenerate into a series of vorticity concentrations if the slightest disturbance is present. Indeed, such a condition is commonly known as Helmholtz instability. Early in the present century Prandtl [16] gave for the occurrence the explanation illustrated in Figure 96. Any perturbation of the sheet, such as the sine wave indicated in the second sketch, will give rise to a force field which tends to amplify the perturbation;

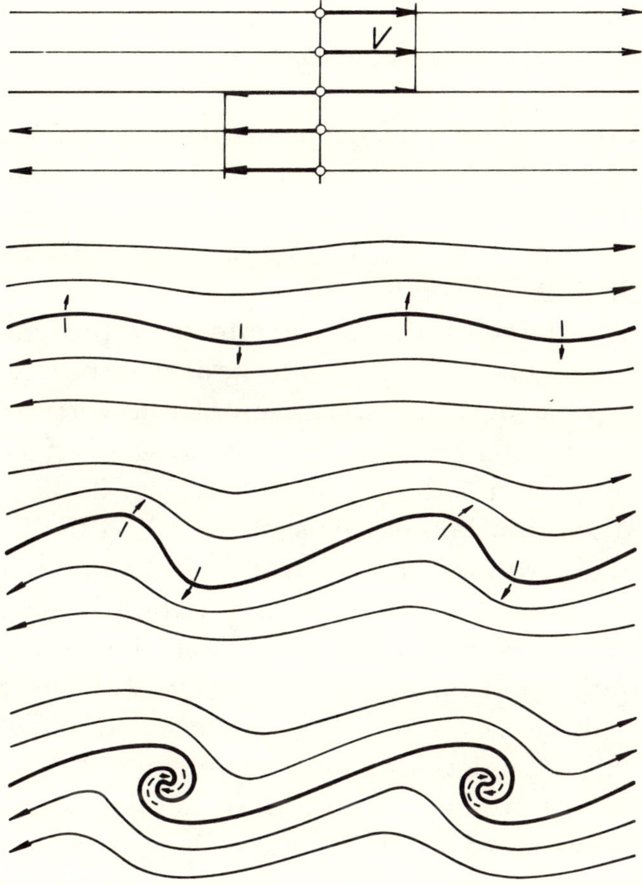

FIG. 96. Concentration of vorticity along a vortex sheet,
after Prandtl.

that is, convergence of the streamlines will produce, according to the Bernoulli theorem, a rise in velocity and a decrease in pressure, and vice versa, a pressure drop thus developing in the direction of the displacement. At the same time that the wave is amplified, it will become asymmetrical and—at least so far as the actual counterpart of the phenomenon is concerned—eventually form a series of vortices as shown.

Prandtl's simple explanation was used as the basis of a numerical solution by Rosenhead,[17] who approximated the continuous distribution of vorticity by a succession of line vortices, and his findings did, in fact, seem to verify the original hypothesis. Some question has since been raised [18] regarding both the validity of the numerical procedure and the correctness of the general hypothesis that the strength of either a vortex sheet or a succession of line vortices would tend to become concentrated in a regular series of localities. Such an objection is, however, of little significance to the present discussion because it is restricted to the idealized condition of zero viscosity, whereas the actual phenomenon must be considered to involve a combination of inertial and viscous effects, each tending to oppose the other. If the viscous qualities of the flow are relatively great, in fact, any initial disturbances of the flow pattern will be damped before eddies can develop. The greater the relative magnitude of the inertial qualities, on the other hand, the greater the tendency toward instability and eddy formation; the role of viscosity then becomes primarily that of vorticity diffusion and energy dissipation in and around the eddy cores.

The independent variables involved in such phenomena of vortex generation are evidently the boundary geometry, the relative velocity (or velocity gradient), the density, the viscosity, and the nature of the disturbance. These can be grouped into two nondimensional parametric ratios, one of the Strouhal type, characterizing the relative aspects of the disturbance, and one of the Reynolds type, characterizing the relative magnitudes of the inertial and viscous effects. From the foregoing discussion, it would appear that each of these parameters must lie within a certain numerical range if vortices are to be generated by the process now under consideration. During the impulsive generation of a separation vortex, for example, there is little tendency for the surface of separation to degenerate into a train of smaller vortices, because the continued acceleration of the motion minimizes the velocity differential across the separation zone. The differential is marked, however, in the case of a quasi-stable eddy, the separation surface of which rolls up into a vortex train—provided that the Reynolds number is sufficiently great and that the proper initial disturbance is present. Like the Kármán trail, once it is formed the vortex train along a separation surface will display a definite mean frequency for a given geometry and relative velocity (much as does a waving flag), unless there exists an extraneous periodic disturbance of sufficient strength to exert control.

The extreme case of this form of vortex generation, necessarily dependent upon the same group of variables and therefore of rather general significance, is represented by flow along a plane boundary of great extent. As has been shown both analytically [19] and experimentally [20] for flow in a boundary layer of thickness δ, below a certain critical value of the Reynolds-number type of parameter any disturbance will be damped. Above this critical value, moreover, disturbances of only an intermediate frequency range will cause a rolling up of the shear zone, those of higher or lower frequency still being damped. Such conditions are illustrated schematically in Figure 97, wherein the respective roles of inertial effects, viscous effects, and the nature of the disturbance are well displayed.

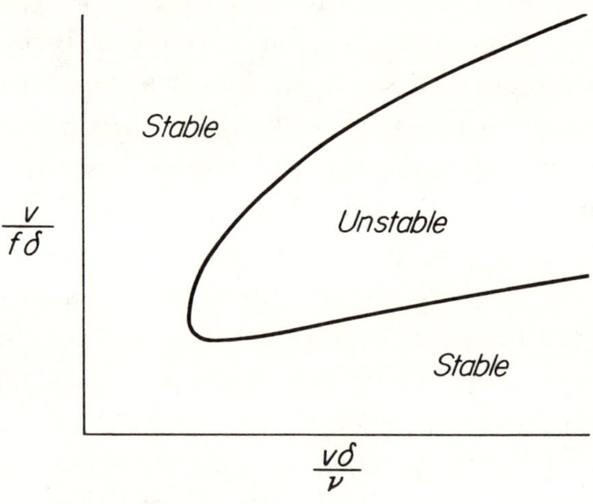

Fig. 97. Susceptibility of boundary-layer flow to
disturbances of various frequencies.

The Eddy Structure of Turbulence

In the wake of a source of eddy motion, one should expect
to observe over an interval of time the passage of eddies
having a considerable variety of sizes, orientations, and
strengths. Each of these characteristics should be statistically
related to the geometry and velocity of the generating mean
flow and to the relative distance between the source and the
point of observation. The direction of rotation of individual eddies, for example, would be predominantly in
accord with the vorticity of the eddies initially formed. The
maximum eddy size would probably be of the same order
as a characteristic dimension of the generating boundary.
The average eddy size—unless the Reynolds number is not
sufficiently large—would doubtless also be governed by the
geometry of the boundary and the point of observation. In
fact, if one were to plot against some measure of the eddy
size either the relative number of eddies, eddy volume, or

kinetic energy per division of the size scale, there should result a distribution curve such as that shown schematically in Figure 98. Because of its likeness to the distribution of light energy as a function of wavelength, a curve of turbulence energy as a function of wave number (which is inversely proportional to the wavelength or eddy scale) is, after Taylor,[21] called the spectrum of turbulence. Eddy scale is used here rather than wave number, because of its pertinence to the present discussion.

The spectral distribution of eddy scale in the wake of such a highly localized source of turbulence as a cylinder or wall must be expected to change considerably in form with both lateral and longitudinal position of the point of observation. This is due in part to the inherent instability of the flow and in part to the essentially nonviscous interaction of the eddies—particularly the large ones—with others in their immediate vicinity. Such interaction has been seen to produce an effective lengthening of the eddy filaments, which has the double result of expanding the eddy zone as a whole and reducing the cross-sectional area of each filament. The expansion of the eddy zone is one aspect of the phenomenon known as turbulent diffusion or mixing. The two effects together yield in turn two related trends: the mark

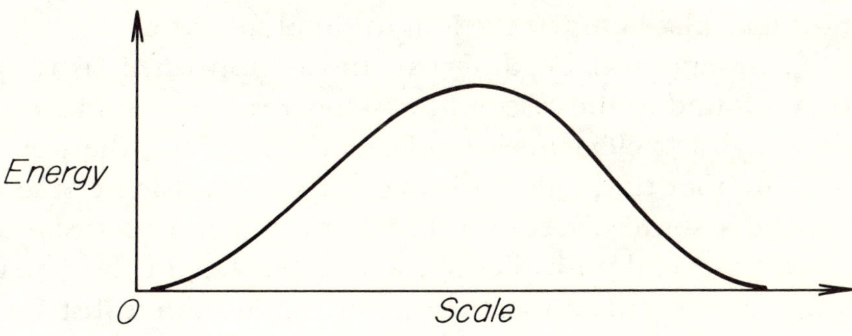

FIG. 98. Schematic representation of turbulence spectrum.

left by the generating body upon the eddies, as evidenced by alignment or sense of rotation, grows steadily less; and their vorticity—and hence the intensity of viscous shear within them—increases, so that their energy is dissipated ever more rapidly. There is, therefore, a continuous flux of kinetic energy from the large-scale toward the increasingly isotropic small-scale portion of the spectrum.[22] Tending to offset this process, on the other hand, is the fact that the rate of dissipation varies with an inverse power of the eddy scale that is well above unity. As a result, downstream from a localized source, the finer-scale eddies tend to decay more rapidly than they are produced, and the average scale of the residual eddies hence gradually increases; the energy which the latter represent, of course, rapidly diminishes.

If, instead of an isolated source of eddies, one considers a regular distribution of sources over a plane normal to the flow (for example, a grid or screen), it is apparent that the individual eddy trails should remain distinct from one another only over a distance equal to several times the spacing of the generating elements. Thereafter, conditions should vary statistically only in the flow direction, so that in any plane at right angles to the flow the statistical characteristics of the turbulence remain both homogeneous (independent of position) and isotropic (independent of direction).[23] The spectral variation of the eddy scale just described should now take place only in the longitudinal direction.

The longitudinal counterpart of the transverse distribution is found in the case of flow along a rough boundary. If the roughness elements are sufficiently large and the Reynolds number of the flow is sufficiently great, each element will be a source of eddies, but beyond a lateral distance equal to several multiples of the element height the generating surface will appear to be a continuous source distribution. A finite length of surface will necessarily produce a nonuniform state of flow (known as a boundary layer), the turbulence characteristics of which vary both longitudinally and laterally. If, however, the boundary is long in comparison with other dimensions—for example, the wall of a pipe or channel—then the flow will vary statistically only at right angles to the bounding surface. Though it is thus homogeneous in planes parallel to the surface, it is now definitely not isotropic,[24] for the eddies lose their boundary bias only as the zone of symmetry is reached. Evidently, since the region of fully developed turbulence does not include the zone of turbulence generation, it matters little how the eddies are originally produced. The bounding surface, in other words, could just as well be smooth and the eddies generated by the rolling up of the vortex layers in the region of maximum susceptibility to the disturbances that are invariably present in actual states of flow. Whether produced by smooth-wall or rough-wall action, the resulting state of statistically steady, uniform flow will now be used as a simple frame of reference in relating the primary turbulance parameters to the eddy characteristics already discussed.

Interpretation of Turbulence Parameters

If a turbulent flow is recognized as a system of heterogeneous eddies superposed upon a fluid stream, then it is obviously an extremely complex state of unsteady, nonuniform motion even though the main stream is invariable with time at each point and invariable with distance along each streamline. The determination of the characteristics of such motion, even statistically, is at best a difficult process, in which one or more of three essentially different procedures must be followed. Thus, an instantaneous picture of conditions in a representative longitudinal plane can be obtained by a photograph of highly illuminated particles suspended in the fluid, the length of exposure being such as to yield short streaks showing the speed and direction of the motion at each point, and the field being large enough to yield a sufficient number of comparable points for statistical analysis. Again, observations can be made as functions of time at various typical points over a given cross section. Finally, marked foreign particles can be introduced, or fluid particles can be tagged by dye or heat, and their courses followed statistically. For experimental reasons the second of these methods, called the Eulerian, is the most common and will serve as the basis of further discussion.

At a fixed point of observation in a statistically steady uniform stream one should expect to measure not only the constant characteristics of the primary flow but also the instantaneous departure therefrom due to the passage of parts of successive as well as superposed (i.e. small on large) eddies past the observation point. This should yield for each of the characteristics (restricted now to the three velocity components and pressure) a continuous record of values fluctuating about fixed means, as illustrated in Figure 99. The root-mean-square deviation for a particular direction, caused as it is by the motion within the eddies, must be indicative of the average tangential velocity in that direction

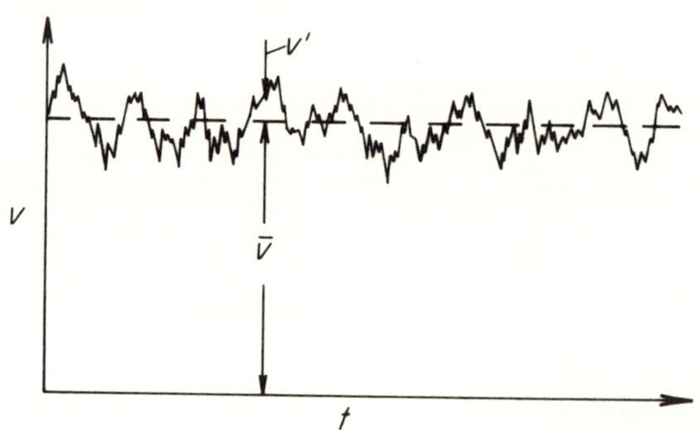

Fɪɢ. 99. Temporal record of velocity fluctuations.

—a factor of proportionality assuredly embodying such complexities as orientation and eccentricity with regard to the direction of mean motion and point of observation, not to mention the departure of each eddy from the symmetry of the vortex model originally discussed. Equidistant from parallel boundaries, or at the axis of a pipe, one would expect the eddies to lack preferred orientation, with the result that all three root-mean-square velocities should then approach the same value. Close to the boundary, however, eddies would have to be so oriented that the normal component of velocity would be very small compared to the lateral and the longitudinal. Whereas vortices with normal axes would permit the remaining components to have comparable magnitudes, the gradient of mean velocity would tend to give newly formed eddies lateral axes, or a longitudinal component greater than the lateral. From the measured values shown in Figure 100,[25] the longitudinal component is, in

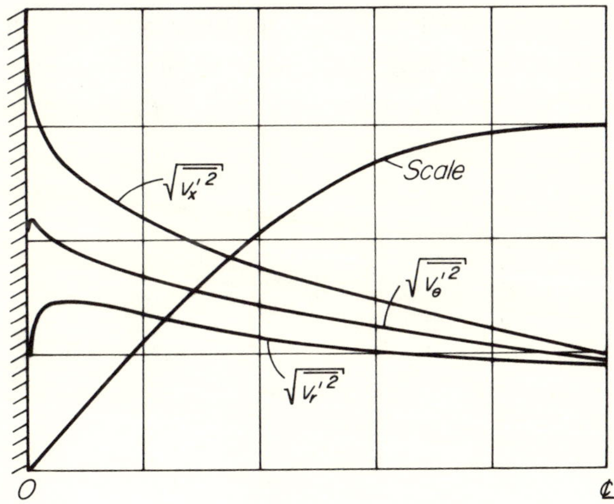

Fig. 100. Variation of eddy characteristics across
a pipe or channel.

fact, seen to remain larger than the lateral, and the lateral larger than the normal, the difference decreasing almost linearly as the central region is approached.

From the type of record shown in Figure 98 one might also expect to obtain some indication of the average eddy size. In fact, if one evaluate the correlation of values at two successive instants [21] in the form

$$R_t = \frac{\overline{(v_x')_t (v_x')_{t + \delta t}}}{(v_x')_t^2} \tag{18}$$

and if one plot the correlation magnitude as a function of the time interval δt, as shown in Figure 101, the interval at which the correlation approaches zero should be proportional to the time required for the average eddy to pass the point of observation. At points within the average eddy, in other words, there should be some degree of velocity dependence, whereas velocities at points in different eddies should be essentially unrelated. Multiplication of the time read on the abscissa scale by the mean velocity thus gives a length L to which the maximum eddy size is proportional; similar treatment of the area under the curve (also a time) would

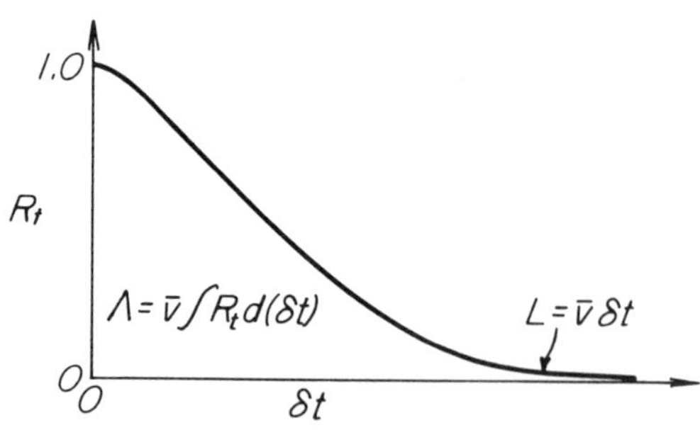

Fig. 101. Determination of mean and maximum eddy scales.

yield a length Λ more nearly indicative of the average eddy size.

A somewhat different type of correlation is that between two parallel components the distance δy apart in the normal direction:

$$R_y = \frac{\overline{(v_x')_y (v_x')_{y+\delta y}}}{(v_x')_y{}^2} \tag{19}$$

A similar plot of this quantity as a function of distance yields a curve like that in Figure 102, from which a length (or an area with the dimension of length) proportional to the maximum (or mean) normal dimension of the eddies can be found directly. The intersection of this curve with the abscissa axis and its final approach from the negative rather than the positive side are due to the fact that a negative correlation must exist between longitudinal velocities on opposite sides of an eddy. (The reverse situations should prevail if the two types of correlation are evaluated in terms of the normal or lateral rather than the longitudinal components.)

Since the correlation coefficient in either instance can approach unity only as $v\,\delta t$ or δy approaches the scale of the

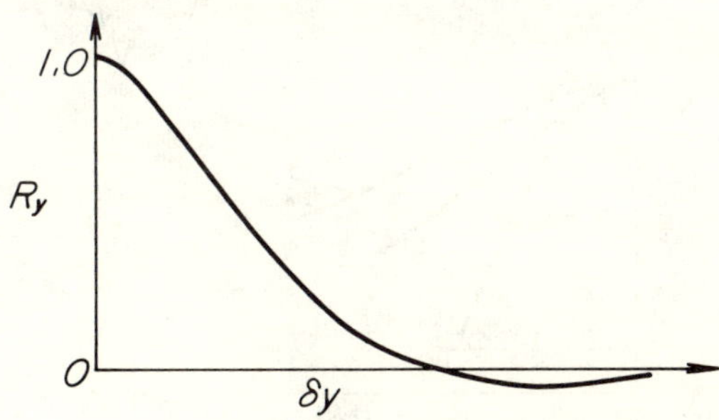

FIG. 102. Characteristics of lateral correlation function.

smallest eddies that are present in any appreciable proportion, the form of the respective correlation curve should
reflect the distribution of eddy sizes. The wider the range in
size, in other words, the sharper the peak of the curve relative to its width or area. Mathematical analysis of the correlation functions shows, in fact, that the initial radius of
curvature of the longitudinal function (or the intercept of
a parabola fitted to the peak of the lateral function) is related
to a length λ (Fig. 103) which characterizes the fine end of
the spectrum much as the rather arbitrary upper limit of the
curve characterizes the coarse end. If it is realized, moreover,
that the superposition of finer eddies upon coarser ones results in higher local velocity gradients in records like the
inset of Figure 103, it will be seen that the mean-square
value of the gradient will be affected most strongly by the
velocity variation within the smallest eddies. Since it is the
smallest eddies which make the greatest contribution to the
dissipation process, it is not surprising that, under conditions
of complete isotropy, toward which the process of eddy dissolution leads, the dissipation rate can be expressed in the
alternative forms:[5]

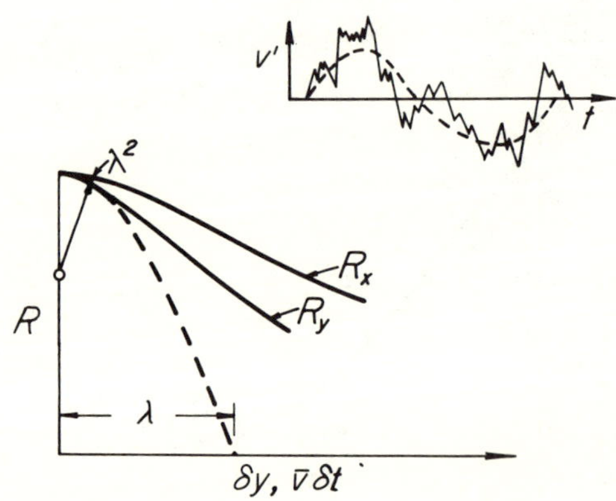

Fig. 103. Determination of the dissipation scale.

$$\epsilon = 15\,\mu\,\frac{\overline{v_x'^2}}{\lambda^2} = 15\,\frac{\mu}{\overline{v}^2}\,\overline{\left(\frac{\partial v_x'}{\partial t}\right)^2} \tag{20}$$

Eddy Diffusion

In order to adapt the basic equations of motion to conditions of fluid turbulence, Boussinesq [26] assumed that the Saint-Venant viscosity coefficient [27] could be considered to have two parts: one depending upon the molecular motion and one upon the eddy motion. For flow in the x-direction, the shear thus becomes

$$\bar{\tau} = (\mu + \eta)\frac{\partial \bar{v}}{\partial y} \tag{21}$$

Just as the ratio of μ and ρ is a kinematic factor μ depending upon a molecular velocity and scale of motion, the ratio of η and ρ is logically a kinematic factor ϵ reflecting an eddy velocity and scale. Reynolds' handling of the same equations gave rise, instead, to a series of fictitious stresses of the form $\rho v_x' v_y'$. Equating the two alternative expressions for the turbulent shear in the direction of flow thus yields

$$\epsilon = \frac{\overline{v_x' v_y'}}{\partial \bar{v}/\partial y} \tag{22}$$

By means of one or another phenomenological approach, efforts were made by Prandtl,[28] Taylor,[29] and others to relate both ϵ and $\overline{v_x' v_y'}$ to the mean velocity gradient through introduction of the so-called mixing length—i.e. the lateral distance over which a small fluid mass would be carried by the mixing process before losing its identity in the new region. Prandtl's hypothesis involved the assumption of constant momentum during lateral displacement, whereas Taylor assumed constant vorticity to prevail. Prandtl's approach is usually regarded as rather artificial in comparison with Taylor's, because it is difficult to conceive of momentum first remaining constant and then changing abruptly, whereas the vorticity of an element in two-dimensional flow is perforce unchanging. As a matter of fact, although both two-dimensional mixing-length hypotheses have provided useful simplifications, a two-dimensional state of turbulence (on which the Taylor hypothesis in particular depends) also departs considerably from reality. As Taylor himself emphasized, an eddy that does not change vorticity because of this two-dimensional restriction cannot produce any of the occurrences characterizing true, three-dimensional turbulence.

Further light upon the subject is obtained by adopting for a moment the Lagrangian rather than the Eulerian point of view.[30] If the paths of individual particles that pass a fixed point are followed, each will be seen to deviate from the linear course of the stream in a manner that changes from instant to instant. From the mathematical point of view, this results from the fact that the velocity is a function of time and space. From the physical point of view it is because the motion of any particle must vary with the motion of the eddy of which it is a part, and because that eddy in turn is subject to the motion of the larger eddy or eddies upon which it is superposed and to the influence of the other eddies in its immediate vicinity. The irregularities of the individual paths thus reflect the scale and velocity characteristics of the eddy structure. Though the paths chosen for examination all originate at a single point, they will generally diverge more and more with distance downstream. In other words, a group of particles that are close together at any instant will gradually be diffused through the eddy action. A diffusion coefficient that measures the rate of spread will necessarily be proportional to the eddy size and velocity, just like the mixing coefficient of Boussinesq.

The extent to which such factors are interchangeable is

somewhat akin to the relationship between the parts played by the molar and the molecular viscosities. The role of the eddies in turbulent flow is essentially threefold: diffusion, shear, and dissipation. Each of these occurrences increases with the velocity characteristics of the eddy motion. With increasing eddy scale, on the other hand, the diffusion increases whereas the dissipation decreases. The mixing characteristics are therefore dependent primarily on the upper end of the eddy spectrum, and the dissipation characteristics on the lower end. The shear, which is independent of scale, involves eddies over the whole range, but particularly in the vicinity of the greatest energy concentration. So far as the diffusion mechanism is concerned, it must be recalled that the process of eddy stretching in itself provides no interchange of fluid with neighboring eddies no matter how complex their interlacing. The intermixing that actually occurs is on the molecular scale. However, the more an eddy has become elongated, the smaller is its diameter, the more pronounced the transverse velocity gradient, and the greater the relative role of viscosity. The part played by molecular diffusion is thus comparable to that played by molecular dissipation: of little moment in the formation and large-scale behavior of the eddies, but soon becoming very essential to their ultimate role.

Conclusion

Each of the foregoing sections contributed at least one fact or concept about the part that is played by eddies in fluid motion, and a recapitulation of the major points should now lend perspective to the general picture.

For the simple Rankine model, combining a central filament of constant vorticity and a surrounding field of constant circulation, all characteristics of the pattern—including the drop in pressure at the axis—are describable in terms of filament density, diameter, and peripheral speed.

Whereas spiral vortices of the cyclonic type exemplify the velocity field and extremely large energies of the Rankine model, they give no clue as to the formation and behavior of the more prevalent type of eddy.

Vortex generation and decay can be calculated in terms of a single filament expanding through the process of viscous shear; this adds viscosity and time to the essential variables, and permits the variation in core pressure, filament diameter, and energy of the motion to be estimated as time passes. On the other hand, the impulsive formation of two counter-rotating filaments by the manipulation of a paddle yields an initially irrotational field without the infinite-energy drawback of the Rankine model; the viscous model then points to the sequence of events in the vicinity of each axis, and the interaction of the essentially irrotational fields gives a glimpse of the complexity of flows upon which numerous eddies are superposed.

Generation of the Kármán vortex trail, in addition to clarifying the fluctuation of pressure behind the generating body, graphically illustrates the increasing rapidity of breakdown of vortices through instability and interaction as the Reynolds number increases. Prevention of the alternate formation and detachment of eddies by means of a splitter plate yields, on either side, a quasi-stable eddy typical of those in all zones of separation in contact with a longitudinal boundary. The production of vortices in the shear zone along such an eddy provides the mechanism for transmitting energy from the ambient flow to the eddy and for dissipating energy in both regions.

The further behavior of turbulent eddies generated in a shear zone exemplifies the sequence of events in any turbulent flow: the gradual reduction in eddy scale and increase in eddy velocity through mutual interaction, and the resulting increase in rate of viscous dissipation. The various turbulence characteristics, usually analyzed statistically for quantitative results, are seen to be subject to significant qualitative analysis in terms of eddy characteristics. In particular it is seen why the largest eddies play the major role in diffusion, the smallest eddies the major role in energy dissipation, and the common middle-size eddies the major role in their own gradual dissolution.

Acknowledgments

Certain of the equations appearing in the foregoing pages were derived by Mrs. Matilde Macagno, and the various parts of the text were critically reviewed by Drs. Enzo O. Macagno and Eduard Naudascher, all staff members of the Iowa Institute of Hydraulic Research. The Office of Naval Research, under Contract Nonr 1509 (03) with the Institute, and the Army Research Office at Durham, under Contract No. 2871-E, contributed financially to a number of the investigations of which results are described.

REFERENCES

1. H. Lamb, *Hydrodynamics,* 6th ed., Cambridge Univ. Press (1932).

2. H. L. F. von Helmholtz, Über Integrale der hydrodynamischen Gleichungen welche den Wirbelbewegungen entsprechen, *J. reine angew. Math., 55,* 25 (1858).

3. W. Thompson, On Vortex Motion, *Trans. Roy. Soc. Edin., 25,* 217 (1889).

4. W. M. J. Rankine, *A Manual of Applied Mechanics,* Griffin, London (1858).

5. H. Rouse, *Elementary Mechanics of Fluids,* Wiley, New York (1959).

6. H. Rouse, ed., *Advanced Mechanics of Fluids,* Wiley, New York (1959).

7. H. Rouse and H. C. Hsu, On the Growth and Decay of a Vortex Filament, *Proc. First Nat. Cong. Appl. Mechanics, ASME,* 741 (1951).

8. F. Hama, Progressive Deformation of a Curved Vortex Filament by Its Own Induction, *Phys. Fluids,* 5 1156 (1962).

9. D. B. Steinman, Problems, of Aerodynamic and Hydrodynamic Stability, *Proc. Third Hydraulics Conf.,* Iowa City (1946), p. 136.

10. V. Strouhal, Über eine besondere Art der Tonerregung, *Ann. Physik. Chem., 5,* 216 (1878).

11. T. von Karman, Über den Mechanismus des Widerstandes den ein bewegter Körper in einer Flüssigkeit erfährt, *Gött. Nach.* 509 (1911) p. 509; (1912) p. 546.

12. S. Goldstein, ed., *Modern Developments in Fluid Dynamics,* Clarendon Press, Oxford (1938).

13. W. M. J. Rankine, On the Mathematical Theory of Stream Lines, *Phil. Trans., 161, 267* (1871).

14. H. Rouse, Energy Transformation in Zones of Separation, *Ninth Convention, Int. Assoc. Hydraulic Res.,* Dubrovnik, Yugoslavia (1938).

15. A. D. Newsham, "Cavitation and Pressure Fluctuation Behind a Bluff Body With and Without a Trailing Splitter Plate," M.S. Thesis, University of Iowa (1963).

16. L. Prandtl, Über Flüssigkeitsgewegung bei sehr kleiner Reibung, *Verhandl. III, Int. Math.-Kong.,* Heidelberg (1904).

17. L. Rosenhead, The Formation of Vortices from a Surface of Discontinuity, *Proc. Roy. Soc. A, 134, 170* (1931).

18. G. Birkhoff, Helmholtz and Taylor Instability, *Proc. Symp. Appl. Math.* (Providence), *13, 55* (1962).

19. W. Tollmien, Über die Entstehung der Turbulenz, *Gött. Nach.,* (1929), p. 21.

20. G. B. Schubauer and H. K. Skramstad, Laminar Boundary Layer Oscillations and Stability of Laminar Flow, *J. Aeronaut. Sci., 14,* 69 (1947).

21. G. I. Taylor, The Spectrum of Turbulence, *Proc. Roy. Soc. A, 164, 476* (1938).

22. J. O. Hinze, *Turbulence,* McGraw-Hill, New York (1959).

23. G. K. Batchelor, *The Theory of Homogeneous Turbulence,* Cambridge Univ. Press (1955).

24. A. A. Townsend, *The Structure of Turbulent Shear Flow,* Cambridge Univ. Press (1956).

25. J. Laufer, The Structure of Turbulence in Fully Developed Pipe Flow, *Nat. Adv. Com. Aeron. Rept.,* (1954), p. 1174.

26. J. Boussinesq, Essai sur la théorie des eaux courantes, Paris (1877).

27. B. de Saint-Venant, Note à joindre au mémoire sur la dynamique des fluides, *Comp. Rend., 17,* 1240 (1843).

28. L. Prandtl., Beriche über Untersuchungen zur ausgebildeten Turbulenz, *Z. angew. Math. Mech., 5,* 136 (1925).

29. G. I. Taylor, The Transport of Vorticity and Heat through Fluids in Turbulent Motion, *Proc. Roy. Soc. A, 135,* 685 (1932).

30. G. I. Taylor, Diffusion by Continuous Movements, *Proc. Lond. Math. Soc., 20,* 196 (1922).

ON THE ART OF ADVANCING THE SCIENCE OF HYDRAULICS

Hunter Rouse

Institute of Hydraulic Research
University of Iowa
Iowa City

My great pleasure over the invitation to participate in this first Australian Conference on Hydraulics and Fluid Mechanics can be fully appreciated only by one who has been in a similar situation. To visit a part of the world that is almost diametrically opposite to one's own state is in itself an extraordinary experience. An opportunity of repaying professional calls that have been made by other world travellers, moreover, serves not only to give such intercourse a proper balance but also to stimulate it further. Finally, the growth of perspective that accompanies the exchange of views at such conferences is, I have found, proportional to at least the first power of the relative displacement of the participants.

We in hydraulics at Iowa have now had the satisfaction of welcoming students — not to mention passing visitors — from thirty-nine different countries. About half of our four-hundred-odd post-graduate alumni have come from abroad. This in itself is indicative of our contact with the profession at large. What is even more pertinent to my subsequent remarks is the fact that two-thirds of our foreign students have been from the countries of the eastern hemisphere represented at this congress. At the moment, for example, we have eleven from Nationalist China and six from India. Although there are only two from Australia, they happen to be very close to the head of our fifteen doctoral candidates, thus ably maintaining the standard set by the four Australians who preceded them. For what significance it may have, I cannot refrain from adding that, to the best of my knowledge, the senior research staff at the Peking hydraulics laboratory in Communist China is composed of five post-war "Iowans".

When our own research institute first became active in 1931, American hydraulics was in the process of receiving a considerable transfusion of new professional blood through a succession of young engineers sent for study to European countries, primarily Germany. This was done at the instigation and largely with the financial support of John R. Freeman, himself a hydraulic engineer of considerable repute and even greater foresight. Like the members of our Institute staff, those now responsible for the progress of hydraulics in the United States are a healthy combination of native-born people educated at home and abroad—with, I might add, a considerable admixture of foreign born, educated either in their own countries or in America, who have chosen to remain with us. It is the breadth of viewpoint that is produced by mingling people of various backgrounds in this manner that I consider to be America's technical as well as social and economic strength.

You will surely have noted my consistent use of the word "hydraulics" to the apparent exclusion of "fluid mechanics", even though it has been my endeavor for the past thirty years to give increasingly greater emphasis to principles of mechanics in hydraulic instruction and research. At the time of my first efforts in this direction, the term "mechanics of fluids" had only recently been introduced into the States by my superior, Boris A. Bakhmeteff, and a considerable amount of missionary work had to be done before American hydraulicians began to recognize the advantages to be gained by the combination of analytical and experimental methods. Those of the Freeman scholars who had studied with members of the Ludwig Prandtl school in Germany, and in particular those who came under Theodor von Kármán's influence later in America, were most effective in this regard. Today the methods of fluid mechanics are embodied in essentially every science in any way involving fluid motion, from astronomy to zoology, and in its modern applications the subject encompasses such a vast field that hydraulics, which once played a parental role, is now very small in comparison with the far more responsible offspring that it helped to beget. I thus adhere to the term "hydraulics" to indicate the specific branch of fluid mechanics that is of primary interest to most of the participants in this conference. Note, if you please, that I also distinguish between hydraulics as an engineering science and hydraulic engineering as its field of application. We are here concerned with the art of advancing the science rather than the art of utilizing it.

I believe that those who organized the conference program thought that I might give some useful advice with regard not only to teaching and research but particularly to the planning of further conferences. They may have felt, perhaps, that certain similarities exist between our situation in America 30 years ago and that here today. I grant that we have had, in the intervening years, a number of successes and failures which might well be considered by any organization before proceeding very far with its own plans. However, there are also certain dissimilarities which should be noted before any advice of mine is either given or taken. Let us first examine the two sides of the picture point by point.

My country is perhaps old as a democracy but still quite young among world civilizations. Here in the eastern hemisphere you have several of the oldest peoples of history as well as some of the youngest independent nations. A bilateral exchange of wisdom would surely be a mutually profitable matter. Indeed, the flow of technical literature between east and west is currently by no means a one-way matter. It is true that since 1939 we at Iowa have held seven triennial hydraulics conferences. But you are also members of the International Association for Hydraulic Research, which has been holding hydraulics conferences since 1937, the fourth having taken place in India in 1951, and a division meeting having been arranged in Japan only this year. Next year the Indian Institute of Science is planning an Asian Conference on Hydraulics and Hydraulic Machinery. Australia may be a beginner in the matter of conferences, but surely not in either hydraulics or the broader mechanics of fluids, if the Australian students that have come to Iowa are any indication; in this regard one might well recall that two of the present international leaders in the field were Australian born and bred. Let me therefore talk as one who is confronted with much the same professional problems as any other conference participant at much the same mean stage of life and time, my advice best being tempered by similar experiences on the part of those interested in heeding it.

So far as hydraulics conferences are concerned, the only affairs of this sort that existed in America at the time ours originated were the annual sessions of the Hydraulics Division of the American

Society of Mechanical Engineers — then a rather conservative group of specialists in hydraulic machinery. Except for the first, in 1939, which offered quite an assortment of topics, each of the seven Iowa programs was built around a central theme such as sediment, instrumentation, or prototype verification. The proceedings of each were published in the course of the following year. The fourth in particular was planned so that its preprinted papers formed, after extensive discussion and revision, the thirteen chapters of the book *Engineering Hydraulics*. Attendance at these conferences rose to a peak of 425 in 1949 and thereafter gradually dwindled to 173 in 1958. The series was then terminated because — far from remaining almost the only thing of its kind — the Iowa conferences had come to be but one of at least ten different American conference series dealing with fluid motion. While the Iowa meetings had been particularly favored for their central theme, careful organization, and faithfully returning alumni, it was felt that national meetings should take precedence over local. My advice, as a result of our own conferences and others that I have attended, is threefold: First, seek to hold neither too many meetings nor yet too few. Second, entrust their planning and execution to an experienced team rather than to a single person or a group of novices. Finally, consider carefully the advantages of the unsolicited-paper type of program in comparison with either round-table discussions or a limited number of invited speakers on well-correlated subjects; decide on (and then expedite) publication of the papers only if their caliber warrants the great expenditure of time involved and the retention of the result. In clarification of this final point, I gladly grant that some outlet should be provided for contributed papers and their discussion, much as I myself favor the prearranged type of program for its unity and general worth; however, I highly recommend use of the round-table device, with selected panels and expert moderators, for the stimulating dissection of a few timely subjects. Large conferences, of course, might well involve a variety of procedures rather than one alone. My feelings about publication warrant still further comment.

At the risk of providing an example of what to avoid, for I understand that my words may eventually appear in print, I should like to voice anew the conviction that far too much of a technical nature is being published today, at least in my own country. This stems in part from the abundance of periodicals that must be filled each month, to some degree from the idea of planned obsolescence that our book publishers seem to have adopted from our automotive industry, and in no small measure from the impatience of many writers to release a manuscript before they really have something to say. Surely such items as progress reports with investigational results that are of immediate interest have their purpose, provided that they do not remain the final form of presentation. The difficulty is that there is no automatic way of extracting the essential for permanent reference and simultaneously discarding the unessential. Instead, our libraries are overflowing with so much material that items of lasting value are frequently hidden by the dross. Effective abstracting services are all too few. To counteract this trend in some small measure, my advice would be to lay extremely great stress upon quality in preference to quantity and upon permanency of value in preference to immediacy of interest – even at the risk of losing the few who are ever loath to publish because their results never reach the state of perfection that they seek.

My views on the scientific development of students in our field are known to some of you through the introductory pages of a recent Iowa bulletin on "Laboratory Instruction in the Mechanics of Fluids". Therein I emphasized the difference between training at the technical-institute level, which must still reflect much of the empirical approach that was long traditional, and education at the university level, which should be more concerned with processes of analysis than with facts and formulas. While it is true that the principles of a science like engineering mechanics remain essentially the same from decade to decade, their application is changing almost from year to year. One need only consider the technical problems introduced by the advent of atomic power and space flight to see my point. We can hence no longer be content to give our engineers a series of procedures for the design of standard products. We must teach them not only how to meet new situations with principles currently at hand but also how to keep pace mentally with world changes that they introduce. In a word, the university must remain at the frontiers of knowledge in every way — which it can do only if the equally essential routine work is undertaken by men trained for this specific purpose: graduates of technical institutes. In America, unfortunately, our engineering colleges are attempting to train the masses as well as educate the leaders, simply because every student wants a college degree regardless of ability or ultimate function. One cannot intelligently depreciate the Soviet goal of having everyone in the university who should be and no one who should not. The distinction between advanced and routine engineering is just as important in hydraulics as in any other field.

This distinction becomes perhaps even more apparent when one considers the related subjects of research, development, and testing —terms that are often confused with one another. Now fundamental research is as essential a function of the university as is higher education. Product or process development also has its creative and instructive sides, but routine testing has little of either. In hydraulics, however, we still find model tests classed not only as development but even as research. This is sometimes the case because the university experiment station is a service department of the state, but perhaps equally often because only in this manner can the laboratory earn money for higher use. If, however, as is all too frequently true, the laboratory program is filled with testing because the staff can do no better, then facilities of the university are being used as improperly as when it functions as a technical institute. I not saying that there is no place in the technical world for model testing, any more than I implied that there is no place for the technical institute. Each has an extremely vital function in our technology. But the last function that either should fill is to crowd out of the university program its most essential features: the discovery and the dissemination of knowledge. Here again the situation is as important in the field of hydraulics as in any other.

If I argue that a university laboratory should stress fundamental research, give due attention to development, and leave routine testing to others, then I must logically concede that there should be laboratories which stress routing testing, give some attention to development, and leave research to the universities. This will surely be the situation in the commercial testing organizations. But a question now arises as to the category in which the laboratory of a state or federal government belongs. In America our federal laboratories dealing with fluid motion tend in both directions, and a comparison of their capabilities is highly relevant to the present discussion. Laboratories of the one type are forbidden by agency policy to spend funds upon anything but solving the specific problem at hand. Accordingly, the same kind of problem is investigated repeatedly, and only through long experience do the underlyng principles of its solution become clear. How much better if a fraction of the cost of the specific tests were put into a general study, thereby reducing expenditures for future tests many-fold! Unfortunately, it is not only a matter of economics: the staffs of such laboratories seldom advance beyond the caliber of testers, in part for lack of incentive and in part for lack of attractiveness to new blood. Laboratories of the other type place quite as much weight upon a general solution as upon a specific one. As a result, the cost of routine testing seems to vary in inverse proportion to some power of the time spent on research. Scientific reputation, in turn, comes to be based upon the analytical power of the staff rather than upon the extent of the experimental facilities (American worship of buildings and equipment to the contrary). I cannot emphasize too strongly the lesson to be gained by such comparison. It is true that experimental hydraulics in the past was greatly furthered by the advent of model testing. Those lessons have long since been learned, however, and continued progress will not come from repeating events of the past.

Ten years ago, when *Applied Mechanics Reviews* first instituted its lead articles summarizing the status of particular fields of investigation, I was invited to prepare several pages on "Present-Day Trends in Hydraulics". The material which was included was used thereafter in slightly augmented form as the last chapter of our *History of Hydraulics* under the title "An Appraisal of the Science at Mid-Century". Less than two years ago I made a related survey in connection with a symposium held by New York University on basic science in France and the United States. A book on the symposium has recently been published, and reprints of my own article, "Current Trends in American Hydraulics", are available on request from the Iowa Institute. I commend the information in both the book and the article to you, because I believe that it is healthy at times to compare one's own research endeavors with those of others. I do not, mind you, hold that one should either follow the leader or purposely choose a diametrically opposite course. Surely duplication of effort is not necessarily something to be avoided at all costs, for complete matching of conclusions is improbable, and partial substantiation (or even disagreement) can serve a very useful purpose. Nor is anything to be gained by insistence upon travelling a totally new road unless it happens to lead in a direction that is in itself attractive. Whether one emulate past and present

masters or not, there is much to be gained by observing their mental
and manual procedures, and a study of both history and current
activity should at once stimulate and lend perspective. The situation
is much like that of an individual who is faced with the need of
developing a new research facility, to whom my advice is invariably
the following: visit as many productive organizations of a related
type as feasible; study carefully their methods and equipment;
then proceed toward your own goal in your own way.

The program that has been devised for this first Australian
conference, consisting as it does of papers submitted by represen-
tatives of organizations from many different countries, quite naturally
illustrates many of the points that I have sought to make both in
this discussion and in the two publications just mentioned. The
contributions range in viewpoint from the traditional to the ultra-
modern; in method, from the empirical to the analytical; in topic,
from electronics to hypersonics; and in substance, from thixotropic fluids to plasmas. Fluid mechanics is indeed a varied field,
and the pursuit of knowledge leads perforce in many directions.
Whether so diverse a group as this can profitably continue to meet
with any great frequency may best be judged at the end of the
conference. In this connection it is significant that at a similar
conference held in Santiago, Chile, just a few months ago it was
decided to have separate meetings for hydraulics and fluid mechanics
in the future Surely one can more profitably discuss particular
problems in closely knit groups, yet one should not ignore the fact
that a great deal can sometimes be gained by noting how the rest
of the world measures and analyzes. In any event, there is much
that is common to all research, particularly that involving fluid
motion. Though I shall continue to deal primarily with hydraulics,
what I still have to say should hence be indirectly significant to
all participants in this conference.

In my earlier survey of hydraulics at mid-century I sought not
only to bring the status of the science up to date but also to project
it a significant distance into the future. I quote the following passage
from the closing pages:

"Even a limited extrapolation beyond the present would be sheer
folly, despite the availability of several centuries of record as a guide,
for the record has been seen to display an appreciable change in
course in the very recent past. However, one can surely hope, if
not predict, that hydraulics will continue to widen its break with

pure empiricism, which long since reached a point of diminishing returns, and to strengthen its tie with physical analysis, for the latter is still unlimited in its possibilities. On the other hand, one could avoid a direct forecast by merely citing... what seem to be the major gaps in hydraulic knowledge, on the assumption that at least some of them will be filled in the near future.... Let us therefore list as problems for the future just four that are typically hydraulic in nature yet little subject to solution by measurement alone. One is the incorporation of gravitational action in the mathematical analysis of flow with respect to fixed boundaries such as gates and spillways. Another follows from the presumption that hydraulic structures will never be fully streamlined: the understanding of the mean and fluctuating patterns of flow in a zone of separation. A third is the statistical evaluation of surface roughness and its effect upon boundary resistance. The last is a problem that is broadly inclusive: the mechanism of turbulent mixing at any boundary, whether the surface of air-entraining water, the bed of sediment-entraining water or air, or the interface between two moving fluids of only slightly different density."

A decade later, in my appraisal of the American situation, I was able to report some progress in each of these four directions but in no case an appreciable reduction of the problem. I can hence still commend each of the avenues to you most emphatically. At the risk of distracting somewhat from these primary goals, in the same paper I named several other matters of immediate concern to American hydraulicians: stability at the interface between air and water or between fluids of only slightly different density; application of the sound-wave analogy with due consideration for boundary resistance; the hydraulic evaluation of natural roughness; shear-flow turbulence; and the measurement of small liquid velocities and velocity fluctuations. Each of these I call especially to your attention, for each remains almost as far from solution today as it has been in the past.

Because of the varied nature of the interests reflected by the conference roster, I would like at this point to illustrate a broader course of advancement by citing three very general problems that we have always had with us but have either set aside for the future or have handled provisionally in a highly simplified manner. Sooner or later each will have to be faced in its entirety, however, and the earlier we begin to give them our serious attention the less remote will be the time at which **we** can consider ourselves reasonably familiar with them. The first is the matter of unsteadiness of flow. The second is the matter of rotationality. The third is the matter of flow measurement as a function of time and space. None of these, you will observe, is at all new, yet I am sure that each of the conference participants can think of some aspects of all three in which he has encountered difficulties, and I am equally sure that these difficulties will differ extensively from one participant to another.

Let us consider first the matter of unsteadiness. Some of the earliest successful analyses of fluid motion dealt with unsteady flow — for example, Franz Josef von Gerstner's wave studies published nearly 160 years ago — but all of these necessarily dealt with conditions in which the variation with time could be eliminated by the principle of relative motion. The transformation of wave shape and celerity by either nonuniformity of boundary geometry or loss of energy through viscous shear is gradually becoming subject to at least approximate analysis, though still far from rigorous treatment. Such inherently unsteady phases as pendulation, flow establishment, water hammer, and surging are likewise receiving ever more accurate evaluation; it must be noted, however, that each of these is very nearly one-dimensional in its aspects, which permits a considerably simplified type of treatment. Great progress has also been made in problems of boundary vibration, in which the departure from irrotationality can be neglected and the added mass calculated with fair precision. Finally, the Tollmien–Schlichting–Lin analyses of flow stability represent a masterpiece of unsteady-flow attack. In contradistinction, the type of unsteadiness that has not yet been treated to any appreciable extent is that involving separation. Prediction of the separation pattern is difficult enough if the flow remains steady; but only experimental evidence can now be trusted to give any idea of the development of this pattern as a function of time. Such knowledge is essential, however, in any analysis of temporally variable fluid resistance, whether it involves the acceleration of submerged bodies or the operation of recoil mechanisms; the action of bilge keels or the suspension of sediment in a turbulent fluid; the oscillation of a penstock gate or the galloping of a suspension bridge. This, then, is the first of the problems that I recommend for your consideration.

The second, that of rotationality, is likewise relatively old, for Helmholtz established its essential character more than a century ago. Various problems involving pronounced rotationality have been solved with different degrees of rigor, as witness the Couette, Stokes, and Poiseuille forms of laminar flow, the Blasius solution for the laminar boundary layer, and the Taylor analysis of flow stability. In general, however, the equations of motion become too unwieldy for more complex boundary forms or more pronounced states of acceleration, and the only means of solution has involved the hypothesis of complete irrotationality. For several generations the apparent unreality of this assumption maintained a rift between theoreticians and engineers, but nowadays not only have the engineers discovered the great applicability of the analytical results to flow past streamlined boundary profiles, but the theoreticians have steadily perfected their methods of analysis. It would next be desirable to begin the extension of such analyses to conditions of appreciable mean-flow vorticity—not only in the boundary layer, as originally hypothesized by Prandtl, but over a considerable portion of the velocity field. Flow around an immersed body, for example, can usually be handled by its segregation after Prandtl into a thin region of rotationality and a surrounding region of irrotationality. Flow through a conduit constriction of similar profile, on the other hand, is entirely rotational, and either the boundary-layer or the irrotational-flow type of analysis will yield only a rough approximation to the actual state of motion. Because of its close relation to the foregoing problem of unsteadiness, the region of separation is a particularly apt instance of rotationality that cannot be handled realistically in any way except through evaluation of the vorticity distibution. The case of the standing eddy within a certain type of separation region is of special importance, because the synthesis of the shear zone through which energy is transferred to the eddy would be a major step in the analysis of rotational flow.

Measuring instruments for fluid motion would seem to be even older than the equations of motion themselves. Henri de Pitot, for example, introduced the stagnation tube, the manometer, and the approximate notion of velocity head in 1732; Daniel Bernoulli, the wall piezometer and the concept of pressure head (though not the Bernoulli theorem) in 1738; Reinhard Woltman, the current meter in 1790; and countless physicists and electrical engineers since Coulomb in the 18th century, the principles on which modern electronic circuits are based. With the resulting means of indicating, recording, and analyzing the pressure and velocity of flow as functions of time and space, one might well wonder why I include flow measurement as a present-day problem. The reason lies in the long-recognized but commonly disregarded effect of fluid turbulence upon the indication of each individual instrument. A current meter calibrated by towing through quiet water will behave differently in a turbulent flow, and differently again in another turbulent flow of unequal intensity or scale. The discrepancy will depend upon the form of the meter as well as the nature of the turbulence, of course, and there are admittedly both meters and states of flow for which the errors are acceptably small. The same is true of stagnation-pressure and velocity tubes: their indication will vary with the geometry of the tube, the intensity of the velocity fluctuations, and the size of the eddies; for normal states of turbulent flow the vagaries of indication can also be estimated with a sufficient degree of approximation. For conditions of excessive turbulence, however, such as that of the shear zone at the edge of a region of separation, not only is it difficult to measure the mean velocity and mean pressure with certainty, but determination of the standard deviations from these means (not to mention such functions as shear and dissipation rate) is still either very inexact or else wholly impossible.

The Bernoulli theorem having already been mentioned, let me still elaborate just a bit upon this topic. As Leonhard Euler first formulated it two hundred and more years ago, and as it is usually used today, no provision is made for either turbulence or energy loss. The latter is frequently added rather arbitrarily, to be sure, and sometimes the former as well, but on the basis of intuition rather than rigorous derivation. Now the study of such matters as separation zones, whether quasi-stable or unsteady, requires the evaluation of energy and work integrals taken either along the streamlines or over a specific region of space. Because the quantities involved cannot yet be predicted analytically, they must be measured. But they still cannot even be measured with certainty. Thus it is that problems as old as rotationality and general unsteadiness of flow remain dependent for their further treatment upon the measurement of such basic quantities as velocity and pressure. Since flow with separation continues to be the rule rather than the exception in hydraulic engineering practice, the hydraulician more than any other should be the one to lead in solving this problem.

Lest either my hosts or my fellow participants in this conference feel that I have been unduly critical, intolerant, or dogmatic in my remarks, I should perhaps confess that it is not really they whom I am judging so much as my own countrymen and perhaps even my own colleagues. In the length of time that I have been concerned with teaching and experimenting, the probability is rather high of having adopted at least one cause—such as research vs. testing, or excellence of staff vs. completeness of equipment. There is undoubtedly also a great deal of personal bias in my selection of problems that warrant consideration, for these are subjects in which we at Iowa are presently interested. Now I would be the last to hold that there is only one right way to accomplish a given end or even that there is a particular major goal to be attained, and I am sure that all would agree with me. I do maintain, nevertheless, that the art of advancing any field of knowledge necessarily involves fundamental research, high-level education, and free intercourse among those who are engaged in such endeavors—each to essentially the same degree. Please believe, with me, that these are equally vital to progress in hydraulics!

Discussion by Hunter Rouse

HUNTER ROUSE,[1] F. ASCE.—This comprehensive treatment of sediment suspension reflects a prevalent dilemma about the source of the "suspended load distribution function," for it states that it "was first presented by Rouse. . .although it was apparently first developed by Ippen. . .at the suggestion of Theodor von Kármán. . ." In view of the fact that this uncertainty has even led sometimes to designation of the function by the anonymous term "classic," a more detailed account of its origin now seems pertinent.

Early in 1936, the writer joined the staff of the cooperative sediment laboratory of the Soil Conservation Service at the California Institute of Technology; this was directed by R. T. Knapp, with von Kármán as consultant and V. A. Vanoni as project supervisor. Ippen had already been on the campus for several years as a pre-doctoral fellow, working at first on the subject of sediment transportation and thereafter on that of high-velocity flow. At about that time, one of von Kármán's papers on turbulence[2] became available to the writer in reprint form, with the indication that the point concentration of sediment in a stream could be predicted from the derived equation

$$\log \frac{n_o}{n} = \rho w \int_0^z \frac{du}{dz} \frac{dz}{\tau} \quad \ldots \ldots \ldots \ldots (2\text{-}D.54)$$

by introducing the velocity gradient from field measurements and the intensity of shear from the point depth and measured stream slope. Because the writer had just published a paper on the Kármán-Prandtl resistance relationships, (2-D.13) it soon occurred to him to substitute these for the field measurements, with the dimensional form of the "suspended load distribution function" as the result.

One day Ippen, in the course of a general conversation, asked the writer what he was doing on the subject of sediment. On being informed of the foregoing accomplishment, Ippen stated that a year or so earlier he had undertaken a similar development in dimensionless form, at the direct suggestion of von Kármán, but had never prepared the derivation for publication. Though enough details were mentioned to make it evident that the same form of rela-

---

[a] September, 1963, Task Committee on Preparation of Sedimentation Manual (Proc. Paper 3636).

[1] Dir., Inst. of Hydr. Research, Univ. of Iowa, Iowa City, Iowa.

[2] Kármán, Th. von, "Some Aspects of the Turbulent Problem," Proceedings of the Fourth International Congress of Applied Mechanics, Cambridge, England, 1934.

tionship was under discussion, at no time was either the equation itself or its plotted form actually demonstrated. In fact, until a recent exchange of correspondence between Ippen and the writer, the latter assumed that the former had also proceeded from the Kármán-Prandtl equation, whereas he had in actuality used that of Krey,[3] with essentially the same results.

In preparing the closure to his ASCE paper on the Kármán-Prandtl resistance analyses for turbulent flow, the writer introduced his own nondimensionalized treatment of the suspension function as a further illustration of the usefulness of their approach (2-D.13). In view of Ippen's previous indication of priority in the derivation, the writer carefully gave him credit for it—so carefully, in fact, that he took no credit whatever for himself, thereby unintentionally implying that he had simply reproduced the Ippen derivation. The resulting situation has, to be sure, been somewhat counter-balanced by a French predilection to call the exponent z in Eq. 2-D.32 the Rouse number. Whether the function itself will come to be regarded generally as "classic," whether it will eventually acquire the names of its independent formulators, or whether it will simply take its place among the many unidentified concepts that the world actually owes to von Kármán is still open to speculation.

Because of Vanoni's prior intention to conduct measurements of suspended load, in 1937 the writer turned instead to the experimental verification of the basic suspension law. The paper under discussion attributes this verification to both Hurst and the writer. Even Hurst himself, however, considered his work as essentially qualitative (2-D.7):

"Experiments have been carried out with three kinds of sand and many different speeds of the propeller used as an agitating device. The results have always been to show that for a given sample of sand and propeller speed, the concentration of sand n at height h is given by

$$n = n_0 \xi^{-\alpha h} \quad\ldots\ldots\ldots\ldots\ldots (2\text{-D}.55)$$

where α is a function of the speed of the propeller and varies with the kind of sand. . .

"Only very general conclusions can be drawn from these curves. These are:

1. For the same velocity of the water the fine sand acquires a greater velocity of agitation than the coarse.

2. The velocity of agitation of the sand increases less rapidly than the velocity of the propeller.

3. The shape of the sand particles must exert considerable influence as two samples of the fine sand from different sources although sieved in the same way gave different results."

The writer's measurements, on the contrary, provided (2-D.8) "definite verification of the suspension theory. . .with regard both to the exponential nature of the distribution and to the role played by the settling velocity of the sediment." Not only were the results quantitatively conclusive even with respect to variation of the turbulence itself, but they also contained the first

---

[3] Krey, H., "Die Quer-Geschwindigkeitskurve bei turbulenter Stromung," Zeit. angew. Math. Mech., Vol. 7, No. 2, 1927.

experimental evidence of a matter which continues to have major importance: "Some question still exists as to the strict similarity between the mixing characteristics of the sediment and those of the fluid, for the factor β appears to vary somewhat with the relative size of the particles in suspension. Moreover, it remains to be seen whether the magnitude of β determined from the velocity gradient in a given turbulent flow will correspond numerically to that for the gradient of sediment concentration."

Discussion by Hunter Rouse

HUNTER ROUSE,[2] F. ASCE.—Nomenclature that is used repeatedly in technical literature becomes ever more difficult to modify—particularly, it would seem, if the term in question is a misnomer. A case in point is found in the title of the paper under discussion. To the writer's best knowledge, the expression "density current" was coined at the Bureau of Standards in the middle 1930's to replace the previous project title "Currents in Lakes and Reservoirs." For over half of the intervening period the writer has sought to limit use of the expression to those conditions which the words really denote; stratified flow occurring as the result of gravitational action is not logically one of these.

The property of either mass or density has two dynamic attributes: Inertia, and mass attraction. The former involves the given substance alone; the latter, its relation to another substance. For the usual engineering condition that the other substance is the earth, the role of mass attraction is embodied in the accelerative effect of the earth's gravitational field. If a third characteristic, viscosity, is added to inertia and gravity then the result is, in essence, a three-dimensional system of independently variable factors controlling fluid motion with density stratification. At the limit, any one of the three alone can govern. For example, shear can be transmitted from one uniform stratum to another by viscous action, much as in simple Couette flow. Similarly, a pressure gradient imposed on a stratified fluid will produce an immediate inertial reaction (i.e., an acceleration) that is inversely proportional to the local density. Gravity, finally, is the factor that is directly responsible for the stratification and for other effects of positive or negative buoyancy. More generally, two of the three will predominate: Inertia and gravity, gravity and viscosity, or viscosity and inertia. The last pair, however, do not represent the conditions of flow now under consideration, for it is gravity rather than inertia which is essential to the discussion. Most generally, of course, all three factors will be involved; however, that is beside the point.

Instead of examining the interplay among the three dimensional characteristics, one can think in terms of the relative importance of each pair as represented by the respective nondimensional parameters. The Froude and Reynolds numbers are commonly associated with such flows, but seldom is it realized that the inertial aspect of the density can be of just as little consequence as it is in the case of Stokes' law, for which neither number is significant. In other words, conditions can prevail in which buoyancy effects are of

---

[a] September, 1963, Task Committee on Preparation of Sedimentation Manual (Proc. Paper 3639).
[2] Dir., Inst. of Hydr. Research, Univ. of Iowa, Iowa City.

appreciable magnitude but are resisted by viscous stresses which are great enough to make inertial resistance to motion relatively negligible. The pertinent nondimensional parameter can be obtained in the usual manner, or (somewhat similar to the author's J) simply by dividing the square of the Froude number by the Reynolds number to eliminate the density, with the following result:

$$G = \frac{\mu V}{\Delta \gamma L^2} \text{ or } \frac{\mu V}{L d\gamma/dz} \quad \ldots \ldots \ldots \ldots (2\text{-}L.15)$$

Resistance to uniform or gradually varied laminar flow, or evenly rapidly varied flow in a porous medium, is a function of this parameter and not of the Froude and Reynolds numbers (unless ρ is willfully or carelessly included in a self-compensating manner).

Because it is gravity rather than inertia that is involved in all of the phenomena under consideration, it is at best ambiguous to call them "density" currents. (From the purist point of view, in fact, the closest approach to a true density current is a submerged jet or wake, for at the limit, it can be regarded as dependent on the constancy of momentum flux alone.) The author of this definitive paper has already shown a clear perception of the connotations involved in the various terms in use. He would be doing the profession a great further favor if he would eliminate the misnomer from the title and—so far as practicable—from the text as well.

Journal of the

HYDRAULICS DIVISION

Proceedings of the American Society of Civil Engineers

CRITICAL ANALYSIS OF OPEN-CHANNEL RESISTANCE

By Hunter Rouse,[1] F. ASCE

INTRODUCTION

For many centuries, if not millenia, the resistance to flow in open channels has claimed the attention of hydraulic engineers. Even the simple question of which way—and about how much—to slope an irrigation ditch involves the basic aspects of channel design, and this matter must already have arisen in pre-historic times. The canals of ancient Egyptian, Sumerian, and Indian civilizations were no small undertakings, moreover, and the aqueducts of Roman times showed considerable planning skill, if only of the rule-of-thumb type. As early as the sixteenth or seventeenth century such present-day correction measures as the straightening of rivers had been attempted, and by the end of the seventeenth century at least one treatise on river hydraulics had appeared in print.

Since that time the literature on open-channel flow has become so replete as to have warranted a considerable historical treatment in its own right.[2] The accomplishments have surely been both numerous and great. As in many another phase of technical endeavor, nevertheless, the hydraulic horizon has advanced quite as rapidly as the state of hydraulic knowledge, and Du Buat's tabulation[3] of gaps that still existed some two hundred years ago might almost be used today as a model for the future. Besides, even historical treatments are of necessity subjective, and what might seem a closed chapter one year could well be reopened for revision the next.

In this period of wide-spread scientific development, a review of existing knowledge with regard to open-channel hydraulics might appear to be as lacking in timeliness as it is in glamour. Yet a glance at publications on the

---

Note.—Discussion open until December 1, 1965. To extend the closing date one month, a written request must be filed with the Executive Secretary, ASCE. This paper is part of the copyrighted Journal of the Hydraulics Division, Proceedings of the American Society of Civil Engineers, Vol. 91, No. HY4, July, 1965.

[1] Dir., Inst. of Hydr. Research, Univ. of Iowa, Iowa City.

[2] Leliavsky, S., "Historic Development of the Theory of the Flow of Water in Canals and Rivers," The Engineer, Vol. 191, April 20 through May 11, 1951.

[3] Rouse, H., and Ince, S., "History of Hydraulics," Dover Publications, Inc., New York, N. Y., 1963.

subject during the past decade or so will reveal many a significant anomaly. Momentum and energy analyses are at the same time over-simplified and confused with one another. Representation of parameters by symbols is mistaken for empirical formulation of functions. Flow formulas are sometimes said to involve the Froude number when they do not, and yet the Froude number is as frequently ignored when it is actually essential. Boundary texture and cross-sectional nonuniformity are often discussed without distinction. Roughness measures are mistaken for r e s i s t a n c e coefficients and are permitted to reflect not only viscous effects but even changes in depth. In a word, principles of mechanics that have proved useful in other phases of fluid motion are still not generally applied to problems of open-channel flow, which perforce remain subject to empirical solution—all too often without even the guidance of physical reasoning.

Certain of the research laboratories in various countries have sought to rectify the situation by approaching the subject in question in as fundamental a manner as possible. The writer's organization, in pursuit of this goal, has maintained a p r o g r a m covering various aspects of free-surface f l o w, endeavoring to correlate these not only with each other but also with comparable aspects of flow in closed conduits and around floating and immersed bodies. Whereas the results are neither more precise nor more numerous than those obtained elsewhere, they are more accessible and familiar to the writer and indeed were in considerable measure obtained for the present paper. He hence asks the reader's indulgence if Iowa material seems to play a preponderant part in the illustration of his remarks.

Whereas the patience of the general reader is invited because of the apparent bias of the illustrative material, that of the design engineer will surely be demanded by the writer's predilection for seemingly esoteric matters of no immediate application. However, it need only be recalled that the Chezy equation was unheard of two centuries ago; the Manning type of formula, one century; the Reynolds-number resistance diagram only a half century; and the Kármán-Prandtl relationships little more than a quarter century—yet all are in regular use today. Application first requires formulation and verification, and it is hence to those engaged in research rather than in design that the following discussion is directed. Only as the many complexities that are discussed become resolved by continued study will the results become readily applicable. The design engineer need only rest assured that progress will continue to be made at least as rapidly as in the past.

The continuing Iowa program can be divided rather arbitrarily into four different periods. First, prewar investigations of a variety of topics: bridge piers, flood waves, the hydraulic jump, and bends, all of which are described elsewhere,[4] the sequence culminating in the work by Powell[5] on bar-type artificial roughness. In the immediate postwar period, attention was focused

---

4 "Two Decades of Hydraulics at the University of Iowa," edited by F. T. Mavis, Studies in Engineering, Bulletin 19, Univ. of Iowa, Iowa City, 1939.

5 Powell, R. W., "Flow in a Channel of Definite Roughness," Transactions, ASCE, Vol. 111, 1946.

on boundary-layer resistance of smooth and rough surfaces,[6,7] in the effort to correlate roughness effects in uniform conduits;[8] attention was also given to high-velocity flow at sills[9] and enlargements.[10] The 1950's saw an intensified attack on matters of open-channel roughness and flow stability.[11] In the 1960's, not only are these studies being extended to both smooth and more naturally roughened channels, but also to wave effects at local constrictions[12] and presently to detailed evaluations of mean-flow and turbulence patterns at bends. Through each of these decades research in sediment transportation was in continuous progress,[13,14] though its bearing upon the open-channel resistance function never received primary emphasis. The related phenomenon of air entrainment has also been disregarded.

Notation.—The letter symbols adopted for use in this paper are defined where they first appear and listed alphabetically in the Appendix.

BASIC PHYSICAL AND DIMENSIONAL CONSIDERATIONS

One may denote by the symbol $\partial F/\partial s$ the local (i.e., cross-sectional) resisting force per unit length of channel. For the case of steady uniform flow, this will be equivalent to the product of the wetted perimeter of the section and the mean intensity of boundary shear. For accelerated flow, however, the forces involved in the acceleration must be distinguished from those involved in the actual resistance. In other words, work is done both conservatively and dissipatively, and it is only the latter which is now under consideration. The force per unit length $\partial F/\partial s$ is therefore that which, when multiplied by the mean velocity of flow, V, yields the rate of energy dissipation. Upon division by the specific weight, γ, and rate of flow, Q, the rate of energy dissipation takes the more familiar form

$$\frac{V \; \partial F/\partial s}{\gamma \; Q} = - \frac{\partial H}{\partial s} \quad \dots \dots \dots \dots \dots \dots \quad (1)$$

in which $-\partial H/\partial s$ is the longitudinal slope of the line of total head in steady

6 Baines, W. D., "An Exploratory Investigation of Boundary-Layer Development on Smooth and Rough Surfaces," thesis presented to the University of Iowa, at Iowa City, in August, 1950, in partial fulfilment of the requirements for the degree of Doctor of Philosophy.

7 Moore, W. L., "An Experimental Investigation of the Boundary-Layer Development along a Rough Surface," thesis presented to the University of Iowa, at Iowa City, in August, 1950, in partial fulfilment of the requirements for the degree of Doctor of Philosophy.

8 Hama, F., "Boundary-Layer Characteristics for Smooth and Rough Surfaces," Transactions, SNAME, Vol. 62, 1954.

9 Forster, J. W., and Skrinde, R. A., "Control of the Hydraulic Jump by Sills," Transactions, ASCE, Vol. 115, 1950.

10 Rouse, H., Bhoota, B. V., and Hsu, E. Y., "Design of Channel Expansions," Transactions, ASCE, Vol. 116, 1951.

11 Koloseus, H. J., "The Effect of Free-Surface Instability on Channel Resistance," thesis presented to the University of Iowa, at Iowa City, Iowa, in August, 1958, in partial fulfilment of the requirements for the degree of Doctor of Philosophy.

12 Hsieh, T., "Resistance of Cylindrical Piers in Open-Channel Flow," Journal of the Hydraulics Division, ASCE, Vol. 90, No. HY1, Proc. Paper 3770, January, 1964.

13 Laursen, E. M., and Toch, A., "Scour Around Bridge Piers and Abutments," Bulletin No. 4, Iowa Highway Research Bd., Ames, May, 1956.

14 Laursen, E. M., "The Total Sediment Load of Streams," Journal of the Hydraulics Division, ASCE, Vol. 84, No. HY1, Proc. Paper 1530, February, 1958.

nonuniform flow, and that part of the slope corresponding to the local dissipation rate in unsteady flow.

Experience indicates that $\partial F/\partial s$ should be a function of the following independent variables: the mean depth, d, as well as the mean velocity, V, at the given section; a length, k, describing the surface roughness; a dimensionless factor, ζ, describing the shape of the cross section, another factor, θ, describing the channel profile, and still another factor, η, describing the channel plan; the fluid density, ρ, specific weight, γ, and viscosity, μ; and the rate of change of either depth with time or discharge with distance in the direction of flow; that is,

$$\frac{\partial F}{\partial s} = F\left(d,\ V,\ k,\ \zeta,\ \theta,\ \eta,\ \rho,\ \gamma,\ \mu,\ \frac{\partial d}{\partial t}\right) \quad \ldots \ldots \ldots \ldots \ldots (2)$$

If the depth, velocity, and density are chosen to appear in combination with each of the remaining variables, the following nondimensional groups will result:

$$\frac{\partial F/\partial s}{\rho\, V^2\, d} = \phi\left(\frac{k}{d},\ \zeta,\ \theta,\ \eta,\ \frac{V}{\sqrt{d\gamma/\rho}},\ \frac{V\, d}{\mu/\rho},\ \frac{\partial d/\partial t}{V}\right) \quad \ldots \ldots (3)$$

The term at the left is of the nature of a local (i.e., cross-sectional) coefficient. So far as its magnitude is concerned, it will be equivalent to one-half the Weisbach resistance coefficient if the section is circular and the flow occurs at a uniform depth equal to the radius. On the assumption of uniformity it can also be related to the Chezy discharge coefficient by the condition of equilibrium that must obtain between the boundary shear and the longitudinal component of weight. However, if it is assumed that the resistance coefficient, f, applies only to the section under consideration, and if the other terms of the nondimensional function are given equally simple symbols, the relationship will take a s c h e m a t i c form that is pertinent to the subsequent general discussion:

$$f = f(\mathbf{R},\ K,\ C,\ N,\ \mathbf{F},\ U) \quad \ldots \ldots \ldots \ldots \ldots \ldots (4)$$

These symbols refer in a descriptive manner to one or more of those in the foregoing function and signify, in the order in which they will be discussed, the Reynolds number, the relative roughness, the cross-sectional shape, the nonuniformity of the channel in both profile and plan, the Froude number, and the degree of unsteadiness.

From a very gross point of view, these six variables might be considered even more significantly in three groups: the first three as the effect of surface resistance (for which the term "friction" has long been a misnomer), the next two as the additional effect of form and wave resistance, and the last simply as the departure therefrom due to local acceleration. In this light, the first three variables can be expected to introduce no principles not common to any conduit flow, whether closed or open; in particular, attention must be restricted to boundary irregularities of the size involved in defining surface texture to the e x c l u s i o n of those producing what is actually cross-sectional nonuniformity. The Froude number likewise plays no part in surface resistance, since no significant change in free-surface configuration takes place in uniform flow. By the same token, however, a proper analysis of non-uniform free-surface flow cannot be made without taking the Froude number, either implicitly or explicitly, into account. If the flow is also unsteady, both surface and form effects must be expected to vary with time.

SURFACE RESISTANCE

Application of modern boundary-layer principles to conduit flow led, in the late thirties,[15] to considerable analytical justification for the form of the Kármán-Prandtl resistance relationships. In brief, the velocity-distribution function can be considered to include a wall zone in which the velocity, v, depends only on the normal distance, y, the wall roughness, k, the kinematic viscosity, ν, and the shear velocity, $\sqrt{\tau/\rho}$,

$$\frac{v}{\sqrt{\tau/\rho}} = \phi_1\left(\frac{y\sqrt{\tau/\rho}}{\nu}, \frac{y}{k}\right) \dots \dots \dots (5)$$

and a central zone which is independent of wall conditions except for $\sqrt{\tau/\rho}$,

$$\frac{v_{max} - v}{\sqrt{\tau/\rho}} = \phi_2\left(\frac{y}{y_{max}}\right) \dots \dots \dots (6)$$

In such circumstances, it is mathematically demonstrable[16] that the functional form of the zone of overlap between them must be logarithmic. This is,[17]

$$\frac{v}{\sqrt{\tau/\rho}} = \frac{1}{\kappa} \ln \frac{y}{y'} \dots \dots \dots (7)$$

in which κ is Kármán's universal constant; though there is experimental evidence that it is what he once termed a "variable constant" with a considerable range of variation, the only value that has been obtained with any consistency is about 0.4. The factor y' is simply the magnitude of y at which $v=0$; this parametric length evidently depends upon the remaining independent variables,

$$y' = \phi_3\left(\frac{\nu}{\sqrt{\tau/\rho}}, k\right) \dots \dots \dots (8)$$

the ratio of the viscosity to the shear velocity being proportional to the thickness of the wall zone of predominantly viscous action. The simplest possible relationship between y' and the other two lengths is

$$y' = C \frac{\nu}{\sqrt{\tau/\rho}} + D k \dots \dots \dots (9)$$

Assuming such a relationship to be valid, one must still expect the numerical

---

15 Millikan, C. B., "A Critical Discussion of Turbulent Flows in Channels and Circular Tubes," Proceedings, Fifth Internatl. Congress for Applied Mechanics, John Wiley & Sons, Inc., New York, N. Y., 1939.

16 "Advanced Mechanics of Fluids," edited by H. Rouse, John Wiley & Sons, Inc., New York, N. Y., 1959.

17 Rouse, H., "Elementary Mechanics of Fluids," John Wiley & Sons, Inc., New York, N. Y., 1946.

coefficients C and D to depend (like κ) upon experimental measurement for their evaluation, although it should be noted that D also involves the somewhat arbitrary definition of k. Thanks to the systematic experiments of Nikuradse[18,19] and Laufer[20,21] not only is Eq. 7 (with $\kappa = 0.4$) known to apply with good approximation over the major portion of both pipes and very wide rectangular conduits, but—with k defined as the diameter, k_s, of the uniform sand grains that Nikuradse cemented to the boundary—for the limiting conditions in which either viscous or roughness influence completely outweighs the other,

$$C = 0.108, \quad D = 0.033 \ldots \ldots \ldots \ldots (10)$$

As will be noted directly, there is also good justification for the form of Eq. 9 in many circumstances.

Although in some respects such a logarithmic relationship is cumbersome, it has the very definite advantage of considerable analytical justification, which is not shared by any purely empirical formulation. Moreover, it readily lends itself to integration over at least elementary types of cross section to yield an expression for the mean velocity

$$V = \frac{1}{A} \int_A v \, dA \ldots \ldots \ldots \ldots \ldots (11)$$

which, through introduction of the definition equation

$$f = 8 \frac{\tau/\rho}{V^2} \ldots \ldots \ldots \ldots \ldots (12)$$

is directly transformable into an expression for the resistance coefficient, or

$$\frac{1}{\sqrt{f}} = A \log \frac{R}{y} + B \ldots \ldots \ldots \ldots \ldots (13)$$

in which R is the hydraulic radius. (The quantity $1/\sqrt{f}$, it should be noted, is the Chezy discharge coefficient divided by $\sqrt{8g}$.) If the cross section is circular, the coefficients (with a slight empirical correction) take the form

$$A = 2, \quad B = -0.62 \ldots \ldots \ldots \ldots \ldots (14)$$

and, if rectangular with great width-depth ratio (with a comparable correction), the form

$$A = 2, \quad B = -0.79 \ldots \ldots \ldots \ldots \ldots (15)$$

The numerical factors distinguishing the two relationships are evidently constants of integration reflecting the shape of the cross section. Except for a slight adjustment in all three factors to yield closer accord with experiment,

---

[18] Nikuradse, J., "Gesetzmässigkeiten der turbulenten Strömung in glatten Rohren," VDI Forschungsheft 356, Berlin (Düsseldorf) Germany, 1932.

[19] Nikuradse, J., "Strömungsgesetze in rauhen Rohren," VDI Forschungsheft 361, 1933.

[20] Laufer, J., "Investigation of Turbulent Flow in a Two-Dimensional Channel," Report No. 1053, NACA, Washington, D. C., 1951.

[21] Laufer, J., "The Structure of Turbulence in Fully Developed Pipe Flow," Report No. 1174, NACA, Washington, D. C., 1954.

the usual forms of the resistance equations (including that of Colebrook[22] for combined viscous and roughness action) result from the introduction of Eq. 9, with the designated values of A, B, C, and D, into Eq. 13. Its general form is retained for the present because of the greater clarity that this will yield.

Though measurements for wide conduits with smooth boundaries are rather sparse, those for rough boundaries are surely profuse—but, by and large, so poorly correlated with the essential parameters as to be of relatively little general use. Since the time of Nikuradse's experiments, it has remained customary to designate roughness in terms of the equivalent sand-grain dimension, k_s. For its proper description, however, a statistical characteristic such as surface texture requires a series of lengths or length derivatives, though the significance of successive terms in the series rapidly approaches a minimum. The eventual analysis of natural roughness will probably proceed in terms of the mean displacement of a stylus[23] and the mean of its first and second derivatives with distance traversed.[24] For the present, however, the nature of the roughness effects is still best seen in terms of the size, shape, and spacing of artificial roughness elements. To date the most significant study of this sort was that undertaken by Schlichting in 1936.[25] The element studied over the greatest range of spacing was the sphere, the results of which are reproduced in Fig. 1 with the ratio of k_s to height, h, plotted against the areal concentration, λ. Under the assumed conditions of a sufficiently high Reynolds number for the resistance of a smooth surface to be negligible in comparison, the two ends of the curve can readily be extrapolated to zero at the concentration limits of zero and unity. Superposed upon the diagram are points determined by Koloseus[11] and by Koloseus and Davidian[26] for cubical elements in a symmetrical diamond-shaped pattern, and other points recently obtained by O'Loughlin, and Macdonald[27] for cubes arranged in the Schlichting pattern and in the same pattern turned through 90°. The diagram also contains the preliminary results of measurements by O'Loughlin and Macdonald[27] on 1/10-in. sand grains cemented to the channel bed in controlled concentrations.

Several significant conclusions can be drawn from this family of curves. Immediately evident is the fact that there is an optimum concentration of 15 to 25% which produces the greatest relative resistance. Below a value of about 15% the resistance varies in direct proportion to the concentration (see Fig. 1, or the logarithmic plot of Fig. 2), and the proportionality coefficient varies approximately with the relative drag of the individual elements. Indeed, unpublished air-tunnel measurements by B. W. Hunt of the drag of a cube with slightly rounded edges, a sphere, and a simulated sand grain, all in contact with a surface, yielded 0.9, 0.7, and 0.6 as drag coefficients, the slopes of the

---

22 Colebrook, C. F., "Turbulent Flow in Pipes with Particular Reference to the Transition Region between the Smooth and Rough Pipe Laws," <u>Journal of the Institution of Civil Engineers</u>, London, Vol. 10, 1939.

23 Todd, F. H., "Skin Friction Resistance and the Effects of Surface Roughness," <u>Transactions</u>, SNAME, Vol. 59, 1951.

24 Posey, C. J., "Measurement of Surface Roughness," <u>Mechanical Engineering</u>, ASME, New York, Vol. 68, No. 4, 1946.

25 Schlichting, H., "Experimentelle Untersuchungen zum Rauhigkeitsproblem," <u>Ingenieur Archiv</u>, Vol. 7, 1936; see also <u>Boundary Layer Theory</u>, McGraw-Hill, 1955.

26 Koloseus, H. J., and Davidian, J., "Roughness-Concentration Effects on Flow over Hydrodynamically Rough Surfaces," <u>U.S.G.S. Water-Supply Paper</u>, (in press).

27 O'Loughlin, E. M., and Macdonald, E. G., "Some Roughness-Concentration Effects on Boundary Resistance," <u>La Houille Blanche</u>, No. 7, 1964.

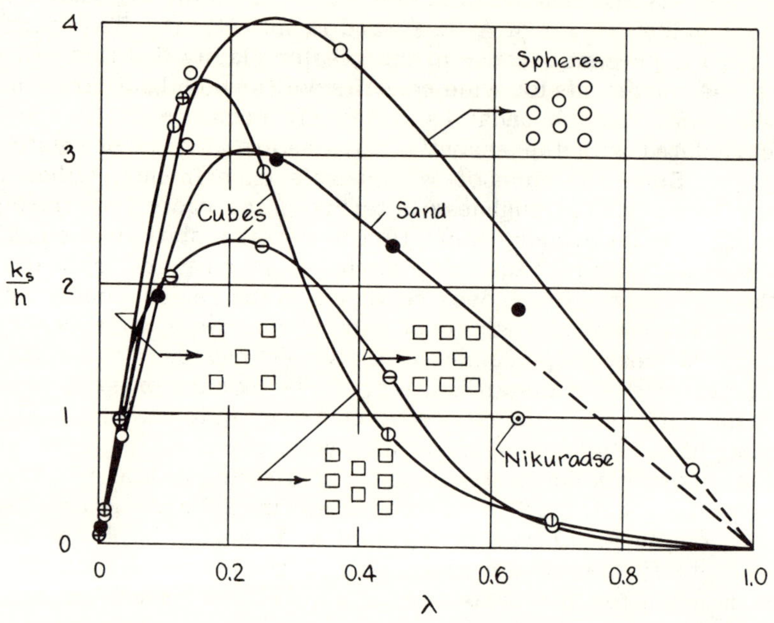

FIG. 1.—EFFECTIVE ROUGHNESS AS A FUNCTION OF FORM, PATTERN, AND CONCENTRATION OF ROUGHNESS ELEMENTS[25,26,27]

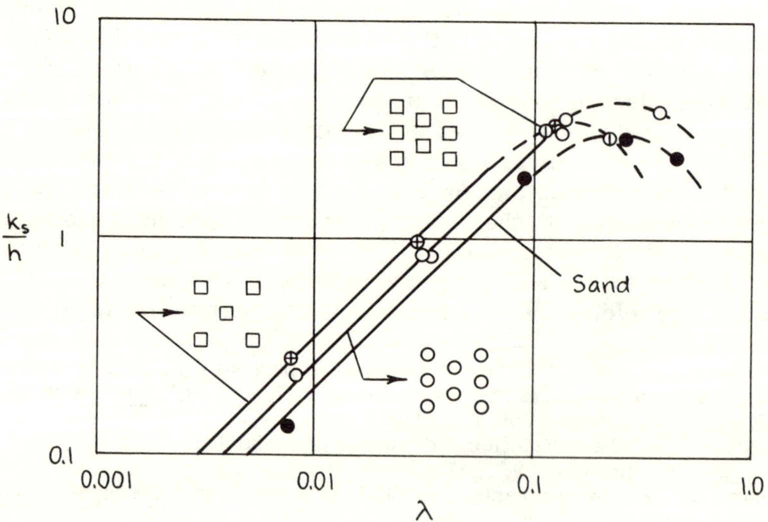

FIG. 2.—LOGARITHMIC PLOT OF DATA FROM FIG. 1 AT LOW CONCENTRATIONS

respective curves in Fig. 1 being 32, 26, and 21. In other words, the resistance function for fully developed roughness of height, h, and concentration, λ, is, for $\lambda < 0.15$, of the form

$$\frac{1}{\sqrt{f}} = A \log \frac{R}{D\ h\ \lambda} + B \ldots\ldots\ldots\ldots\ (16)$$

in which D is now a function of both the shape and the arrangement of the roughness elements. The constant of proportionality may, of course, be removed from the logarithmic term, B then being affected accordingly. For the sanded surface, for example,

$$D = 21, \quad B = 2.17 \quad \text{or} \quad D = 1, \quad B = -0.46 \ldots\ldots\ (17)$$

The influence of element arrangement is evidently negligible at low concentrations, whereas the shadow effect that is especially pronounced when the elements are directly in line is readily apparent from the early deviation of two of the three curves for the different patterns of cubes. Both the curvilinearity and the reversal in position of the cube curves at higher concentrations render them too complex for ready interpretation. On the other hand, the linear nature of both the sphere and the sand curves is conveniently simple, and for at least these two roughness types, the resistance function is of the form

$$\frac{1}{\sqrt{f}} = A \log \frac{R}{D\ h(1-\lambda)} + B \ldots\ldots\ldots\ldots\ (18)$$

For the sanded surface,

$$D = 4.2, \quad B = 2.17 \quad \text{or} \quad D = 1, \quad B = 0.92 \ldots\ldots\ (19)$$

Because the concentration of Nikuradse's sanded surfaces is not known, it can only be assumed to be the same as the maximum attained by O'Loughlin and Macdonald. Its failure to agree with the O'Loughlin-Macdonald curve is surely attributable not only to shape effects but also to the impossibility of either duplicating or estimating the concentration that Nikuradse actually realized. In such circumstances, one is inclined to question the continued desirability of expressing roughness magnitudes in terms of the Nikuradse value.

However extreme the artificiality of the cube as a roughness element actually is, it had three advantages to recommend it for the series of studies still in progress at Iowa: its convenience of manufacture, its comparative freedom from viscous effects, and the continuous approach of an array toward a limiting smooth surface as the concentration approached unity. Whatever the form of the discrete element, however, its use for purposes of both experimental and analytical exploration of the roughness function is well justified by the extreme complexity of natural roughness. One numerical approach to synthesis of the function for low concentrations was undertaken by Roberson[28] with the cube as the elementary form. Air-tunnel measurements were first made of the drag of a cube as a function of both the velocity and the velocity

---

28 Roberson, J. A., "Surface Resistance as a Function of the Concentration and Size of Roughness Elements," thesis presented to the University of Iowa, at Iowa City, Iowa, in August, 1961, in partial fulfilment of the requirements for the degree of Doctor of Philosophy.

gradient. It was then assumed that the drag of discrete elements attached to a plane boundary could be considered to augment the mean shear and thus control the mean slope and axis intercept of the logarithmic velocity curve for the otherwise smooth surface. Roberson was able, in this manner, to effect a computer evaluation of the resistance coefficient for any Reynolds number above 500 and any concentration of cubes below 10%. Typical curves are reproduced in Fig. 3. The characteristic Colebrook type of transition has obviously been realized. Moreover, the computer program readily lends itself to evaluation of the resistance for a statistical distribution of element sizes about the mean. Quantitatively, however, there are still discrepancies between the predicted resistance and that actually measured as a function of concentration. Together with the fact that no prediction whatever can yet be made for the shadow effect at greater concentrations, this means that studies involving drag measurements not only on single elements but on arrays of elements coupled with the intervening boundary are still essential to the clarification and extension of Roberson's analysis. Such an investigation is currently being conducted by O'Loughlin.

So far as the matter of cross-sectional shape is concerned, reference to the derivation of Eq. 13 will reveal that the effect of a change in shape upon the resistance function is actually twofold. On the one hand, it produces a change in the wetted perimeter, P, per unit cross-sectional area, A, the reciprocal of which is designated by the hydraulic radius, R. On the other hand, it produces a change in the distribution of velocity and shear; as a result, the shear will generally vary from point to point of the perimeter, and the ratio $\partial F / P \partial x$ will correspond to its mean value τ_m. Both effects are thus involved in the equilibrium relationship between the gravitational motive force and the surface resistance which the flow entails:

$$\gamma \, A \, S = \tau_m \, P \quad \dotfill \quad (20a)$$

whence
$$\tau_m = \gamma \, R \, S \quad \dotfill \quad (20b)$$

Since Eq. 20 is an exact one for the conditions of uniform flow under consideration, the frequent remark that the hydraulic-radius concept is merely a rough approximation is in no way valid. The fallacy therein results from either ignoring the second aspect of the shape effect or expecting the hydraulic radius to compensate in some way for it. Even if the velocity distribution remained logarithmic in all zones with simply a local change in boundary shear, the numerical constants resulting from integration over the flow section would necessarily change with its shape. It is hence only logical to expect the coefficients in the generalized form of Eq. 13 to vary with the shape of section, whether or not the velocity distribution is logarithmic. Whereas the two effects may well offset one-another to some degree, only by considering them as distinct will their interdependence ever be discerned.

From the considerable amount of rather conflicting information that exists on the effect of cross-sectional shape, only a few generally significant conclusions can be drawn with any confidence. First, not only can the resistance of uniform laminar flow in a duct of any shape be determined either analytically or numerically, but experiments seem to confirm the validity of the results for either confined or free-surface flow. Second, for closed conduits of a given cross-sectional shape, the isovels of turbulent flow tend to differ in

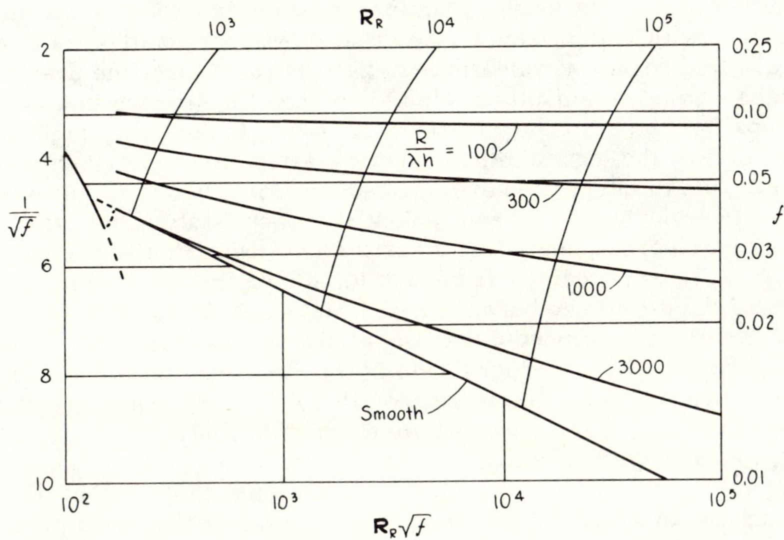

FIG. 3.—COMPUTER SYNTHESIS OF THE RESISTANCE FUNCTION FOR CUBICAL ROUGHNESS ELEMENTS[28]

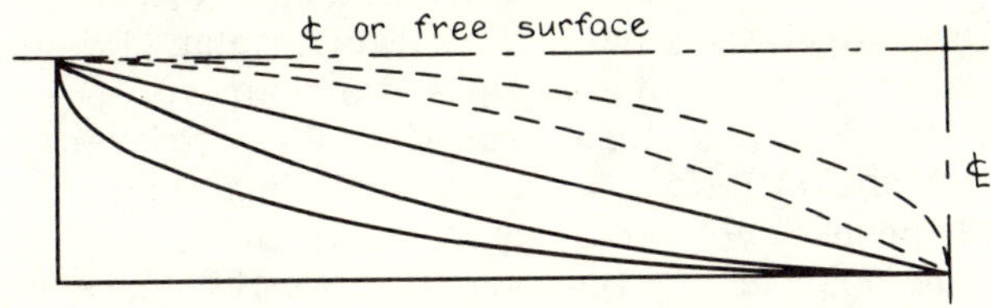

FIG. 4.—DEFINITION SKETCH FOR RECTANGULAR, ELLIPTICAL, CIRCULAR-ARC, AND TRIANGULAR CROSS SECTIONS

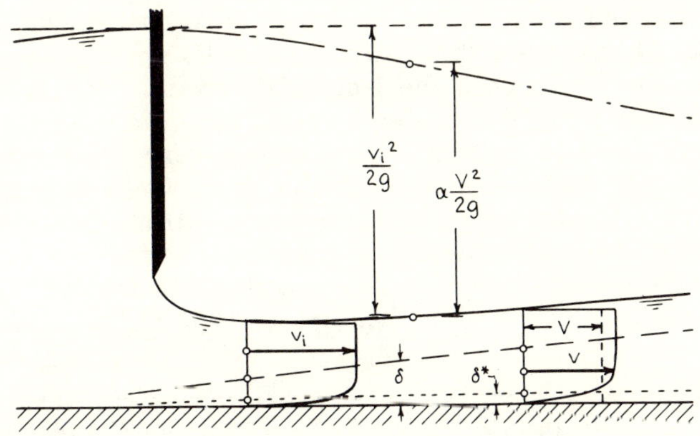

FIG. 5.—ESTABLISHMENT OF FLOW DOWNSTREAM FROM A SLUICE GATE

basic configuration from those of laminar flow because of the fact that secondary motion with components in the cross-sectional plane tends to develop in zones of rapid boundary variation—i.e., corners. Third, the presence of a free surface at what would otherwise be a plane of symmetry in a continuous medium modifies the secondary pattern and hence the primary as well. As a result, not only is the logarithmic velocity-distribution relationship not generally applicable to integration over closed sections, but the additional complication of free-surface influence makes the open section one step further removed from rigorous analytical investigation. Nevertheless, certain indications can still be perceived from application of the foregoing knowledge.

In spite of the difference between the laminar and the turbulent patterns of motion, it is still to be expected that the effect of cross-sectional shape will be of comparable order of progression in the two, even if not of comparable magnitude. For example, conversion of the Poiseuille equations[16] for the very wide and circular sections will yield for the quantity $\mathbf{R}_R f$ the numerical magnitudes 24 and 16, respectively. After an exact analysis by Boussinesq,[29] the value 14.2 is obtained for the square section and 13.3 for that of an equilateral triangle; an exact evaluation for the 2:1-elliptical section provides the value 16.9, evidently somewhat on the other side of that for the circular. Although exact analyses can be performed for laminar flow through conduits of any cross section,[16] an approximation can readily be obtained for sections that are relatively wide compared with their height by assuming the velocity variation in the vertical to be independent of that in the horizontal and proceeding by simple application of the distribution for plane Poiseuille flow. The results become exact as the width-height ratio approaches infinity, but even for ratios of 10:1 the error is less than 1%. With reference to Fig. 4, the limiting values for the elliptical, circular-arc, and triangular sections are 19.8, 13.3, and 12, respectively—evidently again either side of the circular-arc section. (Values approaching the zero limit for the continuing trend shown in broken lines are possible, of course, but without present significance.)

Although the statement is often made that the logarithmic velocity curve for turbulent flow can be integrated over only the circular and wide rectangular sections, in actuality it can also be used to approximate the resistance of any wide section such as those of Fig. 4. Not only should the relative accuracy be comparable to that for laminar flow, but the effect of secondary motion should still be neglibible for both closed and open conduits. However, reference to Eq. 13 will show that different results must be expected as the channel surface varies between the limits of smooth and rough; in a very smooth channel, that is, y' will reflect the variation in viscous effects from one part of the section to another, whereas in a very rough channel y' will vary only with k. On the other hand, it has been shown by Macagno[30] that the process of integration over an irregular section will of necessity change the magnitude of both coefficients of Eq. 13—contrary to the writer's oft-stated belief that the second one alone embodies the shape factor. For a rough chan-

[29] Boussinesq, J., "Essai sur la Théorie des Eaux Courantes," Mémoires présentées par divers Savants à l'Académie des Sciences, Vol. 23, 1877.

[30] Macagno, E. O., "Resistance to Flow in Channels of Large Aspect Ratio," Journal of Hydraulic Research, (in press).

nel of circular-arc form, Macagno and Hung have determined the coefficients to be[31]

$$A = 2.2, \quad B = -0.78 \quad \dots\dots\dots\dots\dots\dots (19)$$

and, for the triangular,

$$A = 2.3, \quad B = -0.70 \quad \dots\dots\dots\dots\dots\dots (20)$$

(The latter values, it should be noted, have been verified experimentally.[32]) For the circular-arc section, moreover, their analysis of the effect of form without regard to secondary flow has shown that A and B will decrease only about 1% as the width-height ratio is reduced as much as from ∞ to 3. As a rough approximation, the effect of secondary flow might be assumed to be of comparable magnitude.

EFFECT OF BOUNDARY NONUNIFORMITY

Any change in shape or size of the flow section that occurs with distance along the channel axis will necessarily be accompanied by a change in the velocity distribution in accordance with the basic equations of fluid motion. These equations are, at present, more significant qualitatively than they are useful quantitatively, for their numerical application is still limited to relatively simple flow conditions. In the case of gradually varied flow represented by backwater analysis it remains customary to ignore such niceties on the assumption that the resistance at any section is equal to what it would be if the same rate of flow took place past the same section under conditions of uniformity. In most instances the error involved in this assumption is probably negligible. There are, nevertheless, at least two circumstances in which the acceleration is sufficiently small for the pressure distribution an any section to be essentially hydrostatic and yet sufficiently large for the boundary-layer development to play an important role. The one is flow down a chute beyond a low spillway,[33] and the other is flow on the apron of a gate.[34] Because of its greater simplicity, the latter will be the one used for illustration.

As indicated in Fig. 5, the flow emerging from the pool behind the gate is very nearly irrotational. As long as any part of it remains irrotational, moreover, that part will be characterized by the initial total head. In the vicinity of the gate section a boundary layer begins to develop, and its continued growth with distance downstream entails the deceleration of a steadily increasing portion of the flow. For reasons of continuity, such modification of the velocity distribution is tantamount to an upward displacement of the flow a distance equal to

---

31 Macagno, E. O., and Hung, T. K., discussion of "Shear and Velocity Distribution in Shallow Channels," Journal of the Hydraulics Division, ASCE, Vol. 90, No. HY5, Proc. Paper 4050, September, 1964, p. 170.

32 Engelund, F., "Flow Resistance and Hydraulic Radius," Acta Polytechnica Scandanavica, 1964.

33 Bauer, W. J., "Turbulent Boundary Layers on Steep Slopes," Transactions, ASCE, Vol. 119, 1954.

34 Rouse, H., "Laboratory Instruction in the Mechanics of Fluids," Studies in Engineering, Bulletin 41, Univ. of Iowa, Iowa City, 1961.

$$\delta^* = \int_0^d \left(1 - \frac{v}{v_i}\right) dy \quad \cdots \cdots \cdots \cdots \cdots \cdots (21)$$

which is aptly called the displacement thickness of the boundary layer. The increase in surface elevation evidently causes a decrease in velocity head (and hence a still further rise in surface), so that even the irrotational flow outside the boundary layer is decelerated, but without loss in total head. The actual loss up to any section is determined by integration of the velocity-distribution curve over the section, yielding, in effect, the quantity

$$\frac{1}{q} \int_0^d \frac{v^2}{2\,g} \, v \, dy = \alpha \, \frac{v^2}{2\,g} \cdot \quad \cdots \cdots \cdots \cdots \cdots (22)$$

which, added to the depth, gives the weighted mean total head for the section, for comparison with the total head of the originally irrotational flow.

For the simple conditions of this example, boundary-layer theory will permit satisfactory prediction of the variation in velocity—and hence in surface profile—with distance downstream. In the general case, on the other hand, not only must the velocity be expected to vary in a different manner over different parts of the cross section, but in few instances can the flow be assumed even approximately irrotational at any section. In a word, at least for the present, the change in surface resistance occasioned by nonuniformity of the flow section can be assessed only qualitatively. In many cases of boundary nonuniformity, however, surface effects play a role that is minor compared with those of shape and wave formation.

These three aspects of channel resistance have much in common with those of ship resistance, which to date have been studied much more systematically in both the laboratory and the field. The drag of a ship is known to be a function of its geometry (including surface texture) and the Froude and Reynolds numbers of its relative motion. While viscous effects may be predominant at model scale, at large scale the hull geometry tends to govern so far as surface and form drag are concerned, and these can then be considered essentially independent of the Reynolds number. At very low Froude numbers, free-surface disturbances are negligible, but a pattern of surface waves gradually develops and thereafter changes continuously in scale. At the higher Froude numbers, the wave drag (the force required to counterbalance the resulting variation in boundary pressure or, alternatively, to maintain the pattern of surface undulations against internal stresses) may completely dominate the picture.

There is no fundamental difference between the drag of a ship moving through still water and the resistance of a stationary pier to flowing water. If the drag per unit projected area in its ratio to the kinetic energy per unit volume is used as a resistance coefficient and plotted against the Froude number, a curve such as that of Fig. 6 will result. (The supporting observations, it should be noted, were made in the Iowa towing tank by H. E. Kobus and A. D. Newsham, specifically for this paper.) The irregularity of the resistance curve is evidently due to the progression of wave crests and troughs along the pier, the difference between the fore and aft pressure loads changing accordingly. A most significant fact about such a curve is that for a channel of un-

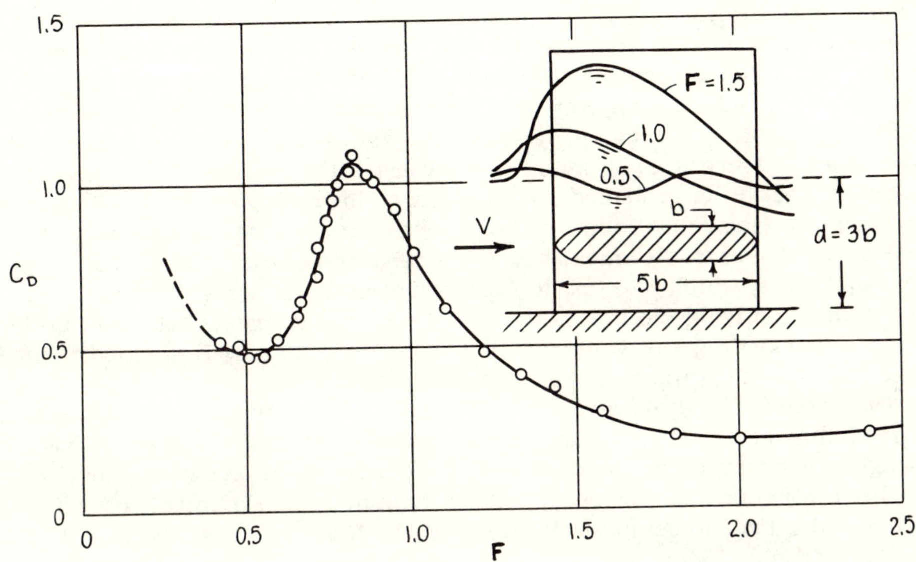

FIG. 6.—RESISTANCE OF A BRIDGE PIER IN A WIDE CHANNEL, AFTER KOBUS AND NEWSHAM

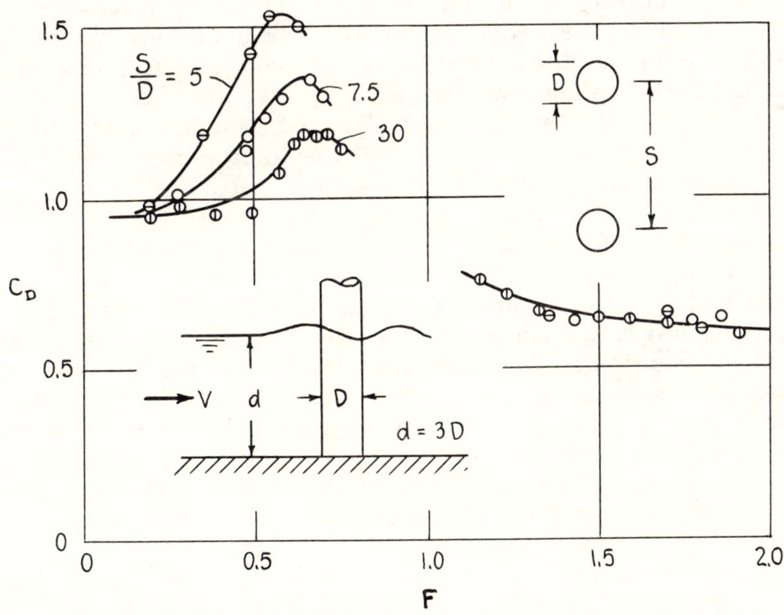

FIG. 7.—VARIATION OF PIER RESISTANCE WITH LATERAL SPACING[12]

limited width it is continuous through the critical Froude number of unity into the supercritical regime. This will not be the case, on the other hand, if the channel is of limited width, for in the vicinity of the critical stage, the pier will tend to choke the channel—that is, the wave formation will be such that flow at the original upstream depth and velocity will become impossible. From the reproduction in Fig. 7 of measurements by Hsieh,[12] the resulting gap in the function is clearly seen to increase in size as the channel width or pier spacing decreases. It actually matters little whether the nonuniformity of cross section is produced by a centrally located pier, by a sidewall contraction of comparable form—or even by a bottom rise. Such conditions are ordinarily evaluated by means of the one-dimensional simplification of the flow equations, which—except for the effect of local wave formation—is just as readily applicable to the case of the pier.[12] In any event, the gap in the curve representing the range of physically impossible flow conditions is a very real aspect of the resistance phenomenon.

The extreme irregularity of a boulder-strewn channel is sometimes simulated schematically by means of a succession of discrete weirs—a reasonable enough procedure if one realizes that the result must eventually become not merely a form of flow impedance but a type of nonuniformity that inherently involves the Froude number. Whereas K. K. Rao, in an extension of the work by O'Loughlin and Macdonald, found the resistance of 1.5-in. cubes at optimum concentration to follow Eq. 16 to R/h values as low as 3,[27] this lack of dependence upon the Froude number was presumably due to the extreme degree of wake interference. For depths greater than the boulder size and spacing, similar field conditions should prevail. At depths comparable to the size and spacing, on the contrary, each boulder should begin to act independently as a local nonuniformity. The valid argument has been advanced[35] that in a considerable reach it is not the individual nonuniformities but the overall channel slope that determines the cumulative rate of loss, since a local increase in resistance will simply cause a local change in surface configuration. Obviously, however, local effects can be evaluated only in terms of local causes, and any local constriction (submerged weirs included) will produce a resistance that varies with the Froude number.

Supercritical flow in bends is now commonly accepted as dependent upon the Froude number, but this has not yet come to pass for subcritical flow, despite the fact that wave resistance is just as important at changes in channel alignment as it is at changes in cross section. It is true that the Froude number for a specific bend will not vary greatly with the rate of flow, so that the resistance is commonly linked only with the bend geometry. However, in comparing one bend with another that is similar, if the Froude numbers are different, one should also anticipate a difference in the loss coefficient. Because the loss at a single bend is difficult to evaluate with any precision, six successive 90° bends of rectangular cross section were built into a tilting flume at Iowa, and the head loss per bend in its ratio to the velocity head was determined by Hayat[36] as a function of F at three different width-depth

---

[35] Ippen, A. T., and Drinker, P. A., "Boundary Shear Stresses in Curved Trapezoidal Channels," Journal of the Hydraulics Division, ASCE, Vol. 88, No. HY5, Proc. Paper 3273, 1962.

[36] Hayat, S., "The Variation of Loss Coefficient with Froude Number in an Open-Channel Bend," thesis presented to the University of Iowa, at Iowa City, in January, 1965, in partial fulfilment of the requirements for the degree of Master of Science.

ratios. The results, shown in Fig. 8, display both the anticipated variation with $\mathbf{F}$ at values below as well as above unity, and the characteristic gap due to choking in the critical region, varying in the expected manner with the width-depth ratio. Such loss of head and the resistance to flow which it represents are obviously related through the rate of energy dissipation which each entails;

$$H_L \, \gamma \, Q = F_D \, V \dots\dots\dots\dots\dots\dots (23)$$

EFFECT OF UNSTEADINESS

Because of the fact that free-surface flow which is unsteady is generally also nonuniform, there is considerable similarity between the arguments of

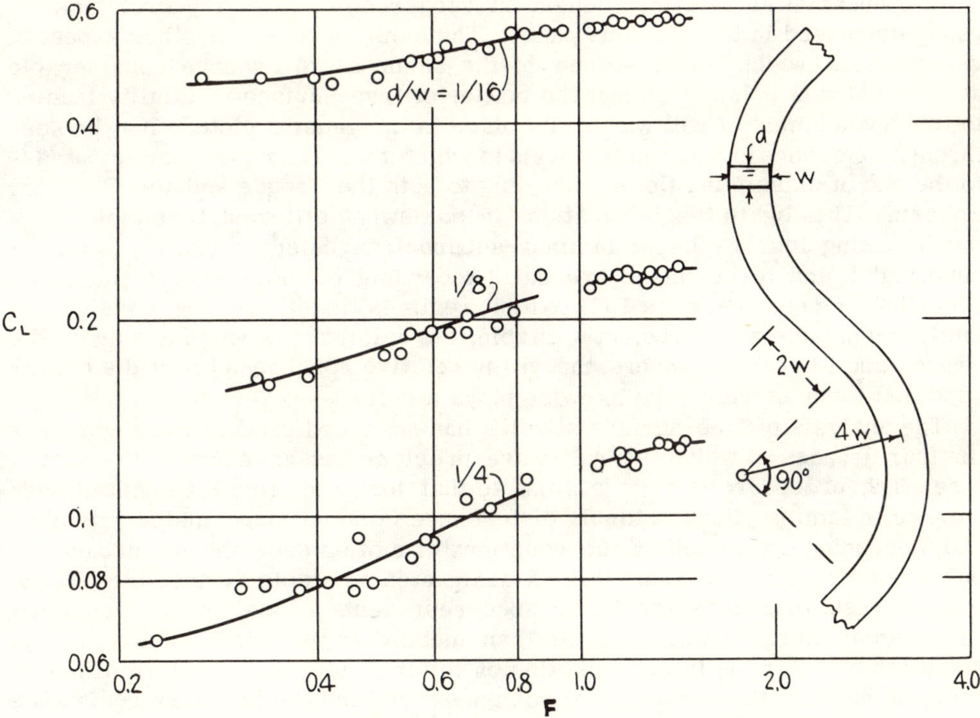

FIG. 8.—LOSS AT ONE OF A SERIES OF CHANNEL BENDS[36]

this section and the previous one. A change in depth, whether as a function of time or a function of distance, must bring with it a change in the velocity distribution in accordance with general boundary-layer principles. One extreme of such variation with respect to time is the relatively slight departure from irrotationality which a solitary wave undergoes as it is propagated through a body of otherwise still water. This phase of the general problem is proving amenable to approximate if not exact solution; unfortunately, the result is a mark of progress in wave analysis rather than in open-channel hydraulics, since a zero ambient flow is inherent to the solution. In other words, the analysis can by no means yet be generalized to wave propagation in a moving fluid, any more than can the application of steady-state boundary-layer theory to the effect of a cross-sectional change on a velocity distribution already established upstream. The other limit of the problem, comparable to gradually varied steady flow, has long since proved even more amenable to treatment. This is the propagation of true flood waves, which takes place in such a manner that inertial effects are small in comparison with resistance. Under these conditions the resistance can be assumed to have essentially the same magnitude as that of steady uniform flow at the same depth and velocity. The unsolved problem in this case is rather the extent of bed variation during the rising and falling stages, which is obviously a matter of sediment transportation in its full complexity.

Progress in the analysis of resistance to unsteady open-channel flow with fixed boundaries is evidently dependent upon resolution of the problems already discussed in the previous pages. The same is true of another aspect of unsteadiness, which is still so new to the scene as to deserve considerable attention at this point.[37] This is the matter of free-surface instability leading to the development of roll waves. Its place in the general picture may be seen through reference to a simple analysis in which the writer once participated[38] to the end of classifying flows according to both the Froude and the Reynolds criteria. This led to the designation (in somewhat different terminology) of the following four regimes: laminar-subundal, turbulent-subundal, laminar-superundal, and turbulent-superundal. According to more recent findings—even then foreseen—two additional regimes must also be considered: laminar-unstable and turbulent-unstable, the criteria for which are again the Froude and Reynolds numbers, though the relative roughness (if not the cross-sectional form as well) now likewise plays a role.

The analysis of free-surface stability has advanced gradually in rigor over the four decades in which the roll-wave problem has appeared in the literature. The procedure is quite similar to that for predicting the onset of turbulence in laminar flow: a small disturbance is superposed mathematically, and, depending upon whether the equations of motion show the disturbance to be damped or amplified, the flow is designated as stable or unstable. In the present case, of course, the disturbance represents a local change in depth—i.e., a small surface wave—rather than merely an internal displacement of streamlines. For the initial condition of uniform flow, a state of equilibrium prevails between the gravitational component γdS and the boundary resistance $\tau = \rho V^2 f/8$. If the flow is stable, the imposed change in depth will produce a resistance change that is greater than the change in gravitational component, and equilibrium will be restored. If the flow is unstable, on the contrary, the increase of gravitational effect will be greater than that of resistance, and the wave will grow as it progresses down the channel.

Any such analysis is perforce approximate rather than rigorous, because of the extreme complexity that rigor would involve, and the results vary con-

---

[37] Rouse, H., Koloseus, H. J., and Davidian, J., "The Role of the Froude Number in Open-Channel Resistance," Hydraulic Research, IAHR, Delft, Holland, Vol. 1, No. 1, 1963.

[38] Robertson, J. M., and Rouse, H., "On the Four Regimes of Open-Channel Flow," Civil Engineering, Vol. 11, No. 3, 1941.

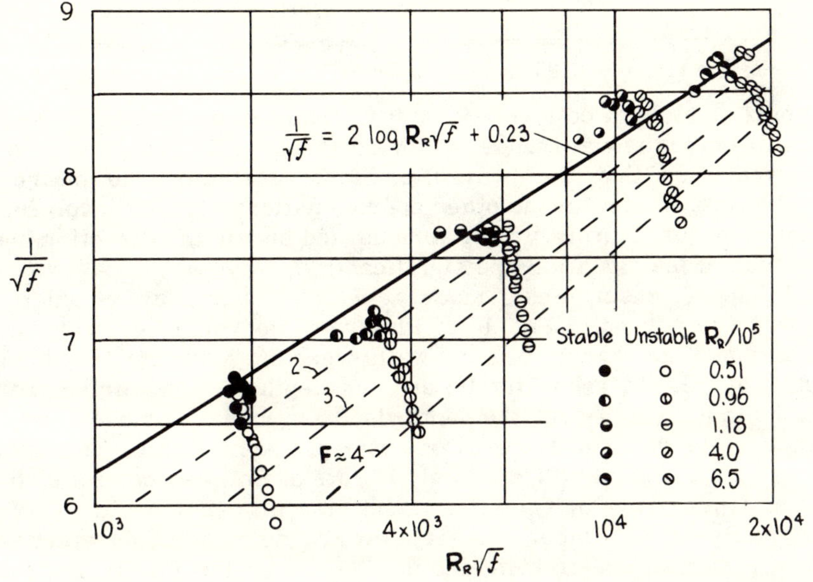

FIG. 10.—EFFECTS OF INSTABILITY ON SMOOTH-CHANNEL RESISTANCE, AFTER NEMEC

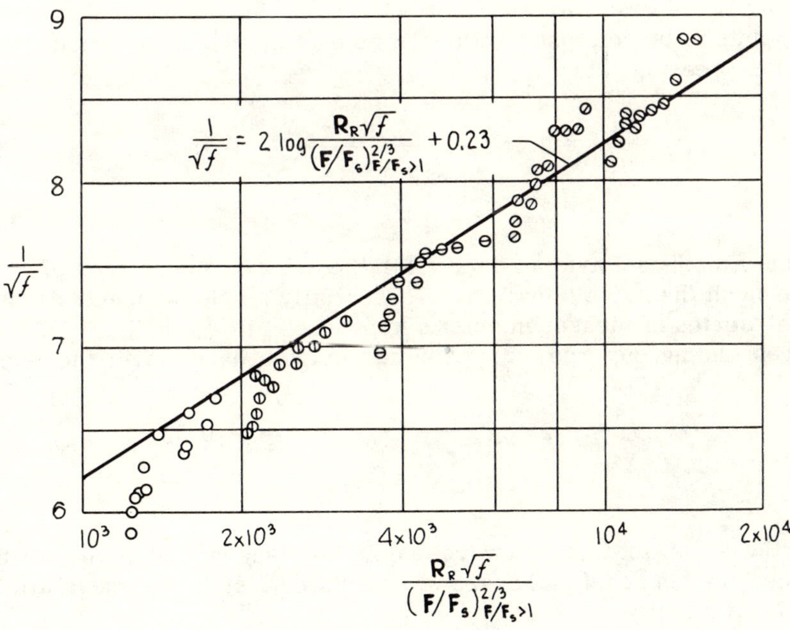

FIG. 11.—SUPERPOSITION OF DATA FROM FIG. 10

and, for a rough boundary,

$$\mathbf{F}_s = \frac{1}{\sqrt{(0.5 + 0.87\ f^{1/2} - 0.78\ f)^2 - 0.78\ f\ (1 + 0.78\ f)}} \quad \dots \ (30)$$

Fig. 9 contains plots of both relationships. Shown in broken lines are approximations of the individual trends with width-depth ratio, based upon the usual faulty assumption that the hydraulic radius embodies the entire cross-sectional effect. In view of the other approximations already involved, neglect of the latter error is probably not serious, and surely the direction and order of magnitude of the variation are significant.

Unpublished measurements made at Iowa in a smooth channel by J. G. Nemec are reproduced in Fig. 10 to illustrate the role of stability in resistance analysis. The channel bed and walls were of glass, the length and width were 85 ft and 2.5 ft, respectively, and the depth of flow varied from a few hundredths to 0.3 ft. Though the width-depth ratio was obviously not infinity, the effect of its departure therefrom was probably small. The successive families of plotted points differed only in rate of flow; since this involved the product of depth and velocity, for constant viscosity they could be considered of constant Reynolds number but variable Froude number, constant values of the latter parameter being indicated on Fig. 10 by the family of transverse lines. As is apparent from the Froude numbers distinguishing the lines, deviation from the heavy line corresponding to the resistance function for smooth boundaries begins approximately with the Froude number marking the border between stable and unstable flow. At least as a first approximation, moreover, use of the ratio between the actual Froude number and that corresponding to the limit of stable flow will, for values greater than unity, permit all points to be collapsed with fair approximation in accordance with the expression (see Fig. 11)

$$\frac{1}{\sqrt{f}} = 2\ \log\ \frac{\mathbf{R}_R\ \sqrt{f}}{\left[\mathbf{F/F}_s\right]^{2/3}_{\mathbf{F/F}_s > 1}} + 0.23 \ \dots \dots \ (31)$$

Use of the Koloseus-Davidian data[43] for the cubic-element type of roughness revealed much the same situation for essentially all the concentrations tested. A typical series of measurements is reproduced in Fig. 12, Fig. 13 showing the corresponding superposition of points in accordance with the expression

$$\frac{1}{\sqrt{f}} = 2\ \log\ \frac{R/\lambda\ h}{\left[\mathbf{F/F}_s\right]^{2/3}_{\mathbf{F/F}_s > 1}} - 0.82 \ \dots \dots \ (32)$$

Noteworthy is the fact that roll waves could be discerned near the downstream end of the flume in nearly all cases of calculated instability and in almost none of calculated stability.

Casual thought about the foregoing analysis will very likely lead to one aspect or another of the following apparent paradox. What is ostensibly a problem of surface resistance is seen to depend primarily upon the Froude number, contrary to a general statement made at the beginning of the paper. Alterna-

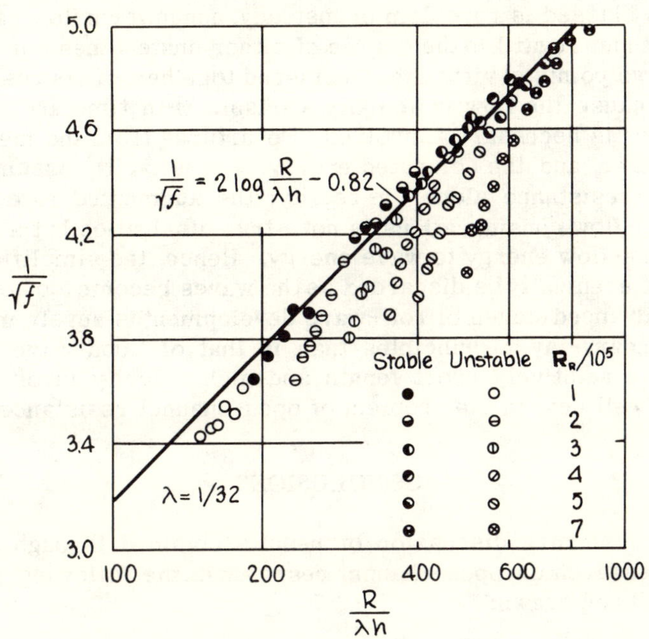

FIG. 12.—EFFECTS OF INSTABILITY ON ROUGH-CHANNEL RESISTANCE[37]

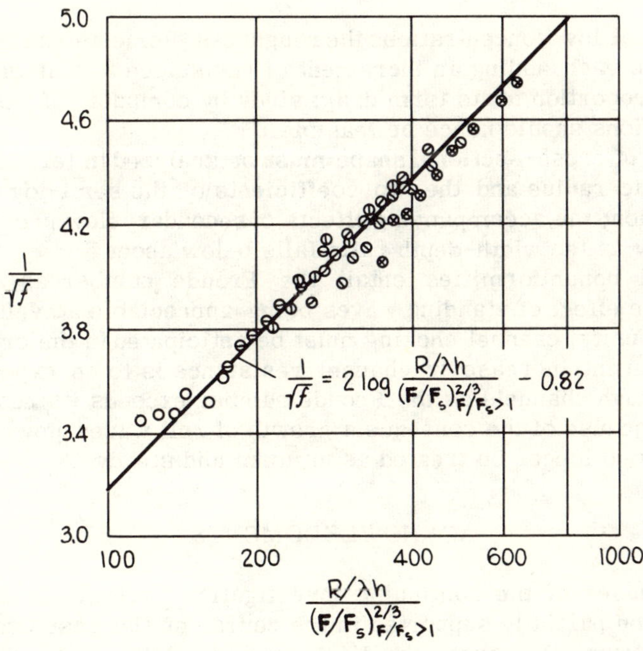

FIG. 13.—SUPERPOSITION OF DATA FROM FIG. 12

tively, what is classed as a problem of unsteady, nonuniform flow can seemingly be treated without regard to the degree of either unsteadiness or nonuniformity. If these two points of view are considered together, of course, the paradox vanishes. Because the flow is actually variable with time and distance, the Froude number is necessarily involved. Departures from the mean depth are ignored, however, and the computed energy loss is tacitly assumed to result from surface resistance alone. In reality, the augmented rate of loss indicated by mean-flow considerations is not a true dissipation but a transformation from mean-flow energy to wave energy. Hence, the simplified method of analysis must eventually be discarded as the waves become too large to ignore. Analysis of advanced stages of roll-wave development is surely more dependent upon boundary-layer principles than is that of flood-wave propagation, because of the relatively short length and high celerity involved, but it is nevertheless well beyond the problem of open-channel resistance itself.

CONCLUSIONS

From the foregoing discussion of results obtained through a continuing program of research on open-channel resistance, the following general conclusions have been drawn:

1. A valid basis for the analysis of wide channels as well as circular pipes is provided by the semi-logarithmic velocity-distribution and resistance laws.

2. The concentration of roughness elements is as essential as their size in determining surface resistance; an optimum concentration of 15% to 25%, depending upon element shape and arrangement, yields a maximum roughness effect.

3. At least at low concentrations the roughness elements can be considered as individuals, each adding an increment of resistance to that of the smooth boundary in proportion to its form drag; study by computer of various statistical distributions should hence be feasible.

4. Effects of cross-sectional shape must be analyzed in terms of variation in the hydraulic radius and the two coefficients of the semi-logarithmic resistance function; the accompanying effects of secondary flow probably become important only as the width-depth ratio falls below about 5.

5. Channel nonuniformities entail the Froude number as a resistance parameter, the effect of standing waves being appreciable at values below as well as above unity; channel choking must be anticipated in the critical range.

6. An apparent increase in channel resistance is to be expected in both smooth and rough channels as the Froude number exceeds its respective stability limit; because of the consequent growth of roll waves, however, the flow can eventually no longer be treated as uniform and steady.

ACKNOWLEDGMENTS

Various phases of the continuing investigation reviewed in the foregoing pages have been partially supported in the course of the past two decades by the Office of Naval Research, the U. S. Geological Survey, and the National Science Foundation under contract or agreement with the Iowa Institute of Hydraulic Research. The manuscript of the present paper was critically reviewed by E. O. Macagno, E. Naudascher, and E. M. O'Loughlin, all Research Engineers of the Institute.

APPENDIX. —NOTATION

The following symbols have been adopted for use in this paper:

A = coefficient (locally, cross-sectional area);
B = coefficient;
C = coefficient (locally, Chezy discharge coefficient);
D = coefficient;
d = depth;
$\mathbf{F}$ = Froude number;
f = resistance coefficient;
g = acceleration of gravity;
H = total head;
h = roughness height;
k = equivalent roughness;
P = wetted perimeter;
$\mathbf{R}$ = Reynolds number;
R = hydraulic radius;
S = slope ($\sin \theta$);
t = time;
V = mean velocity;
v = local velocity;
y = wall distance (locally, depth);
y' = zero-velocity intercept;
γ = weight density;
λ = roughness concentration;
ν = kinematic viscosity;
ρ = mass density; and
τ = boundary shear.

International Association for Hydraulic Research

CAVITATION AND ENERGY DISSIPATION IN CONDUIT EXPANSIONS

by

Hunter Rouse

Director, Institute of Hydraulic Research, University of Iowa
Iowa City, Iowa, USA

and

Vladimir Jezdinsky

Research Engineer, Hydrodynamics Institute, Czechoslovakian
Academy of Sciences, Prague, Czechoslovakia

<u>Synopsis</u>

A laboratory study is described of the onset and development of cavitation in the turbulent eddies produced by the diffusing stream at an abrupt pipe enlargement. Evaluations of incipient cavitation, pressure distribution, head loss, and incipient boundary damage are presented for use in the design of high-head relief conduits.

<u>Résumé</u>

Cette étude de laboratoire traite de la naissance et du développement de la cavitation dans la zone tourbillonnaire provoquée par la diffusion d'un écoulement lors du soudain élargissement d'une conduite. Il y est présenté des évaluations du début de cavitation, de la distribution de pression, de la perte de charge et du début d'érosion des parois dans le but de réaliser des conduites d'évacuation sous forte charge.

Аннотация

В докладе описаны лабораторные исследования по возникновению и развитию кавитации в турбулентных вихрях, возникающих при расширении потока на участке резкого расширения трубопровода. Приведена оценка начальной кавитации, распределения потока, потерь напора и начальной эрозии для использования их в проектировании высоконапорных водосбросных трубопроводов.

<u>Introduction</u>

As modern hydroelectric installations steadily increase in capacity and head, new methods for the dissipation of energy in flood-relief conduits become essential. One of these now finding application is the insertion of gated plugs which direct high-velocity jets into subsequent expansion chambers, the limiting performance criterion being not that of incipient cavitation but that of incipient damage to the walls of the expansion chamber. The intelligent design of such expansions evidently requires greater knowledge of the cavitation phenomenon than is now at hand.

In 1953 the first author[1] published the findings of Whitehouse[2] in a study of eddy cavitation in the shear zone around a submerged jet in an effectively unlimited expansion chamber. The magnitude of the cavitation index

$$\sigma = \frac{h_0 - h_v}{V_0^2 / 2g}$$

for incipient cavitation was found to be about 0.6. More recent tests conducted by Appel[3] in the same cavitation tank but using a focused hydrophone emphasized a number of points evident from Whitehouse's exploratory work and revealed a number of others. Cavitation becomes observable sonically well in advance of its visual appearance. The sonic indications are different for different frequency bands. The bubble collapse which produces the sound (as well as the possible cavitation damage) is a random process requiring statistical analysis, and designation of the state of incipiency is rather arbitrary. The apparent violence of cavitation depends upon the frequency rather than the amplitude of the sound peaks. Appel found an average of over a thousand sound peaks per second to occur at a cavitation index of 0.30 and about 10 at 0.35, and only an indefinite limit of no peaks whatever appeared to exist. Moreover, these values varied considerably with time, air content, and ambient pressure.

More recent tests by the Bureau of Reclamation[4] have indicated that a jet issuing into an expansion chamber would have a comparable incipient-cavitation index, but that it would not begin to produce boundary damage until the index had been considerably reduced. From knowledge of the characteristics of submerged jets,[5] one would expect the index for incipient damage - necessarily 0 at the expansion limit 1:∞ - to increase with increasing expansion ratio. The eddy zone of the submerged jet, in other words, expands linearly, and the intensity of the turbulence varies inversely, with distance from the efflux section; the length of the zone of eddy cavitation should therefore increase with decreasing cavitation index, as should the diameter of an expansion chamber barely subject to damage from the collapsing cavitation pockets.

---

[1] Rouse, H., "Cavitation in the Mixing Zone of a Submerged Jet," <u>La Houille Blanche</u>, January-February, 1953.

[2] Whitehouse, J. P., "An Investigation into the Point of Incipient Cavitation of Submerged Jets," M.S. Thesis, University of Iowa, February 1952.

[3] Appel, D. W., "An Experimental Study of the Cavitation of Submerged Jets," ONR Report, Iowa Institute of Hydraulic Research, June 1956.

[4] Ball, J. W., and Simmons, W. P., "Hydraulic Characteristics of Pipeline Orifices and Sudden Enlargements used for Energy Dissipation," Hydraulics Branch Report No. Hyd-519, U.S. Bureau of Reclamation, December 1963.

[5] Albertson, M. L., Dai, Y. B., Jensen, R. A., and Rouse, H., "Diffusion of Submerged Jets," <u>Transactions, ASCE</u>, Vol. 115, 1950.

Enclosure of a submerged jet by such a chamber will, of course, modify
the mean-flow and turbulence patterns to a varying degree. Recent studies by
Chaturvedi[6] revealed the respective patterns as functions of the expansion angle
for a particular (i.e., 1:2) expansion ratio. Essentially the only guide that
it provides to the effect of this ratio upon variation of the incipient-cavitation index is one of extrapolation according to the elementary Borda loss coefficient, and it is hence the purpose of this paper to supply more significant
information on both cavitation and the likelihood of cavitation damage.

Experimental Facility and Procedure

The special cavitation tank already used by Whitehouse and Appel was
adapted to the present experiments. It is of stainless-clad steel 5 feet in diameter, 10 feet in length, and provided with three observation windows on each of
two 90° axial planes. Flow through a 6-inch recirculating pipe system is produced
by a centrifugal pump, the valved inlet to and outlet from the tank being located
on its axis at either end. A supplementary vacuum pump permits arbitrary control
of the tank pressure. For the present tests (see Fig. 1) a faired reducer to a
1-inch parallel section 1.5 inches long was mounted at the end of a 4-foot uniform
approach of 3-inch diameter, at the normal end face of which could be mounted interchangeably three transparent plastic pipes of 1.6-inch, 2.0-inch, and 3.5-inch
diameters, each 10 diameters in length. All were provided with 6 wall piezometers at suitable intervals. Although an effort was made to reduce the effect of
air in suspension to a minimum by preliminary de-aeration runs, it was again found
that at low velocities the influence of air upon the effective magnitude of the
vapor head h_v was considerable. All test runs were hence made at a sufficiently high velocity to render the magnitude of h_v negligible in comparison with
h_0 in the definition equation for σ - at least in the range of primary interest.

Whereas Appel had utilized a sound-absorbent tank lining so that the
intensity of cavitation noise could be studied as a function of position, the
present tests were conducted without lining, and as a result the location of the
hydrophone made very little difference in its indication. The hydrophone used
was a Massa model M-115, and the sound level was read on a Ballantine voltmeter,
model 861, using the hydrophone manufacturer's calibration and a General Radio
filter, type 1231-P5. As in the previous tests, the indication varied considerably with the frequency band utilized. Typical results are shown in Fig. 2 for
the 1:2 expansion. Use of the entire frequency range of the hydrophone is seen
to result in a fairly constant background-noise level at higher values of the
cavitation index, the sound intensity p_s rising slowly with decreasing σ until rather abruptly a pronounced slope is attained. A plateau is eventually
reached, whereafter another and considerably steeper rise occurs, ending in a
sharp peak. Successive elimination of the lower frequency bands evidently reduces first the sharp peak and then the plateau; narrowing the frequency band
about an intermediate value has an even more pronounced effect.

Discussion of Results

Correlations of sound measurements such as those of Fig. 2 with visual
observations of the cavitation phenomenon indicated that the initial upward trend
corresponds to the development of cavitation (still invisible to the naked eye)
within the small-scale eddies of the diffusing jet. The beginning of the slope
thus represents the point of incipient cavitation, whereas the plateau marks the
convection of the uncollapsed cavitation bubbles as far as the centerline, on the
one hand, and the wall, on the other. The subsequent rapid increase in sound intensity, on the contrary, is produced by rapidly repeated vaporization and collapse of increasingly more extensive portions of the annular eddy enclosing the

---

[6] Chaturvedi, M. C., "Flow Characteristics of Axisymmetric Expansions," Proceedings, ASCE, HY3, May 1963.

central stream. After the final sound peak is reached, this vapor pocket no longer collapses in its entirety but continuously surrounds what has become, at least
at the outset, a free jet.

Measurements of the distribution of piezometric head along the wall of
each expansion showed essentially no variation in form between the point of incipiency and the plateau. Thus, in Fig. 3, the solid points represent conditions
of incipiency, the lined points the vicinity of the plateau, and the semi-solid
ones intermediate values. In all cases the change in head is essentially complete within 4 or 5 diameters from the section of abrupt enlargement. Owing to
the limited length (10 diameters) of the enlargement, the variation in head at
lower values of the cavitation index was not investigated in detail. The single
curve shown as a broken line through open points indicates for the intermediate
expansion the beginning of the expected shift downstream as the primary eddy com-
mences to cavitate. Again because of the relative shortness of the enlargement,
as well as the lack of data on the velocity distribution, the loss of head could
not be determined with precision. However, a fair approximation was obtained, as
shown in the figure, by adding the relative velocity head $(V_1/V_0)^2$ to the relative piezometric head $2gh_1/V_0^2$ and subtracting the sum from the relative total
head of 1.0.

The resulting values of the loss coefficient $C_L = 2gH_L/V_0^2$ are plotted in Fig. 4 against the expansion ratio D_0/D_1 , together with the curve corresponding to the elementary Borda loss

$$C_L = \frac{(V_0 - V_1)^2}{V_0^2} = \left(\frac{D_1}{D_0}\right)^4 - 2\left(\frac{D_1}{D_0}\right)^2 + 1$$

Also plotted as a function of the expansion ratio is the observed index of incipient cavitation σ_i . How this curve actually approaches the limit of zero for the
expansion ratio of unity is of academic interest only, for the range of practical
importance is the left-hand part of the diagram. As a matter of fact, even this
portion of the function is of only indirect concern, for it is not the cavitation
itself which is to be avoided but the structural damage which it can do, and this
should decrease with increasing expansion. At the moment, unfortunately, the only
pertinent information available on boundary damage is that obtained at the Bureau
of Reclamation[4] by discharging water under pressure through a partially (16%) open
3-inch gate valve into concrete-lined pipe of 4.5-inch, 5.2-inch, and 6-inch diameters for 6-inch periods. Values of the cavitation index for conditions of incipient damage (σ_{id}) have been estimated therefrom in terms of the variables used
in the present study and are plotted at the left of Fig. 4. On the surmise that
the index for the beginning of the sound-level plateau would show some degree of
correlation with that for incipient damage, the values of σ_p (see Fig. 2) for
the two expansions tested at this extreme stage were also plotted. However fortu-
itous the consistency of the results may be, the single curve through the two sets
of points provides at least a fair indication of the safety factor that is involved in whatever criterion one actually chooses for design. It should, in other
words, serve as a significant guide in the selection of expansion ratios and suitable ranges of operation, till such time as more extensive field experience has
been acquired.

<u>Acknowledgments</u>

All measurements were made by the second author, in his role as Research
Assistant at the Iowa Institute of Hydraulic Research, under the supervision of
the first author, by whom the paper was written. Like the previous studies of
jet cavitation, the project was supported in large part by the Office of Naval Research under Contract Nonr 1509(03) with the Institute.

503

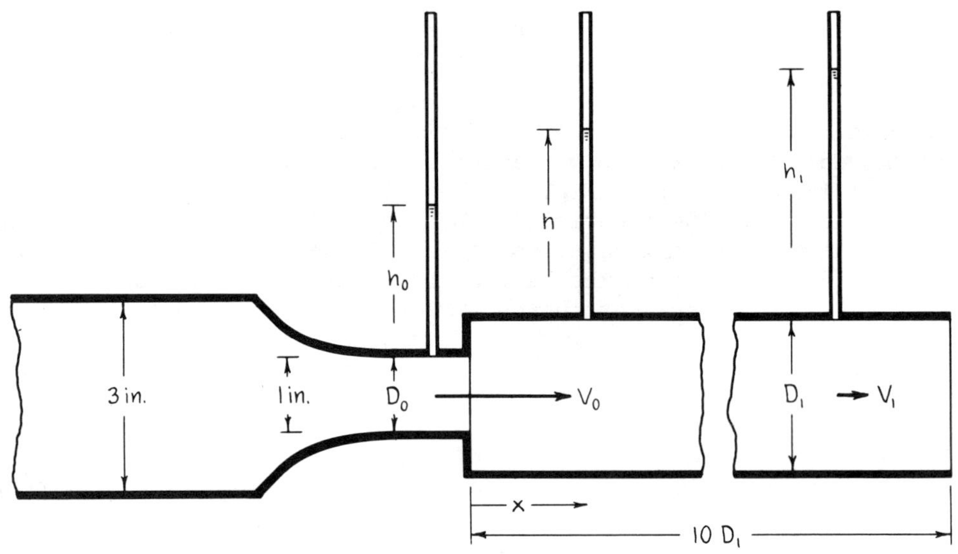

Fig. 1 Details of test expansion.

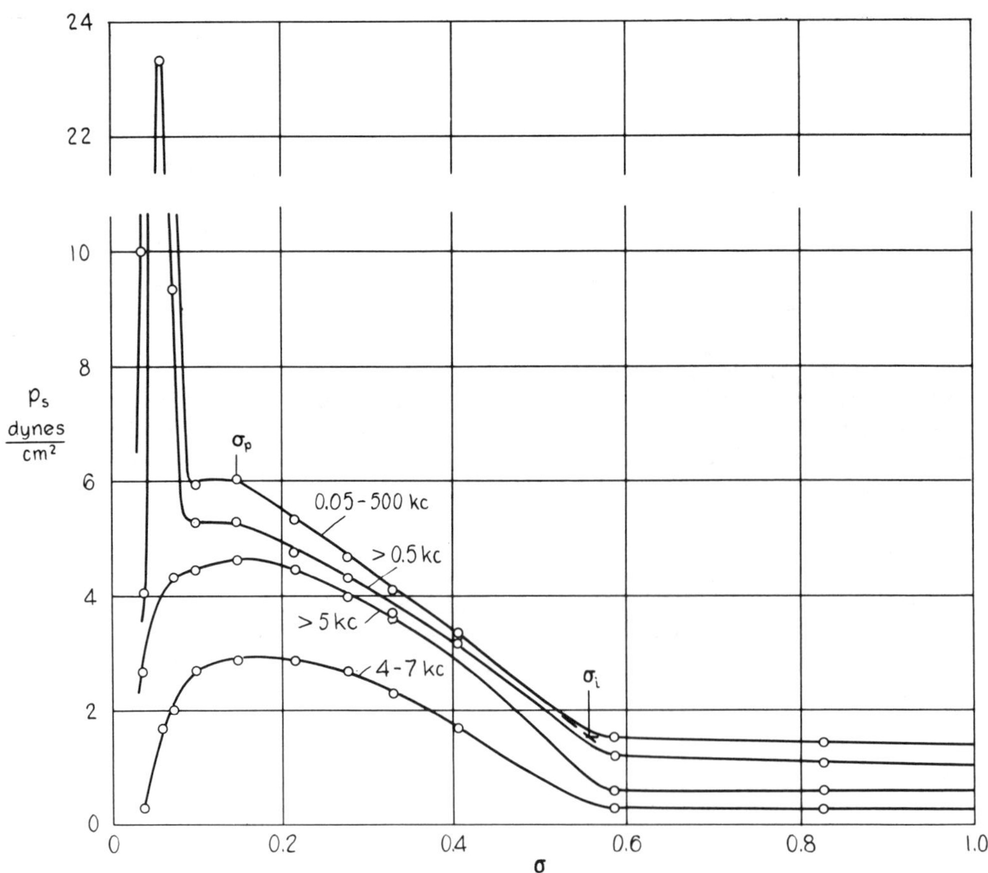

Fig. 2 Typical records of sound intensity for a 1:2 expansion.

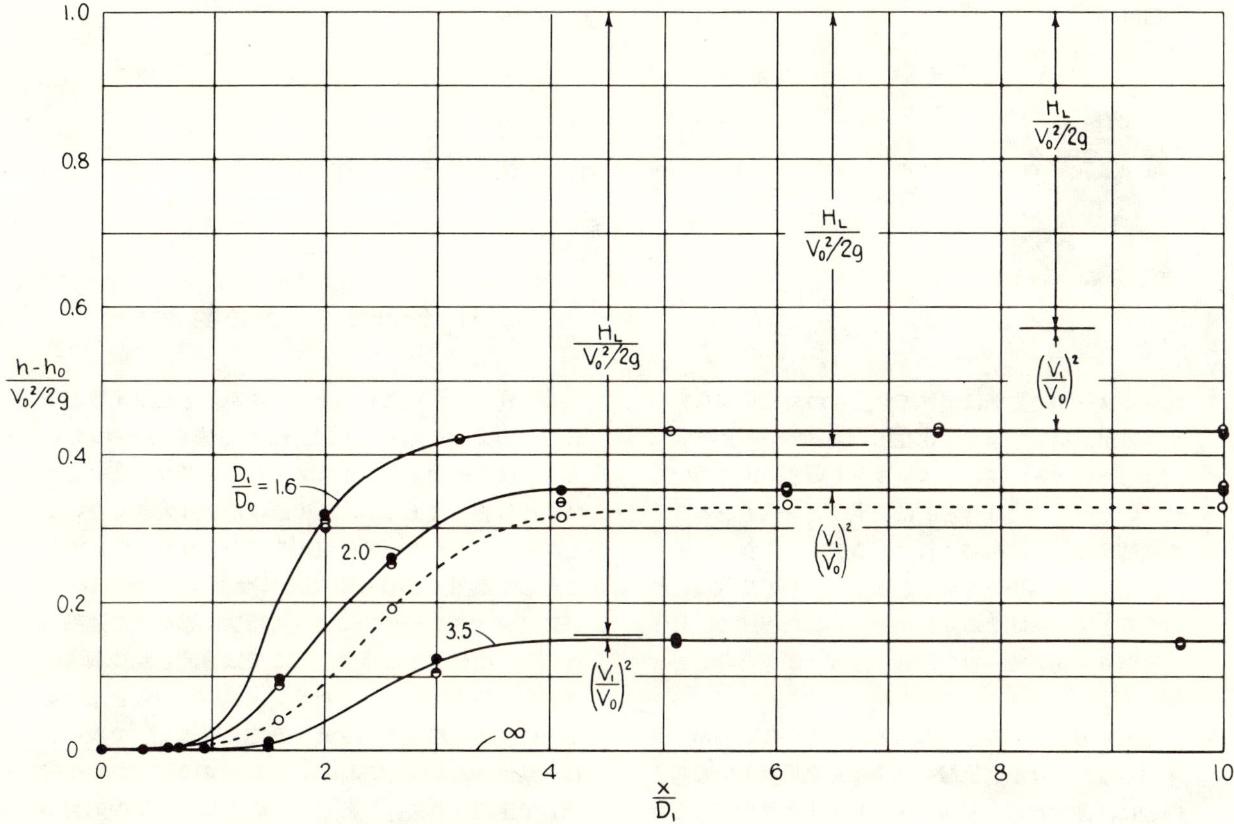

Fig. 3 Distribution of head for various expansion ratios.

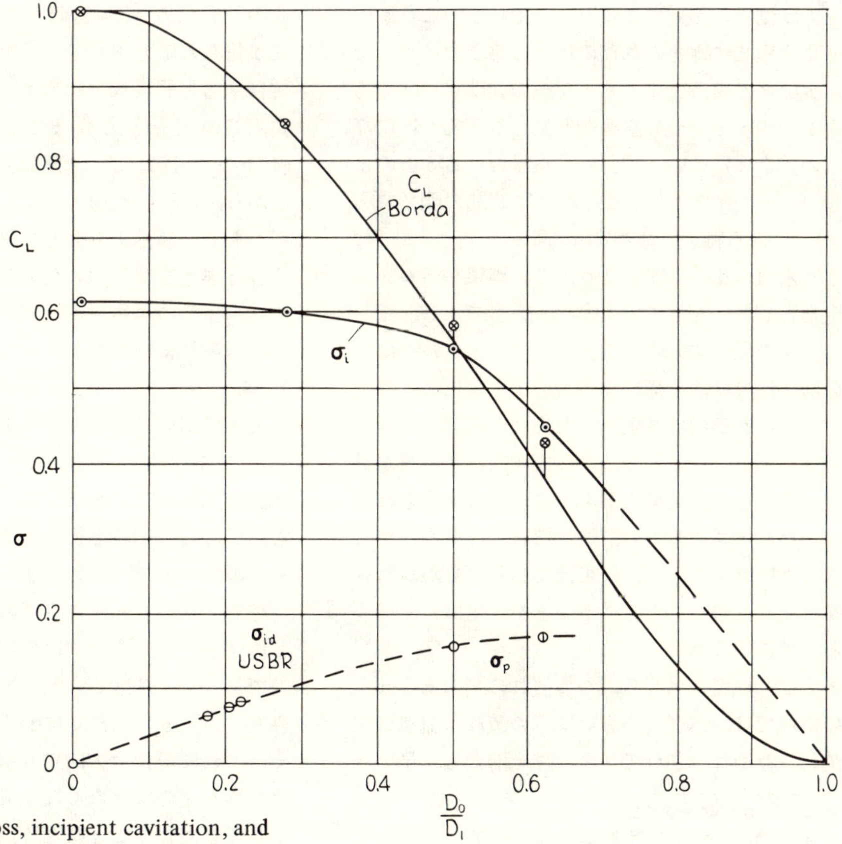

Fig. 4 Coefficients of head loss, incipient cavitation, and
incipient cavitation damage versus expansion ratio.

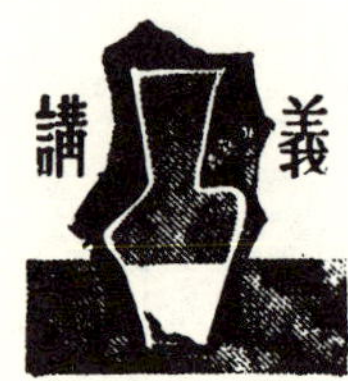

ベルヌーイの定理*

―和文概要―

H. Rouse, 有江幹男訳**

　ベルヌーイの定理は流体工学上きわめて重要なものであるが，果して学生は現在用いられている形の式に到達した経緯と，流れのエネルギをいまだ広く用いられていない形でより合理的に表現できる式もあることを理解するように指導されているであろうか．

　ガリレオの弟子カステリとトリチェリの時代は連続の定理と噴流に関する知識が確立された程度で，ビトーに至って初めて一般の流れに速度水頭があることを認めるようになった．有能な実験家であったベルヌーイは水頭の和が一定となることに気づいていたが，ライプニッツが運動と位置のエネルギを用いていたのと同様に，速度水頭とピエゾメトリック水頭の二項しか採用していない．流体内の圧力については，ベルヌーイの父が息子よりも理解していたものと伝えられるが，圧力は位置の関数であることを知り圧力こう配の形で運動方程式に導き入れたのはその弟子オイラーである．テンソル表示によるオイラーの式は式（2）であり，運動方向 s と一つの座標軸を重ねると式（3）になる．この式を s について積分の結果得られる3個の水頭の和 B は定数ではなくて一般には時間の関数である．しかし，ある特定の時刻では，同一流線上で式（4）のように水頭の和は一定となり，これが現在ベルヌーイの定理として知られている式の一般形である．ところが，流体の保有する単位質量当たりのエネルギは実際の流動条件では同一流線上においても変化するのであって，これを適当に修正するためには式（5）の形にする必要がある．この式の ΔB は機械的エネルギが熱に変換するための損失であるものと一般に考えられているが，かかるエネルギの散逸は機械的エネルギから熱になるという一方向のみに可能であるにもかかわらず ΔB は正にも負にもなり得ることを学生は良く知っているのであろうか．

　たとえば，速度の異なる二つの流れの間にあるせん断流の領域では，速度の小さいほうの流れは流速の大きい流れのため加速されてエネルギを増大し，早いほうの流れはこれに反してそのエネルギを減少する．一般にこのようなエネルギ伝達には乱流が伴われていて，乱流の運動エネルギも考えなければならないように思われ勝ちであるが，この値はあまり大きくないことが多い．ベルヌーイの定理をこのような条件に対して正しく適用する唯一の方法は，ナビエー，コーシー，ポアソン，ストークスによって順次に導かれた運動方程式から出発するこほで，ここでは層流と乱流両者に適用できる式（6）に示したサン・ブナンの式を用いることにする．層流の場合せん断応力は式（7）で表わされ，乱流せん断応力はレイノルズの解析により式（8）で定義されて，一般にはこれら二つの形の応力が同時に作用するものと考えなければならない．サン・ブナンの式に平均流速 $\bar{u}_i$ を乗じて各軸方向に対する式の和を求め $\bar{V}^2 = \sum \bar{u}_i^2$ とすれば式（9）のエネルギ式が得られる．各項を $\bar{V}$ で割って流れの方向のみに着目すると式（10）になり，$-\Delta B$ に相当する項は正負両者の値をとり得ることがめいりょうである．式（10）の右辺第二項は式（11）のようにも書きかえられるから，ΔB は流線上の二点の間で拡散されたエネルギからせん断応力を介して周囲の流体によってなされる全仕事量を差し引いたものとなる．このように考えると，運動のエネルギ，ピエゾメトリック水頭，伝達エネルギおよび損失エネルギの各項をそれぞれ K, P, T, L とすればベルヌーイの式を式（12）のように書くことが適切である．このように拡張した形のベルヌーイの式について有江，チャタベディ，カーモディ，チブレイらがアイオワにおいて行なった研究のうちから，ここには円盤を横切る流れと急拡大管におけるせん断領域の結果を示してある．

　結局，ベルヌーイの定理の適用は流線に対する場合と流路全体を一次元的に取扱う場合とで少しく異なるのであって，速度こう配があってせん断応力の作用が存在するときには流線に沿う方向についてばかりでなく，これに直角な方向にも積分するときにはじめて伝達エネルギの項はなくなって ΔB は二断面間の損失エネルギを表わすことになる．

---

　* 原稿受付　昭和40年9月27日．
　** 正員，北海道大学工学部（札幌市北十二条西8丁目）．

The Bernoulli Theorem*

by Hunter Rouse**
(Japanese Abstract: by Mikio Arie)

Probably no name plays a more prominent part in the mechanics of fluids from the very outset of its study than that of Daniel Bernoulli. The Bernoulli theorem is usually introduced in the first course in high school physics, and the underlying principle that it represents is reflected to some degree in almost every flow phenomenon subsequently treated. Only rarely, however, is the student led to suspect that the principle as it appeared in Bernoulli's Hydrodynamica of 1738 was any different from that of today—and even less often, for that matter, is he given to understand that there is also a far more rigorous form that is still only seldom used.

By Bernoulli's time[1] the principle of continuity and the principle of free fall as applied to jets were both generally accepted, thanks to two of Galileo's pupils, Castelli (1628) and Torricelli (1644). Moreover, a 1732 paper by Pitot had shown that the concept of velocity head for conduit flow was also in the air. Bernoulli, an able experimenter, was obviously aware of the tendency for the sum of the heads to remain constant as the flow section changed in size, and his papers show many perceptive—though not invariably correct—illustrations of manometry for various flow passages. There were only two terms (which we now call velocity head and piezometric head) in the sum that he treated, however, just as there were only two (now known as kinetic energy and potential energy) in the primitive relationship of Leibniz that Bernoulli used when he sought to derive an expression for the pressure. On the basis of earlier writings, one must give him credit for a less primitive conception of fluid pressure than

is reflected in his Hydrodynamica. There he resorted to the device of imagining the end of the passage to be suddenly removed and applying an equation of momentum to express the pressure that had existed just beforehand; the result was far from convincing.

Daniel's father Johann is said to have understood fluid pressure better than his son, as evidenced in his Hydraulica of 1743—which he predated ten years to establish priority[2]! Were Johann's analysis really much superior to Daniel's, the name Bernoulli might still logically remain attached to the theorem. But it was actually a student of Johann's and a colleague of Daniel's—Leonhard Euler—who first conceived of pressure as a function of space and incorporated the pressure gradient in his equations of motion. In 1755, for the particular condition of steady motion in which the velocity and the body force were derivable from potential functions (i.e., irrotational flow in a gravitational field), Euler showed that his equations could be combined to yield

$$\frac{p}{g} = V - \frac{1}{2}\,\mathit{v}\,\mathit{v} + D \quad\cdots\cdots\cdots\cdots\cdots(1)$$

which (with the nomenclature changes $g \rightarrow \gamma\rho$, $V \rightarrow -\Omega$, and $\mathit{v} \rightarrow V$) becomes the first recognizable form of the so-called Bernoulli equation to appear in the literature.

If the Eulerian equations are written in present-day tensor notation as

$$\frac{\partial u_i}{\partial t} + u_j \frac{\partial u_i}{\partial x_j} = -\frac{\partial}{\partial x_i}\left(\Omega + \frac{p}{\rho}\right) \cdots(2)$$

then a considerably simpler derivation can be effected by orienting one of the coordinate axes so that it coincides with the direction of motion s at the point in question, one velocity component thereby acquiring the vector magnitude and the other two becoming equal to zero. For steady

---

* Recieved 30th June, 1964.
** Institute of Hydraulic Research, University of Iowa, Iowa City.

flow the equation in the s-direction reduces to

$$\frac{\partial}{\partial s}\left(\frac{1}{2}V^2+\Omega+\frac{p}{\rho}\right)=0 \quad\cdots\cdots\cdots(3)$$

For purposes of generality we note that the Bernoulli sum $\frac{1}{2}V^2+\Omega+\frac{p}{\rho}=B$ is not a constant but a function of time, in that either the pressure or the gravitational potential can change at the same rate at all points without affecting the flow pattern. At any instant, however, at points along the same streamline,

$$B_1=B_2=\cdots\cdots B_n \quad\cdots\cdots\cdots\cdots\cdots\cdots(4)$$

This is the most general form of the Bernoulli theorem as we encounter it in textbooks today. It differs from the one-dimensional Bernoulli equation in that it recognizes a possibility of variation in the Bernoulli sum from one streamline to the next. The student soon becomes aware, moreover, that the energy per unit mass can also vary from point to point along a streamline, and soon a correction term ΔB is added arbitrarily to the right-hand side of the equation to maintain a proper balance:

$$B_1=B_2+\Delta B \quad\cdots\cdots\cdots\cdots\cdots\cdots(5)$$

This term is commonly considered to represent the loss which results from the transformation of mechanical energy into heat. Only rarely is a student sufficiently astute to appreciate the fact that ΔB is sometimes positive and sometimes negative—this despite the fact that energy dissipation is always a one-way affair. In the shear zone between two neighboring streams of different velocity, for example, or in a submerged jet or wake, the faster stream will accelerate the slower and thereby contribute to its energy per unit mass. Whereas the stream that gives up energy surely displays a decrease in the Bernoulli sum, the gain experienced by the other, must just as surely appear as an increase in B. Since turbulence is frequently involved in such energy transfers, one might suspect that additional terms for the kinetic energy of the turbulence should be introduced. However, this suspicion seldom comes to much, nor would it be of any great aid if it did.

The only way to adapt the Bernoulli theorem correctly to such conditions is to proceed from the pertinent equations of motion, much as Euler did in his time. Those that now come into consideration were derived in turn by Navier (1822), Cauchy (1828), Poisson (1829), and Stokes (1845), with successively greater understanding of what was involved, and the result is customarily named after the first and last together: the equations of Navier-Stokes. The most perceptive derivation, however, was that of Saint-Venant (1843), and we shall use his result here because it is adaptable to both laminar and turbulent flow:

$$\frac{\partial u_i}{\partial t}+u_j\frac{\partial u_i}{\partial x_j}=-\frac{\partial}{\partial x_i}\left(\Omega+\frac{p}{\rho}\right)$$
$$+\frac{1}{\rho}\frac{\partial \tau_{ij}}{\partial x_j} \quad\cdots\cdots\cdots\cdots\cdots(6)$$

Thus, so long as the flow is laminar, the stresses can be expressed in terms of the dynamic viscosity and the rates of deformation:

$$\tau_{ij}=\mu\left(\frac{\partial u_i}{\partial x_j}+\frac{\partial u_j}{\partial x_i}\right) \quad\cdots\cdots\cdots\cdots(7)$$

Likewise, for turbulent flow, use of the nomenclature $u_i=\bar{u}_i+u_i'$ introduced by Reynolds in 1894 permits the fictitious stresses known by his name to be written in terms of the mean products of the velocities of fluctuation:

$$\tau_{ij}=-\overline{\rho u_i'u_j'} \quad\cdots\cdots\cdots\cdots\cdots\cdots(8)$$

As a general rule the acting stresses should be considered as the sum of the two types.

The equations of Saint-Venant can be combined into a single energy relationship by multiplying each one by the corresponding mean component of velocity $\bar{u}_i$, adding the results, and making the substitution $\bar{V}^2=\sum\bar{u}_i^2$:

$$\bar{u}_j\frac{\partial}{\partial x_j}\left(\frac{\bar{V}^2}{2}\right)=-\bar{u}_i\frac{\partial}{\partial x_i}\left(\Omega+\frac{p}{\rho}\right)$$
$$+\frac{\bar{u}_i}{\rho}\frac{\partial \bar{\tau}_{ij}}{\partial x_j} \quad\cdots\cdots\cdots(9)$$

Each of the temporal mean quantities is now indicated by a bar, and the unsteady term has been omitted to avoid ambiguity. If, as before, one axis is made to coincide with the direction of flow, one velocity component will attain the vector magnitude and the other two will become equal to zero (all components in the last term should be retained, however, for reasons that will immediately become clear). Upon division of each term by $\bar{V}$ and integration from s_1 to s_2, the energy equation will reduce to the Bernoulli form:

$$B_1 = B_2 - \frac{1}{\rho} \int_{s_1}^{s_2} \frac{\bar{u}_i}{\bar{V}} \frac{\partial \bar{\tau}_{ij}}{\partial x_j} ds \qquad \cdots\cdots(10)$$

It is obvious herefrom that the integral, equivalent to $-\Delta B$, can be either positive or negative.

By the rules of the calculus, the last term can be replaced by the difference between the following two:

$$-\frac{1}{\rho} \int_{s_1}^{s_2} \frac{\bar{u}_i}{\bar{V}} \frac{\partial \bar{\tau}_{ij}}{\partial x_j} ds = -\frac{1}{\rho} \int_{s_1}^{s_2} \frac{1}{\bar{V}}$$

$$\times \frac{\partial (\bar{u}_i \bar{\tau}_{ij})}{\partial x_j} ds + \frac{1}{\rho} \int_{s_1}^{s_2} \frac{\bar{\tau}_{ij}}{\bar{V}} \frac{\partial \bar{u}_i}{\partial x_j} ds$$

$$\cdots\cdots\cdots\cdots\cdots\cdots(11)$$

The first of these will be found to have the form of total work (per unit mass), whereas the second has the form of dissipative work. If the signs are taken properly into account, the factor ΔB is seen to consist of the energy per unit mass dissipated between points 1 and 2 of the streamline minus the total work per unit mass done by the shearing stresses of the surrounding fluid. Since the negative work done upon a fluid element is equal to the positive work done by the element, the term for total work can also be construed as the energy contributed by the element to its neighbors through the action of the shearing stresses. In other words, if one let the symbols K, P, T, and L denote kinetic-energy, piezometric-pressure, transfer, and loss terms, the complete Bernoulli equation for steady mean flow along a streamline can be written simply but significantly as

$$B_0 = K + P + T + L \qquad \cdots\cdots\cdots\cdots\cdots(12)$$

If the stresses are wholly viscous, the loss term will represent direct transformation into heat. If they are of the Reynolds type, on the contrary, it will represent a loss so far as the mean flow is concerned, but not an immediate change from mechanical to thermal energy. The product of a Reynolds stress and a velocity gradient, in other words, indicates a rate of turbulence production, and only within the turbulent eddies does the viscous dissipation then occur. Although the kinetic energy of the turbulence is still present in the flow, the terms expressing its magnitude evidently do not belong in the Bernoulli equation. By the same token, the equation of energy for the turbulent motion in all its complexity need not be introduced, for energy once lost by the mean motion can never be regained.

Application of the extended form of the Bernoulli theorem, particulary to zones of both positive and negative transfer, is a most informative operation. Needless to say, evaluation of the successive terms for the flow conditions under consideration requires detailed measurements of all the variables involved, and this is no small task. Such measurements for the zone of separation behind a normal wall were obtained at Iowa by Arie[3] some years ago, and they provided the basis for the first analysis that was conducted in the extended Bernoulli form[4]. More recently Chaturvedi[5] carried out an extensive experimental investigation of the flow patterns at a series of pipe enlargements, and Carmody[6] made a similar study of flow in the wake of a disk. Four of these cases have now been analyzed in Bernoulli form by Chevray[7], and two—the disk and the abrupt expansion—are reproduced here for illustration.

For conditions of axial symmetry, and at sufficiently high Reynolds numbers for the viscous stresses to be negligible, the transfer and loss terms appearing in Eq. (11) in tensor notation can be rewritten as

$$T = \int_{s_1}^{s_2} \frac{1}{r \bar{V}} \left[\frac{\partial}{\partial x} r(\bar{u}\overline{u'^2} + \bar{v}\overline{u'v'}) \right.$$

$$\left. + \frac{\partial}{\partial r} r(\bar{u}\overline{u'v'} + \bar{v}\overline{v'^2}) \right] ds \qquad \cdots\cdots(13)$$

$$L = -\int_{s_1}^{s_2} \frac{1}{\bar{V}} \left[\overline{u'^2} \frac{\partial \bar{u}}{\partial x} + \overline{v'^2} \frac{\partial \bar{v}}{\partial r} + \overline{w'^2} \frac{\bar{v}}{r} \right.$$

$$\left. + \overline{u'v'} \left(\frac{\partial \bar{u}}{\partial r} + \frac{\partial \bar{v}}{\partial x} \right) \right] ds \qquad \cdots\cdots(14)$$

At representative sections of the flow pattern each of the terms was evaluated graphically from the measured values of mean velocity, piezometric pressure, shear, and turbulence intensity. The conditions of the experiments were such that the approaching flow was irrotational at the initial section, B_0 thus being constant with r and establishing the constant sum of the three terms at all successive points on the streamlines. For simplicity the reference piezometric pressure was taken as that of the approaching flow, and for generality all terms were made dimensionless through division by the kinetic energy of the approaching flow per unit mass, $V_0^2/2$. The results for the disk are plotted in Fig. 1, with the pattern of streamlines at the top and the distribution curves of K, P, T, and L below the respective sections. The form of the mean pattern is seen at a glance, and the gradual spread of the shear zone and the accompanying turbulence production will be evident from the transfer and loss curves.

Whereas the magnitude of B varies continuously from a maximum in the undisturbed flow to a minimum at the center of the eddy, the sum of B, T, and L depends upon the point chosen as reference. Although $B=K+P$ is invariably positive in the primary flow, at the initial section it has both positive and negative

parts, the piezometric pressure at the edge of the plate being well below zero. Therefore, at the very beginning of the eddy, where the velocity is still zero, the magnitude of B_0 is determined by the negative pressure P, and this in turn determines the sum $K+P+T+L$ for the first streamline. For the remainder of the eddy streamlines, the magnitude of B at the point of zero forward velocity is taken as reference. Along each streamline the magnitude of T now becomes negative, indicating a net transfer toward rather than away from the zero in question. The loss, of course, remains positive. Careful inspection of the eddy region will show that T and L continue to vary around each closed streamline of the eddy, the one negatively and the other positively, T first exceeding L and then vice versa, in such a manner that the two values at the end of a complete circuit are numerically equal and hence cancel each other. The eddy obviously draws its energy wholly from that of the surrounding flow.

The results for the pipe enlargement, plotted in Fig. 2 in a similar manner (but with $2:1$ vertical distortion, because of the length of the separation zone), differ in minor rather

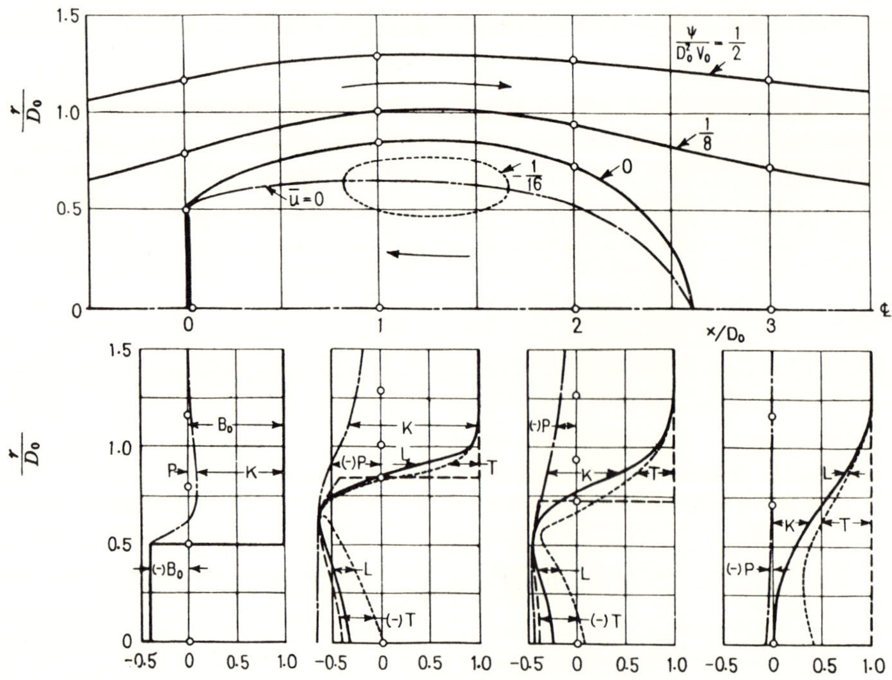

Fig. 1 Interrelationship of the four Bernoulli terms in the lee of a circular disk

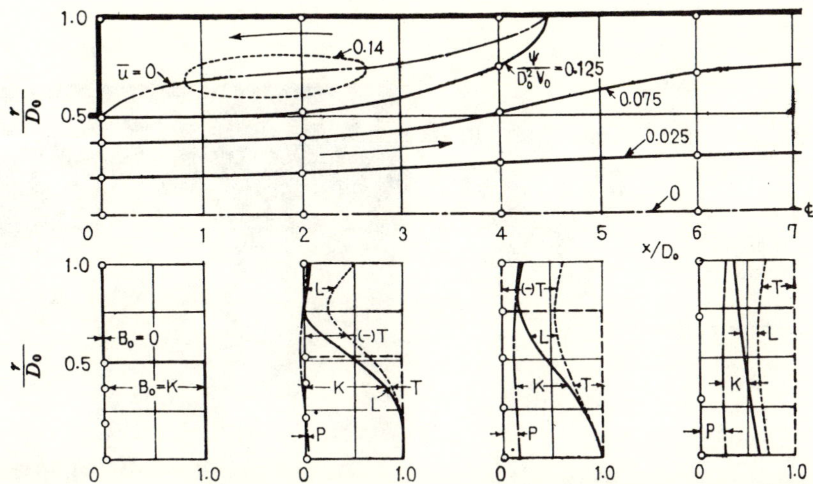

Fig. 2 Variation of the Bernoulli terms at a pipe enlargement (note vertical distortion)

than major detail. The experimental equipment had been so designed as to eliminate the boundary layer of the approaching flow. The magnitude of B_0 was thus constant across the inlet section, and hence the sum of K, P, T, and L is seen to be constant throughout the primary flow. Because the oncoming flow was uniform, the magnitude of B_0 plotted at the very edge of the eddy is zero rather than strongly negative as for the disk. On the other hand, whereas the ambient pressure well away from the disk was practically constant, the enclosed flow in the pipe undergoes an appreciable pressure rise. Through the eddy essentially the same positive and negative variation of L and T occurs as the flow makes a full circuit.

Because primary attention has been given in these analyses to the interaction between the ambient flow and the zone of separation, it has been convenient to ignore both the viscous stresses and the effect of the boundary upon the Reynolds stresses. In a detailed study of the boundary region, of course, neither could be disregarded. In fact, because the velocity gradient usually becomes a maximum, and the mean velocity and turbulence intensity invariably a minimum, at the boundary itself, it is there that the viscous dissipation is the greatest, whereas the limiting value of the transfer there is exactly zero. Herein lies the difference between the Bernoulli equation for the streamline and its one-dimensional counterpart for the flow passage as a whole: if, in addition to integration along a streamline as already performed, the equation is integrated as well across the passage, the transfer term will of necessity vanish (i.e., stationary boundaries absorb no energy and do no work); the result will then be of the form of Eq. (5), the term $\Delta B = L$ representing simply the average energy loss per unit mass between the two limiting sections.

Acknowledgments

The investigations on which this paper was based were partially supported by the AROD and the ONR, respectively, under Contracts 2871-E and Nonr 1509 (03) with the Iowa Institute of Hydraulic Research.

References

（1）H. Rouse and S. Ince, *History of Hydraulics*, (1963), Dover, New York.

（2）D. Bernoulli, *Hydrodynamics*; J. Bernoulli, *Hydraulics*, translated from the Latin by T. Carmody and H. Kobus, Dover, New York, in press.

（3）M. Arie, *Memoirs, Faculty of Engineering, Hokkaido University*, Vol. 10, No. 2 (1956), 211.

（4）H. Rouse, *Miszellaneen der Angewandten Mechanik*, (1962), 267, Akademie-Verlag.

（5）M.C. Chaturvedi, *Proc. ASCE*, Vol. 89, No. HY3 (1963), 61.

（6）T. Carmody, *Trans. ASME*, Ser. D, Vol. 86, No. 4 (1964-12), 869.

（7）R. Chevray, *La Houille Blanche*, No. 6, (1964).

ON A MATTER OF LATITUDE IN PRONUNCIATION

A 'sailor,' with research imbued,
In the lab proved remarkably shrewd.
Yet in spite of his fame
People missay his name:
Is it actually Frowde, Froide, or Froude?

There is surely good reason for the fact that names like Guglielmini, Euler, and Poiseuille vary in pronunciation from one language to another. The German, French, and English versions of Euler, in fact, are scarcely recognizable as having the same spelling. But just as certainly one should expect to find in the language of a single country a reasonable degree of consistency. It is thus the more surprising that within a country the size and age of England the name of an illustrious engineer of the last century—William Froude—should continue to change with latitude! Were the matter purely local, there would be little cause for complaint. However, many other parts of the world seem to have been settled by British emigrants from various latitudes, and the discrepancy is now widespread.

The writer was once told that, as a result of England's past history, the letters 'ou' tend to be pronounced 'oo' in the south and 'ow' farther north. Now Froude came from Devonshire, which is about as far south as England extends. As any naval architect will attest who knew Froude's son Robert (his life extended well into the present century), members of the family used only the sourthern pronunciation of their name.

Unfortunately, not only have erroneous versions become entrenched in various countries (even the farfetched 'Freud' being more than a facetious suggestion among students), but no general effort has ever been made to eliminate them. Indeed, quite the opposite tendency prevails, for in the United States two motion pictures recently appeared with the pronunciation 'Frowde' permanently recorded on their sound tracks. One of these was made by a government agency and the other by an educational organization. In the latter instance, the writer learned, a vote was taken to decide which form to use (though reference need only have been made to a dictionary), and a dilemma now exists

William Froude (1810-1879)

as to whether to correct the error in a subsequent film also using his name or to remain consistently wrong!

Ferdinand Reech is often credited with the first formulation of what is called the Froude criterion for similitude, and Osborne Reynolds with having first applied it in hydraulic research. Yet there is no gainsaying the fact that Froude deserves lasting recognition, not only for contributions to the art and science of ship modeling in a gravitational field, but as well for pioneer boundary-layer research. Whatever may be the logic of the situation, a fundamental flow parameter now universally bears Froude's name; the least we can do in his honor is pronounce it correctly.

Hunter ROUSE.

LA INVESTIGACION Y EL INGENIERO

Por el Prof. Dr. Hunter Rouse
Institute of Hydraulic Research, University of Iowa.
Iowa City, Iowa.

Por varias décadas ya, ha existido un lazo estrecho entre mi organización y varias de las vuestras; es por eso que acepté con gran placer vuestra invitación para tratar un tema que es de extrema importancia en mi vida profesional. Esta no es mi primera visita a vuestra bella y dinámica capital, y no deseo que sea la última. Personalmente, tengo aquí la atracción de una persona de mi familia casada con uno de vuestros ciudadanos. Además, vuestra Universidad Central alberga considerables equipos de laboratorio que fueron ideados en nuestro instituto en la Universidad de Iowa, hace unos pocos años. El hecho más significativo de todos es que doce de vuestros jóvenes siguieron estudios avanzados de hidráulica en nuestra universidad durante los últimos veinte años, y están ahora activamente empleados como ingenieros privados o federales o como educadores en ingeniería. En realidad, para mantener una mutua comprensión en todo el mundo, es mejor apoyarse en este intercambio de profesionales que en el de simples turistas; y vuestra organización debe ser felicitada calurosamente por su contribución a tal fin.

Como mucho de lo que tengo que decir tiene necesariamente que mostrar un tinte objetivo, será conveniente indicar desde un principio varios aspectos de mi educación que influenciaron mi actual punto de vista. Estudié el castellano en la escuela secundaria, preparándome —de acuerdo con mis sueños— para una vida dedicada a construir puentes en América del Sur. Empecé a estudiar ingeniería civil, pero luego escogí el pensum de hidroeléctrica en vez del de estructuras, y después de mi graduación pasé dos años en laboratorios alemanes con una beca de viaje otorgada por el Instituto Tecnológico de Massachusetts. Esos eran todavía los días en los cuales los americanos del Norte y del Sur acostumbraban viajar a Europa para realizar sus estudios de postgrado, y parece que me habitué a mantener un contacto con organizaciones extranjeras de investigación que perdura hasta hoy día. Así, mi perspectiva no es de ninguna manera estrechamente nacionalista, porque un diez por ciento de mi vida lo he pasado en el extranjero, tanto en países latinos como teutónicos.

En cuanto a mi actividad profesional, tiene desde hace mucho tiempo una naturaleza triple. Como profesor universitario soy en primer lugar un educador en ingeniería, y (exceptuando ausencias) he enseñado continuamente a estudiantes universitarios y postgraduados durante los últimos treinta años y he dirigido un número igual de disertaciones doctorales. Como director de laboratorio soy responsable de un considerable programa de investigación que tiene sus aspectos educacionales por el hecho de que unos veinte estudiantes graduados de por lo menos unos doce países son empleados regularmente por nosotros como asistentes de investigación. A este respecto hay que notar que la palabra "Hidráulica" en el nombre de mi organización —"Institute of Hydraulic Research"— refleja solamente una tradición, ya que nuestros estudios abarcan desde crecidas en ríos hasta corrientes en el sistema cardiovascular, desde la resistencia de barcos hasta la carga del viento sobre edificios, y desde la instrumentación más simple hasta intrincados análisis mediante computadores. Por fin, como consultor privado, tengo en cierto grado la oportunidad de aplicar la teoría de la clase y la experiencia del laboratorio a proyectos de ingeniería de una variedad considerable.

Antes de analizar la investigación en sí, tenemos que ponernos de acuerdo sobre el significado de la palabra. En realidad para mucha gente significa muchas cosas diferentes y su connotación en cualquier localidad varía necesariamente con el tiempo y la comprensión. Nosotros, los de habla inglesa, hemos adaptado la palabra francesa "recherche" y hemos adoptado su significado. Los alemanes tienen su propia palabra especial, que es "Forschung". Desgraciadamente, en castellano no existen dos palabras correspondientes a "investigation" y "research" que tanto en inglés, como en francés y alemán, no siempre son equivalentes! El diccionario Webster, que es nuestro árbitro final en los Estados Unidos, da el siguiente significado consagrado por el uso: "Investigación ...indagación crítica y exhaustiva, o experimentación que tiene por fin el descubrimiento de hechos nuevos y su correcta interpretación, la revisión de conclusiones aceptadas, teorías, o leyes, a la luz de los nuevos descubrimientos, o las aplicaciones prácticas de tales conclusiones, nuevas o revisadas". Aun esta definición evidentemente explícita puede estar sujeta a interpretaciones cambiantes con el nivel del desarrollo tecnológico. Así, lo que es investigación este año, puede convertirse en mera repetición de ensayos con el paso de unos años.

Un ejemplo destacado lo da el modelo a escala que se ha demostrado tan esencial para el ingeniero proyectista de más de una estructura prototipo. Cuando, hace dos siglos, John Smeaton, y al fin del siglo pasado, Osborne Reynolds y William Froude inventaron métodos de semejante mecánica para la investigación en laboratorio de ruedas hidráulicas, puertos y barcos, se hablaba de investigación de alto orden porque penetraban en territorio todavía inexplorado. A la vuelta del siglo, la extensión alemana de estas técnicas a modelos de ríos con cauces movibles fue igualmente investigación, como lo fue también la adaptación de estas técnicas a problemas americanos por los becados enviados a estudiar en Alemania justamente para este último fin. Sin embargo, aunque se utilicen los mismos laboratorios y los mismos procedimientos experimentales, los ensayos rutinarios en modelos destinados a verificar proyectos de prototipos no constituyen trabajos de investigación. En una palabra, carecen de la originalidad esencial. Con todo, tan severa fue la lucha inicial para que se aceptara el modelo a escala como un instrumento de diseño, que su reconocimiento final, fue seguido por un oscilar del péndulo hacia el otro extremo: hasta la mera construcción de un modelo parecía conferir al prototipo un mágico factor de seguridad. Desgraciadamente, tanto la lucha preliminar como sus consecuencias se repiten en varios sitios hasta hoy día.

Un título muy frecuente en nuestros presupuestos federales e industriales es "Investigación y Desarrollo" (Research and Development); una parte esencial del mismo, el desarrollo propiamente dicho, evidentemente tampoco es investigación porque de ser así, no llevaría un título distintivo. En realidad, la práctica de ensayos en modelos, que acabamos de discutir, llena una función doble: mientras implique simplemente el clasificar el funcionamiento de un esquema de ingeniería nuevo o propuesto, es tan definitivamente un ensayo como lo es el análisis de una muestra de tierra o de una muestra de sangre; pero una vez que se estudian las formas de mejoramiento, bien sea en economía, seguridad o eficiencia, pertenece tanto a la categoría de desarrollo como el perfeccionamiento de un automóvil de carrera o de un mecanismo para estirar plástico. Desarrollo implica invención y puede llegar hasta requerir investigación, pero en sí mismo no es investigación, porque su fin inmediato es lo que clasificamos generalmente más bien como "ferretería" que como nueva ciencia o nuevo principio.

Aunque algunos (exceptuando quizás los ensayadores mismos) puedan sostener demasiado tenazmente que los ensayos son una forma de investigación, hay que admitir que el desarrollo no está siempre separado de la investigación por una distintiva línea de demarcación. Es quizás por esta razón que los dos son asociados tan metódicamente tanto desde el punto de vista presupuestario como en el taller y el laboratorio. Hay, por otro lado, aquellos que insisten diciendo que el desarrollo pertenece a la tecnología, mientras que la investigación cae en el dominio de la ciencia. Se demostró considerable evidencia de este punto de vista en el planeamiento inicial de nuestra Fundación Nacional de Ciencias, que se organizó sin tomar en cuenta el campo de la ingeniería. La inclusión posterior de un apoyo a la investigación en ingeniería como una ciencia aplicada se efectuó sólo por el esfuerzo combinado de líderes y educadores de la ingeniería. Actualmente, la División de Ingeniería es una de las ramas fundamentales de la Fundación. (Aunque no específicamente relacionado con la investigación, un similar equilibrio de énfasis se demuestra ahora en Washington con la formación de una Academia Nacional de Ingeniería incorporada a la Academia Nacional de Ciencias). Por eso, no es tanto cuestión de si el desarrollo significa una actividad legítima del ingeniero, sino hasta qué punto él puede también lógicamente ocuparse de investigación. Uno bien puede proseguir la cuestión preguntando si el ingeniero está limitado a investigaciones aplicadas o si, ocasionalmente, él puede hasta aventurarse a realizar investigaciones puras.

Probablemente no se resolverá nunca la cuestión de lo que es puro y de lo que es aplicado, pues la decisión es casi invariablemente subjetiva. Cuando yo comencé a enseñar una mecánica de fluidos más general en lugar de la hidráulica tradicional, los ingenieros civiles conservadores consideraron mi orientación ultrateórica y me catalogaron como pseudo-físico, aunque para los físicos mismos aparecía como un ingeniero práctico. Para los científicos, las investigaciones en Ingeniería probablemente parecerán siempre aplicadas y para la mayoría de los ingenieros, las investigaciones científicas verdaderas parecerán conservar su pureza esencial. Pero, como esto es realmente más bien un asunto de punto de vista, y porque los ingenieros están reforzando continuamente su preparación científica, mientras que los científicos asisten cada vez más en la fabricación de "ferretería", se nota una difusión creciente de cada profesión dentro de la otra. En la universidad, por ejemplo, a continuación de las ciencias "básicas" ponemos las que acostumbramos a llamar ciencias de "ingeniería". Algunos de los descubrimientos científicos fundamentales en la actualidad —los rayos laser, entre otros— fueron realizados por ingenieros. Y algunas de las más puras investigaciones —las del campo nuclear— han tenido los más profundos efectos tecnológicos. Pongámonos, pues, de acuerdo sobre el hecho de que investigación en ingeniería puede recorrer toda la escala de la investigación desde la pura hasta la aplicada y que, a lo sumo, sólo puede definir su tipo en términos vagos. Por lo menos deberíamos concordar en que investigación no es ensayo ni desarrollo, sino la búsqueda de nuevos conocimientos y su correlación con los ya disponibles.

Es bastante evidente el por qué tenemos que realizar investigaciones, ya que la humanidad siempre dependerá de nuevos conocimientos para el progreso continuo de la civilización, bien sea en su totalidad o en cada una de sus muchas ramificaciones. Una pregunta a la cual se responde menos automáticamente es por qué un individuo cualquiera debe dedicar tiempo a la investigación si hay otros que pueden ciertamente hacerlo por él; una actitud popular y a primera vista lógica de la pequeña organización o país comparado con los grandes. Sin embargo, la actitud a tomar es sólo un asunto de tamaño relativo. Por ejemplo, ciertos laboratorios federales en los Estados Unidos sostienen que existen específicamente para propósitos de ensayos y que por eso no se justifica gastar dinero de los contribuyentes para investigaciones. Sea grande o pequeña la institución, la situación es efectivamente la misma: ninguna organización puede permitirse carecer de espíritu de investigación. Por lo menos algunos miembros de su personal deben observar, criticar, correlacionar constantemente y tratar de comprender y mejorar lo que ocurre en ellas. Aun el personal de un laboratorio de ensayos puede, sin violar sus funciones, realizar investigaciones de sus procedimientos de ensayos. En realidad, a menos que las realice, gastará probablemente muchas veces el costo de las investigaciones en repetidas operaciones inútiles; hechos significativos inherentes en los especímenes pueden no descubrirse nunca en los ensayos; y la organización no crecerá en habilidad técnica. Si el crecimiento es la única alternativa frente a la decadencia —y hay buena evidencia de que es así— el dinero y el esfuerzo puestos en las investigaciones representan el seguro más económico para una subsistencia continuada. Esto vale tanto para una nación como para una industria.

Hasta ahora hemos hablado de investigación en términos generales. Permítanme ahora examinar su posición particular en la ingeniería. Esta última está ligada en un grado considerable con ciertos aspectos de la historia de la ingeniería que vemos acaecer repetidamente en nuevos países que crecen entre los viejos y en aquellos de los viejos que se rejuvenecen. Al principio la ingeniería consistió fundamentalmente de actividades militares, porque las fuerzas de un comandante contaban mucho con cosas tales como fortificaciones, cruces de ríos, explosivos e instrumentos de asedio. La profesión de ingeniería civil apareció al principio en contraposición a la ingeniería militar y por muchas generaciones involucró toda actividad tecnológica no militar. La separación eventual de subdivisiones específicas —primeramente minera, mecánica y naval, seguidas por eléctrica y aeronáutica— todavía tiene lugar, y cada uno de nosotros podría nombrar una docena de variedades ultramodernas, empezando con la biológica y terminando con la nuclear. No obstante, queda el hecho de que la ingeniería civil continúa encontrándose en la base de la civilización, porque la existencia urbana ha dependido siempre del planeamiento de la ciudad, de la construcción, del suministro de agua, del sistema sanitario, y del transporte y muy probablemente esto será siempre así. Queda también el hecho de que, cuanto más joven y más vital sea un país, más prominentes serán los varios aspectos de la ingeniería civil en su desarrollo. Es de veras apropiado que los tópicos principales de este Simposio sean aquellos en los cuales el ingeniero civil está fundamentalmente comprometido. Hoy tropezamos esencialmente con la misma situación en el continente de Australia, a través de la Unión Soviética y en muchos de los estados de Africa, incluyendo la punta sur del mismo.

Un sitio que ahora se está convirtiendo de un desierto o una selva en área habitada aunque sea en forma esparcida, ciertamente no tiene que pasar por una secuencia idéntica de los hechos que fueron necesarios para una conversión semejante hace una década o un siglo. Tampoco debería un país, que está pasando por etapas de desarrollo ya experimentadas por otro país, repetir los mismos errores. La historia de tales empresas se encuentra, para provecho de todos, en la literatura. Pero nunca se presentan dos casos iguales. Aparecen invariablemente nuevos problemas que necesitan sus soluciones especiales y también se hallan mejores métodos para resolver los viejos problemas. El país que considera como política nacional el resolver sus problemas en vez de ignorarlos y el mejorar siempre lo que aprende de otros, se encontrará pronto entre los líderes, cualesquiera sean su tamaño y edad.

Si una nación contribuye al resto del mundo tanto como recibe de él, adquirirá un espíritu creador. Desde el punto de vista del ingeniero, esto significa que debe avanzar no sólo desde la ingeniería civil fundamental hacia las varias orientaciones apropiadas a su situación especial, sino también de los ensayos fundamentales hacia las innovaciones tecnológicas y luego hacia las investigaciones aplicadas, y finalmente las puras. Los ensayos implican asuntos de normas físicas, solidez de los materiales, modelos hidráulicos, mecánica de suelos y ciertos requerimientos biológicos de sanidad. He mostrado que los ensayos no son ni desarrollo ni investigación; pero son fundamentales para los dos y su práctica inteligente puede muy bien guiar hacia ambos. La innovación tecnológica y el desarrollo aparecen cuando el técnico que hace los ensayos no se conforma con reunir datos solamente, sino que también intenta mejorar sus medios de medición o la calidad del producto que se ensaya. Si su curiosidad le empuja a interpretar sus datos de una manera más significativa de lo necesario, entonces está ya acercándose a la investigación. Es obvio que lo primero viene en primer lugar y que se debe poner la base antes de erigir la superestructura. Pero si un técnico es así atraído hacia una nueva actividad, digamos, la física sólida, cuando tendría que hacer una tarea de rutina, sus superiores deberían felicitarse más bien que reprocharle. El grado en el cual una empresa puede avanzar y sobrepasar su meta original está bien ilustrado por el "U. S. Bureau of Standards". Naturalmente, su personal sigue siendo activo en ensayos de rutina, pero algunas de las más notables investigaciones en la historia de la tecnología se efectuaron en sus laboratorios.

Puesto que es generalmente cierto que el número de personas (así como también el número de máquinas) involucradas tiende a variar inversamente con la pureza de la investigación, una organización o estado que intenta fundar un centro de ensayos, de desarrollo o de investigación aplicada debe pensar en términos de una empresa de un tamaño apreciable más bien que en un grupo pequeño de hombres. Debe disponer de un departamento de administración, de sus ingenieros, de sus técnicos, de su personal mecánico y sus secretarias y debe tener su propia planta. La actitud estadounidense ha sido por mucho tiempo que lo más importante es la planta, lo que muchas veces determinó la exclusión efectiva del resto. Así, más de una instalación fue provista con excelentes equipos pero con muy poco para hacerlos andar. Más de un subsidio para investigación se ha dado para edificios y equipos, sin nada para su personal y, por fin, nada surgió de la inversión. Más de un administrador se ha vanagloriado con el brillo de los instrumentos de su laboratorio, sólo para maravillarse después porque el rendimiento de sitios más pobres excedía el suyo en calidad y cantidad. La alternativa obvia es de poner más relieve en el personal, pero de ninguna manera esto es el secreto completo. Un grupo de hombres capaces con un equipo pobre ciertamente puede crear algo nuevo y útil mejor que un personal pobre con equipo costoso; pero esto no quiere decir que hombres excelentes con facilidades excelentes rendirán invariablemente excelentes resultados.

En realidad dos factores más —dirección y ambiente— intervienen en una investigación exitosa y uno de éstos, a la larga resulta más esencial que todo el resto. No cabe duda alguna que nadie puede trabajar con resultados positivos en un ambiente de disensión, envidia y descontento. Quizás no es tan evidente al lego que, con pocas excepciones, los investigadores progresan en presencia de colegas de intereses iguales con quienes pueden discutir sus problemas e ideas. Además, sus realizaciones normalmente varían en proporción directa con la libertad de que gozan de buscar apoyo financiero para su trabajo. Esto, sin embargo, no es la razón de la importancia de la dirección, porque la tarea de pedir fondos puede ser tan asfixiante para un director como para cualquier otro investigador. Más bien, es porque un grupo no puede ser más fuerte que la inspiración de su líder. Funcionará efectivamente sólo hasta donde se le persuada a hacerlo por una dirección estimulante. El equipo no procrea personal, ni tampoco provee el personal, necesariamente, su propio ambiente armónico; pero un líder capaz puede crear las tres cosas. Mi consejo meditado a una organización que esté a punto de lanzarse a una expansión de sus investigaciones, es de buscar el director más capaz que se pueda encontrar y después dejar el resto en sus manos. Recuerdo muy bien al presidente de un pequeño colegio etadounidense quien, mientras otros presidentes andaban buscando dotaciones para edificios y equipos, puso casi todos sus fondos en pagar a un núcleo de hombres superiores. Cierto es que le quedó poco para instalaciones y equipos, pero esto no causó ninguna dificultad; debido al calibre de su personal superior pronto hubo una generación casi espontánea de personal y planta.

Existen varios tipos de establecimientos para investigación —es decir, el gubernamental, el industrial y el educacional— y cada uno de éstos puede desplegar varias ramificaciones. En general, uno espera que los establecimientos gubernamentales sirvan al país en conjunto y provean aquellos equipos que son demasiado extensos en construcción y mantenimiento para cualquier organización menor. Los laboratorios industriales (p. ej. del tipo automotriz o eléctrico) serán normalmente destinados a servir las necesidades particulares de empresas o grupos de ellas por los cuales fueron establecidos. Sin embargo, en décadas recientes, ha aparecido en los Estados Unidos un nuevo tipo de laboratorio industrial: el instituto de investigación privado sin fines lucrativos, el cual se encarga a precio de costo de las investigaciones en ingeniería, ciencia y economía para cualquier persona o institución que esté dispuesta a pagarlas; cierto es que los gastos no son bajos, pero por un solo proyecto son todavía mucho más bajos que los gastos ocasionados por la formación de una organización especial. Por tradición, la universidad es el centro de investigación pura y es al grupo de candidatos al doctorado y sus consejeros profesorales, que el mundo debe el avance ininterrumpido de los conocimientos científicos que prevaleció durante siglos. (Es oportuno hacer notar que aproximadamente una quinta parte de las investigaciones universitarias en los Estados Unidos corresponde a las ciencias de ingeniería.) Una universidad adecuada con escuela para graduados, biblioteca y un ambiente erudito es así un centro natural de investigación para una nación que desea avanzar técnicamente. Pero la nación tiene que dar a la universidad fondos y estabilidad, porque ninguna supervisión estimulante de las tesis surgirá de hombres que tienen que ir a otros sitios a buscar la mayor parte de sus entradas, o que temen por lo que pasará con sus empleos durante el próximo disturbio político. Por otra parte, en el curso de la segunda guerra mundial se efectuó un constante cambio, que puede no tener un paralelo en otros países, en las investigaciones universitarias de los Estados Unidos. Es el apoyo del gobierno a ambas investigaciones —la pura y la aplicada— en un grado tal que hace parecer pequeñas las sumas previamente gastadas por las mismas universidades. Más de la mitad de la investigación en las universidades en los Estados Unidos es ahora financiada federalmente— pero no tenemos una sola universidad nacional. La diferencia con la mayoría de los otros países es que en ellos las universidades son ya establecimientos federales; esto puede muy bien significar que la libertad de seleccionar, y quizás, la munificencia del apoyo serán reducidas considerablemente.

No hay una sola respuesta a la pregunta: ¿de dónde debe venir el apoyo general de la investigación? En algunos países (p.ej. en la Unión Soviética) es completamente federal. En los Estados Unidos más o menos la décima parte del presupuesto federal está asignado a investigación y desarrollo. La suma gastada por nuestra industria para un fin semejante es un pequeño porcentaje del total de nuestro producto nacional, pero alcanza a la mitad del expendio federal. En realidad, en algunas industrias (automotriz, química, eléctrica y electrónica) el capitalismo funciona tan vigorosamente que los gastos para investigaciones y las mayores ventas de nuevos productos tienden a formar dos espirales que se sostienen mutuamente. En general, las universidades son notoriamente pobres, pero algunas de los Estados Unidos alojan tales concentraciones de espíritus creadores que el gobierno federal y la industria proveen prácticamente la mitad de sus ingresos totales. El hecho de que la investigación universitaria es el cinco por ciento en costo de la federal, se debe en gran parte al apoyo federal, y representa probablemente el expendio más sabio de fondos públicos.

Antes de concluir mis observaciones quisiera comentar sobre dos instrumentos significativos de investigación: uno tan antiguo como la civilización y todavía no totalmente apreciado, y el otro novísimo. El último, se entiende, es el computador electrónico, que parece desempeñar el mismo papel en la tecnología moderna que el que tuvo el modelo a escala hace una o dos generaciones. Así uno encuentra la creencia, en algunas partes, que el uso del computador es todo lo que necesitamos para asegurar el éxito de un programa de investigación, y que sin un computador, uno no puede realizar investigación moderna alguna. En realidad es probablemente cierto que se verificará tanta investigación como antes sin computador, o una cantidad igual con el computador que de otra manera nunca podría cumplirse. El computador sencillamente aumenta la velocidad (si no la precisión) de las operaciones matemáticas en muchos órdenes de magnitud. Los computadores no pueden pensar, pero permiten a los investigadores mismos pasar más tiempo elaborando ideas de un alcance mayor. Las bibliotecas, por otro lado, son tanto las fundaciones como los productos finales de la investigación. Más y más frecuentemente se considera que el doble papel tradicional de las universidades —educación e investigación— está aumentando por un tercero: el servicio de conocimiento. La biblioteca universitaria debe comprender toda la ciencia abarcada por las otras dos funciones de la universidad. Cada laboratorio, igualmente, tendría que poseer o tener acceso a una colección de libros para servicio de sus investigadores en su campo particular de indagación. Sin experiencia acumulada en que basarse, la investigación muy probablemente estará mal orientada o será repetitiva; recíprocamente, mientras no contribuya algo nuevo a la biblioteca, ninguna investigación habrá cumplido su fin.

Journal of the

HYDRAULICS DIVISION

Proceedings of the American Society of Civil Engineers

FLUCTUATION OF PRESSURE IN CONDUIT EXPANSIONS

By Hunter Rouse,[1] F. ASCE, and Vladimir Jezdinsky[2]

INTRODUCTION

Interest in the use of large conduit expansions as means of dissipating energy in hydroelectric developments recently led the writers to investigate both the energy loss and the limiting conditions of cavitation as functions of the expansion ratio.[3] There is evidently still another aspect of the phenomenon that is involved in the structural design of the expansion chamber: the magnitude, scale, and frequency of the pressure fluctuations that are produced by the diffusing jet.

Two previous ASCE papers have direct bearing upon this problem. The first deals with the phenomenon of unconfined jet diffusion.[4] Therein it is shown that the intense shear between the central jet and the surrounding fluid gives rise to the generation of turbulence, the subsequent action of which is to entrain the surrounding fluid and decelerate the high-velocity fluid near the axis, thereby causing the jet to expand. The second paper deals with the modification of flow pattern that results from confinement of the expansion zone by conduit walls.[5] Because of these walls, any fluid that is entrained could only come from farther downstream; in other words, the jet must be surrounded by a quasi-

Note: Discussion open until October 1, 1966. To extend the closing date one month, a written request must be filed with the Executive Secretary, ASCE. This paper is part of the copyrighted Journal of the Hydraulics Division, Proceedings of the American Society of Civil Engineers, Vol. 92, No. HY3, May, 1966. Manuscript was submitted for review for possible publication on September 21, 1965.

1 Dir., Inst. of Hydr. Research, Univ. of Iowa, Iowa City, Iowa.

2 Research Engr., Hydrodynamics Inst., Czechoslovakian Academy of Science, Prague, Czechoslovakia.

3 Rouse, H., and Jezdinsky, V., "Cavitation and Energy Dissipation in Conduit Expansions," Proceedings of the Eleventh International Association for Hydraulic Research Congress, Leningrad, USSR, 1965.

4 Albertson, M. L., Dai, Y. B., Jensen, R. A., and Rouse, H., "Diffusion of Submerged Jets," Transactions, ASCE, Vol. 115, 1950.

5 Chaturvedi, M. C., "Flow Characteristics of Axisymmetric Expansions," Journal of the Hydraulics Division, ASCE, Vol. 89, No. HY3, Proc. Paper 3515, May, 1963.

517

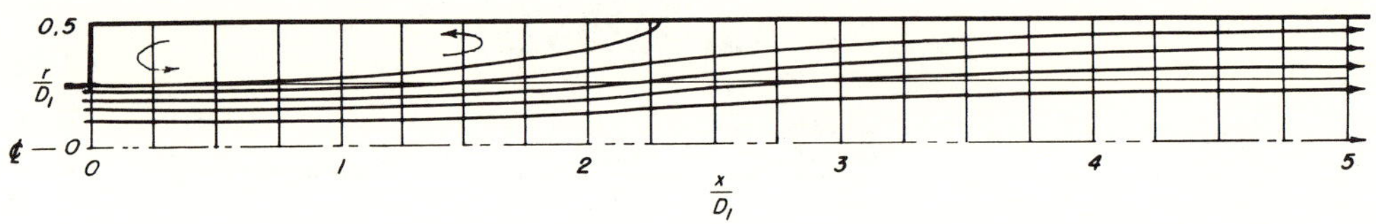

FIG. 1.—GEOMETRY OF MEAN STREAMLINES AND SEPARATION ZONE FOR 2:1 EXPANSION[5]

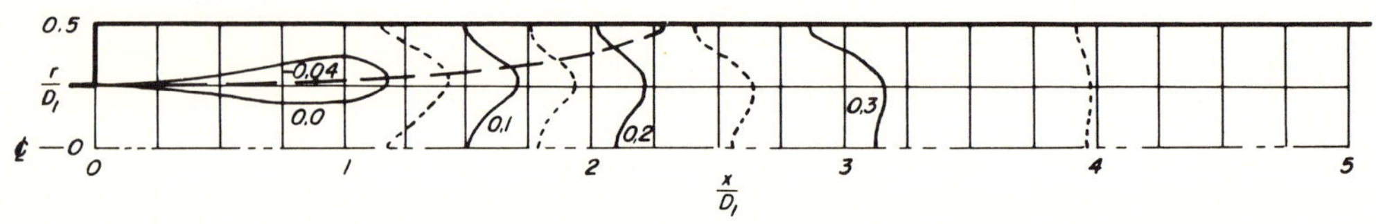

FIG. 2.—DISTRIBUTION OF PIEZOMETRIC HEAD $(h-h_0)/(V_0^2/2g)$ FOR 2:1 EXPANSION[5]

FIG. 3.—DISTRIBUTION OF TURBULENCE ENERGY $\overline{v'^2}/V_0^2$ FOR 2:1 EXPANSION[5]

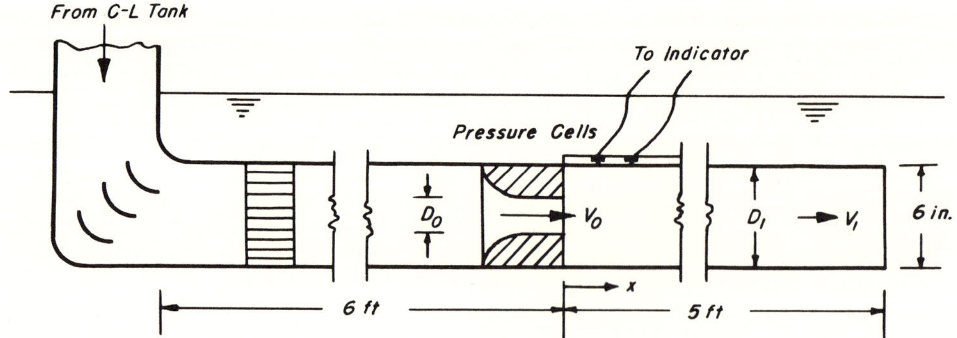

FIG. 4.—SKETCH OF EXPERIMENTAL FACILITY

stable eddy or zone of circulatory flow (see Fig. 1). Whereas the unconfined jet expands in a region of essentially constant mean pressure, confinement produces a longitudinal pressure change, the maximum gradient of which (Fig. 2) occurs shortly before the end of the eddy. The distribution of turbulence seen from the family of contours in Fig. 3 shows the zone of maximum intensity to lie well away from the boundary.

What is not apparent from any of these diagrams is the extent to which the pressure fluctuates at typical points in the field of motion. Turbulence theory[6] relates the fluctuations in velocity and pressure quantitatively for conditions of isotropy, but turbulent shear flow is far from isotropic. Pressure- and velocity-fluctuation measurements in the expanding jet,[7] however, indicate that the two characteristics will vary in much the same manner throughout the region of excessive turbulence, dependent as both are upon the same elements of the flow structure. Moreover, several additional factors controlling the pressure variation are to be gleaned from further experience with jets and expansion chambers. The "roll-up" of the discontinuity surface at the periphery of a submerged jet produces a somewhat irregular series of essentially discrete vortices representing the coarse end of the turbulence spectrum. Similar vortices formed in an expansion chamber at the juncture between the jet and the quasi-stable eddy are carried by the flow in the general direction of the dividing streamline (Fig. 1), some passing off downstream and some being caught in the backflow of the eddy. Whether it is simply the intermittent transit of these vortices, or whether the end of the eddy itself is inherently unstable, conditions in this vicinity are known to be extremely unsteady. The resulting fluctuation in velocity, transmitted upstream and downstream from the shifting point of stagnation, should be anticipated to have the largest-scale effect upon the wall pressure. The onset of cavitation, on the contrary, is found to occur in the spectral elements of smaller scale and higher vorticity produced in the same shear zone.

EXPERIMENTAL PROCEDURE

Whereas the previous expansion study was conducted in a variable-pressure cavitation tank with its own pumping circuit, the necessity of utilizing a vibration-free system for the measurement of pressure fluctuations led to a change of equipment. The test expansions, which previously took place from a single 1-in. approach to interchangeable chambers 1.6 in., 2.0 in., and 3.5 in. in diameter, were enlarged to the constant limiting value of 6 in., with interchangeable approaches 2 in., 3 in., and 4 in. in diameter. The actual approach conduit (see Fig. 4) was a 6-in. steel pipe containing a honeycomb and extending 6 ft beyond a vaned elbow leading from a constant-level tank. The reductions in effective approach diameter were accomplished with well-faired inserts of turned wood painted with epoxy; the parallel sections just prior to the expansion were held to 1 diameter in length in the belief that a minimum degree of boundary-layer development would ensure a maximum degree of fluctuation. The lucite expansion chamber was 10 diameters in length, like that of the previous study. The

---

6 Hinze, J. O., "Turbulence," McGraw-Hill Book Co., Inc., New York, N. Y., 1959.

7 Rouse, H., "Cavitation in the Mixing Zone of a Submerged Jet," <u>La Houille Blanche</u>, Grenoble, France, January-February, 1953.

entire assembly was placed horizontally in a flume, and the free outlet was kept completely submerged during flow.

Along the top of the expansion chamber was cemented a lucite strip containing a series of 41 recesses for pressure cells on 11/16-in. centers over 1/8-in. piezometer openings in the 1/4-in. wall of the chamber. Similar recesses were provided over piezometers spaced at 15° intervals around half the periphery of the chamber 2 diameters from the upstream end; the location of this section could be changed longitudinally through use of short conduit inserts. Two matching Type A-316A (+5 to -10 psi) pressure cells, manufactured by Consolidated Electrodynamics Corporation, were utilized, in combination with an Old Gold Model Type 2 mean-product computer manufactured by the Hubbard Instrument Company, and with a Model HSA-1 spectrum analyzer manufactured by Intercontinental Instruments, Inc. Indications of the root-mean-square pressure fluctuation were first obtained at a sufficient number of points in the longitudinal direction to establish the functional variation for each expansion ratio. Correlation measurements were then made between the two cells spaced at successively greater longitudinal distances either side of several key points chosen on the basis of the root-mean-square fluctuation functions. The peripheral correlation and the spectral distribution of the fluctuations were finally determined at the section of maximum root-mean-square fluctuation for each expansion. Measurements were made at mean velocities of 4.68 fps, 6.08 fps, and 10.7 fps in the expanded section for the 3:1, 2:1, and 1.5:1 expansions, respectively, as indicated by the rate of flow through a calibrated elbow meter in the supply line.

ANALYSIS OF RESULTS

For purposes of clarity, two illustrations from the previous cavitation study are reproduced as Figs. 5 and 6. The first of these is simply the distribution of piezometric head along the wall of each expansion, with an indication of the overall loss in total head. The second involves the following as functions of the inverse expansion ratio: a plot of the Borda loss coefficient, $C_L = \left(V_0 - V_1\right)^2/V_0^2$, together with the coefficient values computed from the measured losses; a plot of the cavitation index, $\sigma = \left(h_0 - h_v\right)/\left(V_0^2/2\,g\right)$, for conditions of incipiency as determined sonically; three points adapted from incipient-damage measurements by the Bureau of Reclamation,[8] and three Iowa Institute points (the lowest of which was determined after submittal of the previous manuscript) representing the sonic "plateau" values of the cavitation index at which the vapor pockets began collapsing in contact with the boundary. Because of possible damage to the boundary surface, no structure should be designed for operation in the vicinity of the bottommost curve. At still lower values of the index, it might be added, dangerously large forces due to the collapse of super-cavitation pockets must also be expected. Not only does the present study not include such forces, but it was conducted under wholly cavitation-free conditions—i.e., above even the curve for incipiency.

---

8 Ball, J. W., and Simmons, W. P., "Hydraulic Characteristics of Pipeline Orifices and Sudden Enlargements Used for Energy Dissipation," Hydraulics Branch Report No. Hyd-519, Bur. of Reclam., U. S. Dept. of the Interior, Washington, D. C., December, 1963.

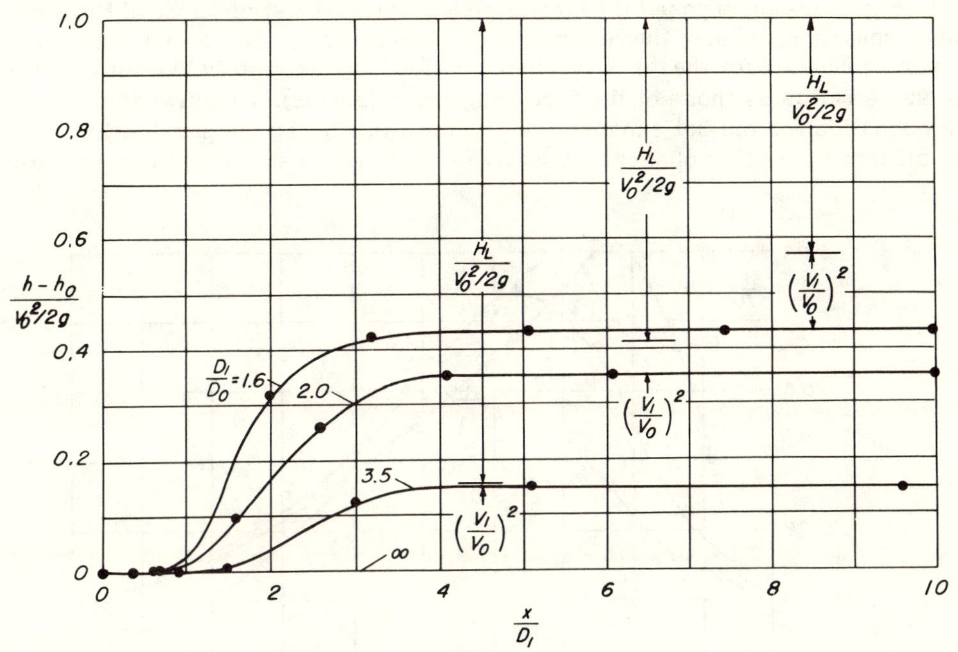

FIG. 5.—DISTRIBUTION OF HEAD FOR VARIOUS EXPANSION RATIOS[3]

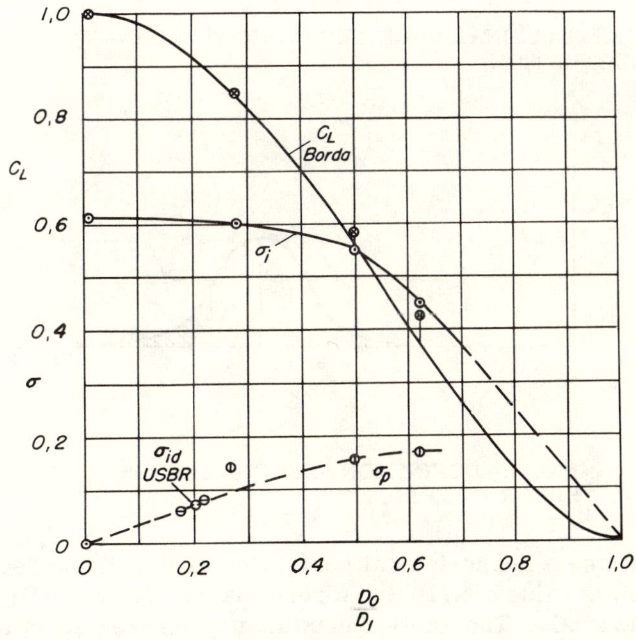

FIG. 6.—COEFFICIENTS OF HEAD LOSS, INCIPIENT
CAVITATION, AND CAVITATION DAMAGE VERSUS
EXPANSION RATIO[3]; $\sigma = (h_0 - h_v)/(V_0^2/2g)$

In Fig. 7 are superposed the measured longitudinal distributions of the root-mean-square pressure fluctuation at the boundary in its ratio to the initial dynamic pressure for the three expansion ratios (approximately, but not exactly, the same ratios as those of the foregoing investigation). Chaturvedi's end-of-eddy location for the 2:1 expansion was shown in Fig. 1; comparable locations for all three cases as observed visually during the present tests are indicated

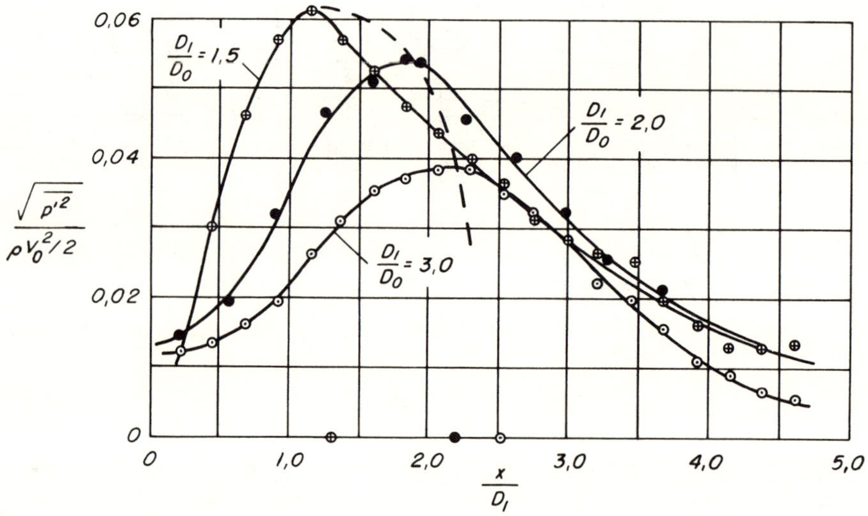

FIG. 7.—LONGITUDINAL VARIATION IN ROOT-MEAN-SQUARE PRES-SURE FLUCTUATION

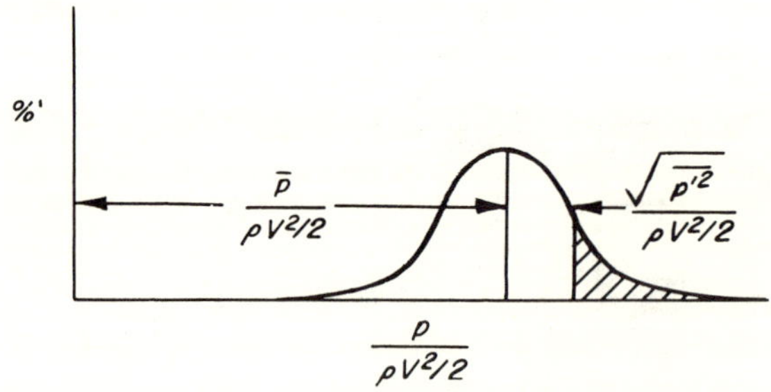

FIG. 8.—DEFINITION SKETCH OF NORMAL ERROR DISTRIBUTION

by the three points on the horizontal scale; the third of these represents very nearly the limiting value of x/D_1 for a zero magnitude of D_0/D_1 and hence an infinite expansion ratio. The points of maximum pressure fluctuation, connected by the broken line, are seen to lie somewhat upstream from the unstable end of the eddy in each instance. As is to be expected, for a given dynamic pressure of the jet, the maximum boundary-pressure fluctuation decreases toward zero with increasing expansion ratio. It should be remarked, however, that the form of the broken line points to the existence of an optimum state of fluctuation at about $D_1/D_0 = 1.5$. (If V_1 rather than V_0 is used as a reference, it should be noted, the pressure ratio increases rapidly with the expansion ratio in the range investigated.) So far as the indicated magnitudes of the fluctuations are concerned, it must be recalled that these are simply statistical measures of quantities having indeterminate upper limits. If it is assumed that their frequency of occurrence approximates a normal error curve (see Fig. 8), as seems to be true of many turbulence characteristics, then the probability (given by the cross-hatched area) of a positive fluctuation exceeding the indicated root-mean-square value is about 16% (of the total area of 100% under the curve). On the other hand, whereas there is no limit to the true error curve, the actual distribution surely does not have an unlimited abscissa range.

Since the foregoing measurements deal only with the fluctuation magnitude, one must turn to the correlation between simultaneous measurements at two different points for an indication of the size of the zone over which individual elements of the fluctuation pattern extend. If the points of measurement were superposed, the correlation R (i.e., degree of agreement) between their indications would be perfect. Likewise, if they were very far apart, the agreement would be negligible. It is the manner in which the correlation function varies from unity toward zero, as shown schematically in Fig. 9, that is thus significant. The abscissa magnitude at which the curve approaches the axis is proportional—if not equal—to the largest distance over which individual fluctuations will probably extend. Since such a measure is at best indefinite, the area of the surface enclosed by the curve is more generally taken as the representative scale of the fluctuation structure. What is sometimes called the microscale is represented by the shape of the correlation function at its peak - that is, the greater the initial radius of curvature, the less important the role played by the small-scale fluctuations, and vice versa. In the case of velocity fluctuations in a uniform flow, the frequency spectrum of their kinetic energy can be determined from the temporal rather than spatial correlation function as a Fourier transform, and the size scale estimated from the time scale through use of the mean velocity of the flow: $\Delta x \approx V \Delta t$. In the case of pressure fluctuations near a zone of stagnation, however, the relation between the correlation function and the size "spectrum" is not so readily approximated. For the moment the latter is simply assumed intuitively to vary with the negative slope of the correlation curve, as sketched in Fig. 10, the ordinate scale (actually, percent per unit of the abscissa scale, just as in Fig. 8) being so chosen as to yield a surface area under the curve equal to 100%. The peak of such a curve evidently indicates the scale of the fluctuations having the greatest probability of occurrence—this scale once again being proportional (perhaps twice) rather than equal to the measured distance Δx.

The five curves superposed in Fig. 11 are the measured correlation functions for five points along the 2:1 expansion: just beyond the expansion; in the zone of rapid increase in fluctuation; at the section of maximum fluctuation; in the zone of rapid decrease in fluctuation; and in the uniform region beyond. The reference magnitude is the mean-square fluctuation at the point under consideration, the points a and b lying the distances $\Delta x/2$ upstream and downstream. The first, third, and fifth curves are seen to indicate that for this expansion ratio the greatest probable longitudinal extent of the individual fluctuations at any section is the order of the conduit diameter. The negative portions of the second and fourth curves are reminiscent of lateral correla-

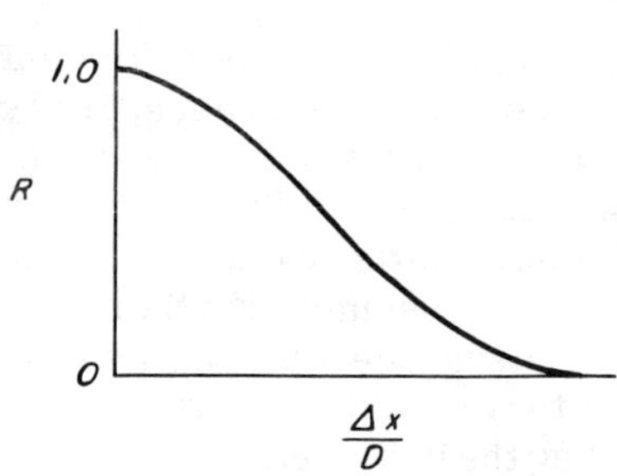

FIG. 9.—DISTRIBUTION SKETCH OF CORRELATION FUNCTION

FIG. 10.—ASSUMED SIZE DISTRIBUTION FOR FIG. 9

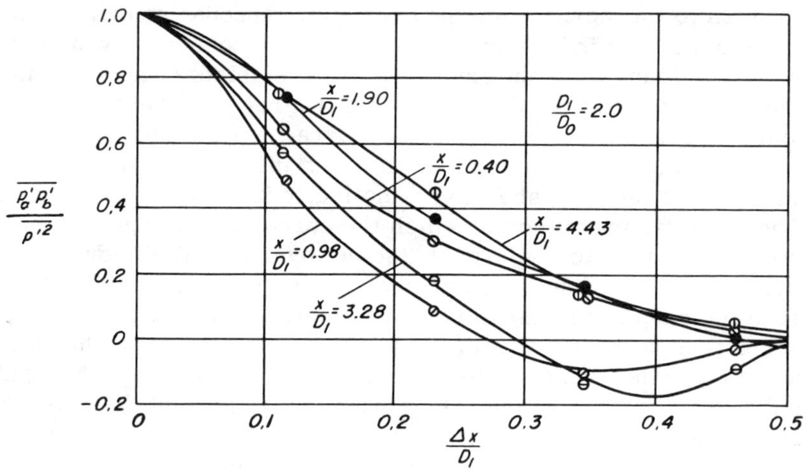

FIG. 11.—LONGITUDINAL-CORRELATION FUNCTIONS FOR VARIOUS SECTIONS OF 2:1 EXPANSION

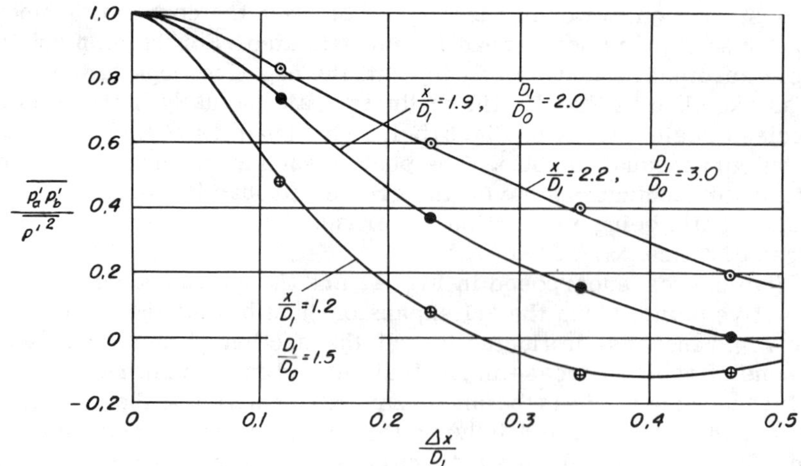

FIG. 12.—LONGITUDINAL-CORRELATION FUNCTIONS AT SECTIONS OF MAXIMUM FLUCTUATION

tions of velocity rather than pressure; they undoubtedly result from a slight regularity of the fluctuation structure, and in all likelihood the other curves also eventually become negative.

Correlation curves for the sections of maximum fluctuation at the three expansion ratios are superposed in Fig. 12. The consistent increase in relative scale of the largest fluctuations with increasing expansion ratio would be somewhat alarming were it not for the fact that the maximum probable fluctuation relative to the initial dynamic pressure $\rho V_o^2/2$ is a reciprocal function of the expansion ratio and hence of steadily decreasing importance. In Fig. 13 will be found the corresponding "spectral" distributions of scale, which roughly depict the probability of occurrence of fluctuations of any particular longitudinal extent. Evidently, the curves peak early, indicating that the vast majority of the fluctuations extend over a relatively limited zone. Around the periphery, on the other hand, the fluctuation scales are greater, as can be seen from the three correlation curves superposed in Fig. 14 for points around the cross

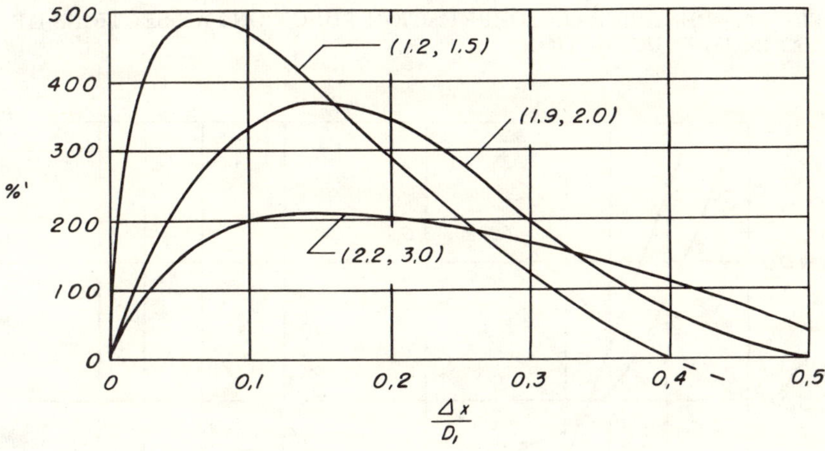

FIG. 13.—APPROXIMATE SIZE DISTRIBUTIONS FOR FIG. 12

sections of maximum fluctuation (the distances are measured as chords rather than circumferentially). Since the correlation is well above zero even between diametrically opposite points, one must conclude that the end of the eddy still fluctuates to some degree in an axisymmetric manner - that is, that its inherent ring-vortex nature has not been completely lost. In observing that the limiting peripheral correlation also increases with expansion ratio, it must be repeated that the maximum fluctuation relative to the initial dynamic pressure simultaneously decreases.

Actual measurements of the pressure-fluctuation spectrum at the section of maximum root-mean-square fluctuation for each of the three expansion ratios are superposed in Fig. 15. The ordinate scale is very nearly equal to $\overline{p_f'^2}\,V_0/\overline{p'^2}\,D_0$, $\overline{p_f'^2}$ being the mean-square fluctuation associated with a particular value f of the frequency range; it is numerically such as to yield an area under each curve, when plotted arithmetically, equal to unity or 100%. The significance of spectral functions of this nature is as much negative as it is

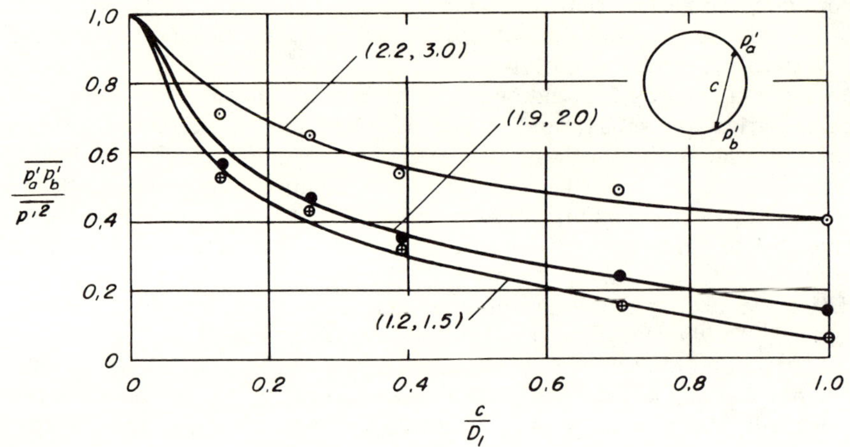

FIG. 14.—PERIPHERAL-CORRELATION FUNCTIONS AT SECTIONS OF
MAXIMUM FLUCTUATION

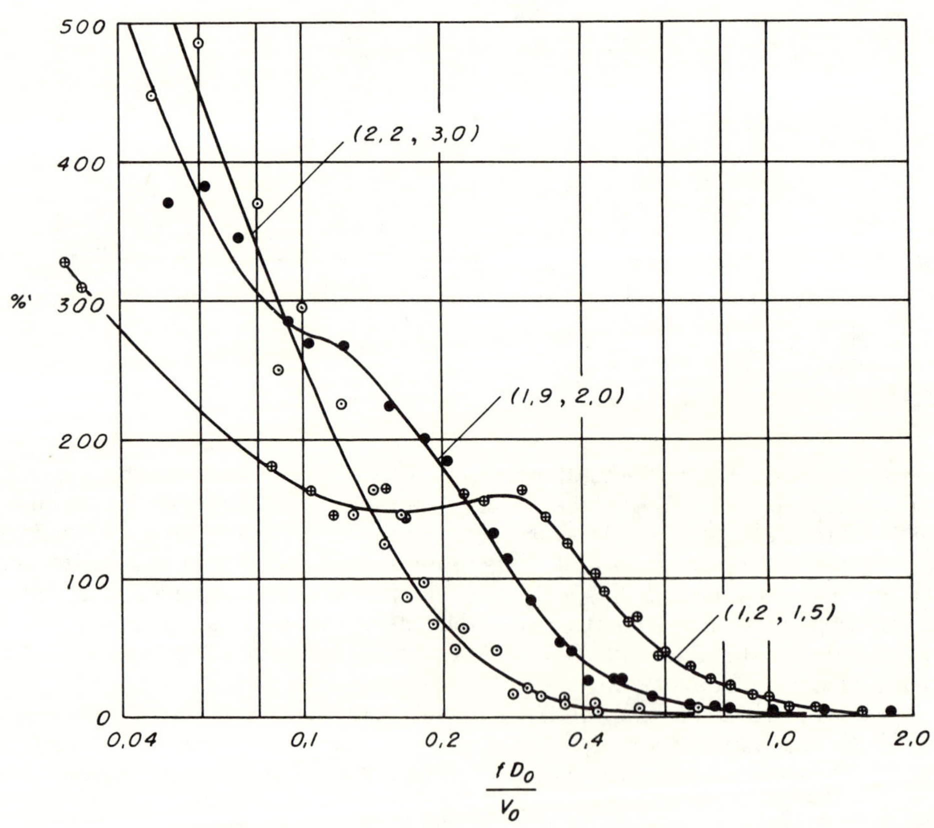

FIG. 15.—MEASURED FREQUENCY SPECTRA AT SECTIONS OF MAXI-
MUM FLUCTUATION

positive. If there is a dominant frequency, such as prevails in the wake of a cylindrical body shedding a regular series of vortices, this will produce a pronounced local rise in the function. For the smallest expansion there appears to be a marked indication of dominance at an abscissa of about 0.3; for the intermediate expansion a slight hump is still evident at an abscissa of about 0.1, but the curve for the 3:1 expansion displays no discontinuity whatever. It is, of course, quite logical that any initial regularity of eddy formation should—like the frequency fluctuation—thus steadily diminish with relative distance from the outlet. So far as the continued rise of each curve with decreasing frequency is concerned, this is presumed (by analogy with velocity-fluctuation spectra[6]) to result from representing a three-dimensional occurrence by a one-dimensional spectrum, rather than from a preponderance of large-scale fluctuations.

In applying these observations to the design of prototype expansion chambers, it must be borne in mind that the experiments were restricted to conditions of axial symmetry. Exploratory studies have shown that a small departure from symmetry can produce relatively large departures from the indicated cavitation characteristics, and presumably comparable departures from the indicated pressure fluctuations. In particular, a deviation in the direction of plane symmetry—i.e., substitution of a slot or several openings abreast of one another— should increase the likelihood of occurrence of a dominant frequency of jet oscillation, both the magnitude and the extent of the resulting fluctuation of pressure being potentially dangerous to the structure.

CONCLUSIONS

In the structural design of expansion chambers as energy dissipators, the following statistical aspects of the fluctuation in pressure should be considered in addition to those involving cavitation:

1. The boundary zone of maximum pressure fluctuation caused by the expanding stream occurs a short distance upstream from the end of the region of separation, within the first two and one-half diameters of the expansion chamber.

2. Whereas the maximum root-mean-square fluctuation in boundary pressure referred to the dynamic pressure of the expanded flow continuously increases with the expansion ratio D_1/D_0, its magnitude referred to the dynamic pressure of the initial jet reaches an optimum and then rapidly decreases.

3. The mean longitudinal extent of the fluctuations is about an order smaller than the diameter of the expansion chamber, but the probable maximum approaches the order of the diameter as the expansion ratio increases.

4. The mean peripheral extent of the fluctuations indicates a tendency toward ring-vortex action that appears to increase with increasing expansion ratio (but hence with decreasing fluctuation magnitude referred to the dynamic pressure of the jet).

5. Although there is indication of a dominant frequency of pressure fluctuation at small expansion ratios, with increasing values of D_1/D_0 the hump in the frequency spectrum decreases in magnitude and shifts in the direction of lower relative frequency.

6. Measurements of cavitation and pressure fluctuation made under conditions of axial symmetry should not be applied quantitatively to axially unsymmetric states of flow.

ACKNOWLEDGMENTS

Most of the measurements presented herein were made by the second author, during his year as Research Assistant with the Iowa Institute of Hydraulic Research, under the supervision of the first author, who wrote the paper. Invaluable aid, particularly in the spectral measurements, was provided by Frederick A. Locher, Research Associate of the Institute. The continuing project on jet cavitation has been supported since its inception by the Office of Naval Research under Contract Nonr 1509(03).

JOURNAL OF THE

BOSTON SOCIETY OF CIVIL ENGINEERS

Volume 53   JULY, 1966   Number 3

JET DIFFUSION AND CAVITATION

BY HUNTER ROUSE*

Presentation of this first John R. Freeman Memorial Lecture has meant a very great deal to the writer, for Dr. Freeman was a man whom he knew and deeply respected. An able experimenter in his own right, Freeman spoke with the voice of a prophet in behalf of experimental hydraulics, and he inspired and supported the study and work of many a young follower. All in all, his influence upon American hydraulics has probably been more far-reaching than that of any predecessor or successor [1].

Freeman himself would probably agree, if he could see things today, that not everything he began had turned out as he originally intended. Of the several dozen Freeman Scholars whom his engineering-society endowments sent to Europe and around the States, some are no longer alive, some were poor choices, and some went into other professions, but many have done their sponsor credit, and a few have even exceeded his expectations. (The writer can speak quite freely, for he was not a Freeman Scholar, but rather one of three M.I.T. students who held in succession an Institute fellowship which Freeman had arranged to be diverted temporarily to hydraulics.) The National Hydraulic Laboratory, founded within the Bureau of Standards at Freeman's instigation, has unfortunately just about reached its last days. For some reason there was little connection between these two undertakings: as far as the writer knows, only two of the forty-two Freeman Scholars have ever been on the Bureau's laboratory staff. Most interesting, indeed, is the fact that the man who made such a

---

* Institute of Hydraulic Research, The University of Iowa, Iowa City.

name as the laboratory acquired was not even an experimental hydraulician, but a mathematical physicist.

So far as Freeman's own investigations are concerned, his finest work—on the resistance of smooth pipes—was conducted years before that of Stanton and Nikuradse and was of fully comparable quality, yet was not published until so many years thereafter [2] that it played essentially no role in the advancement of the subject. For his work on fire streams, however, he received much acclaim, including the highest award of the American Society of Civil Engineers [3]. It is somewhat ironical that the principal factor involved in the poor performance of fire streams—the turbulence of the issuing water—was scarcely suggested in Freeman's papers. Probably the world of fire fighters and their equipment manufacturers is still not ripe for talk of turbulence, for though a paper of the last decade was written to clarify its role in jet diffusion [4], nozzles and monitors are still made much as they were at Freeman's time.

The tremendous impetus that he gave to American hydraulics may not have carried in the direction that he intended, but there is little doubt that he would have approved the healthy condition in which the subject presently finds itself. The writer had his mentor especially in mind, in fact, when he selected the topic of this paper, for it not only deals with jets and turbulence, but it carries their consideration through territory not then foreseen, and it finally brings the subject matter to useful application, just as Freeman believed that research as a whole should do.

A submerged jet differs from a free jet, on the one hand in the lack of gravitational influence, and on the other hand in the interaction between the jet and the surrounding fluid. In fact, from the latter point of view, the submerged jet can be regarded as the limit of the free jet as the density of the fluid stream approaches that of the fluid into which it is ejected. Except at low Reynolds numbers, the intense shear between the two regions results in the formation of turbulence throughout the shear zone, the effect of which is to entrain the surrounding fluid, diffuse the jet, and gradually dissipate its energy. The first mean-flow analysis of the phenomenon was published by Tollmien [5] just a year before the first Freeman Scholar reached Europe, and less than a decade ago Townsend [6] dealt at length with the jet in his general analysis of turbulent shear flow. A good digest of the many intervening papers is given by Hinze [7]. Essentially all of these studies, it should be remarked, were contributed by physicists and aeronautical engineers. While the behavior of a submerged jet surely does not vary with the observer, the point of view and the ultimate use of the observations definitely do. Not only is the writer a hydraulician, but work in certain aspects of jet diffusion has been taking place in his laboratories for the past two decades. He must hence be pardoned for emphasizing his own point of view and observations in the subsequent discussion.

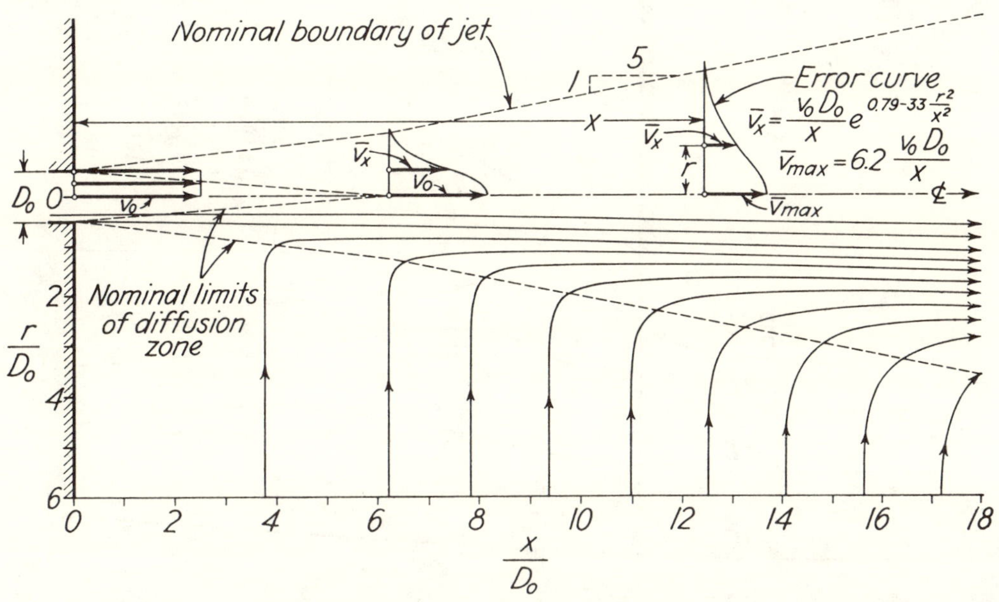

FIG. 1.—MEAN-VELOCITY CHARACTERISTICS OF JET DIFFUSION [8]

If it is assumed that the zone of jet diffusion is one of hydrostatic pressure distribution, then it follows for the boundary geometry shown in Fig. 1 [8] that the same momentum flux must occur past all successive sections, since there is no external force at hand to change it. If it is further assumed that the Reynolds number is so high that at all sections beyond the initial region of establishment a state of similarity exists for both the mean flow and the turbulence, it will follow that the jet must expand linearly and that the velocity along any line will vary inversely with distance from the nozzle [9]. Assumption of a velocity distribution like the error curve shown in the figure will then permit evaluation of the changing rates of volume flux and energy flux past successive sections—except for a single numerical factor, which must be determined empirically. Certain gross aspects of the turbulence can be approximated from the phenomenological relationships of Prandtl; indeed, the reverse process—assumption of the turbulence characteristics and evaluation of the corresponding velocity distribution—was the approach originally used by Tollmien. However, whereas the mean flow is relatively insensitive to the assumed distribu-

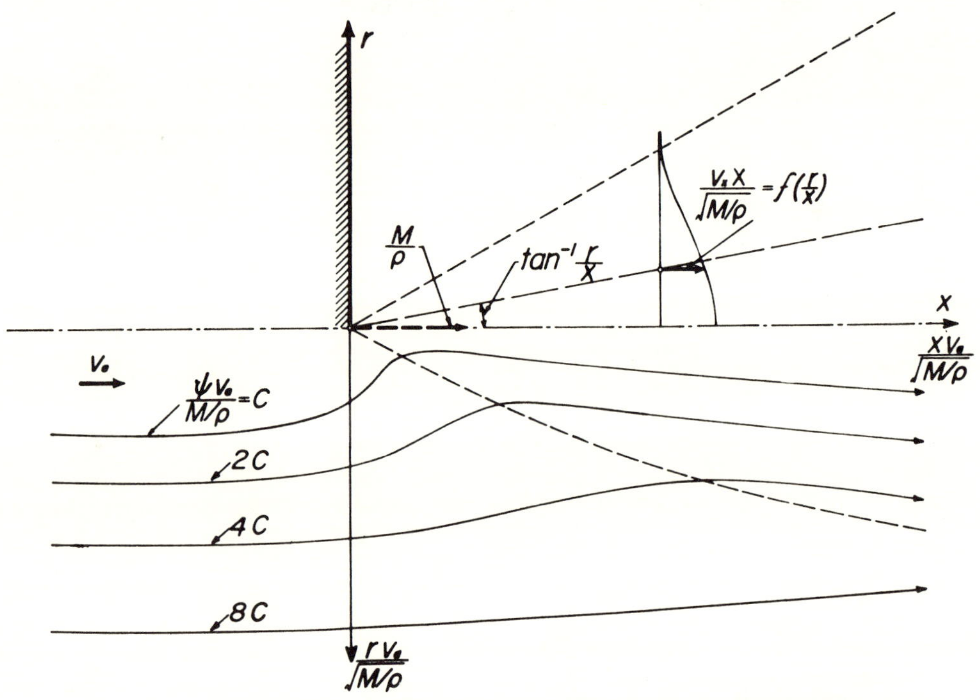

FIG. 2.—POINT SOURCE OF MOMENTUM FLUX: ABOVE, IN A NORMAL WALL; BELOW IN AN AMBIENT FLOW

tion of the turbulence, the reverse is not true, and, as will be illustrated later, information as to the actual turbulence characteristics has continued to depend upon direct measurement.

Consideration of the foregoing discussion will lead to the conclusion that at considerable distances from the efflux section it is neither the efflux velocity nor the outlet diameter which is significant, but the momentum flux $M = \rho v_0^2 \pi D_0^2 / 4$. In other words the quantity M/ρ (with the dimension $[L^4/T^2]$) becomes the sole independent parameter characterizing the jet as a whole, and the efflux section is reduced to a point source of momentum in the axial direction. As with any source flow, there is no reference length, and all cases are perforce dynamically similar and without linear scale. As indicated schematically in the upper half of Fig. 2, every section is similar to every other one in such a rendition, for mean-flow (the actual values of which can be obtained from those of Fig. 1) and turbulence characteristics alike. This intuitive dimensional method of approach can readily be extended to flow from a nozzle in an ambient stream. The normal wall containing the

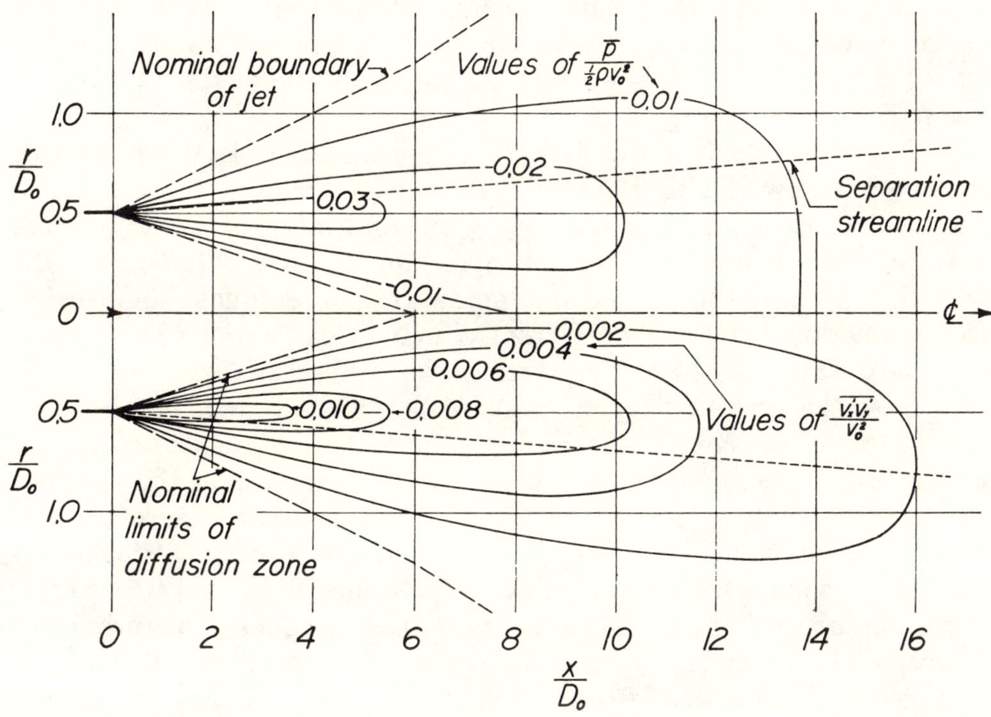

FIG. 3.—DISTRIBUTION OF PRESSURE (ABOVE) AND SHEAR (BELOW) IN A DIFFUSING JET (NOTE SCALE DISTORTION)

outlet must then be eliminated, of course, which necessarily changes the direction of the approaching fluid even for the case of entrainment without ambient flow. If it is now realized that the nondimensional coordinates must incorporate both the kinematic momentum flux M/ρ and the velocity of the ambient flow, they will be found to take the form $r v_a / \sqrt{M/\rho}$ and $x v_a / \sqrt{M/\rho}$. Further thought will show that large values of the radial coordinate will correspond to small rates of momentum flux (or large ambient velocities), and vice versa. The nondimensional pattern of streamlines must hence have a form much like that shown schematically in the lower half of Fig. 2, the streamlines becoming more and more nearly parallel to the axis with increasing relative radius; that is, a single picture will represent all possible conditions ranging from those of relatively high jet strength and weak ambient flow in an enlargement of the pattern (i.e., close to the axis) to those of relatively strong ambient flow and low jet strength in a reduction of the pattern (i.e., far from the axis). To the best knowledge of the writer, this manner of composite presentation of the entire range of conditions from pure jet flow to pure ambient flow is new; he leaves determination of the exact geometry of the pattern to an interested thesis student.

So far as the writer is presently concerned, it is the details of the zone of flow establishment rather than of the zone of established flow which are of special interest. This zone has the distinguishing feature of a central, essentially irrotational core into which the vorticity generated in the surrouding shear zone gradually diffuses. If the distinction between efflux from a cylindrical outlet in a normal boundary and efflux from a nozzle (the difference being restricted largely to the surrounding low-velocity flow) is ignored, the distribution of the mean velocity shown in Fig. 1 can be regarded as typical of either. Although the usual simplified analysis rests upon the assumption of hydrostatic pressure distribution, in actuality there are appreciable departures therefrom: to a minor degree because of the necessary radial change in the velocity of the entrained fluid, and to a greater degree because of the variation in the radial component of turbulence intensity. The two factors can be evaluated through the momentum equation for the radial direction [10]. As-yet-unpublished measurements made at Iowa by Thomas Carmody with a static-pressure probe yield the contours of piezometric head shown in the upper half of Fig. 3 [11]. The head is seen to be lowest in the zones of most-pronounced turbulence, as will immediately be delineated. Far more important than the distribution of pressure or normal stress accomplishing the diffusion of the jet fluid, however, is the distribution of shear or tangential stress. At sufficiently high Reynolds numbers, the viscous contribution to the shear can be neglected in comparison with the so-called Reynolds stresses of the turbulence. Contours of the Reynolds stress $-\rho \overline{v'_x v'_y}$ acting on coaxial cylindrical surfaces, from measurements at Iowa by Sedat Sami, are shown in the lower half of Fig. 3 [11]. In general the maximum intensities of shear would be found to occur at points of maximum mean-velocity gradient and turbulence intensity. Sami's hot-wire data for the several components of the intensity have been combined to yield the contours of $\sqrt{\overline{v'^2}} = \sqrt{\overline{v'_x{}^2} + \overline{v'_y{}^2} + \overline{v'_z{}^2}}$ shown at the top of Fig. 4. The geometrically similar contour plot shown at the bottom of the figure represents Sami's results for the fluctuating pressure obtained by means of a cylindrical piezoelectric crystal flush-mounted at the usual piezometer location on a ⅛-inch-diameter static-pressure

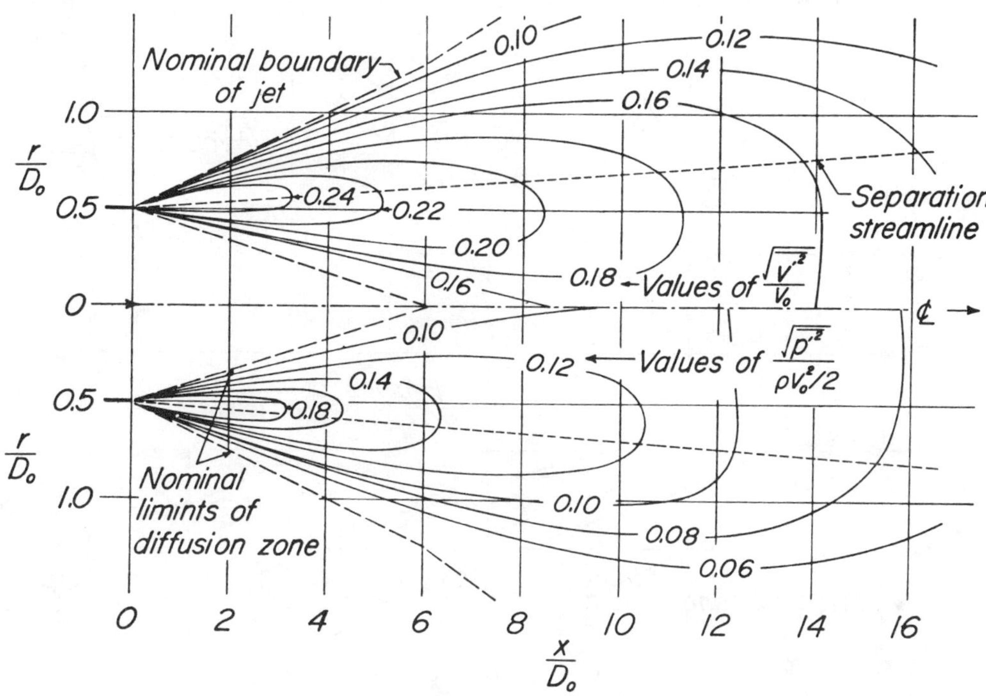

FIG. 4.—DISTRIBUTION OF VELOCITY FLUCTUATION (ABOVE) AND PRESSURE FLUCTUATION (BELOW)

probe. Careful comparison of the two contour families would show that at essentially all points of pronounced turbulence $\sqrt{\overline{p'^2}} = 3.6\,\overline{\rho v'^2}/2$.

Pressure-fluctuation measurements of this sort are relatively new, and the instrumentation involved has not yet been perfected to the extent of matching the hot-wire anemometer in either size or refinement. However, the nearly perfect correlation between the root-mean-square pressure fluctuation and the mean-square velocity fluctuation would lead one to hope that certain pressure characteristics which are not yet measurable may be represented by their velocity counterparts. Principal among these are their linear scales, i.e., the dimensions of the zones over which individual fluctuations extend. Since it is the eddy structure of the turbulence that is responsible for the two types of fluctuation, it is only reasonable that similar spectral distributions should be characteristic of both. In Fig. 5 are shown contour lines of two scales determined from temporal records of the longitudinal velocity fluctuation [11]: above, the size of the average eddies, determined from the correlation (i.e., degree of agreement) between values

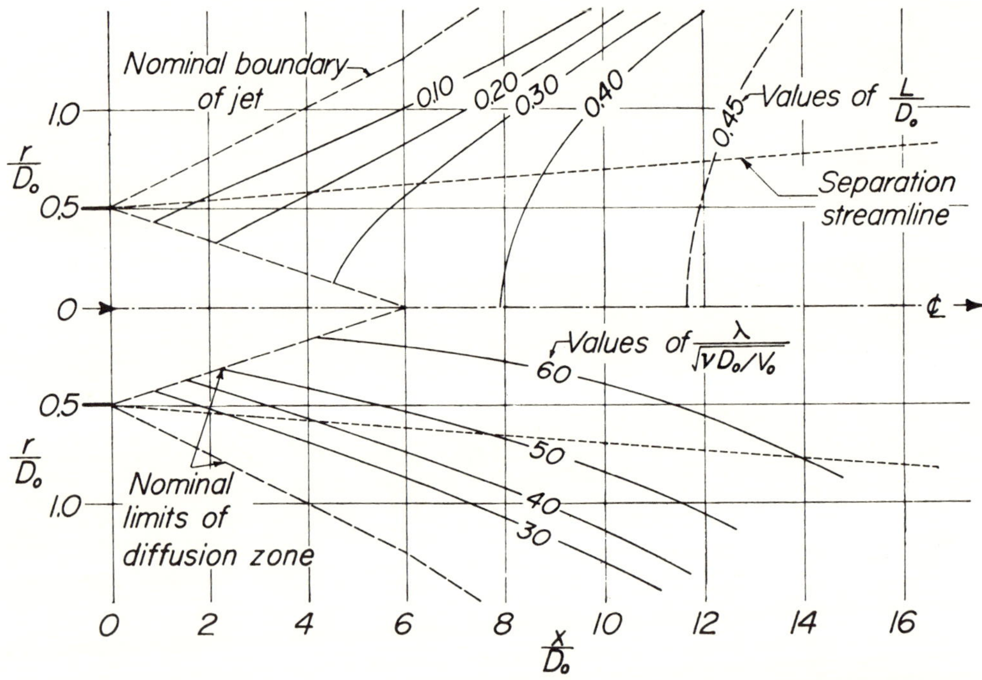

Fig. 5.—Distribution of Mean Eddy Size (Above) and Dissipation Length (Below)

at successive instants a variable time interval apart; and below, the average size of the smaller eddies, known as the dissipation length, determined from the time rate of change of the velocity.

If Freeman could have read the foregoing comments and examined the corresponding illustrations, he might well have lauded the Yankee ingenuity so obviously required by the painstaking and novel experiments that they represent, but he would most certainly have asked what engineering use they could possibly have. As it happens, two primary engineering applications of the information involve phenomena already known at Freeman's time, if not long before. One is the generation of noise. Each pressure fluctuation is a source of sound, whether it occurs in a diffusing jet or along the wall of a conduit, and considerable theoretical study has been given to the phenomenon [12]. However, the fluid compressibility is involved to a considerable degree, and the problem of noise generation is basically one of aerodynamics rather than hydraulics, though it must be granted that plumbing systems are among the most disturbing noise sources in modern civilization.

As a matter of fact, it is cavitation that is usually the cause of noise in plumbing systems, just as it is cavitation that is the other phenomenon involved in applying the foregoing information about pressure fluctuation in submerged jets. The study of jet cavitation as such seems to have originated with the writer some fifteen years ago in connection with the proposed jet propulsion of ships—i.e., the placing of enlarged propellers in pressure passages within the hulls to prevent blade cavitation. It was readily demonstrated that the low-pressure cores of the eddies generated in the shear zone around the emerging jet could well produce cavitation considerably in advance of the ducted propeller. Although the propeller would thus be protected from cavitation damage, the elimination of cavitation as a source of noise and bubbles in the wake would not necessarily be realized. The original studies conducted on the phenomenon at Iowa yielded a magnitude of about 0.6 for the cavitation index $\sigma = (h_0 - h_v)/(V_0^2/2g)$ under conditions of incipiency, as well as the first measurements that appear to have been made of the distribution of pressure fluctuations in the zone of turbulence generation, and the two photographs of vapor bubbles shown in Fig. 6 [13].

Because it was still impossible to predict other than empirically either the magnitude of the cavitation index for incipiency or the position at which collapse of the vapor bubbles could be expected to occur, i.e., the location of the noise source, the two phases of the investigation were continued independently. On the one hand, detailed measurements of the mean-flow and turbulence characteristics of submerged jets without cavitation (for convenience, air into air) were undertaken. These required not only the development of improved instrumentation but the perfection of measurement techniques to the state that the resulting values satisfied the equations of motion in detail as well as in gross. Only now being readied for publication elsewhere [11], the information thus obtained was used by S. T. Hsu of the Iowa Institute staff in preparing Figs. 3, 4, and 5 for the present paper. On the other hand, detailed sonic studies were made by Appel [14] of a cavitating jet at various values of the cavitation index, not only to define the location of the collapse zone but also to clarify a number of apparent anomalies in the earlier study.

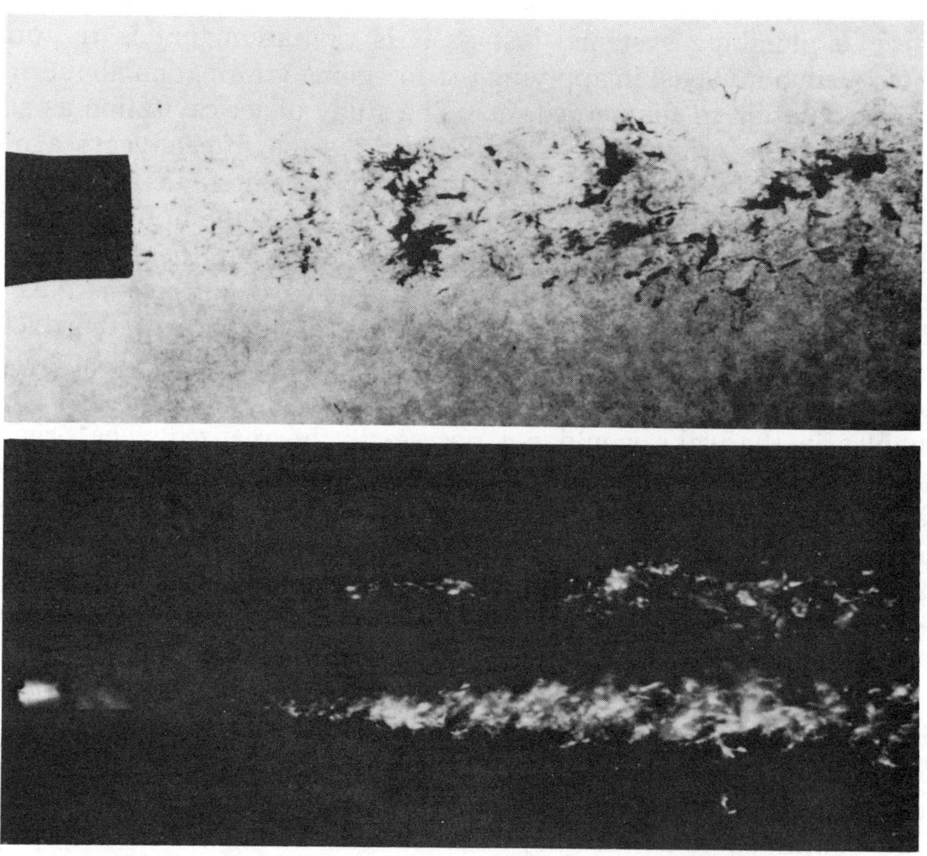

FIG. 6.—HIGH-SPEED PHOTOGRAPHS OF EDDY CAVITATION IN A DIFFUSING JET: ABOVE, SINGLE EXPOSURE, REAR ILLUMINATION; BELOW, 25 EXPOSURES, VERTICAL SHEET ILLUMINATION [13]

As is evident from Fig. 6, the intensity of cavitation-bubble concentration is distributed in much the same manner as the intensity of turbulence, shown by the contours of velocity and pressure fluctuation in Fig. 4. The smaller the cavitation index, moreover, the greater do experiments show the longitudinal and radial range of the visible bubbles to be, as one would expect from the general eddy-cavitation concept. Somewhat surprising, in this regard, was Appel's observation that the maximum source of noise was invariably at the end of the zone of flow establishment, approximately five nozzle diameters downstream from the efflux section. Although bubble collapse must evidently occur beyond any section at which bubbles are still visible, for some reason not yet clear the greatest rate of collapse thus coincides with the zone in which the turbulence reaches the jet axis. Appel's sonic measure-

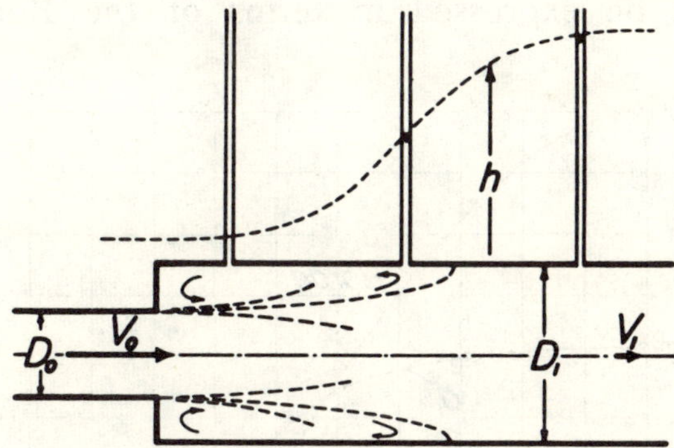

FIG. 7.—JET DIFFUSION AND PRESSURE VARIATION AT A CONDUIT EXPANSION

ments further revealed that at any point the variation in audible noise with the cavitation index involved its frequency but not its amplitude. In other words, each individual bubble collapse produces a pressure fluctuation of essentially the same magnitude, and the relative level of noise is a measure of the relative number of such collapses occurring per unit time. Moreover, the frequency of collapse was found to be quite randomly distributed. Definition of the point of incipiency is thus rather arbitrary. The writer's previous hope [13] of being able to predict at least the point of incipiency from the turbulence data presented herein is now seen to meet yet another obstacle: whereas the distribution of eddy scale shown in Fig. 5 permits the relative magnitude of average eddy size to be predicted at any point (and hence the relative number of cavitation bubbles present at any instant), not only is this a statistical average, but an unknown coefficient of proportionality is also involved. As a result, although two conditions can now be compared quantitatively, one or the other still requires experimental evaluation.

Whereas the direct application of such knowledge about eddy diffusion and cavitation is apparently restricted to the phenomenon of jet propulsion, small changes in boundary geometry can extend its range of interest greatly. Consider, for example, the current use of conduit expansions as a means of energy dissipation in hydroelectric installations. Since these are simply diffusion zones of limited radial extent, all of the foregoing flow characteristics should again be encountered, plus a few more. Principal among the latter (see Fig. 7) are the region of reverse flow enclosing the main stream and the longitudinal rise in pressure as the jet expands. The rate of energy loss at such an expansion can be expressed in terms of the Borda coefficient

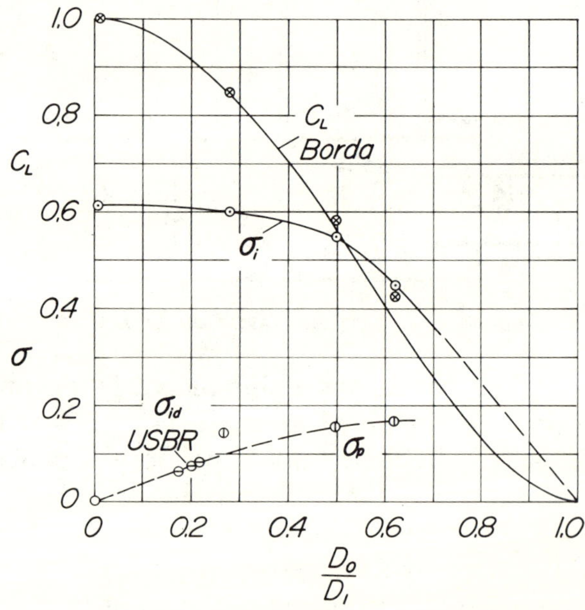

Fig. 8.—Coefficients of Loss, Incipient Cavitation, and Cavitation Damage Versus Expansion Ratio [15]

$C_L = (V_0 - V_1)^2/V_0^2$ as a function of expansion ratio, as plotted in the uppermost curve of Fig. 8 [15]. Conditions of incipient cavitation, determined experimentally, are seen from the middle curve to agree with those for the unconfined jet at the limiting expansion ratio. While even incipient cavitation is to be avoided in most design, it should be noted that actual danger to the structure from cavitation damage will occur only when the bubble collapse takes place in the vicinity of the wall. Such conditions as determined sonically at Iowa (σ_p) are compared near the bottom of the figure with re-evaluated measurements of actual cavitation damage (σ_{id}) conducted at the Bureau of Reclamation [16].

Since cavitation of this nature results from the eddy structure of the turbulence, the pressure fluctuations which the eddies produce cannot be made to disappear, as can the cavitation itself, simply by raising the ambient pressure. Hence, not only must an energy dissipator of the expansion type be designed to perform without cavitation, but the pressure fluctuations must also be considered in the structural design. Such considerations involve not only the magnitude of the pressure departure from the mean but also the size of the region over which the

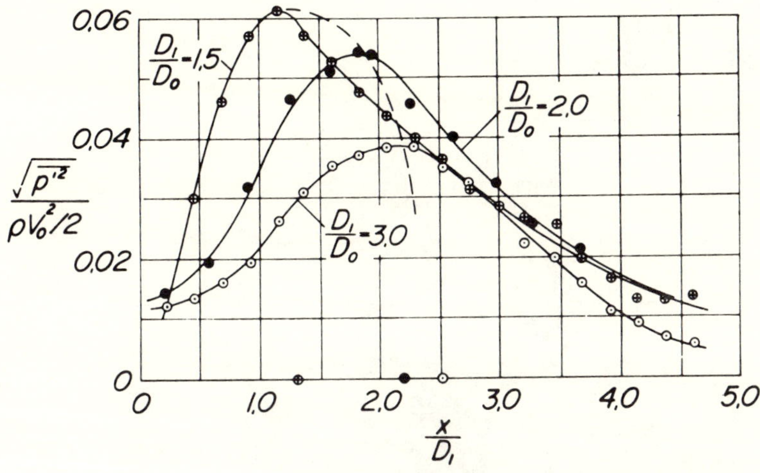

Fig 9.—Longitudinal Distributions of Wall-Pressure Fluctation [17]

average fluctuation extends and its degree of vibration-producing regularity. In a much more recent study [17], an initial exploration of these characteristics was made for the three expansions involved in the previous experiments (which, in turn, had been made with the same equipment used in the original investigations of the submerged jet itself).

Figure 9 reproduces the distribution of root-mean-square pressure fluctuation along the wall of each expansion. Since the fluctuation of pressure, like that of velocity, approximates the Gaussian probability function, magnitudes greater than the root-mean-square can be expected some 16 percent of the time. The root-mean-square values themselves are seen to increase between the initial section and one slightly upstream from that at which the separation streamline reaches the wall (marked by one of the three points along the abscissa scale), and thereafter to decrease. As must be concluded from the broken curve connecting the maximum values, an optimum condition is reached at about the smallest expansion ratio tested; it is, of course, only logical that this curve should approach zero as the expansion ratio becomes either negligible or very large.

In order to obtain an indication of the average size of the zone over which individual fluctuations extend, measurements were made of the correlation (degree of agreement) between pressures at points a and b varying distances $\Delta x/2$ either side of the point of maximum root-

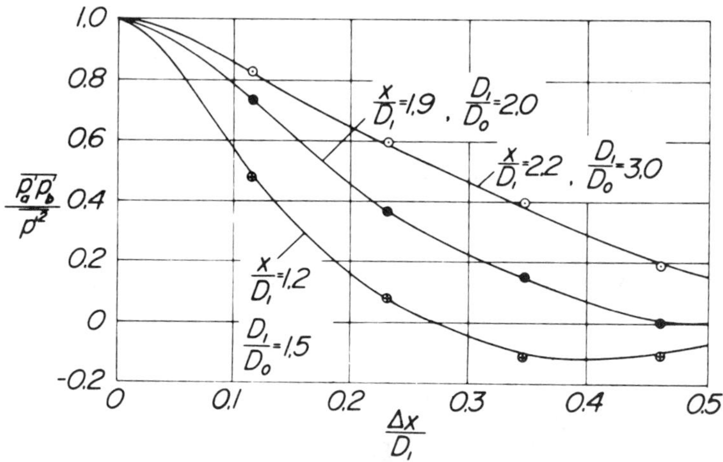

FIG. 10.—LONGITUDINAL WALL-PRESSURE CORRELATIONS [17]

mean-square fluctuation. With a negligible separation, the two values should be almost identical, and the correlation coefficient $\overline{p_a' p_b'}/\overline{p'^2}$ (see Fig. 10) should then be very close to unity; if, on the contrary, the separation is greater than the distance over which the individual fluctuation is likely to extend, the correlation should be essentially zero (the tendency of such a curve to dip below the axis is a sign of some degree of regularity of the fluctuations). As is seen from Fig. 10, the size of the fluid elements involved in the individual fluctuations is generally a small fraction of the conduit diameter.

Finally, the spectral analyses of the fluctuations, again at the sections of maximum root-mean-square fluctuation, are shown in Fig. 11.

The ordinate scale is such as to yield an area of 100 percent under each curve when plotted arithmetically, and the abscissa scale includes the frequency of the fluctuations passed by the filter. It is actually the form of the curves rather than their elevation which is significant, for local humps (such as that in the curve for the 1.5 expansion ratio) indicate a tendency for fluctuations of a particular frequency to predominate and possibly excite elastic vibrations in the structure. It is evident from the figure that this tendency rapidly diminishes with increasing expansion ratio.

The points that the writer has emphasized are but a few of the conclusions that can be reached from the experimental data that have

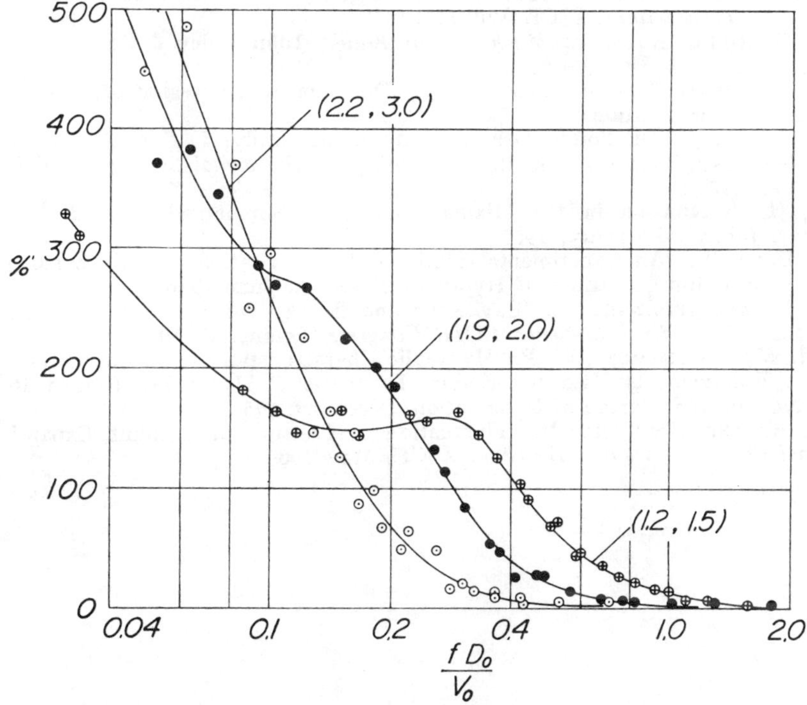

FIG. 11.—FREQUENCY SPECTRA OF WALL-PRESSURE FLUCTUATION [17]

been presented. Likewise, the boundary forms used for illustration are probably but a few of the many which will eventually be investigated in a comparable manner. Phenomena of turbulence and pressure fluctuation are in the air, hydraulically speaking, both here and abroad, particularly, it should be noted, in the Soviet Union. Dr. Freeman would certainly have relished both the situation and the challenge that it brings.

REFERENCES

1. ROUSE, H., AND INCE, S. *History of Hydraulics,* Dover Publications, New York, 1963.
2. FREEMAN, J. R. *Experiments upon the Flow of Water in Pipes and Pipe Fittings,* American Society of Mechanical Engineers, New York, 1941.
3. FREEMAN, JOHN R. "Experiments Relating to Hydraulics of Fire Streams," *Transactions,* ASCE, Vol. 21, 1889.
4. ROUSE, H., HOWE, J. W., AND METZLER, D. E. "Experimental Investigation of Fire Monitors and Nozzles," *Transactions,* ASCE, Vol. 117, 1952.
5. TOLLMIEN, W. "Berechnung turbulenter Ausbreitungsvorgänge," *Zeitschrift für angewandte Mathematik und Mechanik,* Vol. 22, 1926.
6. TOWNSEND, A. A. *The Structure of Turbulent Shear Flow,* University Press, Cambridge, 1956.
7. HINZE, J. O. *Turbulence,* McGraw-Hill Book Company, Inc., New York, 1959.
8. ROUSE, H., editor. *Engineering Hydraulics,* John Wiley & Sons, Inc., New York, 1950.
9. CORRSIN, S. Discussion of "Diffusion of Submerged Jets" by Albertson, Dai, Jensen, and Rouse, *Transactions,* ASCE, Vol. 115, 1950.
10. ROUSE, H., editor. *Advanced Mechanics of Fluids,* John Wiley & Sons, Inc., New York, 1963.
11. SAMI, S., CARMODY, T., AND ROUSE, H. "Jet Diffusion in the Region of Flow Establishment," in preparation.
12. LIGHTHILL, M. J. "On Sound Generated Aerodynamically. I. General Theory, II. Turbulence as a Source of Sound," *Proceedings of the Royal Society,* A, Vol. 221, 1952, Vol. 222, 1954.
13. ROUSE, H. "Cavitation in the Mixing Zone of a Submerged Jet," *La Houille Blanche,* January-February, 1953.
14. APPEL, DAVID W. "An Experimental Study of the Cavitation of Submerged Jets," ONR Report, Iowa Institute of Hydraulic Research, June 1956.
15. ROUSE, H., AND JEZDINSKY, V. "Cavitation and Energy Dissipation in Conduit Expansions," *Proceedings Eleventh IAHR Congress,* Leningrad, 1965.
16. BALL, J. W., AND SIMMONS, W. P. "Hydraulic Characteristics of Pipeline Orifices and Sudden Enlargements Used for Energy Dissipation," Hydraulics Branch Report No. Hyd-519, U.S. Bureau of Reclamation, December, 1963.
17. ROUSE, H., AND JEZDINSKY, V. "Fluctuation of Pressure in Conduit Expansions," *Journal of the Hydraulics Division,* ASCE, May 1966.

German-American Observations on Educational Reform

Hunter Rouse and **Eduard Naudascher**, Institute of Hydraulic Research, University of Iowa[1]

SOME 35 YEARS AGO, the first author was sent to Germany by the Massachusetts Institute of Technology for postgraduate study of experimental hydraulics at the Institute of Technology in Karlsruhe. Thirty years later, the second author came to America for the same purpose—by coincidence, directly from Karlsruhe and eventually to the first author's present institution. Here, perhaps, the similarity reaches an end, for the earlier traveler was quite young, inexperienced, and with scarcely a word of German in his vocabulary, whereas the more recent traveler had already received his doctorate and assisted in the Karlsruhe laboratory for several years, and was fluent in English.

There were lessons to be learned by each country from the other 35 years ago, and the same is true today in terms of the past as well as the present. The two authors have correlated their varied experiences to ascertain what may be learned from past and present educational trends in the two countries that may now be of reciprocal benefit.

An American Student in Pre-Nazi Germany

To the first author as an American hydraulics student in pre-Nazi Germany, many things were deeply impressive. First, the institutes of technology were few and uniformly excellent. Their express purpose was to educate engineering leaders. Training for routine engineering work was the task of the numerous technical institutes. The German engineer's inherent feeling of inferiority alongside other professions did not reach the student's ears till his return to America. Second, the hydraulics departments shared the relative prestige of the institutes of technology as a whole. It was not apparent to the visitor that the civil engineers were still moving with momentum acquired a generation or more before, whereas it was the mechanical and aeronautical engineers who were really making the forward strides. Third, each incumbent of a professorial chair seemed to be world-renowned, and the absence of associate and assistant professors was not remarked amid the many willing hands always in evidence. Finally, the extreme academic and personal freedom of all students was a heady draft for even a young adult to experience for the first time. True, many a lecture was regularly cut, and many a design problem was purchased from the campus draftsman, but the biennial examinations seemed to be passed, and the average end-product—the *Diplom-Ingenieur*—somehow seemed to remain on the highest level.

It was not long before the possibility of earning the German doctorate attracted the American student's attention, but he was told that his bachelor's degree in no way compared with the German engineering diploma and that he must enroll as an undergraduate student. It was never finally decided, however, whether he would have to pass examinations covering the first two or the last two years of prescribed courses. Many of the lectures which he attended were repetitions of those he had heard at home, and others seemed more appropriate to technical institutes. Ultimately he dropped the formal lectures, spent a year in laying the groundwork for a doctoral dissertation, and returned to MIT to get the master's degree, which was reluctantly considered acceptable by the Germans as prerequisite for their doctorate, and to complete the dissertation. He then went back to Karlsruhe for the doctoral examination.

What neither the American nor the Karlsruhe civil engineering faculty realized was that any difference between his preparation for the doctorate and that of his German contemporaries did not lie at the university level but at the level of the secondary schools. The German secondary education of that period was clearly superior to that in America. The student entering a German institute of technology was already well grounded in mathematics, the basic sciences, and the humanities and was capable of independent study; hence, he was certain to complete his student years with what is now considered by many to be the most essential part of engineering education.

A German Educator in America Today

The German author, who came to America in 1959, realizes that he did not have a very broad insight into engineering education in his home country. Having studied civil engineering, he had been affiliated only with the one field in which the need for educational reform had become most evident. His postgraduate experience, moreover, was limited to a few years' assistantship in a

<hr>

[1] Hunter Rouse, formerly Director, Institute of Hydraulic Research, is now Dean of Engineering, University of Iowa. He is the third recipient of the George Westinghouse Award (1948) and the third recipient of the Vincent Bendix Award (1958). Dr. Naudascher is Research Engineer at the Institute of Hydraulic Research and Associate Professor of Mechanics and Hydraulics, University of Iowa. A somewhat longer German-language version of their article appeared in *Mitteilungen des Hochschulverbandes* (Hamburg), Vol. 14, No. 3, p. 79, May 1966.

research institute with a reputation based on the pioneer days of experimental hydraulics. Nevertheless, it seems more than mere coincidence that his trip across the ocean for advanced study was in the opposite direction from that of his coauthor a generation before. The American author's institute is by no means the only one for hydraulic research and instruction in the United States which has gained worldwide repute within the last generation, nor has such development been restricted in any sense to the one field. More than 400 German science and engineering researchers have been registered yearly by the U. S. immigration service. The lack of opportunity for independent research and the lack of responsible positions at an early age are cited as the primary reasons why they go abroad.

Curiosity actually led the German author to come to America. He had been advised against it. Anyone who is attracted to engineering education and research, he was told, should seek to acquire practical experience as rapidly as possible; once one had proved himself in field or office and reached the necessary age and maturity, nothing further would stand in the way of a call to the university. As to the future of an engineering scientist in the United States, opinions were divided. Technological advances in the fundamental engineering fields were difficult to reconcile with the popular view that the American engineer is after immediate practical success. As a matter of fact, the German author needed considerable time to adjust himself to many new impressions. If he expected emphasis on preparation for engineering practice, he soon learned that under conditions of rapid technological change it is realized even in industrial circles that a strong basic curriculum is prerequisite to professional breadth and adaptability. He was surprised to hear heated discussions of curricular reforms, and he was impressed by the relative absence of obstacles to the implementation of new ideas and by the availability of funds for even conflicting proposals. Most of all, he was impressed by the manifold nature of the educational possibilities, which were adaptable in level, substance, and method to a variety of talents, and also by the quantity and the quality of course material available to advanced research students. In view of the many avenues of engineering endeavor, he found especially enlightening the conscious departure from the national uniformity of curriculum so often pictured as an asset in Germany.

For the activity sought by the German author in research and instruction, a number of institutes were available to satisfy his needs. In the single field of fluid mechanics, which is the foundation of all hydraulic engineering, his chosen institute employs six men of faculty rank, each surrounded by a group of postgraduate students and doctoral candidates. Since all of the staff members have slightly different fields of interest, the students are given an extraordinary opportunity to strengthen themselves in almost any aspect of the science, special seminars being arranged if the curriculum should not offer the particular knowledge desired. The staff's diversified interests also ensure that research is conducted under active guidance on a great variety of basic problems. Each staff member is provided with essentially full responsibility in his work and has, within reason, the necessary funds at his disposal. Advice or criticism is not imposed on them by the director; rather, they seek and appreciate it. Exchange of knowledge is always encouraged, within the institute as well as with other departments. In this stimulating atmosphere, theses and dissertations become more than requirements for degrees; they form elements in the solution of complex problems, amenable only to such group endeavor.

Recent Trends in American Education

Much of what impressed the German author in postwar United States had actually developed since the American author's initial experience abroad. The successful collaboration between science and technology during the war years led engineering educators to lay ever greater emphasis upon the engineering sciences. There was, unfortunately, a counterbalancing trend in the high schools in the opposite direction, so that engineering colleges were forced to spend considerable time making up for the weakness in background of entering students. But after the first Sputnik, secondary school training in mathematics and the basic sciences became steadily better. At the same time, the heightened demands for joint efforts on the part of scientists and engineers caused by developments in space and atomic energy were reflected in the engineering colleges. Many of the best students were lured away from engineering by the new glamor of mathematics and physics. Engineering courses placed greater emphasis upon science, even to the extent of arousing widespread criticism for their apparent research orientation.

However, the scientific trend was by no means the only one in engineering education methods. Several others were in conflict. One was based on the logical belief that the difference between pure science and applied science, i.e., engineering, lay in the necessary admixture of judgment and experience with the formalized knowledge involved in solving any technological problem. Rather than supporting the German belief that the graduate engineer must seek practical experience after graduation to balance his classroom education, it was argued that the fostering of uninhibited originality must begin in the first university year, lest the creative spirit of the student be stifled by the formality of science. However, the trend toward a more scientific engineering education involved greater similarity of program for students of all engineering branches: mathematics and basic sciences the first year, mathematics and engineering sciences the second, more engineering sciences the third, and specialization only in the fourth. Proponents of the art as a complement of the science held that this multilayered system of courses should be replaced by one in which originality of planning and design, as well as theoretical analysis, would be stressed in every year, particularly the first.

According to many American educators, the three parallel sequences of mathematics, science, and engineering design would be incomplete unless accompanied by a fourth, the humanities. In this category are such subjects as the social sciences, literature, languages, and the fine and performing arts. Some prefer a series of comprehensive courses specially arranged for engineering students, while others recommend that engineers mix with other university groups in the free selection of subjects. The American author cannot help recalling two aspects of his postgraduate years in Germany. First, music, the theater, opera, art, language, and history did not have to be presented in an introductory course to become real to a student; they were all around him. Second, the greatest possible humanistic awareness in a student is that produced by displacement into cultural surroundings essentially different from his own. Actually, the importance of the humanities to engineering education is appreciated on both continents. It is recognized that the engineer should play a greater part in meeting world changes and crises and that emphasis on writing and speaking is essential to this end.

Amid all the pedagogical ferment sensed by the German author in the United States, one heritage of the American genius for mass accomplishment continues to have an adverse effect. In most countries the student enrolls in a university specifically for professional education. In the United States, however, schooling beyond the secondary level has long since become a status symbol. A majority of the students in college are there because it is a part of the pattern of middle class life.

Although the college of engineering has retained its professional aspect, it cannot help being subject to the student population explosion. Just as educational institutions now seek to upgrade themselves, i.e., from junior college to college and from college to university, so too does every technical student wish to have a university level degree. The result is that every institution teaching engineering seeks the same standard mark of accreditation. It can only follow that this minimum level must not be too high, or there would be many a dissatisfied municipal and state administration. Rather, the differentiation becomes a matter of the reputation of the school and the grades of the students. In other words, the better students of the better institutions become the graduates of true university level, and the poorer students or those of the poorer institutions become the technicians.

If it is true that ten times as many men should be trained for routine engineering tasks as for high level creative work, this means that much of the so-called engineering education in America is actually of technical-institute grade under the guise of accredited university caliber. Several methods of overcoming this situation exist. There is now a concerted drive to strengthen and multiply the number of technical institutes. This is supported by the large engineering colleges, while at the same time technical institutes strive to attain university standing. Formerly the attempt to encompass within the undergraduate program the semester hours required for the proper level of collegiate instruction led to the adoption of a five-year curriculum by a number of respected institutions. This has since become steadily less popular, owing in part to the availability of baccalaureate degrees in the "glamor" professions in only four years. The tendency at present is to offer the Master of Science degree at the end of five years and to encourage attainment of the doctorate after another two or three years of formal courses and research as a further means of differentiation.

Higher Education in Germany

In the development of German higher education, the conditions under the Third Reich and the material losses and casualties of World War II had a nearly fatal influence. Many scholars and educators chose to emigrate; others retreated from public life, burying themselves in their specialties and isolating themselves from ideas and trends outside their country. After the war only 20% of West Germany's academic institutions were fully available, and almost a generation of scholars had been lost.

Government interference with teaching and research was immediately eliminated after the war, but reconstruction, recruitment, and inner reform could not proceed rapidly enough to keep some of the most productive and progressive men from leaving the country. Those who stayed saw their work jeopardized by a great increase in enrollment, which caused the student population to double between 1950 and 1960, and by the demand that applied research should have priority during economic rehabilitation. However, there were more deep-rooted abuses to be remedied than the mere lack of an adequate number of faculty and an unfavorable policy regarding support for research.

Among the recommendations for reform, which are still being heatedly and publicly discussed, the essential ones are: relaxation of the rigid subdivision of the various disciplines and of the autocratic structure within them; abolition of the traditional concept of "one professor, one institute" and establishment of a mid-level personnel with tenure and independence; provision for nonteaching personnel to join university research teams; promotion of better contact between teachers and students through campus communities and organization of study groups, after the British tutorial system, for the early semesters; development of new and wider avenues of secondary education so that more youths may have a chance to enter the university; and creation of new academic degrees so that students not interested in research may leave after two or three years. It is emphasized that the success of any reform plan will depend largely on a redistribution of the power and duties of the German professor, whose present load simply does not permit him to keep up with his academic and research responsibilities.

In order to understand the situation of the German institutes of technology, one must bear in mind that they all developed outside the university system, and that not until the second half of the 19th century did their degrees become commensurate with those of the university. In fact, only today are the engineering sciences being incorporated in one of the newly founded universities and only now is it under consideration to give the institutes university titles. Whether it is due to this development or to other factors, the German institutes of technology have become strongly oriented toward professional practice. Although the policy of using success in engineering design or industry as the primary criterion in appointing a professor has made teachers more aware of the immediate needs of industry and given graduates better preparation for immediate employment, it is now being realized that a too strict adherence to this policy also has serious drawbacks. If the engineering graduate is to master tasks that cannot yet be foreseen in a world of rapidly changing technology, today's engineering practice will not be of much help to him. What he really needs is a program of continued education that keeps him abreast of new knowledge after he has left the university. If, on the other hand, the engineering teacher is to meet his responsibility of advancing the frontiers of basic knowledge in his field, consulting work for industry will not suffice. Obviously, much more could be done in utilizing the great asset of West Germany's 15 technical institutes for each technical university. All routine training must be left to these institutes and all routine testing and design must be left to laboratories and consulting firms outside the universities. Not until then will the latter be free to concentrate on their principal responsibility of developing the mental resources essential to the discovery and utilization of what lies beyond present knowledge and experience.

Weaknesses and Strengths in Both Systems

Consideration of German and American engineering education topic by topic leads to the conclusion that practically every sign of weakness in the one corresponds to a point of strength in the other or vice versa. They complement each other to the extent of demonstrating most of the goals toward which the educational world is striving. On the assumption that strength in one aspect of a topic does not necessarily entail weakness in a closely related aspect, the authors have undertaken to combine the best parts of the two educational systems in the hope that the result may not only represent a well-balanced general goal but may be of direct worth to the individual countries themselves.

What the American author believes that his country has to learn from Germany in engineering education has not changed appreciably since his postgraduate days, and he now feels that it should have been evident for several generations. First, the secondary schools are the backbone of any educational system. Second, training for routine engineering work is the responsibility of technical institutes. Third, education at university level should be aimed at producing the upper stratum of the engineering profession. There is little reason to presume that the strengthening of American high schools pedagogically would weaken them as practice fields for democratic living. Perhaps the German humanistic schooling based on Latin and Greek is too extreme to emulate, but American classes in driving or basketry are quite as extreme in the opposite direction. Technical institutes, in turn, will be able to function properly only as the universities yield to them a portion of their present prestige and only as trade schools take from them a portion of their present association with purely manual activity. Curricula at true university level can then begin to cover adequately in a four-year period the basic combination of mathematics, science, humanities, and creativity that potential engineering leadership requires, with graduate study devoted to specialization. As for tuition, why should elementary and high schools in America be entirely at state expense and the university only slightly? Is not an educated citizen America's most valuable commodity?

The principal areas in which the German author believes the American university to be leading the way in educational reform are those of postgraduate education and faculty policy and organization. He well recalls his coauthor's story of expressing surprise to his Karlsruhe colleagues that only a dissertation was required of German doctoral candidates in engineering and being told, "But we are *Diplom-Ingenieure;* we have already finished all our course work!" Surely no one would claim today to be completely educated. Yet if advancement of the elite is the primary goal, why should the continental doctoral candidate be left to his own resources? How can the solitary German professor, surrounded as he may be with assistants, hope to match either the efficiency or the productivity of his American colleagues, who as a team share his many duties? If it is granted that the faculty is just as responsible for postgraduate as for undergraduate education and for fundamental research as for education in all its aspects, then the American practice of placing administration in the hands of men permanently selected for this purpose should be recognized as a means of transforming administration from drudgery into an effective aid in the long-range planning of programs for education, research, and faculty development. On the other hand, success of these programs in creating an atmosphere of competitive collaboration may be less a matter of administrative accomplishment than a consequence of American secondary schooling and social environment.

High Schools and Technical Institutes

In the matter of secondary education, the task is evidently the mating of a system that emphasizes the fine cultivation of the mind with one that stresses the individual's relationship with his fellow men. On the one hand, the pupil must learn that even the greatest achievement in a specialized field will not make him cultured if he does not possess a breadth of understanding in other areas. On the other hand, he must acquire a sense of responsibility to the community; if he is to be truly effective on a community-wide basis, this sense must be the stronger, the greater the pupil's mental potentiality. The danger in the German emphasis on cultivation of the mind frequently involves the notion that a subject is the more cultivating, the less it is related to practical life; lack of interest in public affairs and in politics and ignorance of achievements in science and technology are some of the adverse consequences to which this notion may lead. Perhaps the most important service that German high schools are supposed to render is the development of self-discipline, the basis for independent learning and thinking. The rule is to give high school students a minimum of freedom in choice of courses and to spur them on through tough examinations. However, once they have passed the final examination (*Abitur*), they are expected to have acquired the maturity necessary to assume responsibility in learning and to use the generous freedom at the university to their best advantage.

It is at the secondary level, in fact, that a distinction in philosophy between the two systems first becomes evident. Whereas in the United States it is the custom to give everybody the same educational opportunity, in Germany the best are given the greatest opportunity. In the United States more than 75% of a given age group complete a secondary education that entitles them to college admission, as compared to about 10% in Germany. Granted that roughly the same percentages continue with postgraduate study in one country as in the other, the fact remains that in the United States at all educational levels except the highest the majority are in general dragged along and thus retard the abler individuals. In contrast, and despite the public demand for an expansion of college education, a group of German educators once went so far as to demand a drastic reduction of admissions to the German university, in recognition of the fact that those who have scientific and creative abilities are by nature limited in number. Their argument was that one must provide a greater variety of schools so that each intelligence level can be given the best education within its grasp. Although diverse programs at schools of the same type, such as the American honors programs, may have a somewhat equivalent effect—with the advantage that individuals may more easily change groups at any stage of their intellectual growth—these German educators reasoned that everyone would profit most from the existence of different types of schools with curricula, teaching methods, and examination systems specially tailored to the student's ability and interest.

West Germany is quite close to this ideal situation with its abundant vocational and trade schools and its 125 reputable technical institutes, which cover every possible specialization from mining to timber and paper engineering. Nevertheless, much needs to be improved in allocating students to the proper institutions and in planning the teaching programs and methods more appropriately. In view of the rapidly increasing need for engineering manpower without an equivalent increase in enrollment, moreover, it is inevitable that the future engineer of either country must increase his efficiency. This calls not only for a greater use of computers, with a corresponding emphasis on teaching computer-aided design, but it also calls for a more efficient use of technicians to assist the engineers, especially in America. If the United States does not succeed in attracting a sufficient number of young men to its technical institutes, perhaps it will have to make them more attractive to women. On the other hand, the fear of many an American high-school graduate that a good technician will be forever a frustrated engineer must not act as a deterrent; some 5% of the German technicians have successfully continued the pursuit of an engineering degree.

Undergraduate Education in Germany and America

Many a special feature of German or European university tradition, which must still be regarded as setting the most favorable conditions for the education of the elite, has become questionable, mainly because the mid-level educational institutions have not been sufficient in number to free the university from the mass rush and permit the formation of a select and highly motivated student body. This is all the more true in America. The traditional form of academic freedom of the German student which allowed him to schedule his time, select courses, replace lectures by independent study obviously created the best atmosphere for scholarly learning, fostering original thought and skeptical contemplation along with the acquisition of knowledge. In fact, the great promoter of academic freedom, Wilhelm von Humboldt, thought of higher education as a process of learning rather than of teaching: a solitary struggle of each individual with himself. Moreover, the examination system, concentrating all the testing in a few comprehensive sessions, corresponds quite well to the responsibility and scholarly habits the students are expected to develop.

A question arises, however, as to whether it is feasible or even desirable today to restrict admission to the university so drastically that every student can be expected to have enough initiative and maturity to find his way through his course work without permanent supervision and the guiding help of recurrent examinations. In view of a gradual change in the optimum ratio of academically trained engineers to technicians that is resulting from the accelerated increase in availability and capability of computers, universities can no longer afford to impose too strict entrance requirements. It is apparently in recognition of this fact that educational reform in Germany has been moving in the direction of reducing academic freedom and handling the examinations more strictly, at least for the early semesters. The objective is to remedy, without restricting admission, the present situation in which roughly 25% of the students leave the German institutes of technology without diploma, and the average length of study of those who remain amounts to almost 12 instead of the required eight semesters. It has been recognized that without proper guidance not only the lazy or incompetent students but also those with too diversified interests may fail to reach the educational goal. However, the advantages of the American system of frequent examinations—the early elimination of students who retard progress, and the gain in efficiency of the learning process—are regarded as substantially outweighed by the drawbacks of preoccupation with getting grades, interference with an orderly sequence of work and with freedom in the use of time, and confrontation by artificially limited problems that usually entail knowledge specified in advance and seldom require perspective or judgment. In engineering education in particular, it is impossible to ignore these drawbacks, considering that engineering, like no other profession, involves the ability to find optimum solutions on the basis of a great variety of conditions that are seldom clearly predescribed.

It is precisely the fact that most engineering problems do not have a unique solution, derivable through logical thinking alone, which makes cultivation of judgment, intuition, and imagination so important in engineering education. Obviously, success in teaching the art of engineering depends much more upon how a subject is presented than upon the subject matter itself. As a matter of fact, since both the application and the advancement of scientific knowledge are much more an art than a science, the intuitive and imaginative capacities can be exercised perfectly well in a science course—provided that it is presented as a process of discovery, provided that a feeling is developed for possible simplifications and short cuts in the derivation of a theory or for possible alternate approaches to the solution of a problem, and provided that the students are exposed to a diversity of thoughts and viewpoints and given experience in finding or verifying a solution by experimental observation. Design courses are thus in no way the only means for the development of imagination and good judgment; indeed, their usefulness depends on the extent to which this goal is attained. The great danger of design teaching lies in the fact that the most convenient approach of presenting routine procedures and design details is least likely to stimulate the creative and decision-making faculties, as well as the fact that today such routines are subject to rapid change and hence of little use to the engineer of tomorrow.

While in the United States a variety of new ways to promote creativity in design have been suggested and experimented with, in Germany one tends to reform but not abandon the traditional approach. Students are given a series of exercises in which they are to work out a design project with as much of their own initiative as possible, the project being selected to show how engineering sciences and economic considerations are applied to problems of synthesis and optimization. The most extensive project, of course, is the *Diplomarbeit*, which—like the bachelor's thesis in some American colleges—forms a basis for the engineering degree. Although this final project may be a theoretical or experimental study, in general it involves a realistic engineering undertaking in which the student has to cooperate with local authorities in assessing the necessary technical, economic, social, and legal factors that are involved. It is obvious that under appropriate guidance such project work can provide an invaluable experience. In regard to orientation in engineering practice, of which 6-12 months of actual experience prior to graduation play an essential part in Germany, the authors are inclined to agree with the majority of American educators that it is most realistic to let the student gain his practical experience later on the job. In fact, many American industrial enterprises have devised special introductory and orientation programs for the newly employed college graduate. The best the university can do in this regard is to cultivate a receptive frame of mind.

Postgraduate Education

One of the great distinctions of American universities, from the European point of view, is their strong and effective postgraduate education. While opinions may differ as to the amount of control that should be exerted on the selection of postgraduate courses, the fact that such courses are offered in abundance and that regular course work is required of the master and doctoral candidates is undoubtedly an exemplary feature. To expect, as is the practice in Germany, that one who has completed an undergraduate course will be able to proceed solely by self-study appears highly unrealistic in view of today's accelerated increase in the scope of scientific knowledge and the complexity of its application to the various branches of engineering. Self-education may develop self-reliance and independence, but seldom leads to more than superficial learning, restricted to what the candidate with his limited judgment considers absolutely necessary for the performance of his research work. Even if it were the rule rather than the exception, professorial advice could give such advanced study direction but not depth and comprehensiveness.

What the German author considers the weakest part of his college education is the strict separation that prevailed between the teaching of fundamentals and that of specialized engineering subjects. All college courses in mathematics and the sciences were offered in the first and second years, when most of the students enjoyed the release from the rigid discipline in high school and were unable to realize the significance of these courses for their subsequent study. In America, on the contrary, postgraduate courses emphasize advanced mathematics and theoretical aspects of engineering science and their application. As a result, the knowledge is acquired by the student at a time when he can appreciate to the fullest its relevance to his research. From either the American or the German point of view, it must be granted that discontinuing instruction in the science of engineering too early is as undesirable as beginning instruction in the art too late.

The offering of advanced courses is but one of many ways in which the postgraduate student is given special attention in the United States. There are also the colloquia and seminars in which selected topics receive thorough attention and the ability to present and defend one's thoughts is exercised. There is the opportunity of commencing an academic career as a part-time teaching associate, as well as the research assistantships, which place the students in direct contact with the research frontiers of senior faculty members. There is the stimulating atmosphere in which professors and students work as partners of a team toward common goals, and there is the great percentage of staff with faculty rank covering many fields of specialization (in Germany, professors with tenure form less than 3% of the academic community). Last but not least, there is the time and effort the thesis adviser devotes to challenging the candidate by setting high standards, stimulating his progress, and criticizing and guiding his work so that it may contribute not only to his intellectual development but also to the advancement of knowledge. It is this which seems to be the key to the reputation of many American institutions: that the university has to prepare for and to form the future, so that research is not merely a means of fostering creativity and making the young engineer ready to face novel situations but also an obligation enjoined on the university for making possible future progress and innovation in technology. The German author is reminded of Vannevar Bush's analogy between the process by which the boundaries of knowledge are advanced and the building of an edifice; what seems to distinguish the American university is the awareness that this ultimate goal, the edifice, must not be neglected, despite the graduates' tendency to see just the brick under hand and the pressure from outside for specific and short-range fabrications.

Faculty Competence

Because of the vital importance of faculty competence to the success of any educational endeavor, questions about the kind of competence to be looked for in faculty recruitment, about the kind of organization that best puts it to use, and about the kind of policy by which it can be most effectively maintained and improved naturally stand in the foreground of reform considerations in any country. Here obviously the United States can teach a lesson. As a matter of fact, the physics department of the Munich Institute of Technology has already made headlines by drastically changing its structure according to what its director found to be worth adopting during his stay at the California Institute of Technology, and other departments may follow suit.

Most impressive for the German observer is the comparatively smooth and democratic functioning of the American multiprofessor departments. He admires the large number of teaching appointments, ensuring better attention for the students as well as more time for research and personal development through a reduction of individual teaching load, and he is fascinated by the independence that every staff member enjoys. The department head, he learns, is appointed according to his administrative abilities and is charged primarily with the task of creating and maintaining, in cooperation with the dean and the university administration, excellence of departmental programs and faculty. The means by which this task is accomplished may to an American seem simple and obvious; to a German they are worth putting into words. On the one hand, programs and teaching methods are held under continuous review and revised at appropriate intervals; on the other hand, promotion, salary adjustment, and tenure are considered solely on the basis of achievement. This strong adherence to the achievement principle, rather than increasing friction and professional jealousy, tends to produce an atmosphere of healthy competition, so vital a shield against complacency even in the academic world. Not that competition is lacking in German universities. It just seems to come to an end once the coveted and highly respected status of professor is attained.

Many details of the successful American professorial policy warrant mention. One is faculty recruitment on the basis of interest and competence in exploring unknown territory and doing creative work. Another is encouragement of continued learning. Actually, little encouragement is needed. Being active at an American university brings about so many stimulations and opportunities that the competitive system is enough to ensure continuous refinement of teaching and research proficiency. Besides the favorable conditions provided by the welcoming of criticism (even to student evaluation of teaching effectiveness), opportunity for self-improvement is available in many forms: in the promotion of colloquia, symposia, and conferences at departmental, interdepartmental, and national or international level; in the policy of technical journals to subject contributions to a careful review before publication (regardless of the author's reputation) and to invite critical discussions; in the support of travel for the purpose of keeping abreast of new developments in the field; in programs for the exchange of faculty members with sister institutions and for the appointment of visiting professors from abroad; in the provision for sabbatical leaves, financed either by the university or through a fundation grant and intended to free the faculty member for a significant professional activity (research, continued study, or even practical experience), usually in another part of the world. What has been particularly new to the German author is that the American universities have devised regular programs for faculty development and have established, mostly in collaboration with industry, centers for advanced engineering study. As incredible as it first appeared to him that his department head at the age of 60 would take a semester off to bring his knowledge of computer techniques up to date, this now strikes him as an excellent example of the spirit that an educational institution should possess.

Other Influences on Education

American universities enjoy many outside influences which contribute to the strength of their programs and the quality of their faculties. Various educational foundations, by instigating experimentation and supporting basically different approaches to engineering education, help to assist the struggle against stagnation, especially dangerous for a profession of which the very goal is innovation. Governmental and industrial agencies give generous support to research on promising basic problems, and they help establish well equipped laboratories and research centers all over the country, promoting advances into new territory and encouraging new interdisciplinary activities. Above all, the United States has the unique advantage of mingling people of various educational backgrounds, both native and foreign-born, which produces a breadth of viewpoint otherwise unattainable. A small country like Germany can benefit from such influence only through participation in undertakings such as the widely disputed University of Europe.

A Word of Caution

In the reciprocal emulation that the authors are recommending to their respective countries, one danger is that in adopting the best of the other's methods one may ignore the best of one's own. It is just as pointless to emphasize the mental upbringing of a country's youth to the neglect of the social as it is to stress the social and ignore the mental. To educate the masses without special attention to the elite is as unbalanced as to focus attention on the elite and let the masses fend for themselves. The routine training provided by the technical institutes is of little value if not complemented by university-level development of originality in design, and vice versa. The art of engineering can no better be sacrificed to the science than can the science of engineering to the art. Undergraduate education without postgraduate continuation is as unrealistic as seeking to make up by postgraduate study what a weak undergraduate curriculum has failed to' accomplish. And to substitute numbers, freedom, and support of faculty for experience, authority, and prestige would be no more salutary than to do the reverse. In a word, one half of the golden apple of educational wisdom is actually no better than the other, and each is far from complete. Although possession of the whole may be an unrealizable ideal, what better objective could any country seek to attain? ■

ON THE USE OF MODELS IN FLUIDS RESEARCH

Hunter Rouse and Enzo O. Macagno

Institute of Hydraulic Research,
The University of Iowa,
Iowa City, Iowa, U.S.A.

THOUGH crude dimensional and physical bases for the principles of similitude can be traced back to classical times, Galileo appears to have been the first actually to utilize principles of scale-model simulation, in his discussions of structural strength[1]—including, it is pertinent to note, the bone structure of living bodies. When the motion of a fluid was first modeled is more difficult to say[2], for any state of flow can correctly be regarded as the model of another, and even an empirical flow relationship is usable as a law of similarity. However, avowed models of water wheels were built and investigated by John Smeaton as early as 1759. Otto Lilienthal began his whirling-arm studies of miniature lifting vanes in 1866, and in the same period William Froude conducted his first towing-tank studies of model ships. Fargue, in 1875, initiated tests of river models, and a decade later Osborne Reynolds conducted laboratory experiments on estuarial tides.

Some of these investigations were based upon the most primitive principles of similitude, and others were quite sophisticated for their time. But only slowly did it come to be realized that the laws of similarity are simply one outcome of the Vaschy theorem of dimensional analysis (now commonly associated with the name of Buckingham). It is by this theorem that a series of dimensional variables completely describing a flow phenomenon can be organized into a set of non-dimensional groups of variables, generally fewer in number by the difference between the number of variables and the number of dimensional categories involved in their measurement (more specifically, the rank of the dimensional matrix in question[3].

For example, if the flow boundaries are determined by the lengths $L_1, L_2, \ldots, L_n$; if the flow is described by the time t, the velocity V, and the differential pressure Δp or the shear τ; and if the fluid properties involved are the density ρ, the specific weight γ, the viscosity μ, the surface tension σ, and the elasticity E; then the Vaschy theorem will permit the variables to be arranged as follows:

$$\varphi\left(\frac{L_2}{L_1}, \cdots \frac{L_n}{L_1}, \frac{tV}{L_1}, \frac{V}{\sqrt{\Delta p/\rho}}, \frac{V}{\sqrt{L_1\gamma/\rho}}, \frac{\rho V L_1}{\mu}, \frac{V}{\sqrt{\sigma/\rho L_1}}, \frac{V}{\sqrt{E/\rho}}\right) = 0$$

The last five terms have come to be known as the Euler, Froude, Reynolds, Weber, and Mach numbers. Whereas the expression states in general that the several terms vary in an interdependent fashion, similitude-wise it indicates that a model will behave like its geometrically similar prototype (i.e. the Euler numbers will be the same) only if the Froude, Reynolds, Weber, and Mach numbers (or whichever of these are appropriate) are respectively equal in magnitude.

Lilienthal (and Smeaton, in effect) ignored all four of these parameters and considered the Euler number to depend on the geometry alone. Froude and Reynolds utilized the Froude criterion (attributable to Ferdinand Reech in 1831) for wave action, and eventually both also simulated surface resistance, though neither actually used the Reynolds number (first formulated by Lord Rayleigh in 1878[4]) as a resistance parameter. Froude was apparently the first to realize that both wave resistance and surface resistance could not be simulated simultaneously at model scale if the same fluid were used in model and prototype (note the appearance of L in the denominator of one and in the numerator of the other parameter), and he devised a method of simulating the first type of resistance and correcting for the second that is still followed today. By the turn of the century Fargue's and Reynolds' successors in river simulation had also encountered difficulties with surface resistance, though with regard to roughness rather than viscous action. Their solution—exaggeration of the vertical scale in comparison with the horizontal—would have sufficed for either effect, although the extent of the model distortion was somewhat arbitrary and hence required in turn another correction dictated by trial and error.

The fact that such models depend primarily on the Froude criterion for similarity has proved of extra use in connection with the simulation of flow with density stratification—whether due to thermal, saline, or sedimentary effects. In fact, replacement of specific weight in the Froude number by the differential specific weight $\Delta\gamma$ has the effect of reducing the gravitational term $g' = \Delta\gamma/\rho$ in such a manner as to simulate most of the usual free-surface phenomena at a density interface even as g' approaches zero. To be sure, as $\Delta\gamma/\rho$ becomes large, two densities (or a density gradient) and possibly also two viscosities (or a viscosity gradient) must be introduced. In these days of space flight, on the other hand, zero gravity can readily be approached for even the air–water combination, and modeling liquid flow under near-weightless circumstances is no longer a hypothetical condition. Of special interest physiologically is the fact that hydrostatic effects (indicated by what could be called the Archimedes number $p/\gamma h$ then disappear much as they do for conditions of neutral buoyancy—i.e. as $\Delta\gamma/\rho \to 0$.

Flows without a free surface or density interface would seem relatively simple to model were it not for the fact that fluid motion at small scale tends to be laminar and at large scale turbulent, and it is not often feasible to modify velocities and fluid properties sufficiently to yield the same order of Reynolds numbers at widely different scales. Often, to be sure, states of turbulence do not change rapidly with the Reynolds number, and the latter can be "distorted" considerably without serious effect. For extreme changes in linear scale, another type of approximation has suggested itself: to simulate the molar diffusion of a turbulent flow (say in the natural atmosphere or ocean) by the molecular diffusion of the laboratory fluid. In some respects this is satisfactory (indeed, many approximate analyses involve the assumption of a constant diffusion coefficient), but in general it is limited in its usefulness by the fact that the molar action varies with the state of motion and the molecular with the fluid temperature only.

The Weber criterion of similarity is perhaps the least used of any, probably because capillary phenomena are the most infrequently modeled. In fact, except for the formation of sprays, its importance to date has been rather negative in nature, for the action of surface tension makes it almost impossible to simulate many gravitational or viscous effects at reduced scale satisfactorily—for example, the entrainment of air by flowing water or the diffusion of liquid nappes in air: in fact, any mixture of gaseous and liquid media flowing simultaneously. On the other hand, modeling such visco-capillary effects as lubrication of the eye should be quite feasible.

The Mach number, of course, finds its greatest application in the transsonic, super-sonic, and hypersonic flow of gases or the motion of boundaries at comparable velocities. Liquids sometimes actually attain the speed of sound nowadays, and the elastic waves of water hammer have been recognized for almost a century. It is significant to note that the name Cauchy, originally used instead of Mach for such effects in fluids, is now re-stricted to the elastic action of the boundary material alone. Both Mach and Cauchy numbers, for example, are important in the forced vibration of flow boundaries. In blood flow, on the other hand, it is the Cauchy number which is pertinent, for the elastic modulus of the blood is so great compared with that of the blood vessel that only the latter plays a role in its modeling. Simulation of boundary elasticity by fluid elasticity (the "Windkessel" method) is thus an artifice of limited rather than general usefulness.

The purely mechanical variables just discussed are commonly increased by both thermal and electrical variables, especially in present-day investigations of the ionized gases known as plasmas. In modeling the flow of body fluids, at least the last named complexity should remain irrelevant, whereas thermal and electrical effects may some-day play a role. On the other hand, much can be learned by considering the flow pattern as a function of only the boundary geometry—for instance, in the study of eddy for-mation in the larger body passages. In its fundamental aspects, the pattern of such motion is independent of any characteristics beyond the geometric, for fluid viscosity and boundary elasticity often have merely a secondary influence at the most.

As a matter of fact, geometrically similar flows involving only one fluid property are perforce dynamically similar—i.e. the corresponding number is a constant; examples are pressure–velocity changes (the density), hydrostatics (specific weight), Couette or Poiseuille flow (the viscosity), and so on. Two fluid properties yield a number that varies with only one other number—problems of resistance, standing waves, capillary flow in soils—and similarity can readily be obtained by making the independent number the same in both model and prototype, through variation of the velocity scale (i.e. $s_v = v_m/v_p$) in accordance with the length and property scales. It is when three or more properties are involved that no amount of velocity variation will yield the desired result. The classic example of such a situation is the requirement already mentioned of satisfying both the Froude and the Reynolds criteria simultaneously, and the artifice of scale distortion has been shown to provide one means of approximate solution. However, elimination of the velocity scale between the two numbers will show that exact similarity is still possible if the remaining scales are related as follows[5]:

$$\frac{s_l(s_g)^{1/3}}{(s_\nu)^{2/3}} = 1$$

Evidently, operation in another gravitational environment could conceivably com-pensate for variation in one or the other remaining scale ratio. If, on the contrary, the same gravitational conditions prevail in the two systems, the length scale must vary with the $\frac{2}{3}$ power of the scale of kinematic viscosity ν.

Even if a liquid could be found that is an order of magnitude less viscous than water, the cost of supplying it in sufficient quantity for towing tanks or river models would surely be economically prohibitive. Biological models, on the other hand, are relatively limited in volume, even if enlarged considerably relative to their prototype; in fact, since the range of viscosities above that of water is far greater than that below, moderate enlargement of a biological model has much in its favor. The search for fluids with other desired properties is also more practicable, in particular since (contrary to the specific weight) all three properties would probably be independently variable. Consider, for example, the necessity of satisfying the Reynolds and Mach numbers together. Elimination of the velocity scale then yields

$$\frac{s_l s_c}{s_\nu} = 1$$

in which $c = \sqrt{E/\rho}$, the celerity of sound. Again it should be relatively easy to find fluids and materials satisfying the modeling requirements if the model is larger than the prototype, since s_c is likely to be very nearly unity. If, however, it is the Cauchy rather than the Mach number that is involved, as is more apt to be the case, the boundary rather than the fluid elasticity would be involved; the ratio of fluid and boundary densities, moreover, would have to be the same in model and prototype. Because of the latter requirement, a scale relationship of identical form would result,

$$\frac{s_l s_{c'}}{s_\nu} = 1$$

the scale of boundary sonic velocities $s_{c'}$ being subject to comparatively great variation.

Since most biological fluids and boundary materials are non-Newtonian and non-Hookean, respectively, their more sophisticated modeling will necessarily be two degrees higher in complexity than the sort just discussed. It goes without saying that the manner of departure from the Newtonian and the Hookean should be the same in model and prototype. Depending upon the manner of departure, the scale relationship is formulated accordingly. Some biological fluids[6] present the additional complication of being cell-laden, and when passing through very fine passages they involve somewhat the same problem as ship navigation through narrow canals. Basically, of course, the modeling procedure is straightforward, but such matters as those just introduced could well make perfect simulation extremely difficult.

A simulation process quite distinct from the foregoing use of scale models is found in the analog: a device that behaves in certain respects in a similar manner despite operation according to a different physical principle. Typical examples of flow analogs are the Hele Shaw method of showing patterns of irrotational motion, the Prandtl membrane analogy for stress distribution, the electrolytic tank, the electrical resistance circuit for representing conduit resistance, and complex networks thereof. The last named are essentially analog computers.

Also distinct are what have come to be called mathematical models: either simplifications of the equations of motion, or arbitrary formulations which — like the analogs — yield reasonable approximations to the actual occurrences. Whereas the latter process is often in frank recognition of the fact that the actual occurrence is not understood (for example, Prandtl's phenomenological approach to fluid turbulence), a model based on the equations of motion is limited in its simulation only by the mathematical ability of the investigator. In these days of analog and digital computers, the degree of similarity that is attained leaves little to be desired — except, to be sure, greater capacity for detail on the part of the computers. Exact solutions of complex states of unsteady, non-uniform irrotational flow are almost the rule, and two-property problems of various sorts have also given way to such attack. Of particular importance in this regard is the fact that the investigator is in no way limited by the practical choice of fluids having the required properties.

Two types of computer solution directly related to blood flow are under continuing investigation at the Iowa Institute of Hydraulic Research. One is the formation of eddies at abrupt changes in conduit section[7]. Although the ultimate goal of simulating turbulence generation by such means may never be realized, not only is the pattern of laminar flow as a function of the Reynolds number already at hand, but the detachment of eddy subdivisions due to velocity perturbations has also been reproduced. Though this has been accomplished only for Newtonian fluids, there is no reason why it cannot be extended to the non-Newtonian as well. To explore this possibility, the basic equation of motion for a deformable continuum

$$\rho\frac{\partial u}{\partial t} = -\frac{\partial p}{\partial x} + \frac{\partial \tau}{\partial y}$$

has been represented by a finite-difference equation and integrated numerically for different expressions of the shearing stress τ. In this manner the classical problem of flow establishment starting from rest under a constant pressure gradient has been solved at Iowa[8] for both dilatant and pseudo-plastic fluids–indeed, for practically any type of relation $\tau = f(\partial u/\partial y)$. The computational model has already been extended to annular pipes, of which the circular is a special case, and to variable pressure gradients, including pulsating. The computational model is well adapted to the introduction of a time-dependent relationship between stress and rate of deformation, and work is also under way in this direction. The latter stage is particularly significant in view of recent results that show a time-dependent behavior in the measurement of blood viscosity[9].

REFERENCES

[1] GALILEO, G. *Dialogues Concerning Two New Sciences*, translation by H. CREW and A. DE SALVIO, McMillan, New York, 1914.
[2] ROUSE, H. and INCE, S. *History of Hydraulics*, Dover Publications, New York, 1963.
[3] ROUSE, H. (Editor), *Advanced Mechanics of Fluids*, Interscience Publ. John Wiley, New York, 1959.
[4] STRUTT, J. W. (LORD RAYLEIGH) *Theory of Sound*, Macmillan, London, 1877–9.
[5] MACAGNO, E. O. *Ciencia y Técnica* **116**, No. 585, 1951.
[6] MERRILL, E. W. and WELLS, R. E. Jr. *Applied Mechanics Reviews* **14**, No. 9, 1961.
[7] MACAGNO, E. O. *La Houille Blanche*, No. 8, 1965.
[8] MACAGNO, E. O. *C.R. Acad. Sci. Paris* **262**, 1121–4, 1966.
[9] GREGERSEN, M. I. Studies on blood viscosity at low shear rates, Progress Report to U.S. Army Medical Research and Development Command, Columbia University College of Physicians and Surgeons, New York, 1965.

PREFACE
TO THE ENGLISH TRANSLATIONS
OF
HYDRODYNAMICA AND *HYDRAULICA*

In the belief that students attaining the doctoral level should know something about the background of their profession, the writer began in 1960 to offer a graduate course at the University of Iowa on the history of hydraulics. Instead of attending lectures, every student was expected to read the Institute book on the subject [1], select a lesser-known investigator from the past in each of his three required doctoral languages (English, of course, included), and submit original monographs summarizing the respective lives and works. If he so preferred, a student could prepare instead one monograph on a single subject, such as pipe resistance, the roots of which would carry him into several source languages. Furthermore, a student whose language background was sufficiently broad to need no further exercise could concentrate on a single national literature, like Russian hydraulics.

The lack of Latin in the writer's background—which had proved particularly troublesome when he was seeking to digest early treatises on hydraulics—led him to suggest to two students with foreign-language upbringing (the one, Italian; the other, German) who were able as well to read Latin that they compare the works on fluid motion of the two Bernoullis, Johann and Daniel. The idea soon grew to the point of involving the complete translation of the two present books, in part as the regular course requirement, in part as salaried employment with the Iowa Institute of Hydraulic Research, and in no small part as a labor of love. To the American-born Thomas Carmody (who really began the undertaking) fell the task of actual translation; but to the German-born Helmut Kobus (who had studied far more Latin) fell that of checking meticulously every word and thought and of preparing the manuscript for the printer after his fellow translator had left for another university. Kobus, in turn, finally left the country, and it remained to Carmody (and the writer) to read proof.

Why the Bernoullis' works should have been singled out for translation seems at first thought rather obvious, if only because of the frequency with which the name Bernoulli is on a hydraulician's lips. But it is only Daniel to whom one is making reference, and the word is gradually spreading that the theorem bearing his name is nowhere to be found in his habitually cited *Hydrodynamica* [2]. Not until the last few years has mention of either the work *Hydraulica* [3] or its author Johann Bernoulli appeared in fluids literature with any frequency whatever, and this almost exclusively in the writings of C. Truesdell [4]. It is Truesdell's thesis that, whereas Daniel has received too much credit for the formulation of the Bernoulli theorem, Johann has received too little. Readers who have not studied Latin, and who may never have the chance of seeing the original works, can now judge the matter for themselves. They can also marvel at the many familiar concepts which Daniel did originate and for which he has received almost no credit at all.

To understand the rather curious relationship between Johann Bernoulli and his son Daniel, one must know something of the family itself [5]. Basel had become a university town in 1460, a center of early printing that attracted such Renaissance writers as Erasmus and Paracelsus, a refuge for Huguenots during the Reformation of 1530, and finally by the seventeenth century a very literate city of strong family ties. To this city, in 1622, came a Huguenot from Antwerp by the name of Bernoulli. He established himself as a merchant and raised sons who also became merchants. One of these fathered a dozen children, of whom four lived: a mathematician, an artist, another mathematician, and at last a merchant. The oldest was Jakob Bernoulli (1654–1705), who became professor of mathematics at the university and finally rector. The second mathematician was Johann Bernoulli (1667–1748), who was trained by his brother, worked with the French mathematician L'Hôpital at Paris, taught mathematics for ten years in Holland, and then succeeded his brother as professor at Basel. It was he who had as sons Nikolaus, Daniel, and Johann II.

Though Jakob was an extremely able mathematician in his own right, his great contribution to the present history was the education of his younger brother, who in turn taught his own sons. Unfortunately, friction developed and steadily increased between the brothers, and their early collaboration eventually changed to rivalry. Johann's bitterness was increased by L'Hôpital's publication in his own name of various discoveries communicated to him in Johann's many letters—not to mention an entire course of instruction which he had given him in Paris—but he nonetheless became upon Newton's death the foremost mathematician in the world. Because of his close association with Leibniz (it was Johann who first applied the new calculus—and, in fact, introduced the word "integral"), he sided eloquently with him against Newton in the fluxion–calculus controversy, to the chagrin of the Royal Society.

Daniel Bernoulli (1700–1782) was born at Groningen near the middle of his father Johann's ten-year Dutch professorship. He studied under his father at Basel after the latter returned to fill the chair left vacant by Jakob's death. In his twenties Daniel spent seven or eight years as professor of mathematics at St. Petersburg, a period darkened only by the early death there of his older brother Nikolaus, also a mathematician. Daniel won or shared, in the course of his life, ten prizes awarded by the Paris Académie des Sciences for the solution of designated problems. The first of these, received at the age of twenty-four, involved the design of a clepsydra for the exact measurement of time at sea. The third, for a paper on tides, was shared with Euler and Maclaurin. One which he divided with his father dealt with the inclination of the planetary orbits, and another, shared with Johann II, was on the best form of anchors. Still another had to do with the nature and cause of ocean currents. In each of these he displayed considerable mathematical ability, to be sure, but above all a keen physical perception and the ingenuity to produce a solution regardless of the method used [1].

Daniel's *Hydrodynamica* was begun in 1729, during his sojourn in Russia, and an uncompleted manuscript of it was left at St. Petersburg when he returned to Basel four years later. In the course of its revision and completion, he wrote for permission to dedicate it to the Empress of Russia, as an acknowledgment of his debt to that country. When the book was finally published in Germany in 1738, he requested that the Russian manuscript be destroyed, but it is still preserved in the files of the Soviet Academy of Science (as remarked in the Preface of a Russian translation of the German edition that recently appeared [6]).

A younger colleague of Daniel's, Leonhard Euler (1707–1783), had also studied mathematics under Johann and, largely as the result of Daniel's influence, was also invited to St. Petersburg by Catherine I. There he became professor of mathematics when Daniel returned to Basel. Euler was eventually to surpass the great Johann Bernoulli as a mathematician (fifty pages were finally required in his eulogy to list merely the titles of his writings), and hence it is hardly surprising that not only Daniel but also Johann was to defer to his judgment at even so early a date.   Daniel wrote him near the end of 1734 that he had arranged with Dulsecker in Strasbourg to publish his new work, but not for three years could he report that it was nearly finished.   Early in 1738 he sent several copies to St. Petersburg, but by the end of the year Euler wrote that they had not yet arrived.   On March 7 of the following year Daniel again protested the total lack of news as to their fate [7].

On the very same day Daniel's father notified Euler that he was sending him the first part of his own manuscript *Hydraulica*.   Now except for a brief criticism of Newton's cataract hypothesis in 1716, there is no record that Johann had written anything whatever on the subject of hydraulics until some months after his son's treatise was off the press, when he stated in a letter to Euler that he was preparing a manuscript on hydraulics which was already well along.   Nevertheless, he indicated in the first part that it was written in 1732, a full year ahead of the Russian version of his son's!   The second part followed the first to St. Petersburg in 1740, and the two were eventually published, as Johann had requested, in the *Memoirs* of the Imperial Academy of Science for 1737 and 1738 (which were printed, respectively, in 1744 and 1747).   They actually first appeared in his collected works, published in Switzerland in 1743.

As surprising as Johann's obvious attempt at seeming to predate his son in publication was Euler's delay in acknowledging Daniel's book until such time as he could do the same for Johann's.   In fact, of the two letters of acknowledgment and praise, written on the same day, that to Johann was far more flattering.   A subsequent letter that he wrote Johann about the book was quoted in part in the foreword to the Swiss version of *Hydraulica* as translated herein (see p. 347).   Johann's reason for using only the first paragraph of Euler's letter is evident from the following translation of the second:

> Truly, regarding the force by which vessels are driven backward, I certainly do not have the least doubt concerning that very method you use for determining this; but when, for pipes attached horizontally to a vessel, you find that the pressure driving the vessel backward is different from that which agrees with the hypothesis of Your son, that force as it is determined by Your Illustrious Son seems to me certainly to be more suited to the truth than Yours; may I have said this without offense to You.   Indeed, from the formula which you present for retroaction in this case it follows that the retroaction can be indefinitely great, even if the orifice is very small and the motion very slow, and the expression given by Your Son does not contain this inconsistency; but I am convinced that if you will deem it worthy to subject this part to examination once

more, Your theory will agree most perfectly with Your Son's idea; indeed, I suspect that fractions have to be inverted, and, with this done, it will agree most perfectly with the truth and with Your Son's expression.

Apparently Daniel first saw his father's treatise when it appeared in 1743 in the collected works, for he thereafter wrote Euler:

I beg your Excellence to tell me in sincere friendship and confidence your opinion of my father's Opera, particularly of the last volume. I for my part have reason of the highest degree to complain about it: The new mechanics problems stem mostly from me, and my father had even seen my solutions before he solved them in his own manner, and nevertheless I am not acknowledged with even a word, which I find the more annoying as my solution is not yet published. My first solution of rotation around an instantaneous center, found from the nature of the least inertia, he has questioned and also contemned for a long time, and finally he published it as his own. But since by a miraculous hazard I obtained a page from his manuscript in which this, his pretended solution, was written, and I complained about it through my brother, he barely let me pass as a second inventor. The matter is roughly the same with the remaining new problems in mechanics. Of my entire *Hydrodynamics*, of which indeed I in truth need not credit one iota to my father, I am robbed all of a sudden, and therefore in one hour I lose the fruits of a work of ten years. All propositions are taken from my *Hydrodynamics*; nevertheless, my father calls his writings *Hydraulics, now first discovered, anno 1732*, since my *Hydrodynamics* was printed only in 1738. Meanwhile my father has gotten everything from me, except that he thought of a different general method to determine the increment of velocity, which invention consists of some few pages. What my father does not claim completely for himself he contemns, and finally, as the height of my misfortune, he inserts the letter of your Excellence in which you, too, diminish my inventions in a field of which I am fully the first, even the only, author and which I claim to have exhausted completely. Your Excellence says that I have determined the pressure of fluids flowing through a conduit in no other way than for the steady state, whereas I show immediately on page 259 toward the bottom that generally the pressure is $\dfrac{a - vv}{2c}$; and what, on the other hand, has my father done in this important new field? The invention of the argument comes from me; the idea to consider the conduit as cut at the point where the pressure is required comes from me; that one should require the acceleration of the last particle at the first instant of interruption is my idea; finally, that from this very acceleration, hindered either partly or completely, the pressure of the elemental volume determines the pressure of the water in the conduit—this also comes from me; and my father has done absolutely nothing else than determine the velocity in his own manner and by repeated reasoning, which is his only invention in the entire work.   The argument about the reaction of fluids my father does not yet understand today; nevertheless, he refutes me in the corollary on page 488 [page 336 of present translation; see also foregoing paragraph from Euler's letter].   All of this is still the least about which I can complain.   In the beginning it seemed almost unbearable to me; but finally I took everything with resignation; yet I also developed disgust and contempt for my previous studies, so that I would rather have learned the shoemaker's trade than mathematics.   Also, I have no longer been able to persuade myself since then to work out anything mathematical.   My entire remaining pleasure is to work some projects on the blackboard now and then for future oblivion.   I could not accept with a clear conscience the call to Berlin, even if the King should give me the honor to send me one, and I beg you therefore not to think of me any more with respect to this matter.   However, I am strongly obliged to your Excellence for your kind services; your most valuable friendship presents me with an innermost and true pleasure, and I esteem such [friendship] much higher in itself than in the profit which could arise to me from it.   I could not abstain from complaining to your Excellence, as my best friend, seeing that the occasion might well arise that you vindicate me of the unjust suspicion of plagiarism without doing wrong to my father, and also bring it about that the truth, as far as the controversial points between my father and me are concerned, does not suffer any injury.   It does not seem proper to me to defend myself.

Far from being guilty of plagiarism, Daniel had based his treatise on material that was not only original but lasting in its interest.   The reader will find in the following pages of translation the initial appearance of topics that are still prominent in the literature even today—from the kinetic theory of gases to the principle of jet propulsion.   Daniel was also the first to connect manometers to piezometric openings in the walls of vessels, to consider the establishment with time of flow in a long conduit, and to attempt to predict conduit pressure in terms of the velocity.   However, his derivation of what has come to be known as the Bernoulli theorem will hardly satisfy any reader but the most casual.   There is no doubt that Daniel understood the theorem in its two-term (velocity head and piezometric head) form.   However, the simplicity of the relationship in comparison with the cumbersomeness of his analysis leads one to suspect that—despite his claim of invariably reasoning first and experimenting thereafter—he actually knew the answer in advance. The artifice of cutting the conduit and relating the pressure to the assumed acceleration seems forced at best.

According to Truesdell [8], most of Daniel's difficulty lay in his imperfect understanding of fluid pressure. Straub, on the contrary, points out that a paper published by Daniel in the St. Petersburg *Commentarii* of 1729 had already contained "an essentially correct formula for the so-called hydrodynamic pressure" [7]. In any event, though his *Hydrodynamica* divided a moving fluid for convenience into a series of slices normal to the direction of motion, in no case was action stated to occur between them. Pressure was treated rather as a condition existing at the conduit wall which would produce a jet (or manometric column) equal in height to the piezometric head if the wall were pierced. Now Johann also assumed the fluid to move in normal slices, but the concept of pressure as a mutual interaction at their surface of contact was essential in his analysis. Moreover, in order to avoid the anomaly of a discontinuity in pressure and velocity at an abrupt conduit contraction or expansion, he imagined the actual flow section to change gradually before the contraction or after the expansion, in a manner reminiscent of Newton's cataract. He called this transition "gurges," which Truesdell insists should be translated as "eddy," despite the fact that Johann referred to the flow within the imaginary throat-like passage rather than to that around it. In fact, far from visualizing circulatory motion in the zone of separation, he specifically considered it to be occupied by stagnant fluid: "And, accordingly, there is formed along the indefinitely small length HG something like a *throat*, IFGH, contracting from the wide into the narrow, through which the liquid must pass, the acceleration being continuous but nevertheless augmented gradually, with a rather small portion of the liquid (which fills the small space IFD) remaining at perpetual rest" [page 357 of the present volume]. The word "gurges" has hence invariably been translated herein as "throat."

Johann was obviously a stride beyond his son in his analysis of the Bernoulli relationship. Whereas Daniel had treated pressure primarily in terms of height of a manometer column or jet, Johann visualized it as a force, albeit one acting over the area of the slice as a whole. It remained to Euler—under the considerable stimulus of Johann's analysis rather than Daniel's—to originate the concept of pressure at a point and to incorporate the pressure gradient into his equations of acceleration. These he finally integrated for specific conditions [9], thereby first deriving in a rigorous manner what is now known as the Bernoulli equation. As in the case of many another aspect of fluid motion for which he has not received proper credit, it is thus the name of Euler which should be most often upon the hydraulician's lips.

Although Daniel's lengthier treatise is much easier to read than Johann's, each will give the reader some difficulty in accustoming himself to the mathematical and the physical style that the respective author used.   He must hence recall that the calculus was still a relatively new concept; that the energy relationship was not yet correctly written and even less well understood than the momentum relationship, with which it was thought by many to be in conflict; and that the theory of dimensions was wholly in its infancy.   So long as the only dimension was length, little difficulty was encountered, as in Daniel's interrelationship of velocity and piezometric heads.   Kinematics was somewhat more complex, as witness his expression of velocity through either its head or a proportionality with an observed time of fall; in fact, it was Johann who first introduced the coefficient of proportionality g.   Dynamics was the most confusing, for force had not yet been universally defined in terms of length, time, and mass through the Newtonian equation, and their interrelationship was still arbitrary; even Euler, whose equations were in fact dimensionally homogeneous, still defined mass and weight as he saw fit.

Except for calling attention to certain of the anomalies or actual errors, the translators have in large part left the assessment of the authors' analyses to the reader himself.   To be sure, many mannerisms of the times have been eliminated, such as the excessive use of abbreviations and symbols like the ampersand and equality sign as parts of speech, and the mathematical notation has been very slightly modernized.   Translators' insertions are invariably placed within brackets, and the bracketed numbers refer to the lists of cited works at the ends of the two books.   For the convenience of the reader, each illustration has been introduced into the text at its first point of reference.   The original figures have been reproduced in all cases.

This Preface would not be complete without reference to a contemporary undertaking [10] begun in the Bernoulli city of Basel some two decades ago and already assuming tangible form: publication of the collected writings of the Bernoulli family, in particular the members referred to herein.   The project is under the general editorship of Dr. Otto Spiess, and the first of well over a dozen projected volumes (the early letters of Johann) recently appeared under his direct guidance [11].   Another, containing Daniel's correspondence as edited by Dr. Hans Straub, is now in preparation.   The scientific world stands greatly in the debt of Dr. Spiess and his colleagues for the impetus that they have given to the tremendous task of making this material generally available.   The writer acknowledges his particular gratitude for stimulating conferences with Drs. Spiess and Straub, for their willingness to review the foregoing pages, and their kindness in providing portraits of Daniel and Johann from approximately the times their books were published. A final touch was Dr. Spiess' indication, during a visit to his home by the writer, of the impending appearance of Daniel's *Hydrodynamica* as translated into his native tongue [12].

Iowa City, Iowa   HUNTER ROUSE
January, 1967

REFERENCES

[1] ROUSE, H., and INCE, S., *History of Hydraulics*, Iowa Institute of Hydraulic Research, 1957.   Dover reprint, 1963.

[2] BERNOULLI, D., *Hydrodynamica, sive de viribus et motibus fluidorum commentarii*, Dulsecker, Strasbourg, 1738.

[3] BERNOULLI, J., *Hydraulica nunc primum detecta ac demonstrata directe ex fundamentis pure mechanicis. Anno 1732. Opera Omnia*, Vol. 4, Bousquet, Lausanne and Geneva, 1743.

[4] TRUESDELL, C., Editor's Introduction to Vol. II 12 of Euler's *Opera Omnia*, Füssli, Zurich, 1954.

[5] SPIESS, O., "Die Basler Mathematiker Bernoulli," *Bulletin de l'Association Suisse des Électriciens*, Vol. 43, No. 8, 1952.

[6] BERNOULLI, D., *Gidrodinamica ili Zapiski o Silakh i Dvizheniyakh Zhidkostei*, translated by V. S. Gokhman, A. I. Nekrasov, K. K. Vaumgart, and V. I. Smirnov, *Izdatel'stvo Akademii Nauk SSSR*, 1959.

[7] STRAUB, H., private correspondence with the writer, 1964–65.

[8] TRUESDELL, C., "Zur Geschichte des Begriffes 'innerer Druck,'" *Physikalische Blätter*, Vol. 12, No. 7, 1956.

[9] EULER, L., "Principes généraux de l'état d'équilibre des fluides"; "Principes généraux du mouvement des fluides"; "Continuation des recherches sur la théorie du mouvement des fluides"; *Histoire de l'Académie de Berlin*, 1753–55.

[10] TRUESDELL, C., "The New Bernoulli Edition," *Isis*, Vol. 49, Pt. 1, No. 155, 1958.

[11] SPIESS, O., Ed., *Der Briefwechsel von Johann Bernoulli*, Birkhäuser, Basel, 1955.

[12] FLIERL, K., "Des Daniel Bernoulli, Hydrodynamik," *Veröffentlichungen des Forschungsinstituts des Deutschen Museums für die Geschichte der Naturwissenschaften und der Technik*, Reihe C, Nr. 1a, Munich, 1965.

Iowa's Quest for Curricular Balance

HUNTER ROUSE, Dean, College of Engineering, The University of Iowa

ONE OF THE PRINCIPAL STRENGTHS of engineering education in the United States is the great variety of institutions and courses of study. The student may choose a private or a public college or university, one that specializes or generalizes, one that caters to the average or the superior intellect, one that looks far ahead or even a bit behind. The number of engineering colleges is so great—nearly 200—as to yield almost continuous gradation from one extreme to the other of any pedagogical characteristic. At the same time it is possible to distinguish fairly sizable groups of institutions having certain qualities in common. It is to those which are similar in one way or another to the University of Iowa that this article is directed.

The Search for Balance

Iowa is a midwestern university (which is not necessarily a similarity requirement), small for a state institution (about 19,500 students) and close to the average for the country in size of engineering college (some 500 undergraduates, and nearly half as many graduates). The University has ten colleges: a large central group of departments composing the College of Liberal Arts, a Graduate College, and eight professional colleges around them. In the basic sense of the Hutchins philosophy—not to mention that of our Administration—the College of Engineering is an essential part of the University, both as the end application of several stems of learning and as the input mechanism for technology in modern culture. Herein lies the first measure of the balance that is sought by the Iowa College of Engineering: a part in the educational process that is neither overpowering nor overpowered, but in a healthy state of dynamic equilibrium with the rest of the University.

A second aspect of the balance that we seek is the matter of undergraduate versus graduate instruction, neither of which do we consider complete without the other. Peculiarities of the engineering curriculum make it undesirable, if not actually impossible, to relegate engineering education to the graduate school. The art of engineering needs to be nurtured from high school on, and the nurturing process must be handled by mature teachers with experience in the creative phase of their profession rather than by the teaching assistants frequently responsible for basic courses. Our faculty adheres to the policy that every member shall teach both undergraduate and graduate courses, recognizing the importance of the interaction between the two academic levels.

Implicit, moreover, in the balance between undergraduate and graduate instruction is that between the two essential parts of university responsibility: the production and the dissemination of knowledge. A professor who does nothing but teach is as much out of balance as one who does nothing but research and consulting. Although the undergraduate phase of engineering education deals primarily with the application of known principles, whereas the graduate phase emphasizes the search for new principles or new applications, neither should the two processes be wholly segregated as such, nor should the individuals who are involved.

Curriculum Goals

So far as the curriculum itself is concerned, balance of a still broader nature is sought. In the past several decades, three distinct aspects of engineering education have received special attention: social-humanistic studies; mathematics and the sciences; and once again the art. Emphasis on one or another of these aspects, without wholly slighting the other two, can produce useful engineers, but we believe that an effort toward approximate balance of all three will achieve a worthwhile goal that is wholly consistent with the other instances of balance just discussed.

With ASEE's *Final Report: Goals of Engineering Education* now before the engineering colleges of this country, it is incumbent upon any faculty contemplating curricular change to give the recommendations of this report their most serious consideration. Let us grant at the outset, as the Chairman of the Goals Study, Eric Walker, has emphasized, that the trends foreseen (particularly emphasis upon the first graduate degree) are not committee-imposed but the natural tendency of the technical world itself. Their acceptance by the profession is nevertheless as varied as its elements have been noted to be. A few of the leaders were already well ahead of the predicted course when it was first publicized. Many institutions of equal size or weight, whether from conservatism, inertia, or continued opposition, show little inclination to change. Some that have still not accepted previous recommendations for emphasis upon the humanities and science are even further from being moved. Others, finally, are experimenting with one aspect or another of the report, convinced that change is inevitable but uncertain as to its probable form.

The situation at Iowa must be typical of that at many comparable institutions. Some of the faculty members are farsighted planners, and some prefer to wait for others to lead the way. (One, who urged that we " be not the first by whom the new is tried, nor yet the last to put the old aside," unwittingly proclaimed the tried-and-true formula for mediocrity.) There are those who fear loss of departmental strength, and others who see emphasis on a common undergraduate core as the salvation of engineering education. Some favor the science and some the art, and some couldn't care less which is favored, in the belief that it is not what is learned that counts but how it is taught. The temptation is sometimes strong to split off the non-believers of the faculty from the believers (using one's own subjective views as the norm), but it invariably gives way to the realization that each group has much to offer the College, and loss of either would be severe. In fact, a positive accommodation of the several views creates a balance that is very potent educationally, for in the continuing effort to agree on as much as possible lies the viability of a dynamic curriculum.

Setting Up a Realistic Curriculum

The first matter of curricular upgrading on which we sought accord was the question, "What should every engineer know?" In the belief that such common knowledge could be classed in four stems that should begin in the freshman year, a 12-man curriculum committee was divided into four subgroups with the task of specifying material in the stems of mathematics, science, socio-humanistic studies, and the art of engineering that should be generally known to all engineers. Because they were instructed to plan uninhibitedly, it was inevitable that enough material was selected for at least four years of study without any specialization whatever. It then became the task of a six-man committee to combine the material into a realistic curriculum that would (a) not exceed 128 semester hours in length, (b) provide all that every properly balanced engineer should acquire in a four-year program, and (c) permit even the most professional-minded department at least the minimum amount of specialization that it considered consistent with a bachelor's degree with departmental designation. The eventual result is shown in Table 1.

The Curricular Stems

Socio-Humanistic Stem. Although many still believe that the ideal liberal course is best administered by the College of Engineering, the only method consistent with the role of the College in a properly balanced university is the administration of the socio-humanistic stem by the College of Liberal Arts. The proposed 8-semester-hour freshman course, literature and composition, has the triple role of providing contact with good prose and poetry, enforcing continued practice in writing, and redirecting to remedial courses in rhetoric those who do not possess the required background. The remaining 16 semester hours of elective courses in this stem are to be carefully chosen, with the adviser's approval, so as to form a social-science sequence and a historical-cultural sequence of at least 6 hours each. In order that exercise in communication will be continued through all four years, careful attention will be given in engineering education to the presentation and evaluation of written and oral reports.

Mathematics Stem. The 10 hours of freshman mathematics are being specially arranged by the Mathematics Department: on 3 days a week the major concepts of the calculus will be treated in a "spiral" manner through several cycles of increasing depth; on the 2 alternate days the student will be introduced to the elements of analytic geometry and linear algebra. In this way, as sufficient geometric background is developed, steadily greater demand can be placed upon the student's intellectual grasp and use of the calculus. Major emphasis will thus be laid on the understanding and application of the concepts, with secondary emphasis on manipulative skill.

Since introductory computer programming will be taught in a first-semester engineering course, use of the computer can be incorporated into the manipulative exercises. In the first semester of the sophomore year, also by special arrangement, 3 semester hours will be devoted to vector analysis, differential and integral calculus of several variables, and ordinary differential equations. The second-semester course will be the first part of the standard sequence in applied mathematics. The final one of the prescribed five semesters will be the regular course in probability. Because updating the curriculum will decrease the time available for practice and application within each element of the mathematics sequence, this aspect of the instruction is to be supplemented by extensive use of mathematics in the science and design stems.

Basic and Applied Science Stem. Defining the pertinent basic and applied sciences occupied a considerable amount of faculty time. The courses finally decided upon are those containing ideas primarily of a conceptual nature relating to the fundamental behavior of the physical world, with a primary goal of emphasizing common principles of analysis. For those students with credit in high-school chemistry, the second semester of the standard chemistry sequence is presumed to suffice for all but chemical engineers. As a result of time limitations and the constraint of a four-year program, the standard course in introductory physics has been eliminated. This means, to be sure, that the courses in engineering science must begin at a more elementary level. However, time is saved, in that unnecessary repetition is avoided, and careful

TABLE 1

The Proposed Iowa Curriculum

Stem	Semester 1	Semester 2	Semester 3	Semester 4	Semester 5	Semester 6	Semester 7	Semester 8	Hr.
Socio-Humanistic	Literature and Composition I 4	Literature and Composition II 4	Socio-Humanistic Guided-Elective Sequence 16 semester hours							24
Mathematics	Math I 5	Math II 5	Math III 3	Math IV 3	Math V 3					19
Basic and Applied Science	Chemistry 4	Thermo-dynamics 4	Mechanics of Solids 4 Materials Science 3	Mechanics of Fluids and Transfer Processes 4	Electro-magnetic Theory 4	Physics I 3	Physics II 3			29
Analysis and Design	Introduction to Engineering I 4	Introduction to Engineering II 4	Dynamic Systems Analysis I 3	Dynamic Systems Analysis II 3	Principles of Design I 3	Principles of Design II 3	Engineering Design I 3	Engineering Design II 3		26
				Departmental Selections 30 semester hours						30
									Total	128

planning of the course sequence should permit thorough correlation of the successive elements.

The first of the engineering sciences is now thermodynamics, in the belief that of the three directions of approach (mechanical, thermal, and electrical) the thermal could best be adapted to the level of the second-semester freshman. This is followed by courses in the mechanics of solids, fluids, and transfer processes, the necessary linear algebra having already been provided in mathematics, and vector analysis being taken concurrently. Though these courses are unorthodox in both coverage and arrangement, it is believed that purposeful application of their principles in the engineering sequence will achieve their essential rounding out. The course in materials science will develop the principles of molecular structure, environmental influences, and the thermal, mechanical, and electrical behavior of engineering materials. Electromagnetic field theory can thus proceed, without duplication, from observations of a static system to the time-varying system and Maxwell's equations, the solutions of which in particular cases yield easily recognizable formulations of physical phenomena. Finally, an upper-level sequence of two semesters will present the fundamentals of wave mechanics and modern physics.

The science stem, it should be remarked, has a dual role: not only the provision of physical principles for immediate application but also—and in current times almost more important—the preparation of the student for a professional lifetime of continuing education. In this regard it is to be noted that only the common core has been delineated and that one or more additional sciences will be taken as departmental requirements.

Analysis and Design Stem. Since none of the first three curricular stems provides the makings of a true engineer, balance in engineering education can only be attained by ensuring comparable strength in the final stem. Call it creative design, analysis, or simply the art of engineering, its proper planning and implementation represent probably the greatest challenge in engineering pedagogy. At Iowa we begin in the freshman year with an introductory two-semester course that is intended to stimulate the student's interest and nurture his originality through open-ended problems of a challenging nature; the pertinent graphic and computer skills are introduced as supplementary to the main theme. (We hope, it might be noted, that Introduction to Engineering will prove attractive to non-engineers as well, and thus provide part of the salutary influence that our College should have upon students in the others).

The introductory course will be followed in the sophomore year by one in dynamic systems analysis, which will apply a uniform approach to the examination of energy storage, transduction, transfer, transmission, and dissipation in lumped thermal, mechanical, fluid, and electrical systems. Since engineering is in reality an iterative combination of synthesis and analysis, the place of this course in the fourth stem is both natural and essential.

In the third year of the analysis and design stem, general principles and methodology applicable to any design problem are introduced. These include optimization methods—i.e., mathematical programming, etc.—simulation, computer-aided design, statistics, reliability analysis, and design and analysis of experiments; the latter build immediately upon the probability studied in the mathematics stem. All of these topics will be studied and applied within the context of the design problems chosen from the various engineering disciplines. The senior design course will involve the same philosophy and methodology but will deal with problems from the student's elected major discipline. Design projects in these courses will be considerably larger in scope. Close but flexible supervision by senior faculty is anticipated.

Some 30 hours of technical courses are presently foreseen as departmental prerogatives, and it is in these that much of the essential laboratory experience will be obtained. It is believed that this is sufficient specialization to warrant granting a B.S. degree with specification of field; indeed, even the chemical engineers will follow much the same curriculum, required courses in chemistry being taken as departmental selections. On the other hand, there is no reason why the technical courses cannot be chosen in different departments, so that the field of the degree will be simply engineering. This is the very direction predicted in the ASEE Goals Report, and in all likelihood Iowa as well as other institutions will steadily approach an undergraduate curriculum which differs from one case to another only in the electives permitted in each of the four stems. In the latter event, the prescribed number of technical electives may well be reduced.

A Flexible, Developing Curriculum

The Iowa curriculum has purposely been made as flexible as possible so that the transition toward a common program for undergraduates may proceed as smoothly and rapidly as the situation dictates. Because of the inherent differences that seem to characterize the various professions throughout the country, the change will very likely proceed at a different rate in each department. Most electrical and mechanical engineers, for example, would be willing to adopt a single program without further ado, and certain members of other departments are of the same mind. By and large, however, chemical engineering still feels compelled by professional pressures to maintain its identity even in undergraduate years; civil engineering, inherently conservative, holds tenaciously for at least some exposure to most of its ever-diverging branches; and industrial engineering would go even further in its new directions of statistics, human factors, and systems analysis. But whatever the rate of eventual change, it cannot be disputed that the undergraduate years are becoming more and more the concern of the engineering college as a whole, with the graduate years more and more definitely those of departmental specialization.

To date, some 30 months have been spent at Iowa on general planning and the initial implementation of experimental courses. After three semesters of teaching first 20 and finally 100 freshmen introductory material in the humanities within the College of Engineering, it was decided for several reasons that this course should be administered and taught in the College of Liberal Arts. On the other hand, after a year's trial of an introductory course in engineering, the single class of 20 is now being enlarged to the complete freshman enrollment and taught by senior staff members from each of the five departments. Other courses are being planned, reorganized, or given experimentally. By the fall of 1969 it should be possible to offer the entire first year of the new curriculum and add the successive years on an annual basis. Many of the major courses of the latter years are still to be designed, needless to say, and problems of considerable complexity continue to face us. One of these is the question of the many transfer students from community colleges—how can their credits be properly assessed in the light of the new curriculum, and to what extent will courses have to be repeated? Such difficulties notwithstanding, we believe that we are within sight of the curricular balance that we currently seek, and yet free to adjust to new conditions as rapidly as technological and social change demand. △

JET-INDUCED CIRCULATION AND DIFFUSION

By Constantin Iamandi[1] and Hunter Rouse,[2] F. ASCE

SUBMERGED JETS AND BUBBLE SCREENS

Circulation and diffusion are often produced in a body of water for one purpose or another by the release of air from a row of orifices in a manifold some distance below the free surface. The buoyancy of the air-water mixture produces an upward flow in the region above the orifices, and the resulting shear between the rising fluid and the surrounding water eventually brings a considerable portion of it into motion.

Much the same situation would prevail if water instead of air were discharged from the orifices, for the shear between the high-velocity jets and the otherwise stagnant water surrounding them would likewise set the water in motion. In fact, the similarities between the two cases of induced flow are so great as to suggest the possible use of water jets instead of bubble screens—and vice versa—under appropriate conditions. There are essential differences, as well as similarities, however, and all must be carefully assessed before a choice is made.

In the case of the submerged jet in an infinite fluid, the absence of external forces beyond the zone of efflux requires that the flux of momentum be the same at all successive sections; that is, for two-dimensional flow from a slot (see the left-hand side of Fig. 1, in which the streamlines are indicted by constancy of stream function ψ) (1)[3]

$$\frac{m}{\rho} = q_0\, v_0 = b_0\, v_0^2 = \int_{-b/2}^{b/2} v^2\, dx \quad \dots\dots\dots\dots\dots\dots\dots\dots\dots (1)$$

Note.—Discussion open until August 1, 1969. To extend the closing date one month, a written request must be filed with the Executive Secretary, ASCE. This paper is part of the copyrighted Journal of the Hydraulics Division, Proceedings of the American Society of Civil Engineers, Vol. 95, No. HY2, March, 1969. Manuscript was submitted for review for possible publication on July 24, 1968.

[1] Visiting Research Engineer, Institute of Hydraulic Research, Univ. of Iowa, Iowa City, Iowa (1967-68); presently Head, Hydrodynamics Laboratory, Institute of Fluid Mechanics, Romanian Academy, Bucharest, Romania.

[2] Dean, College of Engrg., Univ. of Iowa, Iowa City, Iowa.

[3] Numerals in parentheses refer to corresponding items in Appendix I.—References.

In fact, beyond a vertical distance y that is an order greater than the slot breadth b_0, it matters little just how large b_0 and v_0 actually are, provided only that their combination produce the kinematic momentum flux per unit length of slot m/ρ under consideration. The flow could as well proceed from a line source producing the finite rate of momentum flux m/ρ (see the right-hand side of Fig. 1) with a zero rate of volume flux q_0. The rest of the characteristics of the jet are the same for either condition: the jet breadth b increases linearly with height y; the velocity v along any line through the source varies inversely

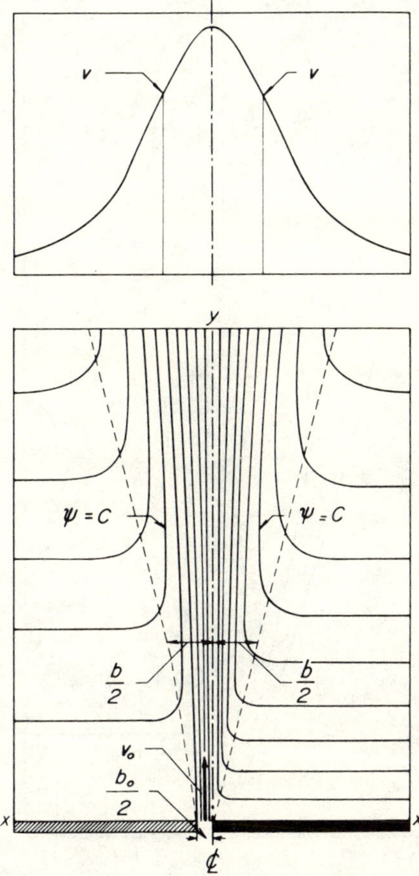

FIG. 1.—PATTERN OF FLOW IN SUBMERGED JET FROM SLOT (LEFT) (1) AND FROM LINE SOURCE OF MOMENTUM FLUX (RIGHT)

with $y^{1/2}$, and the rate of entrainment varies directly with $y^{1/2}$—in fact, the resulting rate of induced flow past any normal section has the magnitude (1)

$$q = \int_{-b/2}^{b/2} v\,dx = 0.62\sqrt{\frac{m}{\rho}}\,y \quad \dotsb \quad (2)$$

All sections of the flow are dynamically similar to all other sections, and the nondimensional parameter for the velocity distribution is evidently of the form $v/\sqrt{m/\rho y}$.

[For the sake of simplicity, the approaching flow in Fig. 1 has been shown parallel to the boundary as in Ref. 1, though this properly notes (see page 663) that the streamlines must initially be asymptotic to parabolas with vertex at the origin if the velocity is to be zero at infinity. In any event, except for the zone of flow establishment at the left, all streamlines should be geometrically similar, differing only in scale. For the present these points are solely of academic interest in connection with Fig. 1 (and Fig. 2), as they have no bearing on the subsequent discussion.]

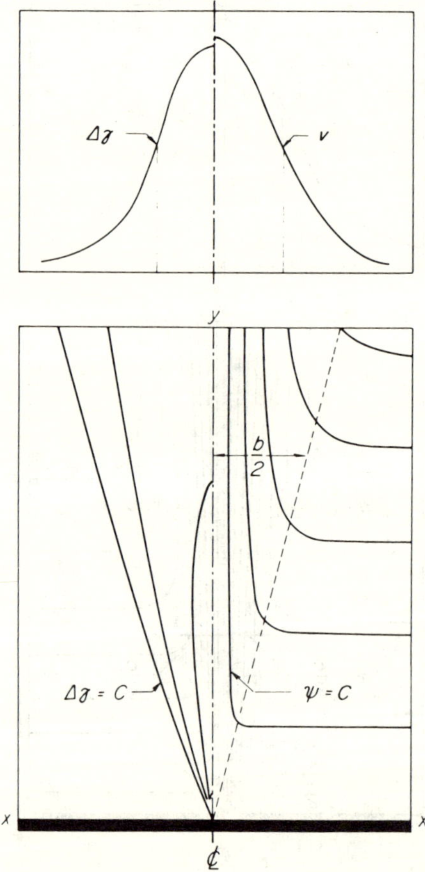

FIG. 2.—PATTERN OF FLOW (RIGHT) AND DISTRIBUTION OF SPECIFIC WEIGHT BY ORDERS OF MAGNITUDE (LEFT) ABOVE LINE SOURCE OF BUOYANCY (6)

In the case of the bubble screen, it might be noted in passing that the release of air under water would have much the same effect (but in the reverse direction) as the release of liquid drops in air, sediment particles in air or water, or a fluid that is colder [or, in the same direction, warmer (6)] than the surrounding fluid. If it is assumed that the effect is essentially one of buoyancy rather than momentum influx, and if the change in density of the rising bubbles is for the moment ignored, the significant parameter in place of m/ρ will be the kinematic weight (i.e., buoyancy) flux per unit length of source

$$\frac{w}{\rho} = q_0\, g = \frac{1}{\rho} \int_{-b/2}^{b/2} v\, \Delta\gamma\, dx \ \ldots\ldots\ldots\ldots\ldots\ldots\ldots\ldots\ldots \ (3)$$

the term $\Delta\gamma$ representing the unit buoyancy of the air-water mixture at any point (see the left-hand side of Fig. 2). For reasons of continuity, the quantity w should as a first approximation be considered constant from section to section, the action of this external force being to increase the rate of momentum flux linearly with distance from the source. Though the breadth b of the flow section again increases linearly with distance from the origin and at about the same rate as the jet, the velocity along any line from the source now remains constant, and the rate of induced flow q past any normal section hence increases linearly with y (6), expressed as

$$q = 0.57 \sqrt[3]{\frac{w}{\rho}}\ y \ \ldots\ldots\ldots\ldots\ldots\ldots\ldots\ldots\ldots\ldots \ (4)$$

The nondimensional parameter for velocity distribution is now $v/\sqrt[3]{w/\rho}$.

Tests have frequently been made of flow patterns produced by jets under various boundary configurations. Two pertinent examples are represented by the impingement of a jet on a normal boundary (4), and the superelevation of that boundary if it is a free surface (3). Though both studies were for conditions of axial symmetry, the conclusions reached are applicable to two-dimensional conditions as well. First, the principal modification of flow pattern is restricted to the zone of impingement. Second, the further modification in the case of the free surface can be approximated by the velocity head of the undisturbed jet at the same distance from the efflux section.

If the jet is also laterally confined by a wall parallel to its axis, a more or less complex pattern of circulation may be expected to develop. So long as the vertical and horizontal dimensions of the fluid body are of comparable magnitude, a single vortex will be produced. Once the length exceeds some multiple of the depth, however, a second, weaker vortex of opposite sense will be formed beyond the first; even a third, still weaker, might be anticipated if the length is sufficiently great.

Essentially the same pattern of induced flow is to be expected in the vicinity of a confined bubble screen. If the submerged jet and the bubble screen are to be compared, it is apparently the rate of flow induction that should be a measure of relative performance. In other words, in water standing at a depth $y = D$, a vertical jet from a bottom slot and a bubble screen from a bottom source might be expected to induce essentially the same motion of the surrounding fluid if, as is indicated by combining Eqs. 2 and 4,

$$\frac{\dfrac{m}{\rho}}{\left(\dfrac{w}{\rho}\right)^{2/3}} \approx D \ \ldots\ldots\ldots\ldots\ldots\ldots\ldots\ldots\ldots\ldots\ldots\ldots\ldots \ (5)$$

The present study has as its purpose the determination of patterns of mean flow and turbulence for various jet: boundary configurations. While a comparable study of the bubble screen must be reserved to a later date, it is anticipated that for the intervening period Eq. 5 may be used as the basis for at least a qualitative adaptation of the one family of flow patterns to that for the other case of flow.

EXPERIMENTAL PROCEDURE

In view of the bubble-screen tests subsequently planned, tests with the submerged jet were initially undertaken in water in a laboratory tank. This had a length of 20 ft, a width of 2.5 ft, and sufficient depth to permit a surface level 2.5 ft over the jet manifold to be maintained. Instead of constructing a slotted efflux section of variable width, the manifold was built to accomodate interchangeable orifices of different diameters and distances center to center. It was also planned to vary the rate of flow, the chamber width, and the chamber

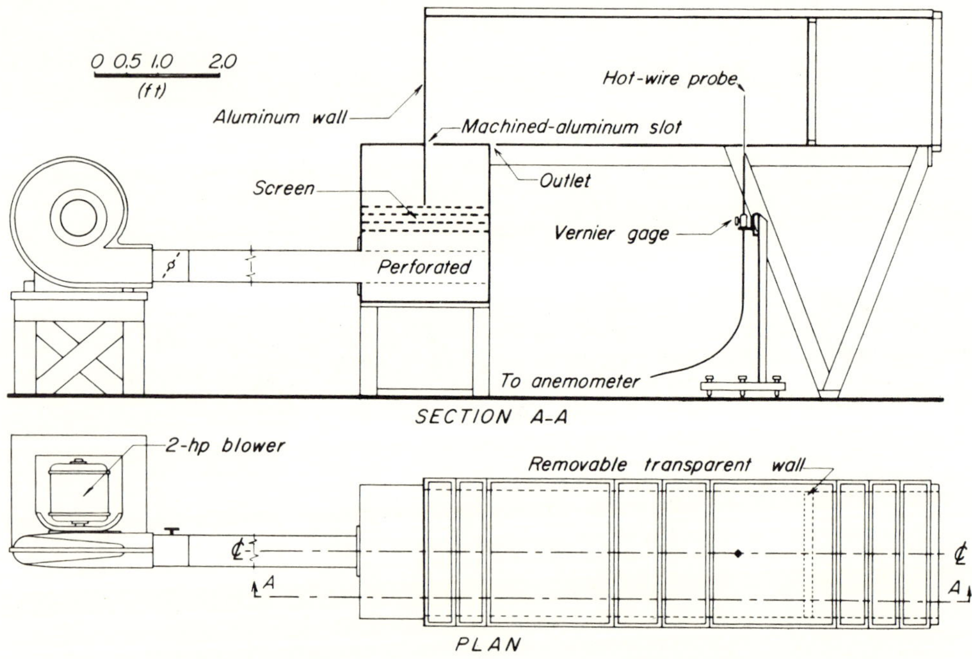

FIG. 3.—DIAGRAM OF EXPERIMENTAL EQUIPMENT

length, and to determine the geometry of the mean flow pattern by analysis of the measured velocity distribution.

Because of the low velocities involved, use of the pitot tube was precluded, and efforts were made to adapt an electrically indicating miniature current meter to the purpose. Such consistent trouble was encountered with variability of the calibration, however, that the project might well have become one of instrument development rather than flow measurement. Shortage of time therefore dictated a change from the use of water to the use of air, with its relatively simple boundary construction and instrumentation.

The flume was thus replaced by a plywood duct 2 ft by 2 ft in cross section and 8 ft long, with a machined-aluminum slot of variable width at the midsection, fed from a plenum chamber connected to a 2-hp blower. A simple traversing mechanism mounted externally permitted placement of a hot-wire probe at any desired point. The wire was normally mounted so as to indicate the magnitude of the velocity vector, regardless of its direction in the plane of motion. The constant-temperature principle of indication was used (2).

Preliminary tests with symmetrical location of the slot permitted qualitative assessment of the effect of rate of flow, slot width, duct width, duct length, and location of waste slot to be made. However, unpredictable instabilities (apparently due to asymmetric formation of secondary eddies) finally led to the decision to deal solely with an end jet for consistency of the results.

The final form of the test equipment was hence as shown in Fig. 3. The end wall at the slot section was of sheet aluminum, to permit the movable edge of the slot to yield an essentially constant breadth of even small openings. The hot-wire probe could be inserted at any of a series of bottom openings and moved vertically to traverse the entire 2-ft height of duct. The wall section at the end opposite to the slot section could be moved to yield L/D ratios of 1, 2, and 4. A longitudinal wall could also be inserted to yield a width-depth ratio of 0.25 instead of the usual 1.0. The waste slot, 1 in. wide, was placed in the floor at the edge of the plenum chamber, since the inlets of a pumping system —whether at the surface or the bottom—would probably be as close to the discharge manifold as feasible. The preliminary tests, moreover, had shown that the change in effect upon the flow pattern decreased as the location of the waste slot proceeded from directly over the inlet, to the upper corner of the end wall, to the lower corner, to the position finally selected.

The hot-wire anemometer was initially calibrated in an air tunnel against a vaned anemometer of known characteristics. However, calibration was eventually carried out by comparison with the indication of a stagnation tube in the irrotational zone of a jet issuing from a streamlined orifice. Although a two-wire probe could have been used to measure the longitudinal and vertical components of the velocity for direct determination of the flow pattern, it was considered simpler to assume the local flow direction, integrate the vector magnitude in the normal direction for a first approximation of the streamline location (i.e., $\psi = \int v \, dn = C$), and then proceed to a second approximation. Usually a single correction of the streamline configuration sufficed. The same wire was used for measuring the intensity of turbulence (actually the root-mean-square of the fluctuating component in the plane of the mean motion).

DISCUSSION OF RESULTS

In view of the introductory discussion, the velocity v at any point in the duct can be expected to depend upon the following variables

$$v = f(x, y, z, D, L, W, m, \rho, \mu) \quad \dots \dots \dots \dots \dots \dots \dots (6)$$

Herein x, y, and z are the coordinates of the point; D, L, and W the depth, length, and width of the duct, m the momentum flux per unit width; and ρ and μ the density and viscosity of the fluid. The variables may be combined to yield the nondimensional function

$$\frac{v}{\sqrt{m/\rho \, D}} = \phi\left(\frac{x}{D}, \frac{y}{D}, \frac{z}{D}, \frac{L}{D}, \frac{W}{D}, \frac{\sqrt{D \, m \, \rho}}{\mu}\right) \quad \dots \dots \dots \dots \dots (7)$$

In order to verify the belief that the last three terms could be considered

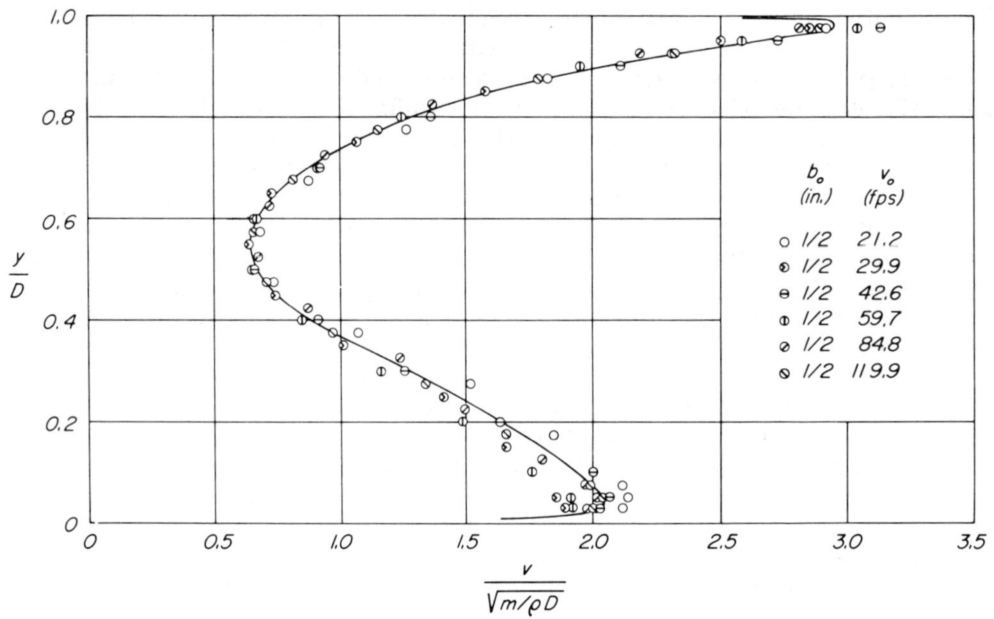

FIG. 4.—VARIATION OF CENTRAL VELOCITY-MAGNITUDE TRAVERSE WITH EFFLUX VELOCITY

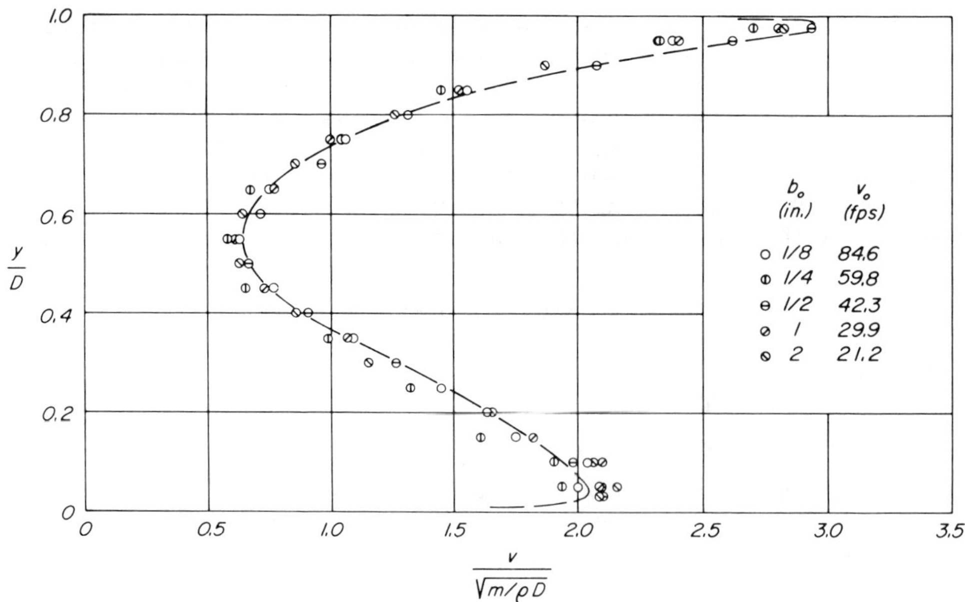

FIG. 5.—VARIATION OF CENTRAL VELOCITY-MAGNITUDE TRAVERSE WITH SLOT BREADTH

of little consequence, preliminary velocity traverses were made at the mid-section of the duct with L = 4 ft and D = 2 ft (except that D was reduced to 1 ft by lowering the ceiling in one run), and v_0, b_0, and W were each varied in turn for constant intermediate values of the other two. Since it was the velocity magnitude that was measured rather than either component, and since the traverse did not pass through the exact center of the eddy, the indications were necessarily positive at every point. Fig. 4 contains the comparative results for various values of v_0 with constant geometry, no systematic displacement of the curve being evident even at the lowest value; it may be concluded that prototype flows above the minimum Reynolds number ($\sqrt{D}\ m\ \rho/\mu \approx 30{,}000$) will be similar. Fig. 5 presents the effect of variation in relative slot width over a 32-fold range, m/ρ being held constant; again no appreciable shift is

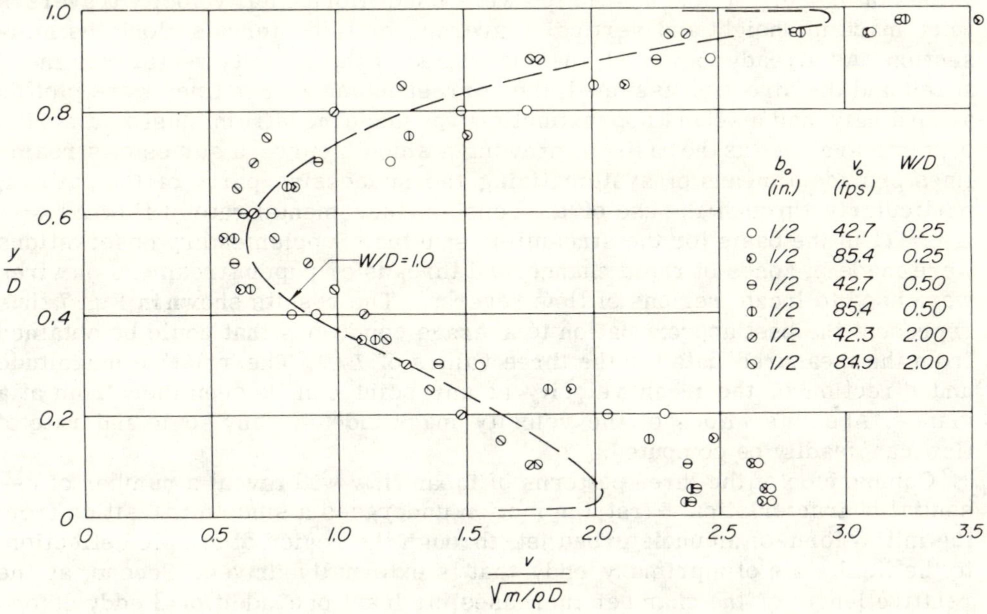

FIG. 6.—VARIATION OF CENTRAL VELOCITY-MAGNITUDE TRAVERSE WITH DUCT WIDTH

apparent. Fig. 6, finally, shows the effect of relative duct width, a systematic displacement now being evident, in particular as W/D becomes small; the latter trend was ignored, however, on the assumption that cross sections would probably be wide rather than narrow. The related effect of z/D, or the wall influence on the velocity distribution, was also ignored, the velocity in the central portion of the duct being accepted as indicative of the average.

The term b_0/D merits further comment. From the introductory discussion it will be recalled that it is the momentum flux $m = \rho\ b_0\ v_0^2$ which characterizes the effectiveness of the jet as a propulsion device, regardless of the slot breadth and efflux velocity, at least for slot breadths that are small compared with other dimensions. Ultimately, of course, the jet flow itself would become preponderant, and the induced flow secondary, as b_0 became large and v_0 small; nevertheless, Fig. 5 indicates that this stage is not yet in sight even for b_0/D =

1/12. In order to maintain the flow as a whole, however, a power input equal to the unit energy flux $e_0 = \rho\, b_0\, v_0^3/2$ is required. Evidently, $e_0 = mv_0/2$. Therefore, whereas a limiting pattern of streamlines is approached as b_0/D becomes small, economy of operation is realized as b_0/D becomes large and appreciable change in pattern becomes evident. On the other hand, under conditions of most economic operation, the flow will be that of the jet itself rather than of the turbulence-induced circulation, and the diffusion will then be minimal; moreover, the initial cost of the pumping system would tend to increase with the pumping capacity. For these reasons it was concluded that the patterns of interest would be those of the type represented by the range of relative slot breadths included in Fig. 5; greater breadths were not investigated.

In conducting the velocity measurement from which the mean patterns of flow were determined for the L/D values of 1.0, 2.0, and 4.0, the intermediate values of $b_0 = 0.5$ in. and $v_0 = 43$ fps were used. Horizontal velocity traverses were made midheight and vertical traverses at 1-ft intervals along the midsection. As already indicated, the magnitude of the velocity vector was measured and the direction assumed, the corresponding streamlines were plotted accordingly, and a second approximation was made therefrom. Just as a series of points are used as the basis for drawing a smooth curve, a series of streamlines provide a means of systematizing the successive parts of the pattern, particularly through the use of the constant incremental rate of flow $\Delta \psi = v\, \Delta n = C$ as the basis for the streamline spacing. Supplementary observations were made in zones of rapid change, and threads or paper streamers on wires were used to locate regions of flow reversal. The results shown in Fig. 7 thus represent the best approximation to average conditions that could be obtained from the measured data for the three values of L/D. The relative magnitude and direction of the mean velocity at any point can be seen therefrom at a glance. Absolute values of the velocity magnitude for any scale and rate of flow can readily be computed.

Comparison of the three patterns of mean flow will reveal a number of essential characteristics. First, the motion undergoes a smooth transition from the initial form of an undisturbed jet, through the region of simple deflection, to the final state of a primary eddy that is externally driven. Second, as the relative length of the chamber increases, at least one additional eddy of opposite rotational sense is formed. Third, the energy level of the motion falls considerably from the jet to the first eddy and then from the first eddy to the second. Fourth, judging by the velocity distribution within the initial jet, the resistance to its motivating action increases with length of chamber. The extent to which these patterns may be combined with their mirror images to predict patterns for symmetrically located jets is problematical. In all likelihood, for small values of L/D, the patterns themselves would be symmetrical. However, in the range of formation of a secondary eddy ($L/D > 2$), it apparently tends to develop first at one end only, with noticeable asymmetry of both jet and surrounding flow as the result.

On the page facing the patterns of mean motion are the corresponding summary plots of turbulence distribution (Fig. 8). The values shown are the indications of the normal wire, which responded to the two components of flow

in the plane of motion—i.e., $\sqrt{\overline{v'^2}} = \sqrt{\overline{v_x'^2} + \overline{v_y'^2}} = \sqrt{\overline{v_s'^2} + \overline{v_n'^2}}$. If it is assumed

(5) that $\overline{v_z'^2} \approx \overline{v_n'^2} \approx (1/2)\,\overline{v_s'^2}$, then the plotted values indicate an intensity $\sqrt{\overline{v'^2}}$ that is roughly 10% below that of the actual turbulence. However, it is the dis-

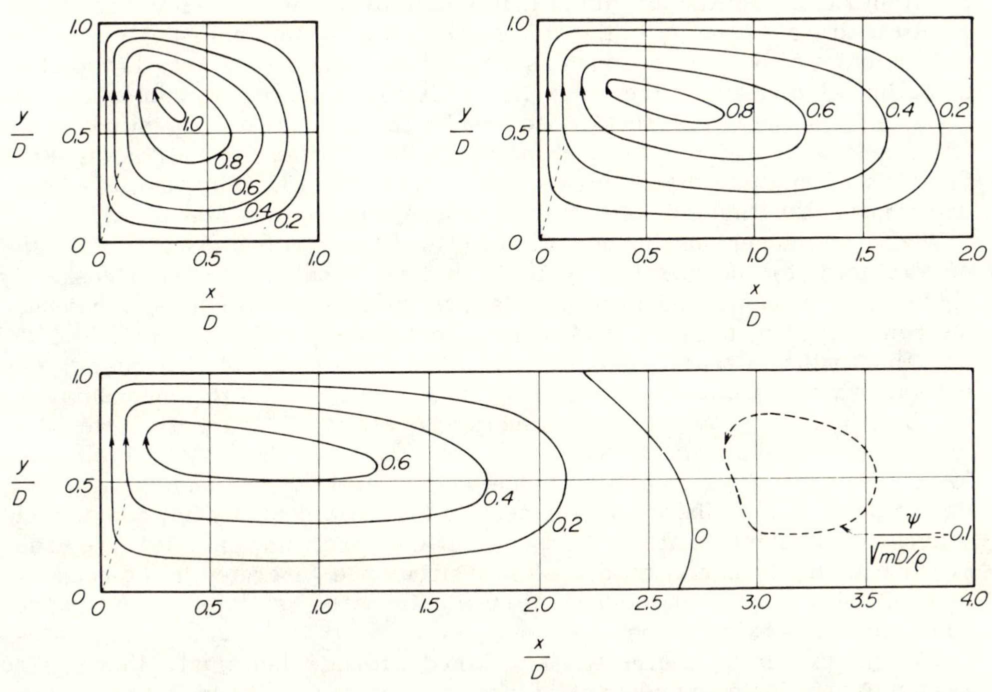

FIG. 7.—PATTERNS OF MEAN STREAMLINES FOR VARIOUS RELATIVE LENGTHS
OF DUCT

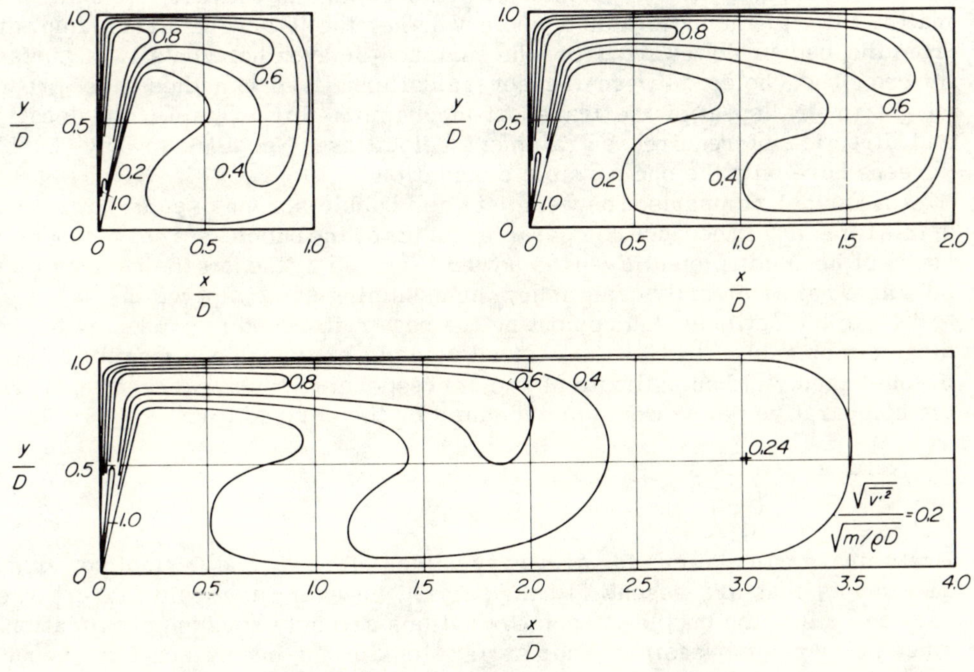

FIG. 8.—DISTRIBUTION OF TURBULENCE INTENSITY FOR PATTERNS OF FIG. 7

tribution rather than the magnitude that is significant. Whereas it is practically identical to that of the jet in the slot vicinity, its variation in the vicinity of the primary eddy is very much as one might expect from the flow pattern throughout the eddy. On the other hand, the rapid decrease in intensity with distance from the initial shear zone results in such a small magnitude in the second eddy, for $L/D = 4$, that the indication there is purely qualitative. Even more significant than the turbulence intensity is the diffusivity, and it is hence unfortunate that time did not permit the measurement of scale as well.

Were the momentum flux actually produced along a line source (i.e., a slot of zero breadth), the spectrum of the resulting turbulence would necessarily extend to zero scale. The effect of enlarging the slot to finite size is to reduce the role played by the very fine-scale turbulence in the initial portion of the jet. Mention has already been made of the fact that continued increase in slot breadth would eventually eliminate the turbulent shear as the mechanism by which the central eddy was kept in motion; in effect, the eddy itself would no longer exist as such, essentially the whole flow being the deflected jet. Even for a value of b_0/D as great as 0.1, however, there would presumably be sufficient shear between the jet and the central flow to generate a full complement of turbulence elements at the coarse end of the spectrum, and it is these which produce the major part of the diffusion. (Turbulence generated in the boundary layer along the walls of the duct, it should be noted, is of minor importance compared with that of the jet.)

Though the flow patterns were obtained with air jets in air, there is no reason to anticipate any essential difference for water jets in water. In fact, it should matter little whether the water were confined or had a free surface, for the slight surface rise in the stagnation zone of the deflected jet would probably be small in comparison with the depth. Moreover, it should also matter little to the pattern as a whole whether the jets were directed upward from the bed or downward from the surface—or even horizontally at surface or bed. The choice between the several alternatives can thus give proper weight to any aspect of the transport mechanism—for example, the location and direction of required entrainment velocities. Needless to say, bubble screens have only the one possible orientation.

Any further comparison between jets and bubble screens seems premature at this time, if only because comparable studies of the bubble-screen mechanism have not been completed. Even the power required by the jets themselves must be somewhat arbitrarily established, by assuming some slot width—say, b_0/D = 0.1—as the optimum and computing the power involved in producing the indicated momentum flux for the desired velocity or turbulence magnitude. Eq. 5 would then yield an estimate of the corresponding bubble-screen input, and the comparative power would be indicated by the derived form

$$\frac{e_{0j}}{e_{0b}} = \frac{1}{2}\frac{v_0}{\left(\dfrac{w}{\rho}\right)^{1/3}} \quad \dots\dots\dots\dots\dots\dots\dots\dots\dots\dots\dots\dots\dots \quad (8)$$

Preliminary calculations for arbitrary values of b_0, v_0, and D yield power requirements that are essentially the same. However, it should be remarked once again that the bubble-screen evaluations are only rough approximations, since neither compressibility nor surface tension has been taken into account. In model studies, to be sure, density changes can probably be ignored for the small depths involved, but an additional head must then be imposed to produce bubbles that are sufficiently small; in the prototype the situation is just the reverse: bubbles that are small relative to the depth are formed without additional work, but considerable compression of the air is necessary at the depths involved. In either case, more power than herein assumed will be required.

CONCLUSIONS

From the foregoing experimental study of jet-induced circulation and diffusion the following conclusions have been reached:

1. Submerged jets provide an alternative (and even more adaptable) means of maintaining the pattern of flow for which a bubble screen is often utilized.

2. The generalized characteristics of the measured mean-flow and turbulence fields are subject to use in design over a considerable range of scales and operational conditions.

3. Until such time as similar studies have been carried out for bubble screens, an approximate method is at hand for comparing their effectiveness in terms of the information for jets presented herein.

ACKNOWLEDGMENTS

The writers of this paper conceived and carried their investigation to completion in the course of exchange visits arranged jointly by the Academies of their respective countries. All experiments were conducted by the first writer (who had already begun bubble-screen studies at Bucharest) during a seven-month period on the staff of the Iowa Institute of Hydraulic Research, under the supervision of the second writer; by whom the paper was written. Financial support was provided by the National Science Foundation through the National Academy of Sciences and the National Academy of Engineering.

APPENDIX I.—REFERENCES

1. Albertson, M. L., Dai, Y. B., Jenson, R. A., and Rouse, H., "Diffusion of Submerged Jets," *Transactions*, ASCE, Vol. 115, 1950.

2. Glover, J. R., "Old Gold Model Type 4-2H Hot-Wire Anemometer and Type 2 Mean-Product Computer," *Report 105*, Iowa Institute of Hydraulic Research, July, 1967.

3. Hunt, B., and Hsu, S. T., "Configuration of the Free Surface above a Vertical Jet," *La Houille Blanche*, No. 6, 1965.

4. Rouse, H., "Seven Exploratory Studies of Hydraulics," *Journal of the Hydraulics Division*, ASCE, Vol. 82, No. HY4, Proc. Paper 1038, August, 1956.

5. Rouse, H., Siao, T. T., and Nagaratnam, S., "Turbulence Characteristics of the Hydraulic Jump," *Transactions*, ASCE, Vol. 124, 1959.

6. Rouse, H., Yih, C. S., and Humphreys, H. W., "Gravitational Convection from a Boundary Source," *Tellus*, Vol. 4, No. 3, 1952.

APPENDIX II.—NOTATION

The following symbols are used in this paper:

C = constant;
D = depth of duct;
L = length of duct;
W = width of duct;
b_0 = breadth of slot;
b = breadth of jet;
e_0 = energy flux per unit width at source;
g = acceleration of gravity;
m = momentum flux per unit width;
n = coordinate normal to streamline;
q_0 = volume flux per unit width at slot;
q = volume flux per unit width;
s = coordinate tangential to streamline;
v = velocity;
v_0 = velocity of efflux;
w = buoyancy flux per unit width;
x = horizontal coordinate;
y = vertical coordinate;
z = coordinate perpendicular to flow;
$\Delta\gamma$ = unit buoyancy of mixture;
μ = viscosity;
ρ = density; and
ψ = stream function.

WORK-ENERGY EQUATION FOR THE STREAMLINE

By Hunter Rouse,[1] F. ASCE

INTRODUCTION

One of the writer's colleagues, in the course of graduate oral examinations over the past decade, frequently asked students to discuss the essential features of the Bernoulli theorem. Most of them usually began (and some ended) by stating that the theorem indicated constancy in the sum of velocity head, pressure head, and elevation. When queried about the conditions under which such constancy prevailed, the majority recalled that the flow had to be steady, the fluid incompressible, and the losses negligible. A few added the further requirement of no change in the overall pressure load with time, since steadiness refers by definition to the velocity only; as a result, constancy of the total head really involved a comparison of points on the same streamline at the same instant. The identical students finally noted that conditions on different streamlines could be compared only if the flow were also irrotational.

If asked how the Bernoulli equation had to be modified to include rotational flow with losses, nearly all the students simply replied that an additional term equal to the loss between sections had to be introduced, the equation presumably being restricted to sections of the same stream tube. That this had to be a gross stream tube (one encompassing the entire flow and hence not readily applicable to flow around a body), was seldom appreciated. Moreover, though a student would stoutly deny that a loss could ever be negative, he would sheepishly grant, when asked, that the increase in head along the central streamline of a wake was a somewhat different matter.

Daniel Bernoulli, of course, probably never puzzled over most of these matters, for this version of what we call the Bernoulli theorem (1) was essentially an application of the then-new Leibniz principle relating potential and kinetic energy. Through his experiments with manometers, Bernoulli had noted that a reduction in flow section produced a decrease in piezometric head

Note.—Discussion open until October 1, 1970. To extend the closing date one month, a written request must be filed with the Executive Secretary, ASCE. This paper is part of the copyrighted Journal of the Hydraulics Division, Proceedings of the American Society of Civil Engineers, Vol. 96, No. HY5, May, 1970. Manuscript was submitted for review for possible publication on July 24, 1969.

[1] Dean, College of Engrg., The Univ. of Iowa, Iowa City, Iowa.

583

proportional to the increase in the square of the velocity, and vice versa. His efforts to incorporate a pressure term in the theorem were hardly satisfactory, however, nor was his father, a far more able mathematician, much more successful (2). It remained to Leonhard Euler (4) to derive the first Bernoulli equation, in the form discussed by the graduate students, by adapting his equations of acceleration to the steady irrotational flow of an incompressible fluid in a gravitational field.

Though the simple Bernoulli relationship has continued to be used throughout the two centuries since its derivation, it has come to be regarded more and more as an over-simplification, with apparently little hope (or desire) for formal rectification. The Euler equations of acceleration have become replaced in the meantime by the more complex Navier-Stokes equations for viscous fluids, and these in turn have been given the form of the Reynolds equations for flow with turbulence. From them, finally, have been derived various forms of the impulse-momentum and work-energy equations, usually as integrals over finite regions of the flow.

For the past twenty years or more the writer has encouraged certain of his graduate students to direct their doctoral studies toward specific cases of turbulent flow with separation, the experimental part of the investigation involving the measurement of the various characteristics of the flow field, and the analytical, the evaluation of the corresponding terms of the momentum and energy relationships. As a general rule the essential balance of the equations themselves was determined in integral form over significant portions of the flow passage. About a decade ago—prompted, in fact, by his colleague's examination questions--the writer prepared a series of invited articles (6,7,8) on adapting the general energy equation to the more familiar Bernoulli form. Because of the comparative clarity that results from visualizing the flow pattern in terms of streamline coordinates rather than their cartesian or polar-cylindrical counterparts, he continues to believe that this point of view has much to recommend it. The articles were published in books or journals of limited American circulation, however, and they have not received very wide attention in this country. Hence, at the risk of seeming either pedantic or repetitious (which he obviously is), the writer now presents his thesis to members of his own Society.

DERIVATION OF EQUATIONS

The Bernoulli theorem in its customary form is derived most simply by relating one of the standard Eulerian equations to the streamline. In modern tensor notation (the convenience of its use will soon become evident) these equations reduce to

$$\frac{\partial u_i}{\partial t} + u_j \frac{\partial u_i}{\partial x_j} = -\frac{\partial}{\partial x_i}\left(\frac{p}{\rho} + \Omega\right) \quad \dots \dots \dots \dots \dots \dots \dots \dots \dots \quad (1)$$

Herein Ω is the gravitational potential, the subscript i refers to a different coordinate direction in three successive equations, and the subscript j refers to the three successive directions in each equation. If one of these is the direction s tangential to the streamline passing through the point under consideration, the other components of the velocity vanish. Thence, for steady flow

$$\frac{\partial}{\partial s}\left(\frac{1}{2}V^2 + \frac{p}{\rho} + \Omega\right) = 0 \quad\cdots\cdots\cdots\cdots\cdots\cdots\cdots\cdots\cdots \quad (2)$$

Evidently, the Bernoulli sum (the quantity with parentheses) must be the same at any instant at all points along the streamline:

$$B_1 = B_2 = \ldots B_n \quad\cdots\cdots\cdots\cdots\cdots\cdots\cdots\cdots\cdots \quad (3)$$

This is the usual form of the Bernoulli relationship for the streamline. If the flow is rotational, B will generally vary from one streamline to the next. If the fluid is viscous, moreover, it is known to vary along the streamline as well, so that an additive term must be introduced to keep the equation in balance:

$$B_1 = B_2 + \Delta B \quad\cdots\cdots\cdots\cdots\cdots\cdots\cdots\cdots\cdots\cdots \quad (4)$$

One might reasonably conclude that the magnitude of ΔB could be determined by proceeding from either the Navier-Stokes or the Reynolds equations in much the same manner as was previously done from the Eulerian. The two together, in fact are embodied in the Saint-Venant form (11):

$$\frac{\partial u_i}{\partial t} + u_j\frac{\partial u_i}{\partial x_j} = -\frac{\partial}{\partial x_i}\left(\frac{p}{\rho} + \Omega\right) + \frac{1}{\rho}\frac{\partial \tau_{ij}}{\partial x_j} \quad\cdots\cdots\cdots\cdots\cdots \quad (5)$$

The intensity of shear τ represents either the product of the dynamic viscosity and the rate of deformation, if the flow is purely laminar, or the negative product of the density and the mean correlation of the components of velocity fluctuation, if the flow is very turbulent, or—in general—the sum of the two together:

$$\overline{\tau}_{ij} = \mu\left(\frac{\partial \overline{u}_i}{\partial x_j} + \frac{\partial \overline{u}_j}{\partial x_i}\right) - \rho\,\overline{u_i' u_j'} \quad\cdots\cdots\cdots\cdots\cdots\cdots \quad (6)$$

(The bars indicate mean values of the quantities involved.)

Before relating the Saint-Venant equations to the streamline, it is necessary to multiply that for each coordinate direction by the respective mean velocity component and add all three together (indeed, a similar procedure is usually followed in deriving the Bernoulli equation from those of Euler); the result for steady flow, after the substitution $\overline{V}^2 = \Sigma\,\overline{u}_i^2$, is

$$\overline{u}_j\frac{\partial}{\partial x_j}\left(\frac{\overline{V}^2}{2}\right) = -\overline{u}_i\frac{\partial}{\partial x_i}\left(\frac{p}{\rho} + \Omega\right) + \frac{\overline{u}_i}{\rho}\frac{\partial \tau_{ij}}{\partial x_j} \quad\cdots\cdots\cdots\cdots \quad (7)$$

If one coordinate direction is that of the streamline at the point under consideration, Eq. 7 will take the form

$$\overline{V}\frac{\partial}{\partial s}\left(\frac{\overline{V}^2}{2}\right) = -\overline{V}\frac{\partial}{\partial s}\left(\frac{p}{\rho} + \Omega\right) + \frac{\overline{u}_i}{\rho}\frac{\partial \tau_{ij}}{\partial x_j} \quad\cdots\cdots\cdots\cdots \quad (8)$$

the last product being left in terms of $\overline{u}_i$, since the indices may now refer to any coordinate system used in obtaining experimental data. Multiplication of each term in Eq. 8 by the volume element $dA\,ds$ places it in a form suitable for integration with respect to each volumetric variable in turn, for the quantity $dQ = \overline{V}\,dA$ remains constant during integration with respect to s:

$$\frac{\partial}{\partial s}\left(\frac{\overline{V}^2}{2}\right)dQ\,ds = -\frac{\partial}{\partial s}\left(\frac{p}{\rho} + \Omega\right)dQ\,ds + \frac{1}{\rho}\frac{\overline{u}_i}{\overline{V}}\frac{\partial \tau_{ij}}{\partial x_j}\,dQ\,ds \quad\cdots\cdots\cdots \quad (9)$$

If integration solely along the streamline is required, the factor dQ can be cancelled throughout, and the integral with respect to s can be written simply as

$$B_1 = B_2 - \frac{1}{\rho} \int_{s_1}^{s_2} \frac{\bar{u}_i}{\bar{V}} \frac{\partial \bar{\tau}_{ij}}{\partial x_j}\, ds \dots\dots\dots\dots\dots\dots\dots\dots (10)$$

Evaluation of the term ΔB of Eq. 4 is now possible, although its significance is not yet clear. By virtue of its derivation, it is a conservative term, which can be either positive or negative. By the rules of the calculus, moreover, it may be related to the two other terms of the following expression:

$$\frac{\partial\left(\bar{u}_i \bar{\tau}_{ij}\right)}{\partial x_j} = \bar{u}_i \frac{\partial \bar{\tau}_{ij}}{\partial x_j} + \bar{\tau}_{ij} \frac{\partial \bar{u}_i}{\partial x_j} \dots\dots\dots\dots\dots\dots\dots (11)$$

The term at the left evidently corresponds to the total rate at which work is done upon a unit volume of the fluid by the shear of the surrounding fluid, the conservative term at the right representing the accompanying rate of change of energy and the second or dissipative term the rate of irrecoverable loss. If Eq. 10 is now written in terms of the integrals for total work and dissipation, it becomes

$$B_1 = B_2 - \frac{1}{\rho} \int_{s_1}^{s_2} \frac{1}{\bar{V}} \frac{\partial\left(\bar{u}_i \bar{\tau}_{ij}\right)}{\partial x_j}\, ds + \frac{1}{\rho} \int_{s_1}^{s_2} \frac{\bar{\tau}_{ij}}{\bar{V}} \frac{\partial \bar{u}_i}{\partial x_j}\, ds \dots\dots\dots (12)$$

which may be given the symbolic form

$$B = K + P + T + L = \text{const}_s \dots\dots\dots\dots\dots\dots (13)$$

Herein K and P are the kinetic-energy and piezometric-pressure terms of the usual Bernoulli sum, T is the transfer of energy through shear to (i.e., work done upon) the surrounding fluid

$$T = -\frac{1}{\rho} \int_{s_1}^{s_2} \frac{1}{\bar{V}} \frac{\partial\left(\bar{u}_i \bar{\tau}_{ij}\right)}{\partial x_j}\, ds \dots\dots\dots\dots\dots\dots (14)$$

and L is the loss suffered by the mean flow through either viscous dissipation or the generation of turbulence:

$$L = \frac{1}{\rho} \int_{s_1}^{s_2} \frac{\bar{\tau}_{ij}}{\bar{V}} \frac{\partial \bar{u}_i}{\partial x_j}\, ds \dots\dots\dots\dots\dots\dots\dots (15)$$

The four quantities of which B is composed must evidently vary along a given streamline in such a manner that their sum is the same at all points at any instant.

If, for the sake of argument, integration of Eq. 9 is performed with respect to Q as well as to s—that is, over the entire region of flow—three significant conditions must be noted. First, the weighted mean value of K for the whole flow section, if written in terms of the mean velocity Q/A, will require introduction of the Coriolis coefficient α:

$$K_m = \frac{A^2}{Q^2} \int_A \frac{\rho \bar{V}^2}{2} \bar{V}\, dA = \alpha \frac{\rho Q^2}{2A^2} \dots\dots\dots\dots\dots (16)$$

Second, unless the pressure distribution is hydrostatic over the flow section, the average magnitude of P will differ from that at the center line by a different weighting factor η:

$$P_m = \frac{1}{Q} \int_A \left(\frac{p}{\rho} + \Omega\right) \bar{V} \, dA = \eta\left(\frac{p}{\rho} + \Omega\right)_{\mathcal{C}} \quad \dots \dots \dots \dots \dots \quad (17)$$

Third, unless the boundary is movable so that work can be done either by or upon it, the integral of the transfer term must be zero except for the usually small longitudinal differential between the end sections:

$$\int_Q T \, dQ = -\frac{1}{\rho} \int_A \int_s \frac{\partial(\overline{u_i \, \tau_{ij}})}{\partial x_j} \, dA \, ds \approx 0 \quad \dots \dots \dots \dots \dots \quad (18)$$

The weighted mean loss is then simply

$$L_m = \frac{1}{\rho Q} \int_A \int_s \bar{\tau}_{ij} \frac{\partial \bar{u}_i}{\partial x_j} \, dA \, ds \quad \dots \dots \dots \dots \dots \dots \quad (19)$$

Under these circumstances (but only then) $\Delta B = L_m$ and the Bernoulli equation can be written in the usually assumed form

$$B_m = K_m + P_m + L_m = \text{const} \quad \dots \dots \dots \dots \dots \dots \quad (20)$$

REPRESENTATIVE APPLICATIONS

Although the foregoing analysis is a rigorous one, and although quantitative evaluation of the several terms from dependable experimental data has proved readily feasible (3), the writer would prefer to use for the present paper both new and significant illustrations for which complete data are not necessarily at hand. Since the purpose is clarification of principle rather than provision of numerical values, qualitative results will hence be presented where quantitative accuracy cannot yet be attained.

The first example is reminiscent of the writer's part in a joint discussion (9) published about a quarter of a century ago: steady, uniform, laminar flow between parallel boundaries (see Fig. 1). For reasons of hydrostatic equilibrium, the shear is linearly distributed, and the velocity distribution takes the usual parabolic form shown at the center:

$$u = -\frac{1}{2\mu} \frac{\partial p}{\partial s} (2by - y^2) \quad \dots \dots \dots \dots \dots \dots \quad (21)$$

The quantity K, which does not vary with s, is likewise plotted at the center. As shown at the left, the pressure is hydrostatically distributed across the section. The transfer term reduces to

$$T = -\nu \int_s \frac{\partial}{\partial y} \left(u \frac{\partial u}{\partial y}\right) ds \quad \dots \dots \dots \dots \dots \dots \quad (22)$$

and the loss term to

$$L = \nu \int_s \frac{1}{u} \frac{\partial u}{\partial y} \frac{\partial u}{\partial y} \, ds \quad \dots \dots \dots \dots \dots \dots \quad (23)$$

each of which is plotted with the reduced value of P at the right of the figure. Evidently, the loss is negligible along a streamline near the center line and rises to a maximum at the wall. The transfer, in turn, is a positive maximum at the center line, zero along some intermediate streamline, and a negative maximum (numerically smaller than $L_{\max}$) at the wall. In other words, whereas the rate of change in pressure is the same along every streamline, energy is transferred from the central region (where T is positive) to the wall region (where T is negative) for dissipation at a far greater rate than the local change

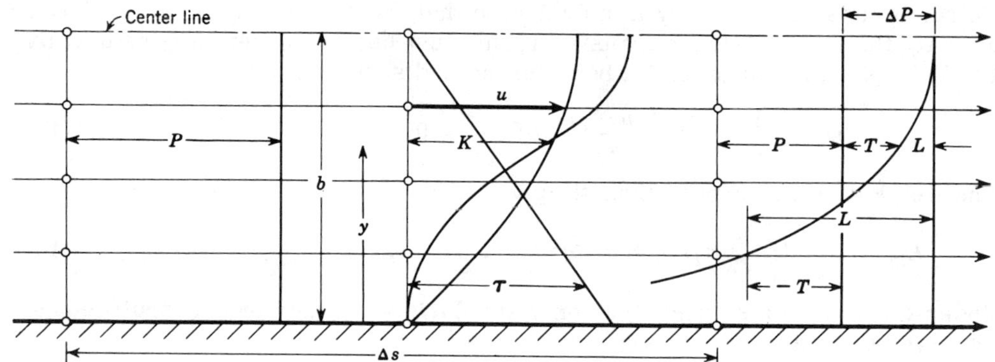

FIG. 1.—TWO-DIMENSIONAL LAMINAR FLOW BETWEEN PARALLEL WALLS

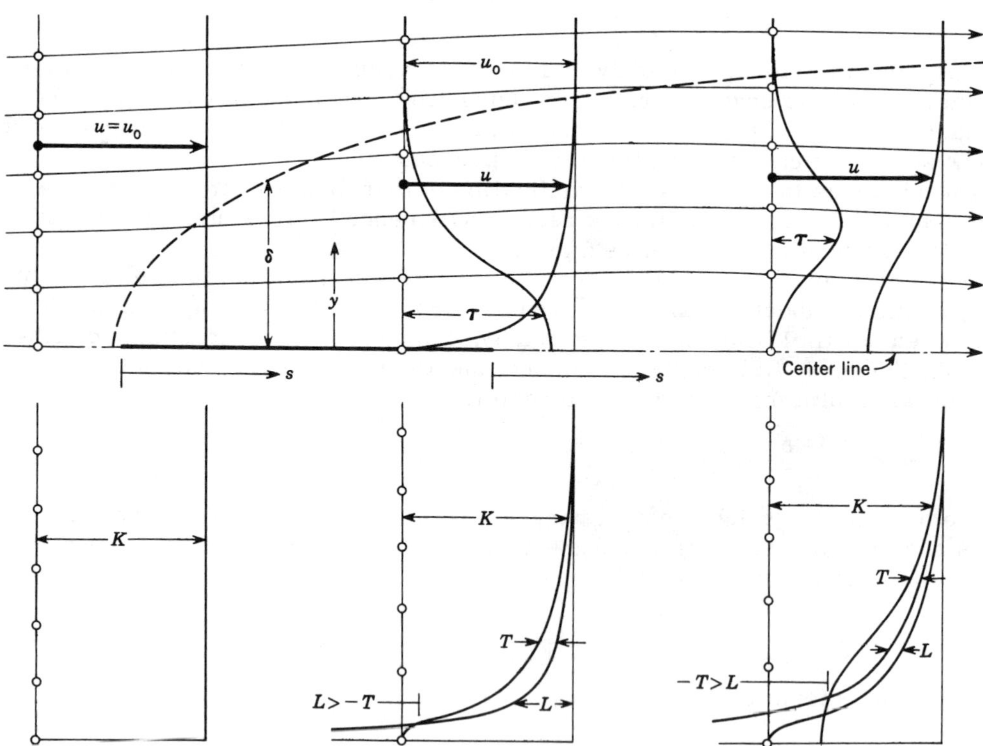

FIG. 2.—TURBULENT FLOW ALONG A PLATE AND IN ITS WAKE (VERTICAL SCALE GREATLY EXAGGERATED)

in P would indicate. From integration of K according to Eq. 16, α will be found to have the value of 1.54. Determination of the total transfer according to Eq. 18 will yield the value of zero. And solution for the net loss according to Eq.

19 will show that $L_m = -\Delta P_m$ as anticipated.

Schematic representation of the wake flow that seemed to belie the students' concept of energy reduction will add a significant variant to the foregoing example. [At the time of writing, only incomplete experimental data were available in the literature for this case; between the dates of manuscript submittal and acceptance, however, an inclusive study was published (5).] The wake under consideration is produced by turbulent flow along a thin, flat plate of finite length, and the primary assumptions involved in elementary boundary-layer theory are presumed to apply to the wake as well—in particular, that of the constancy of pressure at all points. In this connection it must be noted that an extreme degree of scale exaggeration will be used in the normal direction (see Fig. 2) to make the interrelationships graphically clear.

Just ahead of the leading edge the Bernoulli sum can be assumed to consist entirely of the kinetic term K, which is the same at all streamlines for irrotationality of the approaching flow. The linear distribution of shear in the previous example now becomes somewhat S-shaped, the curve of velocity distribution assumes its familiar variation from zero at the plate to nearly a maximum at the nominal boundary-layer limit, and the flow displacement that it indicates grows with distance from the leading edge. The streamlines are deflected accordingly from their parallel course, and for reasons of continuity the nominal limit of the boundary layer, shown as a dashed curve in the figure, must intersect successive streamlines.

Along each streamline the decrease in K (P now remaining constant) represents the sum of the transfer T and the loss L, the effect of viscous shear now being neglected in comparison with the Reynolds stresses of the turbulence. As in the foregoing example, in the boundary vicinity the loss L is greater than the reduction in K, in accord with which the transfer T changes from a positive to a negative value as the boundary is approached. Along the plate L always exceeds T in absolute magnitude, and ΔB is thus invariably positive.

Beyond the end of the plate (located at the midpoint of the figure), the situation is different—at least in the central part of the wake. As before, the change progresses laterally with distance downstream, and it is evident that the effect of the upstream boundary drag is to continue decelerating the outer portions of the flow with distance downstream. The region of deceleration hence continues to expand, but the velocity of the fluid at the axis now increases —at first rapidly and then more slowly—from zero toward its asymptotic limit: that of the originally undisturbed flow. If the distribution of K shown in the middle diagram is assumed to be that of B near the end of the plate and the integration of L and T there begins anew, the result seen in the final diagram will show that near the axis T is now greater than L in absolute magnitude. In other words, in the wake ΔB is negative, for K must gain through transfer from the outer region more rapidly than the local loss would in itself permit. Herein lies the indirect demonstration that ΔB can represent loss alone only if the Bernoulli equation is integrated over the entire region of shear so that T disappears.

The final example of the distribution of the Bernoulli terms is found in an experimental study of the hydraulic jump (10) in which the writer participated. Fig. 3 represents an evaluation of each of the terms in Eq. 13 from the date presented in the cited paper for the Froude number $F = 4$, the results being adjusted for purposes of simplicity to simulate irrotationality of the approaching flow and negligible bed shear. The plotted points are given symbols characteristic of alternate streamlines, so that all distribution curves can be superposed without losing clarity. Though the plot differs somewhat in nature (variation along rather than across the streamlines being shown, and the Bernoulli terms now representing the equivalent heads), the distribution functions for all but the zone of separation will be found to follow the trends discussed in connection with the previous examples. It is specifically the separation zone that introduces a variant of particular significance.

The notation for the first streamline above the bed is indicated in the last section of the grid at the right; it is perfectly normal. The third streamline,

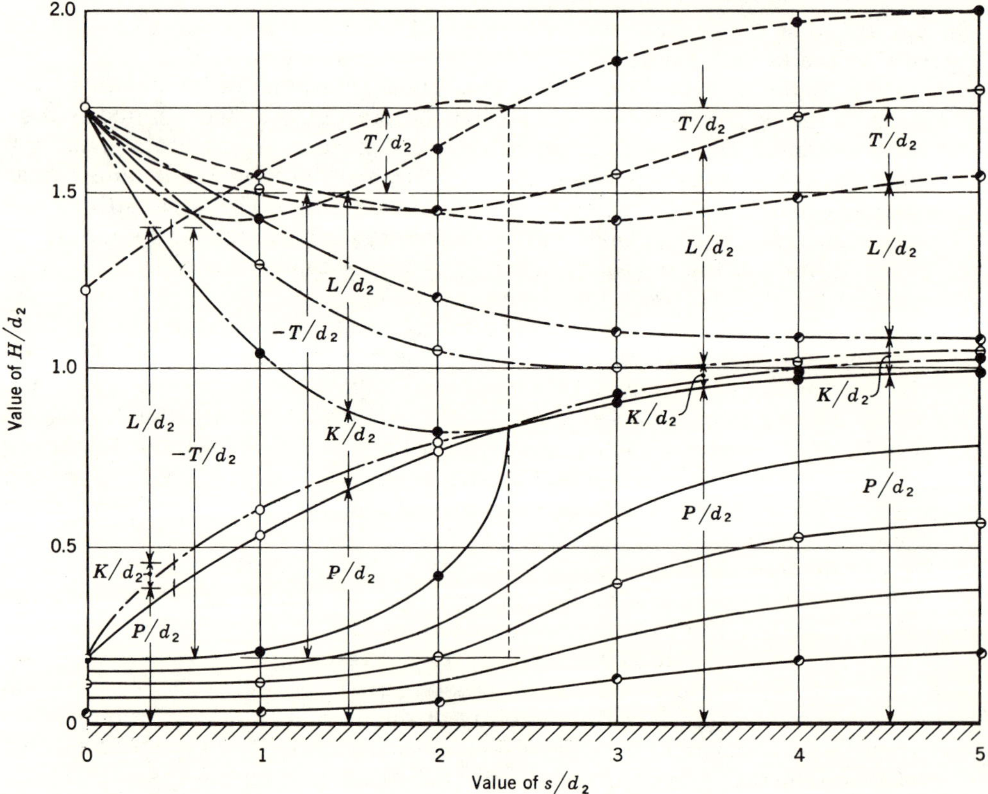

FIG. 3.—DETAILS OF THE MEAN FLOW IN THE MAIN STREAM AND THE ROLLER OF THE HYDRAULIC JUMP FOR **F** = 4

as shown by the notation in the second section from the right, is similar except that the transfer tends to become negative downstream, indicating that positive work must eventually be done upon the flow from below. The situation becomes the more pronounced as one proceeds to the uppermost streamline of the main stream, for the transfer is entirely negative beyond the stagnation point at the end of the separation zone, because of the work required to bring the fluid in this vicinity up to the mean velocity of the flow downstream.

Since the surface roller is not part of the main flow, the energy that is dissipated within it must likewise come through transfer from below. Moreover, since each of its family of streamlines (visible only if the number of divisions in the flow pattern is greatly increased) must be a closed curve, the Bernoulli sum along each one must remain constant through each circuit. However, whereas the total head of the main stream is initially $H_0 = P_0 + K_0$, the fact that the upstream end of the eddy is also a stagnation point of sorts reduces the total head of the outermost streamline of the roller to $(H_0)_r = P_0$. Because the streamline between the roller and the main flow is common to both, the magnitudes of P, K, and L will be the same, as shown in the second grid section from the left (Note that the elevation of the dashed line coincides with the grid line 1.5 by chance only.) On the other hand, because the main flow is providing the energy of the roller, the magnitude of T is positive for the former and negative for the latter, as illustrated, in accord with the difference in energy level. At the downstream end of the roller, the cumulative transfer reaches a maximum value for the roller streamline but a value of zero for that of the main flow, whereafter it remains negative as already discussed.

The free surface of the roller, the notation for which is seen in the first grid section at the left, is characterized by a low velocity, appreciable slope,

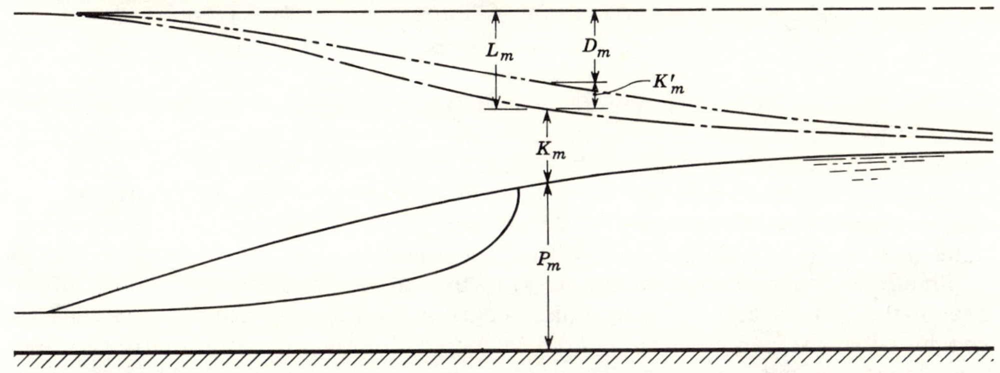

FIG. 4.—CORRECTED FORM OF FIG. 13, $F_1 = 4$, IN REF. 10

and continued increase in the loss term but gradual decrease in the transfer because of lack of direct contact with the main flow. The dashed line, for lack of sufficiently accurate measurements along the roller surface, has been drawn in qualitatively to satisfy these requirements.

If the plotted functions are integrated from bed to free surface in accordance with Eqs. 16 to 20, the magnitude of T_m will be very nearly zero at all sections; this is because the work done by the bed shear is zero, that done at the end sections is generally negligible, and that done by air effects on the free surface is relatively small. Unfortunately, the term was shown schematically in the cited reference as quite appreciable. Fig. 4 is hence presented herewith as a corrected version of the centermost sketch of Fig. 13 (10), the upper and lower sketches of the same figure being subject to similar corrections.

TURBULENCE DIFFUSION AND DISSIPATION

Further reference to Fig. 4 will disclose two terms not previously discussed: (1) The mean kinetic energy of the turbulence K'_m; and (2) the mean dissipation D_m, the sum of which is shown as the mean loss L_m. In the laminar flow of Fig. 1 the energy of the turbulence is zero, and the loss represents direct generation of heat through viscous shear. In Figs. 2 and 3, however, such viscous action is ignored in comparison with the generation of turbulence, the loss term representing at one and the same time the reduction of energy of (or work done by) the mean flow and the increase of energy of the turbulence. Though the generation of turbulence is a dissipative or one-way process, it does not represent true dissipation (i.e., generation of heat), for the latter is a viscous process whereas the production of turbulence at the expense of the mean flow is primarily inertial. The actual dissipation follows thereafter through viscous shear within the turbulent eddies.

There is also a turbulence counterpart of the mean-flow relationship of Eq. 8. It is derived (7) by introducing the Reynolds stresses, performing the operations leading to Eq. 12, and then taking means. It will be seen from the result that the turbulence part of the loss term of Eq. 12 is equivalent to

$$-\rho \int_{S_1}^{S_2} \frac{\overline{u_i' u_j'}}{\overline{V}} \frac{\partial \overline{V}}{\partial x_j}\, ds = \frac{\rho \overline{V_2'^2}}{2} - \frac{\rho \overline{V_1'^2}}{2} + \int_{S_1}^{S_2} \frac{u_j'}{\overline{V}} \frac{\partial}{\partial x_j}\left(\frac{\rho V'^2}{2}\right) ds$$

$$+ \int_{S_1}^{S_2} \frac{\overline{u_i' \frac{\partial p'}{\partial x_i}}}{\overline{V}}\, ds - \mu \int_{S_1}^{S_2} \frac{1}{\overline{V}} \frac{\partial}{\partial x_j}\overline{\left(u_i' \frac{\partial u_i'}{\partial x_j}\right)} ds + \mu \int_{S_1}^{S_2} \frac{1}{\overline{V}} \overline{\frac{\partial u_i'}{\partial x_j} \frac{\partial u_i'}{\partial x_j}}\, ds \quad (24)$$

In other words, the turbulence generated at the expense of the mean flow (the term at the left) is: (1) convected along the streamline by the mean flow (the first two terms on the right); (2) diffused in all directions by the turbulent mixing (the third term); (3) worked upon by the pressure fluctuations (the fourth term) and the viscous stresses within the eddies (the fifth); and (4) caused to decay by viscous shear (the last term).

Since the fourth and fifth terms are very difficult to measure, it is often assumed that they are mutually compensative, and an approximate balance is sought between the production, convection, diffusion, and dissipation of the turbulence without regard to the work otherwise done within the eddies. The plots of the mean Bernoulli terms presented herewith may thus be supplemented by plots showing the corresponding interplay of these four quantities, or the loss term L in Eqs. 12 and 13 may simply be replaced by the sum of turbulent convection C, diffusion by mixing M, and dissipation D to yield

$$B = K + P + T + C + M + D = \text{const}_s \quad \dots\dots\dots\dots\dots\dots \quad (25)$$

quite without regard to the work done by pressure fluctuations and viscous stresses within the eddies.

Though only in this manner can the complete mechanism of flow be investigated in detail, the fact remains that such investigations are extremely laborious and complex. Fortunately, in many cases it is sufficient to follow the energy-conversion process through only the turbulence generation, since the subsequent viscous dissipation generally has little or no effect upon the mean flow pattern. For the time being, therefore, Eqs. 13 and 20 should suffice for the analysis of elementary flow phenomena as illustrated in the four examples.

CONCLUSIONS

1. The equation of energy for shear-free flow may be generalized through

incorporation of the Saint-Venant terms for the work done by tangential as well as normal stresses.

2. Although the resulting equation is conservative, a dissipative term may be introduced by substituting for the work done conservatively the difference between the total work and that done dissipatively.

3. In the case of laminar flow, the usual Bernoulli terms for the streamline become augmented by two: one involving the energy transferred to (or work done upon) the neighboring fluid by viscous shear, and the other the viscous generation of heat.

4. For the mean pattern of turbulent flow, the Reynolds stresses can be used to form two comparable terms; however, the dissipation term now represents the production of turbulence rather than the immediate generation of heat.

5. By means of the transfer mechanism, energy may be dissipated in a given zone at a greater rate than it is lost locally by the flow.

6. In the wake behind a body, energy will be gained through transfer more rapidly than it is dissipated.

7. Since the energy of a standing eddy is supplied entirely by transfer from the main flow, the Bernoulli sum has a much smaller magnitude within the eddy, and the loss and transfer terms undergo no net change around each closed streamline.

8. Though viscous dissipation within the turbulence structure can be described mathematically, for the present the energy transferred from the mean flow to the turbulence may simply be regarded as irrecoverably lost.

ACKNOWLEDGMENT

The computations involved in the analysis of the hydraulic jump were made by Kanwar Sain Satija with the support of the Iowa Institute of Hydraulic Research.

APPENDIX I.—REFERENCES

---

1. Bernoulli, D., *Hydrodynamica*, Dulsecker, Strasbourg, 1738. (See translation by T. Carmody and H. Kobus, Dover Publications, New York, 1968).

2. Bernoulli, J., *Hydraulica*, Bousquet, Lausanne and Geneva, 1743. (See the Carmody-Kobus translation, Dover, 1968.)

3. Chevray, R., "L'application du théorème de Bernoulli dans les zones de décollement," *La Houille Blanche*, No. 6, 1964.

4. Euler, L., "Principes généraux de l'état d'équilibre des fluides"; "Principes généraux du mouvement des fluides"; "Continuation des recherches sur la théorie du mouvement des fluides"; Histoire de l'Académie de Berlin, 1753–55. (See H. Rouse and S. Ince's *History of Hydraulics*, Dover, 1963.)

5. Kovasznay, L. S. G., and Chevray, R., "Turbulence Measurements in the Wake of a Thin Flat Plate," *AIAA Journal*, Vol. 7, No. 8, 1969.

6. Rouse, H., "Energy Transformation in Zones of Separation," General Lecture, *Ninth Convention IAHR*, Dubrovnik, 1961.

7. Rouse, H., "On the Bernoulli Theorem for Turbulent Flow," *Miszellaneen der Angewandten*

Mechanik (Tollmien Anniversary Volume), Akademieverlag, Berlin, 1962.

8. Rouse, H., "The Bernoulli Theorem," *Journal, Japan Society of Mechanical Engineers,* Vol. 68, No. 562, 1965.

9. Rouse, H., and Kalinske, A. A., discussion of Bakhmeteff and Allan's "The Mechanism of Energy Loss in Fluid Friction," *Trans. ASCE,* Vol. 111, 1946.

10. Rouse, H., Siao, T. T., and Nagaratnam, S., "Turbulence Characteristics of the Hydraulic Jump," *Transactions ASCE,* Vol. 124, 1959.

11. Saint-Venant, B. de, "Note à joindre au Mémoire sur la dynamique des fluides," *Comptes Rendus,* Vol. 17, 1843.

APPENDIX II.—NOTATION

The following symbols are used in this paper:

A = cross-sectional area;
B = Bernoulli sum;
C = turbulent convection;
D = dissipation;
d = depth;
F = Froude number;
H = total head;
i, j = tensor subscripts;
K = kinetic energy;
L = energy loss;
M = diffusion due to mixing;
P = piezometric pressure;
p = intensity of pressure;
Q = rate of flow;
s = tangential coordinate;
T = energy transfer;
u = velocity magnitude;
V = vector magnitude of velocity;
x, y = cartesian coordinates;
α = Coriolis coefficient;
μ = dynamic viscosity;
ν = kinematic viscosity;
ρ = mass density;
τ = intensity of shear; and
Ω = gravitational potential.

Mean and instantaneous values are indicated by bars and primes, respectively.

International Heritage

Hunter Rouse

After a century and a half of progress in American engineering education, one tends to regard it as a more or less self-contained development. The sixty-nation UNESCO Conference on Trends in the Teaching of Engineers, held at Paris in December 1968, was a relatively new experience for those who represented the United States; the fact that they seemed in a position to contribute more than they acquired did nothing to rid them of their rather parochial point of view. In actuality, many of America's steps forward have been paralleled and even preceded by those of other countries, yet America's effect upon both older and younger parts of the world has been considerable. Hence, tracing the respective national courses and comparing their influence on one another should be of at least passing interest, and perhaps even salutary in its effect upon the future.

Engineering education of a sort presumably existed in the East or Middle East while the American continent was still relatively uninhabited; traces of ancient civilizations go back many thousands of years, and experience gained in any generation must have been transmitted to the next in one way or another. Much the same informal process of transmission must have occurred in the Inca and Aztec civilizations long before the Europeans reached America. Possible connection between the pyramid builders and irrigation experts of such different localities and epochs is still a moot question. That there was continuity of engineering knowledge, if only rules of thumb passed along by example and word of mouth, is certain. However poorly suited the procedure may have been to mass training, its individual effectiveness is not to be denied; even today, those engineers are lucky indeed who have learned through close association with past masters.

France: Analysis for the Elite

Formal education in engineering is generally agreed to have been established barely two centuries ago by the French, who placed engineering itself on an organized footing. The *Corps des Ingénieurs*, a military subdivision, was founded by Sébastien de Vauban in 1690. Twenty-six years later Jean Perronet established its civil engineering counterpart, the *Corps des Ponts et Chaussées*, and by 1747 Perronet had developed the first engineering school, the *Ecole Nationale des Ponts et Chaussées*, which still exists. This was followed a quarter century later by the *Ecole Nationale Supérieure des Mines*, and in the midst of the Revolution by the *Ecole des Travaux Publics*, a creation of the National Convention. In 1895 the latter school was renamed *Ecole Polytechnique*; such men as Ampère, Carnot, Fourier, Lagrange, and Laplace were on its early staff. It has since become a world-renowned preparatory college in mathematics and science for many of the top professional schools, both civil and military. Similarly, the *Ecole Centrale des Arts et Manufactures*, privately founded in 1829, grew over the years to be the most comprehensive federal engineering institute in the country.

Quite aside from its temporal leadership in establishing engineering education as a formal operation, France deserves the professional world's gratitude on two related counts. First, technical education was given a sound analytical foundation. Second, the engineering profession itself was made attractive, not only to the middle class, but above all to the intellectual elite. The influence of France on surrounding countries was profound; even the United States sent over at least one early observer and brought at least one French authority to the States. Under Napoleon, however, French engineering began to lose its freedom of development, and during the 19th century tended primarily to intensify its analytical approach.

In 1802 the United States Military Academy was founded at West Point under the Corps of Engineers, the American counterpart of the *Corps des Ingénieurs*. The academy was established for three reasons: military experts still had to be imported from Europe; it was believed that more than field experience was needed by engineering officers; and existing colleges were not considered to yield a product conducive to the development of commerce and industry. It is thus noteworthy that when Sylvanus Thayer became superintendent of West Point in 1817, he did not turn for guidance to the *Ecole Spéciale Militaire* (founded by Napoleon at Fontainebleau in 1802 and moved to St. Cyr near Versailles in 1808), but to the *Ecole Polytechnique*.

Germany: Institutes and Complexes

Although France had supplemented its emphasis on top-level engineering education by the formal training of artisans and technical assistants as early as 1788, it was in Germany that the latter class of worker received primary attention. The first *Bergakademie* or school of mining practice was opened at Freiberg, Saxony, in 1766, and the *Bauakademie* for structural practice in 1799 at Berlin. Though the following century saw immense strides in Germanic secondary (*Gymnasium*) and university-level education, for many years these had little effect on the teaching of engineering. Science was stressed in the universities, to be sure, but it was pure rather than applied. The practice academies gradually gave rise to polytechnic schools patterned on the French—in Prague (1806), Vienna (1815), Darmstadt (1822), Munich (1823), and Karlsruhe (1825); the *Oberrealschule* became the strong

Dr. Rouse is dean of the College of Engineering at the University of Iowa. He was a representative of the United States at the UNESCO Conference on Trends in the Teaching of Engineers in December 1968.

secondary-level counterpart of the *Gymnasium;* and the system of ten *Technischen Hochschulen* which developed out of the earlier polytechnic schools became world renowned. Nevertheless, the latter were purposely not accorded university status, and traces of the resulting inferiority complex inherited by most engineers and engineering educators of the intervening period are still in evidence. It was not until 1899 that the first *Technische Hochschule,* that at Charlottenburg, was granted the authority to award the doctoral degree.

However, Germany made at least three distinct contributions to engineering education: great strength of schooling at the preparatory level; emphasis on training in large numbers at the technical institute level; and restriction of top-level education to special institutes of technology for the elite few who were expected to become the creative leaders. A year before the founding of the Karlsruhe institution, a somewhat similar one came into being at Troy, New York: Rensselaer Polytechnic Institute; in 1849 a definite effort was made to pattern its curriculum after that of the Parisian *Ecole Centrale.* As other polytechnic institutes developed in the States, their undergraduate laboratories also grew to be appreciated in Europe; these are believed to have had considerable influence upon German practice. As in France, however, educational research in America remained almost nonexistent well into the twentieth century. In Germany, on the other hand, the professor was definitely of the feudal-lord type and tended to gather around himself a group of advanced workers who assisted him in instruction and research. As a result, the country eventually came to be the same sort of mecca for graduate engineers from all over the world as it had already become for university scientists.

England: Engineering in the University

In England the social status of the technical man was originally even lower than in Germany. Of the three professions acceptable to a gentleman—theology, law, and medicine—only the first received particular attention in the eighteenth-century universities of Oxford and Cambridge. Little science was taught even in medicine, and most scientific achievement occurred outside the university. Toward the end of the century, mechanics institutes were established in the larger cities as an outgrowth of the Industrial Revolution, but these were purely for craftsmen. In 1826, however, dissenters from the Oxbridge classical type of education founded the University of London on the German university pattern, and the teaching of science was gradually strengthened. The University of Glasgow established the first chair of engineering in 1840; University College, London, established its first a year thereafter. The pace increased in the second half of the century, when the British became aware that they were falling behind the continent in industry. Although there were then over 200 mechanics institutes for artisans, a series of civic university colleges was founded; these, with the University of London, assumed the task of training industrial managers as well.

Still, such education—not for culture, but for practical use—was by no means for gentlemen. Though the training of engineers was distinctly second class, as in Germany, the process of its development had made it a responsibility of universities rather than of polytechnic institutes. Even Cambridge established a chair of engineering in 1875, though its technical curriculum was to grow only slowly; Oxford never really followed suit. Whereas the German universities and polytechnics were closely related governmentally, each in its own class, the English universities remained relatively independent of one another. All catered to the few rather than the many, leaving the training of technicians to the trade schools and technical institutes.

The United States: Mass Education

In the United States a series of military schools and polytechnic institutes began to form at much the same time as those in Europe; in 1847 Harvard, through its Lawrence Scientific School, became the first American university to teach engineering. Five years later both Yale and Michigan also offered engineering courses; other universities, both private and public, soon followed suit. Thus there developed in America a close parallel to the English practice of surrounding engineering students with those in liberal arts and sister professions. What first really distinguished American from European technical education was the passage of the Morrill Land Grant Act in 1862 for the specific purpose of expanding efforts in the technological direction. Within ten years 70 engineering colleges had been established, with another 15 in the following decade. Engineering instruction in the land-grant colleges and in the state universities gave American technical education the mark it has borne ever since: attention to the masses rather than to the elite. Engineering students tended to come from blue-collar families; the profession thus acquired somewhat the same complex as in Germany and England relative to medicine and law, signs of which are still evident. Postgraduate study and research were slow to develop, moreover, and little distinction was made between training for routine performance and preparation for leadership; hence lack of technical institutes was long offset by vast productivity of baccalaurate engineers by institutions at many levels. Not till after the mid-twentieth century did the European separation of technologists and professional engineers become widely reflected in American practice; at the same time, the American emphasis on university education for the many rather than the few began to be adopted in other parts of the world.

The Soviet Union: Mass Specialization

Soviet policies in engineering education are of interest less for their past effect upon the world than for their contrast and possibility of influence in years to come. In Czarist Russia the schooling was sound and selective (witness such products as Boris Bakhmeteff and Stephen Timoshenko), yet the best of its graduates had to turn to Germany, France, and Switzerland for advanced study. Under socialist control the educational system has expanded tremendously in scope and productivity, as nearly as practicable in accord with the idealistic slogan, "Everyone in the university who should be; no one who should not." Technical education is seldom provided in the universities, but in highly specialized institutes; for example, there are institutes which teach only civil engineering. In the five to six years required for the diploma (following 10 to 11 years of primary and secondary schooling), the student is well grounded in theory and is given practical training in his special field, followed by a year in industry. Advanced degrees comparable to our doctorate are obtainable in a commensurate time. The true Soviet doctorate, however, is awarded only after many more years of mature productivity. Whereas some research is said to be conducted in the educational institutions, the Soviet system of Science-Academy sponsorship of research in its own laboratories makes this quite different from Western practice. Soviet influence on the rest of the world cannot be considered minimal, as the half dozen principal satellite countries are almost carbon copies scholastically; even mainland China still reflects Soviet practice. In the West its effect has also been great, if only in the sudden upsurge of secondary education produced by the voyage of Sputnik I. Moreover, not only do the Soviets currently produce four times as many engineers per year as the States, but there seems to be no comparable decrease in their production rate. Interestingly enough, they are now giving at least lip service to breadth as well as depth of education, including increased emphasis on the socio-humanistic fields.

The Spread of Student Unrest

Latin American influence upon our teaching would be of little import were it not for the *Reforma,* a movement which began among students and young professors in Córdoba, Argentina, in 1918, just two years after their first one-man/one-vote elected government. Its object was to modernize the extremely conservative educational system then typical of Latin ecclesiastical control. Though the situation in Córdoba was not greatly changed thereby, the drive toward reform was soon extended to La Plata and Buenos Aires. Student participation in university affairs was gradually forced upon the rather impotent administrative bodies, and the students gained not only representation in governance but reduction in costs and the right to new courses parallel to any old ones to which they might object. With the years, the movement spread to other countries—Peru, Venezuela, and Mexico in particular—but at the same time degenerated into misuse by politicians for their own purposes. Student unrest has not diminished, however, and many semesters of instruction continue to be lost by general strike or simple boycott of classes.

Probably more than one North American educator can recall looking southward with disdain until colleges in this country began to display similar student discontent. Military, racial, and other overtones are present here, but pedagogical complaints provide at least the nominal focus of the disturbances. Uprisings in France, Germany, and England are similarly oriented. Perhaps the most severe of the series are those in Japan. At the turn of the century, Japan's university practices were based upon the German feudal system, but under MacArthur the American plan was imposed —without changing the professorial lord-like attitude.

Overtones of Mao, Okinawa, and anti-militarism are evident, to be sure, and the percentage of radicals in the student bodies is probably as low as in the States, but the result is frightening. University buildings are barricaded against assault, professors stand guard duty in their laboratories, and damage is widespread and formidable. The faculties, however, seem obsessed with a spirit of self-reproach rather than exerting any real effort to get at the root of the problem.

Faculties in American engineering colleges pride themselves on the minimal participation of their students in disruptive action. In fact, engineering students are sometimes accused, even by their liberal-arts wives, of scarcely being aware of what is going on around them. How often these or similar anti-societal reproaches have been directed toward engineers! In South America and Japan, on the contrary, engineering students participate fully in the disruptions; their faculties do not hesitate to predict the eventual downfall of their American counterparts by the same process. But is such downfall inevitable? Has emphasis upon socio-humanistic awareness been purely lip service? If so, is it too late to implement the programs about mankind that have been recommended by one ASEE report after another for the past several decades?

American Legacy?

Fortunately, there are small strongholds of awareness in colleges scattered about the country; in many of these localities a start had been made on student participation in college affairs well before unrest brought undue pressure to bear. There are surely limits to realism in such participation, the more radical factions to the contrary, for wisdom and experience do not develop spontaneously as one reaches the age of twenty-one. Like creativity, such matters as tolerance, justice, and understanding of how to live with one's fellow men can be stimulated or stifled. Indeed, it is essential to the education of the engineer that he discover in college what he should appreciate throughout his professional life: the fact that human forces and materials are perhaps the most important of any with which he will have to deal. In addition to extracurricular activities, membership on college committees for which a student's experience has equipped him should be encouraged—and for more than altruistic reasons, as his contributions will not be negligible. Finally, engineers should somehow be persuaded to play a larger role in the life of the university as a whole, particularly in the face of potential disruption, to the end that they may continue to participate effectively in the advancement of mankind.

The innate professional conservatism of engineers is surely not conducive to such action. Nevertheless, the time for action is now especially propitious. On the one hand, the American engineering student is not committed in any fringe direction but is ripe for the acceptance of a progressive yet stabilizing role. On the other hand, international pedagogy is currently looking to the United States for further guidance in the principles of mass education. What better legacy could American engineering educators bestow upon the world? △

Pierre Danel's influence on American hydraulics

MY assigned topic in this well-deserved eulogy of Pierre DANEL might more aptly be worded to highlight his influence on American hydraulicians, for the latter action was as direct and strong as the former was indirect and difficult to pinpoint. But influence us he did, whether by personal charm, perceptive criticism, warmth of hospitality, competitive keenness, brilliance in conversation, or simply purposeful and provocative exaggeration of any point he wished to make. He once told me that he pursued magic as a hobby mainly because of the psychological study of people that it permitted. I am sure that this also played a role in his approach to hydraulics and hydraulicians, particularly those of our country, for he caused lots of healthy thinking—even among those who disagreed with him.

DANEL was a frequent visitor to the States. In the course of his initial visit in 1936, he met Martin A. Mason, who was then on the staff at the National Bureau of Standards, and eventually persuaded him to spend his fifteen months as a Freeman Scholar (1937-38) at Grenoble. There Mason divided his time between the University and Neyrpic (at both of which DANEL was then employed), did research toward and defended his thesis for the engineering doctorate, assisted DANEL and his colleagues in matters of English and American language and contacts, and met many French hydraulicians including Louis and Paul Bergeron.

The elder Bergeron accompanied DANEL in 1937 on one of his many tours of American laboratories and design offices. I can well recall chatting with them both on the sloping platform of Arthur T. Ippen's high-velocity-flow facility during their visit to Robert T. Knapp's laboratories at Caltech in Pasadena. World War II soon intervened, and it was not till a post-war inspection of still-active European establishments in 1946 that I was able to visit him at Grenoble. Socially my reception was a very cordial one, and professionally the tour of the Neyrpic experimental plant was most striking—not, perhaps, for the beauty of the equipment, but surely for the breadth of activity and competence of the staff. In fact, the entire situation there impressed me so favorably that Grenoble was one of three places that I recommended to prospective Freeman Scholars in a post-trip letter (February 1947) to the editor of *Civil Engineering*.

The second Freeman Scholar, George S. Dixon, Jr., of the Corps of Engineers, actually arrived in Grenoble that fall and had received the engineering doctorate by 1949. My enthusiasm next led my colleague, John S. McNown, to spend his year as Fulbright Research Scholar with DANEL in 1950-51; there he wrote a comprehensive paper on fall velocity, did research on seiche, and passed the examination for a doctorate in sciences. During McNown's absence from the Iowa Institute of Hydraulic Research, his chair was filled for a year by Antoine Craya of Grenoble, who later returned to the States to teach at Columbia after the death of Boris A. Bakhmeteff. McNown was followed immediately at Grenoble by Charles W. Thomas, of the Bureau of Reclamation, also a Fulbrighter, as was I in 1952-53. At Grenoble I met Enzo O. Macagno, of Argentina, who was studying there for the doctorate, and in 1956 he moved from La Plata to Iowa City as a member of the Institute staff. My Fulbright successor at Grenoble was Maurice L. Albertson, of Colorado State College, who was the fourth to receive the local doctorate. Ira A. Hunt, of the Corps of Engineers, was the third Freeman Scholar, in 1953-54, and the fifth doctor. The fourth Freeman Scholar at Grenoble, in 1961-62, was Jacques W. Delleur, of Purdue University (and previously one of Craya's students at Columbia). Several other Americans were there for a month or more, and literally hundreds must have visited Neyrpic, and more recently SOGREAH, at one time or another.

DANEL's hospitality to all of these visitors—and, in fact, to any professional friend who came to see him—took many delightful forms. Some of us had Neyrpic cars placed at our continuous disposal, and some even complete apartments. I well remember the cabin that he secured for me on a Turkish steamer bound for the Applied Mechanics Congress at Istanbul, after all other efforts to arrange passage had failed, not to mention a mountain luncheon in which his family and mine consumed twelve dozen snails among the eight of us. I know only by hearsay that he helped Knapp and McNown celebrate Ippen's fiftieth birthday at Sette Ais, Portugal, thirteen years ago, but the tales of the dinner party still echo. The Neyrpic experimental

Il est curieux par contre de constater le mépris et l'ironie ~~de certains~~ de certains théoriciens pour le "Try and see" des américains. Méthode qui demande sans doute plus de bon sens que la résolution d'un système d'équations différentielles.

Le seul moyen de progresser ~~[struck out]~~ est d'interpréter par une théorie, même rudimentaire, les résultats expérimentaux et, guidé par cette théorie, voir dans quel sens il y a lieu de poursuivre les expériences.

facilities were ever at the disposal of those visitors doing their own research, with DANEL's inexhaustible flow of ideas and constructive criticism thrown in. As with his own staff, he preached a judicious blend of theory and experiment, and hammered without mercy at any practice (i.e., American) which tended to separate the two. The beautifully managed Neyrpic library was without peer in the fluids world, and he welcomed its use by the visitor as much as he insisted upon it by his own staff. I had a desk therein for a dozen months, and the result has long since taken a tangible form of appreciation in the dedication to *History of Hydraulics:*

To

PIERRE DANEL

HYDRAULICIAN OF MANY INTERESTS

NOT THE LEAST OF WHICH IS

THE HISTORY OF HIS

PROFESSION

Those who knew DANEL in his own surroundings undoubtedly saw a quite different side from those who received him during his trips through the States. At home he was a lavish host, yet on the road a business man who seemed to have the knack of giving little as he acquired much. I remember telling him once that he was gaining the reputation among Americans of leaving nothing but promises in exchange for the literature and information that were freely afforded him on every hand. This statement (surely more American than French in its bluntness) may well have influenced his subsequent reception of our travellers, or even his publication of *La Houille Blanche* in English as well as French, for Neyrpic was assuredly different enough from a university or governmental establishment to hamper an otherwise completely free exchange. In any event, on his return to France he sent me a copy of Boussinesq's 1877 "Essai sur la théorie des eaux courantes" and what is apparently a life subscription to the magazine. Aside from the bibliographical index cards and laboratory instruments which have a limited sale in the States, perhaps the most tangible mark of his company's influence on this country's hydraulic engineering is found in its ubiquitous tetrapod—the development of which, rumor has it, was prompted by his observation on one visit or another of the vast American need for better shore protection.

No American who ever chatted with DANEL could have failed to be impressed by the strength of his accent and the wealth of his knowledge about esoteric matters normally known only to our native-born. I firmly believe that he held to his accent as assiduously as he sought out the fine points of our culture, if only for the effect that each had in breaking down the artifical barriers that too often obstruct the free flow of ideas. To some extent he did his country a disservice by his insistence on speaking English himself on every possible occasion and by the premium he placed on the learning of English (and other foreign languages) by members of his staff. For a foreigner to speak French to him and receive a French reply was no mean accomplishment, and those of us who finally succeeded felt complimented beyond measure.

Pierre DANEL's greatest influence on the most Americans definitely came from his conviction that the strong German-American liaison in hydraulic research of pre-war days should give way to a post-war Franco-American entente, and he accomplished it almost single-handedly. *La Houille Blanche* is perhaps the continuing symbol of this effort, but the actual vehicle was the now-flourishing International Association for Hydraulic Research. Formed in 1935 largely on German initiative (18 of the 63 charter members were German, 8 American, and only 3 French), the IAHR held but two meetings, at Berlin in 1937 and Liège in 1939, before disruption by the war. With charter members J. Thijsse of Holland and L. G. Straub of the States, DANEL rejuvenated the organization in the post-war years, and the presidency has tended to alternate between the French and the Americans (Straub, DANEL, Ippen, L. Escande, and now J.W. Daily). The 1969 IAHR Register of Members indicates that 636 of the total enrollment of 1,573 are from the Americas, and 348 of these from the United States. For that we can be grateful to Pierre DANEL's initiative, enthusiasm, and originality of belief in what could be accomplished through international collaboration.

Hunter ROUSE.

599

Growth ≠ Progress?

Traffic control, food and drug regulations, laws against murder—these are needed to maintain the cohesion of our social fabric. Add another: A law penalizing excessive births. Otherwise the "population bomb" will go off and smother our best efforts to reverse the tide of pollution. Guiding principle for the future: Growth, technologically and otherwise, of course—but in a qualitative, rather than a quantitative, way.

HUNTER ROUSE[1]

The University of Iowa, Iowa City, Iowa

OUR WORLD would indeed be a different place had science and technology never been applied to its advancement. Whether or not it would be a better place is the question. There are probably few who would willingly do without the comforts of civilization, even were this the only means whereby its discomforts could be eliminated. Nevertheless, many are seeking to allocate responsibility for the degradation of our air, water, and soil, whether to the end of removing the cause or simply of fixing on a scapegoat. The sociologist, for example, is vociferously calling the engineer to task: Technology is the engineer's handiwork, and he has purportedly loosed it upon the world without thought for its consequences. This claim that engineers are not to be trusted with their product naturally calls to mind the equally unrealistic proposal that a moratorium be declared on new technology till man is sociologically mature enough to use it intelligently. In a similar vein, the ecologist recommends that no further technological change be introduced until man is certain of its ultimate consequence. The technologist responds, of course, that being certain about any future action is impossible, particularly when there is no precedent at hand. To wait for a guarantee of performance (or a guarantee of man's reaction to it) before undertaking a project would bring progress to a halt. All of life involves risk; it can be minimized by proper forethought, but never wholly eliminated.

Too Many People

If a scapegoat for the degradation of our world must be named, it is man himself: not the engineer any more than the biologist, ecologist, or sociologist, but

[1] Dean, College of Engineering. Hon. Mem. ASME.
Based on a paper presented under the title "Technology, Growth, and Responsibility" at the Conference on Engineering and the Future of Man held in July 1970 at Chania, Crete, by the International Science Foundation.

every user—or misuser—of a technical product, be it a cigarette, a city, or a supersonic transport. Man is, moreover, not only a polluter but by his very ubiquity a form of contamination. As stated succinctly by Paul Ehrlich in *The Population Bomb*, "Too many cars, too many factories, too much detergent, too much pesticide, multiplying contrails, inadequate sewage treatment plants, too little water, too much carbon dioxide—all can be traced to *too many people*." Indeed, not only are there too many, but their number is increasing at an ever-greater rate.

The rate of population growth depends, of course, upon two essentially independent factors: the birth rate, and the death rate. For millenia the birth rate was wholly without control, for lack not only of contraceptive methods, but even of an understanding of the conception process itself; in most regions, moreover, large families came to be considered desirable for one reason or another: security in old age, religion, fatherly pride, or stigmas associated with barrenness. The death rate, on the contrary, always had controls of a sort, just as in most natural communities of plants or animals. Excessive concentrations of population led intermittently to plague and famine, with sometimes extreme changes in numbers of people left alive. Although hardly intended as population controls, wars had a like effect (offset, to be sure, in the periods of truce immediately following). Relevant in a social sense, though of little weight numerically, is the practice among such peoples as the Eskimos, the Australian aborigines, and peasant families in many lands of terminating the lives of those no longer able to care for themselves because of health or age. Of pronounced contraventive effect, on the other hand, was the role played by the gradual improvement in economic, sanitary, and medical conditions as the centuries passed.

However much the birth and death rates may have changed from one generation to another through such causes, over the last centuries their difference has tended to follow an exponential function, the exponent of which has itself varied exponentially and with very nearly the same doubling period. If one estimates from recorded trends by this rule, at the end of the first millenium A.D. the population of the world would have been about a quarter of a billion and its annual rate of increase about 0.1 percent. Toward the end of the Renaissance both figures would have approximately doubled. By 1830, in the midst of the Industrial Revolution, both would again have doubled, the population according to United Nations studies then being a full billion and the growth rate 0.5 percent (the difference between a birth rate of 3.5 percent and a death rate of 3 percent). A century later the population had reached two billion and the growth rate 1 percent, the death rate having dropped more than the birth rate because of improving conditions. Today the growth rate is some 2 percent and the population is nearing four billion.

The population trend is obviously such that, as emphasized by the National Academy of Sciences, "either the birth rate of the world must come down or the death rate must go back up," whether for reasons of food, health, or simply adequate space to live. Furthermore, the situation varies in gravity from one part of the world to another. In Western Europe both the birth and death rates have dropped markedly in the past century. In Latin America, on the contrary, although the death rate has dropped with that of the rest of the world, the birth rate has not; there the rate of population growth is nearly 3 percent. According to the document "The Population Explosion—a *Present* Danger" assembled for presidential briefing, "The already existing burden of population growth is very great. It actually consumes about two-thirds of the gains in economic growth in the less-developed countries." Not only does a country like India have a productivity per capita only one-fiftieth that of America, but the discrepancy is growing steadily greater. Inside a country, moreover, it is the illiterate poor who show the highest rate of increase, and the educated and well-to-do who show the lowest; the income gap between them thus continues to widen. The parallel with other aspects of pollution is clear: The situation is in good part the result of technological improvement in standard of living; its cure is at hand through contraception, legalized abortion, and sterilization; but it will require human willingness and money to implement.

Two Alternatives

What course civilization takes in the future will lie between the following two extremes. It can continue under the condition of laissez-faire which has prevailed up to the recent past. Without controls any state of pollution tends to grow steadily worse until the local residents become so irritated that they seek by one means or another to remedy the situation. At the moment, the American public is sufficiently incensed against certain types of contamination that nation-wide regulation is in the process of development. Other types, however, are still far from receiving comparable attention. Moreover, it is unpredictable how long the resulting state of arousal will remain effective, and there is no telling whether it would ever pervade the world as a whole and eventually encompass all forms of pollution. In any event, the laissez-faire policy can lead only to continual degradation of the earth; the fact that the engineer could probably continue to keep it habitable almost indefinitely is small justification for permitting this to occur. The second extreme course would proceed from the realization that overpopulation is the primary form of pollution and that other forms will not be amenable to proper control until the population itself becomes stabilized. A start in this direction has already been made in America on the premise that stabilization would be accomplished once and for all if essentially every family limited the number of its offspring to two. Since the practice would have to be absolute and world-wide to be effective, it is obvious that some method of international control never yet realized would have to prevail. Once it did, of course, other aspects of pollution should not long persist, and efforts to upgrade mankind as a whole would certainly follow.

Is Growth Good?

Which of the two extremes is approached by the course actually taken will depend largely upon a single factor: the extent to which it is realized that minimizing the pollution of air, water, and land by man-made contaminants will not be sufficient if man himself continues to multiply at an ever-increasing rate. J. A. Wagner wrote in *Science* of 5 June, "The essential steps ahead are to stabilize human population levels and to learn to recycle as much of our material abundance as possible. . . Unless we can slow the treadmill on which we have been running faster and faster, we may stumble—and find ourselves flung irretrievably into disaster." Many still scoff, of course, at the furor raised by others over the eventual tragedy of the population explosion, maintaining that there will never (i.e., in their lifetime) be cause for alarm. Such contentions lead with increasing frequency to the question, is growth really necessary, or even good? What, indeed, is actually gained by overflowing one neighborhood after another, by proliferation of business and industry, by destruction of a succession of irreplaceable natural resources? Are expansion and improvement truly synonymous? The French claim that a certain amount of growth is essential to the health of a nation regardless of its size, attributing their own economic malaise to their low growth rate. Yet others hold that the moratorium which has been suggested should not be directed against technology but against growth itself—and for an indefinite period.

Optimizing Population Size

Since the earth and its resources are finite in extent, the argument has long been heard that an optimum size of population must exist for each nation and hence for the world as a whole. Such optimum conditions are reached when the criteria that are governed by population attain a maximum (or, conceivably, a minimum) value. Typical criteria are economic welfare, power per capita, health, defensive strength, international harmony, and so on. Evidently, any marked increase or decrease in population from its optimum size will cause the criterion to be reduced. In view of the foregoing discussion, an explicit restriction should now be added: Whatever criterion or combination of criteria is utilized, the optimum state must be attainable without net degradation of the eco-system.

The optimization process has already proved its worth in the engineering analysis of material systems, and it should be readily adaptable to the population problem regardless of the criteria that are involved, once they have been agreed upon. Criteria appropriate for one country and one time may well differ from those for another; some may lead to results that parallel present trends, and some may indicate limits that already have been exceeded. For example, the United States has an average population of about 60 people per square mile; that of the Soviet Union (including Siberia) is only half as great; but China and India have densities that are, respectively, five and seven times this figure, while those of Japan and Holland are fully eleven and sixteen times greater. An optimum state suitable to one might well prove disastrous for another.

Countries which already sense the necessity of planning their population trends before overcrowding has become widespread will surely have less difficulty with the optimization approach than those which do not. Regulation by taxation, subsidy, or reduction of relief would probably have to be imposed upon the recalcitrant. To the question "What right has the government to dictate the number of our children?" there is the obvious answer "The same as the right to regulate the number of your spouses, or the number of your votes in an election, or the number of miles per hour you may drive in urban traffic." Why, indeed, should natural birth not be as thoroughly subject to legal control as is adoption? Once a working model existed, at least a few other countries would surely follow suit in instituting controls, with the expectation of further influence by example after the effectiveness of the procedure had been demonstrated. It goes without saying that migration from countries without to countries with control would have to be prohibited.

What degree of international regulation could be imposed would depend upon how fully the seriousness of the situation became realized, and how persuasive the key countries could be in convincing the others of the fairness of their intentions. The Committee on World Food Supply of the President's Scientific Advisory Board has warned that "Unless the rate of population increase can be sharply diminished, all the efforts to augment agricultural production will merely postpone the time of mass starvation, and increase its agony when it inevitably occurs." Provision of food in return for cooperation would thus be in order, just as would withholding aid for lack of it. Instituting an embargo against uncooperative nations would be the ultimate weapon short of war. Even the latter should not be ruled out, though considered only as a defensive measure against agression, for a country that neither controls its population growth nor possesses sufficient quantities of food is a potential danger in any circumstances. Starvation, quarantine, and war are not matters to be taken lightly, but all are far less serious than permanently damaging a portion of the earth through overpopulation or any other form of contamination.

Were the population problem thus to be brought under control in accordance with the capacity of the earth to maintain it, elimination of the other kinds of pollution would be easy in comparison: The necessary technology is already at hand; the philosophy of concerted effort toward a common goal would by then have been established; and fully half the problem—continued escalation of need because of the population growth—would no longer exist. Once again, moreover, the systems concept of optimization would come into play. Certain aspects of pollution clearly demand all-out action if eventual success is to be achieved. Others may never require complete elimination of the contaminants, but only that fraction which cannot be accommodated by the natural processes available. The air, the water, and the land of the earth have great self-restorative capabilities. On the other hand, essentially all used products should eventually be recycled rather than put to waste. Not only man's numbers, but also his cities, his transportation systems, and his methods of generating power must thus be planned in such a manner as to yield the most efficient combination of natural resources and technological derivatives over an indefinite time.

Quantitative vs. Qualitative Growth

Quantitative control of man's activities by no means implies qualitative limitation. The French premise that growth is essential to the health of a country can be interpreted as increase in excellence as well as change in numbers alone. Indeed, much of the effort that must now go into the correction of past mistakes can eventually be directed toward improving man's general standard of living beyond present possibilities (though it should be remarked that affluence for its own sake is but another sort of pollution). Living standards include, first of all, such material matters as food and housing, urban diffusion, power, and transportation—not for the few, but for all. Quite as essential are health, education, occupation of leisure time, and world understanding. None of these has an evident limit of attainment, but each must be developed in such a way as to require essentially no large-scale corrective measures in the future. Above all, attention should be given sooner or later to the all-important matter of upgrading the human race itself; obviously, were man already perfect, the problems under consideration would not exist.

The biological sciences are rapidly approaching the degree of competence lately attained by the physical. Though the creation of life is still to be accomplished, a gene has now been synthesized, and matters of genetic control and the breeding of plants and animals for the strengthening or reduction of specified characteristics are well advanced. Is man not missing a tremendous opportunity so long as he does not apply the same principles of improvement to himself? Hereditary weaknesses could thereby be eliminated. Disease-resistant traits could be strengthened and susceptibilities diminished. The average level of mankind could be raised by stressing health through preventive medicine rather than merely the extension of the average life span. Physically and mentally, in fact, the human race could be improved without limit.

Unfortunately for his physical and mental progress, man is still adamantly opposed to utilizing the principles of plant and animal improvement for his own advancement. First of all, he doubts his peers' ability to determine the proper course of advancement with any degree of certainty. What traits should be bred into the race and what out, particularly as one progresses from the physical to the mental? Secondly, he is averse on principle to the human control of human life. A judge or jury, to be sure, can imprison or in many states even execute someone considered guilty of a crime; a physician or surgeon surely holds the lives of patients in his hands; a congress can declare war and a commander send his troops into battle. But to regulate another's genes or his right to pass them on is considered akin to playing God. So we let the illiterate have the most children, we lower the average well-being of the race by preserving the ailing and decrepit, and we allow heredity to assume an uncontrolled role.

The issue is clearly an ethical one—to many, even a religious one, though for present purposes the distinction is not significant. Let it be granted that it is the social or behavioral sciences that are involved. We may also assume that the physical and life sciences are at a stage at which they can control technology in its relation to world ecology fairly well, given the human will to do so. Let us further avoid all talk of euthanasia, culling of misfits, and control of another's genes. Surely there are still effective means of self-improvement that can be disseminated by normal educational channels, the use of which would appreciably strengthen the race. For instance, by virtue of world emphasis on athletics, men and women are physically superior on the average to their nineteenth-century counterparts. Appreciation of sound teeth has brought some countries dentally far ahead of others. Mental strengthening has resulted from mass education through succeeding generations. How much more could be accomplished if scientifically accepted principles of improvement were made more widely known, and if people throughout the world could be persuaded, by example, of their worth to the human race! As in the case of population stabilization, the rate of progress would surely increase with the progress itself.

Whether or not mankind actually follows the course that has been outlined will depend in large measure upon a number of factors, each of which is a matter of education toward world understanding of one sort or another. People everywhere must become convinced that neither space nor natural wealth is inexhaustible. Leaders of the developing nations must learn that it is far more effectual to reduce the growth rate of their populations than to increase their gross national product. Engineers must realize that people-problems are a major part of every technical design, and behaviorists, that technology will continue to be essential to human welfare. The mystique of today's youth must be enlisted, not only for its persuasive force but to insure continuity of effort in the future. In a word, all who have any influence upon the gross eco-system must share responsibility for preservation of the system in its many aspects.

COMPLETE LIST OF PUBLICATIONS

This complete list of the publications of Hunter Rouse is arranged by category. A number has been assigned to each publication to indicate chronological order of publication; this number appears in brackets at the end of each entry. An asterisk indicates that the publication is reproduced in this volume.

BOOKS

Fluid Mechanics for Hydraulic Engineers, Engineering Societies Monographs, McGraw-Hill, New York, 1938. *Mekhanika židkosti dlya inženerov-gidrotekhnikov,* Godsudarstvennoe Energeticeskoe Izdatel'stvo, Moscow and Leningrad, 1958. Paperback edition, Dover Publications, New York, 1961. [16]

Elementary Mechanics of Fluids, John Wiley & Sons, New York, 1946. *Hidráulica-Mecánica Elemental de Fluidos,* Editorial Dossat, Madrid, 1951. Chinese translation, Shanghai, 1953. Paperback edition, Toppan Company, Tokyo, 1967. Hindi translation in press. [39]

Engineering Hydraulics (editor and coauthor), John Wiley & Sons, New York, 1950. *Tehnička Hidraulika,* Gradevinska Knjiga, Belgrade, 1969. [54]

Basic Mechanics of Fluids (with J. W. Howe), John Wiley and Sons, New York, 1953. Hindi translation in press. [65]

History of Hydraulics (with S. Ince), English and French fold-in supplements to *La Houille Blanche,* 1954-56. Hard-cover edition, Institute of Hydraulic Research, Iowa City, 1957. Paperback edition, Dover Publications, New York, 1963. [74]

Advanced Mechanics of Fluids (editor and coauthor), John Wiley & Sons, New York, 1959. *Mekhanika židkosti,* Izdatel'stvo literatury po stroitel'stvu, Moscow, 1967. Japanese translation in press. [80]

BOOKLETS

**Verteilung der hydraulischen Energie bei einem lotrechten Absturz* (doctoral dissertation, Technische Hochschule, Karlsruhe), Oldenbourg, Munich and Berlin, 1933. [5]

Cavitation and Pressure Distribution—Head Forms at Zero Angle of Yaw (with J. S. McNown), University of Iowa Studies in Engineering, Bulletin 32, 1948. [48]

Hydraulics of Box Culverts (with D. E. Metzler), University of Iowa Studies in Engineering, Bulletin 38, 1959 [81]

*"Répartition de l'énergie dans des zones de décollement" (doctoral dissertation, University of Paris), *La Houille Blanche,* July 4, 1959. [84]

Laboratory Instruction in the Mechanics of Fluids, University of Iowa Studies in Engineering, Bulletin 41, 1961. [89]

Cavitation and Pressure Distribution—Head Forms at Angles of Yaw, University of Iowa Studies in Engineering, Bulletin 42, 1962. [93]

PAPERS

"Research Institute for Hydraulic Engineering and Water Power," *Transactions ASME Hydraulics Division,* Vol. 54, No. 9, 1932. [3]

*"Model Research on Spillway Crests" (with L. Reid), *Civil Engineering,* Vol. 5, No. 1, 1935. [8]

*"Discharge Characteristics of the Free Overfall," *Civil Engineering,* Vol. 6, No. 4, 1936. [11]

*"Modern Conceptions of the Mechanics of Fluid Turbulence," *Transactions ASCE,* Vol. 102, 1937. (Awarded Norman Medal 1938.) [13]

"Nomogram 'for the Settling Velocity of Spheres," Exhibit D, Annual Report, Committee on Sedimentation, National Research Council, 1936-37. [15]

*"Experiments on the Mechanics of Sediment Suspension," *Proceedings Fifth International Congress of Applied Mechanics* (1938), Cambridge, Massachusetts, 1939. [20]

"Wind-Tunnel Classifier for Sand and Silt" (with G. H. Otto), *Civil Engineering,* Vol. 9, No. 7, 1939. [21]

"An Analysis of Sediment Transportation in the Light of Fluid Turbulence," U. S. Dept. of Agriculture, SCS-TP-25, July 1939. [22]

"Laws of Transportation of Sediment by Streams: Suspended Load," Role of Hydraulic Laboratories in Geophysical Research, National Bureau of Standards, September 1939. [23]

*"Criteria for Similarity in the Transportation of Sediment," *Proceedings Hydraulics Conference,* University of Iowa Studies in Engineering, Bulletin 20, 1940. [25]

"On the Four Regimes of Open-Channel Flow" (with J. M. Robertson), *Civil Engineering,* Vol. 11, No. 3, 1941. [27]

*"Suspension of Sediment in Upward Flow," *Investigations of the Iowa Institute of Hydraulic Research 1939-1940,* University of Iowa Studies in Engineering, Bulletin 26, 1941. [28]

*"Evaluation of Boundary Roughness," *Proceedings Second Hydraulics Conference,* University of Iowa Studies in Engineering, Bulletin 27, 1943. [32]

*"A General Stability Index for Flow Near Plane Boundaries," *Journal of the Aeronautical Sciences,* Vol. 12, No. 4, 1945. [37]

*"Gravitational Diffusion from a Boundary Source in Two-Dimensional Flow" (Sixth International Congress for Applied Mechanics, Paris, 1946), *Journal of Applied Mechanics,* Vol. 14, No. 3, 1947. [40]

"Wind-Tunnel Studies of Fog Dispersal, Gas Diffusion, and Flow over Mountainous Terrain," *NDRC Summary Technical Report,* Chapter 43, Division 10, 1947. [42]

*"Use of the Low-Velocity Air Tunnel in Hydraulic Research," *Proceedings Third Hydraulics Conference,* University of Iowa Studies in Engineering, Bulletin 31, 1947. [44]

"Fundamental Aspects of Cavitation," *Proceedings National Conference on Industrial Hydraulics,* Chicago, October 1947. [47]

*"Diffusion of Submerged Jets" (with M. L. Albertson, Y. B. Dai, and R. A. Jensen), *Transactions ASCE,* Vol. 115, 1950. (Awarded Hilgard Prize, 1951.) [49]

"Cavitation-Free Inlets and Contractions" (With M. M. Hassan), *Mechanical Engineering,* Vol. 71, No. 3, 1949. (Contains fallacy in measurements and conclusions.) [50]

"Characteristics of Irrotational Flow through Axially Symmetric Orifices" (with A. H. Abul-Fetouh), *Journal of Applied Mechanics,* Vol. 17, No. 4, 1950. (Superseded by B. W. Hunt's 1968 paper in *Journal of Fluid Mechanics*). [51]

*"High-Velocity Flow in Open Channels—Design of Channel Expansions" (with B. V. Bhoota and E. Y. Hsu), *Transactions ASCE,* Vol. 116, 1951. [55]

"Model Techniques in Meteorological Research," *Compendium of Meteorology*, American Meteorological Society, 1951. [56]

*"On the Growth and Decay of a Vortex Filament" (with H.-C. Hsu), *Proceedings First National Congress of Applied Mechanics*, 1951. [57]

*"Experimental Investigation of Fire Monitors and Nozzles" (with J. W. Howe and D. E. Metzler), *Transactions ASCE*, Vol. 117, 1952. [58]

"Air-Tunnel Studies of Diffusion in Urban Areas," *Meteorological Monographs*, Vol. 1, No. 4, 1951. [60]

*"Gravitational Convection from a Boundary Source" (with C. S. Yih and H. W. Humphreys), *Tellus*, Vol. 4, No. 3, 1952. [62]

*"Free Convection over Parallel Sources of Heat" (with W. D. Baines and H. W. Humphreys), *Proceedings of the Physical Society*, B, Vol. 66, 1953. [63]

*"Cavitation in the Mixing Zone of a Submerged Jet" (Eighth International Congress for Applied Mechanics, Instanbul, 1952), *La Houille Blanche*, Vol. 8, No. 1, 1953. [64]

*"Measurement of Velocity and Pressure Fluctuations in the Turbulent Flow of Air and Water," *Riabouchinsky Anniversary Volume*, Ministère de l'Air, 1954. [66]

*"Form Drag of Composite Surfaces" (with T. T. Siao), *Proceedings Second U. S. National Congress of Applied Mechanics* (1954), ASME, 1955. [68]

*"Turbulent Diffusion across a Density Discontinuity" (with J. Dodu), *La Houille Blanche*, Vol. 10, No. 4, 1955. [70]

"Seven Exploratory Studies in Hydraulics," *Journal of the Hydraulics Division*, ASCE, Vol. 82, No. HY 4, 1956. [72]

*"Experiments on Two-Dimensional Flow over a Normal Wall" (with M. Arie), *Journal of Fluid Mechanics*, Vol. 1, Part 2, 1956. [73]

*"Diffusion in the Lee of a Two-Dimensional Jet," *Proceedings 9th International Congress of Applied Mechanics* (Brussels, 1956), Vol. 1, 1957. [75]

*"Characteristics of Flow over Terminal Weirs and Sills" (with P. K. Kandaswamy), *Journal of the Hydraulics Division*, ASCE, Vol. 83, No. Hy 4, 1957. [77]

*"Turbulence Characteristics of the Hydraulic Jump" (with T. T. Siao and S. Nagaratnam), *Transactions ASCE*, Vol. 124, 1959. (Awarded Hilgard Prize 1961.) [78]

"Characteristics of Flow with Density Stratification," *Proceedings Second Hydraulics Conference*, Washington State University, October, 1959. [83]

"Distribution of Energy in Regions of Separation," *La Houille Blanche*, Vol. 15, Nos. 3 and 4, 1960. [84]

*"On the Bernoulli Theorem for Turbulent Flow," *Tollmien Festschrift—Miszellaneen der Angewandten Mechanik*, Akademie-Verlag, Berlin, 1962. [86]

*"Interfacial Mixing in Stratified Flow" (with E. O. Macagno), *Transactions ASCE*, Vol. 127, 1962. [91]

*"Energy Transformation in Zones of Separation," General Lecture, *Proceedings IAHR 9th Convention*, Dubrovnik, 1961. [94]

*"The Role of the Froude Number in Open Channel Resistance" (With H. J. Koloseus and J. Davidian), *Journal of Hydraulic Research*, Vol. 1, No. 1, 1963. [96]

*"On the Role of Eddies in Fluid Motion" (a Sigma Xi national lecture, 1965), *American Scientist*, Vol. 51, No. 3, 1966. *Science in Progress*, Yale University Press, 1966. [97]

*"Critical Analysis of Open-Channel Resistance," *Journal of the Hydraulics Division*, ASCE, Vol. 91, No. HY 4, July 1965. [105]

*"Cavitation and Energy Dissipation in Conduit Expansions" (with V. Jezdinsky), *Proceedings of IAHR 11th Congress*, Leningrad, 1965. [106]

*"The Bernoulli Theorem," *Proceedings JSME*, Vol. 86, No. 562, 1965. [107]

*"Fluctuation of Pressure in Conduit Expansions" (with V. Jezdinsky), *Journal of the Hydraulics Division*, ASCE, Vol. 92, No. HY 3, 1966. [110]

*"Jet Diffusion and Cavitation," *Journal of the Boston Society of Civil Engineers*, Vol. 53, No. 3, 1966. [111]

"Jet Diffusion in the Region of Flow Establishment" (with S. Sami and T. Carmody), *Journal of Fluid Mechanics*, Vol. 27, Pt. 2, 1967. [114]

*"On the Use of Models in Fluids Research" (with E. O. Macagno), *Proceedings First International Conference on Hemorheology* (Iceland, 1966), Pergamon Press, London, 1968. [118]

*"Jet-Induced Circulation and Diffusion" (with C. Iamandi), *Journal of the Hydraulics Division,* ASCE, Vol. 95, No. HY 2, 1969. [121]

"Characteristics of Interfacial Turbulence," *Proceedings NATO Advanced Study Institute on Surface Hydro-dynamics* (Bressanone, Aug.-Sept. 1966), Padua, 1969. [123]

*"Work-Energy Equation for the Streamline," *Journal of the Hydraulics Division*, ASCE, Vol. 96, No. HY 5, 1970. [125]

"The Bernoulli Theorem" (in Japanese; transcript of 1969 JSCE lecture, translated by T. Hayashi), *JSCE Journal*, Vol. 55, Nos. 10 and 11, 1970. [129]

DISCUSSIONS

"The Hydraulic Jump in Terms of Dynamic Similarity" (Bakhmeteff-Matzke), *Transactions ASCE*, Vol. 101, 1936. [9]

"Adaptation of Venturi Flumes to Flow Measurements in Conduits" (Palmer-Bowlus), *Transactions ASCE*, Vol. 101, 1936. [10]

*"Pressure Distribution and Acceleration at the Free Overfall" (Langbein), *Civil Engineering*, Vol. 7, No. 7, 1937. [12]

"Varied Flow in Open Channels of Adverse Slope" (Matzke), *Transactions ASCE*, Vol. 102, 1937. (With M. P. White.) [14]

"Laboratory Investigation of Flume Traction and Transportation" (Chang), *Transactions ASCE*, Vol. 104, 1939. [17]

"A Theory of Silt Transportation" (Griffith), *Transactions ASCE*, Vol. 104, 1939. [18]

"Investigation of Errors of Pitot Tubes" (Hubbard), *Transactions ASME,* Vol. 61, 1939. [19]

"Relation of the Statistical Theory of Turbulence to Hydraulics" (Kalinske), *Transactions ASCE*, Vol. 105, 1940. [24]

"Turbulence and Energy Dissipation" (Kalinske), *Transactions ASME*, Vol. 63, 1941. [26]

"Cavitation in Outlet Conduits of High Dams" (Thomas-Schuleen), *Transactions ASCE*, Vol. 107, 1942. [29]

"Profile Curves for Open-Channel Flow" (Gunder), *Transactions ASCE*, Vol. 108, 1943. [30]

"Energy Loss at the Base of a Free Overfall" (Moore), *Transactions ASCE*, Vol. 108, 1943. [31]

"The Circulating Water Channel of the David W. Taylor Model Basin" (Saunders-Hubbard), *Transactions SNAME*, Vol. 52, 1944. (With A. A. Kalinske.) [33]

"Coefficients for Velocity Distribution in Open-Channel Flow" (Eisenlohr), *Transactions ASCE*, Vol. 110, 1945. (With J. S. McNown.) [34]

"A Pattern for Research in Naval Architecture" (Wright), *Transactions SNAME,* Vol. 54, 1946. [35]

*"Friction Factors for Pipe Flow" (Moody), *Transactions ASME*, Vol. 66, No. 8, 1944. [36]

*"The Mechanism of Energy Loss in Fluid Friction" (Bakhmeteff-Allan), *Transactions ASCE*, Vol. III, 1946. (With A. A. Kalinske.) [38]

"Friction Coefficients in a Large Tunnel" (Hickox-Peterka-Elder), *Transactions ASCE,* Vol. 113, 1948. [43]

"An Investigation of Flow Through Screens" (Baines-Peterson), *Transactions ASME,* Vol. 73, 1951. [59]

*"A Note on the Manning Formula" (Chow), *Transactions AGU,* Vol. 37, No. 3, 1956. [71]

"Calculation of Potential Flows with Free Streamlines" (Birkhoff), *Journal of the Hydraulics Division,* ASCE, Vol. 88, No. HY 2, 1962. [92]

"Theory, Experiments, and Philosophy of Hydraulics" (Le Méhauté), *Journal of the Hydraulics Division,* ASCE, Vol. 89, No. HY 1, 1963. (With P. G. Hubbard and E. O. Macagno.) [95]

*"Sediment Transportation Mechanics: Suspension of Sediment, Density Currents," *Journal of the Hydraulics Division,* ASCE, Vol. 90, No. HY 1, 1964. [100]

"Solution of Highly Curvilinear Gravity Flows" (Strelkoff), *Journal of the Engineering Mechanics Division,* ASCE, Vol. 90, No. EM 5, 1964. [101]

"Some Observations on the Undular Jump" (Jones), *Journal of the Hydraulics Division,* ASCE, Vol. 90, No. HY 6, 1964. [102]

"Simulation Techniques in Water Resources Systems," *Proceedings IAHR 13th Congress,* Kyoto, Vol. 5-2, 1969. [124]

ARTICLES

*"Amerikanismus," *The Technology Review,* December 1930. [1]

"Great German Hydraulic Laboratory," *Civil Engineering,* Vol. 1, No. 5, 1931. [2]

*"Night Watch at Obernach," *The Technology Review,* December 1932. [4]

"On the Use of Dimensionless Numbers," *Civil Engineering,* Vol. 4, No. 11, 1934. [6]

"Whirling Enemies of Motion," *The Technology Review,* December 1934. [7]

"Fluid Mechanics," *National Encyclopedia,* P. F. Collier & Son Corporation, New York, 1947. [41]

"Civil Engineers Share Knowledge of Fluid Mechanics with Many Related Professions," *Civil Engineering,* Vol. 17, No. 12, 1947. [45]

"Vaeskemekanikk og Bygningsingeniører," *Teknisk Ukeblad,* Vol. 95, No. 14, 1948. [46]

"Fluid Mechanics," "Hydrodynamics," "Hydraulics," *Encyclopedia Americana,* New York, 1950. [53]

*"Present-Day Trends in Hydraulics," *Applied Mechanics Reviews,* Vol. 5, No. 2, 1952. [61]

"O papel de mêcanica dos fluidos na engenharia hidráulica," *Engenharia,* September 1954. [67]

"Les échanges internationaux de personnes," *Annales de l'Institut Polytechnique de Grenoble,* Vol. 1, No. 1, 1955. [69]

"Estado actual de la hidráulica en los Estados Unidos," *Revista de Ingeniería,* Vol. 9, 1957. [76]

"Una appreciación de la hidráulica al promediar el siglo," *Ciencia y Técnica,* Vol. 125, No. 626, 1958. [79]

"Una valutazione dell'Idraulica verso la metà del secolo ventesimo," *L'Acqua,* Vol. 37, No. 2, 1959. [82]

"Weir," "Flowmeter," "Turbulent Flow in Smooth and Rough Pipes," *Encyclopaedic Dictionary of Physics,* Pergamon Press, Oxford, 1961-64. [85]

*"Current Trends in American Hydraulics," *Physical Sciences, Some Recent Advances in France and the United States,* New York University Press, 1962. [87]

"Hydraulics and Hydraulic Engineering," *The Harper Encyclopedia of Science,* Harper & Row, New York, 1963. [88]

"Nuevos horizontes de la mecánica de fluidos," *Energía Industrial,* Vol. 4, No. 17, 1961. [90]

*"On the Art of Advancing the Science of Hydraulics," *Proceedings 1st Australasian Conference* (1962), Pergamon Press, 1963. [98]

"Hidraulikai kutatások az Iowa-i egyetemen," *Hidrológiai Közlöny*, 1964, 5. sz. [103]

*"On a Matter of Latitude in Pronunciation," *La Houille Blanche*, Vol. 20, No. 6, 1965. [108]

*"La Investigación y el Ingeniero," *Simposio El Ingeniero ante la Ciencia y la Tecnología Contemporáneas*, Colegio de Ingenieros Venezuela, Caracas, 21-28 March, 1965. [109]

*"German-American Observations on Educational Reform" (with E. Naudascher), *Journal of Engineering Education*, Vol. 57, No. 1, 1966. "Deutsch-amerikanische Beobachtungen zur Bildungsreform," *Mitteilungen des Hochschulverbandes*, Vol. 14, No. 3, 1966. [112]

"What Makes an Engineer?" *Iowa Transit*, Vol. 71, No. 8, 1966. [113]

"Teaching *and* Research," *Journal of Engineering Education*, Vol. 58, No. 6, 1968. [116]

"Engineering Education in the Mechanics of Fluids," *La Houille Blanche*, Vol. 23, No. 1, 1968. [117]

*Preface to *Hydrodynamics by Daniel Bernoulli and Hydraulics by Johann Bernoulli*, translated from the Latin by C. Carmody and H. Kobus, Dover, New York, 1968. [119]

*"Iowa's Quest for Curricular Balance," *Engineering Education*, Vol. 59, No. 3, 1968. [120]

*"International Heritage," *Engineering Education*, Vol. 61, No. 3, 1970. [126]

"Du Buat," "Fourneyron," "Froude," "Girard," "Reech," "Weisbach," *Dictionary of Scientific Biography*, Charles Scribner's Sons, New York, 1970- —. [127]

*"Pierre Danel's Influence on American Hydraulics," *La Houille Blanche*, Vol. 25, No. 6, 1970. [128]

*"Growth $\neq$ Progress?" *Mechanical Engineering*, Vol. 93, No. 4, 1971. [130]

REVIEWS

Technische Hydraulik (Jaeger), *Zeitschrift für angewandte Mechanik und Physik*, Vol. 1, No. 5, 1950. [52]

Mécanique Expérimentale des Fluides (Comolet), *Journal of Fluid Mechanics*, Vol. 17, Pt. 2, 1963. [99]

Incompressible Fluid Dynamics (Hunt), *Science*, Vol. 146, 30 October 1964. [104]

Open Channel Flow (Henderson), *Journal of Fluid Mechanics*, Vol. 29, Pt. 2, 1967. [115]

Education for Innovation (De Simone), *Science*, Vol. 164, May 1969. [122]

FILM SCRIPTS

"The Iowa Institute of Hydraulic Research," 1952.

"Introduction to the Study of Fluid Motion," 1961.

"Fundamental Principles of Flow," 1962.

"Fluid Motion in a Gravitational Field," 1963.

"Characteristics of Laminar and Turbulent Flow," 1965.

"Form Drag, Lift, and Propulsion," 1966.

"Effects of Fluid Compressibility," 1970.

UNPUBLISHED WRITINGS

To make certain unpublished writings available to any who may wish to refer to them, copies are being deposited with the Library Archives of the University of Iowa at Iowa City and with the Engineering Societies Library at New York. The collections include reports on the M.I.T., Fulbright, and NSF years in Europe, and on the consulting trips to Egypt and Thailand; twenty annual reports on the Iowa Institute of Hydraulic Research; and the 1970 report to the Iowa Board of Regents on the College of Engineering as reproduced in one of a half dozen alumni newsletters.

COAUTHOR INDEX

Numbers refer to the consecutive pagination of this volume.

Abul-Fetouh, A.-H., 350, 606
Albertson, M. L., 139–164, 242, 285, 517, 581, 606
Arie, M., 264–276, 285, 325, 350, 356, 403, 404, 407, 509, 511, 607
Baines, W. D., 229–235, 607
Bhoota, B. V., 165–181, 477, 606
Carmody, T., 608
Dai, Y. B., 139–164, 242, 285, 517, 581, 606
Davidian, J., 408–410, 481, 492, 607
Dodu, J., 254–261, 371, 607
Hassan, M. M., 606
Howe, J. W., 188–216, 350, 544, 605, 607
Hsu, E.-Y., 165–181, 293, 477, 606
Hsu, H.-C., 182–187, 607
Hubbard, P. G., 609
Humphreys, H. W., 219–228, 229–235, 581, 607
Iamandi, C., 570–582, 608
Ince, S., 475, 544, 556, 565, 605

Jensen, R. A., 139–164, 242, 285, 517, 581, 606
Jezdinsky, V., 500–505, 517–528, 544, 607, 608
Kalinske, A. A., 110–119, 608
Kandaswamy, P. K., 286–298, 607
Koloseus, H. J., 408–410, 477, 481, 490, 493, 607
Macagno, E. O., 369–395, 552–556, 607, 608, 609
McNown, J. S., 605, 608
Metzler, D. E., 188–216, 544, 605, 607
Nagaratnam, S., 299–323, 345, 350, 356, 581, 594, 607
Naudascher, E., 545–551, 610
Otto, G. H., 606
Reid, L., 47–51, 606
Robertson, J. M., 492, 606
Sami, S., 534, 535, 544, 608
Siao, T. T., 248–253, 299–323, 325, 350, 366, 581, 594, 607
White, M. P., 608
Yih, C. S., 219–228, 230, 235, 581, 607

AUTHOR INDEX

Numbers refer to the consecutive pagination of this volume.

Allan, W., 110
Angus, R. W., 350
Appel, D. W., 511, 512, 538, 544
Baines, W. D., 208, 477, 607
Bakhmeteff, B. A., 110, 191, 299, 608
Ball, J. W., 511, 520, 544
Batchelor, G. K., 242, 350, 456
Bauer, W. J., 487
Bazin, H. E., 11, 287, 292
Bernoulli, D., 507, 511, 557–565, 583, 593
Bernoulli, J., 507, 511, 557–565, 594
Bhoota, B. V., 167
Bidone, G., 287, 292
Birkhoff, G., 275, 276, 285, 350, 456, 609
Blasius, H., 109
Bourot, J. M., 186, 187
Böss, P., 12, 54, 288, 292
Boussinesq, J., 385, 451, 452, 456, 486
Bowlus, F. D., 608
Boyer, M. C., 262, 263
Carmody, T., 509, 511, 534, 544, 557, 593
Carstanjen, M., 11
Chang, Y. L., 608
Chaturvedi, M. C., 502, 509, 511, 517, 522
Chevray, R., 509, 511, 593
Chow, V. T., 262, 263
Colebrook, C. F., 96, 98, 99, 101–104, 481
Comings, E. W., 407
Comolet, R., 610
Corrsin, S., 142, 350, 544
Creager, W. P., 47, 50
Darcy, H., 104
Davidian, J., 409
DeLapp, W., 88
De Simone, D. V., 610
Dillmann, O., 51
Dressler, R. F., 493
Drinker, P. A., 490
Du Buat, P. L. G., 287, 292
Ehrenberger, R., 12
Ehrlich, P., 600
Eisenlohr, W. S., 608

Eisner, F., 48
Elder, R. A., 609
Engelund, F., 487
Escande, L., 48, 51
Euler, L., 559–561, 563, 565, 584, 593
Fage, A., 264, 276
Feldman, S., 370
Flierl, K., 565
Folsom, R. G., 99, 104
Forster, J. W., 477
Freeman, J. R., 104, 188, 529, 530, 536, 543, 544
Goldstein, S., 183, 187, 264, 276, 308, 456
Galileo, G., 552, 556
Graebel, W. P., 370
Gregerson, M. I., 556
Greve, F. W., 104
Griffith, W. M., 608
Gunder, D. F., 608
Hama, F., 455, 477
Harleman, D. R., 370
Hayat, S., 490
Helmholtz, H. L. F. von, 416, 423, 426, 430, 455
Henderson, F. M., 610
Heywood, F., 104
Hickox, G. H., 609
Hinze, J. O., 407, 456, 519, 530, 544
Homma, M., 409
Hooker, S. G., 183, 187
Hoskins, L. M., 104
Hotes, F. L., 104
Howarth, L., 141, 350
Howe, J. W., 200
Hsieh, T., 477
Hsu, E.-Y., 167
Hsu, H.-C., 274, 276, 325, 350
Hsu, S. T., 581
Hubbard, C. W., 608, 609
Hubbard, P. G., 242, 270, 285, 310, 350
Humphreys, H. W., 228, 230, 235
Hung, T. K., 487
Hunt, B. W., 581
Hunt, J. N., 610

Hurst, H. E., 63
Ippen, A. T., 292, 370, 490
Iwasa, Y., 409, 493, 494
Jaeger, C., 218, 610
Jeffreys, H., 409, 493
Jegorow, S. A., 409
Johansen, F. C., 264, 276
Jonassen, F., 99, 104
Jones, L. E., 609
Kalinske, A. A., 95, 104, 594, 608
Kampé de Feriet, J., 115
Kandaswamy, P. K., 292
Kármán, T. von, 59–62, 93, 101, 103, 109, 114, 350, 432, 433, 455, 470, 471
Kessler, L. H., 95, 104
Keulegan, G. H., 103, 104, 370, 371, 375, 493
Keuthe, A. M., 141
Kirchhoff, G. R., 253
Klebanoff, P. S., 350
Kobus, H., 511, 557, 593
Koch, A., 11
Koloseus, H. J., 409
Kolupaila, S., 262, 263
Kovasznay, L. S. G., 593
Krey, H., 471
Krumbein, W. C., 49
Lamb, H., 109, 183, 187, 253, 303, 350, 455
Landweber, L., 271, 276
Langbein, W. B., 56
Lauck, A., 288, 292
Laufer, J., 142, 350, 356, 456, 480
Laursen, E. M., 477
Leliavsky, S., 475
Le Méhauté, B., 609
Liepmann, H. W., 142
Lighthill, M. J., 544
Lofquist, K., 371
Lüddecke, T., 4, 5
Macagno, E. O., 486, 487, 556
Macdonald, E. G., 481, 483, 490
Martin, R. R., 104
Martinot-Lagarde, A., 115
Marx, C. D., 104
Matzke, A. E., 299, 300, 608
Mavis, F. T., 476
McNown, J. S., 193, 242, 293, 350
Merrill, E. W., 556
Miller, D. R., 402, 407
Millikan, C. B., 479
Mills, H. F., 104
Mises, R. von, 11, 288, 292, 327, 350
Mittendorf, G. H., 380
Moody, L. F., 105, 106
Moore, W. L., 477, 608
Musterle, T., 12
Nagaratnam, S., 322
Newsham, A. D., 456

Nikuradse, J., 95, 97, 103, 104, 109, 350, 480, 483
O'Brien, M. P., 99, 104
O'Loughlin, E. M., 481, 483, 484, 492
Pai, S. I., 350
Palmer, H. K., 608
Patterson, G. W., 374, 493
Peterka, A. J., 609
Peters, H., 350
Peterson, E. G., 208, 609
Pettijohn, F. J., 49
Pigott, R. J. S., 104
Plessett, M. S., 253
Poleni, G., 286, 292
Posey, C. J., 200, 481
Powell, R. W., 409, 476
Prandtl, L., 187, 438, 451, 456
Prentice, T. H., 55
Preston, J. H., 310
Rankine, A. O., 230, 235
Rankine, W. M. J., 418, 423, 435, 455, 456
Rao, T. R. K., 345, 350
Raynaud, J. P., 370
Rehbock, T., 12, 27, 288, 292
Reichardt, H., 163
Reid, L., 47
Reynolds, O., 331, 350, 451
Riabouchinsky, D., 327
Roberson, J. A., 483, 484
Robertson, J. M., 109
Roesler, F. C., 242
Rosenhead, L., 440, 456
Roshko, A., 264, 276
Ruden, P., 141
Saint-Venant, B. de, 451, 456, 594
Saph, A. V., 104
Saunders, H. E., 608
Schadt, C. F., 285
Schlichting, H., 350, 481
Schmidt, W., 228
Schoder, E. W., 104
Schubauer, G. B., 456
Schuleen, E. P., 608
Shaffer, P. A., Jr., 253
Shields, A., 72, 73, 83
Siao, T. T., 253, 322
Simmons, W. P., 501, 520, 544
Skramstad, H. K., 456
Skrinde, R. A., 477
Spannhake, W., 20, 21, 22
Spengo, A. C., 242
Spiess, O., 564, 565
Squire, H. B., 375
Steinman, D. B., 455
Stokes, G. G., 330. 350
Straub, H., 564, 565
Straub, L. G., 61
Strouhal, V., 433, 455

Strutt, J. W. (Lord Rayleigh), 553, 556
Taylor, G. I., 109, 114, 285, 337, 350, 370, 443, 452, 456
Thomas, H. A., 493, 608
Thompson, W. (Lord Kelvin), 416, 455
Tietjens, O. G., 187
Toch, A., 477
Todd, F. H., 481
Tollmien, W., 141, 456, 530, 532, 544
Tomotika, S., 141
Toussaint, A., 193
Townsend, A. A., 306, 333, 350, 356, 402, 407, 456, 530, 544

Truesdell, C., 558, 563, 565
Vedernikov, V. V., 409, 493
Wagner, J. A., 601
Weisbach, J., 287, 292
Wells, R. E., Jr., 556
White, C. M., 98, 103, 104
Whitehouse, J. P., 242, 501
Wing, C. B., 104
Wright, E. A., 608
Yih, C.-S., 228, 293, 373, 375
Yu, Y.-S., 350

SUBJECT INDEX

Numbers refer to the consecutive pagination of this volume.

Abrupt drop, 178, 179
Abrupt expansion, 169–172
Acceleration, 21–25, 56
Air tunnel, low-velocity, 124–138
American engineering education, 566–569, 596, 597
American hydraulics, 357–368
Amerikanismus, 2–5
Anemometer, hot-wire, 238, 270, 310, 575
    spoke-vane, 223, 232

Bed load, 73
Bed shear, 72, 73
Bends, open-channel, 490, 491
Bernoulli equation, 190, 348, 351–356, 405–407, 419,
    468, 507–511, 562, 563, 583–593
Birth rate, 600, 601
Blasius diagram, 105, 106
Blunt cylinder, flow past, 344–348
Borda loss, 329, 503, 505, 540
Boundary layer, 72, 107, 115, 217, 230, 361, 485, 588
Bridge-pier resistance, 488–490
Broad-crested weirs, 54, 55
Bubble screens, 570–573, 580, 581
Buoyancy, 121, 220, 221, 572, 573

Cavitation, 134, 236–242, 360, 412, 434, 438, 500–505,
    529–544
Cavitation index, 236, 239–242, 402, 501–503, 505, 537
Cavitation measurement, 240, 501, 502
Cavitation noise, 240, 501, 502, 504
Channel expansions, 165–181
Characteristics, method of, 170, 176
Chezy coefficient, 93, 96, 101, 106, 262, 263
Circulation, 416, 570–582
Commercial boundary roughness, 98
Conduit resistance, 105, 106, 217
Continuity equation, 203, 230, 304, 330, 331
Contraction coefficient, 54, 327
Convection, 120, 317–319, 346–348, 404
Coriolis coefficient, 586
Correlation, 149, 336, 448–450, 523–526, 542
Crest depth, 18, 53–56

Critical depth, 15, 53–56
Curricular balance, 566–569
Curvilinear flow, 11–24
Cyclone, 412

Danel's influence on Americans, 598, 599
Darcy equation, 84, 85, 88
Darcy-Weisbach relationship, 93, 101, 105
Death rate, 600, 601
Density-current misnomer, 473
Density interface, 254–261
Development and testing, 514–516
Diffusion coefficient, 149, 451
Diffusion of heat, 120–123, 231
Dimensional analysis, 72–75, 386, 408, 478, 552
Direction measurement, 132, 151, 244, 245, 339
Discharge coefficient, 51, 54, 56, 188–190, 204, 205,
    287–289, 292, 294, 295
Disk, 251, 510
Displacement thickness, 270–272
Dissipation, 113, 114, 116, 316–318, 320, 321, 337, 347,
    348, 353, 355, 404, 450, 500–505, 592
Doublet distribution, 271, 272
Drag coefficient, 248–252, 274, 275, 329
Drag measurement, 250

Eco-system, 601, 603
Eddy motion, 133, 134, 268–275, 277–285, 402, 411–456,
    509–511, 578
Educational reform, 545–551
Energy distribution, 10–45, 325–350, 418–421, 518
Energy equation, 221, 305–308, 318, 333–335, 346, 347,
    403, 404, 508, 583–594
Energy flux, 143—160, 222, 225, 315, 346, 578
Energy line, 18, 32, 36–38
Energy of turbulence, 114, 336, 337, 492
Energy transformation, 396–407
Engineering curriculum, 566–569, 595–597
English engineering education, 596
Euler equations, 20, 352, 507, 584
Evaporation, 138
Expansion, 500–505, 510, 511

Fall velocity, 59, 61, 73–82, 86–92
FIDO, 129
Fire monitors, 188–216
Flow net, 43
Flow patterns, 132–134, 162, 224, 225, 234, 268, 273, 284, 312, 343, 344, 397, 400, 435, 510, 511, 518, 531, 571, 572, 579, 588, 590
Fog dispersal, 120
Form drag, 248–253
Free convection, 229–235
French engineering education, 595
Froude number, 106, 121–123, 131, 166–181, 254, 258–261, 301, 303, 311–323, 377, 378, 384–387, 389, 391–394, 408–410, 478, 488–498, 512, 552–554

Galaxy, 412
Gas diffusion, 128, 129
Gate slot, 135
Geometric mean, 69–71
German engineering education, 547–551, 595, 596
Germany, 2–9
Goals of engineering education, 566, 569
Gradual expansion, 173–180
Gravitational diffusion, 120–123, 129–131, 219–228
Growth rate, population, 600, 601
Guide vanes, 193, 194

Heat diffusion, 120–123, 129–131, 223
High-speed photography, 241
Honeycombs, 193, 194
Humanities, 546, 547, 567
Hydraulic jump, 167, 178, 179, 299–323, 589–591

Initial sediment movement, 72, 73
Inlet, flow at, 136, 343–348
Innovation in research, 516
Interfacial mixing, 369–395
International Association for Hydraulic Research, 599

Jet cavitation, 236–242, 537–541
Jet concentration, 198, 201, 202
Jet-concentration measurement, 197, 198
Jet deflection, 77–81, 277–285
Jet diffusion, 139–164, 192, 203, 277–285, 529–544, 570–582
Jet trajectories, 206–212

Kármán vortex trail, 431–434, 454
Kármán-Prandtl relationship, 94, 95, 105, 106, 471
Kinetic energy of vortex, 186, 419
Kultur, 2–5

La Houille Blanche, 599
Laminar flow, 113, 371–379
Length of jump, 311
Line sources of heat, 120–123, 219–228, 229–235
Loss coefficient, 204, 205
Loss of energy, 137, 354, 405, 406, 426, 509–511, 586, 587

Mach number, 552, 554, 555
Manning formula, 96, 102, 103, 106, 262, 263
Measurement, 364, 468
Mixing, 112–114
Mixing coefficient, 62, 65
Mixing length, 111, 149, 221, 314
Mixing rate, 388, 393
Models, 47–51, 514, 515, 552–556
Momentum equation, 221, 230, 274, 303, 344, 402, 403
Momentum flux, 143–160, 221, 222, 225, 278, 314, 315, 345, 531–533, 570–573, 575, 577

Nappe profiles, 296–298
Navier-Stokes equations, 57, 58, 112, 183, 330–332, 371, 508, 585
Nozzle forms, 195, 213
Nozzles, 188–216

Obernach, 6–9
Open-channel resistance, 475–489
Overfall, 10–45, 52–56

Permeability, 84, 85, 88
Pi theorem, 41, 166, 231, 552
Pipe resistance, 105, 108
Pitot tube, 30, 33, 34, 244
Plates, drag of parallel, 248–251, 266–274
Point source of heat, 219–228
Population density, 602
Population explosion, 601
Population stabilization, 601, 603
Postgraduate education, 550
Potential flow, 217, 359
Pressure distribution, 47–51, 53–56, 134–136, 182–186, 273, 274, 328, 402, 505, 522, 533, 563
Pressure fluctuation, 237–239, 517–528, 535
Pressure-fluctuation measurement, 238, 239, 243–247, 502, 520, 533
Pressure measurement, 30–33, 238, 520
Probability function, 121–123, 143
Production of turbulence, 316, 317, 346–348, 404
Pronunciation of "Froude," 512

Rankine body, 269, 271
Rankine combined vortex, 182, 185, 418–425
Research, 513–516
Research support, 516
Research tools, 517
Resistance coefficient, 383, 408–410, 478–483
Resistance diagram, 100, 105, 408–410
Reynolds equations, 303, 331–333, 401, 585
Reynolds number, 73, 94, 95, 97–103, 105–108, 115–118, 254, 258–261, 275, 337, 338, 376, 378, 384, 386–389, 391–394, 432–434, 478, 486, 488, 495–497, 530, 531, 552–556
Reynolds stresses, 306, 316, 321, 332, 353–355, 508, 532, 533
Rings, drag of multiple, 250

Roll waves, 408, 409
Rotationality, 466, 467
Roughness, 93–104, 106, 263, 361, 367, 408–410, 479–485, 494–497

Saint-Venant equations, 508, 585
Salinity measurement, 256
Scale of cavitation, 242
Scale of turbulence, 114, 336, 337, 442–452, 523–527, 536
Scour, 74–82, 399
Sediment characteristics, 63, 68–71
Sediment concentration, 62–66, 86–92
Sediment suspension, 59–66, 73, 74, 77, 78, 84–92
Sediment transportation, 67–83, 218, 362, 363
Separation, 48, 51, 325–350, 396–407
Shape, cross-sectional, 484–486
Shear, 386, 388, 391, 451
Sills, 286–298
Similarity, 67–83, 144, 203, 217, 363, 364, 552–556
Size-frequency distribution, 69–71, 88, 91, 92
Skewness, 69–71
Slot, flow from, 150, 280, 282
Solidity ratio, 250–252
Soviet engineering education, 596, 597
Specific-energy diagram, 53
Spectrum of turbulence, 336, 337, 443, 444, 526, 527, 543
Spillway crests, 47–51
Square lattice, drag of, 252, 253
Square plates, drag of, 252, 253
Stability, 107–109, 366, 374–377, 408–410, 440–442, 492–495
Stability index, 107–109
Standard deviation, 69–71, 121, 143
Stratified flow, 369–395
Stream function, 267, 268, 271, 310, 328, 329, 342, 343, 359, 575, 578
Streamline location, 267–272, 310, 342–344, 575, 578
Strouhal number, 432
Student unrest, 597

Submerged jet, 139–164, 236–242, 570–582
Supercritical flow, 165–181, 408–410
Surface profiles, 39, 43, 44, 49, 53–55, 296–298
Surface resistance, 479–487
Suspended load, 73, 74
Suspension function, 60–62, 66, 470, 471

Technical institutes, 547–550, 595–597
Temperature distribution, 223, 224, 233
Temperature measurement, 122, 223, 232
Tip vortex, 186
Transfer of energy, 354, 355, 509–511, 586, 587
Transfer of mass, 58, 62, 390
Transfer of momentum, 58
Turbulence, 112, 113, 115, 137, 191, 217, 362, 367, 442–453, 591, 592
Turbulence distribution, 156, 161, 273, 313, 579
Turbulence generator, 63, 255
Turbulence measurement, 243–247, 270, 339, 575
Turbulent diffusion, 120–123, 148, 192, 254–261
Turbulent shear, 148, 149, 220, 314

Unsteadiness, 466

Velocity distribution, 18, 37, 44, 45, 60, 61, 152–159, 182–186, 224, 225, 233, 479, 576, 577
Velocity measurement, 30, 33, 34, 151, 223, 232, 266, 267, 339, 368, 381, 401, 575
Volume flux, 143–160, 222, 225
Vortex growth and decay, 182–187, 423–427, 438–441
Vortex interaction, 428–434
Vorticity, 414–416

Wake oscillation, 275
Wall, flow over, 264–276, 354, 400
Waterspout, 413
Wave theory, 169, 170, 177, 217, 359
Weight flux, 221
Weirs, 47–49, 54, 55, 286–298
Work-energy equation, 601–612